Oceanography and Marine Biology

Oceanography and Marine Biology: An Annual Review remains one of the most cited sources in marine science and oceanography. The ever-increasing interest in work in oceanography and marine biology and its relevance to global environmental issues, especially global climate change and its impacts, creates a demand for authoritative refereed reviews summarising and synthesising the results of both historical and recent research.

Six of the ten peer-reviewed contributions in Volume 62 are available to read Open Access on the OMBAR website (https://www.routledge.com/9781032761961) and on OAPEN.

Supplementary material is provided online (https://www.routledge.com/9781032761961) on the Support Materials tab, for Reviews 2, 5, 6, 7, 8, 9 and 10.

Volume 62 features a review of the biology, ecology and conservation threats to the Iberian harbour porpoise; a review (and translation of) the long documentation of the historical extent and decline of oyster reefs in southern China; a deep-dive into the complexities of coal-macrophyte interactions; consideration of below-ground structures in seagrass restoration; drivers of change in coral ecosystems; an overview of the potential risk to Mediterranean cetaceans and sea turtles from floating marine macro-litter; a history of the social and ecological drivers of the loss of oyster reefs in the Pear River Delta; an exploration of coral-algal interactions; and a review of historical killings of small cetaceans in the Northeast Atlantic and Mediterranean Sea, among others.

An international Editorial Board ensures global relevance and expert peer review, with editors from Australia, Chile, Hong Kong, Ireland, Singapore and the UK. The series volumes find a place in the libraries of not only marine laboratories and oceanographic institutes but also universities worldwide.

If you are interested in submitting a review for consideration for publication in OMBAR, please email the Co-Editors in Chief, Dr Peter Todd (dbspat@nus.edu.sg) and Dr Bayden Russell (brussell@hku.hk).

Oceanography and Marine Biology

An Annual Review, Volume 62

Editors in Chief:

P. A. Todd
Department of Biological Sciences, National University of Singapore, Singapore

B. D. Russell
School of Biological Sciences, University of Hong Kong, Hong Kong, China

Associate Editors

A. J. Lemasson
University of Plymouth, School of Biological and Marine Science, UK

A. L. Allcock
School of Natural Sciences and Ryan Institute, University of Galway, Ireland/UK

M. Byrne
School of Life and Environmental Sciences, University of Sydney, Australia

L. B. Firth
School of Biological and Marine Sciences, Plymouth University, United Kingdom

C. H. Lucas
School of Ocean & Earth Science, University of Southampton, National Oceanography Centre, European Way, Southampton, UK

E. M. Marzinelli
School of Life and Environmental Sciences, The University of Sydney, Australia

P. J. Mumby
Marine Spatial Ecology Lab, School of Biological Sciences, The University of Queensland, Brisbane, Queensland, Australia

J. Sharples
School of Environmental Sciences, University of Liverpool, UK

I. P. Smith
School of Biological Sciences, University of Aberdeen, United Kingdom

S. E. Swearer
School of BioSciences, University of Melbourne, Australia

M. Thiel
Universidad Católica del Norte Chile

CRC Press
Taylor & Francis Group
Boca Raton London New York

CRC Press is an imprint of the
Taylor & Francis Group, an **informa** business

International Standard Serial Number: 0078-3218

Cover credit: Shutterstock

First edition published 2024
by CRC Press
2385 NW Executive Center Drive, Suite 320, Boca Raton FL 33431

and by CRC Press
4 Park Square, Milton Park, Abingdon, Oxon, OX14 4RN

CRC Press is an imprint of Taylor & Francis Group, LLC

ISBN: 9781032761961 (hbk)
ISBN: 9781032832869 (pbk)
ISBN: 9781003477518 (ebk)

DOI: 10.1201/9781003477518

Typeset in Times
by codeMantra

Access the Support Material: https://www.routledge.com/9781032761961

Contents

CONTENTS

Oceanography and Marine Biology: An Annual Review, 2024, **62**, 1-119

Taylor & Francis

AN ENDANGERED POPULATION OF HARBOUR PORPOISE *PHOCOENA PHOCOENA* HIDDEN IN PLAIN SIGHT: BIOLOGY, ECOLOGY AND CONSERVATION OF THE IBERIAN PORPOISE

GRAHAM J. PIERCE[1,2], MARIE A.C. PETITGUYOT[1], PAULA GUTIERREZ-MUÑOZ[1,3], ANDREA FARIÑAS-BERMEJO[1], DIEGO FERNÁNDEZ-FERNÁNDEZ[1], SARAH DOLMAN[4], MICHAEL C. FONTAINE[5,6], ALBERTO HERNÁNDEZ-GONZÁLEZ[1,3], AGATHA GIL[1,7,8], FIONA L. READ[1,3,9], JOSE MARTÍNEZ-CEDEIRA[10], PABLO COVELO[10], BRUNO DÍAZ LÓPEZ[11], SILVINA IVAYLOVA[1], ANGELA LLAVONA[10], ALFREDO LÓPEZ[10,12], MIGUEL LÓPEZ[1], ANA MARÇALO[13], SÉVERINE METHION[11], PAULA MÉNDEZ-FERNANDEZ[14], SINEAD MURPHY[15], BEGOÑA PÉREZ FERNÁNDEZ[3], RAQUEL PUIG LOZANO[1,10,16], EUNICE H. PINN[17], CAMILO SAAVEDRA[3], GREGORY M. VERUTES[18], LUCÍA VIÑAS[3], AND CAROLINE R. WEIR[19]

[1]Instituto de Investigaciones Marinas (CSIC), Eduardo Cabello 6, 36208, Vigo, Spain
[2]Oceanlab, University of Aberdeen, Main Street, Newburgh, Aberdeenshire, AB41 6AA, United Kingdom
[3]Centro Oceanográfico de Vigo, Centro Nacional Instituto Español de Oceanografía (IEO-CSIC), Subida a Radio Faro, 50, 36390, Vigo, Spain
[4]College of Medical, Veterinary & Life Sciences, University of Glasgow, Bearsden Road, G61 1QH, UK
[5]MIVEGEC, Université de Montpellier, CNRS, IRD, Montpellier, France
[6]Groningen Institute for Evolutionary Life Sciences (GELIFES), University of Groningen, Groningen, The Netherlands
[7]Centre for the Research and Technology of Agro-Environmental and Biological Sciences (CITAB), Department of Biology and Environment, University of Trás-os-Montes and Alto Douro. Quinta de Prados, 5000-801 Vila Real, Portugal
[8]Coastal Biodiversity Laboratory (LBC). Interdisciplinary Centre of Marine and Environmental Research – University of Porto (CIIMAR-UP). Terminal de Cruzeiros do Porto de Leixões, Avenida General Norton de Matos, S/N. 4450-208. Matosinhos, Portugal
[9]Life History Studies, UK
[10]Coordinadora para o Estudo dos Mamíferos Mariños (CEMMA), R/Ceán, Nº2, 36350, Nigrán, Spain
[11]Bottlenose Dolphin Research Institute (BDRI), Avenida Beiramar 192, 36980 O Grove, Spain
[12]Departamento de Biología & CESAM, Universidade de Aveiro. Campus de Santiago, Av. João Jacinto de Magalhães, 3810-193 Aveiro, Portugal
[13]Centro de Ciências do Mar (CCMAR), Universidade do Algarve, Campus de Gambelas, Faro, Portugal
[14]Observatoire Pelagis UAR 3462 La Rochelle Université – CNRS, 5 allées de l'Océan, 17000, La Rochelle, France

DOI: 10.1201/9781003477518-1

[15]*Atlantic Technological University, ATU Galway City, Old Dublin Road, Galway, H91 T8NW, Ireland*
[16]*Veterinary Histology and Pathology, Atlantic Center for Cetacean Research, University Institute of Animal Health and Food Safety (IUSA), Veterinary School, University of Las Palmas de Gran Canaria, Calle Transmontaña, s/n, 35416 Arucas, Canary Islands, Spain*
[17]*School of Law, University of Aberdeen, Aberdeen, United Kingdom*
[18]*CRETUS-Department of Applied Economics, Universidad de Santiago de Compostela, Santiago de Compostela, Spain*
[19]*Ketos Ecology, 15 Home Close, Kingsbridge, Devon, United Kingdom*

Abstract The Iberian harbour porpoise (*Phocoena phocoena*) reaches a larger body size than most other harbour porpoise populations and is genetically distinct, albeit closely related to the population in Northwest Africa. Currently comprising an estimated 3000–4000 individuals, genetic evidence and strandings data suggest that the population has declined in recent times, and it is considered to be at risk of extinction. It is distributed all around the Atlantic coast of the Iberian Peninsula, with the highest densities off Galicia in Northwest Spain and Northern and Central Portugal, a highly productive upwelling area characterised by cold-water upwelling. There are occasional reports from the Mediterranean and Macaronesia and some evidence of emigration into the Celtic Sea. It feeds mostly on fish, with pelagic fish being more important than in the diet of porpoises from northern Europe, perhaps due to excursions beyond the narrow continental shelf.

The population faces a number of anthropogenic threats. Historically, porpoises were used for human consumption while current threats include polychlorinated biphenyls (PCBs), with some individuals having concentrations in their blubber above the threshold for impairment of reproduction, and nematode infections, probably also prey depletion, underwater noise and fatal attacks by bottlenose dolphins. The most serious current threat is fishery bycatch mortality. Stranding data suggest that the bycatch mortality increased in the last decade. Although based on information from a small number of documented mortalities (reflecting limited observer coverage especially for small-scale fishing as well as a low number of reported strandings), annual bycatch mortality estimates are in the order of a few hundred animals, which is clearly unsustainable. There is, however, an apparent incompatibility between the high bycatch estimates and the rather similar abundance estimates obtained from large-scale abundance surveys in 2005, 2016 and 2022.

Consistent with population status assessments by Spain and Portugal, OMMEG[1] (Convention for the Protection of the Marine Environment of the North-East Atlantic) concluded that bycatch mortality in Iberian porpoise "is critically exceeding the agreed threshold" of zero. There are several national initiatives in Spain and Portugal including the development of species conservation plans. Continuous reduction of bycatch mortality, preferably until such mortality is eliminated, is a priority to ensure that this population does not disappear in the near future.

Introduction

The Atlantic coast of the Iberian Peninsula (Figure 1) is characterised by a relatively narrow continental shelf, seasonal cold-water coastal upwelling that influences the entire shelf area, equatorward surface circulation in the form of the Canary current, an equatorward Eastern Boundary Current and a poleward flowing slope undercurrent (Fraga 1981, Álvarez-Salgado et al. 2003). Eastern Boundary Currents are shallow, broad equatorward currents, rich in eddies and known for their

[1] OMMEG is the OSPAR Marine Mammal Expert Group, where OSPAR refers to the OSPAR Commission, which administers the Convention for the Protection of the Marine Environment of the North-East Atlantic, better known as the OSPAR Convention (so-called because it originated from a meeting of the Oslo and Paris Commissions), which regulates international cooperation on environmental protection in the North-East Atlantic. It is one of several European Regional Seas Conventions.

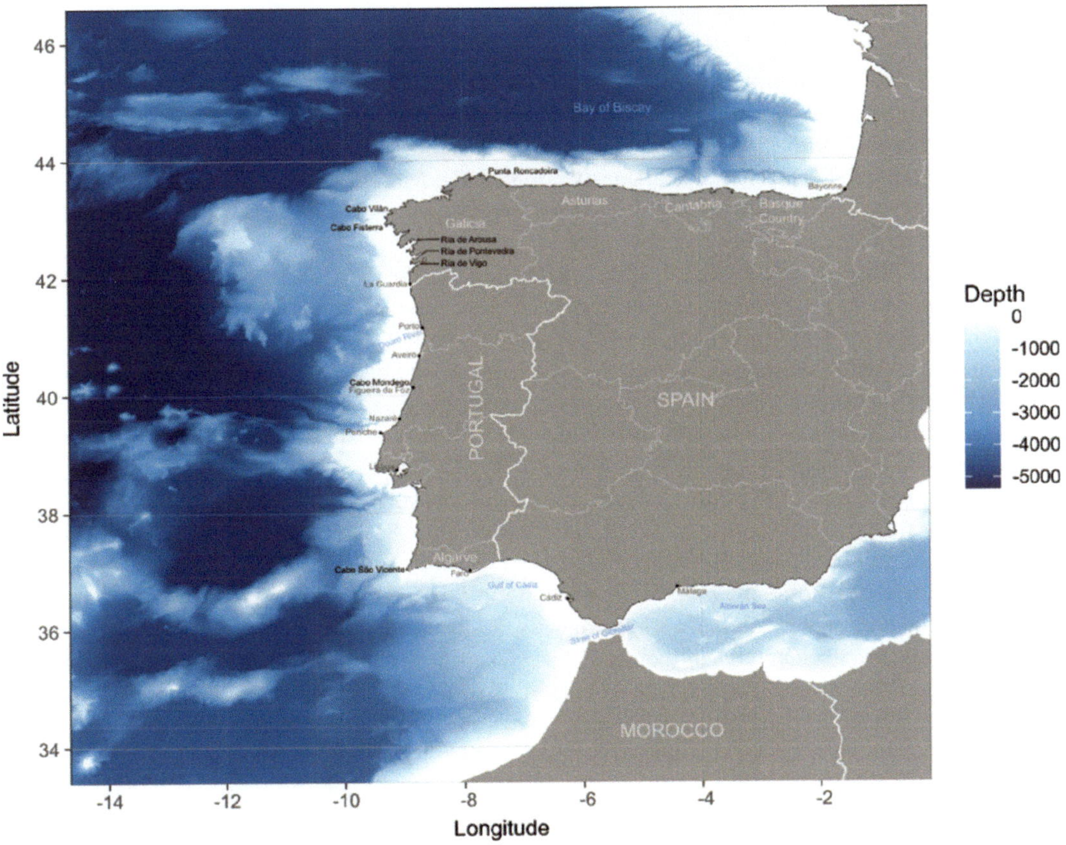

Figure 1 Map of the Iberian Peninsula and adjacent areas showing marine bathymetry data from EMODnet, https://emodnet.ec.europa.eu/en) and the main geographical features mentioned in the text.

upwelling regions along the coast (Moustahfid et al. 2021). The seasonal upwelling supports high productivity, high biodiversity and important fisheries. In Galicia (NW Spain) alone, around 300 species of fish have been recorded (Solórzano et al. 1988), as well as 78 species of cephalopods (Guerra 1992) and at least 22 species of cetaceans and four species of vagrant seals (Fernández de la Cigoña 1990, Penas-Patiño & Piñeiro-Seage 1989, López et al. 2002, López 2017, Covelo & López 2021a). At least two further species are documented from elsewhere on the Spanish mainland Atlantic coast (Fariñas, Petitguyot & Pierce unpublished data). In Portuguese mainland waters, there are at least 150 species of teleost fish (Martins & Carneiro 2018) and 16 species of cephalopods (Jereb et al. 2015) of commercial interest, and 28 species of cetaceans have been reported (Vingada & Eira 2018, Mathias et al. 2023; see also Teixeira 1979, Martin & Walker 1997, Cabral et al. 2006, Wise et al. 2007, Vieira et al. 2009, Brito et al. 2009, Brito & Vieira 2010, Brito & Sousa 2011, Moura et al. 2012, Hammond et al. 2013, Goetz et al. 2015). Mathias et al. (2023) stated that 16 of the 28 cetacean species were commonly found in Portuguese waters while the other 12 were considered as vagrants and that five (vagrant) species of seals are also known from Portugal. Considering records from the extended exclusive economic zone (EEZ) (Correia et al. 2022), the number of cetacean species in Portuguese waters increases to 30.

The harbour porpoise (*Phocoena phocoena*) is one of the most frequently sighted (after the bottlenose dolphin *Tursiops truncatus* and common dolphin *Delphinus delphis*) cetacean species in Atlantic waters of the Iberian Peninsula. In Spanish, the common name for harbour porpoise is *marsopa*; in the Galician language (Galego), it is *toniña* (or *tonina* in López Ferreiro 1895); and in the

Basque Country, it is *moskotxa*. However, in the Canary Islands, *tonina* is a bottlenose dolphin. In Portugal, the harbour porpoise is usually referred to as *boto* (or *bôto*) or *toninha* (or *toninha-comum*). The name *toninha* (sometimes written as *toninha-mansa*) is also used for common dolphin (otherwise *delfim*, *golfinho* or *golfinho-comum*) and (sometimes written as *toninha-brava*) bottlenose dolphin (otherwise *roaz-corvineiro*) (Nobre 1895, Nascimento 1945, Brito et al. 2009, Brito & Vieira 2010, López-Fernández & Martínez-Cedeira 2011, Marçalo et al. 2015, Bencatel et al. 2019). In Porto, fishers used to refer to porpoises as *porcos do rio* (river pigs) (A. Gill, pers. obs.). This list of names and variants may not be exhaustive.

Of the 165 records of *toninhas* in Portugal dating from the thirteenth century onwards compiled by Brito and Vieira (2010), only one can be unambiguously assigned to the harbour porpoise – an animal bycaught and photographed in 1977. It should be noted that the English word 'porpoise' has been used somewhat indiscriminately in the recent past to indicate any small cetacean. For example, Coe et al. (1984) used the word "porpoise" when describing techniques for releasing spotted, spinner and common dolphins (*Stenella attenuata*, *Stenella longirostris*, and *D. delphis*) from purse seine nets in the Eastern Tropical Pacific tuna fishery.

In the mid- to late nineteenth century, several authors referred to the presence of harbour porpoise in Northwest Spain and in Portugal. López Seoane (1861) commented that it could be seen all year round in Galicia, appearing in the ports in summer. Graells (1870) referred to its presence along the whole coast of Galicia and Cantabria, noting that it sometimes entered into the ports and stating that he was unable to obtain a specimen as the fishermen did not target them, only occasionally catching them accidentally in their nets (and resulting in damage to the nets and escape of the catches). Du Bocage (1863) referred to both porpoises ("le Marsouin (*Phocæna communis*)") and common dolphins ("le Dauphin (*Delphinus delphis*, L.)"), especially the former, as usually being seen associated with rivers in Portugal (sometimes at a considerable distance from the river mouth): "se montrent habituellement dans nos principaux fleuves jusqu'à une distance plus ou moins grande de leur embouchure". Nobre (1895) referred to *Phocaena communis* in the Algarve region, southern Portugal, noting that the common name was toninha whereas *Delphinus delphis* was usually referred to as golfinho, and said he saw porpoises frequently while travelling between Lagos and Sines: "je l'ai vu en abondance au nord du cap S. Vincent, pendant la traversée de Lagos à Sines". Both these records clearly distinguished between porpoise and common dolphin. Despite the existence of such records, much of the available information on porpoise occurrence in the Iberian Peninsula derives from the last 50 years, and evidence of significant decline was already available in the 1980s.

Sequeira and Ferreira (1994) noted that while the early naturalists (in the late nineteenth and early twentieth centuries, including the above-mentioned du Bocage and Nobre) considered porpoises to be very common along the Portuguese coast, the situation had "changed drastically" since then, with a reduction in sightings of larger groups and a decline in strandings between 1977 and the late 1980s (see also Sequeira & Teixeira 1988). These authors also noted that many of the stranded animals during this period had net marks around the head and flippers. Perez and Nores (1988) reported that the harbour porpoise declined in Asturias (northern Spain) between 1977–1983 and 1984–1987, based on data from strandings, sightings and catches. They also noted that 7% of 39 bycaught cetaceans obtained during 1977–1987 were porpoises, caught mainly in fixed nets and purse seines.

Lens (1997) summarised almost 200 records from strandings, bycatch and sightings of harbour porpoise, around 100 of which referred to strandings. He observed that porpoise was most abundant along the west coast of Spain, noting that it was absent from the western Mediterranean, and may have declined in the Bay of Biscay and disappeared from the Canary and Azores islands. Over the two and a half decades since that review, new information on Iberian porpoises has emerged from ongoing monitoring of cetacean strandings and fishery bycatch mortality, regional sightings surveys, and various national and international projects. Monitoring of strandings and research on stranded animals have been a major source of information, especially since the establishment of stranding networks in Galicia (run by Coordinadora para o Estudo dos Mamiferos Mariños (CEMMA)) in 1992 and along the central and northern Portuguese coast (run by Sociedade Portuguesa de Vida Selvagem (SPVS) in 2000. Several other networks now cover much of the remaining Atlantic mainland coast and a large part of the coastline of Macaronesian islands (i.e. the Azores, Canaries and Madeira archipelagos). Several large-scale cetacean sightings surveys over the last two decades included Atlantic waters of the Iberian Peninsula, namely, the Small Cetaceans in European Atlantic waters and the North Sea (SCANS II, III and IV) surveys, in 2005, 2016 and 2022, and the Cetacean Offshore Distribution and Abundance in the European Atlantic (CODA) survey in 2007 (Hammond et al. 2009, 2013, 2017, 2021, Gilles et al. 2023). Additional information on distribution and abundance derives from numerous regional and local surveys, conducted from boats, aircraft and land-based observation points (e.g. Pierce et al. 2010, Llavona Vallina 2018, Torres-Pereira et al. 2022).

The biology and ecology of the harbour porpoise are generally well documented on both sides of the North Atlantic (e.g. Sørensen & Kinze 1994, Lockyer 1995a, Read & Hohn 1995, Haug et al. 2003 (and chapters therein), Santos & Pierce 2003, Learmonth et al. 2014; see also numerous reports[2] by the US National Oceanic and Atmospheric Administration). However, the species is less well studied in the Iberian Peninsula, in part because more frequently sighted and stranded species such as common dolphin and bottlenose dolphin (e.g. López et al. 2002, Pierce et al. 2010, Saavedra et al. 2018) have received more attention. Factoring in the small population size, much of the recent evidence available about the biology, ecology and status of Iberian porpoises derives from observations of/on small numbers of animals.

Research on population genetics during the last two decades has revealed that the Iberian porpoise constitutes a genetically distinct population, most closely related to the porpoises of West Africa, and discussions about raising it to subspecies status are ongoing (Fontaine et al. 2007, 2010, 2014, Fontaine 2016, Chehida et al. 2021; see Taxonomy, genetic status, and phylogeography of the Iberian population). The precise limits of its distribution are unknown, and for this reason, within this review we refer to (the small number of) records from the western Mediterranean and Macaronesian islands. The population is small, probably comprising no more than 3000 – 4000 animals and possibly considerably fewer (see Distribution and abundance). Mortality caused by fishery bycatch is of particular concern (Read et al. 2020, Carlén et al. 2021, Pierce et al. 2022, Celemín et al. 2023, Torres-Pereira et al. 2023a). The International Council for the Exploration of the Sea (ICES) Ecosystem Overview for the Bay of Biscay and Iberian Coast states that harbour porpoises "are being caught as bycatch off Iberia, mainly in set nets and beach seines, to the extent that may affect sustainability of the local population" (ICES 2021b).

Despite theoretically strict legal protection (see section 'Conservation of porpoises in Europe: Legal Protection and its implementation'), the conservation of harbour porpoises in Europe is arguably failing due to a combination of "public disinterest, lack of political will to implement conservation measures, and complicated fishing-related issues" (Carlén et al. 2021). A workshop organised by the North Atlantic Marine Mammal Commission (NAMMCO) in 2018 reviewed the status of European porpoise populations, assessed the safe limits for fishery bycatch mortality and compared these with estimated bycatch mortality. The approach taken was based on Potential Biological Removal (PBR, Wade 1998), which is used routinely in the USA for the implementation of the Marine Mammal Protection Act, and fitting logistic population growth models on abundance data within a Bayesian framework (Zerbini et al. 2011). The results highlighted the vulnerable status of the Baltic Proper subpopulation and the Iberian population. In the latter case, based on the best estimates of population size and applying a recovery factor of 0.5, the annual PBR was estimated to be 25 animals (North Atlantic Marine Mammal Commission and the Norwegian Institute of Marine Research 2019). Although a subsequent workshop (NAMMCO 2019) concluded that, due to data limitations, not all assessments were reliable for generating management advice, the small size of the Iberian population and the Baltic Proper subpopulation means that a more optimistic view of their status is unlikely to emerge from future calculations.

[2] See https://www.fisheries.noaa.gov/national/marine-mammal-protection/marine-mammal-stock-assessment-reports-species-stock#cetaceans---porpoises

Concerns about cetacean bycatch in Europe have received greater attention since 2019, when 26 European non-governmental organisations (NGOs) called for the implementation of Fishery Emergency Measures (under the Fisheries Common Policy) to reduce bycatch mortality of common dolphin in the Bay of Biscay and harbour porpoise in the Baltic Sea (Butler-Stroud & Rouley 2019). The European Commission requested advice from ICES concerning the amount of bycatch mortality and appropriate measures to reduce it (see ICES 2020a,b). ICES (2021a) undertook a further assessment of bycatch mortality in response to a request from the OSPAR Commission. The NGOs had considered submitting an Emergency Measures document for the Iberian porpoise but concluded there was insufficient information available (S.J. Dolman, Pers. Comm.).

The Convention on Migratory Species adopted a Concerted Action for the harbour porpoise in the Baltic Sea and the Iberian Peninsula, as proposed by Coalition Clean Baltic, Whale and Dolphin Conservation, Humane Society International and ORCA, referring to the Iberian porpoise as a "critically isolated population", in 2020 (CMS 2020). The Agreement on the Conservation of Small Cetaceans of the Baltic, North East Atlantic, Irish and North Seas (ASCOBANS), through its Secretariat, and the International Whaling Commission's Scientific Committee (IWC SC) (IWC Scientific Committee 2021, 2022, ASCOBANS 2021) have both recently expressed concerns about the status of the Iberian harbour porpoise population. The 26th Meeting of the Advisory Committee of ASCOBANS in 2021 discussed a draft proposal to include the Iberian harbour porpoise population in the Appendices of the UN Convention on the Conservation of Migratory Species of Wild Animals (CMS or the Bonn Convention), noting that this population "has been identified as a high priority for conservation due to its low abundance, genetic distinctiveness, low genetic diversity, and evidence of unsustainably high levels of bycatch mortality in fishing gear" (ASCOBANS 2021). In its 2022 report, the IWC SC recommended "immediate actions to effectively reduce, and where possible eliminate, bycatch of harbour porpoise throughout Iberian Peninsula waters", noting that "measures are urgently needed for fisheries using gillnets and trammel nets but also for beach seines along the Portuguese coast which are used in some areas of high porpoise density" (IWC Scientific Committee 2022). The OSPAR Marine Mammal Expert Group (OMMEG) suggested that a bycatch limit of zero should be applied to the Iberian porpoise population (see ICES 2021a). This threshold was agreed by contracting parties at the Biodiversity Committee of OSPAR in autumn 2021.

The Baltic Proper subpopulation of *Phocoena phocoena*, of which fewer than 500 survive (SAMBAH 2016, North Atlantic Marine Mammal Commission and the Norwegian Institute of Marine Research 2019, Siebert et al. 2020, Owen et al. 2021, Amundin et al. 2022), and the vaquita (*Phocoena sinus*), of which fewer than 10 survive (Jaramillo-Legorreta et al. 2019, Robinson et al. 2022), are arguably in a more perilous state. However, it seems clear that urgent management action is needed to secure the long-term viability of the Iberian porpoise population. The present review encompasses the taxonomic status of the Iberian porpoise, its biology and ecology (highlighting differences from other harbour porpoise populations where applicable), evidence about the status of the population, the various natural and anthropogenic threats it faces – notably mortality due to fishery bycatch, the legal framework under which it is protected and current conservation management. It also aims to identify knowledge gaps and actions necessary to protect this vulnerable population.

Taxonomy, genetic status and phylogeography of the Iberian population

Porpoises from the Iberian Peninsula reach larger sizes than those from further north within Europe (e.g. Smeenk et al. 1992, Donovan & Bjørge 1995, Sequeira 1996, López Fernández 2003, López-Fernández & Martínez-Cedeira 2011). As genetic evidence was amassed, it became clear that Iberian porpoises form a morphologically and genetically distinct, largely isolated, population (Fontaine et al. 2007, 2010, 2014, Celemín et al. 2023) closely related to the population in Northwest Africa (Fontaine et al. 2014). Fontaine et al. (2014; see also the review by Fontaine 2016) proposed that the Iberian and Northwest African porpoise populations together represent a distinct ecotype adapted to upwelling systems. Considering their phylogenetic divergence from the subspecies described in the North Atlantic (*P. p. phocoena*) and Black Sea (*P. p. relicta*), their allopatric distribution, and their morphological and ecological distinctiveness, it was proposed to raise this distinct ecotype as a separate subspecies with the name *P. p. meridionalis* (Fontaine et al. 2014, Fontaine 2016) (Figures 2 and 3). Ben Chehida et al. (2021) and Olsen et al. (2022) identified a fourth porpoise lineage from West Greenland, an area where an offshore ecotype is known to exist (Nielsen et al. 2018). They argued that this fourth lineage probably split off at the same time the other lineages emerged and speculated that it emerged in oceanic waters around the Azores. However, there are very few porpoise records from Macaronesia, and there have been no studies of their genetic makeup.

Within the Iberian Peninsula, porpoises from Spain and Portugal are genetically similar and form part of a single population; no differences in mitochondrial and microsatellite genetic diversity, or in genetic ancestry, were detected between samples from the two countries (Fontaine et al. 2007, 2010, 2014, Llavona Vallina 2018, Ben Chehida et al. 2021, 2023). It should be noted that samples usually come from stranded animals, and given the prevailing currents, it is possible that some porpoises stranded along the northwest coast of Spain had died in Portuguese waters. Genetic diversity of the Iberian population was lower than that of porpoises from the Bay of Biscay, although comparable with that in Mauritania and the Black Sea. Despite similar genetic diversity, the abundance of the Iberian porpoise is around 30 times lower than that of the Black Sea population (see ACCOBAMS 2021 and section 'Distribution and abundance' of this chapter). Ben Chehida et al. (2023) reported that a sharp decline in the genetic diversity of Iberian porpoises occurred between 1990 and 2015, caused by genetic drift and/or a decline in the effective population size, possibly driven by environmental stochasticity, prey depletion or bycatch mortality. Low genetic diversity was also identified in Iberian porpoises by Celemín et al. (2023), in addition to high levels of inbreeding and a low effective population size.

Phylogeographic analyses by Fontaine et al. (2014) suggested that the upwelling ecotype of harbour porpoise present in the Iberian Peninsula and Mauritania descended from a now extinct paleo-population living in the Western Mediterranean Sea during the last glacial maximum (~20,000 years before present) (Figure 3). Porpoises likely entered the Mediterranean Sea from the Northeast Atlantic and split from the Atlantic populations within the past ~30,000 years. Porpoises subsequently disappeared from the Mediterranean during the postglacial warming period, but the Western and Eastern Mediterranean lineages, respectively, gave rise to the 'upwelling' and Black Sea groups, around 15,000 years ago, with the former giving rise to the Iberian and North West African groups (Figure 3).

Population genetic analyses to date suggest that porpoises from the Iberian population mix asymmetrically northwards with those from the Bay of Biscay and the Celtic Sea, with individuals migrating away from the Iberian unit but not into it (Alfonsi et al. 2012, Fontaine et al. 2014, 2017, Ben Chehida et al. 2021). Porpoises found stranded on the northern side of the Bay of Biscay and the Celtic Sea clearly displayed an admixed genetic ancestry, defining a geographically localised tension (or hybrid) zone where most porpoises not only show mixed genetic ancestry between Iberian porpoises and those further north but also exhibit the larger body size typical of the Iberian porpoises (Fontaine et al. 2017). Fontaine et al. (2010, 2014; reviewed in Fontaine 2016) estimated that porpoises from Iberian waters and those further north came back into contact and established this hybrid zone ~300 years ago. Mitochondrial evidence also suggested that individuals from the Iberian population interbreed with the population in Northwest African waters (or with an unknown population in between) (Fontaine et al. 2014, Ben Chehida et al. 2023). Northwest African porpoises are most closely related to Iberian animals, both being descended from the extinct western Mediterranean population and diverging from each other around 3000 years ago. Overall, it seems clear that the porpoises of the Iberian Peninsula should be treated as a distinct unit.

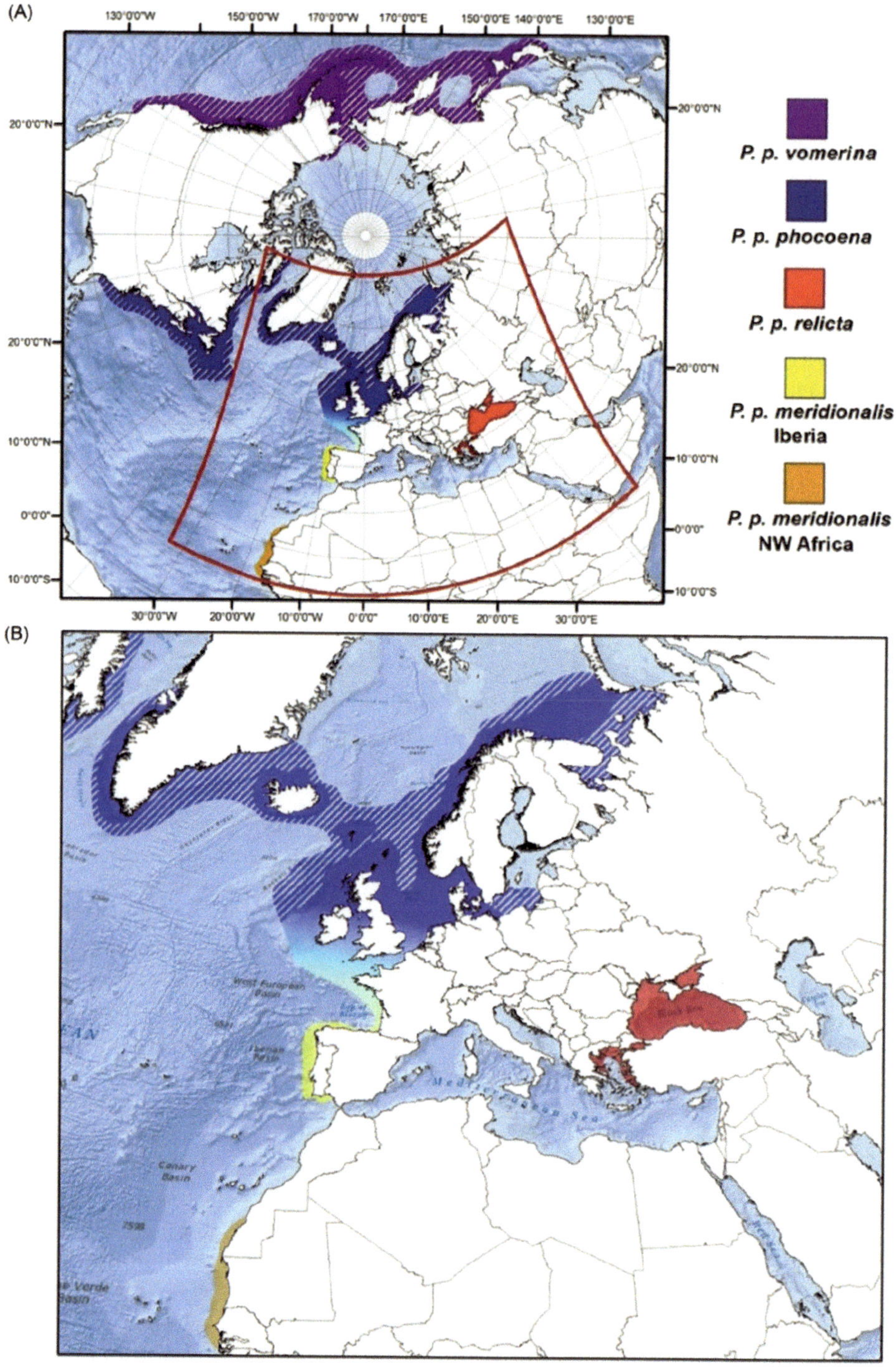

Figure 2 Distribution of the harbour porpoise (*Phocoena phocoena*), showing ranges for the different subspecies, including the proposed *P. p. meridionalis*, and genetically differentiated populations. (A) The global circumpolar distribution of the harbour porpoise. Polygon (in red) indicates the enlarged view displayed in (b). (B) The Northeast Atlantic distribution of the species with the different subspecies and genetically distinct

(*Continued*)

(Continued)

populations bordering the Mediterranean Sea. The four subspecies are displayed with different colours. The known and possible species distribution is shown with plain and hashed surfaces, respectively. A coloured gradient (yellow to dark blue) in the south of the British Isles and Bay of Biscay shows the approximate geographic distribution of the admixture zone between the Iberian population of *P. p. meridionalis* and the population of *P. p. phocoena* north of the Bay of Biscay. The Figure was prepared using ESRI ArcGIS v.10.3 and is based on maps in Gaskin (1984), IWC (1996) and Read (1999), updated with recent observations from the Black Sea and the northern Aegean Sea. (Reprinted from *Advances in Marine Biology*, Vol 75, Michael C. Fontaine, Harbour Porpoises, *Phocoena phocoena*, in the Mediterranean Sea and Adjacent Regions: Biogeographic Relicts of the Last Glacial Period, Page 335, Copyright (2016), with permission from Elsevier.)

Additional evidence about population structure and movements in cetaceans can be derived from ecological tracers such as stable isotope signatures, contaminant concentrations and parasites. Although differences in stable isotope profiles have been used to infer the existence of distinct groups of harbour porpoises within an area (e.g. Jansen et al. 2012) and the absence of differences has been used to infer movements of porpoises between areas (e.g. Angerbjörn et al. 2006), to date such studies have not included Iberian animals. Studies of concentrations of metals and/or organic contaminants in porpoise tissues can provide similar kinds of insights (e.g. Das et al. 2004, Pierce et al. 2008) but again have little to say about the distinctiveness of Iberian porpoise. We revisit this topic in section 'Threats to the Iberian harbour porpoise - Pollutants and other harmful substances', where the role of pollutants as a threat to Iberian porpoise is discussed.

Parasites, including larval nematodes, are routinely used in stock identification for fish (e.g. MacKenzie et al. 2008, MacKenzie & Hemmingsen 2015). In principle, the same approach is applicable to cetaceans, which are the final hosts of *Anisakis* spp. (MacKenzie 2002). Nematodes found in the digestive tract of cetaceans derive from their prey and hence may be useful to identify "ecological stocks" of cetaceans with different habitats and feeding habits, perhaps also true (reproductively isolated) stocks. Gomes et al. (2021) recently found a strong correlation between the proportions of mature nematodes belonging to the sibling species *Anisakis pegreffii* and *Anisakis simplex* (*s.s.*) and stock identity of minke whales (*Balaenoptera acutorostrata*) in the waters around Japan. Results of molecular analysis of *Anisakis* parasites from a small sample of harbour porpoises provided a preliminary indication that parasitological analysis could be used as a complementary method supporting the proposed *P. phocoena meridionalis* subspecies (Cipriani et al. 2022). In one porpoise from Scotland, all the *Anisakis* identified were *A. simplex* (*s.s.*), while three porpoises from Galicia all had mixed infections, with 19% overall of the adult *Anisakis* worms being *A. pegreffi*. This result was consistent with the known distribution of the different *Anisakis* species in the Northeast Atlantic, with *A. pegreffi* being absent from northern waters, while the two species occur sympatrically in fish inhabiting Atlantic waters of the Iberian Peninsula (Abollo et al. 2003, Levsen et al. 2018). Both *Anisakis* species were also present in similar proportions in blue whiting *Micromesistius poutassou* (Roca-Geronès et al. 2020), which is an important prey species for porpoises in the area (Hernandez-Gonzalez et al. 2024).

For the assessment of the status of harbour porpoise in European Atlantic waters in relation to various indicators under the EU Marine Strategy Framework Directive (MSFD), ICES (2014a,b) proposed the use of five Assessment Units, with the Iberia Peninsula being treated as a separate Assessment Unit (Figure 4A). It should be noted that these Units did not include the Baltic Proper sub-population. The OSPAR Commission (2023) subsequently proposed nine Assessment Units for the MSFD bycatch indicator (M6), extending the Iberian porpoise AU into offshore waters, changing the AUs west of the UK and including additional AUs in the north (but still not including the Baltic Sea proper sub-population) (Figure 4B).

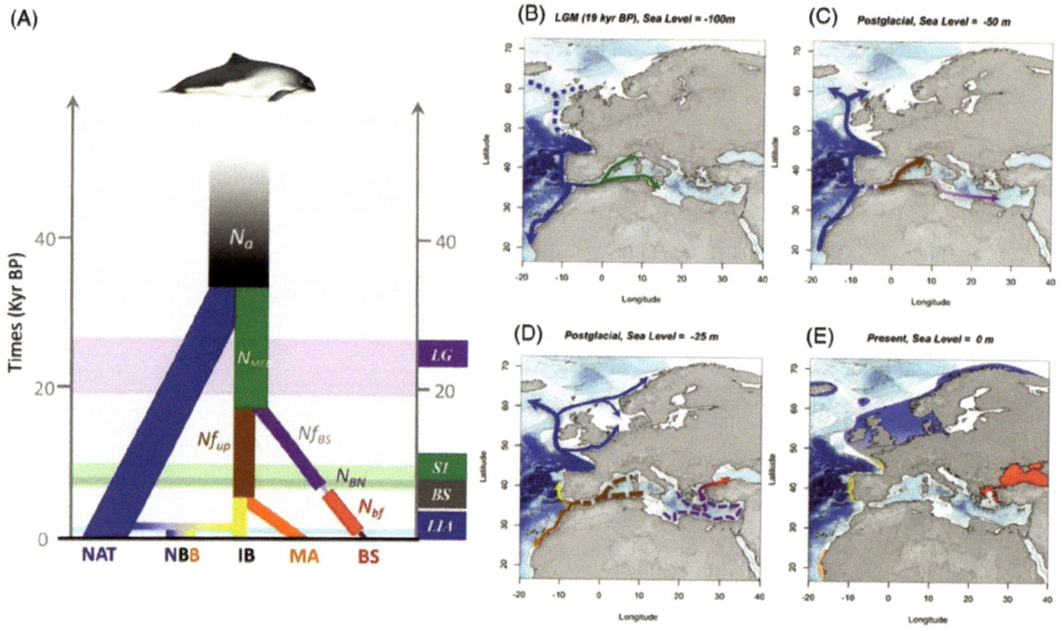

Figure 3 Evolutionary history of the harbour porpoise (*Phocoena phocoena*) as determined from genetic analyses. (A) The evolutionary history of the differentiated groups with a schematic population tree. Each group is shown at the bottom with a colour coding following Figure 1 (BS, population from the Black Sea *P. p. relicta*; IB, Iberian population of the southern ecotype, the proposed *P. p. meridionalis*; MA, Mauritanian–Northwest African population of the southern ecotype; NAT, Northeast Atlantic ecotype (*P. p. phocoena*) inhabiting the European continental shelf north of the Bay of Biscay; NBB, admixed population between IB and NAT in the northern side of the Bay of Biscay). Each group coalesces backwards in time (upward) as indicated by the *Y*-axis, which provides the timeline in thousands of years before present (kyr BP). Changes in population size are depicted by changing line width. Major environmental changes related to the demographic history of harbour porpoises are also plotted: Last Glacial Maximum (LGM period, ca. 23–19 kyr BP) (Clark et al. 2009), Mediterranean Sapropel S1 period (ca. 9.5–6.5 kyr BP) (Spötl et al. 2010, Roberts et al. 2011), flooding of the Black Sea (BS, ca. 8.4–9.4 kyr BP) (Major et al. 2006, Giosan et al. 2009) and Little Ice Age (LIA, ca. 250–700 yr BP) (Osborn & Briffa 2006). (B–E) The biogeographic landscape of porpoise evolutionary history at four time-steps determined by historical changes in sea level: (B) During the LGM (glacial sea level 100 m lower than today), porpoises from the Atlantic likely colonised the Mediterranean Sea, resulting in a divergent group that formed the ancestral Mediterranean population(s) (N_{MED} in A); (C) During the post-glacial Holocene warming and the sea level rise, these ancestral populations in the Mediterranean Sea split into eastern (purple, Nf_{BS} in A) and western (brown, Nf_{UP}) lineages, from which descended the porpoise of the Black Sea and the two populations of the upwelling waters of Iberia and Mauritania–Northwest Africa; (D) Mediterranean conditions became unsuitable for the harbour porpoise at the end of the African Humid Period and Mediterranean Sapropel episodes. Porpoises were thus forced out of the Mediterranean, taking refuge where conditions were still suitable for the species: the Eastern Mediterranean lineage moved into the Black Sea, reconnected to the Mediterranean ca. 8.4 kyr BP, and the Western Mediterranean lineage moved back into the Atlantic waters of Iberia and Northwest Africa; (E) The present distribution. Maps were drawn using MARMAP v0.9.5 package (Pante & Simon-Bouhet 2013) for R (R Core Team 2016) using the ETOPO1 dataset available on the United States National Geophysical Data Center (Amante & Eakins 2009). Sea level was adjusted to account for the historical variation during the LGM and post-LGM period. Panel (A) was based on Fontaine et al. (2012, 2014). (Reprinted from *Advances in Marine Biology*, Vol 75, Michael C. Fontaine, Harbour Porpoises, *Phocoena phocoena*, in the Mediterranean Sea and Adjacent Regions: Biogeographic Relicts of the Last Glacial Period, Page 341, Copyright (2016), with permission from Elsevier.)

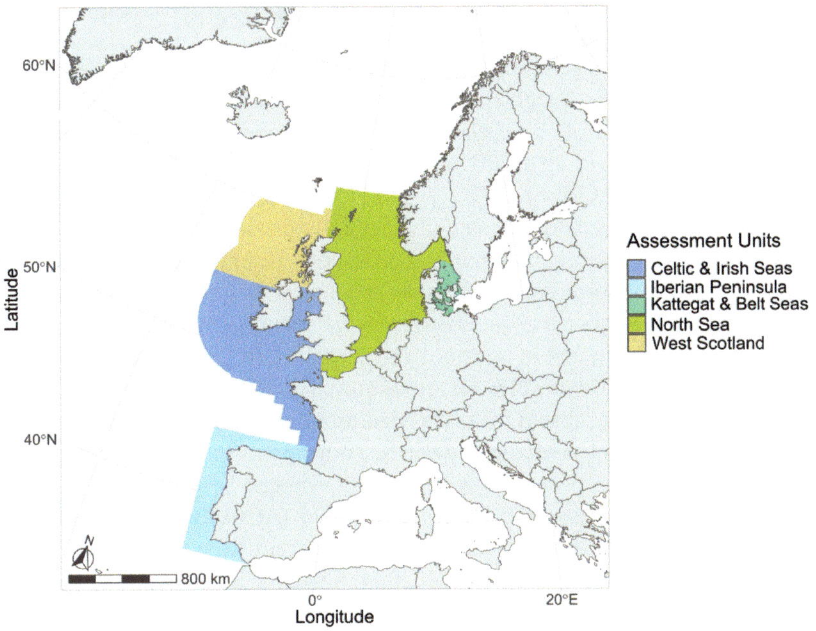

(A)

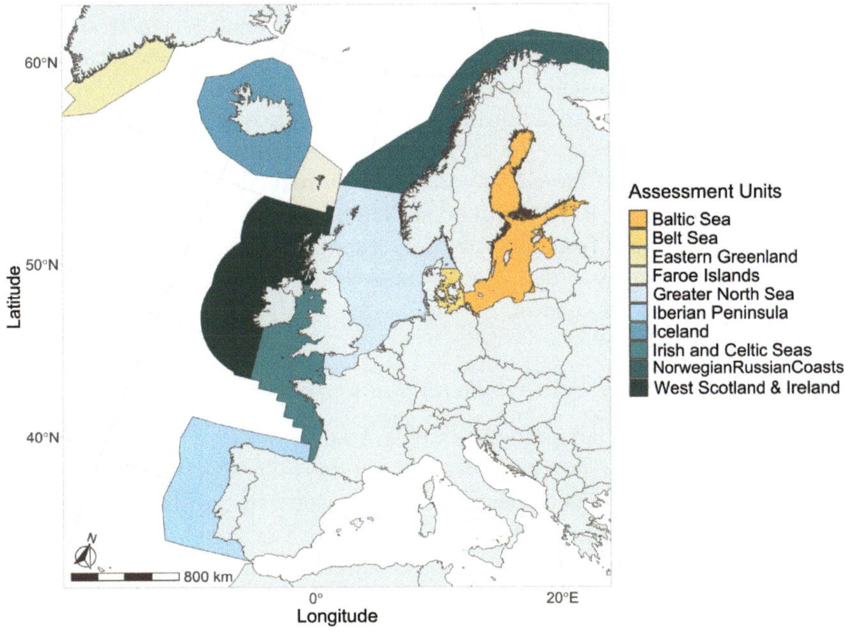

(B)

Figure 4 (A) ICES Assessment Units for harbour porpoise. (Recreated from information in Figure 1.6.6.1.2 in ICES (2014a) and Figure 1 in Appendix 1 of ICES (2014b).) (B) OSPAR Commission Assessment Units for M6 assessment units for harbour porpoise, based on Figure c in Geelhoed et al. (2022) and Figure 1 in OSPAR Commission (2023), which in turn were based on a proposal by North Atlantic Marine Mammal Commission and the Norwegian Institute of Marine Research (2019), plus the Baltic Sea proper Assessment Unit used by HELCOM (2023). The figure was reconstructed using the geopackage https://github.com/osparcomm/Abundance-and-Distribution-of-Cetaceans.

Distribution and abundance

Globally, the harbour porpoise is generally considered to be a coastal species preferring continental shelf waters (e.g. Evans 1987) although there is evidence from satellite tagged animals of movements into deep oceanic waters off Greenland (which has a very narrow continental shelf), with porpoises there diving to depths of up to 410 m (Nielsen et al. 2018, Olsen et al. 2022). As mentioned in the previous section, Ben Chehida et al. (2021) and Olsen et al. (2022) referred to these animals as representing an offshore ecotype and detected a genetically distinct lineage in the region.

The Iberian harbour porpoise population inhabits the cold-water upwelling zone along the Atlantic coasts of Spain and Portugal, from the south Biscay coast to (at least) the Algarve coast of Portugal, bordering the Gulf of Cádiz (Sequeira 1996, Castro 2010). Recent occurrence records are most numerous in the Galicia region of Spain and in northern and central Portugal (e.g., Donovan & Bjørge 1995, Sequeira 1996, Fontaine 2016, Read 2016, Hammond et al. 2017, Torres-Pereira et al. 2022, Gilles et al. 2023). Porpoises are largely absent from the Mediterranean Sea (Frantzis et al. 2001), and there were no sightings of porpoises in the Mediterranean Sea during the Survey Initiative of the Agreement on the Conservation of Cetaceans of the Black Sea, Mediterranean Sea and Contiguous Atlantic Area (ACCOBAMS) in the Mediterranean and Black Sea in 2018 (ACCOBAMS Survey Initiative 2018). However, there is evidence of occasional incursions from Iberian Atlantic waters into the contiguous Mediterranean Sea (e.g. Lens 1997, Cabezón et al. 2004, Sociedad Española de Cetáceos 2006, Anon. 2021). Cabezón et al. (2004) referred to a porpoise stranded in Andalucía (Spanish Mediterranean coast), and Lens (1997) mentioned two records from the Strait of Gibraltar in the early 1980s. Porpoises are also known from the northern Aegean Sea where, for example, a number of sightings and acoustic detections were recorded in both Greek and Turkish waters during a survey in 2013 (Cucknell et al. 2016). Genetic analyses of stranded porpoises from this region indicated that these animals belong to the same genetic pool as Black Sea porpoises (Fontaine et al. 2012).

There are also isolated records from the Azores (Barreiros et al. 2006) and Canary Islands (Díaz-Delgado et al. 2018), although the affiliation of these animals cannot be confirmed, and at least in the Canary Islands, it is more likely that such animals derive from the northwest African population. It is worth noting that the continental shelf of the Iberian Peninsula is quite narrow but to date almost all sightings are from shelf waters, despite several surveys extending into offshore waters. However, the probability of detecting porpoises during visual surveys in oceanic waters, in all but the calmest conditions, is modest.

Large-scale dedicated abundance surveys

The only surveys that have covered almost the whole Atlantic Iberian coast were SCANS-II in 2005 (with the CODA survey in 2007 covering offshore waters), SCANS-III in 2016 and SCANS-IV in 2022, although these surveys did not extend into the interior waters of the Galician Rías. All three surveys took place in summer (June–July). The 2005 SCANS-II survey (Figure 5A and C) covered shelf waters of the Iberian Peninsula and the southern and central Bay of Biscay (Block W) and produced abundance and density estimates for this area of 2357 animals (CV=0.92) and 0.017 animals km^{-2} (CV=0.92), respectively (Hammond et al. 2013). These estimates were revised by Hammond et al. (2021) to 2880 animals (CV=0.72) and 0.021 (CV=0.72) animals km^{-2}, respectively. A complementary survey of adjacent offshore waters (CODA) was carried out in 2007, comprising shipboard visual and acoustic surveys. No porpoises were detected visually and only one acoustically (off southwest of Britain) (Hammond et al. 2009). For the 2016 SCANS-III survey (Figure 5B and D), the survey area was amended to correspond to the Iberian Peninsula Management Unit (IPMU) as proposed by ASCOBANS, the Baltic Marine Environment Protection Commission (better known as the Helsinki Commission or HELCOM) and ICES (Evans & Teilmann 2009, ICES 2009, 2013b, 2014a), which in turn reflected the recent genetics studies on harbour porpoises. The resulting survey

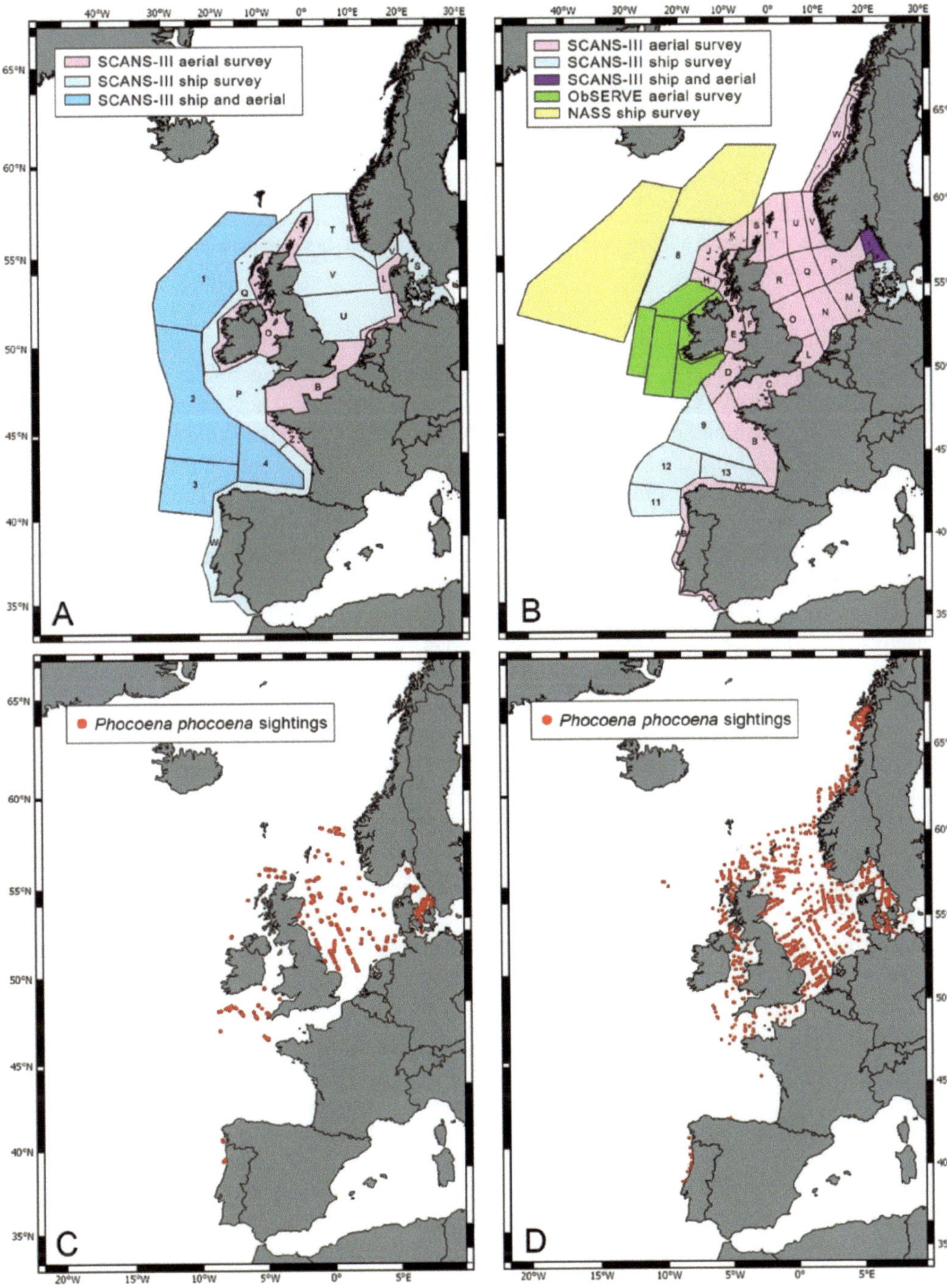

Figure 5 Porpoise sightings from the SCANS II and III surveys: (A) SCANS-II (2005) and CODA (2007) survey blocks, (B) SCANS-III (2016) and related survey blocks, (C) porpoise sightings from SCANS II, (D) porpoise sightings from SCANS III survey. For harbour porpoise, the usable effort comprised aerial surveys carried out in good and moderate conditions and ship surveys undertaken in Beaufort 0–2. (Panels A and B are adapted from Lacey et al. (2022), panel C is based on SCANS-II (2008), recreated using SCANS-II data from OBIS. Panel D is modified from Hammond et al. (2021). (Figures reproduced with permission from Phil Hammond.)

Table 1 Harbour Porpoise Abundance and Density (animals km^{-2}) Estimates from the Iberian Peninsula Block A of the SCANS-III Survey in 2016 (Hammond et al. 2017)

Block	Geographic Region	Abundance	Density	CV	CL Low	CL High
AA	Strait of Gibraltar to Cabo de São Vicente	0	0	0	0	0
AB	Cabo de São Vicente to Cape Finisterre	2715	0.102	0.308	1350	4737
AC	Cape Finisterre to Bayonne (France), including the southern Bay of Biscay	183	0.005	1.020	0	669

CV is the coefficient of variation of abundance and density. CL low and CL high are the estimated lower and upper 95% confidence limits of abundance.

block (Block A) was further divided into three sub-blocks spanning the Atlantic and Bay of Biscay coasts of Portugal and Spain. SCANS-III generated an abundance estimate of 2715 individuals (CV=0.31) for sub-block AB, from Cabo de São Vicente in Portugal northwards to Cape Finisterre in Galicia, which encompasses the core range of the Iberian population (Table 1). No porpoises were seen in sub-block AA from the Strait of Gibraltar to Cabo de São Vicente in Portugal, and few were seen in block AC from Cape Finisterre to Bayonne (France) (Figure 5D, Table 1). The combined abundance estimate for the IPMU was 2898 animals (CV=0.32) (Hammond et al. 2017, 2021).

The porpoise densities recorded for the Iberian Peninsula SCANS blocks during the 2005 and 2016 surveys were among the lowest over the entire European continental shelf. The abundance estimates for the IPMU were also very similar, 2880 (CV=0.72) and 2900 (CV=0.32), respectively, once the former had been corrected for likely negative bias (Hammond et al. 2017, 2021). The similarity of the two abundance estimates suggests that there was no upward or downward trend over this period, although the estimates of porpoise abundance for the Iberian Peninsula are based on a small number of sightings, and such trends could be accommodated within the 95% confidence intervals. It is worth noting that almost all the porpoise sightings along the Iberian coast during the SCANS-III survey were between Viana do Castelo (near the Spanish border) and Peniche (north of Lisbon) along the west coast of Portugal (Figure 5D). SCANS-IV took place in summer 2022. The survey area off the Iberian Peninsula extended into offshore waters, although all porpoise sightings were in inshore waters, and a majority of them were along the western coast of Galicia. The resulting abundance estimate for the Iberian AU was somewhat higher (4043, with a CV of 0.35% and 95% confidence limits of 1842–7309) (Gilles et al. 2023). Note that all calculations of the potential impact of fishery bycatch in section 'Threats to the Iberian harbour porpoise' were based on the previous population estimate of approximately 2900 animals.

National and regional surveys plus opportunistic sightings

Numerous national and regional surveys have recorded cetacean distribution along the Spanish and Portuguese Atlantic coasts (Figure 6) but only a few have generated abundance estimates. The relatively low numbers of porpoise sightings resulted in wide confidence limits around the abundance estimates, and estimates from Portuguese surveys also varied markedly between years (ICES 2014a). The results seem to be consistent with the conclusion from the SCANS surveys that porpoise density in Iberian waters is lower than in most European Atlantic shelf waters. A low sighting rate is also reported for the adjacent porpoise population in Northwest Africa. It was not until 2007 that Boisseau et al. (2007) reported the first sightings of live harbour porpoises (three sightings in Agadir Bay (Morocco) during almost 2000km of effort). Consistent with the results of large-scale surveys (e.g. Figure 5), surveys carried out both in Portugal and northern Spain suggest that the majority of the Iberian porpoise population inhabits Portuguese waters.

The general picture that emerges from sightings in Spanish waters is of small numbers of records along most of the coast and a few hotspots in Galicia, notably over the continental shelf outside the Ría of Arousa, although the high number of sightings here reflects high survey effort,

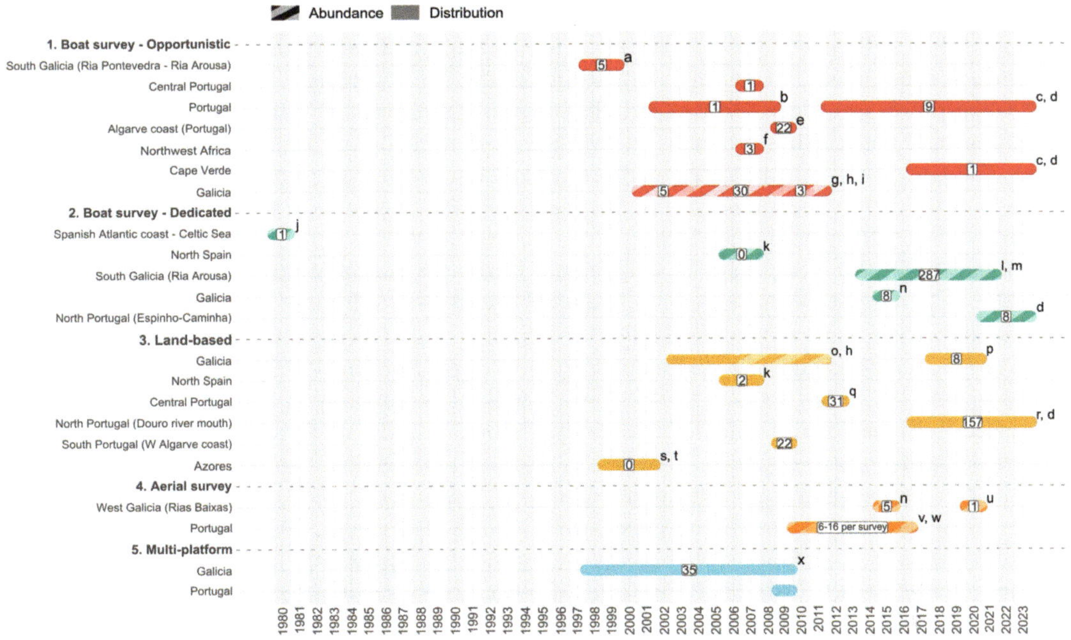

Figure 6 Graphical representation of the different types of surveys, regional and national, carried out in Iberian Peninsula waters between 1980 and 2023 that have collected harbour porpoise sightings. The bars represent the years during which surveys were carried out, plain coloured bars represent surveys that allow determination of porpoise distribution and bars with diagonal line pattern represent surveys from which abundance estimates can be inferred. The number inside each white box represents the number of sightings of harbour porpoise registered (when available) and the letters at the end of the bars indicate the source of the data (when available), as cited in the main text, and detailed below: (A) López et al. 2004; (B) Brito et al. 2009; (C) Correia et al. 2019; (D) A. Gil, pers. comm.; (E) Castro 2010; (F) Boisseau et al. 2007; (G) Spyrakos et al. 2011; (H) Llavona Vallina 2018; (I) López et al. 2011; (J) Aguilar et al. 1983; (K) López et al. 2013; (L) Díaz López and Methion 2018; (M) Díaz López et al. 2022; (N) Martínez-Cedeira et al. 2015; (O) Pierce et al. 2010; (P) López, pers. comm.; (Q) Pereira 2015; (R) Gil et al. 2019; (S) Silva et al. 2011; (T) Barreiros et al. 2006; (U) Martínez-Cedeira et al. 2021; (V) Vingada et al. 2011; (W) Torres-Pereira et al. 2022; (X) Fernández et al. 2013.

since the Bottlenose Dolphin Research Institute (BDRI) is based in the area, coupled with an almost complete absence in deeper offshore waters. In Portugal, there have been more sightings, although with considerable variation between years and between locations.

Spain

Based on the data collected between 2003 and 2011 from multiple sources, López et al. (2013) produced an abundance estimate for harbour porpoises in the Spanish Galician and Bay of Biscay waters of 683 animals (CV=0.63, 95% CI: 345–951, N=40), with a density estimate of 0.0008 animals per km^2. This estimate did not account for availability, perception or responsive movement biases and may therefore have been negatively biased. Summing the Portuguese and northern Spanish estimates, the total estimated abundance for Atlantic waters of the Iberian Peninsula excluding the Gulf of Cadiz (2254+683=2937 animals) is similar to the SCANS II and III abundance estimates. López et al. (2012) provided a separate abundance estimate of 386 (CV=0.71) porpoises in Galician waters (see also CEMMA 2018).

The French SAMM (Suivi Aérien de la Mégafaune Marine, Aerial Census of Marine Megafauna) summer and winter aerial surveys extended to the eastern part of the North coast of Spain. All the porpoise sightings in the southern part of the Bay of Biscay were along the French coast. It was apparent that sightings of this species extended further south in winter (to within 50 km of the Spanish border), mainly along the coast, while in summer occurrences extended further offshore but no porpoises were seen further south than Royan (Lambert et al. 2017 (in their Appendix S4)). This seasonal pattern thus probably relates mainly to animals from the main European population. The SAMM II winter survey involved a higher survey effort (more closely spaced transect lines), and there were more on-effort sightings on porpoises, again including some very close to the Spanish border (Blanchard et al. 2021).

Lens (1997) noted that there were few records of porpoise sightings in the Iberian Peninsula, mostly comprising opportunistic sightings along the west and north coasts. Cabrera (1914) referred to porpoises being present around the Strait of Gibraltar, and there were two further records from this area in 1981, comprising one stranding in Málaga (Rey & Cendrero 1982) and one sighting (Duguy & Desportes 1983). The Ballena surveys, which took place annually in the first half of the 1980s, extended from the Spanish Atlantic coast to the Celtic Sea and focused on (the then still exploited) fin whale *Balaenoptera physalus*. The Ballena 1 survey also yielded a single porpoise sighting, well off the continental shelf, 140 miles west of Lisbon (Aguilar et al. 1983), while surveys 2, 3 and 4 reported small cetaceans but no porpoises were seen (Sanpera et al. 1984, 1985, Sanpera & Jover 1986). McBrearty et al. (1986) mentioned sightings of porpoises along the Spanish and Portuguese Atlantic coasts from British Royal Navy and Merchant Navy vessels during 1978–1982.

During 4 years of shore-based monitoring of the Galician coast (September 2003–October 2007) by CEMMA, involving monthly visits to at least 30 sites, porpoises sightings comprised 10.4% of all coastal cetacean sightings (3.4% of the estimated total number of individuals) and were distributed all along the Galician coast, with the highest sighting frequencies recorded from Faro Punta Roncadoira on the north coast of Galicia, Faro Cabo Vilán near Cabo Fisterra (the westernmost point of Galicia) and A Guarda (on the border with Portugal) (Pierce et al. 2010). Llavona Vallina (2018) repeated the analysis using data up to the end of 2011: over the whole period, porpoise sightings comprised 12.3% of all cetacean sightings. Average annual encounter rates showed a slight overall upward trend (Figure 7). The relative frequency of porpoise sightings was almost unchanged when the study concluded at the end of 2013 (12.1%; A. López, Pers. Comm.)

Combining all boat survey data for Galicia during 1998–2009, there were 35 porpoise sightings, all recorded between Cabo Fisterra and the Portuguese border, suggesting that the south-west region of Galicia is of particular importance for this species (Fernández et al. 2013). Boat-based surveys in Galician waters during 2003–2010 ($N = 111$), using a variety of vessels (including fishing vessels), resulted in a total of 30 porpoise sightings during approximately 844 hours of survey effort, mostly in shelf waters up to 200 m deep (Llavona Vallina 2018). Several of the surveys included in the above-mentioned totals recorded few porpoises, including those using observers onboard fishing vessels in Galician coastal waters during 1998–1999 (López et al. 2004) and 2001–2003 (Spyrakos et al. 2011). The 1998–1999 surveys covered approximately 20,000 km^2 and recorded five porpoise sightings in shallow waters adjacent to the Rías of Pontevedra and Arousa in southern Galicia (López et al. 2004). Several boat-based surveys recorded no porpoises, including surveys along the entire northern Spanish coast in 2006 and 2007 (López et al. 2013), also dedicated surveys of the Galician Bank (180 km off the Galician coast in water depths of 650–1500 m in 2006–2007 and 2009–2011) and the Avilés Canyon (north of the coast of Asturias in waters with depths between 140 m and 4700 m in 2009–2011) (Llavona et al. 2011).

The NGO CEMMA has continued carrying out opportunistic and dedicated surveys. Observations onboard fishing vessels in 2009–2011 in Northwest Spanish shelf waters resulted in three harbour porpoise sightings among 80 cetacean sightings (López et al. 2011, CEMMA 2012).

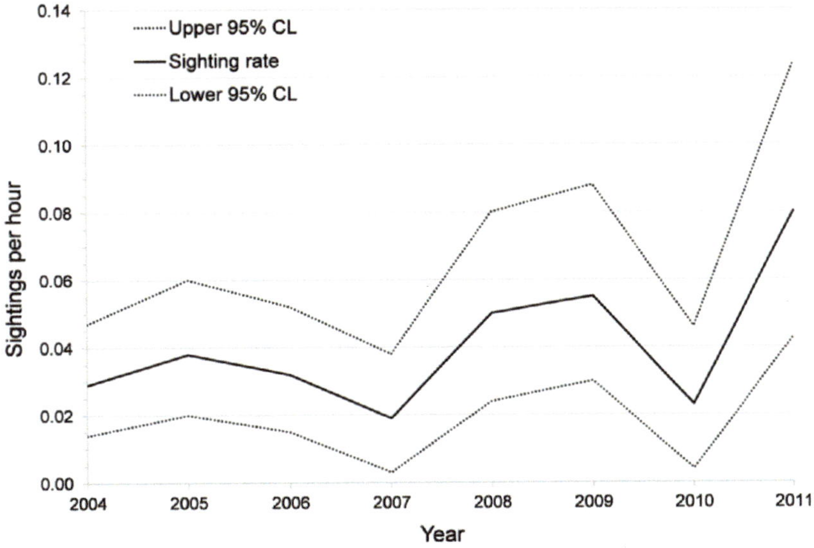

Figure 7 Annual average encounter rates (porpoise sightings per hour) from land-based surveys along the Galician (Northwest Spain) coast between January 2004 and December 2011, across all observation points, with 95% confidence limits (CL, dashed lines). Data collection took place monthly from at least 30 sites. (Adapted from Llavona Vallina 2018.)

Dedicated boat surveys in Galicia in September 2015, which covered an area of nearly 4000 km^2 from the coast (including interior waters of the Rías) to the 200 m isobath, registered eight harbour porpoise sightings among 133 cetacean sightings (Martínez-Cedeira et al. 2015). Aerial surveys carried out in September in 2015 and 2020 in the Rías Baixas (southwest Galicia), again extending to the 200 m isobath, registered five sightings and one sighting of harbour porpoise, respectively, out of 111 and 93 total cetacean sightings (Martínez-Cedeira et al. 2015, 2021). Surveys of coastal waters from the Ría de Vigo to Fisterra during 2018–2020 resulted in eight porpoise sightings among 216 cetacean sightings (A. López, Pers. Comm.).

Boat-based surveys carried out by the Bottlenose Dolphin Research Institute (BDRI) between 2014 and 2021 highlighted that the continental shelf of southern Galicia is a hotspot for harbour porpoises (Díaz López & Methion 2018, Díaz López et al. 2022). During 195 daily dedicated surveys extending to the 1400 m isobath, 287 groups of harbour porpoise were recorded throughout the southern Galician continental shelf, including the Ría de Arousa (Figure 8). All harbour porpoises were observed within the Ría de Arousa and outside the Ria over the continental shelf, in waters with a minimum depth of 2 m and a maximum depth of 231 m (mean depth = 84 m, SE = 2.8 m). The encounter rate inside the Ría de Arousa (0.05 sightings per hour from 1055 hours of observation) was markedly lower than the sighting rate from continental shelf waters outside the Ria (0.42 sighting per hour from 562 hours of observation). A small proportion of survey effort (13 hours) took place beyond the shelf edge but there were no porpoise sightings in these offshore waters.

Along the southern Galician continental shelf, porpoise group size ranged from single individuals to groups of 46 (mean = 4.8 individuals per group, SE = 0.3). The sighting rates were notably higher in summer and autumn than in winter or spring, which may be related to the calving period (see section 'Life history') as well as to high seasonal productivity due to coastal upwelling (Díaz López & Methion 2018) (see Table 2).

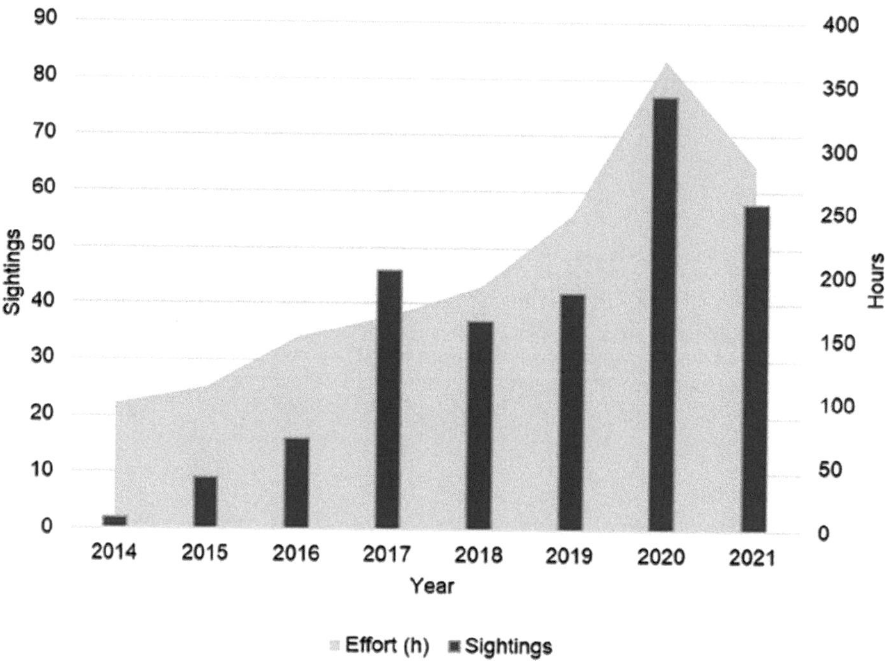

Figure 8 Annual numbers of sightings of groups of porpoises (histogram bars) and survey effort in hours (shaded area) from BDRI boat-based surveys along the Southern Galician coast (Northwest Spain) between April 2014 and November 2021. Data collection took place monthly and year-round. (Adapted from Díaz López et al. 2022.)

Table 2 Seasonal Distribution of Harbour Porpoise Sightings and Group Size

Season	Effort (hours)	No. of Sightings	SPUE (sightings/hour)	Mean Group Size±s.e	No of Individuals	SPUE (adults/hour)
Winter	235	11	0.05	4.1±1.9	45	0.19
Spring	414	42	0.10	3.3±0.4	140	0.34
Summer	736	176	0.24	5.1±0.4	889	1.21
Autumn	245	58	0.24	5.2±0.7	309	1.26
Whole year	1630	287	0.18	5.5±0.3	1421	0.87

Source: Adapted from Díaz López et al. (2022).

Porpoise sightings appear to be rare off southern Spain in the Gulf of Cádiz, with only seven sightings registered in the area (Sociedad Española de Cetáceos 2006), although strandings data confirm the regular presence of porpoises in the Gulf of Cádiz (see section 'Distribution and abundance - Strandings'). Porpoises are generally absent from the Strait of Gibraltar and the western Mediterranean Sea (Frantzis et al. 2001), although two opportunistic sightings were recorded in 2006 in coastal waters off Málaga (Sociedad Española de Cetáceos 2006). There is evidence of increased porpoise presence in the Málaga area since 2011, with a total of over 30 sightings of porpoises from tourist vessels (Samantha Blakeman, Pers. Comm.; see also Pierce et al. 2024).

Portugal

Annual aerial surveys of Portuguese coastal waters were carried out during 2011–2015, in September or October, leading to an overall abundance estimate of 2254 (95% CI = 1287–3949) (Torres-Pereira et al. 2022). There were between 6 and 16 sightings (8–25 individuals) per survey, and the resulting annual abundance estimates ranged from 1196 animals (95% CI = 456–3135) in 2011 to 3207 animals (95% CI = 1531–6718) in 2013. The authors highlighted a sharp drop in estimated abundance between 2013 and 2014 (a 48% decrease to 1653 animals (95% CI = 717–3809)). However, marked increases were seen from 2014 to 2015 (a 30% increase to 2147 (95% CI = 923–4997)) and from 2011 to 2012 (a 150% increase to 2416 (95% CI = 1338–4363)). Considering the small numbers of sightings on which these estimates are based, these changes are unlikely to reflect real changes in abundance and may, for example, indicate changes in distribution or simply stochastic variation in the number of animals seen at the sea surface. Partial results from these surveys were previously reported by Santos et al. (2012) and ICES (2014b).

Aerial surveys between 2008 and 2011 confirmed an important area of porpoise occurrence between Porto and Peniche, especially between Figueira da Foz and Nazaré, i.e., immediately to the south of the area of peak strandings (Vingada et al. 2011). Predicted occurrence maps generated using data from aerial surveys along the Portuguese coastline during 2010–2015 suggested that there was greater year-to-year fluctuation in occurrence in southern Portugal than in the central and northern areas (Araújo et al. 2015; see also Torres-Pereira et al. 2022).

Brito et al. (2009) compiled information on opportunistic cetacean sightings and sightings during boat-based surveys off west-central Portugal during 2002–2008. One porpoise sighting was reported among 45 opportunistic sightings of cetaceans, and no porpoises were reported among a further 45 cetacean sightings during surveys. Wise et al. (2007) recorded the presence of cetaceans in the vicinity of purse seine fishing operations and mentioned a single sighting of two porpoises off Figueira da Foz (south of Porto). During 2009, 22 porpoise sightings were recorded from a whale-watching boat along the western Algarve coast of southern Portugal (Cape São Vicente to Lagos) (Castro 2010). The Cetus project has recorded cetacean sightings from cargo ships between mainland Portugal (and occasionally mainland Spain), the Macaronesian archipelagos and West Africa since 2012. Among over 4200 cetacean sightings logged between 2012 and 2022, based on approximately 166,600 km of survey effort, there were nine sightings of harbour porpoise (Correia et al. 2019, A. Gil Pers. Comm.). Eight of these nine sightings, involving 22 individuals, were over the continental shelf of the mainland coast: one off Caracavelos at the mouth of the Tagus estuary (near Lisbon), six in the vicinity of Porto and one around 50 km northwest of Fisterra on the Galician coast.

A year of shore-based monitoring (2012) from Cabo Mondego in central Portugal produced 31 porpoise sightings, comprising 103 animals (Pereira 2015). Land-based surveys at the Douro River mouth (Porto) in northern Portugal during 2017 resulted in 22 porpoise sightings, with a maximum group size of 4 animals, and included repeated sightings of a leucistic animal (Figure 9A), both alone and in a group; on several occasions, the animals were apparently feeding (Gil et al. 2019). The above-mentioned leucistic individual was spotted with a calf on 28 July 2019 (Figure 9B) but on 20 August of the same year, it was sighted alone. It was last seen in Porto in August 2020, and it was found dead near Aveiro at the end of the month (and found to have been pregnant) (A. Torres-Pereira, Pers. Comm.). There is thus strong evidence that the porpoises that occur at the mouth of the Douro River show some site fidelity and use the area as a breeding and feeding ground. These land-based surveys have continued, involving a total of over 745 hours of observation, with 157 porpoise sightings, up to July 2023. Another leucistic individual was sighted and photographed near the Berlengas islands (off Peniche) (E. Silva, Pers. Comm.) (Figure 9C). A dataset on mammal occurrences in Portugal published in 2022 included 284 opportunistic sightings of harbour porpoise along the southern Portuguese coast during 2009–2019 by Associação para a Investigação do Meio Marinho (Grilo et al. 2022).

Figure 9 (A) The leucistic porpoise seen regularly around the mouth of the river Douro, Porto, Portugal (photograph by Tara Callahan, © the Cetus project), (B) Mother-calf pair of harbour porpoises sighted in the mouth of Douro river. Leucistic adult and normally-coloured calf (photography by Daniela Castelo), and (C) two porpoises, including a leucistic individual, photographed near the Berlengas Islands, off Peniche in Portugal. (Photograph by Elisabete Silva. Photographs reproduced with permission.)

There have been very few porpoise sightings reported from the Portuguese Macaronesian islands. The occurrence of harbour porpoise in Madeiran waters is based on a questionable record of a small cetacean caught in tuna fishing gear in 1905 for which no detailed description is available (Freitas et al. 2012). A total of 3564 hours of observation during 1999–2001 in the Azores, from land stations and during cruises covering the entire archipelago, coupled with use of a porpoise click detector, resulted in no visual or acoustic evidence of the presence of porpoises (Silva et al. 2001, Barreiros et al. 2006). The above-mentioned Cetus project recorded no porpoise sightings in the vicinity of Madeira or the Azores, although there was a sighting of between 20 and 40 individuals 15.7 km north of the island of Sal in the Cape Verde archipelago in waters of 2150 m depth. These were presumably animals from the Northwest African population (Correia et al. 2019, A. Gil, Pers. Comm.). Correia (2020) commented that the species is difficult to spot from a high observation platform like a cargo ship, especially in oceanic waters. Reiner et al. (1996) compiled published sightings records and their own sightings during the early 1990s for the Cape Verde Islands reporting no sightings of porpoises (although there were published records for this species in Mauritania and Senegal).

Strandings

The distribution of reported cetacean strandings in space and time will reflect the distribution of deaths at sea (presumably closely related to the distribution of living animals), as well as the prevailing winds and currents, carcass integrity and buoyancy, the presence of reporting networks and the distribution of search effort along the coast. Carcass drift models have been developed for the French coast in the Bay of Biscay (Peltier et al. 2012), in principle allowing the origin of stranded cetaceans to be inferred but this approach is not yet routinely available for the Iberian Peninsula. In Galicia, there is evidence that some carcasses may have originated in Portuguese waters, reflecting the prevalence of southwesterly winds (Saavedra et al. 2017). While porpoise strandings are reasonably common along the mainland Atlantic coast of the Iberian Peninsula, records of stranded porpoises are extremely rare in the Macaronesian archipelagos.

Spain

Lens (1997) included a table of documented porpoise strandings in Spain during 1978–1994. The great majority (86 out of 100) occurred along the western Galician coast, with a further 12 along the Biscay coasts of Galicia, Asturias, Cantabria and the Basque country. Note that four out of five records of porpoises stranded in the vicinity of A Coruña in Galicia in April 1981 (Rey & Cendrero 1982) were missed out. From 1990 to 2013, porpoises were the third most frequently recorded cetacean species in strandings in Galicia, with a total of 193 individuals recorded. There was a clear seasonal pattern, with strandings being most numerous in late winter to early spring, but no apparent long-term trend (López et al. 2002, Saavedra et al. 2017).

During 2008–2020, 55 stranded porpoises were recorded along the Atlantic coast of Andalucía. Of these, 13 were stranded during 2019–2020 (Anon. 2019, 2020, 2021). Rojo-Nieto et al. (2011) mentioned two harbour porpoises stranded in Cádiz (in 2004 and 2008) among a total of 1198 marine mammal strandings along the southern coast of Spain and the North African coast (between the Spanish enclaves of Ceuta and Melilla) during 1991–2008. Cabezón et al. (2004) reported on the parasites of a porpoise stranded on the Mediterranean coast of Andalucía. Bellido et al. (2006) reported a porpoise floating (alive) off the coast and later stranded (dead) at Benalmádena (Málaga), noting that this was the first porpoise reported stranded in Alboran Sea since the individual stranded in 1981 and reported by Rey and Cendrero (1982). Three stranded porpoises were reported from the Mediterranean coast between Gibraltar and the Alboran Sea between 2008 and 2022 (Anon. 2022a)[3], although the only individual stranding records from this area during that period appear to be from 2009 and 2010 (Anon. 2009a, 2010). There were no records of strandings of harbour porpoises along the French Mediterranean coast during 1970–1994 (Collet 1995). Kaddouri et al. (2023) reported three porpoise strandings on the northwest coast of Morocco during 2016–2021, including one from the Strait of Gibraltar. It is unclear whether these animals would have been from the Iberian population or the African population. The only known stranding of a porpoise in the Canary Islands occurred in 2006 on the Island of Fuerteventura (V. Martín, Pers. Comm., Díaz-Delgado et al. 2018, Puig-Lozano et al. 2020).

Portugal

As previously mentioned, Sequeira and Ferreira (1994) reported a decline in porpoise strandings between 1977 (when strandings monitoring began in Portugal) and the late 1980s. According to Sequeira (1996; see also Sequeira et al. 1996, 1997), 86% (69) of the 80 porpoise strandings in

[3] The same source gives the total for the Atlantic and Mediterranean coasts together as 57, so the numbers do not add up.

Portugal recorded between the nineteenth century and 1994 occurred along the northern and central zones of the west coast, especially around Aveiro and Figueira da Foz (where 67% of strandings occurred). Strandings were thus relatively infrequent south of Lisbon and along the south coast. There was a clear seasonal pattern with more strandings during January to April. Between 2000 (when a new strandings network was established in central and northern Portugal) and 2016, 4021 stranded cetaceans were reported from the Portuguese coast. On average, 17 porpoises were stranded on Portuguese mainland coasts each year, with a generally increasing trend between 2002 and 2014 (43 stranded individuals were recorded in the latter year) (Vingada & Eira 2018). We have no information on search effort, but a gradual increase in effort since the establishment of the network is plausible and might explain the trend in numbers reported. Strandings data on porpoises were also compiled by Gonçalves de Sousa (2010) for the entire mainland coast of Portugal during 1979–2009, who noted that there were 251 porpoise strandings recorded along the Portuguese mainland coast, making up 7.13% of cetacean strandings during that period. Ângelo (2020) compiled strandings data for the west coast of Portugal during 2010–2019 and referred to 278 porpoise strandings during this period. Torres-Pereira et al. (2023a) summarised porpoise strandings in Portugal during 2000–2020, noting an increase in numbers stranded over the years, especially in 2015–2020, and that most strandings occurred in April to June.

In southern Portugal, the regional network established in 2010 recorded on average three to four porpoises strandings per year up to 2019 and since then a drastic decrease in numbers (a single-stranded porpoise was reported in the Algarve from 2020 to 2022) (A. Marçalo, Pers. Obs.). Compiling this information, there are some minor discrepancies in annual totals (so we used the highest values) but the time series clearly illustrates a marked increase in the number of strandings during the 2010s (Figure 10).

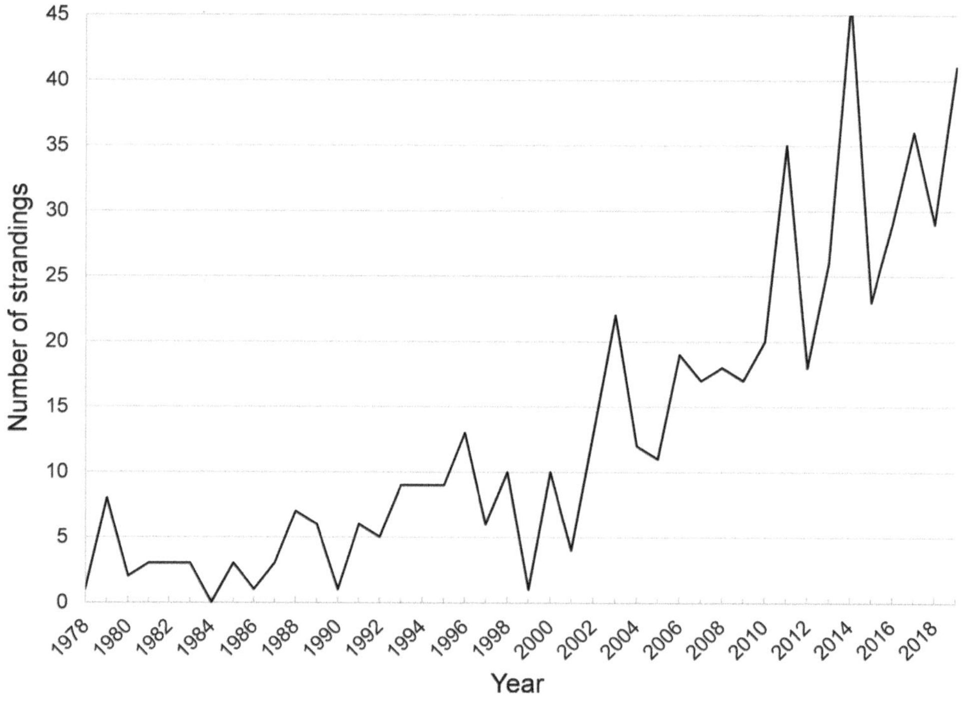

Figure 10 Number of porpoise strandings per year registered along the mainland coast of Portugal from 1978 to 2019. Compiled from Gonçalves de Sousa (2010), Vingada and Eira (2018), Ângelo (2020) and unpublished data from the southern strandings network (A. Marçalo pers. comm.).

In 2004, the stranding of an adult female was recorded on the island of Terceira in the Azores (Barreiros et al. 2006). The origin of this animal remains unknown.

Life history

The most up-to-date life history parameter data for porpoises from the Iberian Peninsula concern the area from the northern limit of Galicia (Northwest Spain) (43°3′N, 7°2′W) to north-central Portugal as far south as Nazaré (39°5′N, 9°2′W) and derive mainly from individuals stranded between 1990 and 2010. A few bycaught animals brought to land during this period were also included (Read et al. 2013, 2020, Read 2016). We refer to this region throughout this section as the Northwest Iberian Peninsula (NWIP). Some historical information for Portugal is available in Sequeira (1996) for the period 1981–1994. In Tables 3A and 3B, published information on the life history parameters of both sexes is summarised for various populations of harbour porpoises in the North Atlantic. Similar (but less extensive) tables may be found in Lockyer and Kinze (2003) and Read (2016). Information from the latter tables was included only when it could be checked against the original source.

A degree of caution is needed when interpreting these life history results. The maximum age of animals in a population will be inversely related to the mortality rate but the likelihood of detecting the oldest animals will be lower if sample size is smaller. Most recent studies on the life history of small cetaceans are based on stranded animals. Given that strandings monitoring networks usually work with limited resources, the oldest age that is determined may also depend on the sample selection strategy. Determining sexual maturity in females should be straightforward although it is important to examine both ovaries for the presence of a *corpus albicans* or *corpus luteum* (even if in harbour porpoise, ovulation and embryonic implantation are normally associated with the left ovary and left uterine horn, respectively) (Harrison 1971, Karakosta et al. 1999). In males, the assessment of maturity is normally based on the histological examination of the testes although Lockyer (1995b) proposed the use of a combined testis weight >200 g as an indicator of maturity.

Age estimation is usually based on counting growth layer groups in teeth, and it is usual for tooth sections to be read by at least two persons (Read et al. 2018) but some studies have used body length as a proxy for age and their conclusions about age at maturity should be treated with caution. To reliably estimate the age and length at sexual maturity (normally taken to be the age and length at which 50% of individuals are mature), a good sample size is needed for animals of ages and sizes closest to the values at which they reach maturity. Evidently, this may be difficult to achieve, because age and size at maturity are initially unknown and, in any case, sample size may be limited by the availability of stranded animals and/or the resources needed to process them. Hence, the overall sample size is not a good indicator of the accuracy and precision of the estimates that can be obtained. In addition, the resulting estimates may depend on what model (if any) is fitted (e.g. logistic regression has an inbuilt symmetry assumption which can drive the resulting estimate) and whether it is fitted to individual data or data that have been grouped by age or length class. Confidence limits and sample sizes are not always reported. Some authors (e.g. Ólafsdóttir et al. 2003) have reported several alternative metrics for Length and Age at Sexual Maturity (LSM and ASM, respectively), including use of the mean age (plus one) of pubertal animals (both sexes), the mean length and age of first-time ovulators, the estimated age at which the number of ovarian scars (corpora albicantia plus corpus luteum) first exceeds zero (based on a linear regression of the total number of scars versus age), and the DeMaster method (DeMaster 1978), which uses the mean proportion of mature animals across all age classes, corrected for the decline in abundance of older age classes (see also DeMaster 1984 for a review of methods).

Assessment of age at physical maturity in cetaceans should ideally be based on the fusion of epiphyseal plates in the vertebrae (e.g. Ridgway & Harrison 1999), but in practice it is often based on fitting a growth curve to determine the asymptotic body length (e.g. Murphy et al. 2009). While it is clear from some studies (e.g. Mead et al. 1982) that growth in cetaceans continues beyond physical

Table 3A Life History Parameters of Female Harbour Porpoises in the Atlantic Ocean

Area and Time Period	Source	Maximum Length (cm)	Maximum Age (yrs)	Sexual Maturity Length (cm)	Sexual Maturity Age (yrs)	Asymptotic Length (cm)	Age at Asymptotic Length (yrs)	Pregnancy Rate (Presence of Foetus)
NWIP (1990–2010)	1	202 (N=125)	18 (N=71)	169 (N=60)	5.5 (N=60)	185 (N=60)	10 (N=60)	0.54 (N=13)
Atlantic Spain (1978–1994)	2	202 (N=32)	n/a	n/a	n/a	n/a	n/a	n/a
Galicia, NW Spain (1990–1999)	3	202 (N=35)	9	166 (N=35)	3	n/a	n/a	n/a
Portugal (1981–1994)	4	208 (N=22)	n/a	n/a	n/a	n/a	n/a	n/a
Portugal (2002–2016)	25	219 (77)	–	168.9 (N=77)	–	–	–	0.68[b]
Portugal (2017–2019)	26	199 (30)	–	–	–	–	–	
Scotland (1992–2005)	5	173 (N=289)	20 (N=170)	138.8 (CI=135.9–141.6) (N=190)	4.35 (CI=3.93–4.71) (N=144)	GAM: 158.4 (SE=2.69, CI=153.1–163.8) Gompertz: 160.7 (CI=157.7–163.8cm) (N=170)	12 (N=170)	i. 0.34 (CI=0.17–0.52) (N=29) ii. 0.40 (CI=0.26–0.55) (N=42)
British Isles (1985–1994)	6	189 (N=128)	22 (N=96)	n/a	n/a	160 (N=96)	8 (N=96)	n/a
England (North Sea MU) (1990–1999)	7	172 (N=79)	22 (N=72)	138.90 (SE=1.46) (N=75)	3.8 (SE=0.23) (N=68)	155.37 (SE=1.95) (N=72)	7.21 (N=72)	0.26 (N=19)
England (North Sea MU) (2000–2013) [a](2000–2012)	7	180 (N=109)	15 (N=51)	139.18 (SE=1.44) (N=90)	4.8 (SE=0.31) (N=49)[a]	155.37 (SE=1.95) (N=51)[a]	11.66 (N=51)[a]	0.30 (N=23)
England and Wales (Celtic and Irish Seas MU (1990–1999)	7	191 (N=127)	15 (N=107)	146.56 (SE=1.71) (N=121)	3.8 (SE=0.23) (N=102)	162.94 (SE=1.95) (N=104)	7.21 (N=104)	0.68 (N=25)
England and Wales (Celtic and Irish Seas MU) (2000–2013) [a](2000–2012)	7	189 (N=269)	21 (N=88)	146.94 (SE=1.32) (N=199)	4.8 (SE=0.31) (N=86)[a]	162.94 (SE=1.95) (N=87)[a]	11.66 (N=87)[a]	0.54 (N=35)
Ireland (2001–2003)	8	175 (N=27)	11 (N=21)	S coast: >150 Irish Sea: >140	S coast: >5 Irish Sea: 3.5–5	n/a	n/a	0.4 (N=5)

Table 3A (*Continued*) Life History Parameters of Female Harbour Porpoises in the Atlantic Ocean

Area and Time Period	Source	Maximum Length (cm)	Maximum Age (yrs)	Sexual Maturity Length (cm)	Sexual Maturity Age (yrs)	Asymptotic Length (cm)	Age at Asymptotic Length (yrs)	Pregnancy Rate (Presence of Foetus)
Denmark (1838–1998) [a](1996–1998)	9	189	23	143[a]	3.5 (*N*=25)[a]	160	n/a	n/a
Germany (Schleswig Holstein - North Sea) (1990–2016)	10	n/a	22 (*N*=311)	n/a	4.95 (95% CI= 4.2–4.8) (*N*=311)	n/a	n/a	n/a
Germany (Schleswig Holstein - western Baltic Sea) (1990–2016)	10	n/a	22 (*N*=215)	n/a	4.95 (CI= 4.2–4.8) (*N*=215)	n/a	n/a	n/a
The Netherlands (1955–1976)	11	186 (*N*=62)	12 (*N*=34)	150 (*N*=19)	6 (*N*=18)	~150 (*N*=34)	~6 (*N*=34)	n/a
The Netherlands (2001–2003)	8	160 (*N*=19)	12 (*N*=14)	>130 (*N*=19)	5 (*N*=14)	n/a	n/a	0.11 (*N*=9)
The Netherlands (2006–2019)	12	175 (*N*=337)	24 (*N*=154)	n/a	4 (CI=3.47–4.48) (*N*=154)	n/a	n/a	0.34 (CI= 0.26–0.43) (*N*=119)
France (Bay of Biscay) (1970–1994)	13	186	n/a	n/a	n/a	n/a	n/a	n/a
France (French English Channel) (1970–1994)	13	190	n/a	n/a	n/a	n/a	n/a	n/a
France (S North Sea, W and E Channel, and Bay of Biscay combined) (2001–2003)	8	192 (*N*=14)	24 (*N*=9)	n/a	n/a	n/a	n/a	n/a
France (North Sea MU) (1990–2015)	14	168	20 (*N*=130)	n/a	3.4 (*N*=130)	n/a	n/a	n/a
France (Celtic Sea MU) (1990–2015)	14	196	24 (*N*=81)	n/a	2.5 (*N*=81)	n/a	n/a	n/a

(*Continued*)

Table 3A (*Continued*) Life History Parameters of Female Harbour Porpoises in the Atlantic Ocean

Area and Time Period	Source	Maximum Length (cm)	Maximum Age (yrs)	Sexual Maturity Length (cm)	Sexual Maturity Age (yrs)	Asymptotic Length (cm)	Age at Asymptotic Length (yrs)	Pregnancy Rate (Presence of Foetus)
West Greenland (1988–89, 1995)	15	166 (*N*=71)	14 (*N*=84)	142 (*N*=71)	3.6	154±2.6	n/a	n/a
West Greenland (2009)	16	n/a	n/a	n/a	3.5 (SE=0.03) (*N*=60)	n/a	n/a	n/a
Norway (1988–1989)	17	168 (*N*=35)	>8 (*N*=35)	n/a	4 (*N*=35)	155.9 (*N*=35)	n/a	n/a
Norway (mainly northern Norway) (2016–2017)	18	173 (*N*=58)	7 (*N*=48)	n/a	4.3 (CI±0.6) (*N*=48)	165.2 (CI= 155.7–176.9) (*N*=58)	n/a	0.85 (CI±0.16) (*N*=20)
Iceland (1991–1997)	19	174 (*N*=474)	20 (*N*=354)	1. 138 (*N*=72) 2. 146 cm 3. 147.6 cm (*N*=30)	1. 2.5 (*N*=62) 2. 3.2 (*N*=269) 3. 2.81 (*N*=21) 4. 2.1 (*N*=51) 5. 4.4 (*N*=293)	160.1 (*N*=314)	n/a	0.98 (*N*=74)
Gulf of Maine/Bay of Fundy (1989–1993) [a](1985–1993)	20, 21[a]	168 (*N*=203)[a]	17 (*N*=109)	n/a	3.36 (SD=0.13) (*N*=99)	158 (CI±1.56) (*N*=203)[a]	~7 (*N*=203)[a]	0.93 (*N*=14)
Bay of Fundy (1969–1973)	22	>165 (*N*=44)	9 (*N*=44)	147 (CI±1.7) (*N*=46)	3.97 (CI± 0.49) (*N*=37)	163 (CI±8.4) (*N*=44)	n/a	0.89 (CI± 0.14) (*N*=19)
Bay of Fundy (1985–1988)	22	~163 (*N*=116)	10 (*N*=116)	143 (CI±0.83) (*N*=108)	3.44 (CI±0.36) (*N*=108)	155 (CI±3.5) (*N*=116)	~6–7 (*N*=116)	0.86 (CI±0.09) (*N*=50)
Newfoundland (1990–1991)	23, [a]24	162 (*N*=35)[a]	9 (*N*=35)	146.4 (SE=0.03) (*N*=32)[a]	3.1 (SE=0.07) (*N*=32)[a]	156.3 (SE=2.9) (*N*=33)	>4 (*N*=33)	0.76 (SE=0.1) (*N*=17)[a]

Table 3B Life History Parameters of Male Harbour Porpoises in the Atlantic Ocean

Area and Time Period	Source	Maximum Length (cm)	Maximum Age (yrs)	Sexual Maturity Length (cm)	Sexual Maturity Age (yrs)	Asymptotic Length (cm)	Age at Asymptotic Length (yrs)
NWIP (1990–2010)	1	189 (*N*=135)	19 (*N*=77)	151 (*N*=47)	3.8 (*N*=47)	162 (*N*=47)	10 (*N*=47)
Atlantic Spain (1978–1994)	2	176 (*N*=27)	n/a	n/a	n/a	n/a	n/a
Galicia, NW Spain (1990–1999)	3	180 (*N*=42)	9	155 (*N*=9)	5	n/a	n/a
Portugal (1981–1994)	4	175 (*N*=15)	n/a	n/a	n/a	n/a	n/a
Scotland (1992–2005)	5	170 (*N*=320)	20 (*N*=176)	132.2 (CI=129.1–135.6) (*N*=145)	5 (CI=4.03–5.88) (*N*=105)	GAM: 147.2 (SE=1.66, CI=143.95–150.5) Gompertz: 148.3 (CI=146.3–150.0 (*N*=180)	12 (*N*=180)
British Isles (1985–1994)	6	163 (*N*=144)	24 (*N*=114)	>130 (*N*=114)	>3 (*N*=114)	145 (*N*=114)	8 (*N*=114)
England (North Sea MU) (1990–1999)	7	153 (*N*=89)	18 (*N*=83)	133.27 (SE=1.33) (*N*=64)	3.56 (SE=0.25) (*N*=62)	140.94 (SE=1.64) (*N*=83)	5.72 (*N*=83)
England (North Sea MU) (2000–2013) [a](2000–2012)	7	161 (*N*=146)	16 (*N*=50)	129.47 (SE=1.29) (*N*=97)	3.62 (SE=0.26) (*N*=45)[a]	140.94 (SE=1.64) (*N*=49)[a]	7.62 (*N*=49)[a]
England and Wales (Celtic and Irish Seas Management Unit) (1990–1999)	7	171 (*N*=129)	18 (*N*=110)	138.73 (SE=1.50) (*N*=92)	3.56 (SE=0.25) (*N*=78)	146.50 (SE=1.60) (*N*=109)	5.72 (*N*=109)
England and Wales (Celtic and Irish Seas Management Unit) (2000–2013) [a](2000–2012)	7	181 (*N*=271)	15 (*N*=84)	133.46 (SE=1.24) (*N*=164)	3.62 (SE=0.26) (*N*=66)[a]	146.50 (SE=1.60) (*N*=83)[a]	7.62 (*N*=83)[a]
Ireland (2001–2003)	8	157 (*N*=17)	7.5 (*N*=14)	S and W coasts: 134–144 Irish Sea: 131–146	S and W coasts: 3–7 Irish Sea: 4–8	n/a	n/a
Denmark (1838–1998) [a](1996–1998)	9	167	23	135 (*N*=135)[a]	3–4 (*N*=135)[a]	145	n/a
Germany (Schleswig Holstein - North Sea) (1990–2016)	10	n/a	n/a	n/a	n/a	n/a	n/a

(*Continued*)

Table 3B *(Continued)* Life History Parameters of Male Harbour Porpoises in the Atlantic Ocean

Area and Time Period	Source	Maximum Length (cm)	Maximum Age (yrs)	Sexual Maturity Length (cm)	Sexual Maturity Age (yrs)	Asymptotic Length (cm)	Age at Asymptotic Length (yrs)
Germany (Schleswig Holstein - western Baltic Sea) (1990–2016)	10	n/a	n/a	n/a	n/a	n/a	n/a
The Netherlands (1955–1976)	11	151 (*N*=43)	12 (*N*=20)	135 (*N*=46)	5 (*N*=20)	~130 (*N*=20)	~4 (*N*=20)
The Netherlands (2001–2003)	8	n/a	n/a	n/a	n/a	n/a	n/a
The Netherlands (2006–2019)	12	n/a	n/a	n/a	n/a	n/a	n/a
France (Bay of Biscay) (1970–1994)	13	168	n/a	n/a	n/a	n/a	n/a
France (French English Channel) (1970–1994)	13	183	n/a	n/a	n/a	n/a	n/a
France (S North Sea, W and E Channel, and Bay of Biscay combined) (2001–2003)	8	165 (*N*=17)	14 (*N*=12)	n/a	n/a	n/a	n/a
France (North Sea MU) (1990–2015)	14	170	14 (*N*=164)	n/a	n/a	n/a	n/a
France (Celtic Sea MU) (1990–2015)	14	183	11.5 (*N*=89)	n/a	n/a	n/a	n/a
West Greenland (1988–89, 1995)	15	158 (*N*=81)	17 (*N*=94)	127 (123–130)	2.45	143±1.7	n/a
West Greenland (2009)	16	n/a	n/a	n/a	3.1 (SE=0.08) (*N*=29)	n/a	n/a
Norway (1988–1989)	17	147 (*N*=41)	>8 (*N*=41)	n/a	3 (*N*=41)	142.3 (*N*=41)	n/a
Norway (mainly northern Norway) (2016–2017)	18	158 (*N*=75)	12 (*N*=75)	n/a	2–3 (*N*=75)	149.0 (CI=145.4–152.8) (*N*=75)	n/a
Iceland (1991–1997)	19	165 (*N*=794)	16 (*N*=615)	1. 135.6 (*N*=33) 2. 135	1.2.9 (*N*=21) 2. 1.9 (*N*=493) 3. 2.6 (*N*=526)	149.6 (*N*=497)	n/a

Table 3B (*Continued*) Life History Parameters of Male Harbour Porpoises in the Atlantic Ocean

Area and Time Period	Source	Maximum Length (cm)	Maximum Age (yrs)	Sexual Maturity Length (cm)	Sexual Maturity Age (yrs)	Asymptotic Length (cm)	Age at Asymptotic Length (yrs)
Gulf of Maine/Bay of Fundy (1989–1993) [a](1985–1993)	20, 21[a]	157 (*N*=198) [a]	15 (*N*=123)	n/a	>3 (3–4) (*N*=31)	143 (CI±1.25) (*N*=198)[a]	~5 (*N*=198)[a]
Bay of Fundy (1969–1973)	22	~150 (*N*=56)	10 (*N*=56)	n/a	n/a	146 (CI±4.1) (*N*=56)	~7–8 (*N*=56)
Bay of Fundy (1985–1988)	22	~153 (*N*=121)	10 (*N*=121)	n/a	n/a	144 (CI±3.5) (*N*=121)	~6–7 (*N*=121)
Newfoundland (1990–1991)	23, 24[a]	155.5 (*N*=59)[a]	12 (*N*=59)	135.1 (SE=0.02) (*N*=59)[a]	3 (*N*=59)[a]	142.9 (SE=1.2) (*N*=59)	>4 (*N*=59)

Sources: 1: Read (2016), 2: Lens (1997), 3: López Fernández (2003), 4: Sequeira (1996), 5: Learmonth et al. (2014), 6: Lockyer (1995a), 7: Murphy et al. (2020), 8: Learmonth et al. (2004), 9: Lockyer and Kinze (2003), 10: Kesselring et al. (2017), 11: van Utrecht (1978), 12: IJsseldijk et al. (2021), 13: Collet (1995), 14: Rouby (2018), 15: Lockyer et al. (2001), 16: Heide-Jørgensen et al. (2011), 17: Bjørge et al. (1991), 18: Unpublished data cited in NAMMCO/IMR 2019, 19: Ólafsdóttir et al. (2003), 20: Read and Hohn (1995), 21: Read and Tolley (1997), 22: Read and Gaskin (1990), 23: Richardson et al. (2003), 24: Richardson (1992), 25: Camarão (2017), 26: Oliveira (2020).

Notes: Camarão (2017) weighed and measured ovaries of 77 porpoises but it is not stated how many of these animals were mature and the pregnancy rate may have been based on a sub-sample of the mature animals. Learmonth et al. (2014) used two different models to estimate asymptotic length, a Generalised Additive Model (GAM) and a Gompertz model. Lockyer and Kinze (2003) assembled a range of different data sources, the earliest dating back to 1838, but it is not always clear which animals contributed to life history parameter estimates. Rouby (2018) used a regression model with Weibull distribution to estimate ASM. The two pregnancy rates cited were based on excluding mature females stranded during (i) April to September and (ii) 26 May to 12 September. Ólafsdóttir et al. (2003) reported several estimates of LSM and/or ASM, some applicable to both sexes: (1) based on the mean age plus one of pubertal animals, (2) the length or age at which 50% of animals are mature, (3) mean length at first ovulation, (4) linear regression method to estimate the age at which the number of ovarian scars first exceeds zero, (5) the DeMaster method (DeMaster 1978), based on the mean of the proportion of mature animals in each age class, corrected for mortality. For some estimates, standard deviation (SD), standard error (SE) and/or 95% confidence intervals (CI) were available and are reported here. Sample sizes (*N*) are reported when available.

NWIP, northwest Iberian Peninsula; S, south or southern; W, west; E, East; MU, management unit; n/a, not available.

[a] Results deriving from a different time period or separate study as detailed in columns 1 or 2.

[b] This is the proportion of mature females with a corpus luteum in the ovaries. The sample size is unknown.

maturity, there appears to be no consensus in the literature about the relationship between asymptotic size and size at physical maturity, with the latter being variously equated to 90%, 95% or (by default) 100% of the former (Aguilar & Lockyer 1987, Pribanić et al. 2000, Murphy et al. 2020, Betty et al. 2022). In addition, fitting different types of growth curve may result in different estimates of asymptotic length (e.g. Richardson et al. 2003, Betty et al. 2022). Most authors used a Gompertz model (although there are different variants of this) and a few used a von Bertalanffy model (e.g. Ólafsdóttir et al. 2003). Where age at physical maturity is reported, it is usually age at asymptotic length; where it was not reported but the growth curve was provided, we have estimated it from the growth curve.

Pregnancy rate may be underestimated if animals that died during the conception period are included in the calculations, because the foetus, if present in such animals, will be very small and hence difficult to detect. A further issue with using data from stranded animals is that pregnancy rate is likely to be underestimated due to the high proportion of sampled females that were in a poor state of health. Where sample sizes permit, this bias can be avoided by basing estimates of pregnancy rate on animals that died due to physical trauma (e.g. fishery bycatch, collisions with boats or fatal interactions with bottlenose dolphins), on the assumption that their health state is more likely to have been representative of the living population. An additional issue with annual pregnancy rate values is that some authors divide the proportion of mature females that are pregnant by the length of the gestation period, typically 10–12 months in porpoises, although where the seasonality of breeding is consistent (as found by Learmonth et al. 2014 in Scotland), implying that the breeding cycle effectively lasts 1 year, such an adjustment is not needed (Read 2016). One possible means to ensure that observed differences are not artefacts of differing methodologies would be to undertake new combined analyses of raw data from different Assessment Units (see ICES 2014a), if available, as done by Murphy et al. (2020) to compare North Sea and Celtic Sea porpoises.

Size, age, growth and maturation

Maximum reported length in animals for the Iberian Peninsula is greater than that in the other populations, as are the lengths at sexual and physical maturity (see Table 3). Vaz (2015) noted that in addition to a larger body length, the cranium of porpoises from Portuguese waters was longer and wider than that of porpoises from the north of the Bay of Biscay. The large body size of Iberian porpoises is well known (e.g. Smeenk et al. 1992, Donovan & Bjørge 1995, Sequeira 1996, López Fernández 2003, Murphy et al. 2020). Maximum lengths recorded were 219 cm for females and 189 cm for males (Sequeira 1996, Read et al. 2013, Camarão 2017). Two of the three porpoises reported stranded on the northwest coast of Morocco during 2016–2021 were measured, a female of 150 cm and a male of 210 cm (Kaddouri et al. 2023). It is not clear whether these would have been African or Iberian animals and the length reported for the male is unusually large. At the opposite extreme, the maximum length for females in the Netherlands reported by Learmonth et al. (2004) was 160 cm and the maximum length reported for males in Norway by Bjørge et al. (1991) was 147 cm.

Various factors operating over different time-scales are expected to influence body size and other life history characteristics in harbour porpoises. The reproductive strategy of the males involves allocating a high proportion of available energy resources to sexual maturation and reproduction, maturing at a relatively small body size and maintaining a large testicular mass during the breeding period, while females need to reach a certain size to be able to carry their offspring. Males retain more paedomorphic features than females (Murphy et al. 2020). While the explanation for the larger body size of Iberian porpoise remains unclear, it may be relevant that their upwelling habitat is characterised by both high productivity and high variability in productivity. Murphy et al. (2020) proposed that the larger body size of Iberian porpoise could be an adaptation to periods of shortage of resources (after Perrin 1989, Ferguson & Larivière 2008, Ferguson et al. 2018). In their review of interspecific variation in body size in terrestrial mammalian carnivores, Ferguson and Larivière (2008) observed that:

species living in highly seasonal environments were associated with larger home ranges and low density that in turn selected for larger body mass and greater sexual size dimorphism. Thus, spacing behaviour provides an important evolutionary link explaining interspecific body size variation.

The maximum age recorded for porpoises in the NWIP was 18 years for females and 19 years for males (Read 2016). One animal of undetermined sex reached 21 years of age. These ages fall within the (admittedly wide) range of maximum ages reported for other areas of the North Atlantic. Maximum ages for females in the Northeast Atlantic ranged from 7 in Norway (North Atlantic Marine Mammal Commission and the Norwegian Institute of Marine Research 2019) to 24 in France (Learmonth et al. 2004), while those for males ranged from 7.5 in Ireland (Learmonth et al. 2004) to 24 in the British Isles (Lockyer 1995a). As noted above, estimated maximum age may depend on the sample size and sample selection strategy, and several of the above-mentioned studies had small sample sizes (Tables 3A and B).

As stated above, in harbour porpoise, ovulation normally occurs only in the left ovary, which tends to be larger and heavier and contain more ova (e.g. Karaskota et al. 1999, Learmonth et al. 2014). This difference has been confirmed in Iberian porpoises (Camarão 2017, Oliveira 2020). Camarão (2017) also reported that corpora albicantia were found four times as frequently in the left ovaries of mature females as in the right ovaries (and never only in the right ovary).

Female Iberian porpoises reached sexual maturity at a length of 169 cm (range of lengths for mature females: 161–202 cm; range for pregnant females: 176–202 cm) (Read et al. 2013, Read 2016). Camarão 2017 reported that mature females ranged in length from 159 cm to 219 cm, with 50% of females reaching sexual maturity at a length of 168.9 cm. The length at maturity was the highest reported for the North Atlantic, the lowest values being those reported for Greenland and various sites along the Atlantic coast of North America (all less than 170 cm). This excludes the 130 cm maximum length reported for the Netherlands by Learmonth et al. (2004), which was based on a very small sample size. Both previous and subsequent studies in the Netherlands based on larger sample sizes gave maximum length values more in line with those for nearby countries such as France, the UK and Denmark (Table 3A).

ASM for female Iberian porpoises was estimated to be around 5.5 years old (Read 2016), among the highest values reported for the North Atlantic and similar to results for the Netherlands: 5 years old according to Addink et al. (1995) and 6 years old according to van Utrecht (1978). The lowest reported estimate of ASM was 3.1 years old in Newfoundland (Richardson et al. 2003). It is not clear why maturation should be relatively late in females from the NWIP; there is no obvious selective advantage. Ideally, the veracity of this finding requires confirmation using larger sample sizes. It should be noted that, based on a much smaller sample size, López Fernández (2003) estimated the age at maturity for females in Galicia to be 3 years old. The multiple estimates of ASM (and of LSM) presented by Ólafsdóttir et al. (2003) for porpoises of both sexes in Iceland illustrate the sensitivity of estimates to the methodology used. The lack of confidence limits for many estimates is also an issue when evaluating differences between porpoises from different areas.

Length at sexual maturity for male Iberian porpoises was estimated to be 151 cm (Read 2016), with mature males ranging in length from 154 cm to 171 cm (Read et al. 2013). This is the highest value reported for the North Atlantic, with the lowest being for Greenland (127 cm, Lockyer et al. 2001). Males mature at 3.8 years old (Read 2016). This falls within the range reported elsewhere in the North Atlantic, where values range from 2.45 in west Greenland (Lockyer et al. 2001) to 5.7 in Scotland (Learmonth et al. 2014).

Growth models indicated that the asymptotic length of Iberian males was 162 cm and that of females was 185 cm (Read 2016), higher than that reported elsewhere in the North Atlantic, with the lowest values being those for the Netherlands (150 cm for females and 130 cm for males; van Utrecht 1978). The age at asymptotic length for Iberian porpoises (approximately 10 years old) was also among the highest (Read 2016; Table 3). Values for both sexes ranged from around 4 years old in the

Netherlands and Newfoundland (van Utrecht 1978, Richardson et al. 2003) to around 12 years old in Scotland (Learmonth et al. 2014); none of the studies reviewed reported true physical maturity.

Over 85% of stranded (and a few bycaught) animals in the NWIP for which age was determined were ≤10 years old, and over 60% were ≤3 years old (Read et al. 2013). Similar age distribution patterns have been observed in the Northeast Atlantic. Based on a large sample size of 645 necropsied porpoises stranded and bycaught along the English and Welsh coasts between 1990 and 2012, Murphy et al. (2020) found that around 80% were ≤5 years old and only 5% were aged 12 years or older. The relative rarity of old animals was also apparent in German waters of the North Sea and Baltic Sea, where 90% of porpoises did not live longer than 12 and 9.6 years, respectively (Kesselring et al. 2017); Danish waters, where less than 5% of porpoises lived to ages beyond 12 years (Lockyer & Kinze 2003); Dutch waters, where 75% of mature females did not live longer than 8.5 years (IJsseldijk et al. 2021a); and Icelandic waters, where up to 90% of the porpoises examined were less than 6 years old (Ólafsdóttir et al. 2003).

Life history traits can vary both over time and geographically, due to both intrinsic and extrinsic factors (e.g. Murphy et al. 2020). Thus, by obtaining more recent life history data from stranding and bycatch since the 2010s, evidence of changes in life history parameters due to pressures such as anthropogenic mortality (e.g. bycatch) may be detectable. Work is still needed to analyse more recent samples from the NWIP.

Murphy et al. (2020) compared length-related life history traits of porpoises from the Celtic and Irish Seas Management Unit (CIS MU) during 1990–1999 and 2000–2013. Porpoises from the CIS MU have a larger body size than animals from other areas in the Northeast Atlantic, which may be inherited from large-sized Iberian porpoises. The admixture zone between the two ecotypes is restricted to the northern part of the Bay of Biscay and southern parts of the Celtic Sea and English Channel (Fontaine et al. 2017). At any given age, porpoises were larger in the 1990s compared to the 2000s and 2010s, and there was also a decline in the Gompertz growth rate parameter, both trends being more evident in females. Declines were also seen in the length at sexual maturity of males and in the occurrence of the "larger-sized morphotype". The authors suggested that these changes may reflect a decrease in movements of Iberian porpoises towards the CIS MU, although acknowledging that other factors could also be at play. If correct, this could be related to the apparent decline in the abundance of Iberian porpoises between 1990 and 2015, as proposed by Ben Chehida et al. (2023) to explain the observed reduction of genetic diversity. Such a decline could have resulted in fewer animals migrating towards northern waters and hence a reduced contribution to the genetic pool of animals in the CIS MU.

Causes of death, sex ratio, age structure and mortality rate

The quality of information available on causes of death depends on several factors, including whether strandings networks have funding for necropsies, access to veterinary expertise and the frequency with which a species is stranded. Ideally, cause of death should be interpreted in the context of a complete gross, histological and histopathological examination of the carcass and sampled tissues. Publications providing guidelines for necropsy protocols and diagnosis of cause of death in cetaceans include Kuiken (1996), IJsseldijk et al. (2019) and Jauniaux et al. (2019).

Among the most important reported causes of death in porpoises in the Northeast Atlantic are bacterial and parasitic pneumonia and fishery bycatch. Typically, bycatch mortality in porpoises can be detected based on the presence of superficial incisions (net cuts), encircling imprints (mainly on the head but also on the flippers and body) and changes in the lungs (pulmonary oedema, emphysema and atelectasis), while circumstantial evidence includes recently ingested prey, reddish or bulging eyes, congestion and disseminated gas bubbles. Post-mortem injuries including stab wounds, amputations and abdominal cuts may have been the result of freeing the animal from a net or trying to sink the carcass. It is also important to be able to exclude other possible causes of death (as determined from other significant pathological findings) (Puig-Lozano et al. 2020, IJsseldijk et al. 2021b). IJsseldijk et al. (2022) reported that infectious diseases (32%) were the largest cause of death category in porpoises stranded in the Netherlands, while bycatch accounted for a further 17% of mortalities. In Sweden, fishery bycatch was the most important cause of death (31%) with diseases (mostly pneumonia) accounting for a further 21% (Neimanis et al. 2022). Cases of starvation in adults and in neonates separated from their mothers, as well as porpoise deaths caused by non-infectious diseases such as congenital anomalies, perinatal complications (dystocia and stillbirths), gastritis and neoplasms, have been described in the UK (Kirkwood et al. 1997), the Netherlands, Belgium, Germany (van Elk et al. 2019) and Sweden (Neimanis et al. 2022). Deaths due to bottlenose dolphin attacks are common in coastal waters of northeast Scotland, where there is a resident bottlenose dolphin population (Ross & Wilson 1996, Davison & ten Doeschate 2020). Acoustic monitoring of porpoise distribution in this area suggested that they tended to avoid spatio-temporal overlap with bottlenose dolphins (Williamson et al. 2022).

In the NWIP, bycatch is a commonly diagnosed cause of death in harbour porpoise (Sequeira & Ferreira 1994, Read et al. 2013, Read 2016). In a compilation of findings from necropsies of cetaceans stranded in Galicia during 2002–2003, results from three porpoises were described. One was diagnosed as bycaught, and no additional findings were mentioned. The second presented parasitic pneumonia, and the third animal had severe pulmonary congestion and pulmonary oedema with lungworms present, also several gastric ulcers and lesions consistent with bycatch (Alonso et al. 2004).

There is little published information on the sex ratio in the Iberian porpoise population. Read et al. (2013) found the sex ratio in stranded animals from the NWIP to be close to parity at 1:1.07 females to males (not significantly different from 1:1). Lens (1997) reported a slightly less even sex ratio in stranded and bycaught animals from Spain, 1.17:1, in this case with more females than males.

Examination of age data from porpoises stranded in the NWIP during 1990–2010 suggests the occurrence of more animals aged 12+ among the strandings towards the end of the study period, implying a decline in mortality rate in porpoises in the NWIP between 2000 and 2010 (Read et al. 2013; see Figure 11A). Read et al. (2013, 2020) used the age-at-death data derived from porpoises stranded in Galicia (Northwest Spain) and north plus central Portugal during 1990–2010 to construct a life table (Table 4A) and to estimate the mortality rate, as well as survivorship and life expectancy at each age (Figure 11B), following the non-parametric Kaplan-Meier approach to estimating survivorship as described in Krebs (1989). The life table also includes calculations for reproductive output, assuming a sex ratio of 1:1 and using estimated values for the proportion of females mature at each age and average pregnancy rate (see 'Life history – Reproductive seasonality, gestation period, pregnancy rate and population reproductive rate'). The estimated annual mortality rate for the entire period was 18%, which is very similar to an estimate for porpoises in Scotland during 1992–2013 (Graham Pierce and Fiona Read, unpublished data). It should be noted that the mortality rate calculated for porpoises in Scotland, also based on stranded animals, may be biased upwards because bottlenose dolphin kills are probably overrepresented in the strandings (since many such mortalities occur close to the coast in northeast Scotland, within the distribution range of the resident bottlenose dolphin population, there is a high likelihood of these animals being stranded). Considering that bottlenose dolphin kills are rarely reported in the Iberian Peninsula, despite the presence of a resident bottlenose dolphin population in Galicia, it is likely that the mortality rate for Iberian porpoises is in fact higher than that in Scotland. The use of the life table approach to estimate mortality rate is, strictly speaking, valid only if population age structure is stable, which can only occur in a stationary population (one in which deaths exactly balance births). Thus, in a declining population (which may or may not be the case here), in which the mortality rate exceeds the birth rate, the age distribution will tend to contract over time with the oldest age classes being lost. Combining data from across many years will then lead to an underestimate of the true mortality rate.

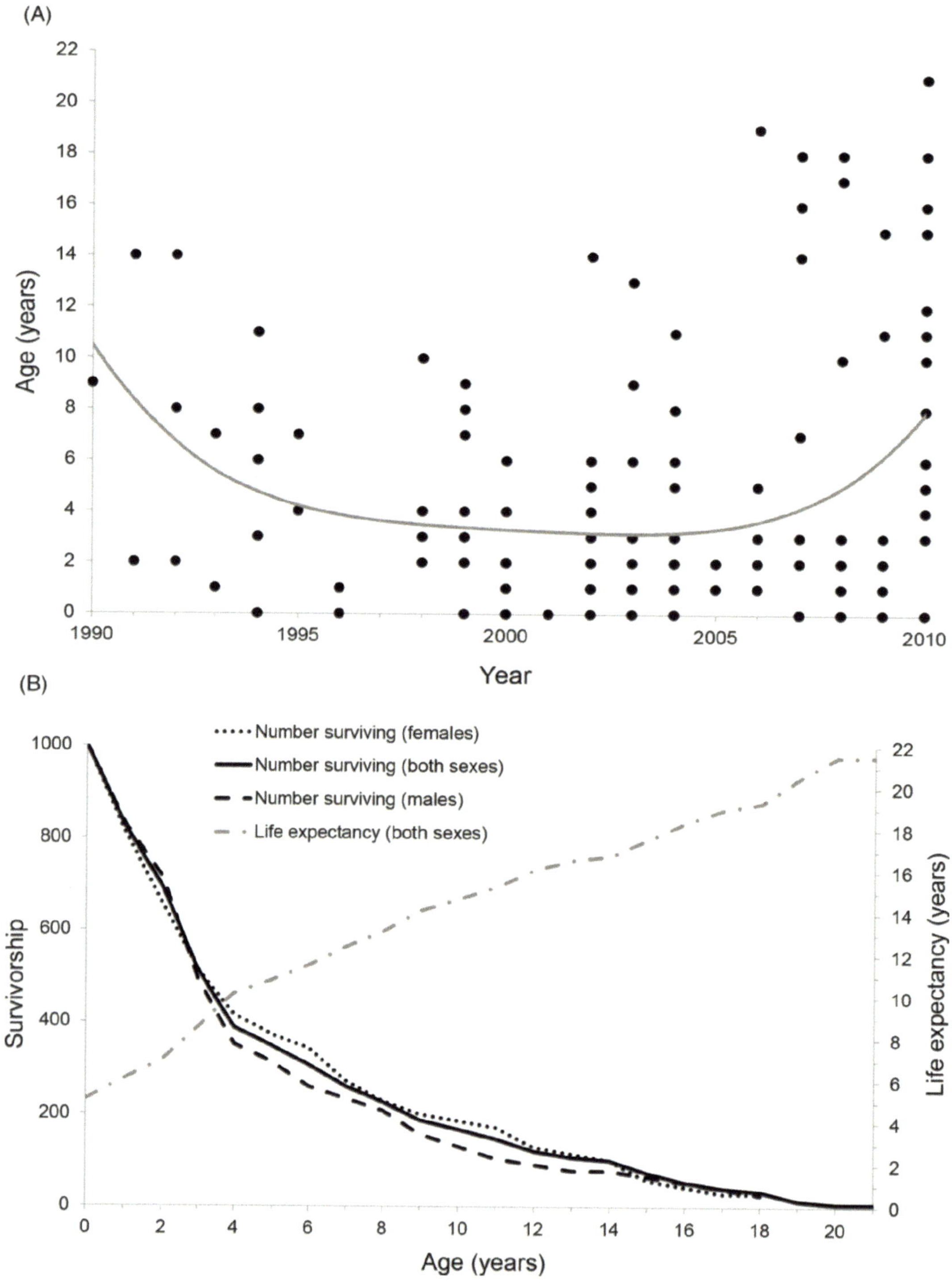

Figure 11 (A) Estimated ages of stranded porpoises from the Northwest Atlantic Iberian Peninsula 1990–2010 (N=151). The fitted smoother represents the underlying trend in average age, which is an indication of the (inverse of the) underlying mortality rate; (B) Estimated survivorship and life expectancy at age for these porpoises.

Table 4A Life Table for Iberian Porpoise, Based on Read et al. (2013) and Following Methodology Described in Krebs (1989)

								APR =	0.540
Age (x)	l_x	d_x	q_x	L_x	T_x	e_x	fem_x	$mat_{f,x}$	b_x
0	151	27	0.179	138	768	5.1	0.5	0	0.0
1	124	20	0.161	114	630	6.1	0.5	0	0.0
2	104	27	0.260	90.5	516	7.0	0.5	0	0.0
3	77	19	0.247	67.5	426	8.5	0.5	0	0.0
4	58	6	0.103	55	358	10.2	0.5	0	0.0
5	52	6	0.115	49	303	10.8	0.5	0	0.0
6	46	7	0.152	42.5	254	11.5	0.5	0.746	9.3
7	39	5	0.128	36.5	212	12.4	0.5	1	10.5
8	34	6	0.176	31	175	13.1	0.5	1	9.2
9	28	3	0.107	26.5	144	14.1	0.5	1	7.6
10	25	3	0.120	23.5	118	14.7	0.5	1	6.8
11	22	4	0.182	20	94	15.3	0.5	1	5.9
12	18	2	0.111	17	74	16.1	0.5	1	4.9
13	16	1	0.063	15.5	57	16.6	0.5	1	4.3
14	15	4	0.267	13	42	16.8	0.5	1	4.1
15	11	3	0.273	9.5	29	17.6	0.5	1	3.0
16	8	2	0.250	7	19	18.4	0.5	1	2.2
17	6	1	0.167	5.5	12	19.0	0.5	1	1.6
18	5	3	0.600	3.5	7	19.3	0.5	1	1.4
19	2	1	0.500	1.5	3	20.5	0.5	1	0.5
20	1	0	0.000	1	2	21.5	0.5	1	0.3
21	1	1		0.5	1		0.5	1	0.3
Sum	843	151							71.6
	M =	0.179						R =	0.085

The table is built from the the number of dead animals recorded (d_x) in each age (year) class (x), from which the notional number of survivors at each age (l_x) can be derived. From these two figures are derived the mortality rate ($q_x = d_x / l_x$)), the average number of animals alive ($L_x = (l_x + l_{x+1})/2$) and total life expectancy ($e_x = \left(\sum_x^{\infty} L_x\right) / l_x$) for each year class. The overall average annual mortality rate (AMR) is given by $\Sigma d_x / \Sigma l_x$. Assuming a sex ratio (proportion of females fem_x) of 0.5 and a single calf being born to each mother, applying a logistic regression describing the proportion of females that are mature in each age class ($mat_{f,x}$), and the estimated (annual) pregnancy rate in mature females (APR), we estimated the number of births annually per age class ($b_x = l_x \times fem_x \times mat_{f,x} \times APR$). The annual reproductive rate is given by $ARR = \Sigma b_x / \Sigma l_x$

Table 4B Proportion of Females Mature at Age for Scotland and the Iberian Peninsula

Age	Scotland	Iberia
2	0	0
3	0.18	0
4	0.25	0
5	0.86	0
6	1	0.746
7	1	1

Dividing the study period into four 5-year blocks and comparing survivorship curves for each period, using the non-parametric Kaplan-Meier method as implemented in Minitab (Minitab Ltd), indicated statistically significant variation in mortality rate over time (Read et al. 2013). It should be noted that the overall sample size was 151 animals, so the number of animals per 5-year block was rather low for applying this methodology, and the analysis should be treated as exploratory. Ideally, more age data are needed. To the best of our knowledge, no equivalent analysis is available for other harbour porpoise populations in the Northeast Atlantic. We used a non-parametric bootstrap with 10,000 repeats to estimate the 95% confidence limits of the annual mortality rate, each time sampling with replacement to obtain 151 ages from the original set of 151 ages, then sorted the 10,000 mortality rates and extracted the values from the 251st and 9750th values. The 95% confidence interval for the estimated annual mortality rate of 18% is from 15.3% to 20.5%.

Estimates of mortality rate and survivorship from strandings are potentially subject to various biases. Thus, whether carcasses reach the shore will depend on factors such as the distance from the coast when the animals die, their buoyancy (related to size and body condition) and prevailing currents. Saavedra Penas (2017) found that the youngest age classes of common dolphin were under-represented in strandings data from Galicia (biasing mortality estimates downwards). In addition, mortality in coastal waters will be over-represented compared to mortality in offshore waters, and different causes of death may have different spatial and temporal distributions, hence potentially being over- or under-represented in the subset of animals reaching the shore. In order to address this heterogeneity, a larger sample size (potentially achievable via improved resourcing of some of the strandings monitoring networks in the area and better collaboration between them), the use of drift modelling (supported by carcass release experiments) to estimate both the likely areas of origin of stranded carcasses and the proportion which reach the coast, and better information on the distribution of porpoises at sea, could all be useful.

Reproductive seasonality, gestation period, pregnancy rate and population reproductive rate

The limited evidence available until recently suggested that calving usually takes place in summer in the NWIP. Read (2016) recorded the presence of seven foetuses (in females aged between 6 and 16 years) in Galicia, with the largest (85 cm length) found in July and the smallest (10 cm) in August. Four neonates were found, three in May and one in August, ranging in size from 84.5 cm to 90 cm.

During monthly surveys in southern Galicia by the BDRI between April 2014 and November 2021, calves (animals less than 1 year old) were present in 55 of the porpoise groups sighted (19.2% of the total). The sightings rate for calves (calves seen per hour of effort) increased from winter to autumn (Table 5) (Díaz López & Methion 2018). The marked increase from spring to summer is consistent with most calves being born in summer.

No estimate is available for the duration of gestation period for porpoises in the NWIP but it is presumably within the range reported in the literature; estimates for Europe range from 10 to 12 months (Møhl-Hansen 1954, van Utrecht 1978, Sørensen & Kinze 1994, Bandomir-Krischack

Table 5 Seasonal Distribution of Harbour Porpoise Sightings and Numbers of Adults and Calves

Season	Survey effort (hours)	No of Sightings	No of Adults	No of Calves	SPUE (calves seen/hour)
Winter	235	11	43	2	0.009
Spring	414	42	130	10	0.024
Summer	176	176	832	57	0.077
Autumn	58	58	285	24	0.098
Overall	287	287	1328	93	0.057

Source: Adapted from Díaz López et al. (2022).

1996, Börjesson & Read 2003, Learmonth et al. 2014). The calving interval, during which gestation, lactation and reproductive resting occur, was estimated to be 1.89 years, and females appeared to remain reproductively active until at least 16–18 years old (Read et al. 2013, Read 2016).

Lockyer (2003) stated that pregnancy rates for this species were generally in the range 0.74–0.988. Most of the recent estimates were based on stranded animals and almost all were considerably lower. As noted previously, pregnancy rate tends to be underestimated in samples from strandings although the downward bias can be reduced by considering only trauma deaths and excluding animals stranded during the conception period. As such, pregnancy rate data derived from strandings are difficult to interpret, and it is not clear to what extent the lower rates reported in many more recent studies reflect real changes over time and/or geographical differences and to what extent they are a consequence of the selection of the samples. The annual pregnancy rate (APR; estimated from the proportion of mature females with a foetus between September and May) for Iberian porpoises was 0.54 ($N=13$, Read 2016), which falls in the mid-range of values reported for the North Atlantic. Camarão (2017) reported that 68% of mature females from Portugal (2002–2016) had a corpus luteum in their ovaries but this is also based of a small sample size (although data are presented for a total of 77 females, the analysis of ovarian scars appears to have been based on a subsample of the mature females). The lowest reported values are from North Sea coasts, including 0.34 for the Netherlands (based on the most recent study) and 0.26–0.30 for the English North Sea coast, and the highest was 0.98 in Iceland, where the sample consisted entirely of bycaught animals (Ólafsdóttir et al. 2003); values for the Atlantic coast of North America were also high (0.76–0.93). Learmonth et al. (2014) reported a pregnancy rate of between 0.34 and 0.40 for Scotland, depending on the range of dates selected to reduce error due to missing early term foetuses, but noted that the rate calculated for females from all year round would have been 0.28. Murphy et al. (2015) reported an overall pregnancy rate of 0.29 for the UK (i.e. using combined data from Scotland, England and Wales), noting that this increased to 0.5 among females which had died due to trauma. Murphy et al. (2020) reported an overall pregnancy rate of 0.47 for porpoises from England and Wales, also observing that there were significant regional differences (from 0.29 in the North Sea to 0.6 in the Celtic and Irish Seas). The low pregnancy rate in the North Sea was related to a high proportion of deaths in the infectious disease and "other" (i.e. live stranding, starvation, neoplasia or undetermined) categories, emphasising the relevance of the health status of the sampled animals (and noting that dead animals will almost inevitably include a higher proportion of animals that were in poor health compared to the living population).

Assuming a 1:1 sex ratio and adding data on female maturity (proportion of females mature at each age), plus the estimated pregnancy rate for Iberian porpoises (54%), to the life table for porpoises in the NWIP (Table 4A) results in an overall annual reproductive rate of only 0.085 (assuming that all foetuses survive to full-term), much too low to balance mortality. Even a 100% pregnancy rate in mature females would generate an annual reproductive rate of only 0.157. A pregnancy rate of almost 114% (i.e. with some mature females carrying twins) would be needed to balance the estimated mortality rate. Twins are extremely rare in harbour porpoises. Kompanje et al. (2017) documented a case of conjoined twins and noted that it was only the second known case of twinning in this species (the first having been reported by IJsseldijk et al. 2014). It is true that the estimated pregnancy rate for the NWIP is based on a very small sample of mature females ($N=13$), and, as previously noted, estimates from strandings tend to be strongly biased downwards unless based only on trauma deaths, and the sample size is evidently insufficient to permit a calculation based only on trauma deaths. However, for the population to be maintained, either both the ASM in females or the mortality rate estimates would need to be adjusted. López Fernández (2003) suggested an ASM of 3 years, also based on a small sample of animals. Applying female maturity at age data from a larger sample for Scotland (Table 4B), where 50% of females are mature at 4.35 years of age (Learmonth et al. 2014) to the NWIP population, a pregnancy rate of 0.54 in mature females results in a birth rate of 0.112, still considerably lower than the mortality rate. A pregnancy rate of 86.3% would then be needed to balance the estimated mortality rate (Figure 12).

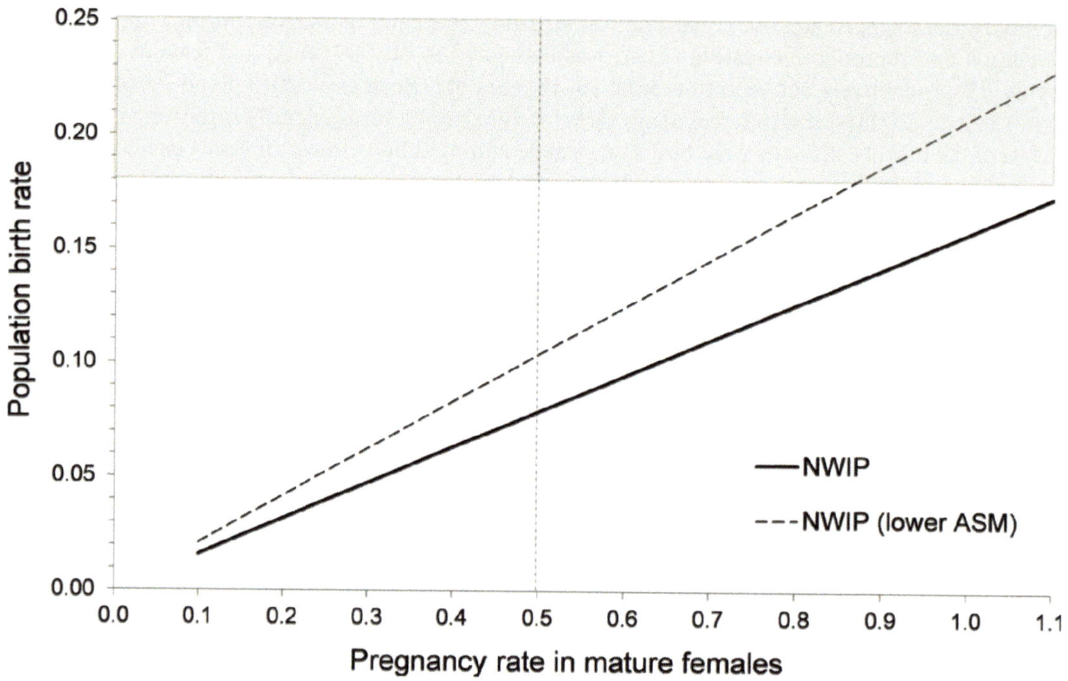

Figure 12 Annual reproductive (birth) rate for the NWIP population as a function of the pregnancy rate in mature females, based on the life table and female age at maturity data from Read et al. (2013). The (grey) shaded area indicates where the annual birth rate equals or exceeds the estimated annual mortality rate. A pregnancy rate exceeding 1.0 would imply that some mature females are carrying twins, which has not been documented in the NWIP. Applying data on Age at Sexual Maturity (ASM) for female porpoises from Scotland from Learmonth et al. (2014) to the NWIP population results in the relationship between pregnancy rate and birth rate indicated by the dashed line.

Feeding ecology

Foraging behaviour

The Iberian porpoise presumably forages mainly on prey distributed along the shelf and continental slopes (i.e. in waters of <200 m depth) since (as mentioned in section 'Distribution and abundance') there are a very few sightings of individuals in the open ocean. The diet of Iberian porpoise has been described based on stomach contents and stable isotope analyses of stranded animals but there have been no specific studies of foraging behaviour.

Harbour porpoises are often described as opportunistic feeders, implying that their prey selection is influenced by availability and not (as optimal foraging theory would predict) by energetic profitability, of which availability is just one component. In practice, we rarely have sufficient knowledge of the factors determining diet selection to justify the use of the word "opportunistic". The apparent large-scale southwards movement of porpoises in the North Sea between 1994 and 2005 may have been due to changes in the distribution and/or availability of prey (Hammond et al. 2013). At a smaller scale, porpoises show seasonal variation in occupancy of foraging grounds (e.g. Nuuttila et al. 2018, Todd et al. 2022), and Williamson et al. (2022) suggested that seasonal shifts in porpoise in distribution in the Moray Firth (northeast Scotland) were related to seasonal prey availability.

Porpoises spend a high proportion of time foraging, "hunting small fish nearly continuously day and night with extreme capture rates" according to Wisniewska et al. (2016), although several studies suggest they are more active at night (e.g. Carlström 2005, V. Todd et al. 2009, Brandt et al. 2014,

Osieka et al. 2020, Todd et al. 2022). The field metabolic rate of harbour porpoises may be as much as twice that of similar-sized terrestrial mammals (Rojano-Doñate et al. 2018). Such high energy requirements potentially make them especially susceptible to disturbance and prey depletion. Food consumption, respiration rate and body condition all vary seasonally (e.g. Kastelein et al. 2018), presumably related to both (temperature-dependent) thermoregulatory costs and the energetic demands of the reproductive cycle. Porpoises are more susceptible to disturbance when accumulating body fat before the winter, when their food intake is higher (Gallagher et al. 2021).

Porpoises forage both in the water column for pelagic prey and close to the seafloor for benthic prey. Porpoises have been observed 'bottom grubbing', whereby the animal forages head down, either vertically or at a near-vertical angle, on the seafloor (Desportes et al. 2000, Lockyer et al. 2003). Porpoises use echolocation (biosonar) to help locate and capture their prey, emitting 'foraging buzzes', i.e., click sequences with short inter-click intervals, when hunting (Verfuß et al. 2009). They usually swallow the prey head first, producing suction using their tongue to move it down the throat (Kastelein et al. 1997).

Although often considered to be solitary hunters, porpoises may aggregate in large numbers when foraging in tide race habitats (Pierpoint 2008, Benjamins et al. 2015), and a recent drone-based study in Denmark showed that they may collaboratively hunt in groups of up to six individuals, within which each animal had one or two primary roles, e.g., swimming around the school or herding fish (Torres Ortiz et al. 2021). Díaz López & Methion (2024) observed group sizes of 1 - 46 porpoises in Galician waters, with a mean group size of 5.5, although 62% groups comprised fewer than 5 animals. Porpoises sometimes forage in close proximity to man-made structures such as bridge pillars that attract aggregations of prey (Brandt et al. 2014) as well as around gillnets (Macaulay et al. 2022).

Diet and prey consumption

Harbour porpoise diets in the Northeast Atlantic include species found in all parts of the water column, including demersal and benthopelagic species like European hake *Merluccius merluccius*, Atlantic cod *Gadus morhua* and whiting *Merlangius merlangus* (typically fished with demersal trawl gear), as well as bathypelagic and pelagic-neritic species like blue whiting, sardine (*Sardina pilchardus*), scad (Atlantic horse mackerel, *Trachurus trachurus*) and mackerels (*Scomber* sp.) (typically fished using pelagic gears). Some prey species are relatively large, while others like gobies (Gobiidae), sandeels (Ammodytidae) and bobtail squids (Sepiolidae) (all of which could be considered demersal and spend at least some of the time partially or wholly buried in the substrate of sandy and muddy seabeds) are much smaller. Very often, a few species dominate the diet. Studies of the diet in UK waters from the 1960s and 1970s suggest that small energy-rich pelagic fish like herring (*Clupea harengus*) used to be an important component of the diet (see Santos & Pierce 2003 for a review) but more recent data suggest that it is no longer the case (Santos et al. 2004, G. Pierce Unpubl. data). Of the most important species in harbour porpoise diet in Galicia, blue whiting is generally found on the continental slope, while other important prey, notably pouting *Trisopterus luscus*, scad and European hake, live mainly in shelf waters (see Pierce et al. 2010). Of the minor prey, garfish (*Belone belone*) is pelagic-oceanic, spotted lanternfish (*Myctophum punctatum*) is bathypelagic and argentine (*Argentina* sp.) could be bathydemersal (*Argentina sphyraena*) or bathypelagic (*Argentina silus*) (remains could be identified only to the genus *Argentina*). All are found over the continental shelf but also in much deeper waters, and the occasional presence of remains of such species would be consistent with some foraging occurring in offshore waters.

Spain

Based on the analysis of stomach contents of 72 harbour porpoises stranded on the Galician coast during 1991–2018 (see Table 6), the main prey consumed in terms of biomass were *Trisopterus* spp. (presumably mainly pouting), scad, blue whiting and European hake. The order of importance

Table 6 The Main Prey in the Diet of Harbour Porpoises Stranded in Galicia (1990–2018)

Prey Taxon	%F	N	%N	W	%W
Fish					
Clupeidae					
Sardina pilchardus	12.5	141	5.4	1113.4	0.9
Gadidae					
Gadiculus argenteus	13.9	278	10.7	1245.5	1.1
Trisopterus spp.	43.1	427	16.5	28,377.3	23.9
Micromesistius poutassou	36.1	517	19.9	22,744.4	19.2
Merlucciidae					
Merluccius merluccius	31.9	140	5.4	18,321.7	15.5
Argentinidae	6.9	21	0.8	415.5	0.4
Sparidae	9.7	13	0.5	588.3	0.5
Gobiidae	22.2	228	8.8	362.7	0.3
Carangidae					
Trachurus trachurus	36.1	209	8.1	23,840.7	20.1
Callionymidae					
Callionymus lyra	9.7	49	1.9	4539.4	3.8
Ammodytidae	5.6	71	2.7	1962.7	1.7
Cephalopod					
Sepiidae					
Sepia officinalis	6.9	12	0.5	410.0	0.4
Sepiolidae					
Sepiola atlantica	18.1	37	1.4	22.0	0.0
Loliginidae					
Alloteuthis spp.	18.1	27	1.0	161.1	0.1
Crustacean					
Decapoda	26.4	–	–	–	–

The importance of each prey taxon is shown as: frequency of occurrence (%*F*), number of prey (*N*), numerical percentage (%*N*), weight of prey (*W*, g) and percentage of reconstructed prey weight (%*W*). Estimated weight of prey is provided in grams (g). Prey taxa with <5% frequency of occurrence are not listed.

of these species changes if the diet is expressed in terms of numbers of prey eaten (blue whiting becomes the most important, followed by *Trisopterus* spp.) or frequency of occurrence (*Trisopterus* spp. remains most important with blue whiting and scad tied for second place). Other important prey included sardine, silvery pout (*Gadiculus argenteus*), gobies and dragonets (Callionymidae). Cephalopods were also present in the diet, especially Atlantic bobtail (*Sepiola atlantica*) and mid-size squid (*Alloteuthis* spp.), but comprised only around 0.9% of prey biomass. Remains of various other invertebrates were found, generally in a small percentage (<5%) of stomachs, although remains of (usually unidentified) decapod crustaceans were found in 26.4% of stomachs. Crustaceans are often assumed to be secondary prey (i.e. from the stomachs of the primary prey) but it is difficult to rule out at least some direct consumption. Several other fish species were found in fewer than 5% of the stomachs, including garfish, Atlantic mackerel *Scomber scombrus* and spotted lanternfish (unpubl. data; see also González et al. 1994, Santos Vázquez 1998, Santos & Pierce 2003, Read et al. 2013, Hernandez-Gonzalez et al. 2024).

Stable isotope data from stranded animals suggest that porpoises in Galicia mainly feed on inshore prey and have a relatively high trophic level, when compared to other common odontocetes in the region, with the exception of coastal bottlenose dolphins (Méndez-Fernandez et al. 2012). Differences in foraging niches between porpoises and other odontocetes inhabiting the NWIP are also revealed by studies that include other 'ecological tracers' such as cadmium and PCB concentrations in body tissues (Méndez-Fernandez et al. 2013, 2017).

Inferences on diet may ultimately be possible from studies on macroparasites of the digestive tract but there is so far only one preliminary study for Galicia (Abollo et al. 1998) which examined parasites in four porpoises from Galicia, recording the presence of two species of nematodes, *Anisakis simplex* (which is present in a wide range of fish species) in the digestive tracts and *Halocercus invaginatus* in the lungs. Evidence of trophic transmission of *Halocercus* in cetaceans is circumstantial (based on its absence in unweaned hosts and increased prevalence in older hosts), and in some cetacean species, it appears to be passed from mother to calf (Pool et al. 2021).

Portugal

Aguiar (2013) reported on the diet of harbour porpoise in Portugal based on the examination of stomach contents from 60 harbour porpoises obtained by the stranding network during 1998–2013 (57 from the north, 3 from the south). As in Galicia, the majority of the diet comprised of mainly fish with a small number of cephalopods. Dietary importance was expressed in terms of occurrence and frequency, with common dragonet (*Callionymus lyra*) being the most important species, followed by *Trisopterus* spp. and mullet (*Liza* spp.). Diet composition varied over the years and between the sexes, size classes and cause of death categories (bycaught versus others). Dragonets were markedly more important in the diet of males, porpoises of intermediate size, bycaught individuals and during the period 2006–2009. Bycaught animals might typically be assumed to have eaten prey species targeted by the fisheries but the common dragonet is a benthic inshore fish of no commercial value which is taken as a bycatch in demersal fisheries (e.g. Chang 1951, King et al. 1994), suggesting that bycatch might also arise due to porpoises feeding on non-target species associated with the target catch. A further study based on the stomach contents of 35 porpoises stranded along the Portuguese coast between 2014 and 2016 found that hake, *Trisopterus* spp. and mullet (*Liza* spp.) were the most important prey categories (Pinheiro 2017). The author noted that the diet included pelagic, mesopelagic and demersal species with pelagic species being most important numerically and of similar importance to demersal species in terms of biomass.

Comparing the diet composition for Galicia and Portugal with diets of harbour porpoise populations from other European regions, some differences in main prey can be observed. Fish species of the family Gadidae are often among the most important prey in many areas (e.g. in the Iberian Peninsula, Bay of Biscay, Celtic Sea, North Sea and Baltic Sea; Desportes 1985, Lick 1991, Malinga et al. 1996, Rogan & Berrow 1996, Santos Vázquez 1998, Víkingsson et al. 2003, Santos et al. 2004, De Pierrepont et al. 2005, Spitz et al. 2006, Jansen et al. 2013, Schelling et al. 2014, Andreasen et al. 2017) but the species involved vary regionally. Compared to (say) the diet in Scotland, the importance of pelagic species seems to be greater, and some of the species eaten could have been taken from waters beyond the Continental Shelf.

Santos et al. (2014) used three approaches to estimate food consumption by porpoises in Atlantic waters of the Iberian Peninsula based on data on diet from Galicia, published information on energy requirements and the most recent estimate of population size (SCANS-II 2008). The latter estimate was adjusted *pro rata* considering that the Iberian Peninsula represented approximately 80% of the relevant survey block. It should be noted that the resulting abundance estimate, 1115 animals (95% CI 297–3798), was less than half the abundance estimate given later by Hammond et al. (2013) based on the same dataset and only 38% of the most recent estimate based on the same dataset but corrected for likely negative bias (2880, CV=0.72) by Hammond et al. (2017). The 'best' estimate of daily food consumption was 1.96 kg day^{-1}, based on a required daily energy intake of 9292 kJ day^{-1} (Otani et al. 2001), assuming 92.5% assimilation efficiency (Lockyer 2007) and applying the average energy density of prey in the diet of Galician porpoises. This estimate represents approximately 4.4% of the average body weight of stranded porpoises in Galicia, thus falling between values of 3.5% and 4%–9.5% obtained from Yasui and Gaskin (1986) and Kastelein et al. (1997), respectively.

Alternative approaches gave somewhat higher daily food intakes: an estimate of 3.0 kg was obtained based on Kleiber (1947), assuming an active metabolic rate three times the basal metabolic rate and again applying an assimilation efficiency of 92.5%. Applying the Innes et al. (1987) equation for food consumption as a function of body weight to estimated body weights of stranded porpoises, resulted in an estimated 3.45 kg of food eaten daily by an average porpoise. The annual estimates for consumption of fish by the porpoise population, assuming that the diet in Galicia could be applied to the whole area, were 39 t (95% CI 2–251 t) of clupeids, 258 t (95% CI 61–1015) of gadids, 72 t (95% CI 11–333 t) of hake and 140 t (32–658 t) of scad. Scaling up to reflect the most recent abundance calculation based on SCANS II data, the annual consumption figures would be 101, 666, 186 and 362 tonnes, respectively.

Threats to the Iberian harbour porpoise

Given the low population size of Iberian porpoise and the importance of the region for fisheries, fishery bycatch is a relevant concern, the more so given the evidence from strandings of significant bycatch mortality, already apparent in the 1980s, as mentioned by Sequeira and Ferreira (1994), who also suggested that there had been an associated decline in abundance. There was a marked decline in genetic diversity between 1990-2002 and 2012-2015 (Ben Chehida et al. 2023), consistent with a substantial decline in effective population size, although the similarity of abundance estimates from the large-scale surveys in 2005 (SCANS-II) and 2016 (SCANS-III), and the apparent increase in abundance seen in the 2022 (SCANS I IV) survey, suggest that abundance was stable or increasing since 2005 (Hammond et al. 2017, 2021, Gilles et al. 2023).

Other relevant anthropogenic threats include contaminants (notably Persistent Organic Pollutants (POPs)), pathogens (including parasites), underwater noise and other forms of disturbance, and collisions with boats. Going back in time, hunting can be added to this list. In relation to natural threats, porpoises may be predated (e.g. by killer whales) and fatally attacked by bottlenose dolphins. In relation to anthropogenic threats, porpoises in the Iberian Peninsula are protected by several European directives, notably the EU Habitats Directive and MSFD, as well as EU fishery regulations and national laws. As such, the status of porpoise populations is routinely monitored and assessed, and, depending on the outcomes of these assessments, mitigation action may be legally required (see section 'Conservation of porpoises in Europe: legal Protection and its implementation' for details).

Fishery bycatch

Fishery bycatch mortality has long been recognised as one of the most severe threats facing cetacean populations worldwide (IWC 1994). It is the proximate cause of the perilous state of a number of populations, including the vaquita in the Gulf of California and the Baltic Proper subpopulation of harbour porpoise. Bycatch mortality of harbour porpoise has been a cause for concern in many parts of the species' range (Gaskin 1984, Read & Gaskin 1988, Kirkwood et al. 1997, Tregenza et al. 1997, Caswell et al. 1998, Cox et al. 1998). In Europe, porpoise bycatch has been a cause for concern for more than two decades – and bycatch in some fisheries was and/or is alarmingly high. For example, based on observer data, Vinther (1999) reported that annual porpoise bycatch mortality in Danish gillnets in the North Sea was around 7000 animals per year. By 2001, estimated annual bycatch mortality had fallen to between approximately 2900 and 3900, depending on the method used to extrapolate from observed bycatch to total bycatch (Vinther & Larsen 2004), and the most recent estimate (for 2020) was from approximately 2100 to 2600, depending on whether the expected impact of pinger deployment was taken into account (Kindt-Larsen et al. 2023).

In Europe, fishery bycatch mortality of porpoises is monitored under Regulation 2019/1241 (and was previously monitored under Regulation 812/2004) and is assessed (nationally) under the MSFD. In their MSFD assessments of porpoise, both Spain and Portugal assessed its status as "Not good" in relation to bycatch mortality (Anon. 2022b). According to the most recent Indicator

Assessment by OSPAR, bycatch of porpoises in three Assessment Units (Greater North Sea, Irish and Celtic Seas, and West of Scotland and Ireland) exceeded the threshold, while despite the (apparent) absence of records of porpoise bycatch in the Iberian Peninsula during 2015–2020 from the on-board observer programme, other sources of evidence suggest that bycatch is well above the threshold (Taylor et al. 2022).

The small size of the Iberian porpoise population means that it has a limited capacity to withstand high bycatch mortality. Recent evidence of cetacean bycatch mortality in Iberian Peninsula waters has emerged from two main sources, on-board observations and the examination of stranded animals. Some additional information is available from other sources, e.g., interviews with fishers (e.g. López et al. 2003, Vingada et al. 2011, Goetz et al. 2014, Martínez-Cedeira & López 2018) and historical records. Until the late twentieth century, human consumption of small cetaceans, most likely of bycaught animals, was common in both Portugal and Spain (Sequeira 1996, López et al. 2003).

Observer, stranding and interview-based estimates of bycatch mortality are all subject to limitations and biases, which should be borne in mind when reading the remainder of this section. It is notable that the coverage provided by observer data tends to be both low and very patchy: data for some fleets were reported only in some years, and small-scale vessels were often entirely excluded. On the other hand, only observer data can provide reliable and unambiguous information on the precise time and location of the bycatch and the gear involved, and they are more likely to have been collected following a specified sampling strategy. Especially in the case of interviews, there may be doubts about the species involved, and about the accuracy and veracity of information that is provided by interviewee. In the case of strandings, the origin and representativeness of the carcasses examined may be in doubt.

For all data sources, the way in which data are processed and raised to fleet or population level is also relevant. For example, in the case of strandings, it is important to know whether undiagnosed deaths were excluded or considered as non-bycatch, and how "possible" or "probable" bycatch mortalities were treated. When observer data have been used, it is often the case that for fleets where observer coverage was very low, any observed bycatches from these fleets were not included in the estimates of total bycatch.

On-board observations

Since 2011, records of cetacean bycatch recorded by observers on vessels fishing in EU Atlantic waters have been compiled annually by the ICES Working Group on Bycatch of Protected Species (WGBYC) in order to permit estimation of total bycatch mortality. While EU Regulation 812/2004 was in force, these data were collected mainly by dedicated observers (although not by France or Spain) but nowadays, under Regulation 2019/1241, they are mainly from fisheries observers (who had, however, previously provided records of bycatch of other protected, endangered and threatened species (PETS)). For the Iberian porpoise, the most relevant ICES fishery divisions are 8c (northern Spanish Atlantic coast) and 9a (western Spanish Atlantic coast, Portuguese coast and southern Spanish Atlantic coast).

Spain Fernández-Contreras et al. (2010) carried out observations on-board pair trawlers in Galicia in 2001 and 2002 and estimated an annual bycatch of 394 common dolphins but no harbour porpoises. The only monitoring of cetacean bycatch in Spain by dedicated on-board observers under Regulation 812/2004 was during a 12-month pilot project (2008–2009) in ICES Subarea 8 (Anon. 2009b, Lens & Díaz 2009). Coverage in 2008 included only the last quarter of the year and recorded no porpoise bycatch. Thirteen porpoises were bycaught in 2009, leading to an estimated total of 208 bycatch mortalities of porpoises in gillnets over the course of the study. However, all the porpoise bycatch occurred further north in the Bay of Biscay, in ICES Divisions 8a and 8b, and thus outside Iberian waters. Based on the data collected by fishery observers, Spain reported bycatches of protected species in pelagic trawls in Division 8c in 2017, for bottom trawls, nets and pelagic trawls in Division 8c in 2018, for rods and lines in Division 8c and Subdivision 8d2 in 2019, and for bottom trawls and nets in 8c and 8d2 in 2020. In all these years, there was no reported bycatch of harbour porpoise (ICES 2019, 2020c, 2021c).

Portugal Information on fishery bycatches of porpoises in Portugal published in papers and project reports is summarised in Table 7. During July–October 2003, cetacean sightings and bycatch interactions were recorded during on-board monitoring of 48 fishing trips by purse-seine vessels operating from four Portuguese ports (Figueira da Foz, Sesimbra, Setúbal and Sines), and interviews with fishers were conducted during 36 trips. No porpoise bycatch was recorded but porpoises were sighted once (among 31 cetacean sightings), with two individuals being seen near Figueira da Foz (Wise et al. 2007). Monitoring of 292 beach seine (*xávega*) hauls in Portugal between 2008 and 2011 recorded five porpoise mortalities, i.e., a mortality rate of 0.017 animals per haul (Vingada et al. 2011). The authors stated that they observed 3.3% of national fishing activity in this fleet. However, based on broad consistency with annual fishing effort data reported in Oliveira et al. (2015), it appears that 292 hauls represented approximately 3.3% of *annual* fishing effort by the fleet. Thus, the annual number of beach seine hauls would be around 8850, implying a total annual bycatch mortality in this gear of around 150 porpoises. Vingada and Eira (2018) reported results on bycatch mortality from the MARPRO project, indicating that the estimated average annual porpoise bycatch during 2010–2015 comprised 17 animals in purse seines, 203 in the polyvalent fleet, 0 in bottom trawls and long lines and 21 in beach seines. This leads to an annual bycatch estimate of 241 porpoises per year. Marçalo et al. (2015) reported on observations on-board purse seiners in Portuguese waters in 2010–2011 (163 days at sea) which included one porpoise bycatch, an animal which was observed being encircled and subsequently escaping. This record was included within Portugal's submission to the 2011 meeting of the ICES Working Group of Bycatch (A. Marçalo, Pers. Obs.). Although such observations may be rare, this illustrates the point that not all recorded cetacean bycatches lead to immediate mortality.

According to reports submitted to the ICES Study Group on Bycatch of Protected Species (SGBYC) in 2010 and to ICES WGBYC in 2011 (ICES 2010, 2011), in 2007–2009, Portugal had no fisheries covered by EU Regulation 812/2004 (see section 'Conservation of porpoises in Europe: legal Protection and its implementation') and no observer programme. In 2010, Portugal reported on observations of polyvalent and purse seine fisheries in Division 9a, reporting five bycaught porpoises in the polyvalent fishery. Observer data for 2010 from the polyvalent fleet data were reported separately for boats targeting demersal and pelagic fish. Porpoise bycatches came from boats targeting mackerel, horse mackerel and sardine. Scaling up (based on the proportion of effort observed in these fisheries), this leads to an estimate of 150 porpoises bycaught (ICES 2013a). In 2011, Portugal reported on observations of purse seine, demersal and polyvalent trawl fisheries, with one reported porpoise bycatch in the purse seine fishery, leading to an estimate of 103 porpoises bycaught

Table 7 Published Information on Harbour Porpoise Bycatch in Portuguese Waters (ICES Division 9a) Collected by On-board Observers during 2003–2020

(A) Data published in scientific papers and reports (for the years 2003–2015)

Years	Country and Source	Fishing Area	Métier	Observed/ Total Effort (% Coverage)	No. of Incidents	No. of Specimens Bycaught	Extrapolated Annual Bycatch Estimate	Effort Unit
2003	Portugal (1)	9	Purse seine	48	0	0	0	Trips
2008–2011	Portugal (2)	9	Beach seine	292 (3% of annual total) [a]		5	150	Hauls
2010–2011	Portugal (3)	9	Purse seine	163	0	0	0	Days
2010–2015	Portugal (4)	9	All gears				241	

(Continued)

Table 7 (*Continued*) Published Information on Harbour Porpoise Bycatch in Portuguese Waters (ICES Division 9a) Collected by On-board Observers during 2003–2020

(B) Data extracted from ICES WGBYC reports from 2011 to 2021 for individual year and gear combinations during 2010–2020)

Year	Country and Source	Fishing Area	Vessel Size (m)	Métier	Target Species	Season	Observed/ Total Effort (% Coverage)	No. of Incidents	No. of Specimens Bycaught	Extrapolated Annual Bycatch (with 95% CI)	Effort Unit
2010	Portugal (5)	9a	NA	Polyvalent	*Trachurus* spp, sardine, chub mackerel	May–October	80/2400 (3.33%)	NA	5	150[b] (23–277)	NA
2011	Portugal (6)	9a	>15	Purse-seine	NA	NA	110/11,320 (0.97%)	NA	1	[103][c] (0–304)	NA
2012	Portugal (7)	9a	>12	Polyvalent	Mixed demersal	NA	71/63,612 (0.11%)	1	1	896[d] (0–2640)	Haul
2015	Portugal (8)	9a	>12	Polyvalent	NA	NA	245/59,679 (0.41%)	NA	6	1462[d] (306–2617)	Days at sea
2010–2015	Portugal (5–8)	9a	>12	Sum across all monitored gears				NA	13	418[e] (86–922)	
2010–2020	Portugal (5–8)	9a	>12	Sum across all monitored gears				NA	13	228[e] (47–503)	

Sources: Years and gear combinations with no reported bycatch of porpoises or no reported observer data are not included in the table. Combined estimates for all monitored gears are given for 2010–2015 and 2010–2020 (note that no bycatch of porpoise was reported for this area during 2016–2020. Approximate 95% confidence limits based on the normal approximation method are provided. Multiannual estimates and 95% confidence limits were obtained summing annual values and dividing by the number of years. Data sources:(1) Wise et al. (2007), (2) Vingada et al. (2011), (3) Marcalo et al. (2015), (4) Vingada and Eira (2018), (5) ICES (2013a), (6) ICES (2013b), (7) ICES (2014c), (8) ICES (2017).

[a] The source text implies that it is 3.3% of total fishing effort over 4 years but based on the reported amount of fishing effort for this metier at this time, it appears to be 3.3% of the annual effort. If this latter interpretation is incorrect, the annual bycatch would be 37.5.

[b] Portugal reported an extrapolated bycatch of 80 in its annual Reg. 812/2004 report, possibly because it separated observations of demersal and pelagic fishing by the polyvalent fleet, wth porpoises being caught only during the latter.

[c] Although this was a bycatch, Marcalo et al. (2015) commented the animal escaped from the net. It should thus not be counted as bycatch mortality.

[d] ICES WGBYC did not extrapolate from these data to estimate bycatch.

[e] The average values over 2010–2015 (6 years) and 2010–2020 (11 years) were calculated excluding the animal caught in 2011 because it was not a bycatch *mortality*.

(ICES 2013b). As noted above, according to Marçalo et al. (2015), the animal escaped from the net and should thus not be counted as bycatch mortality. In 2012, Portugal reported observations from demersal trawl, purse seine and polyvalent (trammel net) fishing to ICES WGBYC (ICES 2014c). Bycatch of one porpoise was recorded in the polyvalent fleet deploying trammel nets. With 63,612 effort units (probably individual hauls), of which 71 were observed, the extrapolated bycatch would be 896 porpoises. In 2013 and 2014, Portugal reported on cetacean bycatch in polyvalent, seine and bottom trawl fleets, which did not include any porpoises (ICES 2015, 2016). In 2015, only observations from the polyvalent fleet were reported, with a bycatch of six porpoises in set nets (ICES 2017). Extrapolation would give a porpoise bycatch of 1462 animals. A small number of trips by boats in the polyvalent fleet deploying fixed nets was observed in 2016, with no porpoise bycatch (ICES 2018a). In 2017, Portugal reported zero bycatch of harbour porpoises in bottom trawls, nets and seines. In 2018, reporting covered bottom trawls, nets, seines and surrounding nets but again there was no observed bycatch of harbour porpoise (ICES 2020c). In 2019 and 2020, no porpoise bycatch was recorded for longlines, nets and surrounding nets (ICES 2021c). See Table 7(B) for a summary of harbour porpoise bycatch in Division 9a reported by ICES WGBYC.

Based on the Portuguese observer data for 2010–2015 compiled in Table 7B, assuming that all the fishing took place in Portuguese waters and that bycatch in beach seines took place only in the years during which that fleet was observed (2008–2011), the estimated average annual bycatch during 2010–2015 across polyvalent and purse seine fleets (435, or 418 if the porpoise which escaped the purse seine in 2011 is not included in the calculations), plus the bycatches in beach seines (50 per year over 6 years assuming that 150 animals were taken in both 2010 and 2011), would be 485 or 468 animals. Taking the absence of observed bycatches over the following 4 years at face value, the annual average for 2010–2020 (including an average of 300/11 = 27 animals taken in beach seines per year is 264 or 255 porpoises per year. Total bycatch is likely to be underestimated because observer effort focused on larger boats (>12 m), and fishing by Spanish vessels is evidently not included. On the other hand, we have included figures that ICES WGBYC did not use: they did not include project-based data collection and did not extrapolate from the bycatches recorded in 2012 and 2015, presumably due to the very low percentage of effort that was monitored. The 95% confidence intervals for the annual bycatch estimates are also provided in Table 7B, based on the normal approximation method to derive 95% confidence limits for a binomial variable, assuming that a single porpoise was caught during each bycatch event (this information was not always available in the WGBYC reports) and thus treating the number of bycaught animals divided by the number of observed fishing events as a binomial variable. If the lower 95% limit was below zero or the upper limit above 100%, we give the values as 0% and 100%, respectively. The estimates and confidence intervals for bycatch per observed fishing event were extrapolated to give the number of animals bycaught annually, by dividing by the proportion of annual fishing events that were observed. Multiannual estimates and associated 95% confidence limits were obtained summing annual values and dividing by the number of years.

The ICES Workshop on estimation of MOrtality of Marine MAmmals due to Bycatch WKMOMA (ICES 2021c) estimated bycatch rates for marine mammals in the OSPAR area using submitted bycatch data for 2005–2021. For the Iberian coast, these data contained only a single porpoise bycatch record, from trammel nets targeting crustaceans (gear code GTR CRU) and no records of bycatch during 2015–2020. Due to the lack of data, it was not possible to assess bycatch for the Iberian coast. Evidently there was a discrepancy with previously reported data, which included several porpoise bycatch events, including six animals bycaught in 2015.

Some of the estimates of bycatch mortality from the on-board observer data are extremely high. In all cases, the extrapolation was based on a small number of observed bycatches, the proportion of fleet activity observed was low and the estimated 95% confidence intervals are very wide. Where the proportion of effort was very low, ICES WGBYC did not extrapolate from these values. If the bycatch mortality is as high as these estimates suggest, there is an apparent contradiction with the lack of change in population size seen between the 2005 and 2016 SCANS II and III surveys. As will be seen below, this issue is not unique to the observer data on bycatch.

Evidence from strandings

Spain Information on bycatch mortality diagnosed from stranded animals in northern Spain mainly derives from the Galician stranding network (CEMMA), since (to date) the other stranding networks operating in northern Spain rarely report on diagnosed bycatch mortality. Read et al. (2013, 2020) compiled data for 1990–2010. Among the 313 (i.e. approximately 15 per year over the whole period) stranded and (a small number of known bycatches) handed in by fishers, 23% were known or diagnosed bycatches, 17% showed no evidence of bycatch, 11% were examined but the cause of death could not be determined and 47% were not examined due to being too decomposed. Based on these data, we excluded the last category from analysis and calculated the % bycatch mortality among "strandings" both with and without the known bycatches handed in by fishers (Table 8). Data for 1990–2019 were summarised by Pierce et al. (2020): there were 306 strandings of harbour porpoise, i.e., on average around 10 per year, 53 of which (17.3% of the total) showed evidence of fisheries interactions. Of these animals, 120 were in a good state of preservation (decomposition state of 3 or less) of which 41 (34.2%) showed evidence of bycatch. Known bycatches ($N = 7$) handed in by fishers were excluded from the calculations. These compilations of data result in similar estimated percentages of stranded animals dying due to bycatch, 31.0% and 34.2%, respectively, when known bycatches are excluded; when known bycatches were included, the former estimate increased to 34.8%, i.e., an upward bias of almost 4%. As for the observer-based bycatch estimates, confidence

Table 8 Strandings of Harbour Porpoises in Galicia 1990–2021: Incidence of Bycatch Mortality

Period, Subset and Source	No. of Strandings	No. F + SD Carcasses Examined	No. with Bycatch Evidence	No. with Bycatch Evidence per Year	No. Bycaught/No. Examined (%)	95% CI for % Bycaught (%)
1990–2010 (no KB) (1)	208	87	27	1.3	31.0	21.3–40.8
1990–2010 (with KB) (1)	213	92	32	1.5	34.8	25.1–44.5
1990–2019 (2)	306	120	41	1.4	34.2	25.7–42.7
1990–1999 (3)	105	38	14	1.4	36.8	21.5–52.2
2000–2009 (3)	95	20	9	0.9	45.0	23.2–66.8
2010–2021 (3)	152	67	27	2.3	40.3	28.6–52.0
1990–2021 (3)	352	125	50	1.6	40.0	31.4–48.6
2018–2020 (4)	51	18	8	2.7	44	21.5–67.4
2000–2020 (PB→NB) (5)	213	83	34	1.6	41.0	30.4–51.5
2000–2020 (PB→B) (5)	213	83	40	1.9	48.2	37.4–58.9

Number of stranded porpoises, number of fresh and slightly decomposed (F + SD) carcasses examined, number with bycatch evidence, number with bycatch evidence per year, and the percentage with bycatch evidence (with 95% confidence intervals estimated using the normal approximation method). All data were collected by CEMMA. For the 1990–2010 data we present results both without and with known bycaught animals (KB, i.e. animals handed in by fishers) included. The 1990–2019 data exclude the known bycaught (KB) animals. For the 2000–2020 data we present results for both (1) treating probable bycatch (PB) as non-bycatch (NB) and (2) treating probable bycatch as bycatch (B)

Sources: (1) Read et al. (2013, 2020), (2) Pierce et al. (2020), (3) CEMMA, unpublished, (4) ICES (2020c, 2021c), (5) Torres-Pereira et al. (2023a).

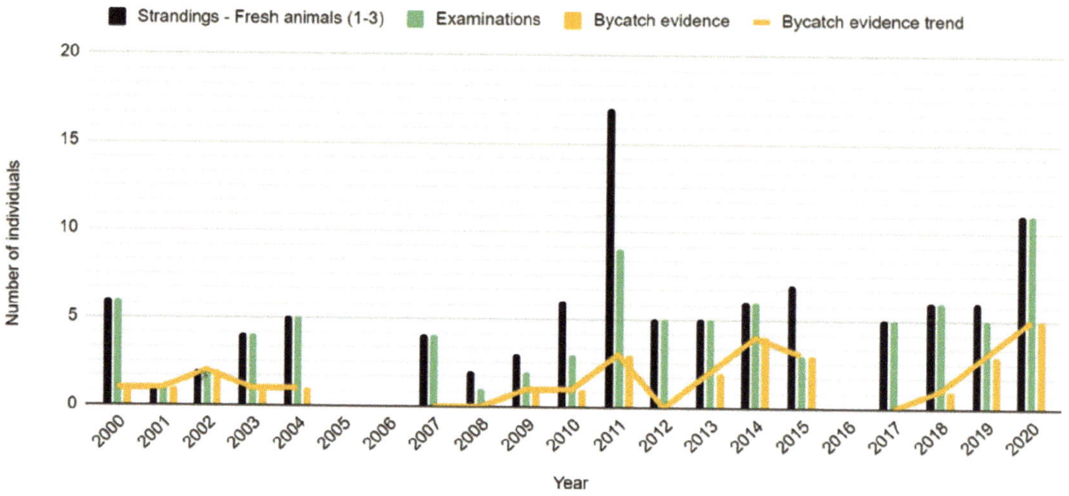

Figure 13 Number of fresh or slightly decomposed harbour porpoises stranded, number of examinations and number of examined stranded animals with evidence of fishery bycatch from 2000 to 2020 in Galicia (NW Spain). In 2005 and in 2006, none of the harbour porpoise strandings was registered as fresh or slightly decomposed. There were no data available for 2016. The trend line refers to the number of bycaught animals recorded. (Data provided by CEMMA.)

limits around the percentage of stranded animals killed by bycatch can be derived using the normal approximation method. Even combining data over two or three decades, the lower and upper confidence intervals span a range of almost 20% (Table 8). Multiplying the 34.2% of strandings that were bycaught by the estimated annual mortality rate of 18% (Read et al. 2013, 2020) gives the annual bycatch mortality rate of 6.2%. Given a population size estimate of 2900 animals, this equates to 178 annual deaths due to bycatch. Finally, the estimate of 34.2% was based on 41 bycaught animals over 30 years, giving a minimum estimate of average bycatch mortality per year of 1.4 animals.

Updated and revised numbers for 1990–2021 provided by CEMMA and a summary for 2018–2020 taken from ICES WGBYC reports (ICES 2020c, 2021c; again based on CEMMA data) also appear in Table 8. The overall percentage of mortality attributed to bycatch across approximately three decades was 40%. The average number of stranded porpoises recorded annually was highest in the most recent period (2010–2021) (12.7) as was the average number of bycaught animals recorded per year (2.3 on average, see also Figure 13). During 1990–2021, the estimated percentage of bycatch mortalities in the strandings was 40% (95% CI 31.4%–48.6%).

Torres-Pereira et al. (2023a) summarised porpoise strandings data from Galicia for the period 2000–2020 (alongside Portuguese data), distinguishing certain and probable bycatch mortalities. In the calculations presented in Table 8, we firstly used only (certain) bycatch and considered probable bycatches as non-bycatch and secondly counted certain and probable bycatches as bycatch. Following the first approach, 41.0% of stranded animals were bycatches, rising to 48.2% when probable bycatches were included.

Portugal As noted previously, since strandings monitoring began in 1977 in Portugal, a decline in strandings was seen up until the late 1980s. Many of the animals showed evidence of bycatch mortality, such as net marks around the head and flippers, presumably due to entanglement in fixed nets (Sequeira & Ferreira 1994). A summary of strandings data from mainland Portugal between 1979 and 2009 suggested that approximately 24% of porpoises strandings were known bycatches or showed indications of bycatch, but no detail was provided on changes over time (Gonçalves de Sousa 2010). Between 2000 and 2016, 4021 stranded cetaceans were reported from the Portuguese coast. Cause of death was determined for 2423 animals, and bycatch was the most frequently diagnosed cause of mortality (44.3%) (Vingada & Eira 2018). Although the report did not provide a full account for each species, it was stated that, of 43 porpoises recorded as stranded in 2014, the most frequent cause of death was bycatch (64%). The southern strandings network recorded 25 porpoise strandings along the south coast of Portugal between 2010 and 2017. Cause of death was recorded in seven cases and two of these were bycatch mortalities (A. Marçalo, Pers. Obs.; see Table 9). Ângelo (2020) compiled strandings data for the west coast of Portugal for 2010–2019, referring to 130 bycaught and probably bycaught animals among 278 strandings. Most bycatches were recorded in spring (54) and summer (37). Based on marks that could be used to identify the gear responsible, bycatches in spring were divided almost equally between those caught in fixed nets (28) and those taken by other and undetermined gears (26). Of the latter, more than half were due to beach seine. In summer, a majority (28) of the bycatches were in the other and undetermined gears category.

Aguiar (2013) reported on the diet of harbour porpoises in Portugal based on the examination of stomach contents from 60 harbour porpoises obtained by the stranding networks during 1998–2013 (57 from the north, 3 from the south). Of these animals, 37 were known or diagnosed bycatch mortalities

Table 9 Strandings of Harbour Porpoises in Portugal 1998–2020: Incidence of Bycatch Mortality

Year, Region, Source, Subset	No. of Strandings	No. F + SD Carcasses Examined	No. with Bycatch Evidence	No. with Bycatch Evidence per Year	No. Bycaught/No. Examined (%)	95% CI for % Bycaught (%)
1990–2010 North (no KB) (1)	97	72	44	2.1	61.1	49.9–72.4
1990–2010 North (with KB) (1)	100	75	47	2.2	62.7	51.7–73.6
1998–2013 (2)	130	60	37	2.3	61.7	49.4–74.0
2014 (3)	43	–	–	–	64	–
2010–2017 South (4)	25	7	2	0.3	28.6	0–62.0
2010–2019 North (7)	278	–	130	13	–	–
2000–2020 North (5) (PB→NB)	453	264	136	6.5	51.5	45.5–57.5
2000–2020 South (5) (PB→NB)	71	17	1	0.0	5.9	5.3–17.1
2000–2020 all (5) (PB→NB)	524	281	137	6.5	48.8	42.9–54.6
2000–2020 North (5) (PB→*B*)	453	264	168	8.0	63.6	57.8–69.4
2000–2020 South (5) (PB→*B*)	71	17	3	0.1	17.6	0–35.8
2000–2020 all (5) (PB*vB*)	524	281	171	8.1	60.9	55.1–66.6
2011–2015 all (5) (PB→NB)		96	49	9.8	51.0	41.0–61.0
2018–2020 (6)	118	72	28	9.1	38.9	27.6–50.1

Number of stranded porpoises, number of fresh and slightly decomposed (F + SD) carcasses examined, number with bycatch evidence, and the percentage with bycatch evidence (with 95% confidence intervals estimated using the normal approximation method). For the 1990–2010 data we present results both without and with known bycaught animals (KB, i.e. animals handed in by fishers) included. For the 2000–2020 data we present results for both (1) treating probable bycatch (PB) as non-bycatch (NB) and (2) treating probable bycatch as bycatch (B).

Sources: (1) Read et al. (2013, 2020), (2) Aguiar (2013) and Vingada and Eira (2018), (3) Vingada and Eira (2018), (4) Ana Marçalo (Pers. Comm.), (5) Torres-Pereira et al. (2023a), (6) ICES (2020c, 2021c), (7) Angelo (2020).

and a further nine were probable bycatch mortalities. Cause of death was undetermined in ten animals. It is not clear if this last category refers to animals in a good state of preservation for which cause of death could not be determined or animals which were not fresh so that diagnosis was not possible. We have assumed the former. Thus, the minimum estimate for the percentage bycatch mortality (assuming the undetermined deaths were not bycatch) would be 61.7% and the maximum (including probable bycatches) would be 76.7%. Since the sample included an unspecified (albeit probably small) number of animals handed in by fishers, the estimated percentage of bycaught animals is likely biased upwards. During the 16-year period covered by the dietary study, a total of 130 porpoises was recorded as stranded in Portugal (Vingada & Eira 2018). Based on the proportion of bycaught animals reported by Aguiar (2013), the total number of porpoise bycatch mortalities recorded by the network during this period would have been between 61.7% and 76.7% of 130, i.e., between 80 and 100 porpoises. If these figures are scaled up to the whole Iberian population, assuming annual mortality of 18% and a population of 2900 animals, the overall annual bycatch mortality rate would be between 11.1% and 13.8%, and the annual bycatch mortality in this population would thus be between 322 and 400 animals.

Torres-Pereira et al. (2023a) presented results from the period 2000–2020, also separately presenting results for 2011–2015 (years for which there was an annual abundance survey). They estimated an average annual bycatch mortality in Portuguese waters of 207 porpoises per year during 2011–2015. They provided separate estimates for north (plus central) and southern Portugal. Based on their data, the estimated bycatch rate among strandings during 2000–2020 was much lower in the south (5.9% or 17.6% depending on whether "probable" bycatches are rated as bycatch or not) than in the north (51.5% or 63.6%), with the combined figures (48.8% or 60.9%) being closer to the latter. Information on numbers of stranded harbour porpoises with evidence of bycatch in Portugal during 2018–2020, as submitted to ICES WGBYC (ICES 2020c, 2021c), suggested a lower percentage of bycaught animals (38.9%) among strandings than most of the estimates from earlier years. The 95% confidence intervals for most of the estimates are wide (spanning 20%+), although narrower (10%–12%) for the large 20-year data sets from the north and all Portugal. Applying the 18% annual mortality rate and population size of 2900, to higher (60.9%) and lower (38.9%) estimates of the percentage of bycatches in Portuguese strandings gives annual population bycatch mortality rates of 10.4% and 7.0%, respectively, equivalent to 303 or 203 animals.

Combined data for Portugal and Galicia As described in section 'Life history – Cause of death, sex ratio, age structure and mortality rate', Read et al. (2013, 2020) used life history data derived from porpoises stranded in Galicia (Northwest Spain) and north-central Portugal during 1990–2010 (plus a few known bycatches handed in by fishers) to estimate the annual mortality rate (18%). Note, however, that the current strandings network in northern Portugal started only in 2010 so the first decade of data are from Galicia alone. These authors then estimated how much of this annual mortality was due to bycatch, based on the proportion of these animals that showed signs of bycatch. Among the 313 stranded and handed-in bycaught porpoises during 1990–2010, 23% were known or diagnosed bycatches, 17% showed no evidence of bycatch, 11% were examined but the cause of death could not be determined and 47% were not examined due to being too decomposed. Evidently, as previously mentioned, including even a small number of animals handed in by fishers will likely result in an upward bias in the estimated bycatch rate. We calculated the percentage of bycaught individuals among all those that were examined as well as repeating the calculations excluding known (handed-in) bycatches. The resulting percentages for the combined dataset were 47.3% (all animals examined) and 44.7% (excluding known bycatches). The 95% confidence intervals span a range of approximately 15% (Table 10). These figures are equivalent to annual bycatch mortality rates in the population of 8.5% (247 deaths annually) or 8.0% (233 deaths annually).

Torres-Pereira et al. (2023a) summarised porpoise strandings data from north Portugal, south Portugal and Galicia for the period 2000–2020. During this period, 756 porpoise strandings were logged. The authors noted that the number stranded per kilometre of coastline in Portugal

Table 10 Strandings of Harbour Porpoises in Galicia and Portugal Combined, 1990–2020: Incidence of Bycatch Mortality

Years, Area, Regions, Subset, Source	No. of Strandings	No. F+SD Carcasses Examined	No. with Bycatch Evidence	No. with Bycatch Evidence per Year	No. Bycaught/ No. Examined (%)	95% CI for % Bycaught (%)
1990–2010, Galicia+North Portugal, no KB (1)	305	159	71	3.4	44.7	36.9–52.4
1990–2010, Galicia+North Portugal, with KB (1)	313	167	79	3.8	47.3	39.7–54.9
2000–2020, Galicia+all Portugal (PB→NB) (2)	756	364	171	8.1	47.0	41.9–52.1
2000–2020, Galicia+all Portugal (PB→*B*) (2)	756	364	211	10.0	58.0	52.9–63.0

Number of stranded porpoises, number of fresh and slightly decomposed (F + SD) carcasses examined, number with bycatch evidence, and the percentage with bycatch evidence (with 95% confidence intervals estimated using the normal approximation method). For the 1990–2010 data we present results both without and with known bycaught animals (KB, i.e. animals handed in by fishers) included. For the 2000–2020 data we present results for both (1) treating probable bycatch (PB) as non-bycatch (NB) and (2) treating probable bycatch as bycatch (B).

Sources: (1) Read et al. (2020), (2) Torres-Pereira et al. (2023a).

was more than double that in Galicia. The animals were examined to determine the cause of death in 364 cases, of which 171 (47.0%) were diagnosed as bycatch mortalities and a further 40 (11.0%) as probable bycatches. Animals for which no cause of death could be determined but which did not present signs of bycatch were considered as non-bycatch. Breaking this down by area, and considering only certain bycatches, the percentage of bycatch mortality was highest in north Portugal (51.5%), somewhat lower in Galicia (41.0%) and lowest (albeit based on a much smaller sample size) in south Portugal (5.9%) (Tables 8 and 9). Combining data from all three networks gives an overall percentage of bycatch mortality among strandings of 47.0%. The 95% confidence intervals for these estimates span a range of approximately 10% (Table 10).

Applying the previously mentioned 18% annual mortality rate and population size of 2900, this translates into an annual population mortality rate due to bycatch of 8.5% or 245 animals per year. If we count probable bycatches as bycatch, these figures increase to 58.0% bycatches among the stranded animals, 10.4% annual mortality due to bycatch in the population or 303 bycatch deaths annually.

As with the observer data, some of the estimates of bycatch mortality from strandings are very high, in the same range as the estimates from observer data. Again the question of their apparent incompatibility with abundance data arises, even if, again, there is substantial uncertainty associated with these estimates. Stranding data could lead to an overestimate of bycatch mortality if bycatch occurred mainly in the coastal zone where carcasses are more likely to wash ashore, although arguably observer data have the opposite bias since they mainly exclude small-scale fishing. It is possible that porpoise abundance has been underestimated due to seasonal movements in and out of the surveyed area, which would probably imply that substantial numbers are present in offshore waters or, perhaps less likely, that there is a substantial amount of movement between Iberia and Africa. Finally, at least until the 2022 (SCANS IV) survey results (Gilles et al. 2023) were published, it seemed plausible that population size had fallen since the 2016 (SCANS III) survey but the abundance estimate from 2022 was around 4000 animals - although, as previously, confidence limits were wide and the estimate was based on a small number of sightings.

The adjacent Northwest African harbour porpoise population in Mauritania presents a low number of strandings due to both the small population size and Mauritanian coastal features that make carcass collection difficult (collection usually involves beach sampling by car), which may produce some biases in the collected stranding data. A coastal monitoring programme carried out along the coast of

Mauritania in 2012 recorded 12 harbour porpoise strandings among other cetacean species strandings. Given the preservation status of the carcasses found, only five could be examined, all of which (100%) showed evidence of bycatch (Mullié et al. 2013). One of three porpoises recorded stranded in Morocco during 2016–2021 by Kaddouri et al. (2023) was bycaught (the cause of death of the other two was unknown) but is unclear whether these would have been Iberian or African porpoises.

Interview surveys

Spain Two interview-based studies in Galicia, by López et al. (2003) and Goetz et al. (2014), generated estimates of total annual cetacean bycatch in Galicia of around 1700 animals. Porpoises were not specifically identified in the earlier study. Goetz et al. (2014) reported that Galician fishers operating fixed gillnets caught an average of 2–3 porpoises per year, with an estimated total annual bycatch by trawl and set gillnet fleets of approximately 40 porpoises, although almost 1300 of the cetaceans bycaught annually were not identified to species. While harbour porpoises were mainly associated with gillnets in Spanish waters, they were also seen close to artisanal gears (those used by the polyvalent fleet), purse seine and beach seine (Goetz et al. 2015).

During 2019 and 2020, the Galician stranding network CEMMA conducted interviews with fishers about their activities in the coastal and adjacent waters from Cape Silleiro to Cape Corrubedo (Galicia) as part of the VIRADA project.[4] The main fishing gears used were trammel nets (*miño*) and single-panel bottom-set gillnet (*beta*), respectively, accounting for 46.7% and 18.1% of all fishing gears used. Of the interviewees, 42.1% reported having had cetaceans, including harbour porpoises, entangled in their nets, mainly in *miños* (Martínez-Cedeira et al. 2021).

Portugal Wise et al. (2007) conducted interviews with skippers during 36 purse seine fishing trips from Portuguese ports to collect information about cetacean sightings and cetacean behaviour around the nets. However, all the reported observations refer to delphinids. Sightings from observer trips carried out in parallel with the interview survey recorded a single porpoise sighting but the only observed bycatch events involved common dolphins.

Face-to-face interviews with skippers were conducted by trained researchers along the whole Portuguese southern coast (Algarve) in the most important landing ports ($n = 19$), between March and November 2018, collecting information on incidents of bycatch of marine megafauna (cetaceans, seabirds and seaturtles) in coastal fisheries which took place during 2017. There were reports of common dolphin and bottlenose dolphin bycatch, but none of porpoises being bycaught (Alexandre et al. 2022).

Historical records

Historical records indicate that fishery bycatch mortality has affected small cetaceans, including harbour porpoises, in European waters for centuries (Petitguyot et al. 2024). As noted above, intense conflicts between fisheries and small cetaceans are documented between the eighteenth and twentieth centuries in the Northeast Atlantic and Mediterranean due to animals being blamed for reducing fish catches and for damaging fishing gear when attempting to take fish from the nets (depredation) and when incidentally getting entangled in the nets (bycatch) (Bearzi et al. 2004, Valdés-Hansen 2004, 2009, Mornet 2008, Petitguyot et al. 2024). By damaging nets, both depredation and bycatch events can result in the loss of catch and force the cessation of fishing activities. In earlier times, nets were made of more delicate materials (e.g. cotton and hemp), probably making damage more likely than in the case of modern nylon (plastic) nets. After a bycatch or depredation event, before fishing could resume, nets needed to be repaired on land, and the consequent loss of time and money led to retaliation in the form of culling campaigns (Bearzi et al. 2004, Petitguyot et al. 2024). Tens of thousands of small cetaceans were killed in European waters during this period, but the extent of bycatch-induced mortality of

[4] https://www.programapleamar.es/proyectos/virada-bases-para-la-reduccion-de-la-mortalidad-por-captura-accidental-en-artes-de-pesca

Iberian harbour porpoises is currently unknown. Given that porpoises are apparently frequently caught in artisanal coastal fisheries in the Iberian Peninsula (see below), and similar fishing activities were taking place between the seventeenth and twentieth centuries (Carmona & López Losa 2009, Amorim 2009, López Losa & Amorim 2012), it is likely that Iberian porpoises were often getting entangled.

Spain Small cetaceans were regularly observed at fish markets in Galicia during the nineteenth century, including "toninas" as mentioned by López Ferreiro (1895). In 1991, a porpoise was auctioned at the fish market in Portonovo in Galicia (A. López, Pers. Obs.).

Portugal As described by Teixeira (1979) and Brito and Vieira (2010), harbour porpoises were bycaught in Portugal in the late 1970s and sold in fish markets. A photograph reproduced by Brito and Vieira (2010) shows a bycaught harbour porpoise at the fish market in Caminha (northern Portugal) in 1977 (Figure 14). The animal had drowned in a bottom-set gillnet and (as permitted by the then existing law) was brought to the fish market and auctioned (A. Teixeira, Pers. Comm.)

Sequeira and Ferreira (1994) summarised available information on bycatch mortality of cetaceans in Portuguese fisheries, reporting on landings and effort and mentioning reported bycatches. Killing and sale of marine mammals were legal in Portugal until 1981, when legislation was passed to protect marine mammals, making killing of cetaceans illegal. Fishers then ceased reporting bycatches, and subsequent official records of bycatches were limited. Official records refer to 18 cetaceans (17 common dolphins and one porpoise) killed in trawl nets. Of these, 12 (including the porpoise) were caught in 1980 and the remainder are reported as drowned in nets, five of them during research cruises. Records since 1977 indicated 132 cases of entanglement and 59 deaths in gillnets, although

Figure 14 A bycaught harbour porpoise at the fish market in Caminha (northern Portugal) in 1977, where it was auctioned, as permitted by Portuguese law at the time. (Photograph taken and provided by Antonio Teixeira.)

the authors commented that most such events were not reported. Most of the deaths were thought to be of common dolphins but striped dolphins and porpoises were also regularly taken.

Overview of uncertainty for porpoise bycatch mortality estimates

Pierce et al. (2020, 2022) summarised the available bycatch mortality estimates for the Iberian porpoise population, and these figures have been revised and updated along with newly published information in the present review (Tables 7–10). In order to provide an indication of the precision of the estimates derived from strandings, we estimated the 95% confidence intervals using a combination of non-parametric and parametric bootstrapping, focusing on multi-year estimates with reasonable sample sizes. For the observer estimates we used a non-parametric bootstrap to resample from observer trips in each year, based on 10,000 runs, to estimate the percentage of observed fishing events with bycatch of porpoise and, multiplying by the total number of fishing events, the estimated number of porpoises that were bycaught during all fishing events. The 95% confidence interval was obtained by sorting the 10,000 estimates and extracting the 251st and 9750th values. Values were then summed across years and divided by the number of years to give average values. The averages were then used to derive the annual mortality rate due to bycatch (number bycaught/population size). We also repeated the latter calculation incorporating a parametric bootstrap to simulate abundance estimation, based on the estimate of 2900 animals and associated coefficient of variation from Hammond et al. (2017) and assuming a log-normal distribution (method 2).

For each relevant strandings dataset, we again carried out 10,000 bootstrap runs as follows: (1) resampling with replacement from the sets of carcases examined to obtain the percentage of bycaught animals; (2) resampling with replacement from the set of ages at death (Read et al. 2013, 2020) to estimate the annual mortality rate in the population; (3) using a parametric bootstrap to simulate abundance estimation, based on the estimate of 2900 animals and associated coefficient of variation from Hammond et al. (2017) and assuming a log-normal distribution; (4) estimating the annual population bycatch mortality rate as the product of values from steps *1* and *2* (method 1); (5) estimating the annual number of bycatch mortalities by multiplying the estimate from step *4* by 2900; and (6) estimating the annual number of bycatch mortalities as the product of the estimates from steps *1*, *2* and *3*. For each estimate, again the 95% confidence interval was obtained by sorting the 10,000 estimates and extracting the 251st and 9750th values. These figures are compiled in Table 11, alongside some of the estimates from observer data. Evidently

Table 11 Summary of Bycatch Mortality Estimates (with 95% Confidence Intervals When Available) for Harbour Porpoises in the Iberian Peninsula 1990–2021, Based on (A) Observer Data and (B) Stranding Data

Areas and Years Sampled, Source, (and Subset)	Type of Data	% Bycatch Mortality in Strandings	Estimated Annual Population Bycatch Mortality Rate (%)	Estimated Annual Bycatch Deaths (Method 1)	Estimated Annual Population Bycatch Mortality Rate (Method 2) (%)
(A) Observer Data					
Portugal, 2010–2015 (1)	Observers (purse seines, polyvalent fleet, bottom trawls, long lines, beach seines)	N/A	8.3	241	N/A
Portugal, 2010–2015 (2)	On-board observation of polyvalent, purse seine and trawl fleets	N/A	14.4 (4.6–28.0)	418 (132–813)	14.4 (3.9–37.9)
Portugal, 2010–2020 (2)	On-board observation of polyvalent, purse seine and trawl fleets	N/A	7.9 (2.5–15.3)	228 (47–503)	7.9 (2.1–20.7)

Table 11 (*Continued*) Summary of Bycatch Mortality Estimates (with 95% Confidence Intervals When Available) for Harbour Porpoises in the Iberian Peninsula 1990–2021

Areas and Years Sampled, Source, (and Subset)	Type of Data	% Bycatch Mortality in Strandings (%)	Estimated Annual Population Bycatch Mortality Rate (%)	Estimated Annual Bycatch Deaths (Method 1)	Estimated Annual Bycatch Deaths (Method 2)
(B) Stranding Data					
Galicia, 1990–2010 (3)	Strandings	31 (21.8–41.4)	5.6 (3.73–7.41)	162 (108–215)	162 (73–300)
Galicia, 1990–2019 (4)	Strandings	34.2 (25.8–42.5)	6.2 (4.4–7.9)	180 (128–230	180 (83–324)
Galicia, 1990–2021 (5)	Strandings	40.0 (31.2–48.8)	7.2 (5.3–9.0)	209 (155–262)	209 (100–379)
Galicia, 2000–2020 (6) (PB → NB)	Strandings	41.0 (30.1–51.8)	7.4 (5.2–9.6)	214 (151–276)	214 (99–392)
Galicia, 2000–2020 (6) (PB → *B*)	Strandings	48.2 (37.3–59)	8.7 (6.4–11.0)	252 (185–319)	252 (118–455)
Galicia, 2018–2020 (7)	Strandings	44.0 (22.2–66.7)	8.0 (3.8–12.2)	232 (110–354)	232 (85–469)
Portugal, 2000–2020 (6) (PB → NB)	Strandings	48.8 (42.7–54.4)	8.8 (7.11–10.4)	255 (206–302)	255 (122–452)
Portugal, 2000–2020 (6) (PB → *B*)	Strandings	60.9 (55.2–66.5)	11 (9.1–12.8)	318 (263–372)	318 (155–561)
Portugal, 2018–2020 (7)	Strandings	38.9 (27.8–50.0)	7.0 (4.8–9.3)	203 (139–269)	203 (92–373)
Galicia+N. Portugal, 1990–2010 (3)	Strandings	44.7 (37.1–52.2)	8.0 (6.2–9.8)	233 (180–283)	233 (110–413)
Galicia+Portugal, 2000–2020 (6) (PB → NB)	Strandings	47.0 (41.8–52.2)	8.5 (6.9–9.9)	245 (200–287)	245 (119–432)
Galicia+Portugal, 2000–2020 (6) (PB → *B*)	Strandings	58.0 (52.7 62.9)	10.4 (8.6–12.1)	302 (251–351)	302 (148–529)

The bootstrap methodology used to derive 95% confidence intervals is described in the text. For data from Read et al. (2013, 2020) and Pierce et al. (2020), known bycatches handed in by fishers were removed from the dataset prior to calculations of bycatch rate. Torres-Pereira et al. (2023a) distinguished (certain) bycatches and probable bycatches. The two sets of estimates refer to when probable bycatch (PB) was treated as (1) non-bycatch (NB) and (2) bycatch (B), respectively. "Method 2" for calculating 95% confidence intervals includes consideration of uncertainty in the population size estimate.

Sources: (1) Vingada and Eira (2018) (MARPRO project), (2) ICES (2013a,b, 2014c, 2017), (3) Read et al. (2013, 2020), Pierce et al. (2020), (5) CEMMA, unpublished data, (6) Torres-Pereira et al. (2023a), (7) ICES (2020c, 2021c), (8) Goetz et al. (2014), (9) Martinez-Cedeira and Lopez (2018), (10) Vingada et al. (2011).

this process does not speak to the accuracy of the estimates, and biases could arise from several sources, including unrepresentative strandings data and changing population size, age structure and mortality rate over time.

As seen in the previous tables and text, the majority of estimates suggest a mortality of a few hundred individuals per year due to bycatch although observer data during 2016–2020 included no records of porpoise bycatch. Strandings continue to include a substantial proportion of bycaught animals, suggesting that on-board observation is not adequately documenting porpoise bycatch, possibly because it mainly occurs as a result of small-scale fisheries. It should be noted that the various strandings datasets all draw on data from the same strandings networks and are thus not all independent estimates. Furthermore, although the estimates from strandings based on data presented by Read et al. (2013, 2020) and Pierce et al. (2020) excluded known bycatches (animals handed in by fishers) from the "strandings" datasets – and there were in any case few such animals – it is not clear whether such animals had been excluded from the other compilations of strandings data. If not, in these cases there would be a small upward bias in the estimated proportion of bycatches among stranded animals.

The 95% confidence limits on the annual bycatch mortality rate and annual number of bycatch deaths are wide (very wide indeed for the observer data, as expected given very low observer coverage) but even the lower 95% confidence limits for the annual mortality rate due to bycatch are always higher and often much higher than the removal threshold of 1.7% mortality due to anthropogenic causes proposed by ASCOBANS (2000) (i.e. 49 animals from a population of 2900), thus also exceeding the annual PBR-based limit of 25 animals calculated during the 2018 IMR/NAMMCO workshop and obviously exceeding the bycatch limit of zero proposed by OSPAR.

Hunting

Harbour porpoises are currently protected by law in both Spain and Portugal, as well as under EU directives, and as such no exploitation is permitted. However, this is a relatively recent development. There is a long history of human exploitation of small cetaceans worldwide, and the Iberian Peninsula is no exception.

Although zooarchaeological remains of harbour porpoise are known from elsewhere in Europe (Evans & Mulville 2018), to our knowledge, there are no published records from the Iberian Peninsula. Porpoise remains dating from the Mediaeval period were recovered in England and the Netherlands (van den Hurk et al. 2020, 2021). In France, remains dating from pre-Roman times (Oueslati 2017, in Bernal-Casasola 2018) and historical records from Mediaeval to modern times attest to long-lasting porpoise exploitation (Berthelot 1840, Musset 1964, Fichou & Levasseur 2004). In the Mediaeval period, cetaceans were "fished" to eat on those days when the church did not permit eating of meat. In England, porpoise with frumenty (a porridge) was served at the wedding feast of Henry IV in 1404; roast porpoise was served at the crowning of Henry V; and porpoise with a sauce made of bread crumbs, vinegar and sugar was one of the favourite foods of Henry XIII, which could be eaten on fast-days because it was considered to be a fish (Fosså 1995). Dillen (2022) describes the annual supply of a porpoise (meerzwijn) for the consumption of the aldermen of Bruges in late Mediaeval times. Records indicate that porpoises were intensively hunted from at least the fourteenth century to the twentieth century in Denmark (Kinze 1995) and that hunting took place in other places in the Northeast Atlantic and Arctic (Stenson 2003) as well as in the Black Sea (Tonay & Öztürk 2012). The sale of a porpoise, alongside two seals and various fish and shellfish, is illustrated in Flemish painter Frans Snyders' painting "The Fish Market" from 1618, while his "Fish Stall" from the same period includes a porpoise, a seal and an otter (Figure 15).

In the Iberian Peninsula, remains of dolphins have been found dating from the Middle and Upper Paleolithic, and remains of other, larger, cetacean species have been found dating from the Late Neolithic/Chalcolithic and Roman periods (Corchón-Rodríguez & Álvarez-Fernández 2008, Mariezkurrena-Gastearena 2011, Álvarez-Fernández 2015, Moreno-García et al. 2017, Benito et al. 2019). Given the rarity of identifiable cetacean remains in the zooarchaeological record (see Speller

Figure 15 Photograph of the oil painting *Fish stall* by Frans Snyders and Jan Wilden, illustrating a stall in the fish market of Antwerp, dated between 1618 and 1621, held at the State Hermitage Museum, St Petersburg, Russia. Among the fish and shellfish for sale are several tortoises, a porpoise, a seal and an otter. Available at: https://commons.wikimedia.org/wiki/File:Frans_Snyders_-_Fish_Stall_-_WGA21521.jpg

et al. 2016, Evans & Mulville 2018), one cannot exclude the possibility of harbour porpoises being exploited in the area during prehistoric and historical times. The coastal distribution of the species would have made it easy for humans to exploit at least beached individuals, and/or those approaching the shoreline. More recent historical information testifies to the exploitation of the species in both Spain and Portugal.

Mediaeval Portuguese historical records from as early as the thirteenth century suggest that *toninhas* were captured and consumed in large numbers for centuries, mainly during the late nineteenth century and much of the twentieth century (Brito & Veira 2010). Although many of these records probably refer to common dolphins, some of the historical accounts cited in Brito and Veira (2010) and Brito (2011) distinguished between *golfinho* (dolphin) and *toninha*, which suggests that porpoises were also exploited. There is also historical evidence of capture of small cetaceans, including harbour porpoises, for human consumption in coastal waters of Northern Spain from as early as the Middle Ages (Valdés Hansen 2004). Records collated by Valdés Hansen (2004, 2009) refer to intense conflicts between fisheries and small cetaceans occurring from Galicia to Cantabria between the seventeenth century and the first half of the twentieth century. Small cetaceans were considered to be direct competitors that depleted fish stocks, damaged the nets by getting themselves entangled and ate the entangled fish. These conflicts resulted in the killing of many cetaceans, including harbour porpoises (Valdés Hansen 2004).

Surveys undertaken along the Portuguese coast between 1976 and 1978 revealed that small cetaceans were captured opportunistically using hand harpoons (and sometimes taken as bycatch) and sold in beach fish markets, as part of a local non-industrial fishery (Teixeira 1979, Brito & Veiera 2010). Most captures were of common dolphin but a harbour porpoise was caught and sold at Cascais market in 1979 (Teixeira 1979, in Brito & Veira 2010). Sequeira (1996) noted that, until

1981, it was legal for porpoises to be caught and sold in Portugal but she considered that they were captured incidentally rather than deliberately targeted.

Examination of carcasses of harbour porpoises stranded in Galicia over the last 30 years has not revealed any evidence of hunting or of deliberate killing by fishers (e.g. perforations in the carcasses) or evidence that flesh was removed for human consumption (e.g. dorsal muscles removed). However, López et al. (2003) reported that 69 fishers interviewed in Galicia in the late 1990s referred to cetaceans being used for human consumption, although the species involved were not identified. Some interviewees said they had eaten cetaceans (fillets or the liver) and others commented that cetaceans were eaten in the Basque country, Portugal and France. The use of cetaceans for bait, animal food and as a source of fat was also mentioned. It is not clear whether these animals were bycatches or were deliberately killed, or whether they included porpoises. In France, it is documented from interviews with retired fishers that they would catch a dolphin or a porpoise after 15–20 days at sea, when meat supplies had run out (Anonymous, Pers. Comm.).

Informal interviews were conducted (by M. Petitguyot) with six retired fishers in August 2022 in the town of A Guarda in the south of Galicia to enquire about consumption of harbour porpoises in the past. The fishers used small boats (*gamelas*) to target various types of fish (e.g. monkfish (*rape*), turbot (*rodaballo*) and shellfish). Although they never hunted porpoises, they would consume porpoises that were bycaught in their gillnets (*trasmallos* and *miños*). Bycaught porpoises represented an easily accessed source of protein that was particularly welcome in periods when food availability was sometimes limited (i.e., during the Spanish Civil War, 1936–1939). The meat was directly consumed by the family or close friends, but not sold to others or in markets. Although one interviewee had eaten porpoises only twice in his life, the others had consumed porpoise regularly when they were actively fishing (one mentioned eating around five porpoises per year). The fishers mentioned that Portuguese fishers also ate porpoises. The loin (*lomo*) was the most commonly eaten part of the animal; the lomo was soaked in water for a few days in order to remove blood and an excessive taste of salt. Once cleaned, it was cooked in various dishes or transformed into cured meat. Boiled lomo was used to make stews, cooked with fish oil (e.g. from sardines) and accompanied by potatoes and peppers. Lomo cured in salt (*salazon*) would be cut into thin slices (similar to *jámon ibérico*). The ribs were left in salt for a whole month to remove the blood and then used to make *cocido* (a stew), eaten with grains or legumes (e.g. chickpeas, haricots, beans, lentils and rice) and other vegetables. The liver was cut into fillets and fried with garlic and salt.

Pollutants and other harmful substances

Cetaceans are exposed to a range of harmful substances of anthropogenic and natural origins. Among the POPs, historically pesticides such as dichlorodiphenyltrichloroethane (DDT) were a major concern but more recently PCBs have attracted most attention, not least due to their known effects on the immune system and on reproduction in mammals. Other potentially harmful organic chemicals include polyaromatic hydrocarbons (PAHs), which are associated with oil pollution, while heavy metals are among the most relevant inorganic chemicals. Aguilar and Borrell (1995) reviewed published information on organic contaminants and metals in harbour porpoise in the eastern North Atlantic (from France to the Baltic Sea, 1967–1991) and noted that levels of organochlorines, especially PCBs, were high enough to cause concern about their effect on population status.

POPs

PCB-induced infertility was demonstrated experimentally in seals (Reijnders 1986). Based on results of studies on mink, otters and seals, a total PCB concentration of 17 μg g^{-1} (i.e. 17 mg kg^{-1}) lipid weight in blubber was estimated to be the threshold level for the onset of physiological effects (including those on reproduction) in aquatic mammals (Kannan et al. 2000), and this value has been applied in various studies on cetaceans (e.g. Schwacke et al. 2002, Pierce et al. 2008, Méndez-Fernandez et al. 2014a). This threshold was based on the commercial PCB mixture Arclor 1254 and was considered to be equivalent to 9 μg g^{-1} lipid weight in blubber when considering the sum of 25 CBs (namely, CBs 18, 28, 31, 44, 47, 49, 52, 66, 101, 105, 110, 118, 128, 138, 141, 149, 151, 153, 156, 158, 170, 180, 183, 187 and 194) (Murphy et al. 2015, Jepson et al. 2016). Jepson et al. (2016) commented that mink are especially sensitive to PCBs, and this threshold may be too low for cetaceans. Based on the analysis of concentrations of 25 PCBs in female harbour porpoises, Murphy et al. (2015) proposed that ΣPCB concentrations over 11 μg g^{-1} in mature females should be considered as indicating infertility or reproductive failure. Jepson et al. (2016) proposed the use of 41 μg g^{-1} as the threshold for reproductive impairment in cetaceans, based on the results for ringed seals in which this concentration was associated with profound reproductive impairment (Helle et al. 1976). In their study of PCB levels in porpoises, Van den Heuvel-Greve et al. (2021) argued that the most relevant congeners are 47, 49, 52, 101, 105, 118, 128, 138, 149, 151, 153, 156, 170, 180, 187, 194 and 202, thus deriving a sum-17PCB, also noting the value of calculating a sum-7PCB for comparison with OSPAR and ICES data, based on CBs 28, 52, 101, 118, 138, 153 and 180 (OSPAR Commission 2016). There has thus been some inconsistency in relation to which CBs have been measured and total concentrations should be treated as thresholds for the onset of reproductive impairment and for complete infertility, and this may of course differ according to species and indeed in relation to other factors such as health status. In relation to effects of POPs on cetacean reproduction, the focus has been almost exclusively on females. However, it has been shown that PCBs also cause reduced testis weight in male harbour porpoises and hence, presumably, also reduces male fertility (Williams et al. 2021).

Results on average PCB concentrations recorded in harbour porpoises in the North Atlantic are compiled in Table 12A. Comparisons are difficult given differences in the number and identities of congeners included in total PCB concentrations (see Table 12B), as well as expected differences related to age, sex and reproductive status of the studied animals. Where possible we included the total concentration of the ICES 7 congeners, thus permitting comparison with the Kannan et al. (2000) threshold of 17 μg g^{-1} lipid, which is equivalent to Σ [ICES 7 CBs] = 5.67 μg g^{-1} lipid (Jepson et al. 2005, Pierce et al. 2008, Murphy et al. 2010). Since the 17 μg g^{-1} threshold based on Aroclor 1254 was considered to be equivalent to 9 μg g^{-1} of the 25 CBs measured by Jepson et al. (2016), it follows that the other thresholds based on 25 CBs suggested by Murphy et al. (2015) and Jepson et al. (2016), i.e., 11 μg g^{-1} and 41 μg g^{-1}, would be equivalent to 6.93 μg g^{-1} and 25.81 μg g^{-1}, respectively, of the ICES 7 CBs.

The EU-funded BIOCET project (2001–2004) surveyed pollutant levels in tissues of small cetaceans on European Atlantic coasts. The animals sampled from Galicia included three porpoises. Total PCB burden was estimated as 3 × Σ [ICES 7 CBs]. One of these individuals had a total PCB concentration in blubber higher than the threshold of 17 μg g^{-1} lipid weight. Of the five cetacean species from the NWIP studied by Méndez-Fernandez et al. (2014a), bottlenose dolphin and harbour porpoise showed the highest concentrations of PCBs in their blubber. In these species, 100% and 75%, respectively, of the individuals analysed exceeded the toxic threshold of 17 μg g^{-1} lipid weight.

Table 12A Mean PCB Concentrations (± Standard Deviation and/or with Range (in Parentheses) When Given) Reported for Harbour Porpoises in the North Atlantic

Area	Time Period	Sex/Maturity	Sum [PCB] (μg.g^{-1} Lipid Weight)	ICES 7 [PCB] (μg.g^{-1} Lipid Weight)	No of Porpoises	No of Congeners	Reference
Canada (Bay of Fundy / Gulf of Maine)	1989–1991	Males	17.28±11.18		55	68	Westgate et al. (1997)
		Females	11.38±4.81		53	68	
Canada (Gulf of St. Lawrence)		Males	10.64±5.43		31	68	
		Females	7.15±3.85		31	68	
Canada (Southeast Newfoundland)		Males	5.24±2.51		18	68	
		Females	5.49±4.37		11	68	
Denmark and Norway	1987–1991	Males	23.27 (3.75–65.26)		34	47	Klevaine et al. (1995)
UK	1990–2012	Males	19.41 (0.44–150.47)	10.14±10.42 (0.25–65.88)	146	23	Jepson et al. (2016)
		Females	13.49 (0.40–159.68)	6.89±8.16 (0.18–71.1)	134	23	
	2000–2003	Females	10.52±13.15	6.77±8.41 31% above threshold	31	16	Pierce et al. (2008)
	2014–2018	Males and Females	16.31 (0.46–159.68)		604	25	Williams (2021)
Ireland	2000–2003	Females	5.35±4.75	3.51±3.15 25% above threshold	12	16	Pierce et al. (2008)
Netherlands and Belgium	2000–2003	Females	15.02±8.57	10.20±6.00 74% above threshold	19	16	Pierce et al. (2008)
Netherlands	1999–2004	Males and Females	12.4		35	21	Weijs et al. (2009)
	2008–2019	Both sexes	12.1±16.42 (0.2–90.2) 38.5% above threshold	8±10.64 (0–55.4)	121	17	Van den Heuvel-Greve et al. (2021)
		Foetuses	3.46±3.10 (0.6–8.8) 0%>above threshold	2.39±2.24 (0–6.3)	9	17	

(Continued)

Table 12A (*Continued*) Mean PCB Concentrations (± Standard Deviation and/or with Range (in Parentheses) When Given) Reported for Harbour Porpoises in the North Atlantic

Area	Time Period	Sex/Maturity	Sum [PCB] (μg.g^{-1} Lipid Weight)	ICES 7 [PCB] (μg.g^{-1} Lipid Weight)	No of Porpoises	No of Congeners	Reference
		Neonates	11.50±7.10 (1.8–24.0)	8.19±5.19 (1.4–17.1)	16	17	
		Juveniles	15.30±20.22 (2.6–90.2) 48.9% above threshold	9.8±12.67 (1.7–55.4)	45	17	
		Mature Males	23.50±15.46 (3.8–62.4) 92.3% above threshold	16.24±10.5 (2.6–42.2)	13	17	
		Mature Females	6.70±13.47 (0.2–76.2) 10.5% above threshold	4.45±8.94 (0.1–49.8)	38	17	
Belgium and France (Southern North Sea)	2010–201	Immature Female	32±21		3	20	Mahfouz et al. (2014)
		Immature Male	20±31		12	20	
		Mature Females	4±1.8		4	20	
		Mature Males	22		1	20	
France	2000–2003	Females	13.81±10.58	9.20±6.96 50% above threshold	2	16	Pierce et al. (2008)
Spain (Northwest)	2000–2003	Females	5.31±4.20	3.42±2.66 33% above threshold	3	16	Pierce et al. (2008)
	2004–2008	Immature Males	9.4±3		3	32	Méndez-Fernandez et al. (2014a)
		Immature Females	10.8±2.8		5	32	
		Mature Males	50.8		1	32	

(*Continued*)

Table 12A (*Continued*) Mean PCB Concentrations (± Standard Deviation and/or with Range (in Parentheses) When Given) Reported for Harbour Porpoises in the North Atlantic

Area	Time Period	Sex/Maturity	Sum [PCB] ($\mu g.g^{-1}$ Lipid Weight)	ICES 7 [PCB] ($\mu g.g^{-1}$ Lipid Weight)	No of Porpoises	No of Congeners	Reference
		Mature Females	37.5±30.8		3	32	
	2015–2019	Immature Males	6.9±6.7	5.1±5.2	4	14	TRANSITION project, unpublished data
		Immature Females	9.1±6.2	6.8±4.7	4	14	
		Mature Males	12.62±9.8	9.2±6.9	4	14	
		Mature Females	14.2±10.6	10.9±8.5	2	14	

We give results in $\mu g.g^{-1}$ (equivalent to $mg.kg^{-1}$). Also indicated are the number, sex and maturity stage of the individuals analysed, number of PCB congeners analysed, and the source of the information. The sum [PCB] columns refers to the set of congeners measures in each study (see Table 8B). To facilitate comparison, where possible we include the summed concentration of the ICES 7 PCBs. Previously reported thresholds correspond to Σ[ICES7 PCBs] of 5.67 $\mu g.g^{-1}$ (Kannan et al. 2000), 6.93 mg kg^{-1} (Murphy et al. 2015) and 25.81 $\mu g.g^{-1}$ (Jepson et al. 2016). Pierce et al. (2008) and Van den Heuvel-Greve et al. (2021) reported on the percentage of animals with PCB concentrations above the Kannan et al. (2000) threshold. Means and standard deviations of total PCB concentration in the latter study were reconstructed from their supplementary data

Table 12B PCB Congeners Measured in the Different Studies Included in Table 12A

Congener	ICES 7	Klevaine et al. (1995)	Westgate et al. (1997)	Pierce et al. (2008)	Weijs et al. (2009)	Méndez-Fernandez et al. (2014a)	Jepson et al. (2016)	Williams (2021)	van den Heuvel-Greve et al. (2021)	Transition project (unpubl.)
18							1	1		
26		1								
28	1			1	1	1	1	1		1
31					1	1	1	1		1
44						1	1	1		
47		1					1	1	1	
49				1		1	1	1	1	
52	1	1	1	1	1	1	1	1	1	1
66		1					1	1		
70		1				1				
74					1	1				
87		1								
91		1								
95		1	1		1					
97						1				
99		1			1	1				
101	1	1	1	1	1	1	1	1	1	1
105					1	1	1	1	1	1
107		1								
110		1			1	1	1	1		
111		1								

(Continued)

Table 12B (*Continued*) PCB Congeners Measured in the Different Studies Included in Table 12A

Congener	ICES 7	Klevaine et al. (1995)	Westgate et al. (1997)	Pierce et al. (2008)	Weijs et al. (2009)	Méndez-Fernandez et al. (2014a)	Jepson et al. (2016)	Williams (2021)	van den Heuvel-Greve et al. (2021)	Transition project (unpubl.)
114		1				1				
118	1	1	1	1	1	1	1	1	1	1
123		1				1				
128		1		1	1	1	1	1	1	
130		1								
132		1				1				
137		1				1				
138	1	1	1	1	1	1	1	1	1	1
141		1		1			1	1		
146		1								
148		1								
149			1	1	1	1	1	1	1	1
151		1	1	1			1	1	1	
153	1	1	1	1	1	1	1	1	1	1
156					1	1	1	1	1	1
157		1				1				
158		1				1	1	1		
167		1				1				
170		1		1	1	1	1	1	1	1
174		1								
177		1								

(Continued)

Table 12B (*Continued*) PCB Congeners Measured in the Different Studies Included in Table 12A

Congener	ICES 7	Klevaine et al. (1995)	Westgate et al. (1997)	Pierce et al. (2008)	Weijs et al. (2009)	Méndez-Fernandez et al. (2014a)	Jepson et al. (2016)	Williams (2021)	van den Heuvel-Greve et al. (2021)	Transition project (unpubl.)
178		1								
179		1								
180	1	1	1	1	1	1	1	1	1	1
183		1		1	1	1	1	1		
185		1								
187		1	1	1	1	1	1	1	1	1
189						1				
192		1								
193		1								
194		1		1	1	1	1	1	1	1
195		1								
196		1								
199					1					
201		1								
202		1							1	
206		1								
209		1				1				
SUM	7	47	68	16	21	32	25	25	17	14

The number 1 indicates that a congener was measured. Note that most studies report the ICES 7 congeners (shaded cells). Westgate et al. (1997) analysed 68 congeners but only mentioned those that together comprised around 50% of the total. Pierce et al. (2008) measured concentrations of 18 congeners but the study included UK porpoise samples for which only 16 congeners had been measured (the two missing CBs were 99 and 177) and, for comparability, results for 16 congeners in porpoises were presented for all regions.

Recent unpublished work conducted in part under the TRANSITION project (Gutiérrez-Muñoz et al. in prep) found somewhat higher total PCB concentrations in porpoises stranded in Galicia during the period 2015–2023 ($N = 19$) than found by Méndez-Fernandez et al. (2014a) during the period 2004–2008. Although further work is needed as sample sizes were small, a decreasing trends over time has been reported elsewhere in Europe (e.g. Williams et al. 2023).

Numerous other Persistent Organic Pollutants are found in cetaceans, including insecticides (e.g. DDTs) brominated flame retardants. The three Galician porpoises included in the Pierce et al. (2008) study had low concentrations of flame retardants in their blubber, namely, polybrominated diphenylethers (PBDEs) (Pierce et al. 2008) and hexabromocyclododecane (HBCD) (Zegers et al. 2005). HBCD concentrations were lower than those in porpoises sampled from Dutch, Irish and Scottish waters. While adverse effects of pesticides on marine animals are well known, the effects of some other POPs remain undescribed. We are not aware of studies on other organic pollutants in Iberian porpoise but it may be noted that in Norwegian and Danish waters, organochlorine pesticide concentrations in porpoises were two to three times higher than those in seals in samples collected during 1987–1991 (Klevaine et al. 1995).

Other contaminants

The proximity of Galicia to one of the world's busiest shipping routes, along with the presence of a refinery located in the port of A Coruña, makes the NWIP coastline particularly vulnerable to oil pollution. Galicia has experienced five out of the eleven major oil spills in Europe in the last three decades (Franco et al. 2006, González et al. 2006, Loureiro et al. 2006). The "Prestige" oil spill in November 2002 released 60,000 metric tonnes of oil into the Atlantic off Galicia and polluted 1300 km of coastline (Loureiro et al. 2006). In the 6 months following the spill, 124 cetaceans were stranded along the Galician coast, of which 35% were oiled and 3% were considered to have died as a direct result of oil. This included the mortality of at least one porpoise (López et al. 2005, Loureiro et al. 2006). It is also likely that there were indirect impacts on porpoises via the food chain. Sánchez et al. (2006) monitored the abundance of four key benthic and demersal species and noted reductions in the abundance of three of them: Pandalid shrimp (*Plesionika heterocarpus*), Norway lobster (*Nephrops norvegicus*) and four-spot megrim (*Lepidorhombus boscii*), although the abundance and distribution of juvenile hake, which are eaten by porpoises, showed no significant changes.

Coastal upwelling systems often experience the occurrence of harmful algal blooms (HABs), e.g., in cultivated mussels (Álvarez-Salgado et al. 2008). The main associated risk to cetaceans relates to the production of natural toxins by various species of phytoplankton and the accumulation of these toxins in food webs (Broadwater et al. 2018, Fire et al. 2021). Domoic acid was found in harbour porpoises (and in a porpoise foetus) from Alaska by Lefebvre et al. (2016), while saxitoxins were reported in porpoises from Canada by Bates et al. (2020). Another toxin associated with HABs is the neurotoxin β-N-methylamino-L-alanine (BMAA), which is produced by both cyanobacteria and diatoms. Soliño et al. (2022) found no evidence of BMAA in stranded porpoises, bottlenose dolphins and common dolphins from Northwest Spain.

Concerning inorganic contaminants, Lahaye et al. (2007) reported low concentrations of the toxic chemical elements mercury (Hg) and cadmium (Cd) in porpoises from Galicia. According to Méndez-Fernandez et al. (2014b), the concentrations of Hg and Cd found in Iberian toothed whales indicate that these populations are not especially threatened by Hg and Cd exposure in the area. Compared with the values reported by Méndez-Fernandez et al. (2014b), 42 harbour porpoises stranded in Portugal from 2005 to 2013 had higher levels of Hg in liver and lower levels of Cd in kidneys (Ferreira et al. 2016). The higher Hg levels may reflect anthropogenic sources, with Portuguese animals inhabiting waters closer to the Mediterranean where high levels of Hg occur in the seawater due to human activities, such as mercury mining, and a high density of sub-marine volcanoes (Bernhard & Renzoni 1977, Cossa et al. 1997, 2022). Nevertheless, the recorded hepatic Hg levels did not exceed the level for toxic thresholds in marine mammals of 60 $\mu g\ g^{-1}$ wet weight (Rawson et al. 1993).

The health implications of the presence of microplastics in cetacean stomachs are presently unknown. Recent studies on plastics in Portugal have shown that the environmental concentration of microplastics is higher in the northern and central coasts of Portugal, although these concentrations are low to moderate compared with other areas of the NE Atlantic (Prata et al. 2020). In the same study, microplastics were found in most of the species of the biota groups investigated (loggerhead turtles, mussels, commercial fish and birds), although with differences in prevalence (Prata et al. 2020). Guilhermino et al. (2021) analysed several species of fish from the estuary of the Minho, on the northern border of Portugal with Spain. Ninety-four percent of the fishes sampled presented microplastics, with a mean of 8 ± 7 microplastic items/fish (among the highest described in the literature at worldwide level). A study of microplastics in anchovy (*Engraulis encrasicolus*), sardine, common dragonet and the mullet *Mullus surmuletus* in Galician waters also showed a high incidence of microplastics (78%) in the fish examined (Filgueiras et al. 2020). Two of these species have been recorded in porpoise diet in Galicia (see section 'Feeding ecology – Diet and prey consumption'). There have been no studies on microplastics in the stomach contents of porpoises in the Iberian Peninsula but a recent study on common dolphin stomach contents in Galicia found microplastics in every stomach ($N=35$) examined (Hernández-González et al. 2018). It would be surprising if this was not also the case for porpoises.

Currently, there is no routine monitoring of contaminants in stranded cetaceans in the Iberian Peninsula. The studies reported above have been based on short-term, usually project-based funding. Consequently, the absence of reports does not necessarily imply the absence of a problem.

Pathogens: viruses, bacteria and parasites

Viruses

Several viruses have been detected in harbour porpoises in Northeast Atlantic waters, including rhabdovirus, papillomavirus, herpesvirus and morbillivirus (Van Bressem et al. 1999, 2014). Both herpesvirus and morbillivirus can cause serious and lethal disease. Cetacean morbillivirus infections result in a high mortality rate and, due to recurrent epizootics, can have long-term effects on population dynamics in cetaceans (Van Bressem et al. 1999). Morbillivirus infections in cetaceans were first described at the time of the 1988 phocine distemper virus (PDV-1) epidemic among harbour seals, when morbillivirus antigens were detected in organs of harbour porpoises that had died on the Irish coast (Kennedy et al. 1988, McCullough et al. 1991), although exposure in cetaceans may have occurred prior to the PDV outbreak (Härkönen et al. 2006). A morbillivirus epidemic was subsequently seen in striped dolphins (*Stenella coeruleoalba*) in the western Mediterranean, from which a dolphin morbillivirus (DMV) was isolated from striped dolphins (Van Bressem et al. 1991). Visser et al. (1993) isolated a (previously unidentified) porpoise morbillivirus (PMV) from two harbour porpoises that died in the Dutch Wadden Sea (North Sea) in 1990. Porpoise and dolphin morbilliviruses represent a distinct lineage closer to ruminant morbilliviruses than to carnivore morbilliviruses (Barrett et al. 1993).

Herpesvirus infections have previously been described in harbour porpoises from Sweden and the Netherlands. Kennedy et al. (1992) reported herpes viral encephalitis in a porpoise stranded in Sweden, while van Elk et al. (2016) examined stranded porpoises in the Netherlands, detecting three types of herpesvirus. A known gammaherpesvirus (PPHV-1) was found in one animal (out of 117 animals), associated with a genital plaque, and two novel alphaherpesviruses. PPHV-2 was found in the brain tissue of one animal (out of 74) associated with encephalitis, while PPHV-3 was found in the brain (4 of 74 animals), blowhole swabs (2 of 43 animals) and genital swabs (2 of 43 animals) but not associated with disease. The authors note that PPHV-2 resemble dolphin herpesviruses, while PPHV-3 was more similar to cervid herpesviruses.

In Portugal, samples from two out of 31 stranded porpoises tested positive for different alphaherpesviruses (representing two out of three monophyletic branches detected among the cetacean samples). In addition, in a phylogenetic analysis, a third porpoise sample clustered with gammaherpesvirus sequences (Bento et al. 2019).

Bacteria

Bacterial species detected in harbour porpoises stranded along the Northeast Atlantic coast include *Actinobacillus delphinicola*, *Actinobacillus scotiae*, *Brucella* spp., *Candida albicanis*, *Cetobacterium ceti*, *Nocardia asteroides*, *Salmonella* and *Streptococcus* spp. (Higgins 2000). *Brucella ceti* appears to be present in cetaceans globally, and distinct dolphin and porpoise types have been identified. Brucellosis in dolphins and porpoises may result in significant clinical and pathological signs related to abortions, male infertility, neuropathogenicity, cardiopathies, bone and skin lesions and death (Guzmán-Verri et al. 2012). Dagleish et al. (2008) and Jauniaux et al. (2010) described pathologies associated with *Brucella* infection in a male porpoise and a female porpoise, respectively, and their findings suggest that infection may result in sterility. Murphy et al. (2015) noted that a porpoise from the UK with a *Brucella ceti* of the mammary glands had lost its calf, possibly due to abortion. *Brucella* was isolated from a striped dolphin live-stranded on the coast of Cantabria (northern Atlantic coast of Spain) in 2004 (Muñoz et al. 2006). Regional variations in pathologies related to bacterial flora were described by Siebert et al. (2009), showing less bacterial growth associated with fewer pathologies in harbour porpoises from Greenlandic, Icelandic and Norwegian waters, compared to individuals from German North and Baltic Seas.

To the authors' knowledge, the only study reporting on bacteria in porpoises stranded along the Iberian coastline was conducted by Soares-Castro et al. (2019) who carried out a comparative analysis of the oral microbiome of the three most common cetacean species stranded on the Northwestern Atlantic Iberian coast (common dolphin, striped dolphin and harbour porpoise). Microbiomes are expected to include both beneficial and potentially harmful microbes. This study revealed significant differences in microbiota between the Phocoenidae and the two Delphinidae species. The Shannon diversity index for harbour porpoise microbiomes was lower than that in common dolphins and higher than that in striped dolphins. The high heterogeneity between species was reflected in the small number (12) of Operational Taxonomic Units (OTU) present in all samples from all three species. Of the 903 OTUs described in the porpoise samples, 18 were present in all porpoise samples (see Table 13). Two bacterial genera were suggested as "fingerprints" for porpoises (*Dethiosulfovibrio* and *Marinicella*), which could be used as bioindicators to develop diagnostic assays and monitoring tools for the assessment of cetacean population and ecosystem health. The microbial community also differed according to age class, with adult and mature animals showing a higher mean number of OTUs and a higher Shannon index. Differences between males and females were detected only at family level. Microbiomes differed depending on the location and health status of the animals, although porpoise samples were insufficiently variable to highlight patterns in this species.

Parasites

Cetaceans host a wide range of parasites. The most visible endoparasites during necropsy include nematodes in the respiratory and gastrointestinal tracts, which are often present in large numbers. Odontocete cetaceans are final hosts of all nine known species of the gastrointestinal parasite *Anisakis* (Mattiucci et al. 2018). Adult *Anisakis* usually live unattached within the forestomach (Smith 1989, Herreras et al. 2004) and feed off the food bolus, something that has been associated with anaemia in heavily infected hosts (Gibson et al. 1998). The prevalence of *Anisakis* spp. is apparently increasing in cetaceans in some areas of the Northeast Atlantic (Lino et al. 2022), where the high abundance of *Anisakis* larvae found in some commercial fish species such as European

Table 13 Taxonomy of the Operational Taxonomic Units (OTUs) Composing the Core Microbiome of the Oral Cavity of the Sampled Harbour Porpoises

Phylum	Class	Order	Family	Genus	Species
Proteobacteria	Gammaproteobacteria	Pasteurellales	Pasteurellaceae	*Phocoenobacter*	Uncultured bacterium
Fusobacteria	Fusobacteriia	Fusobacteriales	Fusobacteriaceae	*Fusobacterium*	Uncultured bacterium
Fusobacteria	Fusobacteriia	Fusobacteriales	Leptotrichiaceae	*Oceanivirga*	Uncultured bacterium
Bacteroidetes	Bacteroidia	Bacteroidales	Porphyromonadaceae	*Porphyromonas*	Uncultured bacterium
Proteobacteria	Gammaproteobacteria	Cardiobacteriales	Cardiobacteriaceae	uncultured	Uncultured bacterium
Proteobacteria	Epsilonproteobacteria	Campylobacterales	Campylobacteraceae	*Arcobacter*	Uncultured bacterium
Firmicutes	Clostridia	Clostridiales	Lachnospiraceae	uncultured	Uncultured bacterium
Bacteroidetes	Flavobacteriia	Flavobacteriales	Flavobacteriaceae	*Maritimimonas*	Uncultured bacterium
Proteobacteria	Gammaproteobacteria	Pasteurellales	Pasteurellaceae	*Actinobacillus*	Uncultured bacterium
Proteobacteria	Gammaproteobacteria	Vibrionales	Vibrionaceae	*Vibrio*	*Vibrio* sp. S-C1-5
Proteobacteria	Epsilonproteobacteria	Campylobacterales	Campylobacteraceae	*Campylobacter*	Uncultured bacterium
Firmicutes	Clostridia	Clostridiales	Family XIII	uncultured	Uncultured bacterium
Proteobacteria	Betaproteobacteria	Burkholderiales	Burkholderiaceae	*Ralstonia*	Uncultured bacterium
Firmicutes	Clostridia	Clostridiales	Family XII	*Fusibacter*	Uncultured bacterium
Proteobacteria	Gammaproteobacteria	Cardiobacteriales	Cardiobacteriaceae	uncultured	Uncultured bacterium
Bacteroidetes	Bacteroidia	Bacteroidales	Bacteroidales S24-7 group	uncultured bacterium	Uncultured bacterium
Bacteroidetes	Bacteroidia	Bacteroidales	Bacteroidales S24-7 group	uncultured bacterium	Uncultured bacterium
Fusobacteria	Fusobacteriia	Fusobacteriales	Fusobacteriaceae	*Fusobacterium*	Uncultured bacterium

Source: From Soares-Castro et al. (2019).

hake and blue whiting is a cause for concern for human consumers (see Levsen et al. 2018, Pascual et al. 2018).

Lungworms (nematodes of the family Pseudaliidae) infect the respiratory tracts of cetaceans, where they may induce parasitic pneumonia, one of the most common causes of death diagnosed in porpoises stranded in the UK (Baker & Martin 1992), and are also found in the cranial sinuses and middle ear. Kuhn (1829) first described the presence of *Strongylus* (now *Pseudalius*) *inflexus* in the lungs, bronchi and typani cavity of a harbour porpoise. Lungworms typically accumulate with age in odontocetes (e.g. Raga & Balbuena 1993; see Measures 2001 for a review of lungworms in marine mammals). They are also vectors of viral and bacterial infections. Davison et al. (2010) isolated *Salmonella enterica* from lungworms *Pseudalius inflexus* removed from a harbour porpoise. It has been suggested that initial infections with the pseudaliid nematode *Stenurus minor* may stimulate protective immunity in porpoises (Faulkner et al. 1998).

The probability of finding parasites in all organs of porpoises increases with body length (ten Doeschete et al. 2017), hence presumably also with age. These authors also reported a higher probability of parasite presence in the ears and stomachs of porpoises in a poorer nutritive condition.

Nine species of parasites have been identified in harbour porpoises bycaught and stranded in Spain (Table 14), as compared to eight species from the Baltic Proper subpopulation. These numbers represent a small proportion of the 55 taxa of parasites that have been described in harbour porpoises worldwide since 1809 (Dzido et al. 2021). Some of the most common species from the global list have been detected in Galician waters, including the respiratory nematodes *Stenurus minor* and *Halocercus invaginatus*, and the gut nematodes *Anisakis simplex* (*s.s.*) and *Anisakis pegreffii*, although several other frequently recorded parasites have not yet been seen in this region, such as the liver fluke *Campula oblonga*, the gut cestode *Diphyllobothrium stemmacephalum* and the respiratory nematodes *Pseudalius inflexus* and *Torynurus convolutus*. Both larvae and adults of *Anisakis simplex* (*s.l.*) (in one case confirmed as *Anisakis simplex* (*s.s.*)) were detected in three out of four porpoises examined in Galicia between 1991 and 1996 (Abollo et al. 1998). Pons-Bordas

Table 14 Parasite Species Detected in Harbour Porpoises from Iberian Waters

Parasite Species	Prevalence	Sample Type	Year	Region	Reference
Anisakis sp.	Unknown	Stranded	2019	Galicia	Covelo and López (2021b)
Anisakis simplex (*s.l.*)[1]	66.7%	Stranded	2017–2018	Galicia	Pons-Bordas et al. (2020)
Anisakis sp.[2]	55.5% (5/9)	Stranded	2019–2021	Galicia	TRANSITION project, unpublished data
Anisakis simplex (*s.l.*)	75%	Stranded	1991–1996	Galicia	Abollo et al. (1998)
Anisakis simplex (*s.s.*)	Unknown	Stranded	2004–2019	Iberian Atlantic coast	Cipriani et al. (2022)
Anisakis pegreffii	Unknown	Stranded	2004–2019	Iberian Atlantic coast	Cipriani et al. (2022)
Anisakis simplex (*s.s.*) × *A. pegreffii* hybrid	Unknown	Stranded	2004–2019	Iberian Atlantic coast	Cipriani et al. (2022)
Crassicauda sp.	Unknown	Stranded	2019	Galicia	Covelo and López (2021b)
Halocercus invaginatus	25%	Stranded	1991–1996	Galicia	Abollo et al. (1998)
Giardia duodenalis[3]	5.9%	Stranded	2008–2012	Galicia	Reboredo-Fernández et al. (2015)
Cryptosporidium sp.	5.9%	Stranded	2008–2012	Galicia	Reboredo-Fernández et al. (2015)
Isocyamus deltobranchium	Unknown	Stranded	2007	Galicia	Martínez et al. (2008)
Toxoplasma gondii	Unknown	Stranded	Unknown	Andalusia[3]	Cabezón et al. (2004)
Stenerus minor	75%	Bycaught	2009–2018	Galicia	Saldaña Ruiz (2021)
Stenerus minor	Unknown	Stranded	2018	Galicia	Díaz Caneiro (2019)

Prevalence is specified when possible. Information about the sample type (known bycatch or stranded (the latter may also include bycaught animals)), date and region of collection is also provided. (Notes: 1. Results based on ulcerative lesions typically associated with the presence of *Anisakis simplex* s.s.; 2. Pending identification to species. 3. *Giardia duodenalis* is also sometimes listed (incorrectly) as *G intestinalis* or *G. lamblia* (Thompson & Monis 2011). 3. Andalusian Mediterranean coast.)

et al. (2020) observed ulcerative lesions associated with the presence of *Anisakis simplex* (*s.l.*) (again in one case confirmed as *Anisakis simplex* (*s.s.*)) in one of three porpoises examined which had stranded in Galicia in 2017 and 2018. Cipriani et al. (2022) identified *Anisakis simplex* (*s.s.*), *Anisakis pegreffii* and a hybrid of both species in harbour porpoises stranded along the Iberian Atlantic coast from samples collected between 2004 and 2019.

Abollo et al. (1998) recorded adult nematodes *Halocercus invaginatus* in the lungs of porpoises from Galicia. *Stenerus minor*, another nematode affecting the respiratory system and the associated cardiovascular organs, was recorded in harbour porpoises in Galician waters by Saldaña Ruiz (2021), who found between 27 and 69 specimens in each of three individuals (see also Saldaña et al. 2022), and Díaz Caneiro (2019), who counted 117 specimens in one stranded porpoise. These numbers are considerably lower than numbers reported from individual porpoises in some other regions: 2928 (Baltic Sea, Dzido et al. 2021), 8920 (Gulf of St. Lawrence, Faulkner et al. 1998) and 11,000 (Black Sea, Biserkov & Dimitrov 1991). Unlike gastrointestinal nematodes, there is a clear link between lungworm infestation and mortality, with parasitic pneumonia being a commonly reported cause of death in other regions (e.g. Neimanis et al. 2022).

Protozoan parasites such as *Toxoplasma gondii* were detected in blood antibodies in one of the few stranded harbour porpoises in the Andalusian Mediterranean coast (Cabezón et al. 2004). Ectoparasites have been observed in porpoises stranded in Galicia: the cyamid *Isocyamus deltobranchium* was present on the caudal peduncle and flukes of one individual (Martínez et al. 2008). A negative relationship between inorganic element concentrations and parasite burdens was reported for porpoises stranded along the Portuguese coast: individuals with high parasite burdens showed a low concentration of zinc and arsenic in all organ systems and nickel in the renal system; it was also noted that such animals tended to be those with better nutritional condition (Ferreira et al. 2016). This may simply imply that animals eating more tend to ingest more parasites.

Other anthropogenic threats

Human activities at sea can cause mortality of porpoises, e.g., due to ship strikes, as well as having sub-lethal effects due to habitat degradation and disturbance, including that caused by underwater noise. Ship strike mortality is perhaps best known in the context of large cetaceans but it also affects small cetaceans and many other marine species. A review by Schoeman et al. (2020) cites published reports of ship strike mortality of harbour porpoises in the UK, the Netherlands and Canada. European marine waters in general have a high density of shipping, with the Bay of Biscay and Iberian coast being one of the areas that accommodates major shipping routes (e.g. Kinneging 2022).

Underwater noise and disturbance

The effect of underwater noise on cetaceans and other marine life has attracted considerable attention (e.g. Thomsen et al. 2021). To date, there has been no published assessment of underwater noise for the Iberian Peninsula (Kinneging 2022) although there have been studies of specific areas, e.g., the Mediterranean port of Cartagena (Rodrigo et al. 2022). In the USA, a framework was developed to assess the Population Consequences of Acoustic Disturbance (PCAD) for marine mammals (NRC 2005), later further developed as more general framework (Population Consequences of Disturbance (PCoD)) (Pirotta et al. 2018). One of the few examples of its application in Europe was a study of the effects of offshore windfarm development in the North Sea on porpoises (King et al. 2015). Avoidance of vessels by porpoises is well known and hence is taken into account when estimating abundance from boat-based surveys (e.g. Palka & Hammond 2001), and there is evidence from both wild and captive studies that porpoises respond to vessel noise. Dyndo et al. (2015) reported that porpoises responded to the noise from passing vessels by displaying stereotypical "porpoising" behaviour, noting that, in the wild, this would not only have an energetic cost (investment in moving, loss of foraging opportunities) but could also lead to abandonment of calves. Wisniewska et al. (2018) demonstrated that vessel noise disrupts foraging in porpoises.

As acoustic deterrents (pingers) on nets have been increasingly deployed in attempts to reduce fishery bycatch of porpoises, the possible effects of the noise they produce, in terms of disruption of behaviour, displacement from foraging areas, stress, hearing loss, and impacts on individual health and reproductive output, and population viability, have also received attention, based on field observations, experimental studies and modelling (e.g. Culik et al. 2001, Kastelein et al. 2000b, Lusseau et al. 2023). This topic is revisited in section 'Current status, knowledge gaps and future research, monitoring and conservation'.

Given the high energetic requirements of the Atlantic harbour porpoises, and hence the large proportion of time they need to spend feeding, it is expected that they will present low resilience to disturbance (Wisniewska et al. 2016), especially if displaced from their feeding grounds due to, for example, continuous underwater sound (IAMMWG et al. 2015). The presence of motor boats and fishing vessels was found to negatively affect the presence and density of porpoises recorded off Galicia (Díaz López & Methion 2018). Similar results were found in central Portugal, where the porpoise sighting rate increased as the number of fishing boats decreased (Pereira 2015). Further work is needed to understand whether disturbance from vessel traffic represents a population-level threat to Iberian porpoises.

Other anthropogenic sources of disturbance documented in other parts of the world but not in the Iberian Peninsula include offshore renewable energy development (e.g. Simmonds & Brown 2010) and seal deterrents used at fish farms (Findlay et al. 2021). Effects may include not only increased stress and displacement from preferred feeding areas but also, in the case of Acoustic Deterrent Devices (ADDs) used in aquaculture, auditory impairment. Despite the increasing importance of aquaculture in Spain, notably in Galicia where species such as turbot, sea bream and mussels are cultured, seals are occasional (vagrant) visitors to most of the Iberian Peninsula (with the exception of a very small resident population of Mediterranean monk *Monachus monachus* seal off Madeira), and we are not aware of the use of acoustic deterrents. The development of offshore renewable energy production has the potential to represent an additional pressure in the future.

Prey depletion

Porpoises are also thus likely to be susceptible to prey depletion due to their high feeding rate (Wisniewska et al. 2016). MacLeod et al. (2007) highlighted the relatively high frequencies of deaths attributed to starvation in porpoises along the Scottish coast at times of low sandeel abundance, although the existence of any causal relationship between the two variables remains speculative: it is sometimes difficult to determine whether weight loss in such animals was due to low food availability or some underlying pathology. Nevertheless, the incidence of starvation/emaciation has been increasing, among necropsied porpoises in UK waters (Deaville et al. 2018), and it was also a leading cause of death among porpoises that died during an unusual mortality event in Dutch waters in 2011 (IJsseldijk et al. 2022).

The main prey of porpoises in Galicia (NWIP) include blue whiting, *Trisopterus*, hake and scad, all of which are commercially important, and at least in terms of numbers eaten, gobies and silvery pout, which are not (Pierce et al. 2010, Read et al. 2013, Hernandez Gonzalez et al. 2024); "see also Feeding ecology". Stocks of scad, European hake and blue whiting are all assessed by ICES. Two stocks of scad occur in the area (see Brunel et al. 2016), one on the north coast (the western stock), which is currently at a historically low level (ICES 2018b), and one on the west coast (the southern stock) which, while less abundant, appears to be increasing (ICES 2018b). The southern stock of European hake is distributed in the Cantabrian Sea and Atlantic Iberian waters while blue whiting forms a single wide-ranging stock, the distribution of which includes the Iberian Peninsula. The abundance of both these stocks is currently above MSY (ICES 2018b). The decline of the western stock of scad in ICES subarea 8 (including the north Spanish coast) is a potential cause for concern. Despite its commercial importance, *Trisopterus luscus* is not assessed by ICES. In general terms, prey depletion is probably not an issue for Iberian porpoise, although low availability of energetically rich species such as sardine and scad could have adverse consequences. The amount of some commercial fish species removed by porpoises was estimated and compared with removal by fisheries (Santos et al. 2014), suggesting that porpoises removed 1% or less of the amount of hake taken by fisheries and less than 0.1% of the amount of sardine taken by fisheries hake. The abundance estimate for Iberian porpoise available at the time was around 1100 porpoises, subsequently revised to approximately 2900, but the impact of porpoises on fishery catches is still likely to be minimal. However, the impact of competition with fisheries on Iberian harbour porpoises is unknown.

Interactions with other species

In the North Atlantic, harbour porpoises are known to be preyed upon by killer whales *Orcinus orca* (e.g. Samarra et al. 2018), grey seals *Halichoerus grypus* (Leopold et al. 2015) and large sharks, e.g., great white shark *Carcharodon carcharias* (Arnold 1972) and Greenland shark *Somniosus microcephalus* (Williamson 2011). The Atlantic coast of Spain is within the known distribution of great white shark, although records of its occurrence in Galicia are considered questionable (the records are of fish that were likely caught elsewhere) (Bañón et al. 2010). According to Fishbase, records of Greenland shark in the eastern Atlantic are rare further south than Ireland and the species is not known from Galician waters (Bañón et al. 2010). Orcas are resident in the Strait of Gibraltar and are sometimes seen along the northwest coast of Spain –– they have recently attracted attention due to interactions with small boats (Esteban et al. 2022). However, this subpopulation of orcas appears to specialise in feeding on tuna (García Tiscar 2010).

Fatal interactions between porpoises and dolphins are well known, mainly involving bottlenose dolphin. Lethal attacks on harbour porpoises by bottlenose dolphins are most common where there is a resident population of the latter, e.g., in northeast Scotland where the phenomenon was described by Ross and Wilson (1996) and remains among the most commonly diagnosed cause of death in porpoises in Scotland (e.g. Davison & ten Doeschate 2020). It has been suggested that such attacks are a result of competition for food resources or that male bottlenose dolphins are practising infanticide (male bottlenose dolphins are known to kill the offspring of other males). Such lethal interactions may also have indirect consequences, including the avoidance by porpoises of areas utilised by bottlenose dolphins. MacLeod et al. (2007) proposed that, where porpoise distribution overlaps with bottlenose dolphin distribution, there is selective pressure for low fat reserves in porpoises because it improves their flight performance and hence increases their ability to avoid fatal bottlenose dolphin attacks.

Although apparently relatively rare, mortalities of porpoises due to bottlenose dolphin aggression are thought to occur in the NWIP (Pierce et al. 2010, López-Fernández & Martínez-Cedeira 2011). No direct observation has been reported to date, but evidence based on examination at necropsy of fresh carcasses shows that fatal interactions with bottlenose dolphins are most likely occurring in Galicia (i.e. presence of rake marks from bottlenose dolphin teeth on the porpoise's skin, fractured scapula and ribs, subcutaneous hematoma) (López & Rodríguez 1995, Alonso et al. 2000). Avoidance behaviour has been observed in common dolphins in the presence of bottlenose dolphins in Ría de Arousa (Galicia), and a fatal attack by a bottlenose dolphin on a solitary common dolphin was reported in the same area (Methion & Díaz-López 2021).

Given that porpoise and bottlenose dolphin distributions also overlap in Portugal, it is likely that such interactions also occur in Portugal. An interspecific interaction was suggested as the possible origin of the trauma that caused the death of the leucistic porpoise had been observed in the Douro River by Gil et al. (2019), which stranded in August 2020 (A. Torres-Pereira, Pers. Comm.).

Conservation of porpoises in Europe: legal protection and its implementation

International agreements

The harbour porpoise is covered by several international agreements, including the Convention of Migratory Species (CMS) and its associated agreements, and the Convention on International Trade in Endangered Species of Wild Fauna and Flora (CITES), as well as European Directives and national legislation. The harbour porpoise is also listed on the IUCN Red List: at global level, the species is of Least Concern (Braulik et al. 2020); in Europe as a whole, its status was updated from Vulnerable to Least Concern in 2023 (Temple & Terry 2007, Sharpe & Berggren 2023); and it is Critically Endangered in the Baltic Sea (Hammond et al. 2008) and Endangered in the Black Sea (Birkun & Frantzis 2008). There is no separate entry for the Iberian porpoise (see Carlén et al. 2021 for a review).

In northern Europe, ASCOBANS, which derives from the CMS and is administered by the United Nations Environment Programme (UNEP), promotes the conservation of small cetaceans and requires Parties to undertake various monitoring and conservation actions. Spain and Portugal are Range States (their seas fall within the area of the Agreement) but neither is a Party to the Agreement. A second CMS-derived agreement, ACCOBAMS, fulfils a similar role for southern Europe. Its current range extends to the Iberian Peninsula and thus overlaps with that of ASCOBANS; both Spain and Portugal are Contracting Parties. ASCOBANS has "long recognized bycatch as the most significant threat to the small cetacean populations in the Agreement Area and resolutions addressing this problem have been passed repeatedly at Meetings of the Parties. These resolutions address the monitoring programmes and mitigation measures required in order to understand the situation and reduce mortality of cetaceans".[5] While ACCOBAMS has arguably placed less emphasis on bycatch in its conservation actions, the Agreement "provides that Parties shall apply, within the limits of their sovereignty and/or jurisdiction and in accordance with their international obligations, appropriate measures for the assessment and management of human-cetacean interactions, stressing that measures concerning fisheries activities shall be applied in all waters under their sovereignty and/or jurisdiction, and outside these waters in respect of any vessel under their flag or registered within their territory".[6] ACCOBAMS and ASCOBANS established a Joint Bycatch Working Group in 2019 and co-organised a workshop on Current cetacean bycatch issues in European waters at the 34th Annual Conference of the European cetacean Society in April 2023.

The systematic recording of strandings in Europe dates from the early twentieth century. The Natural History Museum in the UK began systematic recording of strandings in 1913 and strandings monitoring in France started in 1971. Strandings have been recorded in Galicia since 1973, originally by the Sociedade Galega de Historia Natural. For larger cetaceans such as the sperm whale, historical stranding records going back several centuries can be reconstructed (e.g. Smeenk 1997). Several countries and regions initiated or upgraded strandings monitoring at the start of the 1990s, reflecting commitments formalised under the ASCOBANS Agreement[7] in 1992. The Agreement included a commitment to establish an efficient system for reporting stranded animals; to carry out necropsies to determine the cause of death, study feeding habits and collect tissues for further studies; and to make the information available in an international database. Spain (in 1999) and Portugal (in 2005) joined ACCOBAMS rather than ASCOBANS, but the current Galician

[5] https://www.ascobans.org/en/species/threats/bycatch
[6] https://accobams.org/conservations-action/bycatch-depredation/
[7] https://www.ascobans.org/sites/default/files/basic_page_documents/Ch_XXVII_09_CertifiedTrueCopiesAgreement

strandings network, operated by CEMMA, started in 1990, and the network in northern Portugal dates from 2000.

The 3rd Meeting of Parties of ASCOBANS in 2000 (see ASCOBANS 2000) passed a resolution (Resolution 3.3[8]) about the incidental take of small cetaceans, which noted that "the aim of ASCOBANS can be interpreted as to restore and/or maintain biological or management stocks of small cetaceans at the level they would reach when there is the lowest possible anthropogenic influence" and "the general aim should be to minimise (i.e. to ultimately reduce to zero) anthropogenic removals within some yet-to-be-specified time frame, and that intermediate target levels should be set". 'Unacceptable interactions' were defined as being "in the short term, a total anthropogenic removal above 1.7% of the best available estimate of abundance" and there was an "intermediate precautionary objective to reduce by-catches to less than 1% of the best available population estimate". It was also noted that "if available evidence suggests that a population is severely reduced", then "an anthropogenic removal of much less than 1.7%" could be considered as an "unacceptable interaction".

Applied to the best estimate for the population size of Iberian porpoise (2900), the 1.7% limit suggests an annual removal limit of 49 animals and 1% would be 29 animals. At its COP 13 meeting in 2020, noting the importance of reducing unintentional mortality from fisheries bycatch, under the aforementioned Concerted Action for Baltic and Iberian harbour porpoises, CMS (2020) proposed "developing an action plan for the Iberian harbour porpoise population" before COP 14 (which was scheduled to be held in October 2023). ASCOBANS (2021) developed its aforementioned draft proposal for inclusion of the Iberian porpoise in the Appendices of CMS, citing the benefits to the population by listing it in Appendices I (leading to a collaborative transboundary management plan) and II (offering strict Range State protections, especially in relation to reducing unintentional mortality from fisheries bycatch). During 2023, ASCOBANS held a workshop to recommend new conservation objectives for small cetaceans in relation to anthropogenic removals. Although the outcomes from this workshop are as yet unpublished, it was agreed that the 1% and 1.7% limits were now considered inappropriate as they "would not achieve the desired outcome either when applied to the harbour porpoise or for other species".

European Union Environmental Directives

The EU Habitats Directive (Council Directive 92/43/EEC 1992), which implemented the Convention on the Conservation of European Wildlife and Natural Habitats (Bern Convention), and the Marine Strategy Framework Directive (MSFD; Directive 2008/56/EC 2008) have both resulted in increasing efforts to monitor the status of cetacean populations and to implement conservation measures. Since EU Member States report on the conservation status of cetaceans under both the Habitats Directive and MSFD, the status of porpoises has been assessed at national level. The realisation that the Iberian porpoises represent a genetically isolated population and potentially a distinct subspecies adds urgency to the need for a separate assessment of this population. ICES WGMME (ICES 2013c) recommended that the Iberian population of porpoises should be treated as a separate Management Unit. In practice, the porpoises found in Spanish and Portuguese waters are likely to belong almost exclusively to the Iberian population. Note, however, that Ben Chehida et al. (2023) referred to two porpoises from southern Portugal "carrying a divergent haplotype closely related to those from the Mauritanian population".

The Habitats Directive protects all cetaceans from disturbance and incidental killing and capture, while harbour porpoise is one of two species for which the designation of Special Areas of Conservation (SACs) was proposed. Several European countries have designated SACs for harbour porpoise, including the UK and Ireland. SACs are one of several types of Marine Protected Area falling under the Natura 2000 umbrella (which also includes terrestrial sites). The European Environment

[8] https://www.ascobans.org/sites/default/files/document/MOP3_2000-3_IncidentalTake_1.pdf

Agency maintains a publicly accessible EU database on Natura 2000 that is updated annually, including an online map viewer and a visualisation of area coverage and site numbers by Member State[9,10].

Spain has numerous Natura 2000 marine (or partly marine) sites linked to the Habitats and/or Birds Directives along its Atlantic and Mediterranean mainland coasts, in its north African enclaves and in the Canary Islands. Two SACs (Zonas Especial de Conservación) have been created by Royal Decree, Estrecho Oriental (located to the east of the Strait of Gibraltar; Real Decreto 1620/2012, de 30 de noviembre) and Islas Chafarinas (off the coast of Morocco, to the east of the enclave of Melilla; Real Decreto 190/2018, de 6 de abril). Harbour porpoise is mentioned in the first of these Royal Decrees although in the online map viewer porpoises are not listed under either site. According to the Marine Conservation Institute's Marine Protection Atlas[11], 13% of Spain's total marine area of 1,008,165 km^2 is under some designated protection (10% implemented, 3% unimplemented) although less than 1% of the marine area is highly or fully protected from fishing.

The Spanish Natura 2000 sites that can be viewed in the online map viewer, for which harbour porpoise is included in the associated list of species, are almost all in Galicia and Asturias, on the northwest and north Atlantic coasts. Galicia has nine sites: Complejo Ons-O Grove (7607 ha, i.e., 76 km^2, Habitats Directive), Complexo intermareal Umia – O Grove, A Lanzada, punta Carreirón e lagoa Bodeira (2813 ha, Birds Directive), Costa da Mariña occidental (2169 ha, Birds Directive), Costa da Morte (11,809 ha, Habitats Directive), Costa da Morte (Norte) (7962 ha, Birds Directive), Costa da Vela (1418 ha, Habitats Directive), Illa de Ons (924 ha, Birds Directive), Illas Cies (990 ha, Habitats and Birds Directives) and Illas Estelas (725 ha, Habitats Directives). Asturias has five sites: two named Cabo Busto-Luanco (11,642 ha, Habitats Directive, and 10,043 ha, Bird Directive), Penarrondo-Barayo (4317 ha, Habitats and Birds Directives), Sistema de cañones submarinos de Avilés (338,961 ha, Habitats Directive) and Yacimientos de Icnitas (3560 ha, Asturias, Habitats Directive). The remaining site is Doñana in Andalucia (128,268 ha, Habitats and Birds Directives), a National Park comprising mainly of terrestrial wetlands plus a small area of coastal waters[12].

Portugal has designated 4.5% of its 1,728,718 km^2 marine area, of which less than 1% is highly or fully protected from fishing. Portugal has three Natura 2000 marine sites for which habour porpoise is listed, all designated under the Habitats Directive, although none is an SAC: Maceda – Praia da Vieira to the south of Porto (502,637 ha), Arrábida/Espichel (20433) south of Lisbon and Costa Sudoeste (262,299 ha) in the Algarve[13].

It is apparent that only a small proportion of Iberian Atlantic coastal waters is covered and that, generally, the process of designating sites as SACs has not been completed. In fact, Carlén et al. (2021) argued that, in reality, SACs offered very limited protection (see also Pinn et al. 2021). To date, the Habitats Directive has not demonstrably contributed to the reduction of cetacean bycatch. Firstly, the requirement to reduce incidental killing under this Directive has essentially been ignored until recently with, arguably, too much focus to the establishment of SACs. Secondly, the EC has been slow to challenge Member States in relation to their failure to deliver the required SACs. It can also be argued that declaration of an SAC could benefit a species that has small local resident populations, like bottlenose dolphin, and the benefit for more wide ranging species like harbour porpoise is less obvious.

Article 17 of the Habitats Directive requires Member States to report every 6 years about the progress made with implementing the Habitats Directive. Since the main objective of the Directive is to maintain or restore Favourable Conservation Status for habitats and species of Community interest, reporting under Article 17 is focused on the status and trends of these habitat types and species. Reports for the 2013–2018 period were issued in 2019. In relation to assessment under Article 17 of the Habitats

[9] https://www.eea.europa.eu/themes/biodiversity/natura-2000/the-natura-2000-protected-areas-network
[10] https://natura2000.eea.europa.eu/
[11] https://mpatlas.org/
[12] https://mpatlas.org/countries/ESP/
[13] https://mpatlas.org/countries/PRT/

Directive, Spain assessed the Conservation Status of porpoises to be "Unfavourable-Inadequate" for the period 2013–2018, while Portugal assessed it as "Unfavourable-Bad". For the previous assessment (2007–2012), both countries had assessed it as "Unfavourable-Inadequate", while for 2001–2006, Portugal had rated it as "Unfavourable-Inadequate" and Spain had assessed it as "Unknown".

The MSFD aims to effectively protect the marine environment across Europe, including the marine resource base upon which human economic and social activities depend, through the achievement of Good Environmental Status (GES) of EU waters. Eleven Descriptors of GES have been defined, the first of which is that "Biodiversity is maintained". The criteria to assess the status of cetacean populations under the MSFD in relation to Descriptor 1 (biodiversity) include abundance, distribution, demographic characteristics and the bycatch mortality rate (e.g. Palialexis et al. 2019). Reporting under the MSFD follows a 6-year cycle, the first cycle being completed in 2018. For the first cycle, both Spain and Portugal reported on the status of harbour porpoise in relation to several Criteria for Descriptor 1 (Table 15). While different indicators provided different outcomes, and in most cases, the quantity and/or quality of evidence available was a limiting factor, the overall species assessment in the First Cycle Reports for Spain and Portugal was "Not good".

The Habitats Directive and MSFD do not include specific instructions about how their requirements should be implemented, and the conservation objectives are not quantitatively defined. Developing the necessary practical framework (e.g. the indicators, criteria, baselines, thresholds, assessments, monitoring programmes and programmes of measures) and associated infrastructure has fallen to the Member States, with assistance of European, intergovernmental and international bodies such as the Joint Research Centre (JRC), the International Council for the Exploration of the Sea (ICES) and the (European) Regional Sea Conventions: OSPAR (North East Atlantic), HELCOM (Baltic Sea), Barcelona Convention (Mediterranean) and Bucharest Convention (Black Sea). The EU has also funded numerous projects that aimed to advance and improve the implementation of these directives, notably under the LIFE and MSFD programmes.

In the context of assessing the status of cetaceans in the OSPAR Maritime Area, the OSPAR Marine Mammal Expert Group (OMMEG) has proposed guidelines to set thresholds for anthropic removals, including bycatch, aligning with the conservation objective of maintaining or restoring population to 80% of carrying capacity with probability 0.8 over 100 years (Taylor et al. 2022). These were agreed at the OSPAR Biodiversity Committee meeting in autumn 2021. Given current abundance data on Iberian harbour porpoise, a removal limit of zero should be applied to the Iberian porpoise population (see ICES 2021a, Taylor et al. 2022). This was based on a modified PBR approach (*m*PBR, modified because the underlying conservation objective differs from the original conservation objective of the US Marine Mammal Protection Act), applied to a minimum estimate of population size (N_{min}) and defined as the 20th percentile of the log-normal distribution. If N_{min} was less than 2500 mature animals, *m*PBR was set to zero. Using the Hammond et al. (2021), estimate of the population size for Iberian porpoise, 2898 (CV = 32%) animals, results in an N_{min} value

Table 15 Assessment by Portugal and Spain of the Status of Harbour Porpoise under Descriptor 1 of the MSFD for Each Criterion in the First Cycle Reports

Criterion	Description	Portugal	Spain (North)
D1C1	Bycatch mortality	Not good	Not good
D1C2	Abundance	Not good	Good
D1C3	Demographic characteristics	Not assessed	Not assessed
D1C4	Distribution range and pattern	Good	Good
D1C5	Habitat	Not good	Not assessed
Species assessment		Not good	Not good

Source: Adapted from Anon. (2022).
Porpoise was not assessed for southern Spain.

of 2122, hence a bycatch limit of zero. As previously noted, ICES (2021a) estimated marine mammal bycatch mortality in the OSPAR region but were unable to provide an assessment for Iberian harbour porpoise due to inadequate data on the number of bycaught animals.

It is worth noting that the aspirational approach of European Union conservation Directives concerning conservation contrasts with the detailed prescriptive approach evident in legislation from the USA, for example, the Marine Mammal Protection Act (see Verutes 2022 for a critical review). Having said this, the MMPA remained ineffective until 1994 when PBR was introduced. Previously, management action was to be triggered by the population size falling below the maximum net productivity level (MNPL, similar conceptually and mathematically to Maxium Sustainable Yield (MSY) in fisheries) but, as observed by Taylor et al. (2000), it "failed as a management tool because the parameter that triggered management action was extremely difficult to estimate for the majority of populations". In Europe, it has been necessary not only to develop the management rules but also to define the quantitative conservation objective around which the rules should be designed. Arguably, as mentioned above, there has also been a failure of enforcement by the EC. Furthermore, fishery bycatch of protected species in the EU is covered by both environmental and fisheries law and the overlap in competences and discrepancies in objectives need to be addressed (see, e.g., Fock 2011, Pinn et al. 2021).

European Fisheries Regulations

In addition to the EU's "conservation" Directives, legislation aimed at monitoring and reducing fishery bycatch mortality of cetaceans is also relevant for porpoises. The European Union's Common Fishery Policy originated in 1970 and was reformed in 2002 and 2013, with the latter reform including an increased emphasis on sustainability (e.g. Orach et al. 2017). Monitoring and mitigation of cetacean bycatch was specifically addressed by Council Regulation (EC) No. 812/2004, which mandated the deployment of dedicated observers on-board a sample of vessels in certain fishing fleets to monitor cetacean bycatch, as well as the deployment of acoustic deterrent devices, known as "pingers", on fixed nets in some areas (Kraus et al. 1997, Culik et al. 2001, Pinn 2023). The Regulation stated: "The use of such devices should therefore be required in areas and fisheries with known or foreseeable high levels of by-catch of small cetaceans, and taking into account the cost/efficiency of such requirement". The regulation also encouraged pilot projects and scientific studies on the effects of using pingers. Under the general monitoring obligations of the Regulation, on-board observer coverage aimed to achieve a Coefficient of Variation of 0.30 or less for the estimate of bycatch per unit effort for the most frequently caught species. For pilot projects, minimum observer coverage was also specified, as 10% of fishing effort by pelagic trawlers, covering at least 3 vessels, and 5% or at least 20 vessels for other gears.

As was also the case for several other EU Member States, Spain and Portugal were arguably slow to respond to Regulation 812/2004. Thus, Spain undertook a single Pilot Study and used fishery observers (not dedicated observers) to record cetacean bycatch. In general, across Europe, the success of the Regulation was limited by the fact that it did not apply to all relevant fleets, monitoring intensity was low where it did apply, fisher participation was essentially voluntary and there was little or no monitoring or enforcement of its implementation (e.g. Dolman et al. 2016, 2021, Read et al. 2017).

Regulation 812/2004 was repealed and replaced in 2019 by the Regulation on the Conservation of Fishery Resources and the Protection of Marine Ecosystems through Technical Measures (2019/1241). This offered some advances, notably including an explicit obligation to ensure that bycatch of sensitive species is "minimised and where possible eliminated", a requirement for relevant technical measures to be applied regionally in high risk fisheries, and a requirement for Member States to report on the effectiveness of monitoring of bycatch and measures taken to reduce it. However, the regulation did not clearly define target thresholds for the implementation of mitigation, and the level of monitoring being undertaken was (and is) insufficient to obtain the robust estimates of bycatch rate needed to determine the impact of bycatch at population level (Dolman et al. 2021).

The European Commission adopted the Marine Action Plan in February 2023. The Plan contributes to delivering on the EU Biodiversity Strategy 2030, and under heading "Action to improve fishing selectivity and reduce the impact of fishing on sensitive species", "The Commission calls on Member States to…:

> Adopt national measures or joint recommendations to the Commission to minimise by-catch (or reduce it to the level that enables the full recovery of the populations) of: by the end of 2023, harbour porpoises in the Baltic Proper and the Black Sea, the Iberian Atlantic and the common dolphin in the Bay of Biscay…[14]

National legislation

Spain

Real Decreto 1727/2007, de 21 de diciembre[15], introduced various protection measures for cetaceans in Spain. The Spanish Catalogue of Threatened Species listed the harbour porpoise as "vulnerable" in 2011[16], and this was updated in 2020 to reflect a change in status to "in danger of extinction"[17].

The Ministerial Order APA/1200/2020 of 16th December, published on 18th December 2020 in the Official State Bulletin (Boletín Oficial del Estado, BOE), established new measures for mitigation and monitoring to improve scientific knowledge to reduce cetacean bycatch on fishing activities. Its provisions were described by ICES (2023) and include the following:

- Article 3 of APA/1200/1200 established an on-board observer programme focused on cetacean bycatch. All fishing vessels operating in national fishing grounds of the Bay of Biscay and the Iberian coast, as well as in non-Spanish European waters in the Bay of Biscay, must take scientific observers on board when requested to do so by the General Secretariat for Fisheries. The programme is focused on the national fleet segment that, according to scientific analysis, poses the highest risk of interaction with vulnerable species. The aim is to cover at least those trawling activities involving a major vertical opening (pair bottom trawl), as well as vessels using bottom gillnets or trammel nets with a mesh size equal to or bigger than 80mm. Article 3 also proposed to complement the observer programme with a pilot project on remote electronic monitoring (REM). REM systems have automatic sensors that cannot be manipulated and, as such, should provide reliable records of fishing operations. Currently, only REM systems with associated image analysis software are used. Since they are connected to on-board navigation systems, these images are linked to specific stages of the fishing process. Associated software analysis makes the image processing more efficient and ensures stronger and more reliable data, excluding the possibility of human error.
- Article 4 of APA/1200/2020 established the obligation to use acoustic deterrent devices (pingers) for all Spanish bottom trawlers whose fishing activity is conducted in Cantabrian and Northwest fishing grounds in national waters and in the non-Spanish EU waters of the Bay of Biscay. Currently, 65 trawlers are using pingers in the North Western Cantabrian Sea, plus 12 more in the non-Spanish EU waters of the Bay of Biscay. The devices are used

[14] https://faolex.fao.org/docs/pdf/eur217860.pdf
[15] https://www.boe.es/boe/dias/2008/01/12/pdfs/A02292-02296.pdf
[16] https://www.boe.es/eli/es/rd/2011/02/04/139
[17] https://www.boe.es/eli/es/o/2020/11/20/ted1126

in accordance with the Annex to Commission Implementing Regulation (EU) 2020/967 of 03.07.2020, which lays down rules on the signal and implementation characteristics of acoustic deterrent devices as described in Part A of Annex XIII of Regulation (EU) 2019/1241 of the European Parliament and the Council on the conservation of fisheries resources and the protection of marine ecosystems through technical measures.

- Article 6 established a move-on rule for fishing activities using bottom trawl gear. If more than three cetaceans are caught in the same fishing manoeuvre, or any cetacean is caught in two consecutive hauls, fishing vessels shall move a minimum of 5 miles from the relevant point to continue their fishing activities, at a high navigational speed.
- Article 7 of APA/1200/2020 established the obligation to notify by-catch in logbooks. All fishing vessels, irrespective of the gear used, are obliged to record and transmit information on all bycatch events involving any cetacean species, via their logbook. They should indicate the number of specimens caught, the species, their vital status and relevant morphological characteristics, such as approximate size or whether they show previous marks of possible contact with fishing gear.

On 8th March 2021, the BOE published the Resolution of 2nd March, by the General Secretariat for Fisheries, allocating fishing quotas for scientific purposes in order to implement pilot projects on REM in the context of the mitigation measures for cetacean bycatch. The first stage of this initiative included 13 vessels using bottom trawling and gillnetting. The Resolution of the 15th March 2022, by the General Secretariat for Fisheries provided the basis for continuing the REM project, involving 21 vessels: 61.5% more than those involved in 2021. The project is expected to continue in 2023.

Portugal

Portugal was one of the first European nations to protect cetaceans when it issued Decree-law nº263/1981 of 3rd September. The decree approved the Regulation for the Protection of Marine Mammals in Inland Waters, the Territorial Sea and the Portuguese Continental Exclusive Economic Zone, recognising the high scientific value of cetaceans, the rarity of some species in the seas of mainland Portugal, the alarming decline in the populations of others and the need to start to take conservation measures to protect these populations. Article 2 of this Decree established that it is expressly forbidden, throughout the year, to fish for, catch or kill any species of marine mammal that may occur in estuaries and the Continental Exclusive Economic Zone of Portugal. Article 3 prohibits the sale and commercialisation of marine mammals in fish auctions, markets or anywhere else (even those found dead in nets, fishing gear or on land). Article 4 obliges specialised scientific institutions to assist marine mammals stranded alive. After any necessary assistance, they must be returned to their natural environment as soon as possible. Article 5 of Decree-Law no. 263/1981 established that infringements of the provisions of Articles 2 and 3 would be punishable by seizure and fines of 100,000 escudos (~ €500) for seals, dolphins or porpoises and 900,000 escudos (~ €4500) for sperm whales, rorquals or whales (charged per specimen).

The Decree-Law 140/99 of 24th April[18], later amended by Decree-Law 49/2005 of 24th February, transposed into national law the regulations from EU Habitats Directive (Council Directive 92/43/EEC 1992). Since 2004, following EU Regulation 812/2004, Portugal began using observers to record bycaught cetaceans, using fishery observers (already in place under the EU Data Collection Framework) as well as dedicated observers working on specific projects. Decree-Law no. 9/2006 of 6th January[19] established responsible behaviour during cetacean watching activities on the part of tour operators or audio-visual recorders, sea sports persons, researchers and the general public. Under this regulation, tour operators need prior authorisation to observe cetaceans.

[18] https://faolex.fao.org/docs/pdf/por22472.pdf

[19] https://diariodarepublica.pt/dr/detalhe/decreto-lei/9-2006-168231

The two most recent regulations published are the Ministerial Order 172/2017 of 25th May[20] and the Normative Order no. 19/DG/2020[21] issued by the Portuguese Directorate-General for Natural Resources, Safety and Maritime Services (DGRM).

Article 5 of the Ministerial Order 172/2017 of 25th May stated that (1) Nets used for fishing with traditional beach seines must have acoustic deterrent equipment installed to prevent by-catches of marine mammals, such as harbour porpoises or bottlenose dolphins; (2) If existing data indicate that fishing by traditional beach seines has no impact on cetacean populations, vessels operating in certain zones may be excluded from the first obligation; and (3) the characteristics of the equipment shall be determined by the Order of the DGRM.

Normative Order no. 19/DG/2020 from DGRM drew on the results of the Life+ MARPRO project to set rules for the deployment of pingers. Thus, the use of pingers on traditional beach seines nets is considered unnecessary to protect harbour porpoise and common dolphin but not bottlenose dolphin. Also the lack of reported interactions between marine mammals and the traditional beach seines to the south of Praia da Vieira suggests that the use of pingers in this area is not advised (which will avoid the unnecessary creation of noise pollution). The technical characteristics of the acoustic deterrent equipment were also specified (e.g. in terms of the frequency and pressure level of the emitted sounds (10 kHz and 132 dB for porpoises), the type of battery and the need for it to have a sufficient charge level), and it was specified that one pinger must be installed in each boat and one at the mouth of the net for the duration of the fishing operation and that pingers had to be installed and active in fishing operations within 10 days of the effective date of the order (i.e. by 16 August 2020). Three years on, there has been no routine enforcement, monitoring or assessment of the efficacy of this normative order, although there has been some evaluation of progress, in collaboration with fishers, in the context of two scientific projects (iNOVPESCA and CetAMBICion) in the south of Portugal.

Portaria 201/2019 of 28th June established the management plan for the Sítio de Importância Comunitária (SIC, Site of Community Importance) Maceda-Praia da Vieira, an important area for harbour porpoises. Associated actions include the existing stranding network, a fully equipped rehabilitation centre, abundance surveys and delivery of pingers to fishers, although these have been driven by other obligations; new action has been minimal.

The Red Book of Portuguese Vertebrates assessed the harbour porpoise as "Vulnerable" in 2005 (Cabral et al. 2006) and recently updated this assessment to "Critically Endangered" in the Red Book of Portuguese Mammals (Torres-Pereira et al. 2023b).

Current status, knowledge gaps and future research, monitoring and conservation

Distribution and abundance

Current situation

Information on distribution derives from abundance surveys, opportunistic sightings, strandings and genetic studies. Porpoises seem to be most common on the west coast of the Iberian Peninsula, along the Galician and Portuguese coasts. Almost all sightings are from continental shelf waters and the species is a rare visitor to the Macaronesian islands. There is genetic evidence of past (possibly continuing) emigration into the Celtic Sea. There is insufficient evidence to determine whether the distribution range is expanding or contracting.

The population size of the Iberian porpoise population appears be 3000–4000 animals. Estimates from the SCANS-II (2005) and SCANS-III (2016) surveys were almost identical at

[20] https://dre.tretas.org/dre/2982137/portaria-172-2017-de-25-de-maio

[21] https://www.dgrm.pt/documents/20143/46478/Despacho+19_DG_2020.pdf/6a1bb004-127e-61a4-1adb-257a1225c766

2880 (CV=0.72) and 2900 (CV=0.32), respectively, while the estimate from SCANS-IV (2022) was 4034 (CV=0.35); in all cases, the 95% confidence intervals were wide (e.g. 1842–7309 for SCANS-IV), and the change in the estimate between 2016 and 2022 cannot be assumed to represent a real increase in abundance, although the distribution of sightings was rather different in 2022 (Gilles et al. 2023). The highest published estimate from national surveys was over 3500 (for Portugal) but this is within the 95% CI of the SCANS estimates. This is evidently a small population, and while the large-scale surveys indicate no change in abundance over 11 years, the wide confidence intervals do not preclude upward or downward trends. The population might extend offshore in Portuguese waters beyond the SCANS survey area and the large-scale surveys excluded the Galician rías but it seems unlikely that this has resulted in a serious underestimate. Although the annual abundance estimates for Portugal in 2011–2015 reported by Torres-Pereira et al. (2022) varied considerably between years, they were all arguably broadly consistent with the SCANS results. The low number of porpoise sightings during most surveys (reflecting the low density of porpoise occurrence) is a limitation. Evidence from strandings suggested a decline in porpoise abundance in Portugal in the 1980s. Genetic (Mt DNA) data indicate a loss of genetic diversity (as well as outward movement of animals into the Bay of Biscay and Celtic Sea), and although the research is ongoing, the available genetic results support the idea of a declining population (Ben Chehida et al. 2023).

Future research and monitoring

There remains a need for more frequent large-scale surveys. SCANS IV took place in summer 2022, reducing the time-gap between surveys to 6 years but (assuming this frequency can be maintained) this still means there is only one large-scale survey per reporting cycle for the Habitats Directive and the MSFD.

It would be useful to have better information on movements of porpoises within (and out of) the Iberian Peninsula. Regular surveys of distribution and abundance covering all seasons would be useful, extending into both the Galician rías and offshore waters. Although there are few offshore sightings of porpoises in the Iberian Peninsula (e.g. one mentioned by Aguilar et al. 1983), few offshore surveys have specifically looked for porpoises, and the lack of sightings may be partly due to the difficulty of spotting them. It is possible that increased use of acoustic monitoring and perhaps e-DNA could provide a more complete picture of distribution and occurrence and consequently also assist with the design of future abundance surveys, although the latter is less likely to be useful where density is low. In the Baltic Sea, the Sambah project demonstrated that passive acoustic monitoring can be used to estimate porpoise density and abundance, as well as to map distribution and identify hotspots and preferred habitats and areas with high risks of negative anthropogenic impacts, in a cost-effective way over a large area (SAMBAH 2016). Foote et al. (2012) showed that porpoises could be detected using eDNA within a 10 m distance in a sea pen in a sheltered port area but this is very different to open sea conditions; see Suarez-Bregua et al. (2022) for a review of the utility of the eDNA approach. Another technique based on genetic information, which could provide insights into abundance, is Close Kin Mark Recapture (e.g. Conn et al. 2020, Trenkel et al. 2022).

Demographic characteristics

Current situation

Strandings monitoring provides the main source of demographic data, e.g., age, mortality, reproductive output, health status and potentially indicating changes in distribution and abundance. Both Spain and Portugal have regional strandings networks that have depended heavily on short-term funding and the work of volunteers. They have not always had access to veterinary expertise or sufficient funding to collect and analyse relevant samples. As such, much work has depended on project funding. At the time of writing, national funding has recently been injected into the system in Spain (via *Fundación Biodiversidad*) but coordination across networks still has some way to go.

As seen in section 'Life history', the overall annual mortality rate (including all causes of death) estimated by Read et al. (2013) for the Iberian population from ages of stranded animals during 1990–2010 was 18% (95% CI 5.3%–20.5%). However, the estimated pregnancy rate (54%) from stranded animals, coupled with the estimated age at sexual maturity (ASM) in females, is too low to compensate for 18% annual mortality, and this remains true even if a lower ASM is assumed (see Figure 12). Pregnancy rate is likely to be underestimated from strandings because dead animals inevitably include a higher proportion of animals that were in ill health and hence possibly not able to breed, compared to the living population. The sample sizes here (especially for the estimation of pregnancy rate) were rather small. The validity of the life table-based methodology depends on the stranding data being representative (of deaths) and on population size and age structure being stable, but Read et al. (2013) found evidence that mortality rate varies over time.

Future research and monitoring

The collection of life history, dietary, health and cause of death data from stranded and bycaught animals remains essential to obtain better estimates of demographic parameters and provide a more complete picture of health status and causes of death in this population. They are also essential for genetic studies. One limitation of age-at-death data from strandings is that the youngest animals tend to be underrepresented, possibly because the carcasses are less buoyant and decompose more rapidly. Further work on the population dynamics of Iberian porpoises is needed, for example, using a model-based approach (e.g. Siler 1979) to correct mortality rates (see Saavedra Penas 2017).

As is generally the case in Europe, there is a need for better coordination of the work of strandings networks and sufficient secure funding which covers not only basic monitoring but also full necropsies and the collection and processing of relevant samples. Application of carcass drift models (after Peltier et al. 2012) to the Iberian Peninsula would help to determine the origins of stranded animals and also increase the utility of data collected for estimating bycatch rates (some progress in this direction was made under the recent CetAMBICion project). Most strandings networks depend to a greater or lesser extent on citizen science, which could also be better supported and coordinated. Ongoing initiatives involving ASCOBANS, ACCOBAMS, the IWC and ICES may help to deliver solutions (e.g. ASCOBANS 2023).

Bycatch mortality and other threats

Current situation

The North Atlantic Marine Mammal Commission and the Norwegian Institute of Marine Research (NAMMCO 2019) observed that population size and bycatch data for this population are apparently incompatible – taken at face value the first could not support the second; the population would go extinct in only a few years. Strandings and on-board observations during the last decade suggest total annual bycatch mortalities of between around 200 and 500 animals along the Atlantic coast of the Iberian Peninsula. While estimated bycatch mortality is high, the annual number of *known* bycatch mortalities (i.e. observed bycatch plus bycatch diagnosed in stranded animals) is relatively low (10–15 individuals annually) and thus does not exceed either the ASCOBANS 1.7% threshold (49 animals for this population to achieve 80% of carrying capacity with probability 0.5 (over an indefinite period of time), assuming no biases in abundance or bycatch estimates) or the IMR/NAMMCO Workshop threshold (25 animals for this population to achieve 50% of carrying capacity with probability 0.95 within 100 years). However, it is extremely unlikely that true bycatch mortality is as low as the minimum estimate, and obviously, even the minimum estimate still exceeds the zero limit proposed by OSPAR to achieve 80% of carrying capacity with probability 0.8 within 100 years.

Lack of sufficient reliable data on the amount and distribution of bycatch mortality of porpoises is a key issue. Historically, this reflects the exclusion of the very large number of small (<15 m) fishing vessels from the on-board observer programme and poor implementation of Regulation 812/2004. Although project-based data collection (e.g. from the MARPRO project) filled some gaps, in most cases the proportion of fishing activity monitored was small, and not all fleets were monitored every year, so that the extrapolated bycatch rates were of questionable value. Currently, at a European scale, the increased reliance on fishery observers instead of dedicated observers to collect cetacean bycatch data under Regulation (2019/1241) looks like a backward step. The value of strandings monitoring to provide bycatch data is increasingly recognised and has provided input into recent ICES assessments of cetacean bycatch mortality in the Bay of Biscay (ICES 2020a) but further development of this data source is still needed on the Iberian coast.

Of the other known threats, among the most concerning is the sometimes high PCB concentrations in Iberian porpoises, which are likely to result in at least some reduction in reproductive output. However, as pointed out by Lusseau et al. (2023), porpoises react to underwater noise (to which boat traffic, marine renewable development and the use of pingers on fishing nets, among other sources, all contribute) and their physiological ecology is such that body condition falls rapidly when feeding opportunities are lost. Such effects at individual level can also impact population dynamics. In relation to bycatch mitigation, these authors also warned that noise from pingers can lead to reduced individual reproductive output due to its effects on individual condition and observed that "we do not know how to plan pinger prevalence to minimise bycatch and acoustic impacts" and that there is a need for experimental studies to estimate "dose response relationships" between noise exposure, the resulting behavioural responses and the consequent changes in body condition. In order to achieve this, we also need improved indicators of body condition. Derous et al. (2020) noted that blubber thickness and composition in some parts of the cetacean body does not provide a good measure of health and that more work is needed on the relationships between environmental stressors and blubber biology.

The main measure in place to reduce porpoise bycatch in European waters at present is the use of pingers on fixed nets. To this can be added the recent implementation of 'Move-On' rules and the trialling of pingers and excluder devices for trawl nets in Spanish waters, although these measures are not specifically targeted at porpoises.

In the case of harbour porpoise, many studies, both older and more recent, suggest that pingers can be effective in reducing bycatch mortality in various types of net, increasingly so as pinger design has evolved (e.g. Kraus et al. 1995, 1997, Gearin et al. 1996, Koschinski & Culik 1997, Vingada & Eira 2018, Kindt-Larsen et al. 2019, Brennecke et al. 2022, Königson et al. 2022, Pinn 2023). In Portugal, pingers (F70 and F10) were trialled on several different types of fishing gear within the LIFE+ MarPro project (2011–2017). In purse seines, harbour porpoise mortality was reduced by 75% (based on trials on several vessels over periods of three to 19 months, with a total of 446 hauls with pingers and 551 control hauls); for polyvalent vessels (vessels carrying a range of fixed gears), the reduction was 83% (based on trials with various boats during one to 25 months, with a total of 627 hauls with pingers and 773 control hauls); and for beach seine, the reduction was 55% (based on trials with various vessels over periods of 1–6 months, with a total of 1068 hauls with pingers and 3093 control hauls) (Vingada & Eira 2018).

However, some of the above-mentioned studies also acknowledged negative aspects of pinger deployment: Brennecke et al. (2022) suggested that "pinger use should be limited to critical time periods and regions or that more focus needs to be put on developing acoustic devices which cause less severe behavioral reactions". In general, there is still a need to quantify the effects of disturbance and displacement of animals from their preferred habitat and the consequent effects on body condition and reproductive output (as modelled by Lusseau et al. 2023). This may be most important for small populations and those occupying a confined space (like the Baltic Proper subpopulation) (Sveegaard et al. 2011, Pinn et al. 2021). In Portugal, concern has been expressed that the universal deployment of pingers on beach seines (assuming that they are functioning) may exclude porpoises from important feeding areas.

Perhaps the most glaring issue is the lack of routine bycatch monitoring and mitigation measures (and, where mitigation has been introduced, lack of monitoring of the efficacy of mitigation measures) for many small-scale fisheries in the Iberian Peninsula.

Following the responses by the European Commission and advice from ICES after the 2019 call for Fishery Emergency Measures to reduce bycatch mortality of cetaceans (specifically common dolphins in the Bay of Biscay and harbour porpoise in the Baltic), the Spanish government implemented a strategy to improve knowledge of cetaceans' interactions with fisheries and to reduce bycatch mortality (under Order APA/1200/2020, as detailed in section 'Conservation of porpoises in Europe: legal Protection and its implementation'). The measures included:

- Development of a programme of on-board observation of cetacean bycatch, from September 2020, initially for pair trawl and bottom gillnet vessels using mesh sizes ≥ 80 mm, and more recently extending to (individual) bottom trawls, purse seines and artisanal gears operating in national waters;
- A pilot electronic observation programme for bottom trawl and gillnet vessels;
- Mandatory reporting of incidental catches in the logbook for all fishing vessels;
- Collection of data by stranding networks for collation by the government;
- A pilot study on information collection and harmonisation of protocols for carrying out necropsies.

The European Commission opened an infringement procedure (INFR(2020)4038) against Portugal in November 2023, specifically referring to cetacean bycatch, including that of porpoises[22]. Bycatch in Portugal is being addressed mainly through existing conservation frameworks. Knowledge of the bycatch of protected species in Portugal is based on various previous and ongoing research projects (e.g. SafeSea, MarPro, PescApanha, InovPesca, CetAMBICion), which have provided information on the species and gears involved and bycatch rates, investigated mitigation measures using acoustic devices and developed good practices guides for various fleets. Measures introduced in 2017 and 2019 are described in section 'Conservation of porpoises in Europe: legal Protection and its implementation'.

Under the MSFD programme, several future measures are foreseen, including the establishment of a coordinated strategy (involving Portugal, Spain and France) for the monitoring and assessment of cetacean populations and to address bycatch in the Bay of Biscay and Iberian Coast, recently drafted under the auspices of the CetAMBICion project; formation of a dedicated working group to develop an action plan to reduce bycatch of marine mammals, marine birds and marine turtles; a programme to monitor the use and effectiveness of pingers in the traditional beach seine fishery and to improve control and enforcement; rescue protocols for animals bycaught in traditional beach seines; and an initiative to produce bycatch risk maps and pilot the implementation of spatial and temporal restriction measures for fisheries within the Site of Community Importance (SIC) Maceda-Praia da Vieira (CetAMBICion project partners, Pers. Comm.).

Both Spanish and Portuguese authorities have participated in research projects such as the MSFD 2020 project CetAMBICion, the objectives of which included the determination of abundance and distribution of cetaceans in Spanish waters, the study of the influence of environmental variables and prey availability, calculation of cetacean bycatch mortality and the assessment of the effectiveness of using pingers and other devices for bycatch reduction, and the development of monitoring, evaluation and mitigation measures.

[22] https://ec.europa.eu/commission/presscorner/detail/en/inf_23_5380.

In relation to fishery bycatch and other threats, one of the most significant recent developments was the publication in March 2023 of the OSPAR Quality Status Reports for Indicator Assessments under the MSFD[23]. For the Bycatch Indicator, in relation to bycatch of Iberian porpoise, it is stated that:

> Although it is not possible to quantify this issue using the methods applied to all other AUs given the lack of records through the observer scheme, the evidence suggests a number greater than 200 harbour porpoises are by-caught annually (Vingada and Eira 2018), which, when taken into context with the by-catch threshold of zero, indicates that by-catch in this AU is critically exceeding the agreed threshold.

The estimate from Vingada and Eira (2018) is one of those that appears above in Tables 7 and 10 (Threats to the Iberian harbour porpoise). The reports also include the Pilot Assessment of Status and Trends of Persistent Chemicals in Marine Mammals (Pinzone et al. 2022), and it is expected that a "persistent chemicals indicator" will ultimately be adopted and pave the way for routine monitoring of pollutant levels in cetaceans.

Future research and monitoring

On-board observers and remote electronic monitoring Improved data collection is needed to permit accurate and precise estimates of bycatch rate. As such, ensuring adequate observer coverage of all relevant fleets, including small-scale fleets, preferably using dedicated observers, currently remains a priority. This was demonstrably not been achieved during the last two decades, as evidenced by the extremely patchy observer data presented in section 'Threats to the Iberian harbour porpoise – Fishery bycatch'. Challenges that must be overcome include logistical difficulties with placing observers on small vessels, covering a sufficiently large proportion of fishing activity generally, and the reliance on voluntary collaboration by the fishing sector. It is possible that provision of addition training of fishery observers could help improve the quality of observer data on porpoise bycatch (Scientific, Technical and Economic Committee for Fisheries (STECF) 2019). Increased use of Remote Electronic Monitoring such as on-board cameras should ultimately provide better bycatch estimates (in terms of both accuracy and precision), provided that privacy and data-protection issues can be addressed, and bycaught species can be reliably identified from the images obtained, e.g., using Artificial Intelligence (Ovalle et al. 2021, 2022). Self-reporting of bycatch by fishers could also be useful and can be assisted by new technology, as in the case of the Clean Catch UJ smartphone application (Ryan et al. 2022)[24]. Of course, making it easier to report bycatch does not necessarily mean the frequency of reporting will increase.

Strandings monitoring The value of strandings monitoring to provide bycatch data is increasingly recognised and has provided input into recent ICES assessments of cetacean bycatch mortality in the Bay of Biscay (e.g. ICES 2020a). There are several ways in which this contribution could be enhanced: better resourcing, including availability of veterinary expertise, adoption of common protocols and training in their use, introducing elements of survey design and quantifying the amount and distribution of survey effort, development of drift models, adoption of a common reporting format and construction of a central database (thus also eliminating duplication of calls for data by different organisations). Routine monitoring of pollutant levels in stranded porpoises would be valuable. Even if PCB concentrations in cetacean blubber are generally declining in Europe, many animals continue to have concentrations that represent a toxicological threat (Williams et al. 2023). As noted above, OSPAR has recently published a pilot study on status and trends of PCBs in cetaceans in Europe and proposed the use of PCB concentrations in marine mammal blubber as an indicator (Pinzone et al. 2022).

---

[23] https://oap.ospar.org/en/ospar-assessments/quality-status-reports/qsr-2023/indicator-assessments/

[24] https://www.seafoodinnovation.fund/projects/cetacean%E2%80%91interaction%E2%80%91self%E2%80%91reporting%E2%80%91and%E2%80%91em%E2%80%91verification%E2%80%91sys-tem%E2%80%91rd061/9781032761961_C001.indd

Bycatch mitigation

Globally, a wide range of mitigation measures has been applied to reduce the bycatch of small cetaceans, including acoustic deterrents, gear modifications such as net height, buoy rope modification, weak links and increased rope visibility/reflectivity, acoustic reflectors, time-area closures, visually detectable devices and exclusion devices (e.g. Hamilton & Baker 2019, Kindt-Larsen et al. 2019, Omeyer et al. 2020, Sacchi 2021).

To understand why some mitigation methods work better than others, it is important to understand why cetaceans are entangled and bycaught in the first place, while recognising that the answers may be specific to particular species and particular kinds of gear and may vary over time. Kinze (1994) noted that bycaught porpoises in Denmark tended to include an unexpectedly high proportion of subadults and suggested that individuals may need to learn to avoid entanglement, either by becoming entangled and escaping or through hearing the distress calls of entangled animals. Kastelein et al. (2000a) suggested that echolocating porpoises do not detect gillnets in time to avoid collision. They estimated the target strength of various gillnets and the distance at which the nets would be detected by harbour porpoises and bottlenose dolphins and concluded that a porpoise approaching a net at night at right angles (perpendicular to the net) would detect the nets from a distance of only 3–6 m (depending on the type of net) 90% of the time, whereas a bottlenose dolphin would detect the same nets from 25 m to 55 m away. It has also been suggested that porpoises are distracted due to focusing on the fish trapped in the net (Kastelein et al. 1995) or are aware of the net but consider it not to be a threat (Larsen et al. 2007).

By far the most common approach to date to reduce porpoise bycatch is the use of acoustic deterrents/alarms (pingers) on fixed gear, to date the only bycatch mitigation method for cetaceans that is specifically mandated in certain fisheries in Europe. Pingers actively emit sounds that may both alert porpoises to the presence of a net and have an aversive effect. The first pingers were designed to reduce harbour porpoise mortality in gillnet fisheries (Kraus et al. 1997) and were later developed for use with other cetacean species and other fishing gears (Culik et al. 2001). As previously commented, numerous trials with pingers have suggested that porpoise mortality in fixed nets can be substantially reduced.

In certain areas, e.g., the Inner Danish Sound and, potentially, the Galician rías, porpoises might be constrained to use the area in which nets with the pingers are deployed and thus continue to interact with the nets to feed, despite the sounds emitted by the pingers (Bordino et al. 2002, Forney et al. 2017, Snape et al. 2018, Omeyer et al. 2020). Pingers also may have some adverse effects on porpoises. Displacement could result in stress reactions and reduced food intake, while loud sounds could result in temporary or permanent hearing loss. Captive porpoises avoided active pingers (surfacing farther away from them) but also show an increased respiration rate (Kastelein et al. 2000b). Lusseau et al. (2023) explored the consequences of pingers for a porpoise population using an agent-based model, showing that low rates of pinger deployment (compared to zero deployment) could increase bycatch mortality by leading to movements that increased encounter probability with gillnets not equipped with pingers, while at high rates of pinger deployment, the physiological impacts of acoustic disturbance on reproduction could negatively affect population dynamics. A balance is needed: a relatively low coverage of the fleet may produce the optimum trade-off between positive and negative effects of pinger deployment.

Another solution is increasing the detectability of the nets to echolocation. Metallic-based coatings such as barium sulphate or iron oxide on gillnets have been found to reduce harbour porpoise bycatch (Larsen et al. 2007, Trippel et al. 2008) although this may be due to the increased stiffness of such nets (Larsen et al. 2007, Cox & Read 2004). Gillnets fitted with passive acoustic reflectors (e.g. acrylic glass air-filled "pearls") also reduced porpoise bycatch rates (Gustafsson 2020, Kratzer et al. 2020, 2021). Such nets may be less easy to handle, with more time needed to prepare the nets and to disentangle bycaught animals (Kratzer et al. 2021, Read 2021).

With all such modifications, it is necessary to consider effects on the performance of the gear, e.g., effects on net buoyancy and rigidity, requirements for maintenance and impacts on catches of target species (see Northridge et al. 2013, Mooney et al. 2007, Rowe 2007, Bordino et al. 2013, Knowlton et al. 2016, Kratzer et al. 2020, Read 2021, Sacchi 2021), as well as the potential negative effects of mitigation on the porpoises. As stated by Lusseau et al. (2023), "For marine species that react to noise and are at risk of bycatch, we need to consider the consequences of the interaction of these two stressors more fully".

Alternative approaches include the use of 'move-on' procedures and time/area closures. Time/area closures in the Gulf of Maine proved to be ineffective at reducing bycatch, at least partly due to the (unpredictable) temporal and spatial variation in bycatch rate and partly because fishing effort and the associated bycatch were simply displaced outside the closed area (Murray et al. 2020). Move-on procedures are essentially dynamic closures, and their implementation depends on fishers having access to adequate – and timely – information about the location of cetaceans and bycatch events. The utility of real-time between-vessel data sharing, as used to reduce bycatch of 'choke' species (Marshall et al. 2021) needs to be explored. However, one limitation is the difficulty of observing small species such as harbour porpoise. Possibly porpoise detection could be enhanced by employing acoustic monitoring.

Regardless of the technical solutions, it will be necessary to work with fishers to adapt their fishing practices so that bycatch is reduced (e.g. Ryan et al. 2022). Under the US MMPA, where action on cetacean bycatch is deemed to be required, fishers and scientists are obliged to set up Take Reduction Teams and to devise and follow Take Reduction Plans. Such case-by-case collaborations could be a way forward in Europe.

Part of the problem with implementing these various mitigation methods is that, despite the lethal consequences of bycatch for individual cetaceans and the potentially serious impact at population level, especially for small populations like that of the Iberian porpoise, at the level of individual fishing operations, bycatch of cetaceans is very rare. As such, the costs of mitigation (e.g. installing and maintaining devices, coping with the different handling characteristics of the gear) are arguably much more readily evident than the benefits. This suggests that market incentives (e.g. a premium price for porpoise-safe fish) could make mitigation more acceptable to fishers. Marketing of 'dolphin-safe' tuna, the capture of which did not involve setting nets on dolphins, was part of the solution to reduce bycatch of dolphins in the Eastern Tropical Pacific tuna fishery in the 1980s and 1990s – which is not to undervalue the key roles played by the development of technical measures and training for fishers (National Research Council 1992). Certification of 'dolphin-safe' tuna fishing is currently offered by Friend of the Sea but there is surely scope to develop similar ecolabels for other fisheries.

Finally, in relation to scientific contributions to bycatch reduction, in order to demonstrate a statistically (and practically) significant reduction in bycatch rates, small-scale and short duration trials are often of little value – the performance of the mitigation approach must be monitored over a longer period of time, alongside an equivalent amount of monitored 'control' fishing activity. This in turn implies a need to move away from reliance on short-term project funding for such trials.

Conclusions and recommendations

The Iberian porpoise is an endangered and largely isolated population (and putative subspecies) numbering only a few thousand individuals. Uncertainties notwithstanding, the population evidently suffers high levels of fishery bycatch mortality that may lead to its demise in just a few years. It is essential that we take steps to understand and concurrently facilitate the recovery of this population before it declines to a point at which extinction will become inevitable. If we do not, the case of the vaquita well illustrates the likely outcome (e.g. Rojas-Bracho et al. 2019). Given that the recommended bycatch mortality is zero, the introduction of effective technical measures to reduce bycatch may be insufficient to ensure the long-term survival of the population. However, provided that the measures introduced *are* effective (in terms of both reducing bycatch and avoiding negative consequences like displacement from preferred habitat), slowing the rate of decline would still be a useful first step.

The population occurs mainly along the west coast of the Iberian Peninsula. In terms of its life history and ecology, the most striking distinctive feature of the Iberian porpoise remains its large size, and its diet includes more pelagic fish than is the case for several conspecific populations. Historical records (albeit some of them anecdotal) suggest that the Iberian porpoise was once much more abundant and was also common along the southern coast of Portugal. Today, the population is characterised by a low density and low genetic diversity, with evidence supporting a continuing decline in abundance most likely attributable to fishery bycatch. A comprehensive strategy is needed to save this population, involving effective and well-monitored mitigation actions and supported by increased and well-implemented legal protection (see CMS 2020, ASCOBANS 2021). A joint ASCOBANS-ACCOBAMS conservation plan could be proposed.

In relation to reducing bycatch mortality, the inevitable lack of perfect data cannot be an excuse for inaction. Discarding past data in the hope of collecting more accurate and/or precise data in the future is a fool's errand. Management can be designed to be robust in the face of imperfect data. This rationale underlies the use of Management Strategy Evaluations in fisheries and the use of PBR as a tool under the US MMPA (Wade 1998, Moore et al. 2013). More specifically, management advice can account for imperfect data. However, referring again to fisheries, at least in the European Union, the fact that a particular maximum catch is advised does not necessarily mean that the catch quota which is finally set will follow this advice, nor that once the quota is set, catches will not exceed it. Implementation and compliance/enforcement are not data driven and their success will depend on some combination of (1) the strength (and compatibility) of the relevant legislation related to fisheries (e.g. the Common Fisheries Policy in the EU and the Magnuson–Stevens Fishery Conservation and Management Act (MSFCMA) in the USA) and marine conservation (e.g. the MSFD and MMPA), (2) the willingness of all the stakeholders to work towards achieving the management objectives and (3) public opinion.

Reduction of cetacean bycatch in Europe in general and of porpoise bycatch in the Iberian Peninsula in particular could be facilitated through stronger collaborations with stakeholders (e.g. through Take Reduction Teams) and harnessing public opinion via porpoise-safe ecolabelling.

The European Commission, national governments, international organisations, NGOs, the fishing industry, scientists and (at least to some extent) the general public all realise that action on cetacean bycatch mortality is needed – and that Iberian harbour porpoise is one of the populations for which it is needed most urgently, preferably allowing the recommended bycatch mortality limit of zero to be achieved. A separate assessment of the (Red List) status of this population by the IUCN cetacean group would be welcome.

Acknowledgements

We thank all those colleagues who have contributed directly or indirectly to this work, from NGOs, international organisations, research institutes, universities, national authorities, not least the students and volunteers who contributed to collecting the original data, and M. Begoña Santos Vazquez, the former Head of Marine Mammals at Instituto Español de Oceanografía.

The authors wish to thank an anonymous referee for his/her very useful comments on the submitted manuscript. We are grateful for the help of several retired fishers in conducting informal interviews in the city of A Guarda to obtain historical information on porpoises and to Xabier Pin Vázquez and Marta M. Ortega Cortizo of CEMMA for helping to organise the interviews. We thank the Cetus project and Elisabete Sila for permission to use the photographs reproduced as Figure 9; Antonio Teixeira for permission to use the photo reproduced as Figure 14; Phil Hammond for permission to reproduce figures from some of the SCANS reports; Jenny Renell, Ruth Anderson and Ruth Fernandez for help with citing ASCOBANS and ICES reports; Joana Otero Matias for her input on national surveys and legislation in Portugal; Norma Eréndira García Núñez for her input on Spanish legislation; Arturo Morales and Carlos Nores for confirming that no porpoise remains have been found at archaeological sites in Spain, Catarina Eiras and Andreia Torres-Pereira for information about the death of the leucistic porpoise that had been frequently observed near Porto; Matthieu Authier and Anita Gilles for their assistance with information on Assessment Units; and Florence Caurant, Vincent Ridoux and Cristina Brito for helping us to locate literature from France and Portugal.

The authors from IIM CSIC and IEO CSIC were supported by the following projects: *Coordinated Cetacean Assessment, Monitoring and Management Strategy in the Bay of Biscay and Iberian Coast sub-region* (CetAMBICion, EC DG ENV/MSFD 2020, 110661/2020/839610/SUB/ENV.C.2) (GJP, CS, DFF, PGM); *Monitorización, Evaluación y Reducción de la Mortalidad Accidental de Cetáceos debido a Interacciones con la Flota Española: Revisión y Acción* (MERMA CIFRA, Ministerio de Agricultura, Pesca y Alimentacion (Gobierno de España), MAP2021-02) (GJP, CS, AHG); *Transferencia de Presiones Antropogenicas y Naturales Mediante Relaciones Troficas en el Nekton Oceanico* (Contrato Predoctoral para la Formación de Doctores del Ministerio de Ciencia e Innovación, FPI PRE2019-089740) (AFB); *NUevas Tecnologías moleculares y de control remoto para la Evaluación de las poblaciones de Cetáceos (NuTEC,* Fundacion Biodiversidad) (GJP, CS); *Thresholds in human exploitation of marine vertebrates* (SeaChanges, EU MSCA-ITN-ETN, Grant Agreement number: 813383) (GJP, MACP); and *TRansfers of Anthropogenic and Natural Stressors Involving Trophic Interactions of Ocean Nekton (TRANSITION,* Ministerio de Ciencia, Innovación y Universidades (Proyecto Individual, Plan Nacional, RTI2018-099607-B-I00) (GJP, MBS, PGM, CS, ML, BPF, LV). AL is funded by Portuguese national funds (OE), through FCT – Fundação para a Ciência e a Tecnologia, I.P., in the scope of the framework contract foreseen in the numbers 4, 5 and 6 of the article 23, of the Decree-Law 57/2016, of 29 August, changed by Law 57/2017, of 19 July. The stranding network in Galicia is managed by CEMMA and has the support of the Xunta de Galicia/Dirección General de Patrimonio Natural. AM acknowledges support from the Portuguese Science and Technology Foundation (FCT) through projects UIDB/04326/2020, UIDP/04326/2020 and LA/P/0101/2020 to CCMAR. The authors from the Bottlenose Dolphin Research Institute (BDRI) were supported by the BDRI fundings and the cooperation of all BDRI members and volunteers who generously gave their time to help with field work. Research surveys during 2020 and 2021 were supported by the project "BALAENATUR: seguimiento del rorcual azul dentro de la Red Natura 2000" funded by Fundación Biodiversidad. The research surveys were carried out under an administrative permit from the Spanish Government for the observation of cetaceans for scientific and educational purposes.

References

Abollo, E., López, A., Gestal, C., Benavente, P. & Pascual, S. 1998. Macroparasites in cetaceans stranded on the northwestern Spanish Atlantic coast. *Diseases of Aquatic Organisms* **32**, 227–231.

Abollo, E., Paggi, L., Pascual, S. & D'Amelio, S. 2003. Occurrence of recombinant genotypes of *Anisakis simplex s.s.* and *Anisakis pegreffii* (Nematoda: Anisakidae) in an area of sympatry. *Infection, Genetics and Evolution* **3**(3), 175–181, doi:10.1016/S1567-1348(03)00073-X

ACCOBAMS. 2021. *Estimates of Abundance and Distribution of Cetaceans in the Black Sea from 2019 Surveys*, R.M. Paiu et al. (eds.). Monaco: ACCOBAMS - ACCOBAMS Survey Initiative/CeNoBS Projects, 54 pp.

ACCOBAMS Survey Initiative. 2018. *Technical Report of the Mediterranean Survey*. Monaco: ACCOBAMS.

Addink, M.J., Sørensen, T.B. & Hartmann, M.G. 1995. Aspects of reproduction and seasonality in the harbour porpoise from Dutch waters. *Developments in Marine Biology* **4**, 459–464, doi:10.1016/S0163-6995(06)80046-6

Aguiar, Z. 2013. *Ecologia Alimentar do Bôto (Phocoena phocoena) ao Longo da Costa Continental Portuguesa*. Masters thesis, Universidade de Porto, Portugal.

Aguilar, A. & Borrell, A. 1995. Pollution and harbour porpoises in the eaern North Atlantic: a review. In *Biology of the Phocoenids*, A. Bjørge & G.P. Donovan (eds.). Report of the International Whaling Commission Special Issue 16, 231–242. Online. https://archive.iwc.int/pages/search.php?search=%21collection34

Aguilar, A., Grau, E., Sanpera, C., Jover, L. & Donovan, G. 1983. Report of the Ballena 1 whale marking and sighting cruise in the waters off western Spain. *Report of the International Whaling Commission* **33**, 649–655.

Aguilar, A. & Lockyer, C.H. 1987. Growth, physical maturity, and mortality of fin whales (*Balaenoptera physalus*) inhabiting the temperate waters of the northeast Atlantic. *Canadian Journal of Zoology* **65**(2), 253–264, doi:10.1139/z87-040

Alexandre, S., Marçalo, A., Marques, T.A., Pires, A., Rangel, M., Ressurreição, A., Monteiro, P., Erzini, K. & Gonçalves, J.M.S. 2022. Interactions between air-breathing marine megafauna and artisanal fisheries in Southern Iberian Atlantic waters: results from an interview survey to fishers. *Fisheries Research* **254**, 106430, doi:10.1016/j.fishres.2022.106430

Alfonsi, E., Hassani, S., Carpentier, F.G., Le Clec'h, J.Y., Dabin, W., Van Canneyt, O., Fontaine, M.C. & Jung, J.L. 2012. A European melting pot of harbour porpoise in the French Atlantic Coasts inferred from mitochondrial and nuclear data. *PLoS One* **7**, e44425, doi:10.1371/journal.pone.0044425

Alonso, J.M., López, A. & Gestal, C. 2004. Necropsy reports and summary of causes of death for cetaceans included in BIOCET project, University of Aberdeen, Aberdeen, UK. Annex 6 (Galicia – Spain). In *Necropsy Findings on Cause of Death and Diseases in Small Cetaceans, Including Veterinary Histopathology, Bacteriology and Virology. BIOCET Workpackage 3 Deliverable*, J.A. Learmonth et al. (eds.). In *Bioaccumulation of Persistent Organic Pollutants in Small Cetaceans in European Waters: Transport Pathways and Impact on Reproduction. Final Report to the European Commission's Directorate General for Research on Project EVK3-2000-00027*, G.J. Pierce et al. (eds.).

Alonso, J.M., López, A., González, A.F. & Santos, M.B. 2000. Evidence of violent interactions between bottlenose dolphin (*Tursiops truncatus*) and other cetacean species in NW Spain. *European Research on Cetaceans* **14**, 105–106.

Álvarez-Fernández, E. 2015. L'exploitation des ressources marines au Paléolithique moyen et supérieur initial en Europe: synthèse des données disponibles. *Palethnologie* **7**, 191–209, doi:10.4000/palethnologie.805

Álvarez-Salgado, X.A., Figueira, F.G., Pérez, F.F., Groom, S., Nogueira, E., Borges, A.V., Chou, L., Castro, C.G., Moncoiffé, G., Ríos, A.F., Miller, A.E.J., Frankignoulle, M., Savidge, G. & Wollast, R. 2003. The Portugal coastal counter current off NW Spain: new insights on its biogeochemical variability. *Progress in Oceanography* **56**, 281–321, doi:10.1016/S0079-6611(03)00007-7

Álvarez-Salgado, X.A., Labarta, U., Fernández-Reiriz, M.J., Figueiras, F.G., Rosón, G., Piedracoba, S., Filgueira, R. & Cabanas, J.M. 2008. Renewal time and the impact of harmful algal blooms on the extensive mussel raft culture of the Iberian coastal upwelling system (SW Europe). *Harmful Algae* **7**, 849–855, doi:10.1016/j.hal.2008.04.007

Amante, C. & Eakins, B.W. 2009. *ETOPO1 1 Arc-Minute Global Relief Model: Procedures, Data Sources and Analysis.* NOAA Technical Memorandum NESDIS NGDC-24. National Geophysical Data Center, NOAA, doi:10.7289/V5C8276M

Amorim, I. 2009. Portuguese fisheries, c. 1100–1830. In *A History of the North Atlantic Fisheries. From Early Times to the Mid-Nineteenth Century*, D.J. Starkey, J. Th. Thór & I. Heidbrink (eds.). Bremen: German Maritime Museum of Bremerhaven, Volume **1**, 279–298.

Amundin, M., Carlström, J., Thomas, L., Carlén, I., Koblitz, J., Teilmann, J., Tougaard, J., Tregenza, N., Wennerberg, D., Loisa, O., Brundiers, K., Kosecka, M., Kyhn, L.A., Tiberi Ljungqvist, C., Sveegaard, S., Burt, M.L., Pawliczka, I., Jussi, I., Koza, R., Arciszewski, B., Galatius, A., Jabbusch, M., Laaksonlaita, J., Lyytinen, S., Niemi, J., Šaškov, A., MacAuley, J., Wright, A.J, Gallus, A., Blankett, P., Dähne, M., Acevedo-Gutiérrez, A. & Benke, H. 2022. Estimating the abundance of the critically endangered Baltic Proper harbour porpoise (*Phocoena phocoena*) population using passive acoustic monitoring. *Ecology and Evolution* **12**, e8554, doi:10.1002/ece3.8554

Andreasen, H., Ross, S.D., Siebert, U., Andersen, N.G., Ronnenberg, K. & Gilles, A. 2017. Diet composition and food consumption rate of harbour porpoises (*Phocoena phocoena*) in the western Baltic Sea. *Marine Mammal Science* **33**(4), 1053–1079, doi:10.1111/mms.12421

Ângelo, A.R.M. 2020. *Arrojamentos de cetáceos e interações com as pescas na costa norte de Portugal Continental.* Masters thesis, University of Aveiro.

Angerbjörn, A., Börjesson, P. & Brandberg, K. 2006. Stable isotope analysis of harbour porpoises and their prey from the Baltic and Kattegat/Skagerrak Seas. *Marine Biology Research* **2**(6), 411–419, doi:10.1080/17451000601023896.

Anon. 2009b. *Convenio SGM-IEO para la obtención de datos sobre las capturas accidentales de cetáceos: Final Report.* Spain: Instituto Español de Oceanografía.

Anon. 2009a. APOYO TÉCNICO A LA GESTIÓN SOSTENIBLE DEL MEDIO MARINO. INFORME REGIONAL 2009. Online. https://portalrediam.cica.es/descargas/index.php/s/mxHMWXyHfrCxyNK/download?path=%2F08_AMBITOS_INTERES_AMBIENTAL%2F02_LITORAL_MARINO%2F03_BIOLOGIA%2FInformesMedioMarino%2FDocumentos%2F2009&files=InformeMedioMarino_2009.pdf

Anon. 2010. APOYO TÉCNICO A LA GESTIÓN SOSTENIBLE DEL MEDIO MARINO. INFORME REGIONAL 2010. Online. https://portalrediam.cica.es/descargas/index.php/s/mxHMWXyHfrCxyNK/download?path=%2F08_AMBITOS_INTERES_AMBIENTAL%2F02_LITORAL_MARINO%2F03_BIOLOGIA%2FInformesMedioMarino%2FDocumentos%2F2010&files=InformeMedioMarino_2010.pdf

Anon. 2019. Informe Final de Resultados, Año 2018. Programa de Gestión Sostenible del Medio Marino Andaluz Consejería de Agricultura, Ganadería, Pesca y Desarrollo Sostenible. Junta De Andalucía. Online. https://www.juntadeandalucia.es/medioambiente/portal/landing-page-documento/-/asset_publisher/jXKpcWryrKar/content/informes-regionales-sobre-gesti-c3-b3n-sostenible-del-medio-marino-andaluz-2008-2018-/20151

Anon. 2020. Informe Final de Resultados, Año 2019. Programa de Gestión Sostenible del Medio Marino Andaluz Consejería de Agricultura, Ganadería, Pesca y Desarrollo Sostenible. Junta De Andalucía. Online. https://www.juntadeandalucia.es/medioambiente/portal/landing-page-documento/-/asset_publisher/jXKpcWryrKar/content/informes-regionales-sobre-gesti-c3-b3n-sostenible-del-medio-marino-andaluz-2008-2018-/20151

Anon. 2021. Informe Final de Resultados, Año 2020. Programa de Gestión Sostenible del Medio Marino Andaluz Consejería de Agricultura, Ganadería, Pesca y Desarrollo Sostenible. Junta De Andalucía. Online. https://www.juntadeandalucia.es/medioambiente/portal/landing-page-documento/-/asset_publisher/jXKpcWryrKar/content/informes-regionales-sobre-gesti-c3-b3n-sostenible-del-medio-marino-andaluz-2008-2018-/20151

Anon. 2022a. Informe Final de Resultados, Año 2022. Programa de Gestión Sostenible del Medio Marino Andaluz Consejería de Agricultura, Ganadería, Pesca y Desarrollo Sostenible. Junta De Andalucía. Online. https://portalrediam.cica.es/descargas/index.php/s/mxHMWXyHfrCxyNK/download?path=%2F08_AMBITOS_INTERES_AMBIENTAL%2F02_LITORAL_MARINO%2F03_BIOLOGIA%2FInformesMedioMarino%2FDocumentos%2F2022&files=InformeMedioMarino_2022.pdf

Anon. 2022b. Review of MSFD second cycle reports and state-of-the-art for cetaceans. Deliverable **1**.1. Cetambicion Project.

Araújo, H., Santos, J., Rodrigues, P., Vingada, J., Eira, C., Raínho, A., Arriegas, I., Leonardo, T., Nunes, M. & Sequeira, M. 2015. Proposta técnica de novos Sítios de Interesse Comunitário para a conservação de cetáceos em Portugal Continental para inclusão na Lista Nacional de Sítios. *Anexo do Relatório de Progresso do LIFE+MarPro PT/NAT/00038*, 182.

Arnold, P.W. 1972. Predation on harbour porpoise, *Phocoena phocoena*, by a white shark, *Carcharodon carcharias. Journal of the Fisheries Research Board of Canada* **29**(8), 1213–1214, doi:10.1139/f72-179

ASCOBANS. 2000. Resolution no. 3. Incidental take of small cetaceans. In *3rd Session of the Meeting of Parties*. Bristol, United Kingdom. Online. https://www.ascobans.org/sites/default/files/document/MOP3_2000-3_IncidentalTake_1.pdf

ASCOBANS. 2021. Draft proposal for the inclusion of the Iberian harbour porpoise on the appendices of CMS. In *26th Meeting of the Advisory Committee*, 8–12 November 2021. ASCOBANS/AC26/Doc 4.4b. Online. https://www.ascobans.org/sites/default/files/document/ascobans_ac26_doc4.4b_cms-listing-proposal-iberian-porpoise.pdf

ASCOBANS. 2023. Scoping the development of a European marine strandings database. ASCOBANS/AC28/Doc.8.1. Online. https://www.ascobans.org/sites/default/files/document/ascobans_ac28_doc8.1_scoping-strandings-database.pdf

Baker, J.R. & Martin, A.R. 1992. Causes of mortality and parasites and incidental lesions in harbour porpoises (*Phocoena phocoena*) from British waters. *The Veterinary Record* **130**, 554–558, doi:10.1136/vr.130.25.554

Bandomir-Krischack, B.M. 1996. Preliminary results on reproduction of harbor porpoises in German coastal waters. *European Research on Cetaceans* **9**, 212–214.

Bañón, R., Villegas-Ríos, D., Serrano, A., Mucientes, G. & Arronte, J.C. 2010. Marine fishes from Galicia (NW Spain): an updated checklist. *Zootaxa* **2667**, 1–27.

Barreiros, J.P., Rodeia, J. & Teves, M. 2006. First record of the harbour porpoise *Phocoena phocoena* (Cetacea: Phocoenidae), in the Azores (NE Atlantic). *Aqua – Journal of Ichthyology and Aquatic Biology* **11**(2), 45–46.

Barrett, T., Visser, I.K., Mamaev, L., Goatley, L., van Bressem, M.F. & Osterhaust, A.D. 1993. Dolphin and porpoise morbilliviruses are genetically distinct from phocine distemper virus. *Virology* **193**(2), 1010–1012, doi:10.1006/viro.1993.1217

Bates, S.S., Beach, D.G., Comeau, L.A., Haigh, N., Lewis, N.I., Locke, A., Martin, J.L., McCarron, P., McKenzie, C.H., Michel, C., Miles, C.O., Poulin, M., Quilliam, M.A., Rourke, W.A., Scarratt, M.G., Starr, M. & Wells, T. 2020. Marine harmful algal blooms and phycotoxins of concern to Canada. *Canadian Technical Report of Fisheries and Aquatic Sciences* **3384**, 322.

Bearzi, G., Holcer, D. & Notarbartolo di Sciara, G. 2004. The role of historical dolphin takes and habitat degradation in shaping the present status of northern Adriatic cetaceans. *Aquatic Conservation* **14**, 363–379, doi:10.1002/aqc.626

Bellido, J.J., Castillo, J.J., Farfán, M.A., Martín, J.J., Mons, J.L. & Real, R. 2006. Ejemplar enfermo de marsopa *Phocoena phocoena* (Linnaeus, 1758) varado en las costas de Málaga. *Galemys* **18**(1–2), 37–39.

Bencatel, J., Álvares, F., Moura, A.E. & Barbosa, A.M. (eds.) 2019. *Atlas de Mamíferos de Portugal.* Évora, Portugal: Universidade de Évora, 2ª Edição.

Ben Chehida, Y., Loughnane, R., Thumloup, J., Kaschner, K., Garilao, C., Rosel, P.E. & Fontaine, M.C. 2021. No leading-edge effect in North Atlantic harbor porpoises: evolutionary and conservation implications. *Evolutionary Applications* **14**, 1588–1611, doi:10.1111/eva.13227

Ben Chehida, Y., Stelwagen, T., Hoekendijk, J.P.A., Ferreira, M., Eira, C., Pereira, A.T., Nicolau, L., Thumloup, J. & Fontaine, M.C. 2023. Harbor porpoise losing its edges: genetic time series suggests a rapid population decline in Iberian waters over the last 30 years. *Ecology and Evolution* **13**, e10819.

Benito, J.L., Sanchis, A. & Beneyto, J. 2019. Ballenas y delfines en la prehistoria peninsular. los restos de cetáceo de la Vital (Gandía). In *Recursos Marins En El Passat. IV Jornades D'arqueozoologia*. Valencia: Museu de Prehistoria de Valencia, Diputación de Valencia, 165–192.

Benjamins, S., Dale, A., Hastie, G., Lea, M., Scott, B.E., Waggitt, J.J. & Wilson, B. 2015. Confusion reigns? A review of marine megafauna interactions with energetic tidal features. *Oceanography and Marine Biology: An Annual Review* **53**, 1–54.

Bento, M.C., Canha, R., Eira, C., Vingada, J., Nicolau, L., Ferreira, M., Domingo, M., Tavares, L. & Duarte, A. 2019. Herpesvirus infection in marine mammals: a retrospective molecular survey of stranded cetaceans in the Portuguese coastline. *Infection, Genetics and Evolution* **67**, 222–233, doi:10.1016/j.meegid.2018.11.013

Bernal-Casasola, D. 2018. Whale hunting in the Strait of Gibraltar during the Roman period? *SAA Archaeological Record* **18**(4), 15–22.

Bernhard, M. & Renzoni, A. 1977. Mercury concentration in Mediterranean marine organisms and their environment: natural or anthropogenic origin. *Thalassia Jugosl* **13**, 265–300.

Berthelot, S. 1840. *De la Pêche sur la côte occidentale d'Afrique, et des établissements les plus utiles aux progrès de cette industrie: ouvrage publié sous les auspices de MM. Les ministres de la marine et du commerce.* Paris: Béthune.

Betty, E.B., Stockin, K.A., Hinton, B., Bollard, B.A., Smith, A.N.H., Orams, M.B. & Murphy, S. 2022. Age, growth, and sexual dimorphism of the Southern Hemisphere long-finned pilot whale (*Globicephala melas edwardii*). *Journal of Mammalogy* **103**, 560–575, doi:10.1093/jmammal/gyab165

Birkun, A.A., Jr & Frantzis, A. 2008. *Phocoena phocoena ssp. relicta.* The IUCN Red List of Threatened Species 2008: e.T17030A6737111, doi:10.2305/IUCN.UK.2008.RLTS.T17030A6737111.en

Biserkov, V.Y. & Dimitrov, G.I. 1991. The nematode invasion as a most likely complex lethal factor for *Phocoena phocoena L.* In The Black Sea. In *Second International School: Parasite Host Environment.* Sofía: Publishing House of the Bulgarian Academy of Sciences, 227.

Bjørge, A., Aarefjord, H., Kaarstad, S., Kleivane, L. & Øien, N. 1991. Harbour porpoise *Phocoena phocoena* In Norwegian Waters. ICES Conference and Meeting, Volume **16**, 16–24.

Blanchard, A., Doremus, G., Laran, S., Niviere, M., Sanchez, T., Spitz, J. & Van Canneyt, O. 2021. Distribution et abondance de la mégafaune marine en France métropolitaine. In *SAMM II Atlantique-Manche Hiver 2021. Rapport de campagne août 2021*. Pelagis: La Rochelle Université.

Boisseau, O., Matthews, J., Gillespie, D., Lacey, C., Moscrop, A. & Ouamari, N.E. 2007. A visual and acoustic survey for harbour porpoises off North-West Africa: further evidence of a discrete population. *African Journal of Marine Science* **29**(3), 403–410, doi:10.2989/AJMS.2007.29.3.8.338

Bordino, P., Kraus, S., Albareda, D., Fazio, A., Palmerio, A., Mendez, M. & Botta, S. 2002. Reducing incidental mortality of Franciscana dolphin *Pontoporia blainvillei* with acoustic warning devices attached to fishing nets. *Marine Mammal Science* **18**, 833–842, doi:10.1111/j.1748-7692.2002.tb01076.x

Bordino, P., Mackay, A.I., Werner, T.B., Northridge, S.P. & Read, A.J. 2013. Franciscana bycatch is not reduced by acoustically reflective or physically stiffened gillnets. *Endangered Species Research* **21**(1), 1–12, doi:10.3354/esr00503

Börjesson, P. & Read, A.J. 2003. Variation in timing of conception between populations of the harbor porpoise. *Journal of Mammalogy* **84**, 948–955.

Brandt, M.J., Hansen, S., Diederichs, A. & Nehls, G. 2014. Do man-made structures and water depth affect the diel rhythms in click recordings of harbor porpoises (*Phocoena phocoena*)? *Marine Mammal Science* **30**(3), 1109–1121, doi:10.1111/mms.12112

Braulik, G., Minton, G., Amano, M. & Bjørge, A. 2020. Phocoena phocoena. The IUCN Red List of Threatened Species 2020: e.T17027A50369903, doi:10.2305/IUCN.UK.2020-2.RLTS.T17027A50369903.en

Brennecke, D., Siebert, U., Kindt-Larsen, L., Midtiby, H.S., Egemose, H.D., Torres Ortiz, S., et al. 2022. The fine-scale behavior of harbor porpoises towards pingers. *Fisheries Research* **255**, 106437, doi:10.1016/j.fishres.2022.106437

Brito, C. 2011. Medieval and early modern whaling in Portugal. *Anthrozoos* **24**(3), 287–300, doi:10.2752/175303711X13045914865303

Brito, C. & Sousa, A. 2011. The environmental history of cetaceans in Portugal: ten centuries of whale and dolphin records. *PLoS ONE* **6**(9), e23951, doi:10.1371/journal.pone.0023951

Brito, C. & Vieira, N. 2010. Using historical accounts to assess the occurrence and distribution of small cetaceans in a poorly known area. *Journal of the Marine Biological Association of the United Kingdom* **90**, 1583–1588, doi:10.1017/S0025315410000585

Brito, C., Vieira, N., Sá, E. & Carvalho, I. 2009. Cetaceans' occurrence off the west central Portugal coast: a compilation of data from whaling, observation of opportunity and boat-based surveys. *Journal of Marine Animals and their Ecology* **2**, 10–13.

Broadwater, M.H., Van Dolah, F.M. & Fire, S.E. 2018. Vulnerabilities of marine mammals to harmful algal blooms. In Harmful Algal Blooms, S.E. Shumway et al. (eds.). Chichester, United Kingdom: John Wiley & Sons, 191–222, doi:10.1002/9781118994672.ch5

Brunel, T., Farrell, E.D., Kotterman, M., Kwadijk, C., Verkempynck, R., Chen, C & Miller, D. 2016. *Improving the Knowledge Basis for Advice on North Sea Horse Mackerel. Developing New Methods to Get Insight on Stock Boundaries and Abundance*. Wageningen: IMARES Wageningen UR (University & Research centre), Wageningen Marine Research report **C092/16**, 57 pp.

Butler-Stroud, C. & Rouley, D.A. 2019. Complaint against the breach of species protection obligations under article 12 of the Habitats Directive in relation to cetaceans. Letter to Mr Calleja Crespo (Directorate-General for Environment) and Mr Aguiar Machado (Directorate-General for Maritime Affairs & Fisheries). Online. https://seas-at-risk.org/publications/complaint-against-the-breach-of-species-protection-obligations-under-article-12-of-the-habitats-directive-in-relation-to-cetaceans/

Cabezón, O., Resendes, A.R., Domingo, M., Raga, J.A., Agustí, C., Alegre, F., Mons, J.L., Dubey, J.P. & Almería, S. 2004. Seroprevalence of *Toxoplasma gondii* antibodies in wild dolphins from the Spanish Mediterranean coast. *The Journal of Parasitology* **90**(3), 643–644, doi:10.1645/GE-257R

Cabral, M.J., Almeida, J., Almeida, P.R., Dellinger, T., Ferrand de Almeida, N., Oliveira, M.E., Palmeirim, M.J., Queiroz, A.I., Rogado, L., Santos-Reis, M. (eds.). 2006. *Livro Vermelho dos Vertebrados de Portugal*. Lisboa: Instituto da Conservação da Natureza/Assírio e Alvim, 2a edição.

Cabrera, A. 1914. *Fauna Ibérica. Mamíferos*. Madrid: Museo Nacional de Ciencias Naturales, 441.

Camarão, B.C. 2017. *Estudo da reprodução de pequenos cetáceos através da morfologia do ovário*. Masters thesis, University of Aveiro.

Carlén, I., Nunny, L. & Simmonds, M.P. 2021. Out of sight, out of mind: how conservation is failing european porpoises. *Frontiers in Marine Science* **8**, **617478**, doi:10.3389/fmars.2021.617478

Carlström, J. 2005. Diel variation in echolocation behavior of wild harbor porpoises. *Marine Mammal Science* **21**(1), 1–12, doi:10.1111/j.1748-7692.2005.tb01204.x

Carmona, J. & López Losa, E. 2009. Spain's Atlantic coast fisheries, c. 1100–1880. In *A History of the North Atlantic Fisheries. From Early Times to the Mid-Nineteenth Century*, D.J. Starkey, J.Th. Thór & I. Heidbrink (eds.). Bremen: German Maritime Museum of Bremerhaven, Volume **1**, 250–278.

Castro, J.M. 2010. *Characterization of cetaceans in the South coast of Portugal between Lagos and Cape São Vicente*. MSc thesis, University of Lisbon, Portugal.

Caswell, H., Brault, S., Read, A.J. & Smith, T.D. 1998. Harbor porpoise and fisheries: an uncertainty analysis of incidental mortality. *Ecological Applications* **8**(4), 1226–1238, doi:10.1890/1051-0761(1998)008[1226:HPAFAU]2.0.CO;2

Celemín, E., Autenrieth, M., Roos, A., Pawliczka, I., Quintela, M., Lindstrøm, U., Benke, H., Siebert, U., Lockyer, C., Berggren, P., Özturk, A.A., Özturk, B., Lesage, V. & Tiedemann, R. 2023. Evolutionary history and seascape genomics of harbour porpoises (*Phocoena phocoena*) across environmental gradients in the North Atlantic and adjacent waters. *Molecular Ecology Resources,* doi:10.1111/1755-0998.13860

CEMMA. 2012. Inventario y designación de la Red Natura 2000 en áreas marinas del Estado español. LIFE+INDEMARES. Acciones CEMMA. Informe final.

Chang, H.W. 1951. Age and growth of *Callionymus lyra* L. *Journal of the Marine Biological Association of the United Kingdom* **30**, 281–296, doi:10.1017/S0025315400012777

Cipriani, P., Palomba, M., Giulietti, L., Marcerd, F., Mazzariol, S., Santoro, M., Alburqueque, R.A., Covelo P., López, A., Santos, M.B., Pierce, J.P., Brownlow, A., Davison, N.J., McGovern, B., Frantzis, A., Alexiadou, P., Højgaard, D.P., Mikkelsen, B., Paoletti, M., Nascetti, G., Levsen, A. & Mattiucci, S. 2022. Distribution and genetic diversity of *Anisakis* spp. In cetaceans from the Northeast Atlantic Ocean and the Mediterranean Sea a. *Scientific Reports* **12**, 13664.

Clark, P.U., Dyke, A.S., Shakun, J.D., Carlson, A.E., Clark, J., Wohlfarth, B., Mitrovica, J.X., Hostetler, S.W. & McCabe, A.M. 2009. The last glacial maximum. *Science* **325**, 710–714, doi:10.1126/science.1172873

CMS. 2020. Concerted action for the harbour porpoise (*Phocoena phocoena*) in the Baltic Sea and the Iberian Peninsula. UNEP/CMS/Concerted Action 13.7. Online. https://www.cms.int/sites/default/files/document/cms_cop13_ca.13.7_e.pdf

Coe, J.M., Holts, D.B. & Butler, R.W. 1984. Guidelines for reducing porpoise mortality in Tuna Purse Seining. National oceanic and atmospheric administration technical report, National Marine Fisheries Service, 22. U.S.

Collet, A. 1995. Some data on harbour porpoises off the French Coast. *International Whaling Commission, Scientific Committee* SC/47/SM2.

Conn, P.B., Bravington, M.V., Baylis, S. & Ver Hoef, J.M. 2020. Robustness of close-kin mark-recapture estimators to dispersal limitation and spatially varying sampling probabilities. *Ecology and Evolution* **10**, 5558–5569, doi:10.1002/ece3.6296

Corchón-Rodríguez, M.S. & Álvarez-Fernández, E. 2008. Nuevas evidencias de restos de mamíferos marinos en el Magdaleniense: los datos de La Cueva de Las Caldas (Asturias, España). *Munibe (Antropología-Arkeología)* **59**, 47–66.

Correia, A.M. 2020. *Distribution and habitat modelling for cetacean species in the eastern North Atlantic Ocean*. PhD thesis, University of Porto.

Correia, A.M., Gandra, M., Liberal, M., Valente, R., Gil, A., Rosso, M., Pierce, G.J. & Sousa-Pinto, I. 2019. A dataset of cetacean occurrences in the Eastern North Atlantic. *Scientific Data* **6**, 177, doi:10.1038/s41597-019-0187-2

Correia, A.M., Oliveira-Rodrigues, C., Gandra, M., Liberal, M., Valente, R., Gil, A., Rosso, M., Pierce, G.J. & Sousa-Pinto, I., CIIMAR – UP. 2022. CETUS: cetacean monitoring surveys in the Eastern North Atlantic. *Marine Data Archive*, doi:10.14284/547

Cossa, D., Knoery, J., Bănaru, D., Harmelin-Vivien, M., Sonke, J.E., Hedgecock, I.M., et al. 2022. Mediterranean mercury assessment 2022: an updated budget, health consequences, and research perspectives. *Environmental Science & Technology* **56**(7), 3840–3862.

Cossa, D., Martin, J.M., Takayanagi, K. & Sanjuan, J. 1997. The distribution and cycling of mercury species in the western Mediterranean. *Deep Sea Research Part II: Topical Studies in Oceanography* **44**(3–4), 721–740.

Council Directive 92/43/EEC, of 21 May 1992 on the conservation of natural habitats and of wild fauna and flora. *Official Journal of the European Communities* **L206**, 7–50.

Covelo, P. & López, A. 2021a. First record of dwarf sperm whale (*Kogia sima*) in the north of Spain. *Galemys Spanish Journal of Mammalogy* **33**, 64–69, doi:10.7325/Galemys.2021.N8

Covelo, P. & López, A. 2021b. How to eat a needlefish? The harbour porpoise way. *Marine Mammal Science* **37**, 1109–1115, doi:10.1111/mms.12781

Cox, T.M. & Read, A.J. 2004. Echolocation behaviour of harbour porpoises *Phocoena phocoena* around chemically enhanced gill nets. *Marine Ecology Progress Series* **279**, 275–82, doi:10.3354/meps279275

Cox, T.M., Read, A.J., Barco, S., Evans, J., Gannon, D.P., Koopman, H.N., McLellan, W.A., Murray, K., Nicolas, J., Pabst, D.A., Potter, C.W., Swingle, W.M., Thayer, V.G., Touhey, K.M. & Westgate, A.J. 1998. Documenting the bycatch of harbor porpoises, *Phocoena phocoena*, in coastal gillnet fisheries from stranded carcasses. *Fishery Bulletin* **96**, 727–734.

Cucknell, A.C., Frantzis, A., Boisseau, O., Romagosa, M., Ryan, C., Tonay, A.M., Alexiadou, P., Öztürk, A.A. & Moscrop, A. 2016. Harbour porpoises in the Aegean Sea, Eastern Mediterranean: the species' presence is confirmed. *Marine Biodiversity Records* **9**, 72, doi:10.1186/s41200-016-0050-5

Culik, B.M., Koschinski, S., Tregenza, N. & Ellis, G.M. 2001. Reactions of harbor porpoises *Phocoena phocoena* and herring *Clupea harengus* to acoustic alarms. *Marine Ecology Progress Series* **211**, 255–260, doi:10.3354/meps211255

Dagleish, M.P., Barley, J., Finlayson, J., Reid, R.J. & Foster, G. 2008. *Brucella ceti* Associated Pathology in the Testicle of a Harbour Porpoise (*Phocoena phocoena*). *Journal of Comparative Pathology* **139**, 54–59, doi:10.1016/j.jcpa.2008.03.004

Das, K., Holsbeek, L., Browning, J., Siebert, U., Birkun, A., Jr. & Bouquegneau, J.-M. 2004. Trace metal and stable isotope measurements ($d^{13}C$ and $d^{15}N$) in the harbour porpoise *Phocoena phocoena relicta* from the Black Sea. *Environmental Pollution* **131**, 197–204, doi:10.1016/j.envpol.2004.02.006

Davison, N., Barnett, J., Rule, B., Chappell, S. & Wise, G. 2010. Group B *Salmonella* in lungworms from a harbour porpoise (*Phocoena phocoena*). *Veterinary Record* **167**, 351–352, doi:10.1136/vr.c4495

Davison, N. & ten Doeschate, M. 2020. Scottish marine animals stranding scheme. Annual report 2020, 1 January to 31 December 2020 for Marine Scotland, Scottish Government. Online. https://strandings.org/wp-content/uploads/2022/09/SMASS-AR-2020-final.pdf

Deaville, R., Jepson, P.D., Perkins, M., Brownlow, A., Davison, N., ten Doeschate, M., Smith, B., Laaan, L., Clery, M., Swindells, K., Wilson, S., Sabin, R.C., Penrose, R., Barnett, J.E.F., Astley, K., Clear, N., Crosby, A. & Williams, R. 2018. UK Cetacean Strandings Investigation Programme final contract report to Defra (MB0111 2011-2017). Online. https://randd.defra.gov.uk/ProjectDetails?ProjectId=17835

DeMaster, D.P. 1978. Calculation of the average age of sexual maturity in marine mammals. *Journal of the Fisheries Research Board of Canada* **35**(6), 912–915, doi:10.1139/f78-148

DeMaster, D.P. 1984. Review of techniques used to estimate the average age at attainment of sexual maturity in marine mammals. In *Reproduction of Whales, Dolphins and Porpoises*, W.F. Perrin, J.R. Brownell & D.P. DeMaster (eds.). Report of the International Whaling Commission Special Issue 6, 175–179. Online. https://archive.iwc.int/pages/search.php?search=%21collection34

De Pierrepont, J.F., Dubois, B., Desmonts, S., Santos, M.B. & Robin, J.P. 2005. Stomach contents of English Channel cetaceans stranded on the coast of Normandy. *Journal of the Marine Biological Association of the United Kingdom* **85**(6), 1539–1546, doi:10.1017/S0025315405012762

Derous ten Doeschate, M., Brownlow, A.C., Davison, N.J. & Lusseau, D. 2020. Towards new ecologically relevant markers of health for cetaceans. *Frontiers in Marine Science* **7**, 367, doi:10.3389/fmars.2020.00367

Desportes, G. 1985. *La nutrition des Odontocètes en Atlantique Nord-Est (côtes Françaises-îles Feroë)*. PhD thesis, Université de Poitiers, Portugal.

Desportes, G., Amundin, M. & Goodson, D. 2000. Investigate porpoise foraging behaviour – Task 1. In *The Tail of EPIC: Elimination of Harbour Porpoise Incidental Catch. Final Report to the European Commission of Project No DG XIV 97/00006*, C. Lockyer et al. (eds.), 249.

Díaz Caneiro, C. 2019. *Parásitos en mamíferos marinos varados en el litoral gallego*. BSc thesis, University of Santiago de Compostela, Spain.

Díaz-Delgado, J., Fernández, A., Sierra, E., Sacchini, S., Andrada, M., Vela, A.I., Quesada-Canales, Ó., Paz, Y., Zucca, D., Groch, K. & Arbelo, M. 2018. Pathologic findings and causes of death of stranded cetaceans in the Canary Islands (2006–2012). *PLoS One* **13**(10), e0204444, doi:10.1371/journal.pone.0204444

Díaz López, B. & Methion, S. 2018. Does interspecific competition drive patterns of habitat use and relative density in harbour porpoises? *Marine Biology* **165**, 92, doi:10.1007/s00227-018-3345-8

Díaz López, B., Methion, S., Mosca, O. & Dunel-Roig, N. 2022. *Informe sobre la distribución y dinámica de grupo de la marsopa común en aguas gallegas.* O Grove: Bottlenose Dolphin Research Institute, 39 pp.

Díaz López, B. & Methion, S. 2024. Habitat use by Iberian harbour porpoises: ecological and human factors. *Marine Biology* **171**, 113. doi:10.1007/s00227-024-04438-x"

Dillen, K. 2022. Porpoise, punishment and partnership: the meaning of presenting and consuming a marine mammal in late medieval coastal Flanders. *Urban History* **49**, 568–588, doi:10.1017/S0963926821000225

Directive 2008/56/EC of the European Parliament and of the Council, of 17 June 2008 establishing a framework for community action in the field of marine environmental policy (Marine Strategy Framework Directive). *Official Journal of the European Union* **L164**, 19–40.

Dolman, S.J., Baulch, S.B., Evans, P.G.H., Read, F. & Ritter, F. 2016. Towards an EU action plan on cetacean bycatch. *Marine Policy* **72**, 67–75, doi:10.1016/j.marpol.2016.06.020

Dolman, S.J., Evans, P.G.H., Ritter, F., Simmonds, M.P. & Swabe, J. 2021. Implications of new technical measures regulation for cetacean bycatch in European waters. *Marine Policy* **124**, 104320, doi:10.1016/j.marpol.2020.104320

Donovan, G.P. & Bjørge, A. 1995. Harbour porpoises in the North Atlantic. *Report of the International Whaling Commission* **16**, 3–25.

Du Bocage, M.B. 1863. Liste des mammifères et reptiles observés au Portugal. *Revue et Magasin de Zoologie Pure et Applique* **15**, 329–333.

Duguy, R. & Desportes, G. 1983. Cétacés observés dans le nord-est Atlantique en 1981. *Annals of Biology* **38**, 209–210.

Dyndo, M., Wiśniewska, M., Rojano-Doñate, L. & Teglberg Madsen, P., 2015. Harbour porpoises react to low levels of high frequency vessel noise. *Scientific Reports* **5**, 11083, doi:10.1038/srep11083

Dzido, J., Rolbiecki, L., Izdebska, J.N., Rokicki, J., Kuczkowski, T. & Pawliczka, I. 2021. A global checklist of the parasites of the harbor porpoise *Phocoena phocoena*, a critically-endangered species, including new findings from the Baltic Sea. *International Journal for Parasitology: Parasites and Wildlife* **15**, 290–302, doi:10.1016/j.ijppaw.2021.07.002

Esteban, R., López, A., García de los Ríos, Á., Ferreira, M., Martinho, F., Méndez-Fernandez, P., Andréu, E., García-Gómez, J.C., Olaya-Ponzone, L., Espada-Ruiz, R., Gil-Vera, F.J., Martín Bernal, C., García-Bellido Capdevila, E., Sequeira, M. & Martínez-Cedeira, J.A. 2022. Killer whales of the Strait of Gibraltar, an endangered subpopulation showing a disruptive behaviour. *Marine Mammal Science*, 1–11, doi:10.1111/mms.12947

Evans, P.G.H. 1987. *The Natural History of Whales and Dolphins.* Oxford University Press. ISBN 13: 9780747008002.

Evans, P.G.H. & Teilmann, J. (eds.). 2009. REPORT OF ASCOBANS/HELCOM SMALL CETACEAN POPULATION STRUCTURE WORKSHOP Held on 8–10 October 2007 at UN Campus, Hermann-Ehlers-Str. **10**, 53113. Bonn, Germany: ASCOBANS. Online. https://www.ascobans.org/meeting/download-all-files/1062/all/all

Evans, S. & Mulville, J. 2018. Finding Moby: identifying whales in the archaeological record. A study of the vertebral morphology of cetacean species in the North Eastern Atlantic for the purposes of zooarchaeological analysis. *The SAA Archaeological Record (September 2018)*, 31–37.

Faulkner, J., Measures, L.N. & Whoriskey, F.G. 1998. *Stenurus minor* (Metastrongyloidea: Pseudaliidae) infections of the cranial sinuses of the harbor porpoise, *Phocoena phocoena. Canadian Journal of Zoology* **76**, 1209–1216, doi:10.1139/z98-057

Ferguson, S.H. & Larivière, S. 2008. How social behaviour links environment and body size in mammalian carnivores. *Open Ecology Journal* **1**, 1–7, doi:10.2174/1874213000801010001

Ferguson, S.H., Zhu, X., Young, B.G., Yurkowski, D.J., Thiemann, G.W., Fisk, A.T., et al. 2018. Geographic variation in ringed seal (*Pusa hispida*) growth rate and body size. *Canadian Journal of Zoology* **96**, 649–659, doi:10.1139/cjz-2017-0213

Fernández, R., MacLeod, C.D., Pierce, G.J., Covelo, P., López, A., Torres-Palenzuela, J., Valavanis, V. & Santos, M.B. 2013. Inter-specific and seasonal comparison of the niches occupied by small cetaceans off north-west Iberia. *Continental Shelf Research* **64**, 88–98, doi:10.1016/j.csr.2013.05.008

Fernández-Contreras, M.M., Cardona, L., Lockyer, C.H. & Aguilar, A. 2010. Incidental bycatch of short-beaked common dolphins (*Delphinus delphis*) by pair trawlers off northwestern Spain. *ICES Journal of Marine Science* **67**, 1732–1738, doi:10.1093/icesjms/fsq077

Fernández de la Cigoña, E. 1990. *Os cabaleiros do mar: baleas e golfiños das nosas augas. Guía dos cetáceos de Galicia e Iberia.* Vigo, Asociación Galega para a Cultura e a Ecoloxía (AGCE), 162.

Ferreira, M., Monteiro, S.S., Torres, J., Oliveira, I., Sequeira, M., López, A., Vingada, J. & Eira, C. 2016. Biological variables and health status affecting inorganic element concentrations in harbour porpoises (*Phocoena phocoena*) from Portugal (western Iberian Peninsula). *Environmental Pollution* **210**, 293–302, doi:10.1016/j.envpol.2016.01.027

Fichou, J.C. & Levasseur, O. 2004. Pêcheurs contre Marsouins et Bélougas, XVIe-XXe siècle. *Food & History* **2**(2), 53–86.

Filgueiras, A.V., Preciado, I., Cartón, A. & Gago, J. 2020. Microplastic ingestion by pelagic and benthic fish and diet composition: a case study in the NW Iberian shelf. *Marine Pollution Bulletin* **160**, 111623, doi:10.1016/j.marpolbul.2020.111623

Findlay, C.R., Aleynik, D., Farcas, A., Merchant, N.D., Risch, D. & Wilson, B. 2021. Auditory impairment from acoustic seal deterrents predicted for harbour porpoises in a marine protected area. *Journal of Applied Ecology* **58**, 1631–1642, doi:10.1111/1365-2664.13910

Fire, S.E., Bogomolni, A., Di Giovanni, R.A., Jr, Early, G., Leighfield, T.A., Matassa, K., Miller, G.A., Moore, K.M.T., Moore, M., Niemeyer, M., Pugliares, K., Wang, Z. & Wenzel, F.W. 2021. An assessment of temporal, spatial and taxonomic trends in harmful algal toxin exposure in stranded marine mammals from the U.S. New England coast. *PLoS One* **16**(1), doi:10.1371/journal.pone.0243570

Fock, H.O. 2011. Natura 2000 and the European Common Fisheries Policy. *Marine Policy* **35**, 181–188.

Fontaine, M.C. 2016. Harbour porpoises, *Phocoena phocoena*, in the Mediterranean Sea and adjacent regions: biogeographic relicts of the last glacial period. *Advances in Marine Biology* **75**, 333–358, doi:10.1016/bs.amb.2016.08.006

Fontaine, M.C., Baird, S.J.E., Piry, S., Ray, N., Tolley, K.A. Duke, S., Birkun, A., Jr, Ferreira, M., Jauniaux, T., Llavona, A., Öztürk, B., Öztürk, A.A., Ridoux, V., Rogan, E., Sequeira, M., Siebert, U., Vikingsson, G.A., Bouquegneau, J-M. & Michaux, J.R. 2007. Rise of oceanographic barriers in continuous populations of a cetacean: the genetic structure of harbour porpoises in Old World waters. *BMC Biology* **5**(30), doi:10.1186/1741-7007-5-30

Fontaine, M.C., Roland, K., Calves, I., Austerlitz, F., Palstra, F.P., Tolley, K.A., Ryan, S., Ferreira, M., Jauniaux, T., Llavona, A., Öztürk, B., Öztürk, A.A., Ridoux, V., Rogan, E., Sequeira, M., Siebert, U., Vikingsson, G.A., Borrell, A., Michaux, J.R. & Aguilar, A. 2014. Postglacial climate changes and rise of three ecotypes of harbour porpoises, *Phocoena phocoena*, in western Palearctic waters. *Molecular Ecology* **23**, 3306–3321, doi:10.1111/mec.12817

Fontaine, M.C., Snirc, A., Frantzis, A., Koutrakis, E., Ozturk, B., Ozturk, A.A. & Austerlitz, F. 2012. History of expansion and anthropogenic collapse in a top marine predator of the Black Sea estimated from genetic data. PNAS (Proceedings of the National Academy of Sciences) **109**, E2569–E2576, doi:10.1073/pnas.1201258109

Fontaine, M.C., Thatcher, O., Ray, N., Piry, S., Brownlow, A., Davison, N.J., Jepson, P., Deaville, R. & Goodman, S.J. 2017. Mixing of porpoise ecotypes in southwestern UK waters revealed by genetic profiling. *Royal Society Open Science* **4**(3), 160992, doi:10.1098/rsos.160992

Fontaine, M.C., Tolley, K.A., Michaux, J.R., Birkun, A., Jr, Ferreira, M., Jauniaux, T., Llavona, A., Öztürk, B., Öztürk, A.A., Ridoux, V., Rogan, E., Sequeira, M., Bouquegneau, J-M. & Baird, S.J.E. 2010. Genetic and historic evidence for climate-driven population fragmentation in a top cetacean predator: the harbour porpoises in European water. *Proceedings of the Royal Society B* **277**, 2829–2837, doi:10.1098/rspb.2010.0412

Foote, A.D., Thomsen, P.F., Sveegaard, S., Wahlberg, M., Kielgast, J., Kyhn, L.A., et al. 2012. Investigating the potential use of environmental DNA (eDNA) for genetic monitoring of marine mammals. *PLoS One* **7**, 2–7, doi:10.1371/journal.pone.0041781

Forney, K.A., Southall, B.L., Slooten, E., Dawson, S., Read, A.J., Baird, R.W. & Brownell, R.L. 2017. Nowhere to go: noise impact assessments for marine mammal populations with high site fidelity. *Endangered Species Research* **32**, 391–413, doi:10.3354/esr00820

Fosså, O. 1995. A whale of a dish: whalemeat a food. In *Disappearing Foods: Studies in Food and Dishes at Risk.* Proceedings of the Oxford Symposium on Food and Cookery, H. Walker (ed.). Totnes: Prospect Books, 78–102.

Fraga, F. 1981. Upwelling off the Galician coast, Northwest Spain. In Coastal Upwelling, F.A. Richards (ed.). Washington D.C.: American Geophysical Union, 176–182, doi:10.1029/CO001p0176

Franco, M.Á., Viñas, L., Soriano, J.A., de Armas, D., González, J.J., Beiras, R., Salas, N., Bayona, J.M. & Albaigés, J. 2006. Spatial distribution and ecotoxicity of petroleum hydrocarbons in sediments from the Galicia continental shelf (NW Spain) after the Prestige oil spill. *Marine Pollution Bulletin* **53**, 260–271, doi:10.1016/j.marpolbul.2005.10.004

Frantzis, A., Gordon, J., Hassadis, G. & Komnenou, A. 2001. The enigma of harbor porpoise presence in the Mediterranean Sea. *Marine Mammal Science* **17**, 937–944.

Freitas, L., Dinis, A., Nicolau, C., Ribeiro, C. & Alves, F. 2012. New records of cetacean species for Madeira Archipelago with an updated checklist. *Boletim do Museu Municipal do Funchal* **62**, 25–44.

Gallagher, C.A., Grimm, V., Kyhn, L.A., Kinze, C.C. & Nabe-Nielsen, J. 2021. Movement and seasonal energetics mediate vulnerability to disturbance in marine mammal populations. *American Naturalist* **197**(5), 296–311, doi:10.5061/dryad.80gb5mkpn

García Tíscar, S. 2010. *Interacciones entre delfines mulares (Tursiops truncatus) orcas (Orcinus orca) y pesquerías en el mar de Alborán y estrecho de Gibraltar*. PhD thesis, Universidad Autónoma de Madrid.

Gaskin, D.E. 1984. The harbour porpoise *Phocoena phocoena* (L.): regional populations, status and information on direct and indirect catches. *Report of the International Whaling Commission* **34**, 569–586.

Gearin, P.J., Gosho, M.E., Cooke, L., DeLong, R. & Laake, J. 1996. Acoustic alarm experiment in the 1995 northern Washington marine setnet fishery: methods to reduce by-catch of harbor porpoise. Paper SC/48/SM10. Paper presented at the 48th meeting on the Scientific Committee of the International Whaling Commission.

Geelhoed, S.C.V., Authier, M., Pigeault, R. & Gilles, A. 2022. Abundance and distribution of cetaceans. In *OSPAR, 2023: The 2023 Quality Status Report for the Northeast Atlantic*. London: OSPAR Commission. Online. https://oap.ospar.org/en/ospar-assessments/quality-status-reports/qsr-2023/indicator-assessments/abundance-distribution-cetaceans/

Gibson, D.I., Harris, E.A., Bray, R.A., Jepson, P.D., Kuiken, T., Baker, J.R., & Simpson, V.R. 1998. A survey of the helminth parasites of cetaceans stranded on the coast of England and Wales during the period 1990-1994. *Journal of Zoology*, 244, 563–574. doi:10.1017/S0952836998004099

Gil, Á., Correia, A.M. & Sousa-Pinto, I. 2019. Records of harbour porpoise (*Phocoena phocoena*) in the mouth of the Douro River (northern Portugal) with presence of an anomalous white individual. *Marine Biodiversity Records* **12**, **1**, doi:10.1186/s41200-018-0160-3

Gilles, A., Authier, M., Ramirez-Martinez, N.C., Araújo, H., Blanchard, A., Carlström, J., Eira, C., Dorémus, G., Fernández-Maldonado, C., Geelhoed, S.C.V., Kyhn, L., Laran, S., Nachtsheim, D., Panigada, S., Pigeault, R., Sequeira, M., Sveegaard, S., Taylor, N.L., Owen, K., Saavedra, C., Vázquez-Bonales, J.A., Unger, B. & Hammond, P.S. 2023. Estimates of cetacean abundance in European Atlantic waters in summer 2022 from the SCANS-IV aerial and shipboard surveys. Final report published 29 September 2023. Online. https://tinyurl.com/3ynt6swa

Giosan, L., Filip, F. & Constatinescu, S. 2009. Was the Black Sea catastrophically flooded in the early Holocene? *Quaternary Science Reviews* **28**, 1–6, doi:10.1016/j.quascirev.2008.10.012

Goetz, S., Read, F.L., Ferreira, F., Martínez Portela, J., Santos, M.B., Vingada, J.V., Siebert, U., Marçalo, A., Santos, J., Araújo, H., Monteiro, S., Caldas, M., Riera, M. & Pierce, G.J. 2015. Cetacean occurrence, habitat preferences and potential for cetacean-fishery interactions in Iberian Atlantic waters: results from cooperative research involving local stakeholders. *Aquatic Conservation: Marine and Freshwater Ecosystems* **25**, 138–154, doi:10.1002/aqc.2481

Goetz, S., Read, F.L., Santos, M.B., Pita, C. & Pierce, G.J. 2014. Cetacean-fishery interactions in Galicia (NW Spain): results and management implications of a face-to-face interview survey of local fishers. *ICES Journal of Marine Science* **71**, 604–617, doi:10.1093/icesjms/fst149

Gomes, T.L., Quiazon, K.M., Kotake, M., Fujise, Y., Ohizumi, H., Itoh, N. & Yoshinga, T. 2021. *Anisakis* spp. in toothed and baleen whales from Japanese waters with notes on their potential role as biological tags. *Parasitology International* **80**, **102228**, doi:10.1016/j.parint.2020.102228

Gonçalves de Sousa, A. 2010. *Padrões de arrojamentos de cetáceos na costa continental portuguesa*. Masters thesis, University of Aveiro.

González, A.F., López, A., Guerra, A. & Barreiro, A. 1994. Diets of marine mammals stranded on the northwestern Spanish Atlantic coast with special reference to Cephalopoda. *Fisheries Research* **21**, 179–191, doi:10.1016/0165-7836(94)90103-1

González, J.J., Viñas, L., Franco, M.Á., Fumega, J., Soriano, J.A., Grueiro, G., Muniategui, S., López Mahía, P., Prada, D., Bayona, J.M., Alzaga, R. & Albaigés, J. 2006. Spatial and temporal distribution of dissolved/dispersed aromatic hydrocarbons in seawater in the area affected by the Prestige oil spill. *Marine Pollution Bulletin* **53**, 250–259.

Graells, M. 1870. *Exploración científica de las costas del Departamento Marítimo del Ferrol.* Madrid: Establecimiento Tipográfico de T. Fortanet.

Grilo, C., Afonso, B.C., Afonso, F., Alexandre, M., Aliáacar, S., Almeida, A., et al. 2022. MAMMALS in PORTUGAL: a data set of terrestrial, volant, and marine mammal occurrences in Portugal. *Ecology* **103**(6), e3654, doi:10.1002/ecy.3654

Guerra, A. 1992. Mollusca, Cephalopoda. In *Fauna Ibérica*, M.A. Ramos (ed.). Madrid: Museo Nacional de Ciencias Naturales (CSIC), Volume **1**.

Guilhermino, L., Martins, A., Lopes, C., Raimundo, J., Vieira, L.R., Barboza, L.G.A., Costa, J., Antunes, C., Caetano, M. & Vale, C. 2021. Microplastics in fishes from an estuary (Minho River) ending into the NE Atlantic Ocean. *Marine Pollution Bulletin* **173**, **113008**, doi:10.1016/j.marpolbul.2021.113008

Gustafsson, S. 2020. Pearls as bycatch mitigation strategy for Harbour porpoise (Phocoena phocoena). Second cycle thesis, Swedish University of Agricultural Sciences (SLU), Uppsala, Sweden.

Guzmán-Verri, C., González-Barrientos, R., Hernández-Mora, G., Morales, J.A., Baquero-Calvo, E., Chaves-Olarte, E. & Moreno, E. 2012. *Brucella ceti* and brucellosis in cetaceans. *Frontiers in Cellular and Infection Microbiology* **2**, doi:10.3389/fcimb.2012.00003

Hamilton, S. & Baker, G.B. 2019. Technical mitigation to reduce marine mammal bycatch and entanglement in commercial fishing gear: lessons learnt and future directions. *Reviews in Fish Biology and Fisheries* **29**, 23–247, doi:10.1007/s11160-019-09550-6

Hammond, P.S., Bearzi, G., Bjørge, A., Forney, K.A., Karczmarski, L., Kasuya, T., Perrin, W., Scott, M.D., Wang, J.Y., Wells, R.S. & Wilson, B. 2008. *Phocoena phocoena* (Baltic Sea subpopulation). The IUCN Red List of Threatened Species 2008: e.T17031A98831650, doi:10.2305/IUCN.UK.2008.RLTS.T17031A6739565.en

Hammond, P.S., Lacey, C., Gilles, A., Viquerat, S., Börjesson, P., Herr, H., Macleod, K., Ridoux, V., Santos, M.B., Scheidat, M., Teilmann, J., Vingada, J. & Øien, N. 2017. Estimates of Cetacean Abundance in European Atlantic Waters in Summer 2016 from the SCANS-III Aerial and Shipboard Surveys. *Final Report.* Scotland: University of Saint Andrews.

Hammond, P.S., Lacey, C., Gilles, A., Viquerat, S., Börjesson, P., Herr, H., Macleod, K., Ridoux, V., Santos, M.B., Scheidat, M., Teilmann, J., Vingada, J. & Øien, N. 2021. Estimates of Cetacean Abundance in European Atlantic Waters in Summer 2016 from the SCANS-III Aerial and Shipboard Surveys. *Final Report.* Scotland: University of Saint Andrews.

Hammond, P.S., Macleod, K., Berggren, P., Borchers, D.L., Burt, L., Cañadas, A., Desportes, G., Donovan, G.P., Gilles, A., Gillespie, D., Gordon, J., Hiby, L., Kuklik, I., Leaper, R., Lehnert, K., Leopold, M., Lovell, P., Øien, N., Paxton, C.G.M., Ridoux, V., Rogan, E., Samarra, F., Scheidat, M., Sequeira, M., Siebert, U., Skov, H., Swift, R., Tasker, M.L., Teilmann, J., Van Canneyt, O. & Vázquez, J.A. 2013. Cetacean abundance and distribution in European Atlantic shelf waters to inform conservation and management. *Biological Conservation* **164**, 107–122, doi:10.1016/j.biocon.2013.04.010

Hammond, P.S., Macleod, K., Gillespie, D., Swift, R., Winship, A., Cañadas, A., Vazquez, J.A., Ridoux, V., Certain, G., Van Canneyt, O., Lens, S., Santos, B., Rogan, E., Uriarte A., Hernandez, C. & Castro, R. 2009. Cetacean Offshore Distribution and Abundance in the European Atlantic (CODA*). Final Report.* Scotland: University of Saint Andrews.

Härkönen, T., Dietz, R., Reijnders, P., Teilmann, J., Harding, K., Hall, A., Brasseur, S., Siebert, U., Goodman, S.J., Jepson, P.D., Rasmussen, T.D. & Thompson, P. 2006. A review of the 1988 and 2002 phocine distemper virus epidemics in European harbour seals. *Diseases of Aquatic Organisms* **68**(2), 115–130.

Harrison, R.J. 1971. The life of the common porpoise (*Phocoena phocoena*). *Proceedings of the Royal Institution of Great Britain* **44**, 113–133.

Haug, T., Desportes, G., Vikingson, G.A. & Witting, L. 2003. Harbour porpoises in the North Atlantic. NAMMCO Scientific Publications **5**, doi:10.7557/3.5

Heide-Jørgensen, M.P., Iversen, M., Nielsen, N.H., Lockyer, C., Stern, H. & Ribergaard, M.H. 2011. Harbour porpoises respond to climate change. *Ecology and Evolution* **1**, 579–585, doi:10.1002/ece3.51

HELCOM. 2023. Abundance and population trends of harbour porpoises. HELCOM pre-core indicator report. Online. https://indicators.helcom.fi/indicator/harbour-porpoises-abundance/

Helle, E., Olsson, M. & Jensen, S. 1976. PCB levels correlated with pathological changes in seal Uteri. *Ambio* **5(5/6)**, 261–262.

Hernández-González, A., Saavedra, C., Gago, J., Covelo, P., Santos, M.B. & Pierce, G.J. 2018. Microplastics in the stomach contents of common dolphin (*Delphinus delphis*) stranded on the Galician coasts (NW Spain, 2005–2010). *Marine Pollution Bulletin* **137**, 526–532, doi:10.1016/j.marpolbul.2018.10.026

Hernandez-Gonzalez, A., Saavedra, C., Read, F.L., López, A., Gouveia, A., Covelo, P., Alonso-Fernández, A., Velasco, F., Santos, M.B. & Pierce, G.J. 2024. Feeding ecology of harbour porpoises, *Phocoena phocoena*, stranded on the Galician coast (NW Spain) between 1990 and 2018. *Endangered Species Research* **54**, 105–122. doi:10.3354/esr01328

Herreras, M. V., Balbuena, J.A., Aznar, F.J., Kaarstad, S.E., Fernández, M., & Raga, J.A. 2004. Population structure of Anisakis simplex (Nematoda) in harbor porpoises Phocoena phocoena off Denmark. *Journal of Parasitology*, 90, 933–938. doi:10.1645/GE-188R

Higgins, R. 2000. Bacteria and fungi of marine mammals: a review. *The Canadian Veterinary Journal* **41**(2).

IAMMWG (Inter-Agency Marine Mammal Working Group), Camphuysen, C.J. & Siemensma, M.L. 2015. *A Conservation Literature Review for the Harbour Porpoise (Phocoena phocoena). JNCC Report* No. 566. Peterborough: IAMMWG.

ICES. 2009. ICES working group on marine mammal ecology (WGMME) report 2009. February 2–6 2009, Vigo, Spain. ICES CM 2009/ACOM:21.

ICES. 2010. Report of the study group on bycatch of protected species (SGBYC), 1–4 February 2010, Copenhagen, Denmark. ICES CM 2010/ACOM:25.

ICES. 2011. Report of the working group on bycatch of protected species (WGBYC 2011), 1–4 February 2011, Copenhagen, Denmark. ICES CM 2011/ACOM:26, doi:10.17895/ices.pub.19280807

ICES. 2013a. Report of the working group on bycatch of protected species (WGBYC 2012), Copenhagen, Denmark. ICES CM 2012/ACOM:28, doi:10.17895/ices.pub.19281275

ICES. 2013b. Report of the working group on bycatch of protected species (WGBYC), 4–8 February, Copenhagen, Denmark. ICES CM 2013/ACOM:27.

ICES. 2013c. Report of the working group on marine mammal ecology (WGMME), 4–7 February 2013, Paris, France. ICES CM 2013/ACOM:26.

ICES. 2014a. OSPAR request on implementation of MSFD for marine mammals. In *Report of the ICES Advisory Committee, 2014. ICES Advice 2014, Book 1,* Section 1.6.6.1.

ICES. 2014b. Report of the working group on marine mammal ecology (WGMME), 10–13 March 2014, Woods Hole, Massachusetts, USA. ICES CM 2014/ACOM:27.

ICES. 2014c. Report of the working group on bycatch of protected species (WGBYC), 4–7 February 2014, Copenhagen, Denmark. ICES CM 2014/ACOM:28.

ICES. 2015. Report of the working group on bycatch of protected species (WGBYC), 2–6 February 2015, ICES Headquarters, Copenhagen, Denmark. ICES CM 2015\ACOM:**26**, 82 pp.

ICES. 2016. Working group on bycatch of protected species (WGBYC), 1–5 February 2016, ICES HQ, Copenhagen, Denmark. ICES CM 2016/ACOM:27.

ICES. 2017. Report of the working group on bycatch of protected species (WGBYC), 12–15 June 2017, Woods Hole, Massachusetts, USA. ICES CM 2017/ACOM:24.

ICES. 2018a. Report from the working group on bycatch of protected species (WGBYC), 1–4 May 2018, Reykjavik, Iceland. ICES CM 2018/ACOM:25.

ICES. 2018b. ICES advice on fishing opportunities, catch, and effort ecoregions of the Northeast Atlantic and Arctic Ocean. ICES Advice 2018.

ICES. 2019. Working group on bycatch of protected species (WGBYC). *ICES Scientific Reports* **1**(51), doi:10.17895/ices.pub.5563

ICES. 2020a. Workshop on fisheries Emergency Measures to minimize BYCatch of short-beaked common dolphins in the Bay of Biscay and harbour porpoise in the Baltic Sea (WKEMBYC). *ICES Scientific Reports* **2**(43), doi:10.17895/ices.pub.7472

ICES. 2020b. Bycatch of protected and potentially vulnerable marine vertebrates – review of national reports under Council Regulation (EC) No. 812/2004 and other information. *ICES Advice, Ecoregions in the Northeast Atlantic and adjacent seas.*

ICES. 2020c. Working group on bycatch of protected species (WGBYC). *ICES Scientific Reports* **2**(81), doi:10.17895/ices.pub.7471

ICES. 2021a. Workshop on estimation of MOrtality of Marine MAmmals due to bycatch (WKMOMA). *ICES Scientific Reports* **3**(106), doi:10.17895/ices.pub.9257

ICES. 2021b. ICES ecosystem overviews: Bay of Biscay and the Iberian Coast ecoregion. Published 9 December 2021.

ICES. 2021c. Working group on bycatch of protected species (WGBYC). *ICES Scientific Reports* **3**(107), doi:10.17895/ices.pub.9256

ICES. 2023. Report of the ICES Working Group on Marine Mammal Ecology (WGMME). *ICES Scientific Reports* **5** (88), 123 pp. doi: 10.17895/ices.pub.24131736.

IJsseldijk, L.L., Brownlow, A.C. & Mazzariol, S. (eds.). 2019. Best practice on cetacean post mortem investigation and tissue sampling. Joint ACCOBAMS and ASCOBANS document. Online. https://osf.io/zh4ra/, doi:10.31219/osf.io/zh4ra

IJsseldijk, L.L., Gröne, A., Hiemstra, S., Hoekendijk, J. & Begeman, J. 2014. A record of twin fetuses in a harbor porpoise (Phocoena phocoena) stranded on the Dutch coast. *Aquatic Mammals* **40**(4), 394–397.

IJsseldijk, L.L., Hessing, S., Mairo, A., ten Doeschate, M.T.I., Treep, J., van den Broek, J., Keijl, G.O., Siebert, U., Heesterbeek, H., Gröne, A. & Leopold, M.F. 2021a. Nutritional status and prey energy density govern reproductive success in a small cetacean. *Scientific Reports* **11**, 19201, doi:10.1038/s41598-021-98629-x

IJsseldijk, L.L., Leopold, M.F., Begeman, L., Kik, M.J.L., Wiersma, L., Morell, M., Bravo Rebolledo, E.L., Jauniaux, T., Heesterbeek, H. & Gröne, A. 2022. Pathological findings in stranded harbor porpoises (*Phocoena phocoena*) with special focus on anthropogenic causes. *Frontiers in Marine Science* **9**, 997388, doi:10.3389/fmars.2022.997388

IJsseldijk, L.L., Scheidat, M., Siemensma, M., Couperus, B., Leopold, M.F., Morell, M., Gröne, A. & Kik, M.J.L. 2021b. Challenges in the assessment of bycatch: post-mortem findings in harbor porpoises (*Phocoena phocoena*) retrieved from gillnets. *Veterinary Pathology* **58**(2), 405–415, doi:10.1177/0300985820972454

Innes, S., Lavigne, D.M., Earle, W.M. & Kovacs, K.M. 1987. Feeding rates of seals and whales. *The Journal of Animal Ecology* **56**, 115–130, doi:10.2307/4803

IWC. 1994. Report of the workshop on mortality of cetaceans in passive fishing nets and traps. In Report of the International Whaling Commission Special Issue 15, Gillnets *and* Cetaceans, W.F. Perrin et al. (eds.). Cambridge, United Kingdom: IWC, 1–71.

IWC. 1996. Report of the sub-committee on the small cetaceans. *Report of the International Whaling Commission* **46**, 160–179.

IWC Scientific Committee. 2021. Report of the scientific committee virtual meeting, 11–26 May 2020. *Journal of Cetacean Research and Management* **22**, 1–122.

IWC Scientific Committee. 2022. Report of the Scientific Committee Virtual Meeting, 25 April – 13 May 2022. *Journal of Cetacean Research and Management* **24**, 1–190.

Jansen, O.E., Aarts, G.M., Das, K., Lepoint, G., Michel, L. & Reijnders, P.J.H. 2012. Feeding ecology of harbour porpoises: stable isotope analysis of carbon and nitrogen in muscle and bone. *Marine Biology Research* **8**(**9**), 829–841, doi:10.1080/17451000.2012.692164

Jansen, O.E., Michel, L., Lepoint, G., Das, K., Couperus, A.S. & Reijnders, P.J.H. 2013. Diet of harbour porpoises along the Dutch coast: a combined stable isotope and stomach contents approach. *Marine Mammal Science* **29**(3), doi:10.1111/j.1748-7692.2012.00621.x

Jaramillo-Legorreta, A.M., Cardenas-Hinojosa, G., Nieto-García, E., Rojas-Bracho, L., Thomas, L., Ver Hoef, J.M., Moore, J., Taylor, B., Barlow, J. & Tregenza, N. 2019. Decline towards extinction of Mexico's vaquita porpoise (*Phocoena sinus*). *Royal Society Open Science* **6**, **190590**, doi:10.1098/rsos.190598

Jauniaux, T., André, M., Dabin, W., De Vreese, S, Gillet, A. & Morell, M. 2019. Marine mammals stranding guidelines for post-mortem investigations of cetaceans & pinnipeds & 13th cetacean necropsy workshop special issue on necropsy and samplings (Liege 2019). ACCOBAMS and University of Liege, 45 pp.

Jauniaux, T.P., Brenez, C., Fretin, D., Godfroid, J., Haelters, J., Jacques, T., Kerckhof, F., Mast, J., Sarlet, M. & Coignoul, F.L. 2010. *Brucella ceti* infection in harbor porpoise (*Phocoena phocoena*). *Emerging Infectious Diseases* **16**(12), 1966–1968, doi:10.3201/eid1612.101008

Jepson, P., Deaville, R., Barber, J., Aguilar, À., Borrell, A., Murphy, S., Barry, J., Brownlow, A., Barnett, J., Berrow, S., Cunningham, A.A., Davison, N.J., Ten Doeschate, M., Esteban, R., Ferreira, M., Foote, A.D., Genov, T., Giménez, J., Loveridge, J., Llavona, Á., Martin, V., Maxwell, D.L., Papachlimitzou,

A., Penrose, R., Perkins, M.W., Smith, B., de Stephanis, R., Tregenza, N., Verborgh, P., Fernandez, A. & Law, R.J. 2016. PCB pollution continues to impact populations of orcas and other dolphins in European waters. *Scientific Reports* **6**, 18573, doi:10.1038/srep18573

Jepson, P.D., Bennett, P.M., Deaville, R., Allchin, C.R., Baker, J.R. & Law, R.J. 2005. Relationships between polychlorinated biphenyls and health status in harbor porpoises (*Phocoena phocoena*) stranded in the United Kingdom. *Environmental Toxicology and Chemistry* **24**, 238–248.

Jereb, P., Allcock, A.L., Lefkaditou, E., Piatkowski, U., Hastie, L.C & Pierce, G.J. (eds.). 2015. Cephalopod biology and fisheries in Europe: II. Species accounts. ICES cooperative research report no. 325, 360 pp.

Kaddouri, A., Analla, M., Tamsouri, N., Akissou, M., Awadh, H. & Benchoucha, S. 2023. Evaluation of cetacean strandings on the north-western coast of Morocco from 2016 to 2021. *Mammalian Biology* **103**, 317–328, doi:10.1007/s42991-023-00359-2

Kannan, K., Blankenship, A.L., Jones, P.D. & Giesy, J.P. 2000. Toxicity reference values for the toxic effects of polychlorinated biphenyls to aquatic mammals. *Human and Ecological Risk Assessment* **6**, 181–201, doi:10.1080/10807030091124491

Karakosta, C.V., Jepson, P.D., Ohira, H., Moore, A., Bennett, P.M. & Holt, W.V. 1999. Testicular and ovarian development in the harbour porpoise (*Phocoena phocoena*). *Journal of Zoology, London* **249**, 111–121, doi:10.1111/j.1469-7998.1999.tb01064.x

Kastelein, R.A., Au, W.W.L. & de Haan, D. 2000b. Detection distance of bottom-set gillnets by harbour porpoises (*Phocoena phocoena*) and bottlenose dolphins (*Tursiops truncatus*). *Marine Environmental Research* **49**(4), 359–375, doi:10.1016/S0141-1136(99)00081-1

Kastelein, R.A., de Haan, D., Staal, C., Nieuwstraten, S.H. & Verboom, W.C. 1995. Entanglement of harbour porpoises (*Phocoena phocoena*) in fishing nets. In *Harbour Porpoises – Laboratory Studies to Reduce Bycatch*, P.E. Nachtigall, J. Lien, W.W.L. Au & A.J. Read (eds.). Woerde, The Netherlands: De Spil, 91–156.

Kastelein, R.A., Hardeman, J. & Boer, H. 1997. Food consumption and body weight of harbour porpoises (*Phocoena phocoena*). In *The Biology of the Harbour Porpoise*, W. Read et al. (eds.). Woerden, the Netherlands: De Spil Publishers, 217–233.

Kastelein, R.A., Helder-Hœk, L. & Jennings, N. 2018. Seasonal changes in food consumption, respiration rate, and body condition of a male harbor porpoise (*Phocoena phocoena*). *Aquatic Mammals* **44**(1), 76–91, doi:10.1578/AM.44.1.2018.76

Kastelein, R.A., Rippe, H.T., Vaughan, N., Schooneman, N.M., Verboom, W.C. & de Haan, D. 2000a. The effect of acoustic alarms on the behavior of harbor porpoises (*Phocoena phocoena*) in a floating pen. *Marine Mammal Science* **16**(1), 46–64, doi:10.1111/j.1748-7692.2000.tb00903.x

Kastelein, R.A., Staal, C., Terlouw, A. & Muller, M. 1997. Pressure changes in the mouth of a feeding harbour porpoise (*Phocoena phocoena*). In *The Biology of the Harbour Porpoise*, W. Read et al. (eds.). Woerden, the Netherlands: De Spil Publishers, 279–291.

Kennedy, S., Lindstedt, I.J., McAliskey, M.M., McConnell, S.A. & McCullough, S.J. 1992. Herpesviral encephalitis in a harbour porpoise (*Phocoena phocoena*). *Journal of Zoo and Wildlife medicine* **23**(3), 374–379. Online. https://www.jstor.org/stable/20095242

Kennedy, S., Smyth, J.A., McCullough, S.J., Allan, G.M. & McQuaid, S. 1988. Viral distemper now found in porpoises. *Nature* **336**, **21**, doi:10.1038/336021a0

Kesselring, T., Viquerat, S., Brehm, R. & Siebert, U. 2017. Coming of age: do female harbour porpoises (Phocoena phocoena) from the North Sea and Baltic Sea have sufficient time to reproduce in a human influenced environment? *PLoS One* **12**(10), e0186951, doi:10.1371/journal.pone.0186951

Kindt-Larsen, L., Berg, C.W., Northridge, S. & Larsen, F. 2019. Harbor porpoise (*Phocoena phocoena*) reactions to pingers. *Marine Mammal Science* **35**(2), 552–573.

Kindt-Larsen, L., Glemarec, G., Berg, C.W., Königson, S., Kroner, A.-M., Søgaard, M. & Lusseau D. 2023. Knowing the fishery to know the bycatch: bias-corrected estimates of harbour porpoise bycatch in gillnet fisheries. *Proceedings of the Royal Society B: Biological Sciences 290*, 20222570, doi:10.1098/rspb.2022.2570

King, P.A., Fives, J.M. & McGrath, D. 1994. Reproduction, growth and feeding of the dragonet, *Callionymus lyra* (Teleostei: Callionymidae), in Galway Bay, Ireland. *Journal of the Marine Biological Association of the United Kingdom* **74**(3), 513–526, doi:10.1017/S0025315400047639

King, S.L., Schick, R.S., Donovan, C., Booth, C.G., Burgman, M., Thomas, L. & Harwood, J. 2015. An interim framework for assessing the population consequences of disturbance. *Methods in Ecology and Evolution* **6**(10), 1150–1158, doi:10.1111/2041-210X.12411

Kinneging, N. 2022. Pilot assessment of ambient noise. In *OSPAR, 2023: The 2023 Quality Status Report for the Northeast Atlantic*. London: OSPAR Commission. Online. https://oap.ospar.org/en/ospar-assessments/quality-status-reports/qsr-2023/indicator-assessments/ambient-noise-pilot

Kinze, C.C. 1994. Incidental catches of harbour porpoises (*Phocoena phocoena*) in Danish waters, 1986–1989. *Report of the International Whaling Commission, Special Issue* **15**, 183–187.

Kinze, C.C. 1995. Exploitation of harbour porpoises (*Phocoena phocoena*) in Danish waters: a historical review. *Report of the International Whaling Commission, Special issue* **16**, 141–153.

Kirkwood, J.K., Bennett, P.M., Jepson, P.D., Kuiken, T., Simpson, V.R. & Baker, J.R. 1997. Entanglement in fishing gear and other causes of death in cetaceans stranded on the coasts of England and Wales. *Veterinary Record* **141**(4), 94–98, doi:10.1136/vr.141.4.94

Kleiber, M. 1947. Body size and metabolic rate. *Physiological Reviews* **27**(4), 511–541, doi:10.1152/physrev.1947.27.4.511

Klevaine, L., Skaare, J.U., Björge, A., de Ruiter, E. & Reijnders, P.J.H. 1995. Organochlorine pesticide residue and PCBs in harbour porpoise (*Phocoena phocoena*) incidentally caught in Scandinavian waters. *Environmental Pollution* **89**(2), 137–146, doi:10.1016/0269-7491(94)00066-M.

Knowlton, A.R., Robbins, J., Landry, S., McKenna, H.A., Kraus, S.D. & Werner, T.B. 2016. Effects of fishing rope strength on the severity of large whale entanglements. *Conservation Biology* **30**(2), 318–328, doi:10.1111/cobi.12590

Kompanje, E.J.O., Campphuysen, C.J. & Leopold, M.F. 2017. The first case of conjoined twin harbour porpoises *Phocoena phocoena* (Mammalia, Cetacea). *Deinsea. Online Journal of the Natural History Museum Rotterdam* **17**, 1–5.

Königson, S., Naddafi, R., Hedgärde, M., Pettersson, A., Östman, Ö., Benavente Normann, E. & Amundin, M. 2022. Will harbor porpoises (*Phocoena phocoena*) be deterred by a pinger that cannot be used as a "dinner bell" by seals? *Marine Mammal Science* **38**(2), 469–485, doi:10.1111/mms.12880

Koschinski, S. & Culik, B. 1997. Deterring harbour porpoises (*Phocoena phocoena*) from gillnets: observed reactions to passive reflectors and pingers. *Report of the International Whaling Commission* **47**, 659–668.

Kratzer, I.M.F., Brooks, M.E., Bilgin, S., Özdemir, S., Kindt-Larsen, L., Larsen, F. & Stepputtis, D. 2021. Using acoustically visible gillnets to reduce bycatch of a small cetacean: first pilot trials in a commercial fishery. *Fisheries Research* **243**, **106088**, doi:10.1016/j.fishres.2021.106088

Kratzer, I.M.F., Schäfer, I., Stoltenberg, A., Chladek, J.C., Kindt-Larsen, L., Larsen, F. & Stepputtis, D. 2020. Determination of optimal acoustic passive reflectors to reduce bycatch of odontocetes in gillnets. *Frontiers in Marine Science* **7**, **539**, doi:10.3389/fmars.2020.00539

Kraus, S.D., Read, A.J., Anderson, E., Baldwin, K., Solow, A., Spradlin, T. & Williamson, J. 1995. A field test of acoustic alarms to reduce incidental mortality of harbor porpoises in gill nets. Paper SC/47/SM17 presented at the 47th meeting of the Scientific Committee of the International Whaling Commission.

Kraus, S.D., Read, A.J., Solow, A., Baldwin, K., Spradlin, T., Anderson, E. & Williamson, J. 1997. Acoustic alarms reduce porpoise mortality. *Nature* **388**(6642), 525–526, doi:10.1038/41451

Krebs, C.J. 1989. *Ecological Methodology*. New York: Harper Collins Publishers, 654 pp.

Kuhn, D.M. 1829. Description d'une nouvelle espèce de strongle trouvée dans de marsouin. *Bulletin des sciences naturelles et de géologie, Paris* **17**, 150–153.

Kuiken, T. (ed.). 1996. Diagnosis of by-catch in cetaceans: proceedings of the second ECS workshop on cetacean pathology, Montpellier, France, 2 March 1994. European Cetacean Society Newsletter Special Issue 26.

Lacey, C., Gilles, A., Börjesson, P., Herr, H., Macleod, K., Ridoux, V., Santos, M.B., Scheidat, M., Teilmann, J., Sveegaard, S., Vingada, J., Viquerat, S., Øien, N. & Hammond, P.S. 2022. Modelled density surfaces of cetaceans in European Atlantic waters in summer 2016 from the SCANS-III aerial and shipboard surveys. SCANS-III project report 2, 31pp+Appendices. Online. https://scans3.wp.st-andrews.ac.uk/files/2022/08/SCANS-III_density_surface_modelling_report_final_20220815.pdf

Lahaye, V., Bustamante, P., Law, R.J., Learmonth, J.A., Santos, M.B., Boon, J.P., Rogan, E., Dabin, W., Addink, M.J., López, A., Zuur, A.F., Pierce, G.J. & Caurant, F. 2007. Biological and ecological factors related to trace element levels in harbour porpoises (*Phocoena phocoena*) from European waters. *Marine Environmental Research* **64**(3), 247–266, doi:10.1016/j.marenvres.2007.01.005

Lambert, C., Pettex, E., Dorémus, G., Laran, S., Stéphan, E., Van Canneyt, O. & Ridoux, V. 2017. How does ocean seasonality drive habitat preferences of highly mobile top predators? PartII: the Eastern North-Atlantic. *Deep-Sea Research II* **141**, 133–154, doi:10.1016/j.dsr2.2016.06.011

Larsen, F., Eigarrd, O.R. & Tougaard, J. 2007. Reduction of harbour porpoise (*Phocoena phocoena*) bycatch by iron-oxide gillnets. *Fisheries Research* **85**, 270–278, doi:10.1016/j.fishres.2007.02.011

Learmonth, J.A., Murphy, S., Dabin, W., Addink, M., López, A., Rogan, E., Ridoux, V., Guerra, A. & Pierce, G.J. 2004. Measurement of reproductive output in small cetaceans from the NE Atlantic. BIOCET work-package 5 – final report. Project Reference: EVK3-2000-00027. UK.

Learmonth, J.A., Murphy, S., Luque, P.L., Reid, R.J., Patterson, I.A.P., Brownlow, A., Ross, H.M., Barley, J.P., Santos, M.B. & Pierce, G.J. 2014. Life history of harbor porpoises (*Phocoena phocoena*) in Scottish (UK) waters. *Marine Mammal Science* **30**, 1427–1455, doi:10.1111/mms.12130

Lefebvre, K.A., Quakenbush, L., Frame, E., Huntingdton, K.B., Sheffield, G., Stimmelmayr, R., Bryan, A., Kendrick, P., Ziel, H., Goldstein, T., Snyder, J.A., Gelatt, T., Gulland, F., Dickerson, B. & Gill, V. 2016. Prevalence of algal toxins in Alaskan marine mammals foraging in a changing arctic and subarctic environment. *Harmful Algae* **55**, 13–24, doi:10.1016/j.hal.2016.01.007

Lens, S. 1997. A note on the harbour porpoise (*Phocoena phocoena*) in the coastal waters of Spain. *Reports of the International Whaling Commission* **47**, 841–847.

Lens, S. & Díaz, P. 2009. *Annual Report of Spain to the European Commission on the Implementation of Council Regulation 812/2004 on Cetacean Bycatch. Work conducted during 2008*. Instituto Español de Oceanografía: Spain.

Leopold, M.F., Begeman, L., Heße, E., van der Hiele, J., Hiemstra, S., Keijl, G., Meesters, E.H., Mielke, L., Verheyen, D. & Gröne, A. 2015. Porpoises: from predators to prey. *Journal of Sea Research* **97**, 14–23, doi:10.1016/j.seares.2014.12.005

Levsen, A., Cipriani, P., Mattiucci, S., Gay, M., Hastie, L.C., Mackenzie, K., Pierce, G.J., Svanevik, C.S., Højgaard, D.P., Nascetti, G., González, A.F. & Pascual, S. 2018. *Anisakis* species composition and infection characteristics in Atlantic mackerel, *Scomber scombrus*, from major European fishing grounds - reflecting changing fish host distribution and migration pattern. *Fisheries Research* **202**, 112–121, doi:10.2016/j.fishres.2017.07.030

Lick, R.R. 1991. Parasites from the digestive tract and food analysis of harbour porpoise (*Phocoena phocoena*) from German coastal waters. In *European Research on Cetaceans*, P.G.H. Evans (ed.). Cambridge, United Kingdom: European Cetacean Society, Volume **5**, 65–68.

Lino, L., Ferreira, M., Pereira, A.T., Lopes, A.P., & Pinto, M. d. L. 2022. Ulcerative Lesions Caused by *Anisakis spp* In Stranded Cetaceans along the Portuguese Coast. *Journal of Comparative Pathology* **191**, 56. doi:10.1016/j.jcpa.2021.11.129

Llavona, A., Martínez-Cedeira, J., Covelo, P., Morales, X., Dios, J.J., Bracali, V., Pérez, P., García, J.A., Santos, L. & López, A. 2011. Indemares project: identification of special areas of conservation for cetaceans at "Galician Bank" and "Aviles Canyon" (Atlantic and Cantabrian sea) for Natura 2000. *Presentation at the 25th Annual Conference European Cetacean Society*. Cadiz, Spain.

Llavona Lavina, A. 2018. *Parâmetros populacionais e estrutura genética do bôto (Phocoena phocoena, L. 1758) na costa noroeste da península ibérica*. PhD thesis, University of Aveiro, Portugal.

Lockyer, C. 1995a. Aspects of the biology of the harbour porpoise, *Phocoena phocoena*, from British waters. In *Whales, Seals, Fish and Man*, A.S. Blix et al. (eds.). Amsterdam, Netherlands: Elsevier Science, 443–457, doi:10.1016/S0163-6995(06)80045-4

Lockyer, C. 1995b. Investigation of aspects of the life history of the harbour porpoise, *Phocoena phocoena*, in British waters. *Reports International Whaling Commission* **16**(Special Issue), 190–197.

Lockyer, C. 2003. Harbour porpoises (Phocoena phocoena) in the North Atlantic: Biological parameters. *NAMMCO Scientific. Publications* 5, 71–90.

Lockyer, C., Desportes, G., Hansen, K., Labberté, S. & Siebert, U. 2003. Monitoring growth and energy utilisation of the harbour porpoise (*Phocoena phocoena*) in human care. *NAMMCO Scientific Publications* **5**, 107–120, doi:10.7557/3.2743

Lockyer, C. 2007. All creatures great and smaller: a study in cetacean life history energetics. *Journal of the Marine Biological Association of the United Kingdom* **87**(4), 1035–1045. doi: 10.1017/S0025315407054720.

Lockyer, C., Heide-Jørgensen, M.P., Jensen, J., Kinze, C.C. & Buus Sørensen., T. 2001. Age, length and reproductive parameters of harbour porpoises *Phocoena phocoena*, (L.) from West Greenland. *ICES Journal of Marine Science* **58**(1), 154–162, doi:10.1006/jmsc.2000.0998

Lockyer, C. & Kinze, C. 2003. Status and life history of harbour porpoise (*Phocoena phocoena*) in Danish waters. *NAMMCO Scientific Publications* **5**, 143–176, doi:10.7557/3.2745

López A. 2017. Mammalia. In *Inventario de la biodiversidad marina de Galicia: Proyecto LEMGAL*, R. Bañón (ed.). Consellería do Mar, Xunta de Galicia: Santiago de Compostela, 567–570.

López, A., Covelo, P., Vázquez, J.A., Martínez, J.A., Caldas, M., Lago, R., Llavona, A. & Alonso, J.M. 2011. Bases para a conservación e xestión das especies de cetáceos ameazadas nas augas Atlánticas e Cantábricas. *Eubalaena* **12**, 62.

López, A., Llavona, A., Alonso, J.M., Martínez-Cedeira, J.A., Caldas, M., Valeiras, X. & Covelo, P. 2005. Effects of the "Prestige" oil spill in aquatic mammals and sea turtles of the Galician coast (NW Spain). In *Symposium on Marine Accidental Oil Spills*, Vertimar, Vigo, Spain, 13–16 July 2005.

López, A., Pierce, G.J., Santos, M.B., Gracia, J. & Guerra, A. 2003. Fishery by-catches of marine mammals in Galician waters: results from on-board observations and an interview survey of fishermen. *Biological Conservation* **111**(1), 25–40, doi:10.1016/S0006-3207(02)00244-6

López, A., Pierce, G.J., Valeiras, X., Santos, M.B. & Guerra, A. 2004. Distribution patterns of small cetaceans in Galician waters. *Journal of the Marine Biological Association of the United Kingdom* **84**(1), 283–294, doi:10.1017/S0025315404009166h

López, A. & Rodríguez, A. 1995. Agresión de arroás (*Tursiops truncatus*) a toniña (*Phocoena phocoena*). *Eubalaena* **6**, 23–27.

López, A., Santos, M.B., Pierce, G.J., González, A.F., Valeiras, X. & Guerra, A. 2002. Trends in strandings and by-catch of marine mammals in northwest Spain during the 1990s. *Journal of the Marine Biological Association of the United Kingdom* **82**(3), 513–521, doi:10.1017/S0025315402005805

López, A., Vázquez, J.A., Martínez Cedeira, J.A., Cañadas, A., Marcos, E., Maestre, I., Ruano, A., Larias, L., Llavona, A., MacLeod, K. & Evans, P. 2013. Abundance estimates for harbour porpoise (*Phocoena phocoena*) in the Spanish area of the Iberian Peninsula management unit. Paper SC/65a/SM20. *International Whaling Commission* **64**, 812.

López, A., Vázquez, J.A., Martínez-Cedeira., J.A., Marcos-Ipiña, E., Laria, L., Maestre, I., Carbó, A., Llanova, A., Fernández, M., Díaz, J.I., Santos, L., Ruano, A., Fernández, R. & Méndez, P. 2012. *Bases para el desarrollo de los planes de conservación de las especies de cetáceos protegidas en la Demarcación Marina Noratlántica.* Madrid, Spain: Ministerio de Agricultura, Alimentación y Medio Ambiente - Fundación Biodiversidad.

López Fernández, A. 2003. *Estatus dos pequenos cetáceos da plataforma de galicia.* PhD thesis, University of Santiago de Compostela, Spain.

López-Fernández, A. & Martínez-Cedeira, J.A. 2011. Marsopa - *Phocoena phocoena.* In *Enciclopedia Virtual de los Vertebrados Españoles*, A. Salvador & J. Cassinello (eds.). Madrid, Spain: Museo Nacional de Ciencias Naturales. Online. https://www.vertebradosibericos.org/

López Ferreiro, A., 1895. *Fueros municipales de Santiago y de su tierra.* Santiago de Compostela, Spain: Imprenta y Encuadernación del Seminario Conciliar Central.

López Losa, E. & Amorim, I. 2012. The fisheries of the Iberian Peninsula in modern times. In *The History of the North Atlantic Fisheries. From 1850s to Early Twenty-First Century.* D.J. Starkey & I. Heidbrink (eds.). Bremen: German Maritime Museum of Bremerhaven, Volume **2**, 253–276.

López Seoane, V., 1861. *Fauna Mastológica de Galicia.* Santiago de Compostela: Imprenta de Manuel Mirás.

Loureiro, M.L., Ribas, A., López, E. & Ojea, E. 2006. Estimated costs and admissible claims linked to the Prestige oil spill. *Ecological Economics* **59**(1), 48–63, doi:10.1016/j.ecolecon.2005.10.001

Lusseau, D., Kindt-Larsen, L. & van Beet, F.M. 2023. Emergent interactions in the management of multiple threats to the conservation of harbour porpoises. *Science of the Total Environment* **855**, 158936, doi:10.1016/j.scitotenv.2022.158936

Macaulay, J., Kingston, A., Coram, A., Oswald, M., Swift, R., Gillespie, D. & Northridge, S. 2022. Passive acoustic tracking of the three-dimensional movements and acoustic behaviour of toothed whales in close proximity to static nets. *Methods in Ecology and Evolution* **13**(6), 1250–1264, doi:10.1111/2041-210X.13828

MacKenzie, K. 2002. Parasites as biological tags in population studies of marine organisms: an update. *Parasitology* **124**(7), 153–163, doi:10.1017/S0031182002001518

MacKenzie, K. & Hemmingsen, W. 2015. Parasites as biological tags in marine fisheries research: European Atlantic waters. *Parasitology* **142**(1), 54–67, doi:10.1017/S0031182014000341

MacKenzie, K., Campbell, N., Mattiucci, S., Ramos, P., Pinto, A.L. & Abaunza, P. 2008. Parasites as biological tags for stock identification of Atlantic horse mackerel *Trachurus trachurus* L. *Fisheries Research* **89**(2), 136–145, doi:10.1016/j.fishres.2007.09.031

MacLeod, C.D., Santos, M.B., Reid, R.J., Scott, B. & Pierce, G.J. 2007. Linking sandeel consumption and the likelihood of starvation in harbour porpoises in the Scottish North Sea: could climate change mean more starving porpoises? *Biology Letters* **3**, 185–188, doi:10.1098/rsbl.2006.0588

MacLeod, R., MacLeod, C.D., Learmonth, J.A., Jepson, P.D., Reid, R.J., Deaville, R. & Pierce, G.J. 2007. Mass-dependent predation risk and lethal dolphin-porpoise interactions. *Proceedings of the Royal Society B; Biological Sciences* **274**, 2717–2723, doi:10.1098/rspb.2007.0786

Mahfouz, C., Henry, F., Jauniauz, T., Khalaf, G. & Amara, R. 2014. Organochlorines in harbour porpoises (*Phocoena phocoena*) stranded along the southern North Sea between 2010 and 2013. *Environmental Science: Processes and Impacts* **16**, 2774–2781, doi:10.1039/c4em00490f

Major, C.O., Goldstein, S.L., Ryan, W.B., Lericolais, G., Piotrowski, A.M. & Hajdas, I. 2006. The co-evolution of Black Sea level and composition through the last deglaciation and its paleoclimatic significance. *Quaternary Science Reviews* **25**(17–18), 2031–2047, doi:10.1016/j.quascirev.2006.01.032

Malinga, M., Kuklik, I. & Skóra, K.E. 1996. Food consumption of harbour porpoises (*Phocoena phocoena*) in Polish waters of the Baltic Sea. *European Research on Cetaceans* **10**, 24.

Marçalo, A., Feijó, D., Katara, I., Araújo, H., Oliveira, I., Santos, J., Ferreira, M., Monteiro, S., Pierce, G.J., Silva, A. & Vingada, J. 2015. Quantification of interactions between the Portuguese sardine purse seine fishery and cetaceans. *ICES Journal of Marine Science* **72**(8), 2438–2449, doi:10.1093/icesjms/fsv076

Mariezkurrena-Gastearena, K. 2011. Hallazgos de macromamíferos poco frecuentes en yacimientos arqueológicos y paleontológicos del Pleistoceno de la región Cantábrica. *Kobie serie Paleoanthropología* **30**, 83–110.

Marshall, C.T., Macdonald, P., Torgerson, E., Asare, J.L. & Turner, R. 2021. *Design, Development and Deployment of a Software Platform for Real-Time Reporting in the West of Scotland Demersal Fleet.* A study commissioned by Fisheries Innovation Scotland (FIS): Scotland, United Kingdom.

Martin, A. & Walker, F.J. 1997. Sighting of a right whale (*Eubalaena glacialis*) with calfoff S.W. Portugal. *Marine Mammal Science* **13**(1), 139–140.

Martínez, R., Segade, P., Martínez-Cedeira, J.A., Arias, C., García-Estévez, J.M. & Iglesias, R. 2008. Occurrence of the ectoparasite *Isocyamus deltobranchium* (Amphipoda: Cyamidae) on cetaceans from Atlantic waters. *The Journal of parasitology* **94**(6), 1239–1242, doi:10.1645/GE-1518.1

Martínez-Cedeira, J. & López Fernández, A. 2018. Actualización del estado de conservación de la marsopa en las demarcaciones marinas noratlántica y sudatlántica y elaboración del borrador del plan de conservación. Final Report on the work carried out under the INTEMARES Project. Coodinadora para o Estudios dos Mamíferos Mariños - CEMMA, Gondomar, Spain.

Martínez-Cedeira, J.A., Vázquez Bonales, J.A., Izquierdo Ferreiro, I. & Fernández Acevedo, O. 2021. *Proyecto VIRADA: Bases para la reducción de la mortalidad por captura accidental en artes de pesca de la marsopa (*Phocoena phocoena*) y el delfín mular (*Tursiops truncatus*) en áreas Red Natura 2000 de Galicia.* Informe campaña aérea. CEMMA and OPFF-4: Vigo, Spain.

Martínez-Cedeira, J.A., Vázquez Bonales, J.A. & López Fernández, A. 2015. *PHOCOEVAL: Evaluación y determinación de la categoría de conservación de la marsopa común (*Phocoena phocoena*) en el NW de la Península Ibérica.* Coordinadora para o Estudios dos Mamíferos Mariños (CEMMA): Nigrán, Spain.

Martins, R. & Carneiro, M. 2018. *Manual de identificação de peixes ósseos da costa continental portuguesa - Principais Características Diagnosticantes.* Instituto Português do Mar e da Atmosfera: Lisbon, p. 204.

Mathias, M.L., Fonseca, C., Rodrigues, L., Grilo, C., Lopes-Fernandes, M., Palmeirim, J.M., Santos-Reis, M., Alves, P.C., Cabral, J.A., Ferreira, M., Mira, A., Eira, C., Negrões, N., Paupério, J., Pita, R., Rainho, A., Rosalino, L.M., Tapisso, J.T. & Vingada, J. (eds.) 2023. *Livro Vermelho dos Mamíferos de Portugal continental.* FCiências.ID, ICNF: Lisboa

Mattiucci, S., Cipriani, P., Levsen, A., Paoletti, M., & Nascetti, G. 2018. Molecular Epidemiology of *Anisakis* and Anisakiasis: An Ecological and Evolutionary Road Map. *Advances in Parasitology* **99**, 93–263. doi:10.1016/bs.apar.2017.12.001

McBrearty, D.A., Message, M.A. & King, G.A. 1986. Observations on small cetaceans in the North-east Atlantic Ocean and the Mediterranean Sea: 1978–1982. In *Research on Dolphins*, M.M. Bryden & R. Harrison (eds.). Oxford: Clarendon Press, 225–249.

McCullough, S.J., McNeilly, F., Allan, G.M., Kennedy, S., Smyth, J.A., Cosby, S.L., McQuaid, S. & Rima, B.K. 1991. Isolation and characterization of a porpoise morbillivirus. *Archives of Virology* **118**, 247–252, doi:10.1007/BF01314034

Mead, J.G., Walker, W.A. & Houck, W.J. 1982. Biological observations on *Mesoplodon carlhubbsi* (Cetacea: Ziphiidae). *Smithsonian Contributions to Zoology* **34**, 1–25.

Measures, L.N. 2001. Lungworms of marine mammals. Chapter 10. In *Parasitic Diseases of Wild Mammals*, W.M. Samuel, M.J. Pybus & A.A. Kocan (eds.). Ames, IA: Iowa State University Press, 279–300.

Méndez-Fernandez, P., Bustamante, P., Bode, A., Chouvelon, T., Ferreira, M., López, A., Pierce, G.J., Santos, M.B., Spitz, J., Vingada, J.V. & Caurant, F. 2012. Foraging ecology of five toothed whale species in the Northwest Iberian Peninsula, inferred using carbon and nitrogen isotope ratios. *Journal of Experimental Marine Biology and Ecology* **413**, 150–158, doi:10.1016/j.jembe.2011.12.007

Méndez-Fernandez, P., Pierce, G.J., Bustamante, P., Chouvelon, T., Ferreira, M., González, A.F., López, A., Read, F.L., Santos, M.B., Spitz, J., Vingada, J.V. & Caurant, F. 2013. Ecological niche segregation among five toothed whale species off the NW Iberian Peninsula using ecological tracers as multi-approach. *Marine Biology* **160**, 2825–2840, doi:10.1007/s00227-013-2274-9

Méndez-Fernandez, P., Webster, L., Chouvelon, T., Bustamante, P., Ferreira, M., González, A.F., López, A., Moffat, C.F., Pierce, G.J., Read, F.L., Russell, M., Santos, M.B., Spitz, J., Vingada, J.V. & Caurant, F. 2014a. An assessment of contaminant concentrations in toothed whale species of the NW Iberian Peninsula: Part I. Persistent organic pollutants. *Science of the Total Environment* **484**, 196–205, doi:10.1016/j.scitotenv.2014.02.045

Méndez-Fernandez, P., Webster, L., Chouvelon, T., Bustamante, P., Ferreira, M., González, A.F., López, A., Moffat, C.F., Pierce, G.J., Read, F.L., Russell, M., Santos, M.B., Spitz, J., Vingada, J.V. & Caurant, F. 2014b. An assessment of contaminant concentrations in toothed whale species of the NW Iberian Peninsula: Part II. Trace element concentrations. *Science of the Total Environment* **484**, 206–217, doi:10.1016/j.scitotenv.2014.03.001

Méndez-Fernandez, P., Simon-Bouhet, B., Bustamante, P., Chouvelon, T., Ferreira, M., López, A., Moffat, C.F., Pierce, G.J., Russell, M., Santos, M.B., Spitz, J., Vingada, J.V., Webster, L., Read, F.L., González, A.F. & Caurant, F. 2017. Inter-species differences in polychlorinated biphenyls patterns from five sympatric species of odontocetes: can PCBs be used as tracers of feeding ecology? *Ecological Indicators* **74**, 98–108, doi:10.1016/j.ecolind.2016.11.013

Methion, S. & Díaz López, B. 2021. Spatial segregation and interspecific killing of common dolphins (*Delphinus delphis*) by bottlenose dolphins (*Tursiops truncatus*). *Acta Ethologica* **24**, 95–106, doi:10.1007/s10211-021-00363-0

Møhl-Hansen, U. 1954. Investigations of reproduction and growth for the porpoise (*Phocaena phocaena* (L.)) from the Baltic. *Viddenskaplige Meddelelser fra den Danske Naturhistoriske Forening* **116**, 369–396.

Mooney, T.A., Au, W.W., Nachtigall, P.E. & Trippel, E.A. 2007. Acoustic and stiffness properties of gillnets as they relate to small cetacean bycatch. *ICES Journal of Marine Science* **64**(7), 1324–1332, doi:10.1093/icesjms/fsm135

Moore, J.E., Curtis, K.A., Lewison, R.L., Dillingham, P.W., Cope, J.M., Fordham, S.V., Heppell, S.S., Pardo, S.A., Simpfendorfer, C.A., Tuck, G.N. & Zhou, S. 2013. Evaluating sustainability of fisheries bycatch mortality for marine megafauna: a review of conservation reference points for data-limited populations. *Environmental Conservation* **40**, 329–344, doi:10.1017/S037689291300012X

Moreno-García, M., Pimenta, C.M. & de Fátima Palma, M. 2017. Recursos cinegéticos y ganaderos en Myrtilis (Mértola, Portugal) en los inicios de la Romanización: una aportación desde la Arqueozoología. In *La romanización en la Península Ibérica, una visión desde la Arqueozoología*, S. Valenzuela, L. Colominasy & C. Fernández (eds.). *Archaeofauna* **26**, 179–198.

Mornet, R. 2008. *Les mots de la mer*. Geste: La Crèche.

Moura, A.E., Sillero, N. & Rodrigues, A. 2012. Common dolphin (*Delphinus delphis*) habitat preferences using data from two platforms of opportunity. *Acta Oecologica* **38**, 24–32, doi:10.1016/j.actao.2011.08.006

Moustahfid, H., Hendrickson, L., Arkhipkin, A., Pierce, G.J., Gangopadhyay, A., Kidokoro, H., Markaida, U., Nigmatullin, C., Sauer, W.H.H., Jereb, P., Pecl, G., de la Chesnais, T., Ceriola, L., Lazar, N., Firmin, C. & Laptikhovsky, V. 2021. Ecological-fishery forecasting of squid stock dynamics under climate variability and change: review, challenges and recommendations. *Reviews in Fisheries Science and Aquaculture*, **29**, 682–705.

Mullié, W.C., Wagne, M.M., Elmamy, C.A., Mint Yahya, F., Veen, J. & Van Waerebeek, K. 2013. Large number of stranded harbour porpoises *Phocoena phocoena* as by-catch victims in Mauritania. Paper SC/65a/. Online. https://archive.iwc.int/pages/search.php?search=%21collection149

Muñoz, P.M., García-Castrillo, C., López-Garciá, P., González-Cueli, J.C., De Miguel, M.J., Marín, C.M., Barberán, M. & Blasco, J.M. 2006. Isolation of Brucella species from a live-stranded striped dolphin (*Stenella coeruleoalba*) in Spain. *Veterinariy Record*, **158**(13), 450–451, doi:10.1136/vr.158.13.450

Murphy, S., Barber, J., Learmonth, J.A., Read, F.L., Deaville, R., Perkins, M.W., Brownlow, A., Davison, N., Pierce, G.J., Law, R.J. & Jepson, P.D. 2015. Reproductive failure in UK harbour porpoises *Phocoena phocoena*: legacy of pollutant exposure? *PLoS One* **10**, e0131085, doi:10.1371/journal.pone.0131085

Murphy, S., Petitguyot, M.A.C., Jepson, P.D., Deaville, R., Lockyer, C., Barnett, J., Perkins, M., Penrose, R., Davison, N.J. & Minto, C. 2020. Spatio-temporal variability of harbor porpoise life history parameters in the North-East Atlantic. *Frontiers in Marine Science* **7**, 502352, doi:10.3389/fmars.2020.502352

Murphy, S., Pierce, G.J., Law, R., Bersuder, P., Jepson, P., Learmonth, J., Addink, M., Dabin, W., Santos, M., Deaville, R., Zegers, B., Mets, A., Rogan, E., Ridoux, V., Reid, R., Smeenk, C., Jauniaux, T., López Fernández, A., Farré, A. & Boon, J. 2010. Assessing the effect of persistent organic pollutants on reproductive activity in common dolphins and harbour porpoises. *Journal of Northwest Atlantic Fishery Science* **42**, 153–173, doi:10.2960/J.v42.m658

Murphy, S., Winship, A., Dabin, W., Jepson, P.D., Deaville, R., Read, R.J., Spurrier, C., Rogan, E., López, A., González, A.F., Read, F.L., Addink, M., Silva, M., Ridoux, V., Learmonth, J.A., Pierce, G.J. & Northridge, S.P. 2009. Importance of biological parameters in assessing the status of *Delphinus delphis*. *Marine Ecology Progress Series* **388**, 273–291, doi:10.3354/meps08129

Murray, K.T., Read, A.J. & Solow, A.R. 2020. The use of time/area closures to reduce bycatches of harbour porpoises: lessons from the Gulf of Maine sink gillnet fishery. *Journal of Cetacean Research and Management* **2**(2), 135–141.

Musset, L. 1964. Quelques notes sur les baleiniers normands du Xe au XIIIe siècle. *Revue d'histoire économique et sociale* **42**, 147–161.

NAMMCO. 2019. Report of the NAMMCO Scientific Committee Working Group on Harbour Porpoise, 19-22 March, Copenhagen, Denmark. Online. https://nammco.no/wp-content/uploads/2019/02/final-report_hpwg-2019.pdf

Nascimento, L.G. 1945. O delfim: um inimigo irreconciliável da sardinha. *Boletim da Pesca* **8**, 18–25.

National Research Council. 1992. *Dolphins and the Tuna Industry*. Washington, DC: The National Academies Press, doi:10.17226/1983

Neimanis, A., Stavenow, J., Ågren, E.O., Wikström-Lassa, E. & Roos, A.M. 2022. Causes of death and pathological findings in stranded harbour porpoises (*Phocoena phocoena*) from Swedish waters. *Animals* **12**, 369, doi:10.3390/ani12030369

Nielsen, N.H., Teilmann, J., Sveegard, S., Hansen, R.G., Sinding, M.H.S., Dietz, R. & Heid-Jørgensen, M.P. 2018. Oceanic movements, site fidelity and deep diving in harbour porpoises from Greenland show limited similarities to animals from the North Sea. *Marine Ecology Progress Series* **597**, 259–272, doi:10.3354/meps12588

Nobre, A. 1895. Notes sur les poissons d'Algarve. *Annaes de Sciencias Naturaes: Revista de Historia Natural, Agricultura, Piscicultura, e Pescas Maritimas* **2**, 223–232.

North Atlantic Marine Mammal Commission and the Norwegian Institute of Marine Research. 2019. Report of Joint IMR/NAMMCO International Workshop on the Status of Harbour Porpoises in the North Atlantic, Tromsø, Norway. Online. https://nammco.no/wp-content/uploads/2020/03/final-report_hpws_2018_rev2020.pdf

Northridge, S.M.D.S., Waples, D. & Read, A.J. 2013. To ping or not to ping: the use of active acoustic devices in mitigating interactions between small cetaceans and gillnet fisheries. *Endangered Species Research* **19**(3), 201–221, doi:10.3354/esr00464

NRC. 2005. *Marine Mammal Populations and Ocean Noise: Determining When Noise Causes Biologically Significant Effects*. Committee on Characterizing Biologically Significant Marine Mammal Behavior, National Research Council. National Academies Press. Online. https://www.nrc.gov/docs/ML1434/ML14345A574.pdf

Nuuttila, H.K., Bertelli, C.M., Mendzil, A. & Dearle, N. 2018. Seasonal and diel patterns in cetacean use and foraging at a potential marine renewable energy site. *Marine Pollution Bulletin* **129**, 633–644, doi:10.1016/j.marpolbul.2017.10.051

Ólafsdóttir, D., Víkingsson, G.A., Halldórsson, D.S. & Sigurjónsson, J. 2003. Growth and reproduction in harbour porpoises (*Phocoena phocoena*) in Icelandic waters. *NAMMCO Scientific Publications* **5**, 195–210, doi:10.7557/3.2747

Oliveira, A.F. 2020. *Caracterização do estado reprodutor de fêmeas de pequenos cetáceos na costa norte de Portugal*. Masters thesis, University of Aveiro.

Oliveira, N., Henriques, A., Miodonski, J., Pereira, J., Marujo, D., Almeida, A., Barros, N., Andrade, J., Marçalo, A., Santos, J., Benta Oliveira, I., Ferreira, M., Araújo, H., Monteiro, S., Vingada, J. & Ramírez, I. 2015. Seabird bycatch in Portuguese mainland coastal fisheries: an assessment through on-board observations and fishermen interviews. *Global Ecology and Conservation* **3**, 51–61, doi:10.1016/j.gecco.2014.11.006

Olsen, M.T., Nielsen, N.H., Biard, V., Teilmann, J., Ngô, M.C., Víkingsson, G., Gunnlaugsson, T., Stenson, G., Lawson, J., Lah, L., Tiedemann, R. & Heide-Jørgensen, M.P. 2022. Genetic and behavioural data confirm the existence of a distinct harbour porpoise ecotype in West Greenland. *Ecological Genetics and Genomics* **22**, 100108, doi:10.1016/j.egg.2021.100108

Omeyer, L.C.M., Doherty, P.D., Dolman, S., Enever, R., Reese, A., Tregenza, N., Williams, R. & Godley, B.J. 2020. Assessing the effects of banana pingers as a bycatch mitigation device for harbour porpoises (*Phocoena phocoena*). *Frontiers in Marine Science* **7**, 285, doi:10.3389/fmars.2020.00285

Orach, K., Schlüter, M. & Österblom, H. 2017. Tracing a pathway to success: how competing interest groups influenced the 2013 EU Common Fisheries Policy reform. *Environmental Science and Policy* **76**, 90–102, doi:10.1016/j.envsci.2017.06.010

Osborn, T.J. & Briffa, K.R. 2006. The spatial extent of 20th-century warmth in the context of the past 1200 years. *Science* **311**, 841–844, doi:10.1126/science.1120514

Osieka, A.N., Jones, O. & Wahlberg, M. 2020. The diel pattern in harbour porpoise clicking behaviour is not a response to prey activity. *Scientific Reports* **10**, 14876, doi:10.1038/s41598-020-71957-0

OSPAR Commission. 2016. OSPAR Coordinated Environmental Monitoring Programme (CEMP). OSPAR Agreement 2016-01, OSPAR Commission. Online. https://www.ospar.org/documents?v=32943

OSPAR Commission. 2023. OSPAR CEMP Guideline. QSR23 Common Indicator Assessment: M6 Marine Mammal By-catch (harbour porpoise; common dolphin; grey seal) Region II, III and IV and QSR23 candidate indicator assessment: m6 marine mammal by-catch (harbour porpoise; grey seal) OSPAR Region I. Agreement 2022-03. Online. https://www.ospar.org/documents?v=48551

Otani, S., Naito, Y., Kato, A. & Kawamura, A. 2001. Oxygen consumption and swim speed of the harbor porpoise *Phocoena phocoena*. *Fisheries science* **67**(5), 894–898, doi:10.1046/j.1444-2906.2001.00338.x

Oueslati, T. 2017. Aller à la pêche au gros et produire du garum à l'époque romaine dans le nord de la Gaule. In *L'exploitation des ressources maritimes de l'Antiquité. Activités productives et organisation des territoires*, R. González Villaescusa, K. Schorle, F. Gayet & F. Rechin (eds.). Antibes: APDCA, 41–56.

Ovalle, J.C., Velasco, E., Vilas Fernández, C., Abad, E., Valeiras, J., Pérez Martín, R.I. & Antelo, L.T. 2021. Advances in electronic monitoring of fishing catches based on artificial intelligence. 9th International Workshop on Marine Technology (MARTECH 2021). *Instrumentation Viewpoint* **21**, 37–38. Online. https://hdl.handle.net/2117/360219

Ovalle, J.C., Vilas, C. & Antelo, L.T. 2022. On the use of deep learning for fish species recognition and quantification on board fishing vessels. *Marine Policy* **139**, 105015, doi:10.1016/j.marpol.2022.105015

Owen, K., Sköld, M. & Carlström, J. 2021. An increase in detection rates of the critically endangered Baltic Proper harbor porpoise in Swedish waters in recent years. *Conservation Science and Practice* **3**(8), e468, doi:10.1111/csp2.468

Palialexis, A., Connor, D., Damalas, D., Gonzalvo, D., Micu, D., Mitchel, I., Korpinen, S, Rees, A.F. & Somma, F. 2019. Indicators for status assessment of species, relevant to MSFD Biodiversity Descriptor. Identifying methods to set thresholds for the GES assessment. JRC Technical Report JRC117126, EUR 29820 EN, Publications Office of the European Union, Luxembourg, doi:10.2760/282667.

Palka, D.L. & Hammond, P S. 2001. Accounting for responsive movement in line transect estimates of abundance. *Canadian Journal of Fisheries and Aquatic Sciences* **58**, 777–787.

Pante, E. & Simon-Bouhet, B. 2013. Marmap: a package for importing, plotting and analysing bathymetric and topographic data in R. *PLoS One* **8**, e73051, doi:10.1371/journal.pone.0073051

Pascual, S., Rodríguez, H., Pierce, G.J., Hastie, L.C. & González, A.F. 2018. The NE Atlantic hake: a neglected high exposure risk for zoonotic parasites in European fish markets. *Fisheries Research* **202**, 69–78, doi:10.1016/j.fishres.2017.12.008

Peltier, H., Dabin, W., Daniel, P., Van Canneyt, O., Dorémus, G., Huona, M. & Ridoux, V. 2012. The significance of stranding data as indicators of cetacean populations at sea: modelling the drift of cetacean carcasses. *Ecological Indicators* **18**, 278–290, doi:10.1016/j.ecolind.2011.11.014

Penas-Patiño, X.M. & Piñeiro-Seage, A. 1989. *Cetáceos, focas e tartarugas das costas ibéricas*. Consellería de Pesca, (Xunta de Galicia): Santiago de Compostela.

Pereira, A.T. 2015. *Monitorização do uso costeiro por pequenos cetáceos e avaliação do uso de pingers para mitigação de capturas acidentais na região norte de Portugal*. MSc thesis, Universidade do Porto, Portugal.

Perez, C. & Nores, C. 1988. Cetacean Research in Asturias (North Spanish coast). *European Researh on Cetaceans* **2**, 17–23.

Perrin, M.R. 1989. Alternative life-history styles of small mammals. In *Alternative Life-History Styles of Animals. Perspectives in Vertebrate Science*, M.N. Bruton (ed.). Dordrecht: Springer, Volume **6**.

Petitguyot, M.A.C., Bearzi, G., van den Hurk, Y., Tejedor Fuentes, M. & Pierce, G.J. 2024. Intentional killings and culling of small cetaceans due to perceived competition with fisheries in the Mediterranean Sea and Northeast Atlantic between the Eighteenth and Twentieth centuries. *Oceanography and Marine Biology: An Annual Review* **62**, 120–162.

Pierce, G.J., Caldas, M., Cedeira, J., Santos, M.B., Llavona, A., Covelo, P., Martinez, G., Torres, J., Sacau, M. & López, A. 2010. Trends in cetacean sightings along the Galician coast, north-western Spain, 2003–2007, and inferences about cetacean habitat preferences. *Journal of the Marine Biological Association of the United Kingdom* **90**, 1547–1560.

Pierce, G.J., Petitguyot, M.A.C., Gutiérrez-Muñoz, P., Fernández-Fernández, D., Fariñas-Bermejo, A., Read, F.L., Saavedra, S., López, A. & Martínez-Cedeira, J. 2022. Iberian harbour porpoise - an update on fishery bycatch mortality. Paper SC/68D/HIM/10. Online. https://archive.iwc.int/pages/search.php?search=%21collection30114

Pierce, G.J., Petitguyot, M.A.C., Gutiérrez-Muñoz, P., Fernández-Fernández, D., Marçalo, A., Blakeman, S., Puig-Lozano, R., Saavedra Penas, C., Vázquez Bonales, J.A., Suarez-Bregua, P., Hernandez Gonzalez, A., López, A., Covelo, P., Pin, X., Vázquez, U., González, M. Martínez-Cedeira6, J., Vilas, N., Read, F.L., Leaper, R., Simmonds, M., Pinn, E. & Dolman, S., 2024. Iberian harbour porpoise: an update. Paper SC/69B/HIM/12. Scientific Committee of the International Whaling Commission. https://archive.iwc.int/pages/view.php?ref=22104&search=&offset=240&order_by=field8&sort=DESC&archive=0

Pierce, G.J., Santos, M.B., Murphy, S., Learmonth, J.A., Zuur, A.F., Rogan, E., Bustamante, P., Caurant, F., Lahaye, V., Ridoux, V., Zegers, B.N., Mets, A., Addink, M., Smeenk, C., Jauniaux, T., Law, R.J., Dabin, W., López, A., Alonso Farré, J.M., González, A.F., Guerra, A., García-Hartmann, M., Reid, R.J., Moffat, C.F., Lockyer, C. & Boon, J.P. 2008. Bioaccumulation of persistent organic pollutants in female common dolphins (*Delphinus delphis*) and harbour porpoises (*Phocoena phocoena*) from western European seas: geographical trends, causal factors and effects on reproduction and mortality. *Environmental Pollution* **153**, 401–415, doi:10.1016/j.envpol.2007.08.019

Pierce, G.J., Weir, C., Gutiérrez, P., Verutes, G., Fontaine, M.C., Hernández-González, A., Saavedra, C., Llavona, A., Martínez Cedeira, J., Covelo P., López, A., Read, F.L. & Dolman, S. 2020. Is Iberian harbour porpoise (Phocoena phocoena) threatened by interactions with fisheries? Paper SC/68B/SM04 Rev2. Online. https://archive.iwc.int/pages/search.php?search=%21collection29958

Pierpoint, C. 2008. Harbour porpoise (*Phocoena phocoena*) foraging strategy at a high energy, near-shore site in south-west Wales, UK. *Journal of the Marine Biological Association of the United Kingdom* **88**(6), 1167–1173, doi:10.1017/S0025315408000507

Pinheiro, G.A.J. 2017. *Contribuição para o estudo da dieta de pequenos cetáceos em Portugal continental*. Masters thesis, Univerity of Aveiro.

Pinn, E.H. 2023. Poroises, by-catch and the ‘pinger’ conundrum. *Aquatic Conservation: Marine and Freshwater Ecosystems* **33**(11), 1360–1368, doi:10.1002/aqc.4004

Pinn, E.H., Macleod, K. & Tasker, M.L. 2021. Conservation of transnational species: the tensions between legal requirements and best scientific evidence. *Aquatic Conservation: Marine and Freshwater Ecosystems* **31**(11), 3013–3343, doi:10.1002/aqc.3693

Pinzone, M., Parmentier, K., Siebert, U., Gilles, A., Authier, M., Brownlow, A., Caurant, F., Das, K., Deaville, R., Galatius, A., Geelhoed, S., Hernández Sánchez, M.T., Mendez-Fernandez, P., Murphy, S., Persson, S., Roos, A., van den Heuvel-Greve, M., Vinas, L. & Williams, R. 2022. Pilot Assessment of Status and Trends of Persistent Chemicals in Marine Mammals. In OSPAR. 2023: The 2023 Quality Status Report for the North-East Atlantic. OSPAR Commission, London. Online. https://oap.ospar.org/en/ospar-assessments/quality-status-reports/qsr-2023/indicator-assessments/pcb-marine-mammals-pilot

Pirotta, E., Booth, C.G., Costa, D.P., Fleishman, E., Kraus, S.D., Lusseau, D., Moretti, D., New, L.F., Schik, R.S., Schwrz, L.K., Simmons, S.E., Thomas, L., Tyack, P.L., Weise, M.J., Wells, R.S., Harwood, J. 2018. Understanding the population consequences of disturbance. *Ecology and Evolution* **8**(19), 9934–9946, doi:10.1002/ece3.4458

Pons-Bordas, C., Hazenberg, A., Hernández-González, A., Pool, R.V., Covelo, P., Sánchez-Hermosin, P., López, A., Saavedra, C. & Fraija-Fernández, N., Fernández, M. & Aznar, F.J. 2020. Recent increase of ulcerative lesions caused by *Anisakis* spp. in cetaceans from the north-east Atlantic. *Journal of Helminthology* **94**, e127, doi:10.1017/S0022149X20000115

Pool, R., Romero-Rubira, C., Raga, J.A., Fernandez, M. & Aznar, F.J. 2021. Determinants of lungworm specificity in five cetacean species in the western Mediterranean. *Parasires & Vectors* **14**, 196, doi:10.1186/s13071-021-04629-1

Prata, J.C., da Costa, J.P., Lopes, I., Duarte, A.C. & Rocha-Santos, T. 2020. Environmental status of (micro) plastics contamination in Portugal. *Ecotoxicology and Environmental Safety* **200**, 110753, doi:10.1016/j.ecoenv.2020.110753

Pribanić, S., Mioković, D. & Kovačić, D. 2000. Preliminary growth rate and body lengths of the bottlenose dolphins *Tursiops truncatus* (Montagu, 1821) from the Adriatic Sea. *Natura Croatica* **9**(3), 179–188.

Puig-Lozano, R., Fernández, A., Sierra, E., Saavedra, P., Suárez-Santana, C.M., De la Fuente, J., Díaz-Delgado, J., Godinho, A., García-Álvarez, N., Zucca, D., Xuriach, A., Arregui, M., Felipe-Jiménez, I., Consoli, F., Díaz-Santana, P.J., Segura-Göthlin, S., Câmara, N., Rivero, M.A., Sacchini, S., Bernaldo de Quirós, Y. & Arbelo, M. 2020. Retrospective study of fishery interactions in stranded cetaceans, Canary Islands. *Frontiers in Veterinary Science* **7**, 567258, doi:10.3389/fvets.2020.567258

Raga, J.A. & Balbuena, J.A. 1993. Parasites of the long-finned pilot whale, Globicephala melas (Traill, 1809), in European waters. *Report of the International Whaling Commission* **14**(Special Issue), 391–406.

Rawson, A.L., Patton, G.W., Hofmann, S., Pietra, G.G. & Johns, L. 1993. Liver abnor- malities associated with chronic mercury accumulation in stranded Atlantic bottlenose dolphins. *Ecotoxicology and Environmental Safety* **25**, e41–e47.

R Core Team. 2016. *R: A Language and Environment for Statistical Computing*. Vienna, Austria: R Foundation for Statistical Computing. Online. https://www.R-project.org.

Read, A. J. 1999. Harbour porpoise (*Phocoena phocoena*). In: Ridgway, S., Harrison, R. (Eds.), *Handbook of Marine Mammals*. Academic Press, London, San Diego, pp. 323–350.

Read, A.J. & Gaskin, D.E. 1988. Incidental catch of harbor porpoises by gill nets. *Journal of Wildlife Management* **52**(3), 517–523, doi:10.2307/3801603

Read, A.J. & Gaskin, D.E. 1990. Changes in growth and reproduction of harbour porpoises, *Phocoena phocoena*, from the Bay of Fundy. *Canadian Journal of Fisheries and Aquatic Sciences* **47**, 2158–2163, doi:10.1139/f90-240

Read, A.J. & Hohn, A.A. 1995. Life in the fast lane: the life history of harbour porpoises from the Gulf of Maine. *Marine Mammal Science* **11**, 423–440, doi:10.1111/j.1748-7692.1995.tb00667.x

Read, A.J. & Tolley, K.A. 1997. Postnatal growth and allometry of harbour porpoises from the Bay of Fundy. *Canadian Journal of Zoology* **75**, 122–130, doi:10.1139/z97-016

Read, F.L. 2016. *Understanding Cetacean and Fisheries Interactions in the North-West Iberian Peninsula.* PhD thesis, University of Vigo, Spain.

Read, F.L. 2021. Cost-benefit Analysis for Mitigation Measures in Fisheries with High Bycatch. ASCOBANS Technical Series No. 2. ASCOBANS Secretariat, Bonn, Germany.

Read, F.L., Evans, P.G.H & Dolman, S.J. 2017. Cetacean Bycatch Monitoring and Mitigation under EC Regulation 812/2004 in the Northeast Atlantic, North Sea and Baltic Sea from 2006 to 2014. A WDC Report, Whale and Dolphin Conservation. Chippenham, UK. Online. https://www.wdcs.org/submissions_bin/EU-Cetacean-Bycatch-Monitoring-Mitigation-Report.pdf

Read, F.L., Hohn, A.A. & Lockyer, C.H. 2018. A review of age estimation methods in marine mammals with special reference to monodontids. *NAMMCO Scientific Publications* **10**, 1–67, doi:10.7557/3.4474

Read, F.L., Santos, M.B, Ferreira, M., López, A. & Pierce, G.J. 2020. Harbour porpoise (*Phocoena phocoena*) and fisheries interactions in the north-west Iberian Peninsula. Paper SC/68B/SM/06. Online. https://archive.iwc.int/pages/search.php?search=%21collection29958#

Read, F.L., Santos, M.B., González, A.F., López, A., Ferreira, M., Vingada, J. & Pierce, G.J. 2013. Understanding harbour porpoise (*Phocoena phocoena*) and fishery interactions in the north-west Iberian Peninsula. Final Report to the 20th ASCOBANS Advisory Committee Meeting. Warsaw, Poland, 27–29 August 2013.

Reboredo-Fernández, A., Ares-Mazás, E., Martínez-Cedeira, J.A., Romero-Suances, R., Cacciò, S.M. & Gómez-Couso, H. 2015. *Giardia* and *Cryptosporidium* in cetaceans on the European Atlantic coast. *Parasitology Research* **114**, 693–698, doi:10.1007/s00436-014-4235-8

Reijnders, P.J.H. 1986. Reproductive failure in common seals feeding on fish from polluted coastal waters. *Nature* **324**, 456–457, doi:10.1038/324456a0

Reiner, F., Dos Santos, M.E. & Wenzel, F.W. 1996. Cetaceans of the Cape Verde archipelago. *Marine Mammal Science* **12**, 434–443.

Rey, J.C. & Cendrero, O. 1982. Les mammifères marins trouvés sur les côtes espagnoles en 1981. Paper CM 1982/N:6. *International Council for the Exploration of the Sea, Council Meeting.*

Richardson, S.F. 1992. *Growth and reproduction of the harbour porpoise, Phocoena phocoena, from the eastern Newfoundland.* MSc thesis, Memorial University of Newfoundland, Canada.

Richardson, S.F., Stenson, G.B. & Hood, C. 2003. Growth of the harbour porpoise (*Phocoena phocoena*) in Eastern Newfoundland, Canada. *NAMMCO Scientific Publication Series*, **5**, 271–302, doi:10.7557/3.2748

Ridgway, S.H. & Harrison, R. 1999. *Handbook of Marine Mammals. Volume 6. The Second Book of Dolphins and Porpoises.* London: Academic Press.

Roberts, N., Brayshaw, D., Kuzucuoglu, C., Perez, R. & Sadori, L. 2011. The mid-Holocene climatic transition in the Mediterranean: causes and consequences. *The Holocene* **21**, 3–13, doi:10.1177/0959683610388058

Robinson, J.A., Kyriazis, C.C., Nigenda-Morales, S.F., Beichman, A.C., Rojas-Bracho, L., Robertson, K.M., Fontaine, M.C., Wayne, R.K., Lhmueller, K.E., Taylor, B.L. & Morin, P.A. 2022. The critically endangered vaquita is not doomed to extinction by inbreeding depression. *Science* **376**, 635–639, doi:10.1126/science.abm1742

Roca-Geronès, X., Segovia, M., Godínez-González, C., Fisa, R. & Montoliu, I. 2020. *Anisakis* and *Hysterothylacium* species in Mediterranean and North-East Atlantic fishes commonly consumed in Spain: Epidemiological, molecular and morphometric discriminant analysis. *International Journal of Food Microbiology* **325**, 108642, doi:10.1016/j.ijfoodmicro.2020.108642

Rodrigo, F.J., Ramis, J., Carbajo, J. & Poveda, P. 2022. Underwater anthropogenic noise pollution assessment in shallow waters on the South-Eastern Coast of Spain. *Journal of Marine Science and Engineering* **10**, 1311, doi:10.3390/jmse10091311

Rogan, E. & Berrow, S.D. 1996. A review of harbour porpoises, *Phocoena phocoena* L., in Irish waters. *Report of the International Whaling Commission* **46**, 595–606.

Rojano-Doñate, L., McDonald, B.I., Wisniewska, D.N., Johnson, M., Teilmann, J., Wahlberg, M., Højer-Kristensen, J. & Madsen, P. 2018. High field metabolic rates of wild harbour porpoises. *Journal of Experimental Biology* **221**, jeb185827, doi:10.1242/jeb.185827

Rojas-Bracho, L., Gulland, F.M.D., Smith, C.R., Taylor, B., Wells, R.S., Thomas, P.O., Bauer, B., Heide-Jørgensen, M.P., Teilmann, J., Dietz, R., Balle, J.D.,, Jensen, M.V., Sinding, M.H.S., Jaramillo-Legorreta, A., Abel, G., Read, A.J., Westgate, A.J., Colegrove, K., Gomez, F., Martz, K., Rebolledo, R., Ridgway, S., Rowles, T., van Elk, C.E., Boehm, J., Cardenas-Hinojosa, G., Constandse, R., Nieto-Garcia, E., Phillips, W., Sabio, D., Sanchez, R.J., Sweeney, J., Townsend, F., Vivanco, J., Vivanco, J.C. & Walker, S. 2019. A field effort to capture critically endangered vaquitas *Phocoena sinus* for protection from entanglement in illegal gillnets. *Endangered Species Research* **38**, 11–27.

Rojo-Nieto, E., Álvarez-Díaz, P.D., Morote, E., Burgos-Martín, M., Montoto-Martínez, T., Sáez-Jiménez, J. & Toledano, F. 2011. Strandings of cetaceans and sea turtles in the Alboran Sea and Strait of Gibraltar: a long-time glimpse of the north coast (Spain) and the south coast (Morocco). *Animal Biodiversity and Conservation* **34**, 151–163.

Ross, H.M. & Wilson, B. 1996. Violent interactions between bottlenose dolphins and harbour porpoises. *Proceedings of the Royal Society B* **263**, 283–286, doi:10.1098/rspb.1996.0043

Rouby, E., 2018. *Démographie du Marsouin commun (Phocoena phocoena, Linné 1758) au sein des eaux Atlantiques françaises.* Masters thesis, University of Sorbonne (Paris) and La Rochelle University, France.

Rowe, S.J. 2007. *A Review of Methodologies for Mitigating Incidental Catch of Protected Marine Mammals.* Dunedin, New Zealand: Science & Technical Publishing, Department of Conservation.

Ryan, K., Mynott, S., Lyons, C., Clare, T., Day, E., Bell, C., Wild, M., Engelhard, G. & Hetherington, S.J. (eds.) 2022. Hauling Up Solutions 2: Exploring new ways to expand the wildlife bycatch reduction toolkit. Final Workshop Report. 27 pp. Online. https://www.cleancatchuk.com/wp-content/uploads/2022/09/HUS2-Report.pdf

Saavedra, C., Gerrodette, T., Louzao, M., Valeiras, J., García, S., Cerviño, S., Pierce, G.J. & Santos, M.B. 2018. Assessing the environmental status of the short-beaked common dolphin (*Delphinus delphis*) in North-western Spanish waters using abundance trends and safe removal limits. *Progress in Oceanography* **166**, 66–75, doi:10.1016/j.pocean.2017.08.006

Saavedra, C., Pierce, G.J., Gago, J., Jusufovski, D., Cabrero, A., Cerviño, S., López, A., Martínez-Cedeira, J.A. & Santos, M.B. 2017. Factors driving patterns and trends in strandings of small cetaceans. *Marine Biology* **164**, 165, doi:10.1007/s00227-017-3200-3

Saavedra Penas, C. 2017. *Multispecies population modelling of the common dolphin (Delphinus delphis), the bottlenose dolphin (Tursiops truncatus) and the southern stock of European hake (Merluccius merluccius), in Atlantic waters of the Iberian Peninsula.* PhD thesis, Universidad de Vigo, Spain.

Sacchi, J. 2021. Overview of mitigation measures to reduce the incidental catch of vulnerable species in fisheries. *Studies and Reviews (General Fisheries Commission for the Mediterranean)* **100**, 137, doi:10.4060/cb5049en

Saldaña, A., López, C.M., López, A., Covelo, P., Remesar, S., Martínez-Calabuig, N., García-Dios, D., Díaz, P., Morrondo, P., Díez-Baños, P. & Panadero, R. 2022. Specificity of *Stenurus* (Metastrongyloidea: Pseudaliidae) infections in odontocetes stranded along the north-west Spanish coast. *International Journal for Parasitology: Parasites and Wildlife* **19**, 148–154, doi:10.1016/j.ijppaw.2022.09.002

Saldaña Ruiz, A. 2021. *Infección por Stenurus spp. En odontocetos varados en las costas gallegas.* BSc thesis, University of Santiago de Compostela, Spain.

Samarra, F.I.P., Bassoi, M., Béesau, J., Elíasdóttir, M.Ó., Gunnarsson, K., Mrusczok, M.T., Rasmussen, M., Rempel, J.N., Thorvaldsson, B. & Vikingsson, G.A. 2018. Prey of killer whales (Orcinus orca) in Iceland. *PloS One* **13**(12), e0207287, doi:10.1371/journal.pone.0207287

SAMBAH. 2016. LIFE Project NumberLIFE08 NAT/S/000261. Final Report: Covering the project activities from 01/01/2010 to 30/09/2015. Online. https://www.sambah.org/SAMBAH-Final-Report-FINAL-for-website-April-2017.pdf

Sánchez, F., Velasco, F., Cartes, J.E., Olaso, I., Preciado, I., Fanelli, E., Serrano, A. & Gutierrez-Zabala, J.L. 2006. Monitoring the Prestige oil spill impacts on some key species of the Northern Iberian Shelf. *Marine Pollution Bulletin* **53**, 332–349.

Sanpera, C., Aguilar, A., Grau, E. & Jover, E. 1984. Report of the 'Ballena 2' whale marking and sighting cruise in the Atlantic waters off Spain. Paper SC/35/O 12. *Report of the International Whaling Commission* **34**, 663–666.

Sanpera, C., Grau, E., Jover, E., Recasens, A., Aguilar, A. & Olmos, M. 1985. Report of the 'Ballena 3' fin whale marking and sighting cruise off Spain, 1983. Paper SC/36/O 15. *Report of the International Whaling Commission* **35**, 495–497.

Sanpera, C. & Jover, E. 1986. Report of the 'Ballena 4' fin whale sighting cruise. Paper SC/37/Ba2. *Report of the International Whaling Commission* **36**, 253–255.

Santos, J., Araújo, H., Ferreira, M., Henriques, A., Miodonski, J., Monteiro, S., Oliveira, I., Rodrigues, P., Duro, G., Oliveira, F., Pinto, N., Sequeira, M., Eira, C.E. & Vingada, J. 2012. Chapter I: Baseline estimates of abundance and distribution of target species. Annex to the Midterm Report of project LIFE+MarPro PT/NAT/00038

Santos, M.B. & Pierce, G.J. 2003. The diet of harbour porpoise (*Phocoena phocoena*) in the Northeast Atlantic. *Oceanography & Marine Biology: An Annual Review* **41**, 355–390.

Santos, M.B., Pierce, G.J., Learmonth, J.A., Reid, R.J., Ross, H.M., Patterson, I.A.P., Reid, D.G. & Beare, D. 2004. Variability in the diet of harbour porpoises (*Phocoena phocoena*) in Scottish waters 1992–2003. *Marine Mammal Science* **20**, 1–27, doi:10.1111/j.1748-7692.2004.tb01138.x

Santos, M.B., Saavedra, C. & Pierce, G.J. 2014. Quantifying the predation on sardine and hake by cetaceans in the Atlantic waters of the Iberian Peninsula. *Deep-Sea Research II* **106**, 232–244, doi:10.1016/j.dsr2.2013.09.040

Santos Vázquez, M.B. 1998. *Feeding ecology of harbour porpoises, common and bottlenose dolphins and sperm whales in the Northeast Atlantic.* PhD thesis, University of Aberdeen, Scotland.

SCANS-II. 2008. Small Cetaceans in the European Atlantic and North Sea. Final Report to the European Commission under project LIFE04NAT/GB/000245. Available from SMRU, Gatty Marine Laboratory, University of St Andrews, St Andrews, Fife, United Kingdom.

Schelling, T., Van der Steeg, L.J. & Leopold, M.F. 2014. The diet of harbour porpoises *Phocoena phocoena* in Dutch waters: 2003-2014. IMARES Report (No. C136/14).

Schoeman, R.P., Patterson-Abrolat, C. & Plön, S. 2020. A global review of vessel collisions with marine animals. *Frontiers in Marine Science* **7**, 292, doi:10.3389/fmars.2020.00292

Schwacke, L. H., Voit, E. O., Hansen, L. J., Wells, R. S., Mitchum, G. B., Hohn, A. A., & Fair, P. A. (2002). Probabilistic risk assessment of reproductive effects of polychlorinated biphenyls on bottlenose dolphins (*Tursiops truncatus*) from the southeast United States coast. *Environmental Toxicology and Chemistry: An International Journal* **21**(12), 2752–2764. doi: 10.1002/etc.5620211232

Scientific, Technical and Economic Committee for Fisheries (STECF). 2019. *Review of the implementation of the EU regulation on the incidental catches of cetaceans (STECF-19-07).* Luxembourg: Publications Office of the European Union. Online. https://stecf.jrc.ec.europa.eu/documents/43805/2489016/STECF+19-07+-+Incid+catches+cetaceans.pdf/3485bafd-4350-40af-8d72-0226a68cb86e

Sequeira, M. 1996. Harbour porpoises (*Phocoena phocoena*) in Portuguese waters. *Report of the International Whaling Commission* **46**, 583–586.

Sequeira, M. & Ferreira, C. 1994. Coastal fisheries and cetacean mortality in Portugal. *Report of the International Whaling Commission* **15**(Special Issue), 165–181.

Sequeira, M., Inácio, A. & Silva, M.A. 1997. Cetacean strandings in Portugal: 1993–1995. *European Research on Cetaceans* **10**, 136–140.

Sequeira, M., Inácio, A., Silva, M.A. & Reiner, F. 1996. *Arrojamentos de mamíferos marinhos na costa continental portuguesa entre 1989 e 1994.* Estudos de Biologia e Conservação da Natureza, Lisbon: ICE.

Sequeira, M. & Teixeira, A.M. 1988. Marine mammal surveys in Portugal. *European Research on Cetaceans* **2**, 9–12.

Sharpe, M. & Berggren, P. 2023. *Phocoena phocoena* (Europe assessment). The IUCN Red List of Threatened Species 2023: e.T17027A219010660. https://www.iucnredlist.org/species/17027/219010660. Accessed on 13 August 2024.

Siebert, U., Pawliczka, I., Benke, H., von Vietinghoff, V., Wolf, P., Pilats, V., Kesselring, T., Lehnert, K., Prenger-Berninghoff, E., Galatius, A., Anker Kyhn, L., Teilmann, J., Hansen, M.S., Sonne, C. & Wohlsein, P. 2020. Health assessment of harbour porpoises (*Phocoena phocoena*) from Baltic area of Denmark, Germany, Poland and Latvia. *Environment International* **143**, 105904, doi:10.1016/j.envint.2020.105904

Siebert, U., Prenger-Berninghoff, E. & Weiss, R. 2009. Regional differences in bacterial flora in harbour porpoises from the North Atlantic: environmental effects? *Journal of Applied Microbiology* **106**, 329–337, doi:10.1111/j.1365-2672.2008.04006.x

Siler, W. 1979. A competing-risk model for animal mortality. *Ecology* **60**(4), 750–757, doi:10.2307/1936612

Silva, M.A., Prieto, R., Magalhães, S., Cabecinhas, R. & Cruz, A. 2001. Estatuto da conservação do boto, *Phocoena phocoena*, nos Açores. Documento preparado no âmbito do Projecto Life - Gestão Integrada de Zonas Costeiras e Marinhas nos Açores. Contrato LIFE B4-3200/98-509 (Unpublished Report). Portugal.

Simmonds, M.P. & Brown, V.C. 2010. Is there a conflict between cetacean conservation and marine renewable-energy developments? *Wildlife Research* **37**, 688–694, doi:10.1071/WR10020

Smeenk, C. 1997. Strandings of sperm whales *Physeter macrocephalus* in the North Sea: history and patterns. *Bulletin de l'Institut royal des Sciences naturelles de Belgique, Biologie* **67**(Suppl.), 15–28.

Smeenk, C., Leopold, M.F. & Addink, M.J. 1992. Note on the harbour porpoise *Phocoena phocoena* in Mauritania, West Africa. *Lutra*, **35**, 98–104.

Smith, J. W. 1989. Ulcers associated with larval *Anisakis simplex* B (Nematoda: Ascaridoidea) in the forestomach of harbour porpoises *Phocoena phocoena* (L.). *Canadian Journal of Zoology*, **67**, 2270–2276. doi:10.1139/z89-319

Snape, R., Broderick, A., Çiçek, B., Fuller, W., Tregenza, N., Witt, M. & Godley, B. 2018. Conflict between dolphins and a data-scarce fishery of the European Union. *Human Ecology* **46**, 423–433, doi:10.1007/s10745-018-9989-7

Soares-Castro, P., Araújo-Rodrigues, H., Godoy-Vitorino, F., Ferreira, M., Covelo, P., López, A., Vingada, J., Eira, C. & Santos, P.M. 2019. Microbiota fingerprints within the oral cavity of cetaceans as indicators for population biomonitoring. *Scientific Reports* **9**(1), 1–15, doi:10.1038/s41598-019-50139-7

Sociedad Española de Cetáceos. 2006. Plan de monitorizaci6n de la marsopa (Phocoena phocoena) en Andalucía. Primera fase 2007–2010. LIFE02NAT/E/8610. Spain.

Soliño, L., Kim, S.Y., López, A., Covelo, P., Rydberg, S., Costa, P.R. & Lage, S. 2022. No β-N-Methylamino-L-alanine (BMAA) was detected in stranded cetaceans from Galicia (North-West Spain). *Journal of Marine Science and Engineering* **10**, 314, doi:10.3390/jmse10030314

Solórzano, M.R., Rodríguez, J.L., Iglesias, J., Pereiro, F.X. & Alvarez, F. 1988. Inventario dos peixes do litoral galego. (Pisces: Cyclostomata, Chondrichthyes, Osteichthyes). Cadernos da Área de Ciencias Biolóxicas. *(Inventarios) Seminario de Estudios Galegos* **4**, 69.
Sørensen, T.B. & Kinze, C.C. 1994. Reproduction and reproductive seasonality in Danish harbour porpoises, *Phocoena phocoena. Ophelia* **39**(3), 159–176, doi:10.1080/00785326.1994.10429541
Speller, C., van den Hurk, Y., Charpentier, A., Rodrigues, A., Gardeisen, A., Wilkens, B., McGrath, K., Rowsell, K., Spindler, L., Collins, M. & Hofreiter, M. 2016. Barcoding the largest animals on earth: ongoing challenges and molecular solutions in the taxonomic identification of ancient cetaceans. *Philosophical Transactions of the Royal Society B: Biological Sciences* **371**, 20150332, doi:10.1098/rstb.2015.0332
Spitz, J., Rousseau, Y. & Ridoux, V. 2006. Diet overlap between harbour porpoise and bottlenose dolphin: an argument in favour of interference competition for food? *Estuarine, Coastal and Shelf Science* **70**, 259–270, doi:10.1016/j.ecss.2006.04.020
Spötl, C., Nicolussi, K., Patzelt, G., Boch, R. & Team, D. 2010. Humid climate during deposition of sapropel 1 in the Mediterranean Sea: assessing the influence on the Alps. *Global and Planetary Change* **71,** 242–248, doi:10.1016/j.gloplacha.2009.10.003
Spyrakos, E., Santos-Diniz, T.C., Martínez-Iglesias, G., Torres-Palenzuela, J.M. & Pierce, G.J. 2011. Spatiotemporal patterns of marine mammal distribution in coastal waters of Galicia, NW Spain. *Hydrobiologia* **670**, 87–109, doi:10.1007/s10750-011-0722-4
Stenson, G.B. 2003. Harbour porpoise (*Phocoena phocoena*) in the North Atlantic: abundance, removals, and sustainability of removals. *NAMMCO Scientific Publications* **5**, 271–302, doi:10.7557/3.2830
Suarez-Bregua, P., Álvarez-González, M., Parsons, K.M., Rotllant, J., Pierce, G.J. & Saavedra, C. 2022. Environmental DNA (eDNA) for monitoring marine mammals: challenges and opportunities. *Frontiers in Marine Science* **9**, 987774, doi:10.3389/fmars.2022.987774
Sveegaard, S., Teilmann, J., Tougaard, J., Dietz, R., Mouritsen, K.N., Desportes, G. et al. 2011. High density areas for harbor porpoises (*Phocoena phocoena*) identified by satellite tracking. *Marine Mammal Science* **27**, 230–246, doi:10.1111/j.1748-7692.2010.00379.x
Taylor, B.A., Wade, P.R., De Master, B.P. & Barlow, J. 2000. Incorporating uncertainty into management models for marine mammals. *Conservation Biology* **14**(5), 1243–1252.
Taylor, N., Authier, M., Banga, R., Genu, M., Macleod, K. & Gilles, A. 2022. Marine mammal by-catch. In *OSPAR, 2023: The 2023 Quality Status Report for the Northeast Atlantic*. London: OSPAR Commission. Online. https://oap.ospar.org/en/ospar-assessments/quality-status-reports/qsr-2023/indicator-assessments/marine-mammal-bycatch
Teixeira, A.M. 1979. Marine mammals of the Portuguese coast. *Zeitschrift für Säugetierkunde* **44**(4), 221–238.
Temple, H.J. & Terry, A. 2007. *The Status and Distribution of European Mammals.* Luxembourg: Office for Official Publications of the European Communities.
ten Doeschete, M.T.I., IJsseldiijk, L.L., Hiemstra, S., de Jong, E.A., Strijkstra, A., Gröne, A. & Begeman, L. 2017. Quantifying parasite presence in relation to biological parameters of harbour porpoises *Phocoena phocoena* stranded on the Dutch coast. *Diseases of Aquatic Organisms* **127**, 49–56, doi:10.3354/dao03182
Thompson, R.C.A. & Monis, P.T. 2011. Taxonomy of *Giardia* species. In *Giardia. A Model Organism*, H.D. Lujá & S. Svärd (eds.). Vienna: Springer, doi:10.1007/978-3-7091-0198-8_1
Thomsen, F., Mendes, S., Bertucci, F., Breitzke, M., Ciappi, E., Cresci, A., Debusschere, E., Ducatel, C., Folegot, F., Juretzek, C., Lam, F-P., O'Brien, J. & dos Santos, M.E. 2021. Addressing underwater noise in Europe: current state of knowledge and future priorities. In *2021 Future Science Brief 7 of the European Marine Board*, P. Kellett (eds.). Belgium: Ostend, doi:10.5281/zenodo.5534224.
Todd, N.R., Jessopp, M., Rogan, E. & Kavanagh, A.S. 2022. Extracting foraging behavior from passive acoustic monitoring data to better understand harbor porpoise (*Phocoena phocoena*) foraging habitat use. *Marine Mammal Science* **38**, 1–20, doi:10.1111/mms.12951
Todd, V.L.G., Pearse, W.D., Tregenza, N.C., Lepper, P.A. & Todd, I.B. 2009. Diel echolocation activity of harbor porpoises (*Phocoena phocoena*) around North Sea offshore gas installations. *ICES Journal of Marine Science* **66**(4), 734–745, doi:10.1093/icesjms/fsp035
Tonay, A.M. & Öztürk, A.A. 2012. Historical records of cetacean fishery in the Turkish seas. *Journal of the Black Sea/Mediterranean Environment* **18**(3), 388–399.
Torres Ortiz, S., Stedt, J., Midtiby, H.S., Egemose, H.D. & Wahlberg, M. 2021. Group hunting in harbour porpoises (*Phocoena phocoena*). *Canadian Journal of Zoology* **99**(6), 511–520, doi:10.1139/cjz-2020-0289

Torres-Pereira, A., Araújo, H., Matos, F.L., Bastos-Santos, J., Sá, S., Ferreira, M., Martínez-Cedeira, J., López, A., Sequeira, M., Vingada, J. & Eira, C. 2022. Harbour Porpoise abundance in Portugal over a 5-year period and estimates of potential distribution. *Animals* **12**, 1935, doi:10.3390/ani12151935

Torres-Pereira, A., Araújo, H., Monteiro, S.S., Ferreira, M., Bastos-Santos, J., Sá, S., Nicolau, L., Marçalo, A., Marques, C., Tavares, A.S., De Bonis, M., Covelo, P., Martínez-Cedeira, J., López, A., Sequeira, M., Vingada, J. & Eira, C. 2023a. Assessment of harbour porpoise bycatch along the Portuguese and Galician coast: insights from strandings over two decades. *Animals* **3**, 2632, doi:10.3390/ani13162632

Torres-Pereira, A., Ferreira, M., Eira, C., López, A. & Sequeira, M. 2023b. *Phocoena phocoena* boto. In *Livro Vermelho dos Mamíferos de Portugal Continental*, M.L. Mathias et al. (eds.). Lisboa: FCiências.ID, ICNF.

Tregenza, N.J.C., Berrow, S.D., Hammond, P.S. & Leaper, R. 1997. Harbour porpoise (*Phocoena phocoena* L.) by-catch in set gillnets in the Celtic Sea. *ICES Journal of Marine Science* **54**, 896–904, doi:10.1006/jmsc.1996.0212

Trenkel, V.M., Charrier, G., Lorance, P. & Bravington, M.V. 2022. Close-kin mark-recapture abundance estimation: practical insights and lessons learned. *ICES Journal of Marine Science*, **79**, 413–422, doi:10.1093/icesjms/fsac002

Trippel, E.A., Holy, N.L. & Shepherd, T.D. 2008. Barium sulphate modified fishing gear as a mitigative measure for cetacean incidental mortalities. *Journal of Cetacean Research and Management* **10**, 235–246.

Valdés-Hansen, F. 2004. El problema pesquero con los delfines y su persecución en Galicia (siglos XIII al XX). *Cuadernos de Estudios Gallegos, Tomo LI* **117**, 313–362.

Valdés-Hansen, F. 2009. Pescadores y delfines en el norte de España. Historia de su interacción desde la Edad Media hasta el siglo XX. *Itsas memoria: revista de estudios marítimos del País Vasco* **6**, 629–641.

Van Bressem, M.F., Duignan, P.J., Banyard, A., Barbieri, M., Colegrove, K.M., De Guise, S., Di Guardo, G., Dobson, A., Domingo, M., Fauquier, D., Fernandez, A., Goldstein, T., Grenfell, B., Groch, K.R., Gulland, F., Jensen, B.A., Jepson, P.D., Hall, A., Kuiken, T., Mazzariol, S., Morris, S.E., Nielsen, O., Raga, J.A., Rowles, T.K., Saliki, J., Sierra, E., Stephens, N., Stone, B., Tomo, I., Wang, J., Waltzek, T. & Wellehan, J.F. 2014. Cetacean morbillivirus: current knowledge and future directions. *Viruses* **6**(12), 5145–5181, doi:10.3390/v6125145

Van Bressem, M.F., Van Waerebeek, K. & Raga, J.A. 1999. A review of virus infections of cetaceans and the potential impact of morbilliviruses, poxviruses and papillomaviruses on host population dynamics. *Diseases of Aquatic Organisms* **38**(1), 53–65, doi:10.3354/dao038053

Van Bressem, M.F., Visser, I.K.G., Van de Bildt, M.W.G., Teppema, J.S., Raga, J.A. & Osterhaus, A.D.M.E. 1991. Morbillivirus infection in Mediterranean striped dolphins (*Stenella coeruleoalba*). *Veterinary Record* **129**, 471–472.

van den Heuvel-Greve, M.J., van den Brink, A.M., Kotterman, M.J.J., Kwadijk, C., Geelhoed, S.C.V., Murphy, S., van den Broek, J., Heesterbeek, H., Grone, A. & Ijsseldijk, L.L. 2021. Polluted porpoises: generational transfer of organic contaminants in harbour porpoises from the southern North Sea. *Science of the Total Environment* **796**, 148936, doi:10.1016/j.scitotenv.2021.148936

van den Hurk, Y., Rielly, K. & Buckley, M. 2021. Cetacean exploitation in Roman and medieval London: reconstructing whaling activities by applying zooarchaeological, historical, and biomolecular analysis. *Journal of Archaeological Science: Reports* **36**, 102795, doi:10.1016/j.jasrep.2021.102795

van den Hurk, Y., Spindler, L., McGrath, K. & Speller, C. 2020. Medieval whalers in The Netherlands and Flanders: zooarchaeological analysis of medieval cetacean remains. *Environmental Archaeology* **27**(3), 243–257, doi:10.1080/14614103.2020.1829296

van Elk, C.E., van de Bildt, M.W., van Run, P.R., Bunskoek, P., Meerbeek, J., Foster, G., Osterhaus, A.D.M.E. & Kuiken, T. 2019. Clinical, pathological, and laboratory diagnoses of diseases of harbour porpoises (*Phocoena phocoena*), live stranded on the Dutch and adjacent coasts from 2003 to 2016. *Veterinary Research* **50**, 88, doi:10.1186/s13567-019-0706-3

van Elk, C.E., van de Bildt, M.W., van Run, P.R., de Jong, A., Getu, S., Verjans, G., Osterhaus, A. & Kuiken, T. 2016. Central nervous system disease and genital disease of harbour porpoises (*Phocoena phocoena*) are associated with different herpesviruses. *Veterinary Research* **47**, 28, doi:10.1186/s13567-016-0310-8

van Utrecht, W.L. 1978. Age and growth in *Phocoena phocoena* Linnaeus, 1758 (Cetacea, Odontoceti) from the North Sea. *Bijdragen tot de Dierkunde* **48**(1), 16–28.

Vaz, A.R.C. 2015. *Morfometria craniana do boto (*Phocoena phocoena) *na costa portuguesa.* Masters thesis, Universidade de Aveiro.

Verfuß, U. K., Miller, L. A., Pilz, P. K., & Schnitzler, H. U. 2009. Echolocation by two foraging harbour porpoises (*Phocoena phocoena*). *Journal of Experimental Biology*, **212**(6), 823–834. doi: 10.1242/jeb.022137

Verutes, G. 2022. An Evaluation of International Policies and Local Management Strategies to Reduce Marine Mammal Bycatch. PhD thesis, University of Santiago de Compostela.

Vieira, N., Carvalho, I. & Brito, C. 2009. Occurrence and relative abundance of common dolphins in three sites of the Portuguese shore. Paper SC/61/SM16. Paper presented at the 61st meeting of the Scintific Committee of the International Whaling Commission **16**, 7.

Víkingsson, G.A., Ólafsdóttir, D. & Sigurjónsson, J. 2003. Geographical, and seasonal variation in the diet of harbour porpoises (*Phocoena phocoena*) in Icelandic coastal waters. *NAMMCO Scientific Publications* **5**, 243–270, doi:10.7557/3.2829

Vingada, J. & Eira, C. 2018. Conservation of Cetaceans and Seabirds in Continental Portugal. The LIFE+ MarPro project. Final project report NAT/PT/00038, Portugal.

Vingada, J., Ferreira, M., Marçalo, A., Santos, J., Araújo, H., Oliveira, I., Monteiro, S., Nicolau, L., Gomes, P., Tavares C. & Eira, C. 2011. SAFESEA - Manual de Apoio para a Promoção de uma Pesca Mais Sustentável e de um Mar Seguro para Cetáceos. Programa EEAGrants - EEA Financial Mechanism 2004–2009 (Projecto 0039), Braga, Portugal.

Vinther, M. 1999. Bycatches of harbour porpoises (*Phocoena phocoena* L.) in Danish set-net fisheries. *Journal of Cetacean Research and Management* **1**(2), 123–135.

Vinther, M. & Larsen, F. 2004. Updated estimates of harbour porpoise (*Phocoena phocoena*) bycatch in the Danish North Sea bottom-set gillnet fishery. *Journal of Cetacean Research and Management* **6**(1), 19–24, doi:10.47536/jcrm.v6i1.785

Visser, I.K.G., Van Bressem, M.F., de Swart, R.L., van de Bildt, M.W.G., Vos, H.W., van der Heijden, R.W.J., Saliki, J.T., Orvell, C., Kitching, P., Kuiken, T., Barrett, T. & Osterhaus, A.D.M.E. 1993. Characterization of morbilliviruses isolated from dolphins and porpoises in Europe. *Journal of General Virology* **74**, 631–641, doi:10.1099/0022-1317-74-4-631

Wade, P.R. 1998. Calculating limits to the allowable human-caused mortality of cetaceans and pinnipeds. *Marine Mammal Science* **14**, 1–37, doi:10.1111/j.1748-7692.1998.tb00688.x

Weijs, L., Dirtu, A.C., Das, K., Gheorghe, A., Reijnders, P.J.H., Neels, H., Blust, R. & Covaci, A. 2009. Inter-species differences for polychlorinated biphenyls and polybrominated diphenyl ethers in marine top predators from the Southern North Sea: Part 1. Accumulation patterns in harbour seals and harbour porpoises. *Environmental Pollution* **157**(2), 437–444, doi:10.1016/j.envpol.2008.09.024

Westgate, A.J., Muir, D.C., Gaskin, D.E. & Kingsley, M.C. 1997. Concentrations and accumulation patterns of organochlorine contaminants in the blubber of harbour porpoises, *Phocoena phocoena*, from the coast of Newfoundland, the Gulf of St Lawrence and the Bay of Fundy/Gulf of Maine. *Environmental Pollution* **95**(1), 105–119, doi:10.1016/S0269-7491(96)00073-5

Williams, R. 2021. *Persistent Organic Pollutants: Assessing the threat to cetaceans*. PhD thesis, Brunel University London, United Kingdom.

Williams, R.S., Brownlow, A., Baillie, A., Barber, J.L., Barnett, J., Davison, N.J., Deaville, R., ten Doeschate, M., Penrose, R., Perkins, M., Williams, R., Jepson, P.D., Lyashevska, O. & Murphy, S. 2023. Evaluation of a marine mammal status and trends contaminants indicator for European waters. *Science of the Total Environment* **866**, 161301, doi:10.1016/j.scitotenv.2022.161301

Williams, R.S., Curnick, D.J., Brownlow, A., Barber, J.L., Barnett, J., Davison, N.J., Deaville, R., ten Doeschate, M., Perkins, M., Jepson, P.D. & Jobling S. 2021. Polychlorinated biphenyls are associated with reduced testes weights in harbour porpoises (*Phocoena phocoena*). *Environment International* **150**, 106303.

Williamson, G.M. 2011. Common porpoise from the stomach of a Greenland shark. *Journal of the Fisheries Research Board of Canada* **20**(4), 1085–1086, doi:10.1139/f63-073

Williamson, L.D., Scott, B.E., Laxton, M.R., Bachl, F.E., Illin, J.B., Brookes, K.L. & Thompson, P.M. 2022. Spatiotemporal variation in harbor porpoise distribution and foraging across a landscape of fear. *Marine Mammal Science* **38**, 42–57, doi:10.1111/mms.12839

Wise, L., Silva, A., Ferreira, M., Silva, M.A. & Sequeira, M. 2007. Interactions between small cetaceans and the purse-seine fishery in Western Portuguese waters. *Scientia Marina* **71**, 405–412, doi:10.3989/scimar.2007.71n2405

Wisniewska, D.M., Johnson, M., Teilmann, J., Rojano-Doñate, L., Shearer, J., Sveegaard, S., Miller, L.A., Siebert, U. & Madsen, P.T. 2016. Ultra-high foraging rates of harbour porpoises make them vulnerable to anthropogenic disturbance. *Current Biology* **26**, 1441–1446, doi:10.1016/j.cub.2016.03.069

Wisniewska, D.M., Johnson, M., Teilmann, J., Siebert, U., Galaatius, A., Dietz, R. & Madsen, P.T. 2018. High rates of vessel noise disrupt foraging in wild harbour porpoises (*Phocoena phocoena*). *Proceedings of the Royal Society B: Biological Sciences* **285**, 20172314, doi:10.1098/rspb.2017.2314

Yasui, W.Y. & Gaskin, D.E. 1986. Energy budget of a small cetacean, the harbour porpoise, *Phocoena phocoena* (L.). *Ophelia* **25**(3), 183–197, doi:10.1080/00785326.1986.10429749

Zegers, B.N., Mets, A., Van Bommel, R., Minkenberg, C., Hamers, T., Kamstra, J.H., Pierce, G.J. & Boon, J.P. 2005. Levels of hexabromocyclododecane in harbor porpoises and common dolphins from Western European seas, with evidence for stereoisomer-specific biotransformation by cytochrome P450. *Environmental Science and Technology* **39**, 2095–2100, doi:10.1021/es049209t

Zerbini, A.N., Ward, E., Kinas, P.G., Engel, M. & Andriolo, A. 2011. A Bayesian assessment of the conservation status of humpback whales (*Megaptera novaeangliae*) in the western South Atlantic Ocean. *Journal of Cetacean Research and Management* **3**, 131–144, doi:10.47536/jcrm.vi3.320

Oceanography and Marine Biology: An Annual Review, 2024, **62**, 120-162

Taylor & Francis

INTENTIONAL KILLINGS AND CULLING OF SMALL CETACEANS DUE TO PERCEIVED COMPETITION WITH FISHERIES IN THE MEDITERRANEAN SEA AND NORTHEAST ATLANTIC BETWEEN THE EIGHTEENTH AND TWENTIETH CENTURIES

MARIE A.C. PETITGUYOT[1*], GIOVANNI BEARZI[2,3,4], YOURI VAN DEN HURK[5], MARISA TEJEDOR FUENTES[6], AND GRAHAM J. PIERCE[1]

**mariepetitguyot@palaeome.org*
[1]Instituto de Investigaciones Marinas, Centro Superior de Investigaciones Científicas (IIM-CSIC), Vigo, Pontevedra, Spain
[2]Dolphin Biology and Conservation, Cordenons, Italy
[3]OceanCare, Wädenswil, Switzerland
[4]CNR National Research Council, ISMAR Institute of Marine Sciences, Venice, Italy
[5]Department of Archaeology and Cultural History, NTNU Vitenskapsmuseet, Norwegian University of Science and Technology, Trondheim, Norway
[6]Canary Islands Cetacean Stranding Network, Department of Biological Research, Playa Blanca, Lanzarote

Abstract Between the eighteenth and twentieth centuries, many small cetaceans (particularly dolphins) were killed in large numbers off the coasts of Europe and North Africa, due to perceived competition with fisheries. Dolphins were long considered as pests, responsible for reduced fish catches and damage to fishing gear. This conflict was particularly acute between the second half of the nineteenth century and the first half of the twentieth century, when governments from various countries encouraged and subsidised the extermination of small cetaceans. While the precise number of killings and the effects these had on the past and current status of cetacean populations are largely unknown, historical records and osteological collections can help us appreciate their magnitude. Here we first summarise information available regarding conflict between fishers and small cetaceans in various countries around the Mediterranean and Northeast Atlantic. Then, based on an extensive analysis of historical literature that included landing and bounty reports, and on a review of osteological collections, we provide new and extensive information on conflicts in France and Algeria. For France and Algeria, we (1) provide information regarding the context and chronology of killings and culling campaigns, (2) describe the methods used to deter and kill cetaceans, (3) identify the species affected, and (4) attempt to quantify the number of individuals killed. Our results suggest that tens of thousands of small cetaceans were killed in France and Algeria, primarily between the 1880s and the 1930s. Total mortality of small cetaceans due to human conflict in the Mediterranean Sea and Northeast Atlantic is certainly much higher, as that includes the killings in all other countries where similar conflicts existed. Such a high mortality likely had an important negative impact on dolphin population trajectories, particularly in the Mediterranean Sea. In the second half of the twentieth century, intentional killings largely stopped, but small cetaceans became exposed to a variety of other anthropogenic threats. Reconstructing past mortality can help

 DOI: 10.1201/9781003477518-2

us understand current population trends, viability and potential for recovery, as well as help us to set meaningful baselines for conservation.

Keywords: Cetaceans; Historical Ecology; Human-Wildlife Conflicts; Marine Environmental History; Past Abundance and Distribution; Osteological Collections

Introduction and overview

Killings of small cetaceans: from resources to competitors

The relationship between humans and 'small cetaceans' (hereafter referring to small dolphins of the family Delphinidae and to Phocoenidae) has been evolving through history. Once considered as useful resources, small cetaceans were used as food (primarily when freshly dead or live-stranded) or to make weapons and tools with their bones during prehistoric times. For instance, there is zoo-archaeological evidence of human use in various locations around the Mediterranean and Northeast Atlantic, dating back to the Paleolithic (Pascual Benito et al. 2019), the Lower Mesolithic (Lightfoot et al. 2011, Trantalidou 2008), the Mesolithic-Neolithic Transition (Mannino et al. 2015) and the Late Neolithic (Cauliez et al. 2004). Small cetaceans continued to be used during the Bronze Age (Pascual Benito et al. 2019), and possibly in the Iron Age (Bernal-Casasola et al. 2016), during the Antiquity (Papadopoulos & Paspalas 1999, Pascual Benito et al. 2019) and in pre-Roman times (Bernal-Casasola 2018).

During the Medieval period, all around Europe, small cetaceans were hunted or otherwise used for their meat and blubber. Clear evidence of consumption was reported from the Mediterranean and Atlantic waters of Spain (Garrido 2014, Alcover Cateura 2022, Valdés-Hansen 2004), along the French Atlantic coast (Guizard 2018, Musset 1964, Berthelot 1840), Portugal (Brito 2011), England (Dobney et al. 2007, van den Hurk et al. 2021, 2023), Scotland (Szabo 2008, van den Hurk & McGrath 2021) and the Faroe Islands (Szabo 2008). Interest in the consumption of small cetaceans gradually decreased during the sixteenth century (Fichou & Levasseur 2004, Garrido 2014). In Europe, a change in the perception that humans have of small cetaceans occurred towards the end of the sixteenth century, when these animals were no longer seen predominantly as resources, but rather as competitors for fishery catches. Written records testify to the animosity towards dolphins, which were blamed for reducing fish catches and damaging fishing gear. For instance, a Papal Decree was issued in 1587 "anathematising this vermin", after concerns were raised in France regarding the effect of dolphins on fisheries (Smith 1995). Additionally, in 1612, Pope Paul V recommended to the bishop of Marseille (French Mediterranean) to show the Holy Cross to dolphins at sea in order to scare them away (Faget 2009). In Spain, the "pilot whales conjuration" was an agreement signed in 1624 between Asturian fishers to hire a clergyman of the Holy Office of the Inquisition to chase away cetaceans (Valdés-Hansen 2004, 2009). From then onwards, historical records indicate an intensification of the perceived conflict between small cetaceans and fishers. The conflict was particularly acute between the second half of the nineteenth century and the first half of the twentieth century, when governments from various countries around the Mediterranean Sea and Northeast Atlantic Ocean encouraged and subsidised the extermination of small cetaceans (Bearzi et al. 2004, Fichou & Levasseur 2004, Valdés-Hansen 2004, 2009, Faget 2009, Garrido 2014, Meliadò et al. 2020, Sokou et al. 2022; see Figure 1).

Intentional killings in the Mediterranean Sea and Northeast Atlantic

There is abundant evidence in the literature of small cetaceans being deliberately killed due to the perceived competition with fisheries in the Mediterranean and Northeast Atlantic. A brief overview follows for the countries that have published information:

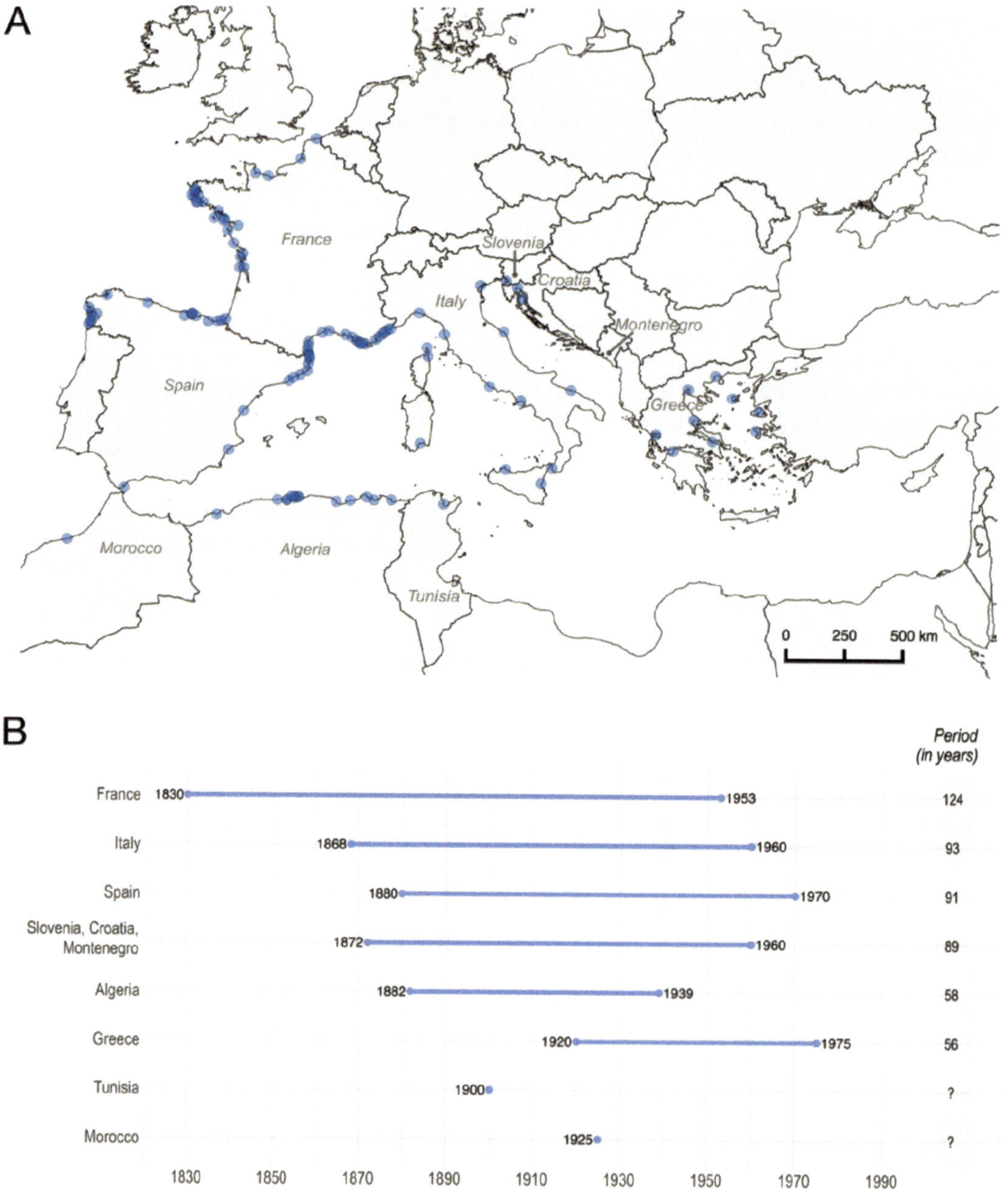

Figure 1 (A) Spatial distribution of the Mediterranean Sea and Northeast Atlantic locations considered in this study, where historical material was available about the conflict between small cetaceans and fisheries (mainly between the eighteenth and the mid-twentieth centuries). (B) Duration of the reported main period of historical conflict between fisheries and small cetaceans around the Mediterranean and Northeast Atlantic. The duration of the conflicts is approximate and is based on the earliest and latest dates in historical documents and published studies, mentioning organised efforts to kill small cetaceans. Information in this figure is based on the following studies – France: Fichou and Levasseur (2004), Faget (2009), this study; Spain: Valdés-Hansen (2004, 2009), Garrido (2014); Slovenia, Croatia, and Montenegro: Bearzi et al. (2004); Italy: Bearzi et al. (2004), Meliadò et al. (2020); Greece: Sokou et al. (2022); Algeria, Tunisia and Morocco: this study.

Italy

In Italy, requests to receive monetary rewards in exchange for dolphin killings occurred as early as in 1868 (in the Gulf of Genoa), and formal bounties started being issued in 1872 (in Trieste; Bearzi et al. 2004). Subsequently, systematic cullings were implemented all over the country until 1942 (Meliadò et al. 2020). During 1928–1938, bounties paid could reach 100 Lire (the equivalent of approximately 45–90 Euros) if a pregnant female was killed (Bearzi et al. 2004). Using historical records of landings of animals and bounties paid for the killings from 13 Compartimenti Marittimi (Marine Departments) in Italy, Meliadò et al. (2020) reported a total of 6785 dolphins killed between 1927 and 1937. The highest total catches were reported in Ancona (Adriatic Sea) and Palermo (Tyrrhenian Sea), each with more than a thousand animals killed during the 11-year period. The authors also report around 40 animals (i.e., 2.7 ± 0.5 dolphins per year) killed in the Chioggia Compartimento (close to Venice) during the period 1946–1960. Meliadò et al. (2020) also noted that additional mortality of wounded animals that died later was not taken into account in these figures. The authors suggest that the most common species killed were common bottlenose dolphins *Tursiops truncatus* and common dolphins *Delphinus delphis*, and possibly striped dolphins *Stenella coeruleoalba* to a lesser extent. Cetaceans became legally protected under Italian law in 1979 (Bearzi et al. 2004).

Slovenia, Croatia, and Montenegro ("former Yugoslavia")

In Croatia, bounties were first offered in 1872 in Rijeka (Bearzi et al. 2004). In the Kvarner Gulf (Croatia), Meliadò et al. (2020) reported that about 222 dolphins were killed between 1914 and 1925, 844 dolphins were killed between 1927 and 1937, and 600 dolphins were killed after the Second World War. Bearzi et al. (2004) reported that 335 dolphins were killed in former Yugoslavia (exact location unknown) between 1933 and 1935. The main culling campaign started in Croatia in 1949 (Bearzi et al. 2004). The authors also reported that 788 dolphins were killed between 1955 and 1960 in former Yugoslavia. The campaign was planned to last until 1965, but there was no record of bounties paid after 1960, possibly because of a decrease in the amount of money received for each dolphin killed in 1959. Common bottlenose dolphins and common dolphins were the main species targeted by these campaigns (Bearzi et al. 2004). While culling campaigns in the eastern part of the northern Adriatic stopped around 1960, killings continued until cetaceans became legally protected under the Croatian law in 1995 (Bearzi et al. 2004). The figures above should be considered minimum estimates and provide only a partial indication of the actual number of animals killed across a century.

Greece

In Greece, conflicts between dolphins and fisheries took place during the early phases of fishery development, from the early twentieth century until the mid-1970s, increasing in intensity from the interwar period (1920–1940) onwards (Sokou et al. 2022). The conflict was mainly reported in coastal areas of the northern and central Aegean Sea, and in the eastern Ionian Sea (Sokou et al. 2022), and it involved common dolphins and common bottlenose dolphins (Gonzalvo et al. 2015, Sokou et al. 2022). Dolphins were blamed for net destruction and reduced fish catches, notably in the purse seine fishery (Sokou et al. 2022). In response, fishers called for the extermination of dolphins (Maynou et al. 2011, Gonzalvo et al. 2015, Foskolos et al. 2020, Sokou et al. 2022). Requests were made to allow the use of harpoons, guns and explosives to kill these animals (Sokou et al. 2022). Culling campaigns were encouraged by the Greek government, including through bounties (e.g., 600 Drachmes, approximately 1.76 Euro, per dolphin killed in 1961; Gonzalvo et al. 2015). Turkish experts in cetacean hunting techniques reportedly travelled to Greece to share their know-how (Tonay & Öztürk 2012, Sokou et al. 2022). Fishers also requested monetary compensation for damaged fishing gear, and there were proposals to make use of the skin and fat of killed animals (Sokou et al. 2022).

Tunisia and Morocco

A few documents from Tunisia and Morocco suggest that in the early twentieth century, small cetaceans could be killed due to conflict with fisheries. A newspaper article from 1900 claimed that small cetaceans were responsible for the decline of fish stocks in the Gulf of Tunis (Tunisia; Anonymous 1900g). A newspaper article from 1950 mentioned that on the 10th of August 1900, the Chamber of Commerce encouraged all fishers in Tunisia to kill small cetaceans to receive a 25 Francs reward (Anonymous 1950). In Morocco, a report from 1925 mentioned that the option of offering bounties to kill small cetaceans was being considered (Pérard 1925). Another Moroccan newspaper article from 1934 mentioned the killing of two animals measuring 2.5 m close to Casablanca: one was harpooned and the other was shot (Anonymous 1934c).

Turkey

In the Turkish portions of the Black Sea and Mediterranean Sea, common bottlenose dolphins, common dolphins and harbour porpoises *Phocoena phocoena* were extensively exploited for oil and meat for more than 2300 years, until they became legally protected in 1983 (Tonay & Öztürk 2012). Birkun et al. (2014) suggested that another reason for killing cetaceans could have been conflict with fisheries. The killings peaked in the 1930s, and later in the 1950s and 1970s. Commercial exploitation was particularly intense in the Black Sea, but records indicate that cetaceans were also extensively killed in the Aegean and Marmara Seas. For instance, Tonay and Öztürk (2012) reported that 84.9 tonnes of dolphins were caught in the Turkish Aegean Sea in 1969, which would roughly correspond to 1700 animals (for calculation details, see Tonay & Öztürk 2012). While it was estimated that a few million cetaceans were killed in the Black Sea, there are no such estimates available for the Turkish Mediterranean Sea (Tonay & Öztürk 2012).

Spain

In the Mediterranean Sea off Spain, dolphin killings attributed to conflict with fisheries occurred between the second half of the nineteenth century and the mid-twentieth century (Garrido 2014). In the Costa Brava region (part of Catalonia), conflict could be traced back to 1880 and lasted until the 1970s (Garrido 2014). Three periods of acute conflict were identified: 1903–1905, 1915 and 1960, which correspond to years when catches of small pelagic fishes decreased drastically. Fishers from this region blamed dolphins for low catches and gear damage (affecting gillnets, trawls and purse seines). Common bottlenose dolphins were identified by fishers from Costa Brava as being the main culprit, followed by common and striped dolphins (Garrido 2014). In retaliation, fishers killed dolphins by using harpoons ("dofinera"), explosives and fishing nets modified with attached blades or wooden sticks (Garrido 2014). Hundreds of dolphins could be caught during a single fishing session, and some would be eaten by the fishing community (Garrido 2014). During the period of intense conflict, Spanish fishers operating in Mediterranean waters repeatedly asked the government to take measures to eradicate dolphins. However, these requests were refused, and the killing of dolphins was prohibited in 1905 (Garrido 2014). One reason for such denial may be that fishers from the Atlantic coast of Spain, particularly the ones in Cantabria using the "manjua" (a sort of purse seine), perceived dolphins as being essential to help them detect shoals of fish (Garrido 2014). However, other historical documents indicate that conflict between fishers and small cetaceans also took place in Cantabria (Valdés-Hansen 2004). In 1913, the Spanish government finally issued bounties of 2.5 Pesetas (approximately 0.012 Euros) per dolphin killed in Costa Brava, and these bounties were later increased to 7.5 Pesetas (approximately 0.042 Euros). However, this measure was rapidly abandoned. A few representatives of the Spanish fishing industry also considered commercialising the killing of dolphins to exploit their blubber and meat, but these initiatives were never implemented (Valdés-Hansen 2004, 2009, Garrido 2014), with the exception of a few unsuccessful trials in waters of Andalusia and the Strait of Gibraltar (Aguilar 2013). Garrido (2014) also mentions that conflict took place in the area of Valencia. While we could find no reports of killings in other regions of the Spanish Mediterranean, we cannot exclude the possibility that such conflict took place in other areas.

Along the Atlantic coast of Spain, the first evidence of dolphin killings dates back to the early seventeenth century in the region of Asturias, with a peak in killings occurring between the eighteenth and mid-twentieth centuries (Valdés-Hansen 2004). Although killings took place along the whole Spanish Atlantic coast, including the regions of Basque Country, Asturias and Cantabria, the most affected region was Galicia. Here, the main species killed was the common bottlenose dolphin, but common dolphins and harbour porpoises were also targeted (Valdés-Hansen 2004). Occasionally, other cetaceans could have been killed, such as striped dolphins and Risso's dolphins *Grampus griseus*, as well as pilot whales *Globicephala* sp. and killer whales *Orcinus orca* (Valdés-Hansen 2004). Fishers killed dolphins using harpoons and guns, and fishing nets were used to trap the animals. In the nineteenth century, fishers in the town of Pontevedra (Galicia) organised common bottlenose dolphin "corridas", which involved trapping dolphins in the town's estuary with large nets and then killing them. These corridas were popular events that took place during festivals (Valdés-Hansen 2004). While fishers operating along the Atlantic coast asked the Spanish government to take measures for the extermination of dolphins, and these requests were usually refused, a few unsuccessful trials were conducted in 1911 by the government, which sent military vessels equipped with torpedoes and machine guns to kill dolphins in unspecified locations. Similar trials took place in Galicia in 1927 (Valdés-Hansen 2004). In addition, the commercial exploitation of small cetaceans was carried out in Galician waters on small whaling boats for a short period around 1925 (Aguilar 2013). Apart from the dolphin "corridas", along the Atlantic coast of Spain there seemed to have been no formal culling campaigns due to perceived competition with fisheries, killings being conducted largely based on the initiative of individual fishers.

Portugal

In mainland Portugal, small cetaceans were killed opportunistically since at least the thirteenth century, with particularly large numbers being killed between the late nineteenth and twentieth centuries (Brito & Vieira 2010, Reiner & Brito 2012). Small cetaceans killed opportunistically (e.g., by using harpoons) by Portuguese fishers could still be found in fish markets around the country in the late 1970s (Teixeira 1979, Brito & Vieira 2010). However, we could find no reports of formal extermination campaigns. Valdés-Hansen (2004) reported that some fishers from Galicia (Spain), who had issues with common bottlenose dolphins in their regions, did not have their fishing gear damaged by dolphins when they fished in Portuguese waters.

British Isles

There seem to be no historical records of culling campaigns of small cetaceans in the British Isles. Of note, Sir Sidney Frederic Harmer, director of the British Natural History Museum between 1919 and 1927, reported that—to his knowledge—there was no conflict between fishers and cetaceans in England (Harmer 1926).

Iceland and Norway

Although they are not considered under our definition of 'small cetaceans', it is interesting to note that in Iceland, hundreds of killer whales were killed in the mid-1950s by the U.S. Navy using machine guns and rockets, after Icelandic fishers complained about these animals damaging fishing gear (Jourdain et al. 2019, Samarra & Esteban 2022). In Norway, fishers complained that killer whales were interfering with the recovery of the Norwegian spring-spawning stock of Atlantic herring (*Clupea harengus*) in the late 1960s, and that may have contributed to the development of killer whale hunting in Lofoten (Øien 1988, Plagányi & Butterworth 2009, Samarra & Esteban 2022).

France

Finally, conflicts were documented on both the Atlantic and Mediterranean shores of France. Fichou and Levasseur (2004) reported dolphin killings in the Atlantic region of Brittany, between the end of the eighteenth and the mid-twentieth centuries, while Faget (2009) described killings in the Mediterranean area of Marseille between the nineteenth and twentieth centuries. These reports are suggestive of intensive culling campaigns that resulted in the killing of many animals over a long period of time. Detailed information about the conflict that took place in France is one focus of the current review (Case study 1).

Scope of the review

Because the present status of small cetaceans may be linked with the levels of mortality they faced in the past (e.g., Bearzi et al. 2004), reconstructing the magnitude of historical killings can help us understand current population demography and trends, including the viability and potential for recovery of populations, as well as set meaningful baselines for conservation.

The practice of killing small cetaceans due to perceived competition with fisheries is well documented in Mediterranean countries like Italy and Croatia, and somewhat consistent information comes from other European areas, particularly Catalonia and the Atlantic coast of Spain, as well as Brittany and the Mediterranean southeast of France. However, there seems to be scant published information on small cetacean killings in several other countries, including North African countries and the Middle East. Filling these gaps and knowing whether, for example, lack of information reflects lack of killings or, alternatively, insufficient research on this subject, would help elucidate the actual magnitude of historical killings.

This study aims to complement the available information with additional investigations intended to assess the magnitude of killings that occurred between the eighteenth and twentieth centuries, with a focus on two particular case studies: France and Algeria. Based on an extensive analysis of historical information that included landings and bounty reports, as well as of osteological specimens from museum collections, we (1) provide information on the context and chronology of killings and culling campaigns, (2) describe the methods used to kill and harm small cetaceans, (3) identify the species affected (when information on morphology, behaviour and feeding ecology is available) and (4) attempt to quantify the number of animals killed.

Records of killings of larger delphinids, such as killer whales, false killer whales and pilot whales, were included, but we did not make a systematic search for these species.

Materials and methods

Search for historical records

Historical sources were accessed through the following open access digitalised repositories: the Biodiversity Heritage Library (https://www.biodiversitylibrary.org), the Bibliothèque Nationale de France (https://gallica.bnf.fr), the Collection Musée de la carte postale de Baud (https://www.cartolis.org), the Ville de Paris/Bibliothèque Forney (https://bibliotheques-specialisees.paris.fr) and the Cnum – Conservatoire numérique des Arts et Métiers (http://cnum.cnam.fr). As the main countries of interest to this review (France and Algeria) used the French language between the eighteenth and twentieth centuries, the search mostly, but not exclusively, targeted sources in French. The main keywords used to search for information were "dauphin", "marsouin" and "béluga" (i.e., "dolphin", "porpoise" and "beluga", respectively, in English). The words "marsouin" and "béluga" were commonly used in France and Algeria to refer to dolphins and did not necessarily refer to members of the family Phocoenidae or to the beluga *Delphinapterus leucas*. A list of all keywords used to search for information is available in Supplementary Material S1 (Supplementary Table 1.1). More than 5000 historical documents (including photographs, illustrations, newspaper articles, governmental and scientific reports or publications, and books) dating from the late sixteenth century (i.e., 1551) to the second half of the twentieth century were accessed through open access digitalised repositories. Of these, in total, 311 documents were considered relevant to this work, with the majority of these dating from the nineteenth and twentieth centuries.

Osteological collections

To better understand which species were killed during the period of conflict between the eighteenth and the twentieth centuries, information on Mediterranean small cetaceans held in osteological collections of museums and universities in France and Monaco was collated through public online databases when possible, or via direct contact with curators and conservators, and used to complement the historical documents. The reason for focusing on osteological specimens of small cetaceans from the French Mediterranean, and not French Atlantic waters, relates to the use of the French common name "marsouin" (porpoise in English) in historical documents referring to small cetaceans in the Mediterranean. Harbour porpoises are currently not present in French Mediterranean waters, and the presence of the species in osteological collections from the French Mediterranean could help us understand whether the species occurred and was killed in the Mediterranean during the eighteenth to twentieth centuries.

A total of 18 institutions (museums, universities and research institutes) provided information on their collections of Delphinidae and Phocoenidae. Of these, 12 collections included at least one specimen that came from the Mediterranean, while six did not have any Mediterranean specimens in their collection (see Figure 2 for the list of institutions). The 12 institutions provided information on 298 specimens. Among these, seven specimens were identified as coming from outside of the Mediterranean (from Africa (exact provenance unknown), Antarctica, the Arctic, North Sea, or the South or Northeast Atlantic) and were thus subsequently removed from the analyses. We also removed from further analyses any specimens that had a known date of acquisition after 1940 ($n = 57$), after the end of the period of conflict between fishers and small cetaceans in France, and specimens that could be identified as duplicates of individuals that were already included ($n = 8$). Among the remaining 226 specimens, 106 came from the Mediterranean and 120 had an unknown provenance. Among the Mediterranean specimens, 63 were labelled as coming from the French Mediterranean and Monaco. The rest of the specimens came from Algeria ($n = 20$), Spain ($n = 5$) and unidentified Mediterranean locations ($n = 18$).

The information collected on each specimen consisted of the species (and, if it had been recently re-identified, the species it was originally thought to be), museum code, collector, type of specimen (e.g., entire skeleton and skull only), method of conservation (e.g., formalin and dry mounted), date and place of provenance, cause of death, sex, age class and length of the specimen. Not all the above information was available for every specimen.

As the species could not be confirmed for some of the specimens, the certainty of species identification was assessed based on the information provided by each institution. Identification was judged reliable when carried out in recent years by experts trained in cetacean osteological identification, and it was judged uncertain when institutions indicated that it was not done recently (e.g., identification dated from the nineteenth century) or when no information on certainty of identification was provided. When possible, photographs and measurements were obtained to confirm that the identification was correct. Four institutions (i.e., Université de Montpellier, Muséum Départemental du Var, Muséum d'Histoire Naturelle de Toulouse, Muséum d'Histoire Naturelle de Nîmes) provided photographs for 49 of the specimens whose identification was not reviewed recently. Institutions were asked to provide photographs showing the underside of skulls, to allow identification based on the absence or presence of palatal grooves (to discriminate between common dolphins and other species, since the former is the only relevant species with this feature).

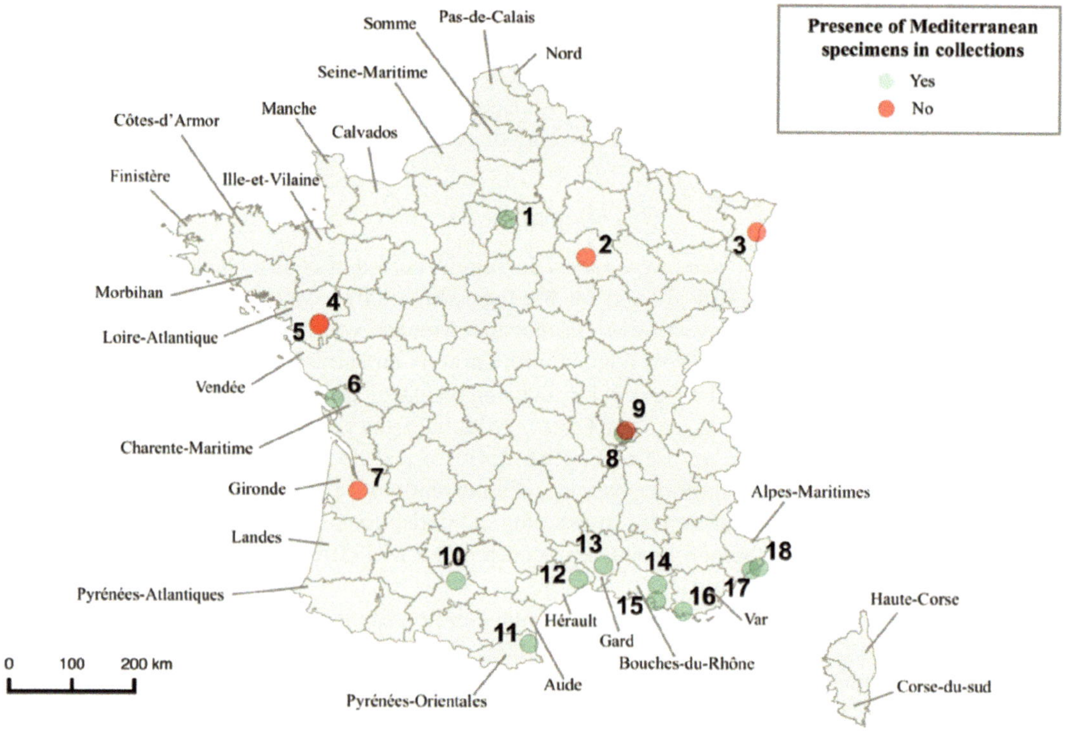

Figure 2 Map of France, with names of coastal "départements". Numbers correspond to the location of institutions that provided information on their odontocete osteological collections. The presence of Mediterranean specimens in their collection is indicated by green dots and the absence is indicated by red dots. In green: (1) Muséum National d'Histoire Naturelle de Paris; (6) Muséum d'Histoire Naturelle de La Rochelle; (8) Musée des Confluences de Lyon; (10) Muséum d'Histoire Naturelle de Toulouse; (11) Muséum d'Histoire Naturelle de Perpignan; (12) Université de Montpellier; (13) Muséum d'Histoire Naturelle de Nîmes; (14) Muséum d'Histoire Naturelle d'Aix en Provence; (15) Muséum d'Histoire Naturelle de Marseille; (16) Muséum Départemental du Var; (17) Muséum d'Histoire Naturelle de Nice; and (18) Institut Océanographique de Monaco. In red: (2) Muséum d'Histoire Naturelle de Troyes; (3) Musée Zoologique de Strasbourg; (4) Université de Nantes (collection de zoologie); (5) Muséum d'Histoire Naturelle de Nantes; (7) Muséum d'Histoire Naturelle de Bordeaux; and (9) Université Claude Bernard Lyon 1 (collection de zoologie).

The Université de Montpellier and the Muséum d'Histoire Naturelle de Nîmes also provided measurements on five different locations of the mandibles for 23 out of the 49 photographed specimens. These measurements were compared with those of 383 modern specimens of 21 known species within the database "Osteological Reference for Cetaceans in Archaeology-Manual (ORCA – Manual)" (van den Hurk 2020). Using principal component analysis and linear discriminant analysis, a species identification was attributed when measurements from the 23 specimens fell within clusters belonging to particular species. Details on the methodology are available in Supplementary Material S1 (Supplementary Figure 1.1 and Supplementary Table 1.2).

As the identifications could not be confirmed by examining the specimens in person, and photographs sometimes did not allow us to assess the presence or absence of palatal grooves, all species identifications based on the analyses of photographs and mandible measurements ($n = 49$) were classified as uncertain. When species identification based on photographs and on mandible measurements did not match, the identification based on mandible measurements was retained as the final species identification. If mandible measurements did not support identification to the species level, specimens were classified only to the family or higher level (e.g., Delphinidae, unknown cetacean and non-cetacean).

Case study 1: France

Context and perceptions

Our review of historical information from France suggests that conflict between humans and small cetaceans encompassed the whole period surveyed, from the sixteenth century to the twentieth century. Records mentioning conflicts were scarce between the sixteenth century and the late eighteenth century. From then onwards, records became abundant. These documents suggest that the intensity of conflict increased starting from the late eighteenth century, reaching a peak between the 1880s and the 1930s, and the conflict ending in the 1950s. Along the Mediterranean and Atlantic coasts of France, but particularly in the Mediterranean, small cetaceans were blamed for scaring fish away, eating large amounts of fish, therefore reducing catches, and damaging the nets when attempting to take fish stuck in meshes (referred to as "depredation" in Anonymous (1903a), but see Bearzi and Reeves (2022) where the use of this word is questioned) and when they accidentally get entangled in the nets (bycatch, which often also resulted in injury or death) (Perrier 1889, Gourret 1894, Léotard 1894, Caffarena 1887, Le Gall 1904, Legendre 1926a, 1929, Fichou & Levasseur 2004, Faget 2009). Although attitudes towards small cetaceans were generally hostile, there apparently was an exception: some fishers in the French Basque country considered small cetaceans to be essential to localise shoals of sardines and had developed fishing techniques to use small cetaceans to detect and catch fish (Legendre 1926a, Anonymous 1895a, 1903b). The main fisheries affected by the conflict were targeting shoaling fish such as sardines, anchovies and, to a lesser extent, mackerels (e.g., Legendre 1926a, 1929, Fichou & Levasseur 2004, Faget 2009). In the case of damage to fishing gear, the fishers often needed days or weeks to repair damaged fishing gear, which reportedly cost up to several hundreds of Francs (Legendre 1926a). Note: the value of the old Franc has fluctuated over time. Conversions into Euros are available for the years 1901–1960 from the currency converter tool of the Institut National de la Statistique et des Études Économiques, taking the inflation rate into account; the conversions from old Francs to Euros provided in this chapter are approximate. The value of the Franc varied between 4.27 and 0.02 Euros over the period 1901–1960 (Institut National de la Statistique et des Études Économiques 2022). As a result of the damage caused by small cetaceans, some fishers were reportedly reluctant to deploy their nets and even considered stopping fishing for commercially important species (Legendre 1926a, Anonymous 1894a, 1898a). Figure 3 shows two examples of fishing nets allegedly damaged by small cetaceans in the French Atlantic and Mediterranean.

Fishers' complaints regarding the negative impact of small cetaceans were amplified by local newspapers through sensational articles. Media propaganda claimed that there was an "open war" between fishers and small cetaceans (Anonymous 1893a) and that the "brave" fishers had to be "protected" (Anonymous 1931a). Articles depicted small cetaceans as vile animals that had to be "annihilated" (Anonymous 1898b). Small cetaceans were considered as pests that should be exterminated in the same way as wolves, bears, foxes, moles, mice and cockchafers (Anonymous 1894b). For instance, one newspaper article referred to these animals as "monsters", "sea pirates", "voracious enemies", "nuisance species" and "brigands" and encouraged the readers to "destroy" and "get rid of" them (Anonymous 1894b). This kind of loaded language was also common among naturalists and scientists. For instance, the President of the Naturalists Association from Nice and the Alpes-Maritimes, Eugène Caziot, wrote in the Association's bulletin:

> The ancients have given us tales for pleasure, marvellous stories that Cuvier was naive enough to suppose true, at least some of them. Dolphins are, in reality, stupid, brutal, voracious animals, having only

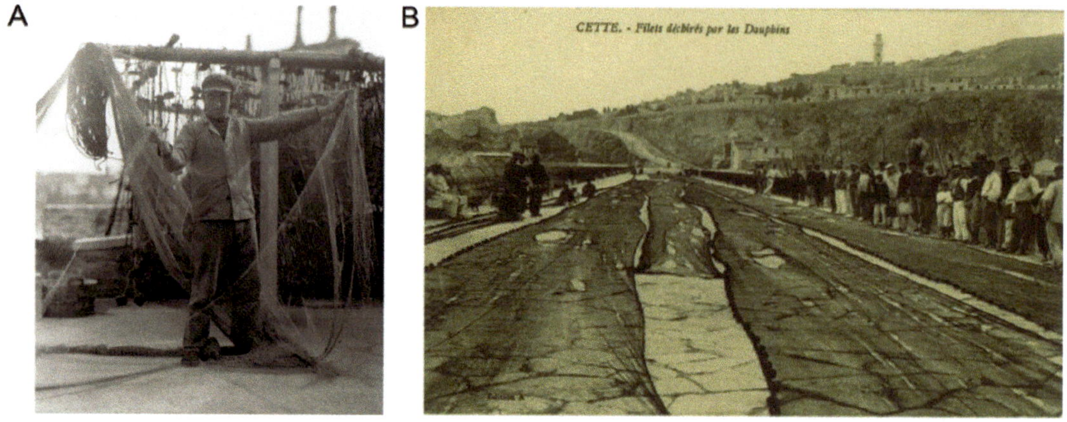

Figure 3 Damage to fishing nets in (A) Brittany (Atlantic) and (B) city of Sète (Mediterranean Sea). Legends indicate: (A) "Audierne – The damages made by the béluga", photograph by François Kollar, entitled "Pêcheurs. Les dégâts du béluga", published between 1931 and 1934; (B) "Cette – Nets damaged by dolphins", published around 1910. Photograph (A) was provided by Ville de Paris / Bibliothèque Forney and is visible at: https://bibliotheques-specialisees.paris.fr; photograph (B) was provided by Collection Musée de la carte postale – Baud and is visible at: https://www.cartolis.org.

> enough intelligence to devour their prey and reproduce their species. They are unfortunately all too often seen in the vicinity of Nice.
>
> —*Caziot (1913)*

Caziot (1913) reused, almost verbatim, the description of dolphins given by Pierre Larousse in his Dictionary of 1870 (Larousse 1870), which highlights the long-standing animosity towards these animals. Both fishers and journalists called on the French government to act to remedy the situation (Anonymous 1893a, Legendre 1926a). Various solutions were suggested to exterminate cetaceans, and military workers from the French Navy, as well as fishers and other seafarers, were encouraged to kill these animals and, in the case of fishers and seafarers, to exploit them commercially, e.g., by marketing their skin to make leather or their meat as food (Anonymous 1893a, 1903a, Legendre 1926a).

Methods to kill small cetaceans

Mobilisation of the fishing community and the military

To encourage the organised killing of small cetaceans by the fishing community, the French government, counties' representatives and Prud'homies (i.e., collectives of fishers from the French Mediterranean; see for instance Decugis 2015) offered monetary rewards, also called bounties, for each animal caught. Details about the information found in historical material regarding these bounties are available in Supplementary Table 2.1. The available information suggests that bounties were issued as early as in the 1830s and that the bounty system stopped in 1927. There was mention of bounties paid in all the Mediterranean "départements" (counties) of France (see Figure 2 for a map of these "départements"), except for the "départements" of Aude and Corse-du-sud. While some accounts referred to France as a whole, we could find no specific mention of bounties in French Atlantic "départements". However, previous work by Fichou and Levasseur (2004) indicates that bounties were also issued in the Atlantic "départements" (e.g., in the region of Brittany). Bounties were awarded to fishers or other seafarers on the condition that the head of the killed "porpoise" (in this case, small cetacean) was shown to the administration in charge of delivering the bounties (Anonymous 1901a). To avoid frauds (e.g., reusing the same head several times to obtain more money), the head was destroyed when handed to the administration (Anonymous 1901a). The value of the bounties varied markedly across years and areas, from a maximum of 35 Francs (approximately 149 Euros) per "porpoise" head in Marseille in 1899 (Anonymous 1900a) to a minimum of 2.50 Francs (approximately 10 Euros) in 1910 (location not provided; Legendre 1926a). However, because catching small cetaceans was no easy task and demanded extra efforts, numerous fishers did not systematically engage in the hunting of small cetaceans, and many complained about the situation not improving. In response to these complaints, the government occasionally offered monetary compensation for damaged nets. For instance, in 1868, 833 Francs were distributed among 13 boat owners (Anonymous 1889).

Fishers were not the only ones involved in small cetacean killings. After the fishing community called for the help of the government, the French Navy deployed several types of military vessels along the Mediterranean and Atlantic coasts. The first request for the use of military vessels dates back to 1872 (Anonymous 1872). The most commonly used vessels were torpedo boats ("torpilleurs", see Figure 4). Other types included counter torpedo boats ("contre-torpilleurs"), fishery control boats ("garde-pêches"), tug boats ("remorqueurs"), cannon ships ("cannonières") and small warships ("avisos").

Military vessels patrolled the French coast in search of small cetaceans. Once these were detected, officers on board would use various methods to kill the animals. Torpedo boats were equipped with cannons of 37 mm and 75 mm (Legendre 1926a), as well as cavalry lances (see Figure 6 and the next section for more details). Supplementary Table 2.2 provides information on the vessels used to kill small cetaceans in France. The first account of military vessels sent by the government to patrol the Atlantic coast of France dates back to 1903, in the region of Brittany, whereas in the Mediterranean, the first account dates back to 1883, in the region of Provence-Alpes-Côte d'Azur. The last accounts are from 1953 in the Atlantic and 1932 in the Mediterranean. Over a period of at least 71 years (1883–1953), at least 27 military vessels (10 in the Atlantic, 16 in the Mediterranean

Figure 4 "The boat Chasseur 115, in charge of killing 'bélugas' (i.e., most likely dolphins) in the waters where sardines are caught". Photograph taken on the 16th of September 1932, Belle-Île (Morbihan, Brittany). Agence de presse Mondial Photo-presse. (Source: Bibliothèque Nationale de France (https://gallica.bnf.fr).)

and 1 in an unknown location) were deployed along the shores of France. Deployment of torpedo boats had particularly high costs. For instance, more than 180,000 Francs (approximately 205,000 Euros) were spent in 1923 for the deployment of the boats MISTRAL and "321" in the French Atlantic and for the boat "339" in the French Mediterranean (Anonymous 1923a).

The scientific community played an important role in the quest for the eradication of small cetaceans. Scientists developed or conceived some of the extermination methods and disseminated information in scientific essays to encourage killings and commercial exploitation (Legendre 1926a, 1929). Boats belonging to scientific institutes were occasionally used to hunt small cetaceans or to test some of the killing methods. Boats employed for these purposes included LE ROLAND in 1901, owned by the laboratory Arago in the Pyrénées Orientales (Anonymous 1901b), and the NÉREIS in 1926, owned by the laboratory of Concarneau (Legendre 1929).

Killing methods and frequency of use

A variety of methods was deployed by the fishing community and the military to kill small cetaceans. Attempts were made to identify and assess efficient and cost-effective ways of eradicating these animals, for instance, through surveys in fishing ports, as done in 1901 by the Minister of the French Navy (Anonymous 1903d), or by creating expert commissions, as done in 1923 (Pérard 1925) and 1925 (Legendre 1929). Often, these enquiries concluded that the methods employed did not produce satisfactory results, that some methods were too expensive or dangerous and that the best way of eradicating small cetaceans was relying on the private initiative of fishers, rather than involving the military or other government entities. A system of bounties was often regarded as the best solution (Anonymous 1903d), but it was also recommended to keep searching for new methods of killing (Legendre 1926a). Supplementary Table 2.3 lists some of the methods employed to kill small cetaceans during the years 1865–1953.

A common method included a "herding and beaching" technique (Legendre 1926a, Anonymous 1873), where fishers would spot a group of animals, herd them towards shallow waters and finally trap them, force them to beach and kill them. Harpoons and fishing nets (used on purpose to catch small cetaceans) were also considered as effective ways of killing the animals (Legendre 1926a; Figure 5A). In addition to fishing nets being used with the purpose of intentionally trapping small cetaceans and then killing them (for instance with harpoons), small cetaceans could also be caught incidentally in fishing gear during normal fishing operations. When live animals were bycaught, fishers would sometimes free them in order to avoid gear damage (in the past, fishing gear was made of relatively fragile fabric such as hemp, cotton or linen). When live animals could not be released, the fishers would pull them onboard to kill them (Garau 1909).

When military vessels were used, cetaceans were shot with cannons, torpedoes and rifles, as well as killed using explosives (Legendre 1926a, Caffarena 1887; Figure 6A). Guns were also distributed by the government directly to the fishers to encourage them to kill more cetaceans (Anonymous 1903e; Figure 6B). Finally, cavalry lances were used by both military personnel and fishers (Anonymous 1930a; Figure 6A).

Besides the methods mentioned above, other techniques developed by scientists, engineers and other workers proved to be difficult to implement due to either high cost, dangerousness or ineffectiveness (Supplementary Table 2.3). Such methods were usually abandoned after a few trials. Some examples are given below.

Bellot's needles This method ("aiguilles de Bellot") was developed in 1892 by Mr. Bellot, a factory worker from Douarnenez (Brittany). Originally used to kill foxes in Provence (Gourret 1894), it consisted of two long needles (8–10 cm long) made of steel, which were attached together centrally by a piece of rubber and tied at the ends using catgut. The needles were inserted into fish bait (sardines or other small fish), which were then thrown into the sea, with the intention that small cetaceans would eat them. Once ingested, the catgut pieces at both ends of the system dissolved, and the two

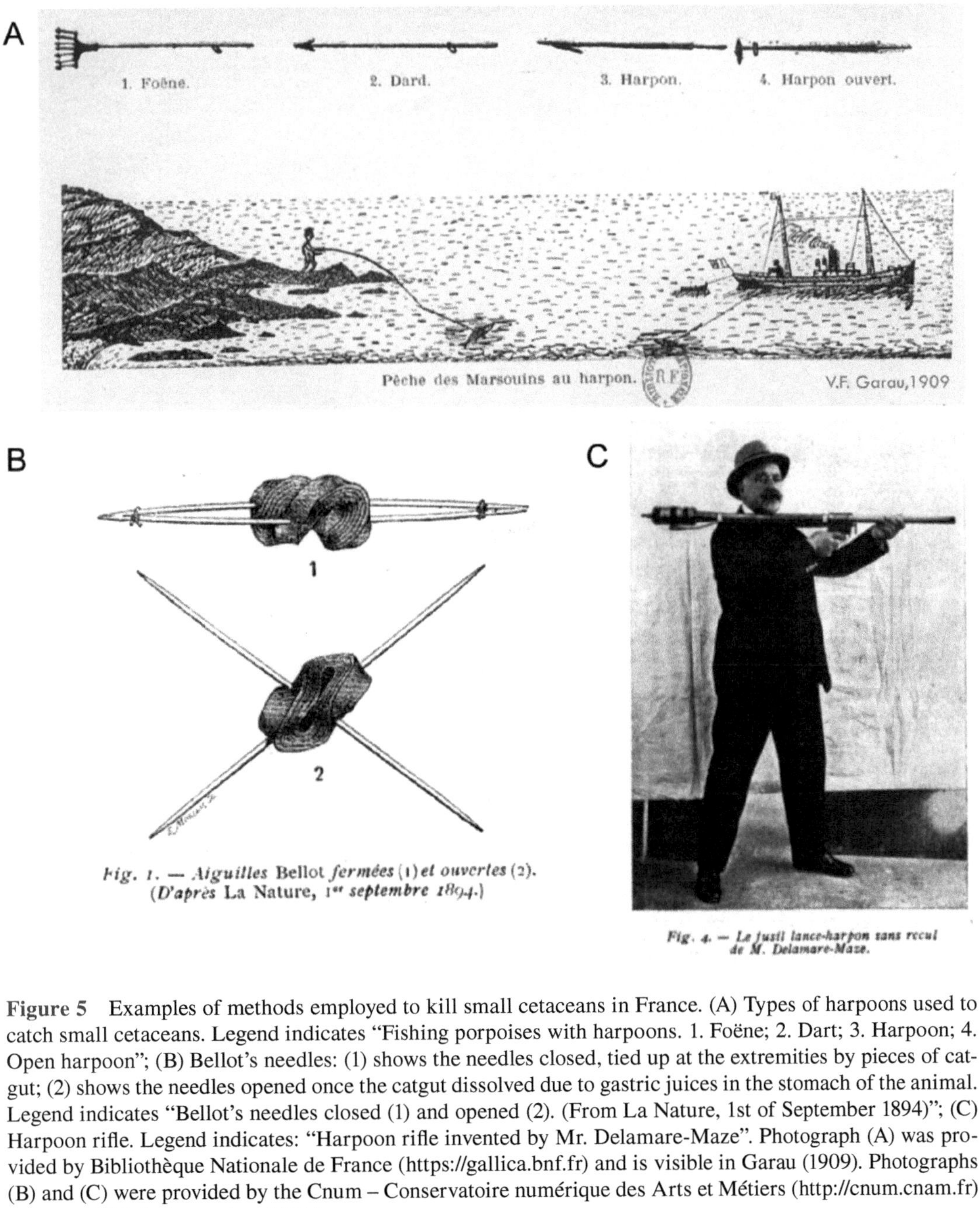

Figure 5 Examples of methods employed to kill small cetaceans in France. (A) Types of harpoons used to catch small cetaceans. Legend indicates "Fishing porpoises with harpoons. 1. Foëne; 2. Dart; 3. Harpoon; 4. Open harpoon"; (B) Bellot's needles: (1) shows the needles closed, tied up at the extremities by pieces of catgut; (2) shows the needles opened once the catgut dissolved due to gastric juices in the stomach of the animal. Legend indicates "Bellot's needles closed (1) and opened (2). (From La Nature, 1st of September 1894)"; (C) Harpoon rifle. Legend indicates: "Harpoon rifle invented by Mr. Delamare-Maze". Photograph (A) was provided by Bibliothèque Nationale de France (https://gallica.bnf.fr) and is visible in Garau (1909). Photographs (B) and (C) were provided by the Cnum – Conservatoire numérique des Arts et Métiers (http://cnum.cnam.fr) and are visible in Legendre (1926a).

needles would pivot so as to be at 90 degrees to each other, thus perforating the wall of the digestive tract and resulting in the death of the animal (Figure 5B). Approximately 35,000 Bellot's needles were distributed in fishing ports along the French Atlantic and Mediterranean shores: 15,000 in 1893, 10,000 in 1894 and 10,000 in 1915. This method apparently gave mixed results: reportedly, it worked in some regions (e.g., in the cities of Auray and Noirmoutier) and had no results in others (e.g., in Nantes and Royan). While the suggested explanation for these mixed results was that small

A

B

Figure 6 Further examples of methods employed to kill small cetaceans in France. (A) The illustration shows a variety of methods to kill small cetaceans onboard warships: handguns, rifles, cannons and cavalry lances. One of the animals caught with a lance is tied to a rope. The legend indicates: "Torpedo boats hunting porpoises" (Anonymous 1903c). (B) The illustration shows fishers hunting for dolphins in the region of Brittany (Atlantic). A warship with seamen shooting is visible in the background. The legend indicates: "Hunting dolphins in the bay of Douarnenez" (Anonymous 1903e). Both illustrations were provided by Bibliothèque Nationale de France (https://gallica.bnf.fr).

cetaceans were reluctant to ingest the dead fish used as baits (Léotard 1894, Anonymous 1895a), it is unclear how that would explain the differences observed among different areas.

The Ocellus method The Ocellus method was developed by engineer Mr. Ocellus and tested between 1893 and 1895 in La Ciotat and the Gulf of Marseille (Mediterranean). The system consisted of an electric cable connected to a battery installed onboard a ship, attached to a long (around 400 m) "sardinal" (a net to catch sardines), deployed in waters up to 100 m deep, with dynamite attached to the net every 15 m (Ropers 1906). The net was deployed in areas where shoals of sardines were detected. When a group of small cetaceans approached the net to feed on sardines, an electric current resulted in the dynamite explosion. The effectiveness of this method is unclear: there are reports mentioning deaths of between 1 and 80 animals (Anonymous 1893a, 1895a). Eventually, the technique was judged inefficient, expensive and complicated and thus abandoned (Legendre 1926a).

Harpoon rifles A harpoon rifle ("fusil lance-harpon sans recul" in French), which is a small version of the whaling harpoon cannon, was developed around 1926 (Legendre 1926a; see Figure 5C). A few trials were made, including some in the area between Concarneau and Camaret (Brittany), onboard the fishery control vessels PÉTREL and CALEBAS, as well as the vessel NÉRÉIS owned by the Maritime Laboratory of Concarneau. After these trials, harpoon rifles were abandoned (Legendre 1929).

Other methods Several other methods were deployed, but their frequency of use or success was not reported. For instance, Anonymous (1908a) reported that priests were being paid to pray, specifically to take away small cetaceans' taste for sardines. Noisy rockets were shot to scare the small cetaceans away (Pérard 1925). Other methods were proposed but never implemented. One of the most bizarre, developed in 1896 in Marseille (Bouches-du-Rhône), aimed to attract small cetaceans by producing a bright underwater light, which was supposed to "hypnotise" the animals and allow for their capture (Anonymous 1895a, Legendre 1926a). There were requests to use explosive balls, gas projectiles and poison (e.g., poisoned tubes and poisoned sardines), as well as proposals to inoculate small cetaceans with rabies (Legendre 1926a). In 1890, it was suggested to use poisoned baits, but this method was judged dangerous and not implemented (Legendre 1926a). In 1891, the Minister of the Navy ordered explosive cartridges to the pyrotechnic factory of Toulon, specifically to kill small cetaceans (Anonymous 1900b). In 1914, a deputy of Morbihan (Brittany) requested permission to put explosive mines in places frequently used by small cetaceans, but this request was also not implemented due to the associated costs and dangers (Legendre 1926a). There was a proposal to build small speedboats and to fly seaplanes and airships dedicated to hunting small cetaceans (Anonymous 1930a,b, 1922a, Legendre 1926a). In the area of Brest (Brittany), it was proposed to deploy the seaplane GOLIATH to kill small cetaceans, with steel arrows, bombs and bullets, but it was never used as fishers were afraid of being harmed during the operations (Fichou & Levasseur 2004).

Regional differences

Although a variety of methods was implemented along the coasts of France, there were differences in frequency among regions. For instance, in 1894, 10,000 Bellot's needles were distributed to fishers in Normandy (northern France), Brittany (north-west), Vendée, Charentes-Maritimes and Gironde (Bay of Biscay), Bouches-du-Rhône and Var (Mediterranean). While Bellot's needles were used by most fishers in these locations, some fishers from the Pyrénées-Atlantiques (French Basque Country) categorically refused to use the needles, arguing that small cetaceans were essential to help them detect shoals of fishes and they had no intention of killing them (Anonymous 1895a). Nonetheless, small cetaceans were still killed in the French Basque Country. For instance, Figure 7G shows fishers from Biarritz (a city in the French Basque Country) posing with a dead harpooned common dolphin, described as a "sardine school destroyer". Some of the killing methods, such as the Ocellus method, were only implemented in the Mediterranean. Moreover, the deployment of military vessels in the Atlantic seems to have occurred more often in the region of Brittany, whereas the entire French Mediterranean coast was patrolled by these vessels in search of small cetaceans to kill.

Effectiveness of killing methods

Government articles, official reports from dedicated commissions and interview surveys emphasised the need to commercialise the exploitation of small cetaceans. Authors often referred to other countries, where the commercial exploitation of cetacean blubber, skin and meat was profitable (e.g., in England and along the Black Sea; Gourret 1894, Le Gall 1904), when calling for increased killings in France. However, such commercialisation never occurred, with the exception of a few unsuccessful trials to market canned cetacean meat prepared in the canning factories of Concarneau, Quiberon and Guilvinec (Finistère and Morbihan, in the region of Brittany; Fichou & Levasseur 2004). Another proposal involved the development of sport hunting for small cetaceans (Pérard 1925).

Reports describing the inefficiency of some methods in certain areas raise the question of whether animals were in fact killed. For instance, René Legendre (1929), the deputy director of the Maritime Laboratory in Concarneau (Finistère, Brittany), suggested that the fishery control vessel PÉTREL had never succeeded in killing small cetaceans and that harpoon rifles were never used after the initial trials in Concarneau (Legendre 1929). One document from 1931 also suggested that the Navy was aware of the ineffectiveness of military vessels, but that the boat CHASSEUR 115 was nevertheless required to patrol the shores of Brittany in order to reassure fishers (Anonymous 1931b). However, other documents suggested that the deployment of military vessels was effective. For instance, Fichou and Levasseur (2004) mentioned a document reporting that the boats CAPITAINE-MEHL and "132" arrived in Port-Vendres in April 1907 to "protect" the sardine fisheries in the Mediterranean, and after patrolling the area between Cap Cerbère and La Nouvelle 277 times, the commander believed that their presence had protected fishers' nets against cetaceans' attacks. While reports tend to be unclear and non-quantitative, at least 27 vessels were deployed along the shores of France for at least 71 years. Considering the presumably significant costs associated with keeping these military vessels engaged in fishery support over several decades, it could be argued that small cetaceans were being killed, although in unknown numbers.

Use of small cetaceans after killing

Many historical records attest to the practice of consuming the meat of small cetaceans in France, particularly during the Middle Ages (Musset 1964) and the Renaissance (Van Beneden & Gervais 1880). Between the tenth and fourteenth centuries, there was an intense fishery in Normandy, with small cetacean meat sold in markets around the country and exported to other countries as well (Fischer 1881). It should be noted that in the Middle Ages, cetaceans were not perceived as mammals but rather as fish, and as such, they were consumed by religious communities during periods of fasting (Fichou & Levasseur 2004, Guizard 2018, van den Hurk 2020, van den Hurk et al. 2021). In French monasteries, the meat was consumed until at least the seventeenth century (Dubuc 1968). The oil extracted from the blubber was also used in monasteries for lighting (Musset 1964, Berthelot 1840, Fischer 1881). The meat of small cetaceans was eaten at the tables of the royals until at least the seventeenth century (Clavel 2001). This taste for dolphin and harbour porpoise meat decreased from the seventeenth century onwards (Fichou & Levasseur 2004). Zoology essays from the late eighteenth century and mid-nineteenth century suggest that small cetaceans were still consumed occasionally at the time (Fichou & Levasseur 2004). For instance, Bonnaterre (1789) wrote about the flesh, blubber and intestines of *Delphinus tursio* (most likely the common bottlenose dolphin) being eaten (no location was provided). However, some accounts from the early nineteenth century questioned the use of this "tough" and "mediocre" meat (Berthelot 1840), with some authors expressing disgust towards dolphin and porpoise meat (Lesson 1838).

Even though the meat of small cetaceans appears to have fallen out of favour, it was still consumed during the period of most acute conflict between fishers and cetaceans (i.e., between the early nineteenth and the mid-twentieth centuries), when it even regained some popularity (Legendre 1926a) and was sold in markets, for example, in the Paris market, where it was sold for 10 Francs kg^{-1} in 1926 (Legendre 1926a). During this period, several authors stated that harbour porpoise or dolphin meat was tasty (e.g., Anonymous 1903d, Legendre 1926a) although others reported that it was sold rarely, as customers did not like it (Anonymous 1929). Some documents refer to sausages made with the meat of small cetaceans ("saucisson de marsouin"), as well as to ventral and dorsal muscles being consumed (Anonymous 1903f). In 1934, small cetacean meat ("exquis suprême de marsouin aux herbes marines") was served at the spring banquet of the Société Nationale d'Acclimatation in Paris (Anonymous 1934a). In the historical material reviewed, there is no mention of small cetaceans being consumed for religious reasons during the period of acute conflict with fisheries.

The carcasses of small cetaceans were also used as agriculture fertilisers, for example, in La Rochelle around 1894 (Anonymous 1894c, Fichou & Levasseur 2004). Some carcasses, however, were simply piled up on the quays of ports, where they were left to rot (Fichou & Levasseur 2004).

Species of small cetaceans killed

Knowing which species of small cetaceans were killed by fishers and navy personnel during the conflict is critical to understand if and how the populations suffered from direct removals. However, species determination is mostly hampered by the lack of photographs in the documents accessed. Thus, historical records (text and other iconography) and osteological collections were used to identify the species, when possible.

Information from historical records

Species identification relying on photographs One way of identifying species reported in historical documents is by using photographs of appropriate quality. Photographs associated with the words "marsouin", "béluga", "dauphin", "souffleur", "squale", "taupe" and "cétacé" helped us to understand the use of these words. Our search yielded a total of 12 photographs, all published in the early twentieth century, which permitted a reliable identification of the species caught or killed by fishers. Of these, seven were of common dolphins (four in the Bay of Biscay (Figure 7F, G, J, K), two in the English Channel (Figure 7E, I) and one in the Mediterranean (Figure 7H)). Figure 7H, taken from Pourcel (1910), shows a pregnant common dolphin and her male foetus caught by a fisherman from Carqueiranne (close to Toulon, Var) in 1910. This animal most likely corresponds to the specimen 1-M-5-DELDEL-(1)/1910, kept by the Muséum Départemental du Var. The text associated with the photograph identifies the animal as *Delphinus delphis mediterraneus*, which was stated to be rare in the region of Toulon (Pourcel 1910). However, Pourcel (1910) differentiated between *Delphinus delphis mediterraneus* and *Delphinus delphis*, which may imply that the latter was not rare in that region. Two of the 12 photographs portrayed harbour porpoises: one caught in the Bay of Biscay (Figure 7L) and one caught in the English Channel (Figure 7D). The search also revealed that species other than cetaceans were also occasionally associated with the name "porpoise" and "dolphin". For instance, three photographs showed different species of sharks. Figure 7A and C shows sharks reported as "porpoises", and 7B shows a shark reported as "dolphin".

Identification of species based on common names Historical records often use common names in French to refer to small cetaceans in general, which prevents the determination of the species involved in the conflict with fishers. Two of the most common terms were "marsouin" and "béluga", which translate as porpoise and beluga in English, although these names did not necessarily refer to members of the family Phocoenidae or to the beluga *Delphinapterus leucas*. Normally, the common names in French referred to dolphins, particularly to small members of the family Delphinidae. The word "marsouin" was used throughout France, whereas "béluga" was mainly used along the Atlantic coast of France and more specifically in Brittany. Other commonly used words included "dauphin", "souffleur", "squale" and (more rarely) "taupe", which can be translated in English as dolphin, blower, shark and mole, respectively. Below, we discuss which species correspond to these common names.

Marsouin The word "*marsouin*" (porpoise) was used to refer to dolphins, without distinguishing between species (e.g., Rolland 1877). To refer to actual harbour porpoises, French fishers used "poursille" or "poursil", or various other names depending on the region (see Supplementary Table 2.4 for a list of common names used to refer to Delphinidae and Phocoenidae between 1798 and 1900). In Brittany, fishers reportedly perceived that the "poursil" was not involved in conflict with fishing activities (Legendre 1926a). Nonetheless, Figure 7 shows photographs of harbour porpoises killed in the English Channel in 1910 (D) and in the Bay of Biscay in 1926 (L), which is indicative of the species sometimes being among those killed in French Atlantic waters. Figure 7 also shows photographs published around 1910 (F) and in the 1930s (J), with legends referring to *D. delphis* as "marsouins", and photographs published in the 1930s (A) and around 1905 (C), with legends referring to sharks as "marsouins". This highlights the variety of species that the term encompassed at that time.

Physical descriptions of "marsouins" in historical documents often did not match those of *P. phocoena*, but rather those of dolphins. For instance, some reports mentioned "marsouins" as having pointy beaks (e.g., Caffarena 1887). In the sixteenth century, Belon (1551) had already noted that dolphins were wrongly called "marsouins" in France. Nonetheless, some of the descriptions, mainly those from zoology textbooks, actually seem to correspond to harbour porpoises. It should be noted that authors during that period tended to quote work from others without verifying the information themselves, which could result in an author describing a species without ever having seen it.

While the presence of *P. phocoena* is established in the Northeast Atlantic, reference to "porpoises" in the vast majority of historical documents from the French Mediterranean Sea raises the question of whether the species was indeed present in the French Mediterranean between the eighteenth and the twentieth centuries. Genetic analyses have shown that harbour porpoises were present in the Mediterranean basin thousands of years ago, but have likely disappeared from it during the second half of the Holocene period (ca. 5.5 ka), after conditions became unsuitable for the species at the end of the African Humid Period (14.8–6 ka) and the Mediterranean Sapropel S1 period (ca. 9.5–6.5 ka) (Fontaine et al. 2014, Fontaine 2016). Nowadays, harbour porpoises are largely absent from the Mediterranean, apart from a few stranding records in the western basin (e.g., Cabezón et al. 2004), presumably vagrant individuals from the Atlantic population, and free-ranging animals in the northern Aegean Sea (Cucknell et al. 2016) which presumably derive from the Black Sea population.

Scientists from the late nineteenth century with experience in cetology did not mention the presence of harbour porpoises in the French Mediterranean. For example, Fischer (1881) did not consider harbour porpoises as being present in the French Mediterranean. According to Van Beneden and Gervais (1880):

> The harbour porpoise does not exist, at least to my knowledge, in the Mediterranean. I did not see any taken on the coasts of France during my long stay in Montpellier, and my excursions along the coast, from Cerbère and Collioure to Menton, have not yet brought me across one. It does not exist either in the Italian collections that I visited, particularly in those of Bologna, Pisa, and Naples, and I do not find it either among the Delphinids that were sent to me from the coasts of Algeria.

Figure 7 Photographs of small cetaceans and sharks caught in France between the 1900s and the 1930s. (A) A shark, probably caught in the "département" of Calvados. The description indicates: "Vierville-sur-Mer – A porpoise. This 80 kg glutton was caught by its tail in a herring net. Shipped to Paris and cut into thin slices, it was a treat for Parisians under a variety of names: sole, tunas, etc.", published in the 1930s; (B) A shark, probably caught in the "département" of Pas-de-Calais. The description indicates: "Étaples-sur-Mer (Pas-de-Calais) – Dolphin, taken offshore Étaples", published around 1907; (C) Small sharks, probably caught in the "département" of Somme. The description indicates: "Cayeux-sur-Mer. – boat bringing back Porpoises", published around 1905; (D) A harbour porpoise, probably caught in the "département" of Calvados. The description indicates: "Saint-Aubin-sur-Mer – A great catch", published around 1910; (E) A common dolphin, probably caught in the "département" of Manche. The description indicates: "Le Grand-Vey – A good fishing trip – capture of a Dolphin", published in the 1930s; (F) Two common dolphins, probably caught in the "département" of Loire-Atlantique. The description indicates: "Piriac – Porpoise hunting", published around 1910; (G) A common dolphin probably caught in the "département" of Pyrénées-Atlantiques. The description indicates: "Biarritz – seamen from Biarritz who just harpooned a destroyer of sardines shoals", published

(Continued)

(Continued)

around 1910; (H) A common dolphin probably caught in the "département" of Var. The description indicates: "*Delphinus delphis L.*, Var. *Mediterraneus* Loche, female, length 2.30 metres. Her male foetus, measuring 30 centimetres", published in 1910; (I) A common dolphin probably caught in the "département" of Somme. The description indicates: "Cayeux-sur-Mer – Catch of a dolphin weighting 200 kg", published around 1910; (J) A common dolphin's head, from an animal probably caught in the "département" of Loire-Atlantique. The description indicates: "Piriac – A porpoise head", published in the 1930s; (K) A common dolphin caught in the "département" of Finistère. The description indicates: "A 2.15 metres long dolphin harpooned by tuna fishers and brought to the laboratory in Concarneau", published in 1926; (L) A harbour porpoise caught in the "département" of Finistère. The description indicates: "A 1.80 metres long porpoise brought to Concarneau", published in 1926. Photographs (A)–(G), (I) and (J) were provided by the Collection Musée de la carte postale – Baud and are visible at: https://www.cartolis.org. Photographs (K) and (L) were provided by the Cnum – Conservatoire numérique des Arts et Métiers (http://cnum.cnam.fr) and are visible in Legendre (1926b). Photograph (H) was provided by Bibliothèque Nationale de France (https://gallica.bnf.fr) and is visible in Pourcel (1910).

Béluga The word "béluga" was used in the region of Brittany to refer to the small cetaceans involved in conflict with fishers targeting small fish (mainly sardines). René Legendre, deputy director of the Maritime Laboratory in Concarneau (Finistère), tried to elucidate the identity of the Breton fishers' "béluga", which was evidently not the actual beluga, *Delphinapterus leucas*.

Legendre (1926a, 1929) suggested that the term was introduced by Breton fishers from Douarnenez (Finistère). Some Breton fishers believed that these "bélugas" could actually be Risso's dolphin *Grampus griseus*, a delphinid that tends to whiten with age. However, Legendre doubted that the species called "béluga" was the Risso's dolphin, as most of the cetaceans killed by fishers (mainly tuna fishers) and brought back to his laboratory were common dolphins. Over the course of his career at the Maritime Laboratory, Legendre (1926b) received hundreds of killed common dolphins, in addition to a few harbour porpoises and only one common bottlenose dolphin. Therefore, the author was convinced that the species called "béluga" were in fact common dolphins and common bottlenose dolphins, and possibly sharks as well. He based his conclusions on observations that the common dolphin was the species most killed by fishers, that a common bottlenose dolphin caught in Audierne (Finistère) had pieces of nets in its stomach and that sharks he had dissected in his laboratory had sardines (without their heads, suggesting they had been taken from a net) in the stomach. In addition, he noted that some fishers reported that the "béluga" turned on its back to chew on the nets, which, according to him, could refer to shark behaviour. Legendre excluded harbour porpoises, since he believed that they were rare in the Atlantic (being mainly present in the English Channel and the North Sea) and that Breton fishers called them "poursil" and not "béluga". He also said that harbour porpoises were "innocents" and too small to be confused with the "béluga". He also excluded Risso's dolphins, considering that this species was rarely stranded or caught and that stomachs of the few specimens that were dissected mainly contained octopuses and "pebbles", very few cod bones and no sardines (which were said to be the main prey of the "béluga") (Legendre 1926a).

Souffleur The word "souffleur" and its variants (soufflur, souflur) were mainly used to refer to common bottlenose dolphins (Supplementary Table 2.4). However, "souffleur" was sometimes used to refer to other members of the family Delphinidae (Bouvier 1891) or even to cetaceans generally (Companyo 1841). "Souffleur" could be used to refer to common dolphins and occasionally to harbour porpoises, sperm whales *Physeter macrocephalus*, pilot whales and Risso's dolphins (Supplementary Table 2.4).

Dauphin This word was principally used to refer to the species common dolphins and, to a lesser extent, to common bottlenose dolphins. The terms "nésarnak" (and its variants nesarnak or nasar nack) and "souffleur" were more often used to refer to common bottlenose dolphins (see Supplementary Table 2.4). A few documents use the word "dauphin" to refer to other members of the family Delphinidae (e.g., harbour porpoises and killer whales, in Companyo 1841).

Squale Some documents mention "squales" as being responsible for damaging fishing gear and removing fish from nets. In modern French, the term "squale" refers to the shark. A newspaper article reported dolphins ("marsouins" in the text) taking fish from nets (Anonymous 1894a) and then used both terms "cetaceans" and "squales" to refer to the same dolphins. Legendre (1929) wrote about fishers potentially confusing cetaceans and "squales" (used here to refer to sharks). Photographs in Figure 7A–C clearly show sharks, and their associated legends refer to them as being "marsouins". A newspaper article reported that "these were not the typical 'marsouins à bec de cane' (porpoises with a long beak), but rather sharks, with a rounded head, a mouth placed ventrally, and a typical way of turning on their back to feed" (Anonymous 1903c). Bounties of 10 or 25 Francs were offered in the Gard "département" in 1900 to capture sharks and "lamies" (a word that currently refers to sharks of the genus *Lamna*) that destroyed nets (Anonymous 1900a).

Taupe The word "taupe" means mole in English. This term was used to refer to common bottlenose dolphins and harbour porpoises (Supplementary Table 2.4), possibly because of the dark grey-brown colour of the mole, which may resemble the colour of both of these cetacean species. In contemporary French, the term "taupe de mer" refers to sharks of the genus *Lamna*.

Identification of species based on scientific names Scientific names of cetacean species have varied greatly over the centuries, sometimes making it difficult to know which species is being described, especially when the reports lack photographs or illustrations. In addition, some of the reports may have been written by individuals with limited knowledge of cetaceans, leading to the dissemination of potentially incorrect information. While the identification of the cetacean species involved in conflict with fishers is often challenging, two species were clearly identified by several authors as competing with fishers in French waters, namely, the common dolphin and the common bottlenose dolphin. Where a different scientific name was used in the past, we refer to the scientific name as given in the cited reference but also mention the contemporary common name.

In the late nineteenth century, common dolphins were believed to be a common species in both the Atlantic and Mediterranean waters of France (e.g., Fischer 1881, Bouvier 1891). Some authors considered the species as being the most abundant cetacean in French waters, particularly in the Mediterranean (Flower 1879, van Beneden 1889). Gervais (1864) differentiated common dolphins (*Delphinus delphis*) from striped dolphins (*Delphinus tethyos*) based on the presence of palatal grooves in the former, and noted that unlike common dolphins, striped dolphins were rare in the French Mediterranean. He also noted that common bottlenose dolphins (*Tursiops tursio*) were far less common than common dolphins in the same area (Gervais 1864). Some reports regarded common dolphins as the main species responsible for conflict with fishers. For instance, Van Beneden and Gervais (1880) wrote that "fishermen have always feared the common dolphin because of the damage it does to their nets; hence it is the only species that was first well known". Caziot (1913) reported that "*Delphinus delphis*, L. is only too well known to Mediterranean tuna fishermen for the great damage it does to their nets called madragues". Madragues were fixed nets acting as traps that were used to fish tunas around the eastern Mediterranean coast of France until the beginning of the twentieth century (Farrugio 2012).

Bouvier (1891) reported that common bottlenose dolphins *Tursiops tursio* (Fabricius) were well known for causing damage to fishing nets around the coasts of France. According to the author, the killing of common bottlenose dolphins gave rise to celebrations in the city of Nice (Alpes-Maritimes). People adorned the killed dolphins with flowers and marched with their catch triumphantly through the streets, stopping in front of the homes of rich people, who could offer monetary compensation for the loss of fish and the damage to the nets caused by these dolphins (Bouvier 1891).

The harbour porpoise *Phocoena communis* was the only species identified as not being involved in the conflict with fishers in Brittany (Legendre 1926a,b). Van Beneden (1889) mentioned that a few harbour porpoises were killed in French Atlantic waters and subsequently donated to museums, but it is not known why these animals were killed. In addition to harbour porpoises, there are a few records from museum collections of other cetacean species, such as striped dolphins, Risso's dolphins, pilot whales and killer whales, killed in the Atlantic and Mediterranean waters of France, but again it is not known why these animals were killed, and there was no indication of these other species being involved in conflict with fishers in the region at that time (Van Beneden 1889).

Identification of species based on morphology, behaviour and feeding ecology The identification of species reported in historical texts can occasionally benefit from information on morphology, behaviour and feeding habits. Historical reports sometimes included contradictory information regarding the morphology of animals. For instance, an article described how the "porpoise" was different from *Delphinus delphis* since it was smaller (1.5 m), but then went on saying that its beak was 15–20 cm long, narrow, cylinder shaped and similar to a goose's beak (Anonymous 1903b). Most morphological descriptions referred to animals 2–2.5 m long, weighing 40–110 kg, with a grey colour, swimming in groups, in pairs or alone (e.g., Anonymous 1894a, 1895b, 1898b,c, 1903b, 1909, Fichou & Levasseur 2004). This information does not allow for species identification, unless accompanied by additional information, for example, about particular behaviours or the prey targeted. For example, one report (Pérard 1925) mentioned that small cetaceans were waiting for and then following trawlers, a common behaviour among common bottlenose dolphins, worldwide (Bonizzoni et al. 2022). There were also records of animals entering rivers to chase fish (particularly mullets) in shallow waters (e.g., Anonymous 1898b), which is another common behaviour among common bottlenose dolphins (Silber & Fertl 1995). Interactions between small cetaceans and fishers mainly occurred in coastal waters, but some killings were also reported "offshore" (i.e., in Brittany, Legendre 1926a). Some sources mentioned the prey species that were targeted by the cetaceans. This type of information sometimes allowed us to infer the predator species, based on comparisons with contemporary diet studies. However, the fish species targeted by the fisheries in the late nineteenth century and early twentieth century are nowadays eaten by several different cetacean species, and in any case, caution is required because cetacean diets may have changed over the last 100 years.

The main reported prey of small cetaceans in French Atlantic waters were sardines and anchovies (e.g., Anonymous 1903b, 1930a). In this area, in the last two decades, sardines are a known prey of common dolphins and common bottlenose dolphins, as well as of harbour porpoises (Spitz et al. 2006a, Meynier et al. 2008, Marçalo et al. 2018). They also constituted a small fraction of striped dolphin diet off Portugal (Marçalo et al. 2021), but not in French waters (Spitz et al. 2006b). Anchovies are prey of common, common bottlenose and striped dolphins, as well as harbour porpoises (Spitz et al. 2006a,b, Meynier et al. 2008). Mullets were also mentioned in some reports (e.g., Fichou & Levasseur 2004), and these are the prey of common bottlenose and occasionally of striped dolphins in French waters (Spitz et al. 2006a,b), and of common bottlenose and common dolphins (although to a lesser extent) off the Iberian Peninsula (Silva 1999, Santos et al. 2007, Marçalo et al. 2018).

In the Mediterranean, the main fisheries reported to be suffering from conflict with small cetaceans were those targeting sardines, anchovies and mackerels (e.g., Anonymous 1894a, 1895b, 1898a,d, 1900a). In this region, sardines are part of the present-day diet of common dolphins (Bearzi et al. 2006, Larbi Doukara 2015, Giménez et al. 2018, Milani et al. 2018) and common bottlenose dolphins (Blanco et al. 2001, Larbi Doukara 2015, Bräger et al. 2016, Giménez et al. 2017, Borrell et al. 2021). They also represent a small fraction of the diet of striped dolphins (Meotti & Podestà 1997, Gómez-Campos et al. 2011, Aznar et al. 2017, Saavedra et al. 2022). Anchovies and mackerels are part of the diet of common, common bottlenose and striped dolphins (Meotti & Podestà 1997, Blanco et al. 2001, Gómez-Campos et al. 2011, Larbi Doukara 2015, Aznar et al. 2017, Giménez et al. 2017, 2018). Conflict with fisheries targeting mullets was also mentioned (e.g., Anonymous 1898b), and mullets are normally targeted by common bottlenose dolphins (Scheinin et al. 2014, Giménez et al. 2017, Milani et al. 2018). Finally, one report mentioned a conflict with fishing for bogues (Anonymous 1898d), which are part of the diet of common, common bottlenose and striped dolphins (Würtz & Marrale 1993, Gómez-Campos et al. 2011, Larbi Doukara 2015, Aznar et al. 2017, Giménez et al. 2017, 2018).

Information from osteological collections

The review of osteological collections of Delphinidae and Phocoenidae held in French institutions revealed that the most common species, in decreasing order of importance, were the common dolphin, long-finned pilot whales *Globicephala melas*, striped dolphins, common bottlenose dolphins and Risso's dolphins (Table 1). Other species present within the collections were killer whales, short-finned pilot whales *Globicephala macrorhynchus*, harbour porpoises, false killer whales *Pseudorca crassidens* and unidentified Delphinidae. However, species identification could not be confirmed for all specimens (see Table 1 and Supplementary Table 2.5). Results from the identification of the 49 specimens whose identification was not reviewed recently, based on the analyses of photographs and mandible measurements, are presented in Supplementary Material S1 (Supplementary Figure 1.2, Supplementary Tables 1.3 and 1.4).

The confirmed species acquired by institutions during the period of acute conflict between fishers and small cetaceans (i.e., during 1841–1940) were long-finned pilot whales ($n = 6$), common dolphins ($n = 5$), common bottlenose dolphins ($n = 2$), striped dolphins ($n = 1$), Risso's dolphins ($n = 1$), killer whales ($n = 1$), short-finned pilot whales ($n = 1$) and harbour porpoises ($n = 1$).

However, there are doubts about the two harbour porpoise specimens included in Table 1. The specimen MNHN-ZM-AC-1993-64, a harbour porpoise with a certain identification held at the Muséum National d'Histoire Naturelle de Paris (MNHN), was collected in Palavas ("département" of Hérault) in 1870. The specimen was previously identified as *Orcinus orca* or *Globicephala melas* and was later re-identified as *P. phocoena* (Céline Bens/MNHN, pers. comm.). The original identification as *O. orca* is still attached to the skull of the specimen. Van Beneden and Gervais (1880) and Van Beneden (1889) mentioned a very young killer whale *Orcinus orca* stranded on the beach of Palavas (Hérault), unknown date, which could correspond to specimen MNHN-ZM-AC-1993-64, implying that the harbour porpoise skull present in the MNHN collection may have been mislabelled. However, this could not be confirmed. The second specimen of harbour porpoise corresponds to a record of an animal caught by fishers in the area of Marseille in 1841, from the Muséum d'Histoire Naturelle de Marseille (MNHM). The specimen was recorded as being "un très gros marsouin indigène" (a very large native porpoise; Borrely 2021). As discussed above, the word "porpoise" may not refer to actual harbour porpoises. In addition, this specimen is currently not present in the zoological collection of the MNHM, and it does not appear in more recent records (many specimens were destroyed and replaced between 1850 and 1870; Christophe Borrely/MNHM, pers. comm.). Therefore, it was impossible to verify whether this specimen was a *Phocoena phocoena*. In summary, the information obtained from French osteological collections cannot confirm the presence of harbour porpoises in the French Mediterranean.

Among the 63 cetacean specimens originating from the French Mediterranean and Monaco, 42 had an unknown cause of death. Thirteen specimens were reportedly killed by humans: six were harpooned, four were "captured" (without further information) and three were "captured in fishing nets". An additional eight specimens were found stranded. Among the 13 specimens killed by

Table 1 Summary of the Specimens from the French Mediterranean and Monaco Held in French Institutions

Species	Period/Year of Acquisition Known (1841–1940)		Period/Year of Acquisition Unknown		Total
	Species Identification Certain	Species Identification Uncertain	Species Identification Certain	Species Identification Uncertain	
Delphinus delphis	5	13	0	0	18
Globicephala melas	6	5	0	0	11
Stenella coeruleoalba	1	1	2	4	8
Tursiops truncatus	2	4	0	2	8
Grampus griseus	1	7	0	0	8
Orcinus orca	1	1	0	0	5
Delphinidae	0	3	0	0	3
Phocoena phocoena	1[a]	1[b]	0	0	2
Globicephala macrorhynchus	1	0	1	0	2
Pseudorca crassidens	0	1	0	0	1
Total	54		12		63

Information on the certainty of species identification is provided.

[a] Indicates specimen MNHN-ZM-AC-1993-64, a harbour porpoise whose provenance could not be confirmed (Muséum National d'Histoire Naturelle de Paris).

[b] Indicates a "harbour porpoise" caught by fishers in the area of Marseille in 1841, whose identification could not be confirmed (Muséum d'Histoire Naturelle de Marseille).

humans, two were common dolphins (both "captured"), four were long-finned pilot whales (three harpooned and one "captured") and three were Risso's dolphins (one "captured", one "captured in fishing nets" and one harpooned). The collections also included one striped dolphin and one killer whale, both harpooned, and one common bottlenose dolphin and one false killer whale, both "captured in fishing nets".

In addition to specimens with a known provenance from the French Mediterranean and Monaco, there were 143 specimens without a known provenance (see Supplementary Table 2.5). Of these, 27 were acquired prior to 1940, and 116 did not have a date of acquisition. The species acquired prior to 1940 were common bottlenose dolphins ($n = 9$), long-finned pilot whales ($n = 3$), killer whales ($n = 3$), common dolphins ($n = 2$) and Risso's dolphins ($n = 1$), all species that are currently present in the Mediterranean. There was also one specimen of spinner dolphin *Stenella longirostris* and one short-finned pilot whale, neither of which is currently known to occur in the Mediterranean. Among the 116 specimens with an unknown date of acquisition, the most common species were the striped dolphin, common bottlenose dolphin, common dolphin, long-finned pilot whale, Risso's dolphin, unidentified *Stenella* sp., harbour porpoise, pygmy killer whale *Feresa attenuata*, false killer whale and rough-toothed dolphin *Steno bredanensis*, all currently present in the Mediterranean. There were also species not presently known to occur in the Mediterranean: the pantropical spotted dolphin *Stenella attenuata*, spinner dolphin, Fraser's dolphin *Lagenodelphis hosei*, Indo-Pacific humpback dolphin *Sousa chinensis* and Clymene dolphin *Stenella clymene*.

Zoological museums tend to hold specimens acquired around the world as well as locally sourced, and the latter may have been particularly important for small museums. Historical documents from the eighteenth to twentieth centuries confirm that curators of zoological collections obtained at least some specimens from their own region. For instance, Louis Companyo (1867), who was the director of the Muséum d'Histoire Naturelle de Perpignan, acquired one specimen (specimen 2008.0.3) of *Globicephala melas* that was part of a large group of which some individuals were caught and other stranded close to Barcarès (Pyrénées-Orientales) in 1864. Companyo (1867) mentioned that Paul Gervais, Dean at the Université de Montpellier, brought four animals from the same group to the collection of his university. The collection of the Université de Montpellier currently holds one specimen of *G. melas* which is probably one of the animals mentioned above. Thus, it is possible that some of the specimens of unknown provenance in Supplementary Table 2.5, which were identified as species known to occur in Mediterranean waters, were sourced locally from the French Mediterranean.

To summarise, based on the evidence from historical documents and osteological collections, the main species involved in conflict with fishers in French Atlantic and Mediterranean waters were common bottlenose dolphins and common dolphins. In addition, other species such as striped dolphins, pilot whales or Risso's dolphins were killed in both the French Atlantic and the Mediterranean, as well as harbour porpoises in the French Atlantic only, but these species were not the main target.

Minimum estimates of numbers of small cetaceans killed

Estimating the total number of small cetaceans killed during the main period of conflict is a difficult task. Firstly, no time series of bounties being actually paid were found in our review of historical literature. Bounty reports are most likely available in non-digitalised archives (e.g., municipal or council archives), and thus, a search in these physical archives could potentially provide comprehensive time series of bounty rewards. For example, by accessing archives of the city of Sète (Mediterranean coast), Faget (2009) collated records of bounties offered to fishers for a total of 254 small cetaceans, between 1904 and 1910. Secondly, even if such time series could be accessed, bounties would only be likely to reflect a fraction of the actual number of animals killed (Bearzi et al. 2004). Furthermore, only fishers and other seafarers claimed bounties, while the mortality caused by the military would not be reflected in bounty reports.

The difficulty of recovering and bringing carcasses (that could weigh up to a few hundred kilograms each) back to the port sometimes forced fishers to abandon the carcasses at sea (Anonymous 1900c). Fishers could also have killed small cetaceans without knowing or caring about monetary rewards. For example, Legendre (1926a) wrote about tuna fishers who had harpooned many dolphins in the offshore waters of Brittany in order to get fresh meat. These fishers reportedly felt that a 10 Francs (approximately 7 Euros) bounty per dolphin head was worthless in comparison to the 500–600 Francs (approximately 356–427 Euros) they received per tuna, and consequently, they did not ask for bounty rewards. In addition, when small cetaceans were killed by torpedoes or bombshells, it was impossible to determine how many were killed, "because bombshells are only leaving behind fragments of skin, flesh, and other body parts" (Anonymous 1909). When shot with guns, animals reportedly tended to sink and rarely floated so that they could be collected (Legendre 1912). Shooting may also fail to kill the animals immediately, but the wounds could lead to a later death. Other killing methods (e.g., Bellot's needles, see Methods to kill small cetaceans) would also have resulted in later deaths and lack of carcass recovery.

While exact numbers cannot be computed, information from historical records and osteological collections allows us to make minimum estimates that help us appreciate the magnitude of killings around the shores of France. Supplementary Table 2.7 collates records of small cetaceans killed between 1826 and 1954, based on the historical material reviewed in this study. When adding up the numbers from Supplementary Table 2.7, a minimum estimate of 8860–8907 animals is obtained, based solely on reports of anecdotal kills mentioned in newspapers articles, a few photographs of killed animals, a few specimens reported from museum collections and a few years of bounty rewards reported by some authors. In summary, these figures clearly substantially underestimate the number of animals killed.

The bounty rewards were reported by (1) Fichou and Levasseur (2004), who accessed historical documents and found that thousands of small cetaceans were killed during a 5-year period between 1894 and 1898 (location unknown, most likely in French Atlantic waters); (2) Legendre (1926a), who reported 5590 animals killed during a 5-year period between 1921 and 1925 (location unknown); and (3) Faget (2009), who counted a total of 254 animals killed during a 7-year period between 1904 and 1910, exclusively by fishers from the city of Sète (French Mediterranean). Sète was just one among many locations in France where fishers were rewarded with bounties, and it was also a small city in comparison with, for instance, Marseille (Mediterranean coast), the latter city being a place where fishers notoriously complained about small cetaceans and where high bounties were being offered (Perrier 1889, Anonymous 1882). In total, between 6660 and 6707 animals killed in France were counted, which included 644 animals in the French Atlantic, between 425 and 472 in the French Mediterranean and 5591 in unknown provenance (i.e., from either the Atlantic or Mediterranean shores of France). In addition to these numbers, "hundreds", "thousands" and "many" more animals killed in the French Atlantic were mentioned in historical reports (see Supplementary Table 2.7; Legendre 1926b, Fichou & Levasseur 2004), which adds up at least 2200 animals (assuming that "hundreds" corresponds to a minimum of 200 individuals, and "thousands" corresponds to a minimum of 2000 individuals), thus increasing this minimal estimated number from Supplementary Table 2.7 to 8860–8907 kills.

Finally, one should consider that (1) the main period of conflict in France lasted for more than a century; (2) bounty rewards were offered for about a century (with the first indication in the 1830s and the last indication in 1927, with variations over time and regions depending on the budget available); and (3) at least some of the methods used to kill small cetaceans were particularly effective (e.g., harpoons and nets, as well as the guns, bombshells and torpedoes used by military vessels for at least 71 years). Taking into consideration the whole historical scenario, it is likely that at least tens of thousands of small cetaceans, mainly common dolphins and common bottlenose dolphins, were killed in the Mediterranean and Atlantic waters of France, particularly in the French Mediterranean, between the eighteenth and the mid-twentieth centuries.

Case study 2: Algeria

Context and perceptions

Algeria was a French territory from 1830 through 1962, and thus, historical documents from the French government included information about this country. These document the existence of conflict between fishers and small cetaceans off the coasts of Algeria between the end of the nineteenth century and the mid-twentieth century. Algerian newspapers provide additional evidence of conflict. The context was the same as in France: small cetaceans being blamed for reducing catches (mainly of sardines), taking fish from the nets, damaging fishing gear and occasionally getting entangled in such gear (Anonymous 1901c).

Methods to kill small cetaceans

Since at least the 1880s, Algerian fishers started asking for bounty rewards for cetacean killings, as happened in France (Anonymous 1882). These requests were approved in 1900, when it was decided to offer a 25 Francs (approximately 107 Euros) bounty per animal killed (Anonymous 1900d). Administrative issues delayed implementation for a few years, and the first indication of bounty payments for killing "porpoises" (most likely dolphins) and seals (most likely the Mediterranean monk seal, *Monachus monachus*) dates back to 1904 (Jonnart 1904, 1908). Before the implementation of bounties, Algerian fishers were granted financial compensation (between 20 and 25 Francs per event) for the damage caused by small cetaceans to their nets (Anonymous 1903g). Bounties were paid for 36 years, from 1904 to 1939 (the latter date being associated with the last indication of an annual budget dedicated to bounty payments; Anonymous 1939c). Monetary compensation for each animal killed ranged between 5 and 10 Francs, depending on the year. In the documents reviewed, we could find no mention of the 25 Francs bounty proposed in the early decision reported by Anonymous (1900d).

The most common method to kill small cetaceans was to use fishing nets (mainly "lamparos" and "bonitières", i.e., surrounding nets) to trap the animals (Steeg 1922). Fishers would then use harpoons to kill the trapped animals. Cetaceans were also incidentally caught in fishing nets, and if fishers could not release them to avoid gear damage, they killed them using harpoons (Garau 1909). In addition, fishers used guns and harpoons to target animals swimming close to their boats. Guns and harpoons were normally considered as the most efficient killing methods and were distributed to fishers by the government (Anonymous 1908b). However, reservations were expressed about the danger of distributing guns to fishers (e.g., in case of altercations), and the use of harpoons was preferred (Anonymous 1908b). Other methods included harpoon rifles (Anonymous 1937) and dynamite (Anonymous 1900e). The use of Bellot's needles was considered (Anonymous 1893b), but we could find no evidence of implementation. As for France, military ships and fishery control vessels were sent to patrol Algerian waters to kill small cetaceans, from at least 1902 through 1913 (Anonymous 1902a, 1903a, 1904a, 1913, Jonnart 1907, 1908, 1910, Lutaud 1911). The names of six of these vessels were identified in the historical material reviewed. These were the boats FLIBUSTIER, FORBAN, MOULOUYA, CHIFFA, CHÉLIFF and DAGUE (Anonymous 1902a, 1913, Révoil 1902).

Algerian fishers regularly consumed the meat of the small cetaceans they killed (Anonymous 1912), and consumption of cetacean meat was further encouraged by the press (Anonymous 1935). The meat was also sold in fish markets. For instance, common bottlenose dolphins and common dolphins, as well as monk seals, could be found in Algier's fish market (Drouet 1887).

Species of small cetaceans killed

Information from historical records

Species identification relying on photographs Three photographs showing dolphins were found in the historical material from Algeria dating back to the late 1930s. Figure 8A shows a dead dolphin (possibly a striped or a common dolphin) killed by seafarers, Figure 8B shows a killed common dolphin and Figure 8C shows a group of common dolphins swimming, which documents their presence in Algerian waters. In all cases, these dolphins are referred to as "marsouins".

Identification of species based on common and scientific names Fewer words were used to refer to dolphins in Algeria than in France, the most common being "marsouin" (porpoise), followed by "dauphin" (dolphin). The animals were also occasionally called "souffleur" and "taupe" (blower and mole, respectively; Anonymous 1902b), "squale" (shark; Jonnart 1908, Lutaud 1911) and "poisson" (fish) in some newspapers (Anonymous 1900f, 1928). In newspapers, the word "marsouin" was often used to refer to common dolphins and possibly also to striped dolphins. The word "souffleur" is still being used in Algeria and other North African countries to refer to common bottlenose dolphins (Larbi Doukara 2015).

In a catalogue of animals from Algeria, Loche (1858) identified the common bottlenose dolphin as one of the species of Delphinidae inhabiting the waters of Algeria. Drouet (1887) listed *Tursiops tursio* (Gerv.), *Delphinus algeriensis* (Loche) and *Delphinus mediterraneus* (Loche) as being sold in fish markets. Whereas the identification of *Tursiops tursio* (Gerv.) is certain, there is confusion regarding the use of *Delphinus algeriensis* (Loche) and *Delphinus mediterraneus* (Loche). *Delphinus algeriensis* was described by Loche (1860) from a specimen caught near Algiers in 1859, and the description corresponds to a striped dolphin. This is confirmed by an illustration of the specimen in Bouvier (1891), which is reproduced from Fischer (1881). The name *Delphinus mediterraneus* was used to refer to both common and striped dolphins. For example, a common dolphin caught close to Toulon (French Mediterranean) was identified as a *Delphinus delphis mediterraneus* by Pourcel (1910; Figure 7H). A striped dolphin caught near Algiers in 1860 (most likely corresponding to the specimen MNHN-ZM-AC-1860-132, currently held in the collection of the Muséum National d'Histoire Naturelle de Paris) was first misidentified as a *Delphinus delphis mediterraneus* and later identified as *Stenella coeruleoalba*. This striped dolphin was also described as being a *Delphinus delphis mediterraneus* in Bouvier (1891), and the illustration shows a striped dolphin.

A newspaper article describing the conflict between fishers and small cetaceans in Algeria mentioned that "three species of ichthyophagous cetaceans inhabit the Algerian coasts: the marsouins (*Delphinus delphis*), the souffleurs (*Nelphinus tursio*), and the taupes (*Delphinus communis*)" (Anonymous 1902b). Descriptions of the appearance and behaviour given by the author suggest that the "marsouins" could be common and/or striped dolphins (described as occurring in large groups, fast swimming and displaying aerial behaviour, not longer than 2.35 m). The "souffleurs" could be common bottlenose dolphins, described as being solitary animals or swimming in pairs, slower and less commonly seen than the "marsouins", and reaching 3.50 m. Finally, the description of the "taupes" (rarely seen, less than 1.60 m long) might refer to harbour porpoises. As discussed earlier, harbour porpoises disappeared from the western Mediterranean thousands of years ago (Fontaine et al. 2014, Fontaine 2016), but it cannot be excluded that the species occurred sporadically in Algerian waters in the nineteenth and twentieth centuries, given that a few stranded individuals were reported in recent decades (e.g., Cabezón et al. 2004).

Identification of species based on morphology, behaviour and feeding ecology Morphological descriptions of small cetaceans killed were available in a few newspaper articles and referred to animals 2–2.50 m long and weighing 90 kg (Anonymous 1934c, 1938, 1939b), as well as to animals of 3–3.50 m long weighing around 250 kg (Anonymous 1923b, 1924, 1934b), and to larger animals

Figure 8 Photographs of dolphins published in Algerian newspapers in the late 1930s. (A) A dolphin (possibly a striped or a common dolphin) killed by seafarers. The legend indicates: "A porpoise hunt offshore of Guyotville. The boat Chéragas is back to the harbour after a successful hunt with the T.C.F crew" (Anonymous 1937); (B) A common dolphin captured by fishers. The legend indicates: "A porpoise captured in waters offshore Lapérouse. A porpoise measuring 2.12 m and weighting 92 kg was captured in waters offshore Lapérouse by Mr. Jean Ganouste, posing in the centre of the picture, surrounded by his friends." (Anonymous 1938); (C) A group of common dolphins. The legend indicates: "A curious picture of porpoises taken by our fishing boxers" (Anonymous 1939a). All photographs were provided by Bibliothèque Nationale de France (https://gallica.bnf.fr).

weighting 400–600 kg (Anonymous 1900c,e,h, 1925). Although this information alone does not allow for identification to the species level, the lengths and weights would match those of several Mediterranean Delphinidae. In the western Mediterranean, common and striped dolphins can reach lengths slightly above 2 m (Di-Méglio et al. 1996, Larbi Doukara 2021), common bottlenose dolphins can reach 3–3.50 m (Sharir et al. 2011) and Risso's dolphins can reach 3–4 m (Bearzi et al. 2011). The reports of animals weighting 400–600 kg might correspond to large Risso's dolphins (animals of 3.2 m can reach 500 kg in the Faroe Islands; Bloch et al. 2012) or to young individuals of larger species of Delphinidae (e.g., pilot whales, killer whales and false killer whales).

Small cetaceans were seen in large groups, in pairs or alone (Anonymous 1900e,h, 1902b, 1904b, 1925, 1939b). Interactions between small cetaceans and fisheries appeared to be mainly coastal (Anonymous 1899a, 1900c,e,h, 1923b, 1924, 1928), but considering that Algeria has a narrow continental shelf, interactions with pelagic species such as striped dolphins would not be rare. Garau (1909) reported various types of interactions with fishing gear. At the time, fishing nets were made of delicate fabrics such as linen, cotton and hemp (Garau 1909), which meant they could easily be damaged. The "sardinal", a type of gillnet used to catch sardines, round sardinellas and anchovies, was the gear reported to be most frequently damaged by small cetaceans when they fed on prey stuck in meshes. According to Garau (1909), the animals also damaged several other types of gear such as gillnets, beach seines, bottom pair trawl nets and tuna traps. Interestingly, the term "déprédation" (depredation) was used in two articles (Imbert 1900, Anonymous 1931c), showing that this loaded word (Bearzi & Reeves 2022) was already in use more than one hundred years ago. Damage could also be caused by small cetaceans that became incidentally entangled in surrounding nets and beach seines (Garau 1909). While Garau (1909) did not report bycatch of small cetaceans in "madragues" (tuna traps), cases of incidental mortality in these nets were reported from other Mediterranean areas (e.g., in Italy; Anonymous 1881). Some animals were reported to feed on fish discarded by bottom pair trawlers (Anonymous 1899b, 1901c, 1922b), which is a common behaviour among common bottlenose dolphins, worldwide (Bonizzoni et al. 2022). Recent work on dolphin interactions with fisheries in Algerian waters indicates that the main species of cetaceans bycaught in drift gillnets and trawlers in recent years are common bottlenose, common and striped dolphins (and occasionally Risso's dolphins, long-finned pilot whales and goose-beaked whales; Larbi Doukara 2015, 2021). Common bottlenose dolphins were also reported to forage in proximity and within fishing nets, including trawl nets (Larbi Doukara 2015).

The main prey species targeted by small cetaceans in Algerian waters were shoaling fish, mainly sardines and anchovies (Jonnart 1906, 1908, Garau 1909, Anonymous 1899a, 1900d,e, 1902c, 1928, 1931c), in addition to round sardinellas and bonitos (Anonymous 1923c, 1931c). Sardines and anchovies are part of the present-day diet of common dolphins and common bottlenose dolphins in Algeria (Larbi Doukara 2015) and in other regions of the western Mediterranean (e.g., Blanco et al. 2001, Giménez et al. 2017, 2018). They are also part of the diet of striped dolphins in the western Mediterranean (Gómez-Campos et al. 2011, Aznar et al. 2017, Meotti & Podestà 1997). Common dolphins feed on round sardinellas in Algerian waters (Larbi Doukara 2015) and in the western Mediterranean (Gimenez et al. 2018). While historical documents also mention bonitos as being targeted by small cetaceans, the bonito species to which they referred is unclear. The Atlantic bonito (*Sarda sarda*) was not found to be part of the diet of common dolphins and common bottlenose dolphins in Algerian waters (Larbi Doukara 2015), but it may be targeted by common bottlenose dolphins in other areas (e.g., in Portugal, Dos Santos et al. 2007, and in the Black Sea, Gladilina & Gol'din 2014). The prey species listed above are normally not targeted by Risso's dolphins and pilot whales in the Mediterranean, both species being mainly teuthophagous (Bearzi et al. 2011, Praca et al. 2011), nor by killer whales (which principally feed on tunas in the Strait of Gibraltar; Esteban et al. 2016). Little is known about the diet of false killer whales in the Mediterranean, but in other areas the species appears to feed mainly on cephalopods and various species of fish, but not on shoaling fish (Zaeschmar & Estrela 2020).

Information from osteological collections

French osteological collections of small cetaceans provided information on four specimens acquired from Algeria during the period of intense conflict with fisheries. These comprised three common dolphins and one striped dolphin from the second half of the nineteenth century (1860–1876). There were also an additional 16 specimens with an unknown date of acquisition: 10 specimens could be common dolphins (species identification is uncertain) most likely acquired in the twentieth century (date unavailable), four were striped dolphins and one was a pilot whale, all acquired prior to 1978–1983 (the dates of entry in the collection of the Muséum National d'Histoire Naturelle de Paris). Finally, there was one common bottlenose dolphin with an unknown date of acquisition. Information on the cause of death was available for only one of the 20 specimens listed: a striped dolphin "captured" in 1860 (specimen MNHN-ZM-AC-1860-132, kept at the Muséum National d'Histoire Naturelle de Paris; Loche 1860).

The limited information available from the osteological collection does not allow us to draw clear conclusions. However, based on the evidence reported above from historical documents, the main species involved in conflict with fishers in the waters of Algeria were common bottlenose dolphins and common dolphins, as well as striped dolphins (probably to a lesser extent). In addition, some larger species (e.g., Risso's dolphins and pilot whales) may have been killed, but were likely not the main target given their feeding preferences (i.e., small shoaling fish are not their main prey).

Minimum estimates of numbers of small cetaceans killed

Several historical documents provided information on small cetaceans killed in Algeria between 1900 and 1939 (Supplementary Table 3.1). This information included (1) the annual budgets available to pay bounties, (2) the budgets that were spent and (3) the number of animals killed annually, which were reported in government documents. Information was not consistently available across years. Additional information on the killing of small cetaceans was found in newspaper articles.

Bounties were paid between 1904 and 1939. Annual budgets varied between 1500 and 2500 Francs per year and were described as "bounties to eradicate porpoises and seals". However, no reference to seals was found in the reports, which only refer to "porpoises". In some years, the entire budget was used to pay bounties, and in some cases, the budget was reportedly insufficient (e.g., in 1909), whereas only a small fraction of the budget was spent in other years (e.g., in 1919 and 1920). Bounties between 5 and 10 Francs were paid per animal killed, depending on the year. The numbers of bounties claimed, corresponding to the number of animals killed, were registered by "département". These included Algiers, Oran, Bône (a name used during the colonisation period for the city of Annaba) and Philippeville (the city of Skikda) and varied between 30 and 434 animals killed per year between 1904 and 1939. Numbers were not available for all years. Minimum estimates, based on the information found in the consulted documents, totalled 5769 animals killed. Of these, 5725 were reported in government documents, and 44 were reported in newspaper articles. It cannot be excluded that these 44 animals killed that were reported in newspapers were duplicates of killing events also reported in government documents. Because the numbers of animals killed per year during the period when bounties were paid were available for only 24 out of 36 years, estimates for the missing years were calculated based on the total budget available, the budget spent and the amount of money given per animal. These estimates yielded 2159–2583 extra animals killed, resulting in 7928–8352 animals killed between 1900 and 1939 (see calculation details in Supplementary Table 3.1). Numbers did not always match: for instance, in 1924, one document indicated 181 animals being killed, but when calculating the numbers killed based on the budget spent (2460 Francs) and the amount of the bounty for that year (10 Francs per animal), 246 animals would have been killed.

Although the numbers above include extrapolations and estimates, the consulted documents provide clues suggesting that even the maximum estimate in Supplementary Table 3.1 is likely to underestimate the actual number of killings. More specifically: (1) numbers of small cetaceans killed by fisheries control vessels and military vessels were not accounted for; (2) bounties were unlikely to be claimed for every animal killed; and (3) some of the methods employed to kill cetaceans, such as dynamite, may destroy the bodies, making it difficult to bring the head back to shore in order to claim a bounty. For instance, one newspaper article mentions that two kilos of dynamite were thrown into the middle of a group of 15–20 cetaceans, which resulted in almost all of them being killed and sinking to the seafloor, with the exception of two animals. However, due to their heavy weight (reportedly 400–500 kg), only one of these two animals was brought to shore (Anonymous 1900c). Another article indicated that three small cetaceans were shot and only two of them subsequently stranded (Anonymous 1925). Finally, another article reported that fishers caught a large animal (3.4 m) and that it required several boats to take it out of the water due to its heavy weight (Anonymous 1923b). Fishers working from small boats or on their own could have difficulty handling heavy animals (such as common bottlenose dolphins) and lifting the body on board. Consequently, some of these small cetaceans could not always be landed. Additionally, wounds caused by guns, harpoons and dynamite could have resulted in animals evading capture and dying later.

Overall, the historical documents reviewed suggest that the number of small cetaceans killed in Algeria between 1900 and 1939 is in the order of 10,000.

Final remarks

Between the eighteenth and twentieth centuries, small cetaceans (particularly dolphins) were killed in large numbers off the coasts of Europe and North Africa, due to perceived competition with fisheries. Dolphins in these areas were long considered as pests responsible for reduced fish catches and damage to fishing gear. Such conflict was particularly acute during the second half of the nineteenth century and the first half of the twentieth century, when governments from various countries encouraged and subsidised the extermination of small cetaceans. In the Mediterranean, intentional killings and culling campaigns took place, with varying intensity, particularly in France, Italy, "former Yugoslavia" (in our study; Slovenia, Croatia and Montenegro), Algeria, Greece, Spain, and possibly in Turkey, Tunisia and Morocco, while in the Northeast Atlantic, they took place in France and Spain (Figure 1).

Our review of historical documents and osteological collections focused on the conflict between fishers and small cetaceans in France and Algeria, and it uncovered exceptionally high levels of animosity and a strong determination to eradicate these animals, which was consistent with information from other areas (e.g., the seas around Italy and "former Yugolsavia"; Bearzi et al. 2004, Meliadò et al. 2020). Therefore, future efforts should aim to uncover information in multiple languages, with emphasis on geographical areas that have not been the focus of historical research. In addition, more comprehensive research should be conducted in physical libraries to reveal the actual extent of mortality of small cetaceans documented by recorded bounties. We acknowledge that our study did not manage to encompass all the potential sources of information, to the extent that the number of killings reported here might be just the tip of the iceberg.

While the precise numbers of killings remain largely unknown, our review provides an appreciation of their possible magnitude. Our results suggest that tens of thousands of small cetaceans were killed in France and around 10,000 were killed in Algeria, primarily between the 1880s and the 1930s. Total mortality of small cetaceans in the Mediterranean Sea and Northeast Atlantic due to human conflict is certainly much higher, as that includes the killings in all other countries where similar conflicts existed. For instance, minimum estimates in Italy and "former Yugoslavia" approached 10,000 dolphins killed (Bearzi et al. 2004, Meliadò et al. 2020). As described in our study, killings also occurred in unknown but possibly significant numbers in Greece, Spain, Tunisia and Morocco. Off the Mediterranean coast of Turkey, small cetaceans were killed primarily for commercial purposes (at least 1700 animals). Finally, little is known about the numbers of small cetaceans killed intentionally in Portugal, and virtually nothing is known about historical killings in parts of the Middle East. By combining all these exceedingly imprecise and partial figures, one may conclude that, at the very minimum, 40,000 small cetaceans were killed. However, for all the reasons explained in our manuscript, the actual numbers were certainly far greater. Such a high mortality most likely had important negative impacts on population trajectories, particularly in the Mediterranean Sea (Bearzi & Reeves 2021). For instance, common dolphins were considered to occur regularly in the Mediterranean before the mid-twentieth century, but are nowadays a rare sight in most of the region (Genov et al. 2021, Bearzi & Genov 2022). In the inner Mediterranean Sea, common dolphins are currently listed as Endangered in the International Union for Conservation of Nature Red List of Threatened Species (Bearzi et al. 2022). Intensive culling and killing campaigns in the past may have contributed to the decline of this species, in combination with more recent impacts such as prey depletion and habitat degradation (Bearzi et al. 2004, 2022, Bearzi & Genov 2022). Furthermore, the species is presently rare in French Mediterranean waters. For instance, there were no common dolphin records during aerial surveys in recent years (Panigada et al. 2024), and only 21 strandings of common dolphins were recorded in the last 30 years (Dars et al. 2023).

In the second half of the twentieth century, intentional killings largely stopped, coinciding with a widespread change in the perception and use of marine megafauna (Mazzoldi et al. 2019), and the beginning of legal protection of all cetaceans in European waters. However, European cetaceans have become increasingly exposed to a variety of other anthropogenic threats (Bearzi & Reeves 2021). While the deliberate killings have largely stopped, conflict between fisheries and small cetaceans is still occurring in European waters (e.g., Snape et al. 2018). In some areas, dolphins continue to be considered as pests, and there are reported cases of fishers still seeing culling as an appropriate measure to mitigate conflicts (Gonzalvo et al. 2015, Pardalou & Tsikliras 2018).

Documenting past events and their consequences can not only deepen our understanding of the long-term relationship between humans and the marine environment but also encourage us to reflect on the lessons learned and strive not to repeat our past mistakes. Given that some populations of small cetaceans have been the target of massive killings for centuries, their past abundance and distribution around the Mediterranean and Northeast Atlantic may be different from what we know today, largely or entirely as a consequence of these killings. Historical ecology and environmental history (Holm et al. 2001) have the potential to provide historical baselines of cetacean diversity, abundance and distribution (Brito & Vieira 2010), while contributing datasets of historical documents and osteological collections that can help us to investigate population trends, viability, and potential for recovery, as well as to set meaningful baselines for conservation (Bearzi & Reeves 2021). This study represents a step in that direction.

Acknowledgements

The authors are grateful to the following people, who provided information, photographs and measurements of specimens of small cetaceans held in French osteological collections: Jérémy Migliore (Muséum Départemental du Var); Mehdi Mouana, Audrey Theron and Anne-Lise Charruault (Université de Montpellier); Laura Morazzani, Agnès Servière and Adeline Rouilly (Muséum d'Histoire Naturelle de Nîmes); Adeline Aumont (Muséum d'Histoire Naturelle de La Rochelle); Michèle Bruni (Institut Océanographique de Monaco); Aude Lalis and Céline Bens (Muséum National d'Histoire Naturelle de Paris); Christophe Borrely (Muséum d'Histoire Naturelle de Marseille); Henri Cap (Muséum d'Histoire Naturelle de Toulouse); Nicolas Vialle (Muséum d'Histoire Naturelle d'Aix en Provence); Olivier Gerriet (Muséum d'Histoire Naturelle de Nice); Didier Mary (Muséum d'Histoire Naturelle de Perpignan); Didier Berthet (Musée des Confluences de Lyon); Marie-Dominique Wandhammer (Musée Zoologique de Strasbourg); Matthieu Landreau (Muséum d'Histoire Naturelle de Bordeaux); Priscilla Decottignes and Bruno Cognie (collection de zoologie de l'Université de Nantes); Blandine Bartschi (collection de zoologie de l'Université Claude Bernard Lyon 1); Eric Guiho (Muséum d'Histoire Naturelle de Nantes); and Ghislain Gregoire (Muséum d'Histoire Naturelle Troyes). We thank Pauline Bisiaux at the Musée de la Carte Postale de Baud for providing information on historical photographs. We thank Ville de Paris/Bibliothèque Forney, gallica.bnf.fr/Bibliothèque Nationale de France, and the Cnum – Conservatoire Numérique des Arts et Métiers – http://cnum.cnam.fr for granting us authorisation to reprint historical photographs and illustrations. We thank the editors and the two referees, Cristina Brito and Aldemaro Romero Jr., for their valuable comments, which improved the content and quality of the article.

MACP, GJP and GB were supported by funding from the European Union's Horizon 2020 research and innovation programme under the Marie Skłodowska-Curie Grant Agreement No. 813383 ("SeaChanges – Thresholds in human exploitation of marine vertebrates"). YvdH was funded by the MSCA-IF project Demise of the Atlantic Grey Whale (DAG) under the European Union's Horizon 2020 research and innovation programme (Grant Agreement No. 101025598).

References

Aguilar, À. 2013. Chimán. In *La pesca ballenera moderna en la península Ibérica*. Barcelona, Spain: Edicions Universitat Barcelona.

Alcover Cateura, P.J. 2022. Aproximación a la pesca, venta y consumo de delfín en la Corona de Aragón (siglos XIV-XV). *Imago Temporis Medium Aevum* **15**, 583–602.

Anonymous. 1872. *Conseil général du Var, procès-verbal des délibérations, session ordinaire du mois d'avril 1872*. Draguignan, France.

Anonymous. 1873. Pêche de marsouins dans le golfe de Marseille. In *La Nature, revue des sciences et de leurs applications aux arts et à l'industrie, Journal hebdomadaire illustré*, n°19, 11th of October 1873. Paris: Masson et Cie.

Anonymous. 1881. *Ministère de la Marine et des Colonies, Revue Maritime et Coloniale, 1881*. Paris: Berger Levrault et Cie, Volume 70.

Anonymous. 1882. *Newspaper Le petit colon Algérien*, n° 1567, 25th of October 1882. Alger, Algeria.

Anonymous. 1889. *Journal Officiel de la République*, n°138, 22nd of May 1889. Paris, France.

Anonymous. 1893a. *Article "La destruction des marsouins". Newspaper Le Courrier de La Rochelle*, n°74, 14th of September 1893. La Rochelle, France.

Anonymous. 1893b. *Newspaper L'impartial*, n°146, 2nd of July 1893. Djidjelli, Algeria.

Anonymous. 1894a. *Article "Encore les marsouins". Newspaper Le Petit Bastiais*, n°9972, 5th and 6th of March 1894. Bastia, France.

Anonymous. 1894b. Article "Les marsouins". *Newspaper Le Petit Bastiais*, n°10017, 26th of April 1894. Bastia, France.

Anonymous. 1894c. *Ministère de la Marine, Revue Maritime et Coloniale*. Paris: Librairie militaire de L. Baudoin, Tome 123.

Anonymous. 1895a. Article "La destruction des marsouins". *Ministère de la Marine, Revue maritime et coloniale*. Paris: Librairie militaire de L. Baudoin, Tome 124.

Anonymous. 1895b. *Ministère de la Marine, Revue maritime et coloniale*. Paris: Librairie militaire de L. Baudoin, Tome 126.

Anonymous. 1898a. Article "Sardines et Marsouins". *Newspaper Le Petit Bastiais*, n°263, 13th of November 1898. Bastia, France.

Anonymous. 1898b. Article "Le Golo envahi par les marsouins !!". *Newspaper Le Petit Bastiais*, n°163, 13th of July 1898. Bastia, France.

Anonymous. 1898c. Article "La cavalcade". *Newspaper Le Petit Bastiais*, n°68, 20th of March 1898. Bastia, France.

Anonymous. 1898d. Article "Boques et marsouins". *Newspaper Le Petit Bastiais*, n°81, 4th and 5th of April 1898. Bastia, France.

Anonymous. 1899a. Article "pêcheurs et marsouins". *Newspaper L'Oued-Sahel*, n°1410, 8th of October 1899. Bougie, Algeria.

Anonymous. 1899b. *Conseil Supérieur de Gouvernement. Procès-verbaux des délibérations et exposé de la situation générale de l'Algérie. Sessions de décembre 1899.* Mustapha, Algeria.

Anonymous. 1900a. *Conseil général du département du Gard, session ordinaire d'avril* 1900. *Procès-verbaux des séances*. Nîmes, France.

Anonymous. 1900b. Article "Les marsouins". *Newspaper La Sardine* n°118, February 1900. Dijon, France.

Anonymous. 1900c. *Newspaper La démocratie algérienne*, n°4394, 21st of April 1900. Bône, Algeria.

Anonymous. 1900d. Article "Pêche maritime". *Newspaper Le Ralliement*, n°45, 22nd of March 1900. Philippeville, Algeria.

Anonymous. 1900e. Article "Marsouins et dynamite". *Newspaper L'impartial*, n°502, 22nd of April 1900. Djidjelli, Algeria.

Anonymous. 1900f. *Newspaper Le Ralliement*, n°24, 8th of January 1900. Philippeville, Algeria.

Anonymous. 1900g. Article "Le dépeuplement des mers". *Newspaper La Dépêche Tunisienne*, n°3575, 22nd of April 1900. Tunis, Tunisia.

Anonymous. 1900h. Article "la capture des marsouins". *Newspaper Le Ralliement*, n°52, 19th of April 1900. Philippeville, Algeria.

Anonymous. 1901a. Article "Chasse aux marsouins". *Newspaper Le Petit Bastiais*, n°272, 22nd of November 1901. Bastia, France.

Anonymous. 1901b. *Conseil Général des Pyrénées-Orientales, 19th of* August 1901. Perpignan, France.

Anonymous. 1901c. *Conseil supérieur de gouvernement. Procès-verbaux des délibérations et exposé de la situation générale de l'Algérie. Session ordinaire de 1901.* Mustapha, Algeria.

Anonymous. 1902a. *Newspaper Les Nouvelles*, n°421, 29th of January 1902. Mustapha, Algeria.

Anonymous. 1902b. Article "Guerre aux marsouins". *Newspaper L'avenir de l'est, gazette Algérienne*, n°1,803, 14th of May 1902. Bône, Algeria.

Anonymous. 1902c. Article "Encore les marsouins". *Newspaper L'avenir de l'est, gazette Algérienne*, n°1,789, 29th of March 1902. Bône, Algeria.

Anonymous. 1903a. Article "Utilisation des marsouins". *Newspaper Le Progrès*, n°867, 7th of February 1903. Mascara, Algeria.

Anonymous. 1903b. Article "La Sardine : Des différentes pêches à la sardines et à l'Anchois. Du marsouin et de son utilité comme auxiliaire pour la pêche des sardines et anchois. De la méthode de pêche des Basques et des Espagnols à l'aide du marsouin". *Newspaper La Conserve Alimentaire*, n°7, July 1903. Meulan, France.

Anonymous. 1903c. Article "Les Marsouins". *Newspaper Le Petit Journal, Supplément illustré*, n°640, 22nd of February 1903. Paris, France.

Anonymous. 1903d. Article "Ces pauvres marsouins". *Newspaper Le Grand Echo du Nord et du Pas-de-Calais*, n° 157, 5th of June 1903. Lille, France.

Anonymous. 1903e. *Newspaper Le Petit Parisien, Supplément Illustré*, n°760, 30th of August 1903. Paris, France.

Anonymous. 1903f. Article "Emploi des chairs de marsouins". *Newspaper La Conserve Alimentaire* n°3, December 1903. Meulan, France.

Anonymous. 1903g. Gouvernement général de l'Algérie. *Délégations financières*, session de mai 1903. Alger, Algeria.

Anonymous. 1904a. *Newspaper L'impartial*, n°715, 5th of June 1904. Djidjelli, Algeria.

Anonymous. 1904b. Article "les marsouins à Djidjelli". *Newspaper Les Nouvelles*, n°1560, 9th of June 1904. Mustapha, Algeria.

Anonymous. 1908a. *Newspaper Le Progrès*, n°887, 27th of June 1908. Mascara, Algeria.

Anonymous. 1908b. *Gouvernement général de l'Algérie. Conseil supérieur de gouvernement, session ordinaire de 1908.* Alger, Algeria.

Anonymous. 1909. Ministère de la Marine. *Revue Maritime*. Paris, France, Tome 180.

Anonymous. 1912. Article "La mer alimentaire". *Newspaper L'Echo d'Alger*, n°219, 20th of October 1912.

Anonymous. 1913. *Newspaper Le petit oranais*, n°6373, 18th of November 1913.

Anonymous. 1922a. Article "Pour détruire les marsouin". *Newspaper Les Nouvelles*, n°7602, 6th–7th of February 1922. Mustapha, Algeria.

Anonymous. 1922b. Article "Pêche aux filets bœufs sur les côtes Algériennes". *Newspaper L'impartial*, n°1.638, 18th November 1922. Djidjelli, Algeria.

Anonymous. 1923a. Article "Nos sénateurs contre les belugas". *Newspaper Le Citoyen*, n°27, 5th of July 1923. Quimper, France.

Anonymous. 1923b. *Newspaper Les Nouvelles*, n°8543, 28th of February 1923. Mustapha, Algeria.

Anonymous. 1923c. Article "Un marsouin de grande taille". *Newspaper Les Nouvelles*, n°8547, 28th of February 1923. Mustapha, Algeria.

Anonymous. 1924. Article "Une belle pêche". *Newspaper L'Écho d'Alger*, n°5502, 5th of November 1924. Alger, Algeria.

Anonymous. 1925. *Newspaper L'Echo d'Alger*, n°5680, 27th of April 1925. Alger, Algeria.

Anonymous. 1928. *Newspaper L'Echo d'Alger*, n°6737, 21st of March 1928. Alger, Algeria.

Anonymous. 1929. Article "Les cétacés dans l'alimentation". *Revue scientifique illustrée*, n°1, January 1929. Paris, France.

Anonymous. 1930a. Article "La sardine - contre les bélugas". *Newspaper Le Breton socialiste*, n°85, 16th of August 1930. Morlaix, France.

Anonymous. 1930b. Article "Chasse aux marsouins". *Newspaper L'Echo d'Alger*, n°7735, 13th of December 1930. Alger, Algeria.

Anonymous. 1931a. Article "La chasse aux Bélugas". *Newspaper Le Citoyen*, n°41, 8th of October 1931. Quimper, France.

Anonymous. 1931b. Article "La chasse aux bélugas". *Newspaper L'Afrique du Nord illustrée : journal hebdomadaire d'actualités nord-africaines : Algérie, Tunisie, Maroc. Nouvelle Série*, n°534, 25th of July 1931. Alger, Algeria.

Anonymous. 1931c. *Newspaper L'Echo d'Alger*, n°7936, 2nd of July 1931. Alger, Algeria.

Anonymous. 1934a. Article "Le banquet de printemps de la Société nationale d'acclimatation". *Newspaper L'Echo d'Alger*, n°8706, 6th of June 1934. Alger, Algeria.

Anonymous. 1934b. *Newspaper L'Echo d'Alger*, n°8835, 13th of October 1934. Alger, Algeria.

Anonymous. 1934c. Article "Capture de deux marsouins". *Newspaper L'Echo d'Alger*, n°8762, 1st of August 1934. Alger, Algeria.

Anonymous. 1935. Article "Mangeons du marsouin". *Newspaper L'Écho d'Alger*, n°8952, 7th of February 1935. Alger, Algeria.

Anonymous. 1937. Article "Une chasse aux marsouins au large de Guyotville". *Newspaper L'Echo d'Alger*, n°9849, 24th of July 1937. Alger, Algeria.

Anonymous. 1938. *Newspaper L'Echo d'Alger*, n°10235, 14th of August 1938. Alger, Algeria.

Anonymous. 1939a. *Newspaper L'Echo d'Alger*, n°10584, 29th of July 1939. Alger, Algeria.

Anonymous. 1939b. *Newspaper L'Echo d'Alger*, n°10387, 13th of January 1939. Alger, Algeria.

Anonymous. 1939c. *Gouvernement général de l'Algérie. Conseil supérieur de gouvernement. Session ordinaire de 1939.* Alger, Algeria.

Anonymous. 1950. Article "Il y a cinquante ans, 10 août 1900". *Newspaper La Dépêche Tunisienne*, n°20658, 10th of August 1950. Alger, Algeria.

Aznar, F.J., Míguez-Lozano, R., Ruiz, B., de Castro, A.B., Raga, J.A. & Blanco, C. 2017. Long-term changes (1990–2012) in the diet of striped dolphins *Stenella coeruleoalba* from the western Mediterranean. *Marine Ecology Progress Series* **568**, 231–247.

Bearzi,, G. & Genov,, T. 2022. Imperiled common dolphins of the Mediterranean Sea. In *Imperiled: The Encyclopedia of Conservation*, D. DellaSala, M. Goldstein & M.J. Costello (eds.). Amsterdam: Elsevier, 837–846, doi:10.1016/B978-0-12-821139-7.00059-3

Bearzi, G., Genov, T., Natoli, A., Gonzalvo, J. & Pierce, G.J. 2022. *Delphinus delphis* (Inner Mediterranean subpopulation) (errata version published in 2022). The IUCN Red List of Threatened Species 2022: e.T189865869A210844387. (Accessed 23 July 2023).

Bearzi, G., Holcer, D. & Notarbartolo di Sciara, G. 2004. The role of historical dolphin takes and habitat degradation in shaping the present status of northern Adriatic cetaceans. *Aquatic Conservation: Marine and Freshwater Ecosystems* **14**(4), 363–379.

Bearzi, G., Politi, E., Agazzi, S. & Azzellino, A. 2006. Prey depletion caused by overfishing and the decline of marine megafauna in eastern Ionian Sea coastal waters (central Mediterranean). *Biological Conservation* **127**(4), 373–382.

Bearzi, G. & Reeves, R.R. 2021. Shifting baselines of cetacean conservation in Europe. *ICES Journal of Marine Science* **78**(7), 2337–2341.

Bearzi, G. & Reeves, R.R. 2022. Marine mammals foraging around fishing gear or preying on fishing catch and bait: it may not be "depredation". *ICES Journal of Marine Science* **79**, 2178–2183.

Bearzi, G., Reeves, R.R., Remonato, E., Pierantonio, N. & Airoldi, S. 2011. Risso's dolphin *Grampus griseus* in the Mediterranean Sea. *Mammalian Biology* **76**(4), 385–400.

Belon, P. 1551. *L'histoire naturelle des estranges poissons marins, avec la vraie peincture et description du Daulphin, et de plusieurs autres de son espèce, observée par Pierre Belon du Mans*. Paris: Imprimerie de R. Chaudière.

Bernal-Casasola, D. 2018. Whale hunting in the Strait of Gibraltar during the Roman period. *The SAA Archaeological Record* **18**, 15–22.

Bernal-Casasola, D., Gardeisen, A., Morgenstern, P., Horwitz, L. K., Piqués, G., Theodoropoulou, T. & Wilkens, B. 2016. Ancient whale exploitation in the Mediterranean: the archaeological record. *Antiquity* **90**(352), 914–927.

Berthelot, S. 1840. *De la pêche sur la côte occidentale d'Afrique et des établissements les plus utiles aux progrès de cette industrie*. Paris: Béthune Éditeur, Arthus Bertrand Libraire-Éditeur.

Birkun, A., Jr, Northridge, S.P., Willsteed, E.A., James, F.A., Kilgour, C., Lander, M. & Fitzgerald, G.D. 2014. *Studies for carrying out the Common Fisheries Policy: adverse fisheries impacts on cetacean populations in the Black Sea*. Final report to the European Commission, Brussels.

Blanco, C., Salomón, O. & Raga, J.A. 2001. Diet of the bottlenose dolphin (*Tursiops truncatus*) in the western Mediterranean Sea. *Journal of the Marine Biological Association of the United Kingdom* **81**(6), 1053–1058.

Bloch, D., Desportes, G., Harvey, P., Lockyer, C. & Mikkelsen, B. 2012. Life history of Risso's dolphin (*Grampus griseus*) (G. Cuvier, 1812) in the Faroe Islands. *Aquatic Mammals* **38**(3), 250–266.

Bonizzoni, S., Hamilton, S., Reeves, R.R., Genov, T. & Bearzi, G. 2022. Odontocete cetaceans foraging behind trawlers, worldwide. *Reviews in Fish Biology and Fisheries* **32**, 827–877.

Bonnaterre, P.J. 1789. *Tableau Encyclopédique et méthodique des trois règnes de la nature. Cétologie*. Paris: Panckoucke Libraire.

Borrell, A., Gazo, M., Aguilar, A., Raga, J.A., Degollada, E., Gozalbes, P. & García-Vernet, R. 2021. Niche partitioning amongst northwestern Mediterranean cetaceans using stable isotopes. *Progress in Oceanography* **193**, 102559.

Borrely, C. 2021. *État des lieux des cétacés conservés au Muséum d'histoire naturelle de Marseille*.

Bouvier, A. 1891. *Les mammifères de France*. Paris: Éditeur Georges Carré.

Bräger, Z., Gonzalvo, J., Agazzi, S. & Bearzi, G. 2016. Identification of bottlenose dolphin (*Tursiops truncatus*) prey using fish scale analysis. *Aquatic Mammals* **42**(1), 63.

Brito, C. 2011. Medieval and early modern whaling in Portugal. *Anthrozoös* **24**(3), 287–300.

Brito, C. & Vieira, N. 2010. Using historical accounts to assess the occurrence and distribution of small cetaceans in a poorly known area. *Journal of the Marine Biological Association of the United Kingdom* **90**(8), 1583–1588.

Cabezón, O., Resendes, A.R., Domingo, M., Raga, J.A., Agustí, C., Alegre, F., Mons, J.L., Dubey, J.P. & Almería, S. 2004. Seroprevalence of *Toxoplasma gondii* antibodies in wild dolphins from the Spanish Mediterranean coast. *The Journal of Parasitology* **90**(3), 643–644.

Caffarena, L. 1887. *Journal officiel de la République Française*, n°126.

Cauliez, J., Blaise, E., Cade, C., Desse, J., Desse-Berset, N., Durrenmath, G., Gilabert, C., Martin, S. & Vella, C. 2004. Paysage et implantations du Néolithique final à l'âge du Bronze ancien au Collet-Redon (Martigues-Bouches-du-Rhône). In *Paysages et peuplements. Aspects culturels et chronologiques en France méridionale*, P. Fouéré et al. (eds.). Périgueux: Actes des VIèmes rencontres méridionales de préhistoire récente, Volume 11, 125–139.

Caziot, E. 1913. *Bulletin de l'Association des naturalistes de Nice et des Alpes-Maritimes, bulletin n°2, liste des cétacés de la mer de Nice, avec indications permettant de les reconnaître*. Nice: Imprimerie Centrale.

Clavel, B. 2001. L'animal dans l'alimentation médiévale et moderne en France du Nord (XIIIe - XVIIe siècles). *Revue archéologique de Picardie* **19**, 9–204.

Companyo, L. 1841. *Catalogue descriptive des mammifères qui ont été observés et qui vivent dans le département des pyrénés-orientales*. Perpignan: Imprimerie J.-B. Alzine.

Companyo, L. 1867. Notice sur des cétacés échoués sur les côtes de la Méditerranée, entre Saint-Laurent-de-la-Salanque et Leucate, en février 1864. In *Société Agricole, Scientifique et Littéraire des Pyrénées-Orientales*. Perpignan: Imprimerie Charles Latrobe, Volume 15, 127–139.

Cucknell, A.-C., Frantzis, A., Boisseau, O., Romagosa, M., Ryan, C., Tonay, A.M., Alexiadou, P., Öztürk, A.A. & Moscrop, A. 2016. Harbour porpoises in the Aegean Sea, Eastern Mediterranean: the species' presence is confirmed. *Marine Biodiversity Records* **9**, 72.

Dars C., Méheust E., Genu M., Méndez-Fernandez P., Peltier H., Wund S., Caurant F., Dabin W., Demaret F., Spitz J. et Van Canneyt O. 2023. Les échouages de mammifères marins sur le littoral français en 2022. Rapport scientifique de l'Observatoire Pelagis, La Rochelle Université et CNRS, 49 pages.

Decugis, C. 2015. Mediterranean Prud'homies. In *Marine Productivity: Perturbations and Resilience of Socio-ecosystems*, H.J. Ceccaldi et al. (eds.). Cham: Springer, 295–298.

Di-Méglio, N., Romero-Alvarez, R. & Collet, A. 1996. Growth comparison in striped dolphins, *Stenella coeruleoalba*, from the Atlantic and Mediterranean coasts of France. *Aquatic Mammals* **22**(1), 11–21.

Dobney, K., Jaques, D., Barrett, J., Johnstone, C., Carrott, J., Hall, A., Herman, J., Nichols, C., Muldner, G. & Grimes, V. 2007. Exploitation of resources and procurement strategies. In *Farmers, Monks, and Aristocrats: The Environmental Archaeology of Anglo-Saxon Flixborough, Excavations at Flixborough 3*, K. Dobney et al. (eds.). Oxford: Oxbow Books, 190–212.

Dos Santos, M.E.D., Coniglione, C. & Louro, S. 2007. Feeding behaviour of the bottlenose dolphin, *Tursiops truncatus* (Montagu, 1821) in the Sado estuary, Portugal, and a review of its prey species. *Revista Brasileira de Zoociências* **9**(1), 31–40.

Drouet, H. 1887. *Alger et le Sahel*. Paris: Librairie Hachette et Cie.

Dubuc, A. 1968. Prix alimentaires et autres à Dieppe de 1690 à 1723 d'après le registre de comptabilité des Minimes. In *93è congrès national des sociétés savantes*. Paris: hist. mod., t. 111, 23–34.

Esteban, R., Verborgh, P., Gauffier, P., Giménez, J., Foote, A.D. & de Stephanis, R. 2016. Maternal kinship and fisheries interaction influence killer whale social structure. *Behavioral Ecology and Sociobiology* **70**(1), 111–122.

Faget, D. 2009. Les tueries de dauphins en Méditerranée ou l'impossible rationalisation d'un massacre (XIXe siècle-milieu XXe siècle). *Provence Historique* **59**(237), 379–396.

Farrugio, H. 2012. Données historiques sur les anciennes madragues françaises de Méditerranée. *Collective Volume of Scientific Papers ICCAT* 67(1), 112–123.

Fichou, J.C. & Levasseur, O. 2004. Pêcheurs contre "Marsouins" et "Belougas" (XVI-XIXème siècles). *Food and History* **2**(2), 53–86.

Fischer, P.H. 1881. *Cétacés du sud-ouest de la France, extrait des Actes de la Société Linnéenne de Bordeaux, t. XXXV*. Paris: F. Savy Libraire-Editeur.

Flower, W.H. 1879. On the common dolphin, *Delphinus delphis*, Linn. In Proceedings of the Zoological Society of London. London: Academic Press, 382–384.

Fontaine, M.C. 2016. Harbour porpoises, *Phocoena phocoena*, in the Mediterranean Sea and adjacent regions: biogeographic relicts of the last glacial period. *Advances in Marine Biology* **75**, 333–358.

Fontaine, M.C., Roland, K., Calves, I., Austerlitz, F., Palstra, F.P., Tolley, K.A., Ryan, S., Ferreira, M., Jauniaux, T., Llavona, A., Öztürk, B., Öztürk, A.A., Ridoux, V., Rogan, E., Sequeira, M., Siebert, U., Vikingsson, G.A., Borrell, A., Michaux, J.R. & Aguilar, A. 2014. Postglacial climate changes and rise of three ecotypes of harbour porpoises, *Phocoena phocoena*, in western Palearctic waters. *Molecular Ecology* **23**, 3306–3321.

Foskolos, I., Gkikopoulou, K. C. & Frantzis, A. 2020. Current state of knowledge and conservation perspectives on the cetaceans of the Aegean Sea. In *The Aegean Sea Environment: The Natural System*, C.L. Anagnostou et al. (eds.). Berlin: Springer, 1–28.

Garau, V.F. 1909. *Traité de pêche maritime pratique et des industries secondaires en Algérie*. Alger: Imprimerie P. Crescenzo.

Garrido, A. 2014. La relación entre pescadores y delfines en la Costa Brava en el discurso de la comunidad pesquera local. In *Conferencia leída en la jornada de estudios "Gestion et exploitation des ressources de la mer de l'époque moderne à nos jours". Marseille (MUCEM)*, 12 febrero, 18 pp.

Genov, T., Kotnjek, P. & Centrih, T. 2021. Occurrence of common dolphins (*Delphinus delphis*) in the Gulf of Trieste and the northern Adriatic Sea. *Aquatic Conservation: Marine and Freshwater Ecosystems* **31**, 69–75.

Gervais, P. 1864. Cétacés des côtes francaises de la Méditerranée. *Comptes Rendus de l'Academie des Sciences, Paris* **2**, 876–881.

Giménez, J., Marçalo, A., García-Polo, M., García-Barón, I., Castillo, J. J., Fernández-Maldonado, C., Saavedra, C., Santos, M.B. & de Stephanis, R. 2018. Feeding ecology of Mediterranean common dolphins: the importance of mesopelagic fish in the diet of an endangered subpopulation. *Marine Mammal Science* **34**(1), 136–154.

Giménez, J., Marçalo, A., Ramirez, F., Verborgh, P., Gauffier, P., Esteban, R., Nicolau, L., González-Ortegón, E., Baldó, F., Vilas, C., Vingada, J., Forero, M.G. & de Stephanis, R. 2017. *Diet of bottlenose dolphins (Tursiops truncatus) from the Gulf of Cadiz: insights from stomach content and stable isotope analyses. PLoS One* **12**(9), e0184673.

Gladilina, E.V. & Gol'Din, P.E. 2014. *New prey fishes in diet of black sea bottlenose dolphins, Tursiops truncatus* (Mammalia, Cetacea). *Vestnik zoologii* **48**(1), 83.

Gómez-Campos, E., Borrell, A., Cardona, L., Forcada, J. & Aguilar, A. 2011. *Overfishing of small pelagic fishes increases trophic overlap between immature and mature striped dolphins in the Mediterranean Sea. PLoS One* **6**, e24554.

Gonzalvo, J., Giovos, I. & Moutopoulos, D.K. 2015. Fishermen's perception on the sustainability of small-scale fisheries and dolphin-fisheries interactions in two increasingly fragile coastal ecosystems in western Greece. *Aquatic Conservation: Marine and Freshwater Ecosystems* **25**(1), 91–106.

Gourret, P. 1894. *Les pêcheries et les poissons de la Méditerranée (Provence).* Paris: Libraire J.-B. Baillière et fils.

Guizard, F. 2018. Delfines nec non et ballenae... Les cétacés de l'Atlantique nord au haut Moyen Âge: représentation, identification et consommation. *Anthropozoologica* **53**(1), 115–123.

Harmer, S.F. 1926. A propos des "bélugas". In *La Nature, revue des sciences et de leurs applications aux arts et à l'industrie, Journal hebdomadaire illustré*, n°2736, 11th of September 1926. Paris: Masson et Cie.

Holm, P., Smith, T.D. & Starkey, D.J. (eds.). 2001. *The Exploited Seas: New Directions for Marine Environmental History.* Liverpool: Liverpool University Press, doi:10.2307/j.ctt21pxj1m

Imbert, A. 1900. *Notice sur les services maritimes de l'Algérie.* Alger: Giralt Imprimeur-Photographeur.

Institut national de la statistique et des études économiques. 2022. *Convertisseur franc-euro; Pouvoir d'achat de l'euro et du franc.* https://www.insee.fr/fr/information/2417794 (Accessed 4 July 2023).

Jonnart, C. 1904. *Exposé de la situation générale de l'Algérie, année 1904.* Alger: Imprimerie Victor Heintz.

Jonnart, C. 1906. *Exposé de la situation générale de l'Algérie, année 1905.* Alger: Imprimerie Victor Heintz.

Jonnart, C. 1907. *Exposé de la situation générale de l'Algérie, année 1906.* Alger: Imprimerie Victor Heintz.

Jonnart, C. 1908. *Exposé de la situation générale de l'Algérie, année 1907.* Alger: Imprimerie Victor Heintz.

Jonnart, C. 1910. *Exposé de la situation générale de l'Algérie, année 1909.* Alger: Imprimerie Victor Heintz.

Jourdain, E., Ugarte, F., Víkingsson, G.A., Samarra, F.I., Ferguson, S.H., Lawson, J., Vongraven, D. & Desportes, G. 2019. North Atlantic killer whale *Orcinus orca* populations: a review of current knowledge and threats to conservation. *Mammal Review* **49**(4), 384–400.

Larbi Doukara, K. 2015. *État de la conservation et évolution de la population des petits delphinidés: Grand dauphin Tursiops truncatus et Dauphin commun Delphinus delphis dans le littoral occidental algérien.* PhD thesis, University of Oran, Algeria.

Larbi Doukara, K. 2021. A stranding record of the short-beaked common dolphin (*Delphinus delphis*) in Algerian West Coast, during 2008–2012. *Aquatic Conservation: Marine and Freshwater Ecosystems* **31**, 8–14.

Larousse, P. 1870. *Grand dictionnaire universel du XIX*è *siècle.* Paris: Administration du Grand Dictionnaire Universel, Tome 6.

Le Gall, T. 1904. *L'industrie de la pêche dans les ports sardiniers bretons, crise et palliatifs.* Rennes: Imprimerie Guillemin et Voisin.

Legendre, R. 1912. Notes sur le système nerveux central d'un dauphin (*Delphinus delphis*). In *Archives d'anatomie microscopique*, L. Ranvier & L.-F. Henneguy (eds.). Paris: Masson et Cie éditeurs, Tome 13.

Legendre, R. 1926a. La chasse aux "bélugas" et l'utilisation des petits cétacés. In *La Nature, revue des sciences et de leurs applications aux arts et à l'industrie, Journal hebdomadaire illustré*, n°2723, 12th of June 1926. Paris: Masson et Cie.

Legendre, R. 1926b. Les cétacés de nos côtes. In *La Nature, revue des sciences et de leurs applications aux arts et à l'industrie, Journal hebdomadaire illustré*, n°2721, 26th of May 1926. Paris: Masson et Cie.

Legendre, R. 1929. Les Belugas. In *Recherches et inventions*, n°180, September 1929. Levallois-Perret, France.

Léotard, J. 1894. La guerre aux marsouins. In *La Nature, revue des sciences et de leurs applications aux arts et à l'industrie, Journal hebdomadaire illustré,* n°*1109, 1st of* September 1894. Paris: Masson et Cie.

Lesson, P. 1838. *Compléments de Bouffon, tome premier, Livre XI.* Paris: Pourrat Frères Eds.

Lightfoot, E., Boneva, B., Miracle, P. T., Šlaus, M. & O'connell, T.C. 2011. Exploring the Mesolithic and Neolithic transition in Croatia through isotopic investigations. *Antiquity* **85**(327), 73–86.

Loche, V.J.-F. 1858. *Catalogue des mammifères et des oiseaux observés en Algérie.* Paris: Librairie Arthus Bertrand.

Loche, V.J.-F. 1860. Travaux inédits, description de deux nouvelles espèces du genre Dauphin. In *Revue et magasin de zoologie pure et appliquée,* F.E. Guérin-Méneville (ed.). Paris: Revue et magasin de zoologie, 473–479.

Lutaud, C. 1911. *Exposé de la situation générale de l'Algérie, année 1910.* Alger: Imprimerie Victor Heintz.

Mannino, M.A., Talamo, S., Tagliacozzo, A., Fiore, I., Nehlich, O., Piperno, M., Tusa, S., Collina, C., Di Salvo, R., Schimmenti, V. & Richards, M.P. 2015. *Climate-driven environmental changes around 8,200 years ago favoured increases in cetacean strandings and Mediterranean hunter-gatherers exploited them. Scientific Reports* **5**(1), 1–12.

Marçalo, A., Giménez, J., Nicolau, L., Frois, J., Ferreira, M., Sequeira, M., Eira, C., Pierce, G.J. & Vingada, J. 2021. Stranding patterns and feeding ecology of striped dolphins, *Stenella coeruleoalba*, in Western Iberia (1981–2014). *Journal of Sea Research* **169**, 101996.

Marçalo, A., Nicolau, L., Giménez, J., Ferreira, M., Santos, J., Araújo, H., Silva, A., Vingada, J. & Pierce, G.J. 2018. *Feeding ecology of the common dolphin (Delphinus delphis) in Western Iberian waters*: has the decline in sardine (*Sardina pilchardus*) affected dolphin diet? *Marine Biology* **165**(3), 1–16.

Maynou, F., Sbrana, M., Sartor, P., Maravelias, C., Kavadas, S., Damalas, D., Cartes, J.E. & Osio, G. 2011. *Estimating trends of population decline in long-lived marine species in the Mediterranean Sea based on fishers' perceptions. PLoS One* **6**(7), e21818.

Mazzoldi, C., Bearzi, G., Brito, C., Carvalho, I., Desiderà, E., Endrizzi, L., Freitas, L., Giacomello, E., Giovos, I., Guidetti, P., Ressurreição, A., Tull, M. & MacDiarmid, A. 2019. *From sea monsters to charismatic megafauna: changes in perception and use of large marine animals. PLoS One* **14**(12), e0226810, doi:10.1371/journal.pone.0226810

Meliadò, E., Bavestrello, G., Gnone, G. & Cattaneo-Vietti, R. 2020. Historical review of dolphin bounty hunting in Italy with a focus on the period 1927–37. *Journal of Cetacean Research and Management* **21**(1), 25–31.

Meotti, C. & Podestà, M. 1997. Stomach contents of striped dolphins, *Stenella coeruleoalba* (Meyen, 1833), from the Western Ligurian, Sea (Cetacea, Delphinidae). *Atti della Società Italiana di Scienze Naturali e del Museo Civico di Storia Naturale di Milano* **137**, 5–15.

Meynier, L., Pusineri, C., Spitz, J., Santos, M.B., Pierce, G.J. & Ridoux, V. 2008. Intraspecific dietary variation in the short-beaked common dolphin *Delphinus delphis* in the Bay of Biscay: importance of fat fish. *Marine Ecology Progress Series* **354**, 277–287.

Milani, C.B., Vella, A., Vidoris, P., Christidis, A., Koutrakis, E., Frantzis, A., Miliou, A. & Kallianiotis, A. 2018. Cetacean stranding and diet analyses in the North Aegean Sea (Greece). *Journal of the Marine Biological Association of the United Kingdom* **98**(5), 1011–1028.

Musset, L. 1964. Quelques notes sur les baleiniers Normands du Xè au XIIIè siècle. *Revue d'histoire économique et sociale* **42**(2), 147–161.

Øien, N. 1988. The distribution of killer whales (*Orcinus orca*) In the North Atlantic based on Norwegian catches, 1938–1981, and incidental sightings, 1967–1987. *Rit Fiskid* **11**, 65–78.

Panigada, S., Pierantonio, N., Araujo, H., David, L., Di-Meglio, N., Doremus, G., Gonzalvo, J., Holcer, D., Laran, S., Lauriano, G., Paiu, R-M., Perri, M., Popov, D., Ridoux, V., Vazquez, J.A., & Canadas, A. 2024. The ACCOBAMS survey initiative: the first synoptic assessment of cetacean abundance in the Mediterranean Sea through aerial surveys. Front. Mar. Sci. **10**, 1270513. doi: 10.3389/fmars.2023.1270513

Papadopoulos, J.K. & Paspalas, S.A. 1999. Mendaian as Chalkidian wine. *Hesperia: The Journal of the American School of Classical Studies at Athens* **68**(2), 161–188.

Pardalou, A. & Tsikliras, A.C. 2018. Anecdotal information on dolphin-fisheries interactions based on empirical knowledge of fishers in the northeastern Mediterranean Sea. *Ethics in Science and Environmental Politics* **18**, 1–8.

Pascual Benito, J.L., Sanchis, A. & Beneyto, J.P. 2019. Ballenas y delfines en la prehistoria peninsular: los restos de cetáceo de La Vital (Gandia). *Recursos marins en el passat, IV Jornades d'arqueozoologia, Museu de Prehistòria de València*, 165–192.

Pérard, J. 1925. *Comptes rendus des séances du IXè congrès national des pêches et industries maritimes tenu à Bordeaux, 14–19th of Septembre 1925*. Orléans: Imprimerie du Loiret, 25–27.

Perrier, E. 1889. Rapport adressé au Ministre de la Marine au nom du comité des pêches maritimes. *Journal Officiel de la République* **138**.

Plagányi, É.E. & Butterworth, D.S. 2009. Competition with fisheries. In *Encyclopedia of Marine Mammals*, W.F. Perrin et al. (eds.). Academic Press, 2nd edition, 269–275.

Pourcel, L. 1910. Le dauphin (*Delphinus delphis mediterraneus*). In *Annales de la Société d'Histoire Naturelle de Toulon, année 1910*. Toulon: Imprimerie Mouton, 55–58.

Praca, E., Laran, S., Lepoint, G., Thomé, J.P., Quetglas, A., Belcari, P., Sartor, P., Dhermain, F., Ody, D., Tapie, N., Budzinski, H. & Das, K. 2011. Toothed whales in the northwestern Mediterranean: insight into their feeding ecology using chemical tracers. *Marine Pollution Bulletin* **62**(5), 1058–1065.

Reiner, F. & Brito, C. 2012. Baleias e golfinhos sob mira: A baleação e outras capturas históricas. In *Cetáceos de Portugal: Passado, Presente e Futuro*, C. Brito & I. Carvalho (eds.). Lisboa: Escola de Mar. Paleta Natura, nº 2, 27–35.

Révoil, P. 1902. *Exposé de la situation Générale de l'Algérie, 1902*. Alger: Imprimerie Victor Heintz.

Rolland, E. 1877. *Faune populaire de la France, Les mammifères sauvages*. Paris: Maisonneuve & Cie, Libraires-Éditeurs.

Ropers, L. 1906. *Exposition de la condition économique et sociale du pêcheur sardinier*. PhD thesis, Faculté de droit de l'Université de Paris, France.

Saavedra, C., García-Polo, M., Giménez, J., Mons, J.L., Castillo, J.J., Fernández-Maldonado, C., de Stephanis, R., Pierce, G.J. & Santos, M.B. 2022. Diet of striped dolphins (*Stenella coeruleoalba*) in southern Spanish waters. *Marine Mammal Science* **38**(4), 1566–1582.

Samarra, F.I. & Esteban, R. 2022. Killer Whale *Orcinus orca* (Linnaeus, 1758). In *Handbook of the Mammals of Europe*, K. Hackländer & F.E. Zachos (eds.). Cham: Springer International Publishing.

Santos, M.B., Fernandez, R., López, A., Martínez, J.A. & Pierce, G.J. 2007. Variability in the diet of bottlenose dolphin, *Tursiops truncatus*, in Galician waters, north-western Spain, 1990–2005. *Journal of the Marine Biological Association of the United Kingdom* **87**(1), 231–241.

Scheinin, A.P., Kerem, D., Lojen, S., Liberzon, J. & Spanier, E. 2014. Resource partitioning between common bottlenose dolphin (*Tursiops truncatus*) and the Israeli bottom trawl fishery? Assessment by stomach contents and tissue stable isotopes analysis. *Journal of the Marine Biological Association of the United Kingdom* **94**(6), 1203–1220.

Sharir, Y., Kerem, D. & Spanier, E. 2011. Small size in the common bottlenose dolphin *Tursiops truncatus* in the eastern Mediterranean: a possible case of Levantine nanism. *Marine Ecology Progress Series* **438**, 241–251.

Silber, G.K. & Fertl, D. 1995. Intentional beaching by bottlenose dolphins (*Tursiops truncatus*) in the Colorado River Delta, Mexico. *Aquatic Mammals* **21**, 183–186.

Silva, M.A. 1999. Diet of common dolphins, *Delphinus delphis*, off the Portuguese continental coast. *Journal of the Marine Biological Association of the United Kingdom* **79**(3), 531–540.

Smith, T.D. 1995. Interactions between marine mammals and fisheries: an unresolved problem for fisheries research. In Whales, Seals, Fish *and* Man, A.S. Blix et al. (eds.). Elsevier Science, Volume **4**, 527–536.

Snape, R.T.E., Broderick, A.C., Çiçek, B.A., Fuller, W.J., Tregenza, N., Witt, M.J. & Godley, B.J. 2018. Conflict between dolphins and a data-scarce fishery of the European Union. *Human Ecology* **46**, 423–433.

Sokou, V., Gonzalvo, J., Giovos, I., Brito, C. & Moutopoulos, D. 2022. Tracing dolphin-fishery interaction in early Greek fisheries. *Annals for Istrian and Mediterranean Studies* **32**(2), 443–450.

Spitz, J., Richard, E., Meynier, L., Pusineri, C. & Ridoux, V. 2006b. Dietary plasticity of the oceanic striped dolphin, *Stenella coeruleoalba*, in the neritic waters of the Bay of Biscay. *Journal of sea Research* **55**(4), 309–320.

Spitz, J., Rousseau, Y. & Ridoux, V. 2006a. Diet overlap between harbour porpoise and bottlenose dolphin: an argument in favour of interference competition for food? *Estuarine, Coastal and Shelf Science* **70**(1–2), 259–270.

Steeg, T. 1922. *Exposé de la situation générale de l'Algérie, année 1921*. Alger: Imprimerie Victor Heintz.

Szabo, V.E. 2008. *Monstrous Fishes and the Mead-Dark Sea: Whaling in the Medieval North Atlantic. The Northern World*. Leiden, the Netherlands: Brill, Volume **35**.

Teixeira, A.M.A.P. 1979. Marine mammals of the Portuguese coast. *Zeitschrift für Säugetierkunde* **44**, 221–238.

Tonay, A.M. & Öztürk, A.A. 2012. Historical records of cetacean fishery in the Turkish seas. *Journal of Black Sea/Mediterranean Environment* **18**(3), 388–399.

Trantalidou, K. 2008. Glimpses of Aegean island communities during the Mesolithic and Neolithic periods: the zooarchaeological point of view. In *A Colloquium on the Prehistory of the Cyclades*, N. Brodie et al. (eds.). Cambridge: McDonald Institute Monographs, 19–27.

Valdés-Hansen, F. 2004. El problema pesquero con los delfines y su persecución en Galicia (siglos XIII al XX). *Cuadernos de Estudios Gallegos* **LI**(117), 313–362.

Valdés-Hansen, F. 2009. Pescadores y delfines en el norte de España. Historia de su interacción desde la Edad Media hasta el siglo XX. *Itsas memoria: revista de estudios marítimos del País Vasco* **6**, 629–641.

Van Beneden, P.J. 1889. *Histoire Naturelle des Delphinidés des mers d'Europe*. Bruxelles: P. Hayes, Académie royale de Belgique.

Van Beneden, P.J. & Gervais, P. 1880. *Ostéographie des cétacés vivants et fossiles, comprenant la description et l'iconographie du squelette et du système dentaire de ces animaux, ainsi que des documents relatifs à leur histoire naturelle*. Paris: Ed. Arthus Bertrand.

van den Hurk, Y. 2020. *Cetacean exploitation in Medieval northern and Western Europe: zooarchaeological, historical, and social approaches*. PhD thesis, University College London, England.

van den Hurk, Y. & McGrath, K. 2021. Whaling in Iron Age to post-medieval Scotland; a zooarchaeological and biomolecular study of cetacean remains from selected sites in Caithness, the Orkney Islands and the Shetland Islands. *Proceedings of the Society of Antiquaries of Scotland* **150**, 451–474, doi:10.9750/PSAS.150.1324

van den Hurk, Y., Riddler, I., McGrath, K. & Speller, C. 2023. Active whaling, opportunistic scavenging or long-distance trading: zooarchaeological, palaeoproteomic, and historical analyses on whale exploitation and bone working in Anglo-Saxon Hamwic. *Medieval Archaeology* **67**(1), 137–158, doi:10.1080/00766097.2023.2204674

van den Hurk, Y., Rielly, K. & Buckley, M. 2021. Cetacean exploitation in Roman and medieval London: reconstructing whaling activities by applying zooarchaeological, historical, and biomolecular analysis. *Journal of Archaeological Science: Reports* **36**, 102795.

Würtz, M. & Marrale, D. 1993. Food of striped dolphin, *Stenella coeruleoalba*, in the Ligurian Sea. *Journal of the Marine Biological Association of the United Kingdom* **73**(3), 571–578.

Zaeschmar, J. & Estrela, G. 2020. False Killer Whale *Pseudorca crassidens* (Owen, 1846). In *Handbook of the Mammals of Europe*, K. Hackländer & F.E. Zachos (eds.). Cham: Springer, doi:10.1007/978-3-319-65038-8_102-1

Supplementary Materials are provided online at: https://www.routledge.com/9781032761961

Oceanography and Marine Biology: An Annual Review, 2024, **62**, 163-191

A LONG HISTORY: SOCIAL-ECOLOGICAL SYSTEMS AS DRIVERS OF OYSTER REEF LOSS IN THE PEARL RIVER DELTA AND THE BROADER ASIAN REGION

JESSICA M. WILLIAMS[1], SALLY C.Y. LAU[2], MARINE THOMAS[3], CHUN-SEE TSAO[3], AND BAYDEN D. RUSSELL[4,5,6,*]

[1]The Centre for Civil Society and Governance, The University of Hong Kong, Hong Kong SAR, China
[2]Centre for Sustainable Tropical Fisheries and Aquaculture and College of Science and Engineering, James Cook University, Townsville, QLD, Australia
[3]The Nature Conservancy Hong Kong, Hong Kong SAR, China
[4]The Swire Institute of Marine Science and Area of Ecology and Biodiversity, School of Biological Sciences, The University of Hong Kong, Hong Kong SAR, China
[5]Institute for Climate and Carbon Neutrality, The University of Hong Kong, Hong Kong SAR, China
[6]The Joint Laboratory for Marine Ecology and Environmental Sciences (JLMEES), The Swire Institute of Marine Science, The University of Hong Kong, Hong Kong SAR, China
**Corresponding author: brussell@hku.hk*

Abstract It is well established that oyster reefs and associated ecosystems provide a range of important ecosystem functions and services beneficial to the human communities. Yet, oyster reefs have been heavily degraded or extirpated by historical and ongoing exploitation and coastal development globally. Consequently, restoration projects are underway across the world to re-establish these ecosystems. Knowing where oyster reefs were historically located, as well as the drivers of their destruction, provides important insights into these projects and increases success. In the Pearl River Delta, and the Asian region in general, there is a paucity of knowledge regarding the historical presence, functioning and societal relationships surrounding oyster reefs. Here we present a historical review of the social and ecological relationship between humans and oysters in the Pearl River Delta region, covering documents spanning 7000 years, and contextualise our findings within the broader Asian region. Our findings reveal evidence of extensive natural oyster habitats in the region, leading to a long history of social and ecological relationships between humans and oysters. Imbalances between the social and ecological systems, however, led to the demise of reefs in the region through overexploitation and hold important lessons for current and future revitalisation attempts.

Keywords: Aquaculture; Asia; Ecosystem Degradation; Fisheries; Historical Ecology; Oyster Reef; Pearl River Delta; Social-Ecological Systems

DOI: 10.1201/9781003477518-3

Introduction

Oyster reefs historically formed significant components of many temperate and subtropical inshore coastal systems but have suffered extensive declines globally (Beck et al. 2011), with a concurrent loss of services and the associated financial benefits (Grabowski et al. 2012). While restoration programmes to re-establish these ecosystems are becoming more common, the extirpation of these systems generally occurred before living memory, creating a 'shifting baseline' (Pauly 1995) and making it difficult to determine target locations for restoration.

Oyster reefs provide a variety of environmental, social and economic benefits, including essential ecosystem services (Chan et al. 2022). The ecosystem functions provided by reefs form the basis of healthy and sustainable subtropical and temperate estuarine waters (Smaal et al. 2019, Chan et al. 2022). They protect shorelines (Ysebaert et al. 2018), shielding valuable infrastructure from extreme waves and storm surge events; prevent erosion; and provide coastal stability, having the ability to grow alongside sea level rise and so provide continuous protection over time (Rodriguez et al. 2014). They also play a large role in pollution amelioration through water filtration and nutrient cycling (Kellogg et al. 2014) and support other important habitats such as mangroves. This benefits society through improved water quality, and so public health (Zu Ermgassen et al. 2016). Oysters also provide an important source of protein for local communities, both directly through oyster meat and indirectly by dramatically enhancing biodiversity, crustacean and fish production (Zu Ermgassen et al. 2016). Economic opportunities for communities are also created through enhanced oyster farming (from increased larval supply) (Schulte & Burke 2014), fishing and ecotourism where healthy oyster reefs exist in coastal areas (Rusher 2003).

Globally, over 85% of oyster reefs have been lost in the last century, the largest documented global loss for any coastal ecosystem. This understanding, however, is largely based on developed nations such as North America, Europe and Australia (Beck et al. 2011, Gillies et al. 2018, Pogoda 2019, Grabowski et al. 2012). The more extensive documentation in these regions of declines in commercial fisheries and oyster reefs in the nineteenth century provides information vital for reef restoration, particularly locations to focus restoration efforts. In Asia, however, the historical extent and current status of oyster reefs have not been assessed, largely because human alteration of coastal environments has been occurring for over 1000 years, making it challenging to map socio-ecological changes with quantitative certainty.

The Pearl River Delta (PRD) is the second largest estuary in China and thirteenth largest estuary in the world (Chen et al. 2008). Understanding the historical status of oyster reefs in the PRD region, and Asia more generally, is pertinent as tropical and subtropical cities in East and Southeast Asia tend to possess high biodiversity, making ecosystem loss especially catastrophic (Tittensor et al. 2010). Such regions are characterised by large, rapidly increasing, human populations and exponential infrastructure development (Yeung 2001). Sustaining ecosystem services, protecting biodiversity and ensuring functional sustainable social-ecological systems require not only preventing these ecosystems from further degradation but also effective restoration (McDonnell & MacGregor-Fors 2016, Lau et al. 2020).

The theory of historical ecology has demonstrated the utility of drawing on historical perspectives to increase understanding of the complex and dynamic nature of landscapes to enable a frame of reference for the assessment of contemporary patterns and processes. Such historical insights can provide reference conditions when setting management goals (Swetnam et al. 1999). In this vein, the relationship between the social and ecological systems embedded within the history of oyster reefs in the PRD region can provide important insights into their cultural and ecological roles as well as aid future restoration efforts to support environmental, societal and cultural goals. Understanding oyster reefs as a social-ecological system (SES) stresses the integrated concepts of humans within nature and that the separation of economic and social systems is contrived and arbitrary (Folke et al. 2005).

Importantly, a SES approach recognises that ecosystem services are generated not solely by ecosystems but also by SES (Carpenter et al. 2012) and so people and their actions are essential to sustaining healthy, functioning ecosystems and their services. The survival of social systems depends on their interrelations with natural resource systems and the environment, and natural resources are conditioned by the actions of human populations (Schouten et al. 2009, Ambrosio-Albala & Delgado 2008). Therefore, drawing on SES and marine historical ecology helps us to address how and when ecosystems changed, as well as the more pertinent questions of why the changes occurred and the socio-ecological outcomes and interactions (Thurstan 2022). This can provide new perspectives on socio-ecological interactions around marine ecosystems as well as the health of such habitats, helping to challenge misconceptions surrounding shifting baselines and provide decision-makers with more accurate frames of reference when establishing goals and policies (Alleway & Connell 2015, Thurstan 2022).

Here, we undertake a review of historical documents to reconstruct the transformations of oyster reef ecosystems in the PRD region with a focus on the changing societal and cultural relationships related to the presence and use of shellfish (particularly oyster) resources. By integrating data from the archaeological, anthropological and geographical literature, government and media accounts, we locate historical shellfish presence and human activities that have led to shellfish habitat changes in the region over the past 7000 years. Firstly, we present a synopsis of the information contained in documents, historical records and first-hand accounts translated from Chinese (both Traditional and Simplified; >130 translations) to make it accessible in the published scientific (English) literature. These documents are contained in historical archives, but until now only in Chinese. We combine this information with a review of the contemporary published literature about the PRD region.

Finally, to aid the evaluation of cultural and natural causes of variability and gain a broader understanding of the dynamic characteristics of ecosystems (Swetnam et al. 1999), we increase the number of comparative historical perspectives from different locations by integrating the review of the PRD with a review of the scientific literature documenting the status (and loss) of oyster reefs across South, Southeast and East Asia. This review is a comprehensive longitudinal synthesis on understanding social and ecological relationships surrounding shellfish, in particular historical oyster reefs, in the southern China region. We also explore the social and ecological relationships between humans and oysters for the wider Asia continent, which provides important insights for restoration efforts going forward.

Background

Oyster reefs in Asia

In Asia, coastal habitat loss is historically driven by economic factors (Williams et al. 2016), while socio-cultural factors account for ongoing loss and a lack of protection or restoration (Lau et al. 2020). The perception in regional Asia tends to be that the value of ecosystems is in their use for local benefits, often for the continuation of 'traditional' harvesting, and requires adequate support and compensation to stop exploitation (Bennett & Dearden 2014). On top of this, over the last several hundred years, coastal cities in the region have undergone dramatic landscape changes, eroding memories of the historical conditions of ecosystems and habitats (Turvey et al. 2010, Lau et al. 2020). As a result, the prevailing assumption is that oyster reefs were not historically extensive, and restoration efforts are perceived as unwarranted in Asia compared to other regions of the world where oyster-related ecological and human histories are more understood.

In contemporary Asia, there are 30 recorded extant species of reef-building oyster species (family Osteridae), including 11 species of *Magallana* (formerly *Crassostrea*). The majority of the known natural oyster populations are reportedly over-harvested, most to the point of functional extinction, and oyster production is maintained by aquaculture. The *Magallana* species have been documented as forming natural beds and reefs along Asian intertidal and shallow subtidal shorelines in northern China (Quan et al. 2012), southern China (Lau et al. 2020), Thailand (Brohmanonda et al. 1988) and Japan (Smith et al. 2018). It is only recently that the extent to which natural oyster habitats have disappeared in these regions is starting to be realised; natural oyster beds and reefs have become increasingly sparse, having effectively become functionally extinct in the region (Lau et al. 2020).

Lack of knowledge as a barrier to restoration

Restoring oyster reefs along urbanised coastlines has come to be recognised as a powerful strategy for mitigating environmental problems typically associated with coastal development, for example, eutrophication (Kellogg et al. 2014), storm surge damage (Walles et al. 2016) and declines in biodiversity (Coleman & Williams 2002). Supporting these natural functions has historically been challenging as until recently ecosystem 'management' focused on reshaping the natural environment to suit human populations and economies. This economic focus did not consider long-term sustainability and served to increase the rate of loss of ecosystems and their associated functions (Crain et al. 2009, Robinson 2012). This anthropocentric view of what is "desirable" or "useful" is a typical example of disconnecting the social system from the ecological, resulting in an unbalanced, and so unsustainable, social-ecological system.

Basing natural resource management decisions on contemporary knowledge can be problematic as it can contribute to misleading management targets and continued patterns of degradation through the presence of shifting baselines (Alleway & Connell 2015). Shifting baseline syndrome (Pauly 1995) is the result of intergenerational shifts in expectations about how a 'natural' ecosystem should function. While this phenomenon can occur in any ecosystem that undergoes change across generations (Hanazaki et al. 2013, Papworth et al. 2009), the inability of the general population to directly observe many marine ecosystems, combined with their largely undocumented history, means shifting baseline syndrome is particularly prevalent in marine systems (Pauly 1995, Thurstan 2022). As such, it is imperative to accurately quantify historical baselines and understand past loss when looking to restore reefs as this will provide more realistic expectations and allow for more accurate conservation modelling (Alleway & Connell 2015). Major programmes to restore oyster reefs in North America (Baggett 2014) and Europe (Laing et al. 2006), some of which have been ongoing for decades, have been extremely successful in restoring the structure and function of oyster reef systems (Hernández et al. 2018) because of the well documented and understood historical status and extent of reefs in these regions. Restoration is still much more limited in the southern hemisphere despite the documented collapse of reefs (Cranfield et al. 2001). The exception here being some recent successful (and still expanding) efforts undertaken in places like Australia (McAfee et al. 2020) where a historical baseline has been established (Alleway & Connell 2015).

While the ecological context of restoration is important, restoration cannot be successful without the support of the local community (McAfee et al. 2021). The socio-economic and cultural facets of such systems need to be properly understood and aligned for successful revitalisation to occur (Williams et al. 2021, McAfee et al. 2020). Generating self-sustaining oyster reefs in heavily developed and populated coastal megacities like Hong Kong and the wider PRD region has been problematic for two major reasons: (1) natural reefs were overexploited to the point of extirpation prior to living memory, leading to a general misconception that they are not naturally extensive on the shores of the region; and (2) there is a general perception that the oyster culture industry is a historical form of economy in decline. Ultimately, it is the lack of knowledge about the historical extent and important services of oyster reefs in Asia, coupled with the perception that their value is in exploitation, which forms the largest barrier to restoration efforts (Caswell et al. 2020). Therefore, here we provide a review of the historical documentation of oyster reefs in Hong Kong and the broader PRD and their importance in SESs in the broader Asian context.

Literature search

A review of historical documents was undertaken focused on the Pearl River Delta (PRD) region, China. Between 2018 and 2023, historical documents were located via extensive searches in multiple online databases and electronic document repositories; the Google Search Engine, Google Scholar, HathiTrust Digitial Library (https://www.hathitrust.org/), Internet Archive (https://archive.org/), JSTOR (https://www.jstor.org/), Biodiversity Heritage Library (https://www.biodiversitylibrary.org/), the Chinese Text Project (https://ctext.org/), Multimedia Information Service of the Hong Kong Library (https://mmis.hkpl.gov.hk/) and the University of Hong Kong library archive using keyword combinations varying between "Zhujiang", "Pearl River Delta", "Pearl River Estuary", "Pearl River", "Kwantung", "Canton", "oyster", "reef", "lime", "shell" and "shellfish" in both English and Chinese (traditional and simplified). Chinese documents were translated into English.

The documents were then coded into themes based on the topics that they discussed. These themes included shoreline change, documented presence of shells, shells as tools, aquaculture, shell-related economy, harvesting and extraction, pearl oysters, decline of habitat, cultural references, shellfish as food, oysters as not being important and maps. If a document dealt with multiple themes, they were coded in each category to ensure all the relevant information was included. These documents were used to construct a timeline of the history of social interactions with oysters and the habitats they formed. Where documents spanned more than one time period, they were included in the counts for both time periods. This allowed a more accurate understanding of the oyster-related activities of the time periods to be built as it enabled a greater level of detail to be harvested from the data and avoided arbitrary pigeonholing of sources into one category. Historical descriptions of currencies were not adjusted to present-day inflation to reflect oyster-related economy largely as some of the documents pre-date accurate inflation records; however, we have standardised the currencies to USD to aid understanding.

To increase the number of comparative histories from different locations, this longitudinal study is accompanied by a more succinct review of the published historical studies of oyster-related SES in the wider Asian region in Google Scholar using the keyword combinations of "history", "oyster" and "[country name]", where the [country name] used were China, Japan, Korea, Thailand, Vietnam, Cambodia, India, Bangladesh, Sri Lanka and Taiwan.

Analysis of the history of oysters in the PRD

Overall, our searches identified 134 sources that fitted the search criteria. These sources included academic journal articles, books and book chapters, newspaper articles, maps, ancient chorography, translations from historic literature and government records and reports. Of these, 61 were in Chinese and were translated into English. From these, we excluded documents with statements or references that we could not ascertain if they were geographically referring to the PRD. This left 51 different documents containing 129 references to oysters, shellfish or lime production in the PRD region. As well as more conventional academic articles and books, the review also located first-hand accounts (27 sources) such as newspaper articles (5), government reports (5), ancient Chinese literature (5 sources) and 12 historical maps. Given that many of these sources are from non-traditional formats (e.g., media) and in Chinese, we have cited the source as a "translated text" (e.g., Translate text 1) where necessary and provided the full reference and translation of the relevant section in the supplementary materials.

The Pearl River Delta region

Oysters historically occurred naturally along the Eastern and Southern coastline of China and provided important resources for human populations and societal development. This study focuses on the Pearl River Delta (PRD) region in southern China, including the regions of Guangdong, Hong Kong and Macau. This geographical focus aligns with an area with a history of aquaculture and oyster shell-related lime industry.

The Pearl River is China's third longest river and traverses from the West to the East, emptying into the South China Sea. The delta region includes the municipalities of Guangzhou, Shenzhen, Zhuhai and Dongguan in Mainland China and the Special Administrative Regions of Hong Kong and Macau (Figure 1). The past three decades have seen the region transform from rural villages to one of China's main economic hubs, with the reputation of being a 'world factory', having large financial and technology sectors, and it is one of the most densely urbanised regions in the world (Bie et al. 2015).

In this region, pressures from increasing populations, urbanisation and shifting socio-economic patterns have contributed to the decline of oyster reefs. More specifically, these factors contributed to unsustainable extractive practices, increasing sedimentation and land reclamation, resulting in the misalignment of coastal social and ecological systems (Weng 2007). The region in general also lacks limestone and so the region historically relied on burning shells and coral to produce lime for construction (Meacham 2009). As a result, natural oyster beds in the region are now rare and, where extant, are extremely degraded and sparse. Consequently, there were no reports of natural oyster reefs or habitats within scientific literature until relatively recently (e.g., Lam & Morton 2004, Lau et al. 2020). Information from the rest of the PRD, and China as a whole, is even more limited. Extrapolating from contemporary Hong Kong would suggest that natural oyster habitats in the coastal areas of the region are extremely degraded, failing to form oyster reefs and instead consisting of sparsely distributed oyster beds on soft sediment shores. Importantly, oyster populations are subject to heavy on-going recreational harvesting pressures and a loss of suitable habitat, limiting any potential for natural recovery (Lau et al. 2020).

Figure 1 A map of southern China showing the municipalities which line the Pearl River and its delta. Delta boundary is indicated by a thick black line. Municipality boundaries surrounding the delta are indicated by dotted lines. The entire map is part of the Province of Guangdong. (Map shapefiles were extracted from Global Administrative Areas (2018).)

Given the suitability of the coastal and estuarine conditions of the PRD for oyster species and the necessary abundance of oysters to support centuries of extractive harvesting/practices for related industry (i.e., lime, food and construction materials), there is reason to believe that oyster reefs were historically much more extensive. The region possesses a subtropical monsoonal climate, creating favourable estuarine conditions for natural oyster reefs and aquaculture (Lau et al. 2020). This is supported by archaeological evidence of a shell-based lime industry (Meacham 1978) and records of deliberate mudflat cultivation of the Hong Kong oyster *Magdallana hongkongensis* (Morton & Wong 1975). It is, therefore, highly likely that extensive loss of historical oyster reefs has occurred and that the region's natural oyster reefs were destroyed before being scientifically recognised and recorded in modern documentation (Lau et al. 2020). While spatially extensive surveys below the sedimentary surface could be conducted to find buried or formerly productive oyster reefs (Zong et al. 2009), sediment cores indicate the PRD basin is heavily reworked by increased sediment discharge linked to human activities (Wei et al. 2016). Additionally, the mass exploitation of reefs across the PRD basin (i.e., physical removal of shells), evident from historical literature, indicates that sub-surface investigation is unlikely to provide an accurate regional picture, and information about where to search for these reefs is scarce.

Pre- and early settlement (pre-Tang, prior to 618 CE)

The shoreline of the PRD during this period was minimally influenced by human activities and changed gradually, allowing oyster populations to continually adapt to ecosystem habitats even as sea levels rose and fell by up to 5 m from the Late Neolithic Age (3250–2250 BCE) to 1 CE (Weng 2002). Given that this period was up to 5000 years ago, there were relatively few resources available (Supplementary Table 1). In total, 29 references were found from a variety of sources: journal articles, book chapters, newspapers and ancient chorography (translated from Chinese). Of these, 22 made specific reference to this time period, while seven made a general reference (i.e., 'first millennium AD' and 'last 2000 years'). Only one document was found that originated from this time period (ancient chorography), with the rest of the data for this period located from documents that referenced oysters in some form during this time period.

Of the 29 documents, 19 referred to oyster reefs in the PRD prior to 1 CE. These stated that prior to 1 CE, naturally occurring oysters were highly abundant along extensive areas of shorelines throughout the PRD. Approximately 70% of the documents (20 sources) provide evidence of humans engaged in the harvesting of oysters for food and shells from as early as 5000 BC. The time period is characterised by minimal human-induced habitat destruction on the shoreline despite the use of oysters and other shellfish as part of a mobile fisher-hunter-forager lifestyle and much smaller local populations (Atha & Yip 2016).

Documentation of shell midden sites throughout the PRD region provides indication of substantial oyster and shellfish habitats that were exploited by humans (Peregrine 2001, Weng 2002, Hingham 1996, Rogers & Widdowson 1996). The largest of these, Hedang in Foshan, recorded 70–86 cm of accumulated shells and covered 10,000 m^2 and was dated back to the late third Millennium BCE (Hingham 1996, Peregrinee 2001). Shell records at a Han period (220 BCE–206 CE) midden site in Hong Kong also feature significant shell deposits including the reef-building species *Magallana (Crassostra) gigas* and possibly *Magallana rivularis*, *Saccostrea cucullata* and *Saccostrea kegaki* (Rogers & Widdowson 1996). Oysters were also consumed as food, with one translated text from 430 to 479 CE specifically referring to the oysters in the area tasting 'quite good' (Translated text 1. 420–479). Accounts from this period also reveal the advancement of technological tools to expedite shellfish harvesting with rudimentary boats used to harvest shellfish, such as the "Ni-Mo-Chuan", in existence as early as 2205 BCE (Worcester 1958).

We found six references to humans utilising oyster shells as tools. Half of these refer to oyster shells being used in buildings or as artefacts (Guangzhou Daily 2018, Atha & Yip 2016, Peregrine

2001). The other half relate to oysters being used to make lime (Carran et al. 2010, Meacham 2009). Oysters were also found to have had uses as weapons, grave goods and in small-scale production of lime for construction. In the Middle Neolithic (4000–2500 BCE), the use of oyster shells as picks was documented in sites across the PRD area including Tung Wan and Sham Wan (Guo 1999). In the late Neolithic (2500–1500 BCE), shell weapons were found at the Hedang shell midden site. Oyster shells were also discovered as part of grave good assemblages in a Bronze Age grave in Sha Po, graves at Tung Wan Tsai North and at a child's burial at the Yinzhou shell midden (Atha & Yip 2016). This reference to the widespread usage of shells, including oysters, is an indicator of early human uses of oysters as food, tools and objects imbued with every day and cultural value.

The located texts make little reference to the period from 700 BCE to 300 CE as there is an unexplained break in human habitation in the PRD area (Meacham 2009). In Hong Kong, Shenzhen and Haifeng, there are very few sites with evidence of human dwellings or activities during this period. The exception is a journal article referencing the use of clam shells in producing lime in the construction of a Duke's tomb in 635 BCE (Carran et al. 2010), meaning that shellfish habitat was being used and that this use likely included oyster habitats. In 400–500 CE, there is evidence that human activities began to modestly return to the PRD region, coinciding with the rise of the first lime industry, which became widespread in the PRD (Meacham 2009). Populations continued to modestly increase until the Tang dynasty, when rapid development in the northern provinces of China created the need for a reliable source of lime for construction, and the PRD became a primary source (Weng 2002).

Settlement in Tang Dynasty and later (618 CE–1800 CE)

The Tang dynasty marked a turning point in the PRD. Political instability and conflict in northern China drove migrants southward into the PRD region, increasing population pressures. Extensive dyke-building occurred to increase land for food production and to control floods, dramatically changing coastlines (Weng 2002). Correspondingly, an increase in the number of texts referencing oysters was recovered (48 making direct reference to this time period and five making general references (e.g., past 2000 years); Supplementary Table 2). Of these, 19 references were found from historic texts written during this time period and which we translated.

During this period, an industry producing lime from oyster shells emerged drawing on the abundant beds of shellfish and oysters. Of the sources from that period, 12 of the 48 make references to oysters in relation to lime and/or kilns that were used to produce lime from shells. While only referenced once, aquaculture is stated as becoming more widespread during this period, further replacing natural oyster habitats with human-made benthic oyster farms (Translated text 4. 1687). These texts are the earliest documentation of how this period is characterised by more rapid modification of shorelines to the detriment of naturally occurring oyster habitats (Huang et al. 1982). These documents also point towards an increasingly extractive relationship between oysters and humans, which lasted until the collapse of the first lime industry in the tenth century. This could have been due to an environmental collapse during this period, potentially caused by increased sedimentation in the PRD or as the lime industry relocated into other areas of the PRD (Meacham 1994, 2009, Huang et al. 1982).

By the late Tang period (867–904), food production from existing lands, including less favourable uphill areas, was insufficient to meet population demands (Weng 2002). Dyke-building to produce agricultural fields became a practice increasingly relied upon by successive dynasties. In particular, the Song (960–1270 CE) and Yuan (1279–1368 CE) dynasties developed extensive large-scale dyke-building projects that were continued through the Ming (1368–1644 CE) and into the Qing dynasty (1368–1644 CE) (Weng 2002). These dykes changed sedimentation patterns and were some of the earliest land reclamations in the region for agricultural purposes, changing shorelines and potentially rapidly removing estuarine areas of oyster habitat.

A lime industry reliant on oyster-shellfish beds as raw materials re-emerged in the second century to ninth century, producing lime to meet various industries along the coastline, including salt production (Atha & Yip 2016, Wong 2008), construction (Wong 2008), shipbuilding (Ting & Donnelly 1925, Wong 2008) and agriculture (Wong 1984). Archaeological records are relatively scant across most of the PRD; however, the abundance of lime-making sites in Hong Kong suggests that similar sites extracting oyster and shellfish shells may have been operating in the PRD. Furthermore, evidence from the gathered documents suggests these production sites were not operating independently but likely functioned within larger imperial networks (Atha & Yip 2016). In Hong Kong alone, dozens of lime production sites have been found with some making concurrent use of 15 open kilns to fire shells. The scale of this indicates both an abundance of shells in locally accessible areas and a thriving market that paid for the employment of fishermen as shell extractors or labourers. It also points towards a larger regional need for lime, with Hong Kong in the PRD being a likely export supplier for areas such as Canton (Meacham 2009).

Of note, oysters also emerged within cultural texts from the Tang (618–906 CE) to Song (960–1279 CE) dynasties and later. One notable example is in a poem by the famous Song dynasty exile, Su Shi. Here, Su describes oysters as especially delicious when paired with wine (Po 1099). Outside of literati circles, oysters were associated with poverty, wildness and lower social classes such as the Tanka boat people or the mythical Lutings (Translated text 2. 867–904, Hansson 1996, Watson & Rawski 1990). The Luting folk were believed to be the descendants of the Peasant Uprising Army of Lu Xun. After the uprising failed, Lu Xun threw himself into the water and the survivors fled to the numerous islands near the estuary of the Pearl River, becoming Haiyi (sea gypsies/barbarian). They lived on oysters and built houses with oyster shell walls (e.g., Figure 2; Translated text 3. 618–907). Villages in the PRD region often had many buildings with oyster shell walls, and larger buildings (e.g., village temples) could reportedly be constructed with up to one million oyster shells (Figure 2).

A translated historical text from 1678 shows the different nature of human interactions with oysters throughout the PRD. They describe how oysters in the Zhong Shan (香山) region were harvested from the natural environment, rather than farmed. In contrast, the text describes how in Dongguan oysters were farmed by placing bamboo sticks in the sea (Translated text 4. 1687). This is the first historical reference located in this review of active oyster farming in the region. Throughout this time period, reference is made to eating oysters (six sources referencing oysters as food), describing how it can be cooked and prepared (Translated text. 2 867–905, Translated text. 4 1687, Translated text. 5 1596) potentially indicating the role of oysters as a relatively easily acquired protein source in the region.

Oysters are recorded in abundance in Macau, with oyster lime being used for fort construction (1500–1700), lime kilns (1789) (Garrett 2010) and glass windows (1695) (Careri 1700/1704). In Guangdong and a few other coastal areas in 1639, oyster ashes are recorded as a slaked lime substitute for boat caulking (Ting & Donnelly 1925). Written accounts of the 1800s describe thriving oyster aquaculture throughout the area, suggesting a continued relationship between humans and oysters from 1000 to 1800 CE (Hayes 1964).

Two references are also made in this time period to oysters being harvested for pearls. This industry would have been in the eastern waters of Hong Kong where the conditions are more oceanic influenced rather than influenced by the Pearl River and so more appropriate for the survival of the naturally occurring pearl oysters (Yang et al. 2022). The documents indicate a relatively small, but likely profitable, industry that later declined. The first describes how the emperor of the Southern Han dynasty established 'Meichuan sector', likely around what is now known as Tolo Harbour, and had it guarded by troops. At the invasion of the Northern Song dynasty, the sector was abolished in 972 CE as it was believed that pearl picking would do the kingdom harm (Wong 2008), though the reason for this belief is not clarified. The second account describes the perils faced by pearl divers in Canton in the 1400s, who were subject to frequent shark attacks. Consequently, dredging was developed to gather the shells, and the harvest of oysters began to be industrialised (Hayes 1856).

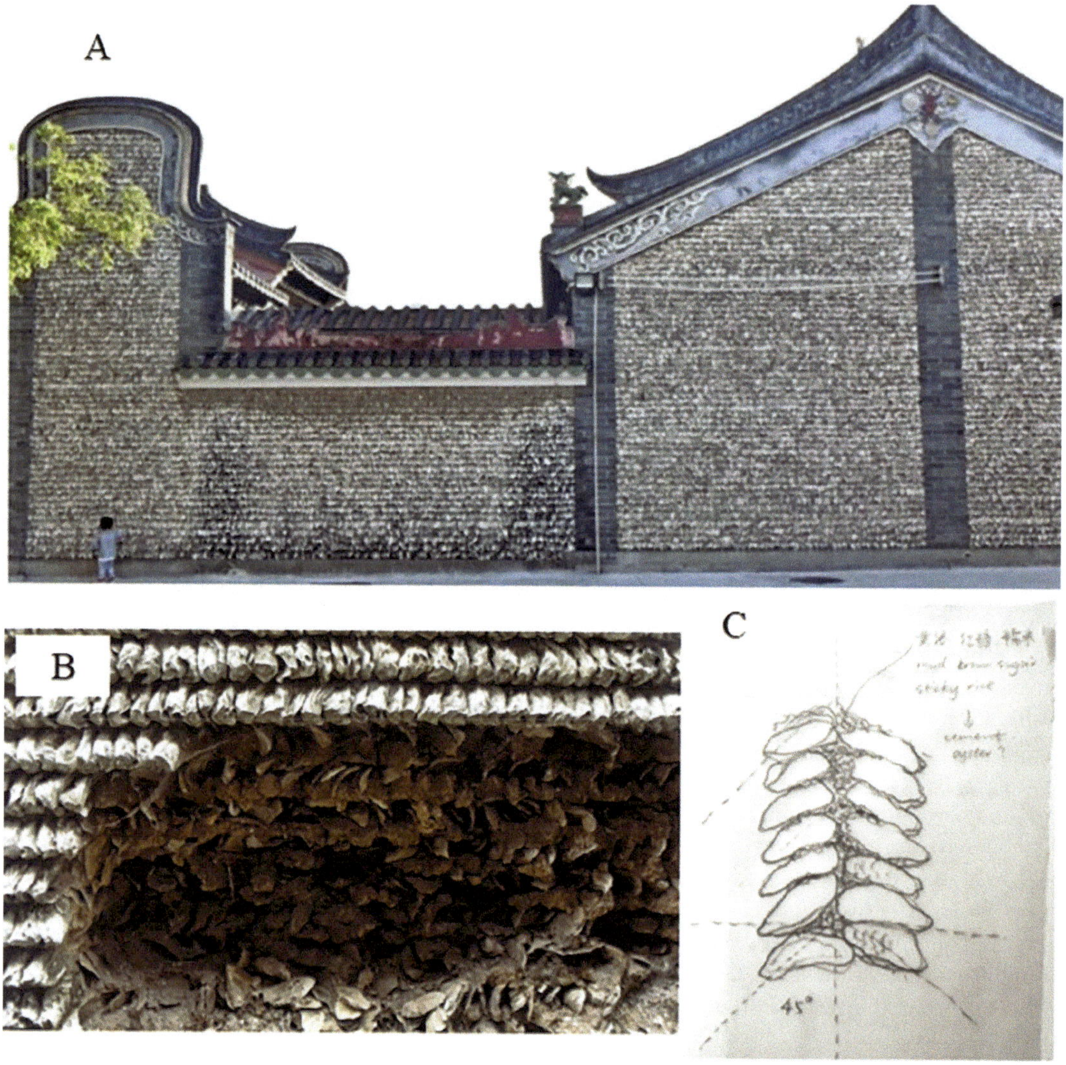

Figure 2 Oyster shells were extensively used for building materials throughout the Pearl River Delta and adjacent provinces. (A) A village temple in Shenzhen, which is on the shores of Deep Bay and historically had extensive natural oyster beds that were later transformed to aquaculture. (B) An excavated village house wall. Foundations and walls for buildings contained several layers of shell cemented together with lime or mud and so required large quantities of oyster shells. (C) A hand-drawn diagram used to illustrate how to make building walls with oyster shells. ((A) and (B) reproduced (with permission) from an online article on oyster shell use in houses in the PRD (https://mp.weixin.qq.com/s/tIqxudus-cAW6W4p_BhnZQ), (C) reproduced with permission from the author, Haihong Huang (Source: https://projects.arch.chalmers.se/haihong-huang/2/).)

1800s to 1960s

From 1800 to 1960, the balance of production of oysters shifted from wild harvest to being dominated by aquaculture production. At the beginning of the period, wild oysters were relatively abundant, and oyster-reliant industries were once again proliferating. From this period, 48 references to oysters were located, seven of these generally referenced this time period (i.e., referring to the past 2000 years) and 25 were from sources written during this time period (Supplementary Table 3). Two major industries characterise this period: a resurgent lime industry (mentioned 23 times) and a rising aquaculture industry for oysters (referred to seven times).

The re-emergence of lime-making industries based on the wild harvest of oysters was well underway by the 1860s (Wong 1984). Records show that the annual wild harvest in Tsing Yi, in the middle of Hong Kong's waters, in ~1915 was >8000 tonnes (Figure 2). Just to the north on another tributary of the Pearl River, landings at Jiangmen were ~1500 tonnes per year (Figure 2). In the New Territories of Hong Kong alone, in 1901, official records show 1435 men and 194 boats employed in the work of "Taking Coral and Shell from the Sea" (Hong Kong Government Gazette 1901). Not only did shell resources drive the lime industry but also they drove secondary industries. In kiln sites, such as those in Ping Chau, selling wood and grass to the lime kilns became profitable (Strickland 2010). Records from 1908 of Jiang Men, Guangdong, noted that dredging of 'large shells' produced 12,000 tonnes of lime annually (worth approximately $40,000) (The Directory & Chronicle for China 1908).

The lime industry was a formative industry of Hong Kong society, with some family-run business such as the San Shing Lei lime kiln factory run by at least five generations. By 1930, Walter Schofield, a governmental officer, recorded a "lime-burning and sand digging boom…in full boom because of the roadbuilding and building then going on in Hong Kong and Kowloon". He noted kilns in the western side of Hong Kong in sites such as Peng Chau, Nei Kwu Chau, Tsing Yi and Hang Hau, attributing their location to market proximity and the comparatively clean water in the western parts of Hong Kong. Schofield also noted the comparative unrest in mainland China during this period as contributing to the industry being located in Hong Kong (Strickland 2010).

Another source records that the strength of the industry could still be found in the 1950s, with the lime kiln industry (and its associated use of shells) producing at least 3000 tonnes of lime monthly (36,000 tonnes of lime annually) in production sites spread across Hong Kong. The key raw material – shells – were in abundant supply, with local boat people providing continued transportation of shells on their boat to the lime sites. Most lime kilns occupied a wide spread of space on coastal areas, with spaces to store prepared lime awaiting transport and private piers to facilitate logistics (Wong 1984).

Despite the relative strength of the lime industry, its demise is also recorded during this period; of the 23 references found to the lime industry, five refer to its decline. Taking Peng Chau as an example, in its heyday, the island was recorded as being the site of 20 kilns; by the 1930s, the industry on the island was showing signs of a demise and only six kilns were operational. Post-war, only four kilns, operating at partial capacity, re-opened but ownership of these kilns was eventually lost to outsiders (Strickland 2010).

There is some disagreement as to when this demise occurred; some sources state that the industry was in 'deep trouble' by the 1930s (Meacham 2009, Strickland 2010, Schofield 1983); another describes the industry as being in 'full boom' by 1930 (Schofield 1983); and a later source details an active industry in the 1950s (Wong 1984). There is a greater consensus as to the causes of decline, namely that the environmental cost of the lime industry was being felt during this period. Two references note that the 'shell and coral resources were dwindling' and that coral and shell were 'almost worked out' of the sea, which at times led to violent altercations (Meacham 2009, Schofield 1930). While oyster aquaculture was quite extensive and increasing in scale by this time, the focus of the aquaculture was oyster meat, and the quantities of shell were much less than that supplied by the wild harvest required in the lime industry. Lime kilns were also forced to close in the face of fierce competition with imported lime from China and Japan (Wong 1984, Strickland 2010). A new government practice of using Japanese lime on its public works sounded the death knell of the lime industry in Hong Kong, with many kilns ceasing operation in the early 1960s (Wong 1984). Ultimately, from the available documents, it seems that the most likely causes of the decline of the lime industry were a combination of the over-extraction of wild oyster reefs and beds to the point of collapse and the shift in focus to importing lime from other regions.

Oyster aquaculture in the PRD from the early 1800s to 1960s was prolific and widespread, with recorded fields in Macau, Hong Kong and the Guangzhou region. In Heukang, a location opposite Macau, while no unit is specified, oyster beds were recorded in 1875 as being '490 in length and 200 in breadth', producing 'magnificently large' oysters. The 'Heng-kang establishment', an oyster aquaculture company staffed by eight men, made an annual profit of 7–8000 taels (approx. 265–302 kg, 1 tael = 37.78 g) after deducting some 1900–2000 taels (approx. 72–76 kg) in expenses (De Thiersant 1875). Oyster aquaculture could also be found in other locations. A newspaper article published in 1883 describes the large-scale aquaculture practices that took place in the Deep Bay areas. The oyster beds in the four villages of Sai Heung, Ku su, Fuk-Wing and Sha Zheng were recorded to be at least 50–60 sq miles in size. Almost all villagers were involved in oyster cultivation, with Sha Zheng alone having at least 10,000 people engaged in cultivating, preparing and selling oysters. As another indicator of the scale of this industry, in one small creek in Hong Kong, 40 oyster junks passed the observer in half an hour. The smallest of the junks carried 10 tonnes, and the majority were junks of about 40–50 feet long by 10 feet wide (Hong Kong daily Press 1883). There is also evidence that habitats made by the Hong Kong oyster, *Magallana hongkongensis*, were manipulated and cultivated for meat and lime on mudflats in Deep Bay, Hong Kong, over the past 150–700 years (Morton & Wong 1975)

By 1875, oyster cultivators in Macau had reached sufficient scale and visibility that the risks faced by oyster aquaculture were well known to the local government, earning them exoneration from government taxes (De Thiersant 1875). Similarly, in 1933 in Hong Kong, records of there being a thriving oyster industry are demonstrated with government leasing of land exclusively for oyster breeding. One such bed with 4.83 acres in Yuen Long Bay, near Lungku Tan, Hong Kong, was secured on a 21-year lease with a Crown rental of $25.50 (presumably HKD; The Hong Kong Telegraph 1933).

On a larger scale in the PRD, a Shanghainese customs report from 1924 records a large flourishing trade in oysters centred on Canton and Hong Kong as destinations for secondary processing. It records "nearly every bay and hamlet has its beds" in the Sunon and Sunning districts, including the "important Shatsing beds". Oyster cultivation was prolific, with great value being placed on the oysters produced, in particular that "oysters of the estuary of the Canton and West Rivers are highly valued". At Tongka village in Heung Shan, the oyster beds owned by the municipality cleared no less than 50,000 Haikwan Tael (approx. $38,000 USD) during a good season. Good, dried oysters were priced at 95 Haikwan Tael (approx. $72 USD) per picul (approx. 60 kg) (Decennial Reports on the Trade Navigation Industries, Etc., of the Ports Open to Foreign Commerce in China and Korea, and on the Conditions and Development of the Treaty Port Provinces 1924).

The sheer size and scale of this industry indicate a healthy and flourishing aquaculture industry supporting tens of thousands of workers across the PRD (Figure 3). Western reports from 1875 to 1883 detail an admiration for the complexity and success of oyster aquaculture methods in the PRD. In 1875, French diplomat Darby de Thiersant commented that "ostriculture gives in China the best results. Local cultivators were keenly aware of ecological and climate-related risks to oyster beds, such as Tchou Mouh (a predatory whelk), chang-yu-yu (a predatory fish called the eagle-head fish), dryness (likely related to rainfall patterns), changes in salinity and typhoons, which are all deeply connected to oyster survival rates" (de Thiersant 1875).

Techniques used to build oyster beds also significantly improved from 1875 to 1958, for instance, transitioning from using stones to seed oysters (de Thiersant 1875) to using old oyster shells as seed materials (Hong Kong Daily Press 1883), then using a split bamboo with two oyster shells of different thicknesses (Worcester 1958). These techniques reduced oyster harvesting periods from 2–3 years to every 8 months (de Thiersant 1875). This well-developed body of knowledge and skill indicates a growing complexity in the human-oyster relationship during this period both through increased ecological knowledge and oyster aquaculture techniques.

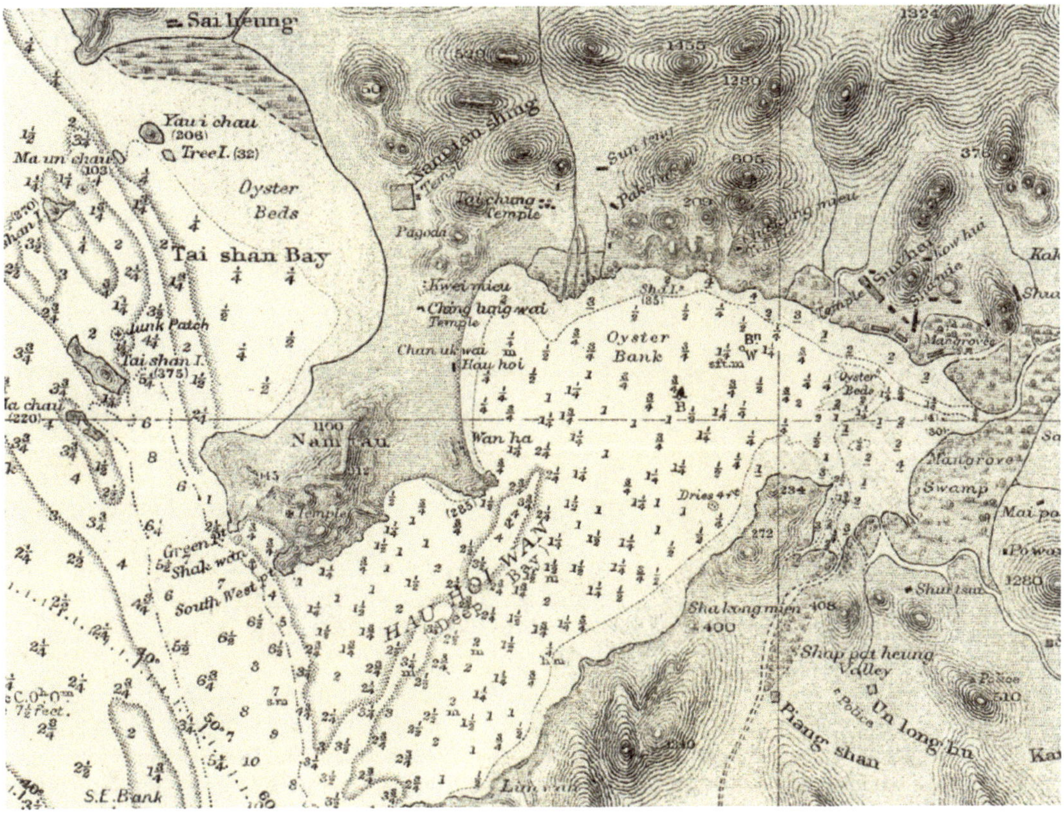

Figure 3 A map of Deep Bay in northern Hong Kong and Shenzhen area, at the end of the Pearl River Delta, showing the extensive oyster beds that were present on the mudflats (circa. 1920s). These mudflats historically had natural oyster reefs which were harvested heavily and then later transformed into benthic oyster leases. (Map produced by The Admiralty, London, 1899, most recently updated 1928. Map courtesy of Geographicus Rare Antique Maps (http://www.geographicus.com).)

1960s to present

From the 1960s to the present, both wild and aquaculture oyster populations had dramatically declined to be essentially non-existent in the PRD by 2020 (Du 2020). While Shajing, just to the north of Hong Kong, was still producing up to 4000 tonnes annually until the 1970s, aquaculture in Hong Kong declined from its peak in the 1950s to relatively low production through this period to the present (Figure 4). Aquaculture industries are now a shadow of their former self, and lime industries are non-existent. Rapid reclamation, dam building and shoreline modification during this period have replaced coastal habitats that could have been inhabited by self-sustaining shellfish habitats and reefs.

References to oysters during this period mostly involve shoreline change (6 out of 11 located references; Supplementary Table 4), mainly pertaining to how human influences have altered the PRD region and subsequently reduced the presence of oyster reefs and suitable habitat (Figure 5, Supplementary Table 5). Navigation charts demonstrate how between 1850 and 2015, the PRD gained 1160 km^2 of new land (Wu et al. 2018), over 230 km^2 of this occurred from 1955 to 2010 (Wu et al. 2016). This change did not occur gradually but rather there was an abrupt increase between 1980 and 2000 (Wu et al. 2016). Consequently, intertidal and subtidal areas in the PRD decreased and the local impacts of construction activities damaged an even greater area. Between 1850 and 2015 tidal flat area decreased by 32%, with nearly all of the 1850 intertidal land being

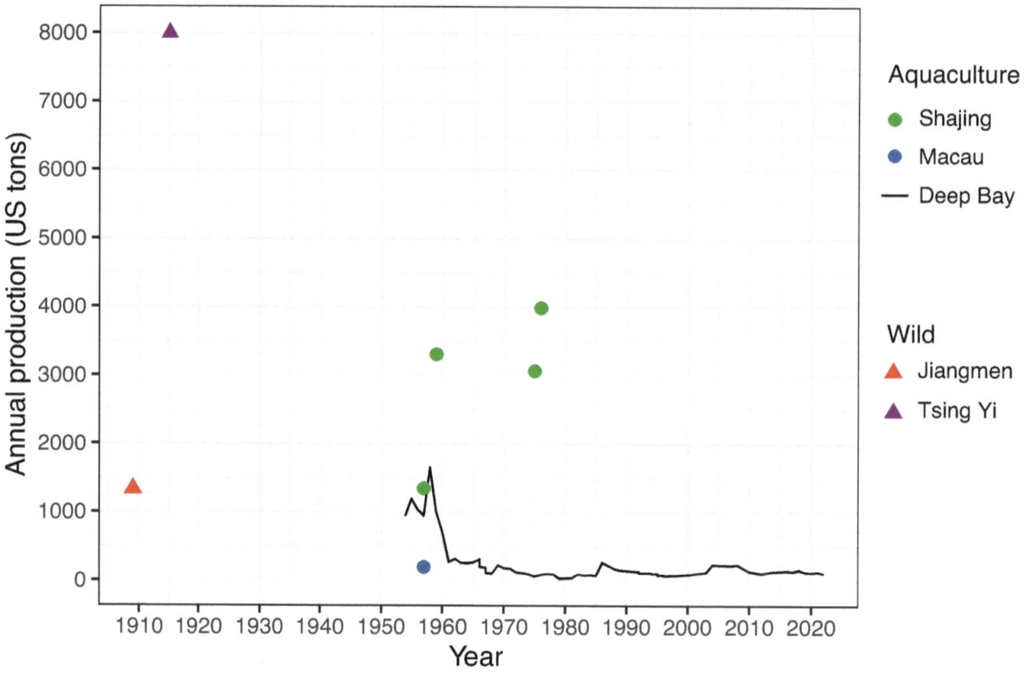

Figure 4 Annual oyster production statistics across the Pearl River Delta since 1909, showing both wild collection (triangles) and aquaculture (circles and line). Wild collection statistics from Jiangmen were related to oyster shell landings from dredging vessels, and statistics from Tsing Yi were related to lime kiln production from wild oyster collection. The most continuous record of documented oyster aquaculture production is observed at Deep Bay, Hong Kong. For Deep Bay, a rapid decline in oyster aquaculture production was recorded from 1960 until the present day (2023). (Deep Bay aquaculture production data were extracted from Caswell et al. (2020) and AFCD (2019–2023))

occupied by reclaimed land by 2015 (Wu et al. 2018). This new land was derived from the disposal of dredge spoils (Wu et al. 2018), sediment from deforestation and, later in the 2000s, dam construction (Tan et al. 2016). Human activities resulted in sediment being trapped in the encircled tidal flats along the front of delta plains (Zong et al. 2009), depriving oysters of the habitat required to form reefs.

Socio-ecological history of oyster reefs in South and East Asia

The lack of data on the history of oyster reefs and their socio-ecological relationships with these systems in the PRD extends to the wider Asian region. A review of English language academic publications faces similar barriers regarding a scarcity of academic resources. What is available is summarised below and shows how in East Asia, China, Japan and Korea, the history of oysters revolves around aquaculture and cultivation for food, while in South Asia, India, Bangladesh, Pakistan and especially Sri Lanka, the harvesting and trade of oyster pearls dominates. Investigation into the Mekong region, Thailand, Cambodia and Vietnam revealed a paucity of data and few insights. The synopsis below is a non-exhaustive synthesis of the historical socio-ecological history of oysters in Asian countries. Our efforts here aim to encourage local researchers to explore similar accounts in multiple languages and to reconstruct their histories related to human livelihoods and fisheries in a global context.

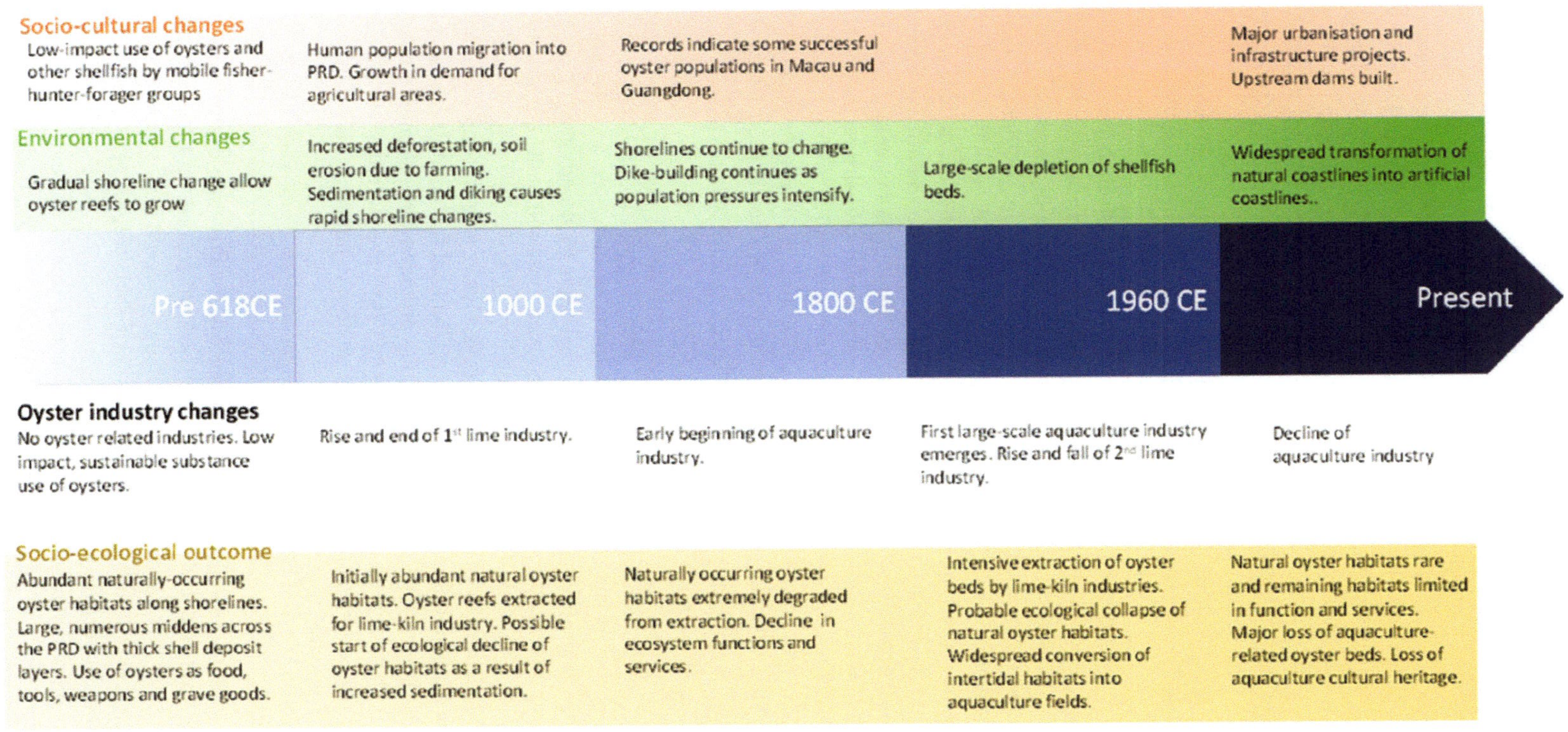

Figure 5 A timeline summarising the socio-cultural, environmental and oyster-related industry changes in the Pearl River Delta, and the resulting socio-ecological outcomes relating to oyster habitats and ecosystems in the region.

China

Regarding China as a whole, a few academic papers make reference to the likelihood of a long history of oyster reefs in Bohai Bay and the coast of Jiangsu (Wang & van Strydonck 1997, Lin et al. 2021). Loss of reefs has been noted in the Bohai Sea, which suffered a 70% loss of reefs, and in Dajiawa, where there is a 90% loss (Fang et al. 2007, Beck et al. 2011). These losses are attributed to several factors, all of which were common in the PRD. For example, in Dajiawa, overfishing of oyster reefs was a significant problem from the 1960s to the 1980s, with oysters being harvested for food and lime. Urbanisation caused direct habitat loss as cities expanded and reefs were destroyed to allow commercial ships access. Pollution from chemical industries and higher salinity due to reduced riverine discharge caused further declines and likely reduced the probability of natural recovery (Deng & Jin 2000, Lin et al. 2001).

Also significant to the China region is the aquaculture of oysters. In temperate Asia, the cultivation of oysters has been practised for at least 2000 years (Guo et al. 1999). The aquaculture of *Magallana* (*Crassastrea*) spp. resulted in major changes to the remnants of the Dajiawa oyster reefs (Beck et al. 2011). Contemporary oyster aquaculture in the region is primarily based on the collection of wild spat (Guo et al. 1999), which along with the growth of oyster farms can negatively impact natural reefs (Beck et al. 2011). In some cases, however, cultivation of oysters can still create oyster habitats that function to some degree as reefs, such as in Southern China (Guangxi and Fujian provinces) where temporary structures placed in intertidal habitats to collect spat loosely resemble natural oyster reefs. Elsewhere in Northern, Central and Southern China, however, oysters are cultivated with suspended rafts in a way that does not function as natural reefs (Beck et al. 2011).

In socio-economic terms, boat people in China, and East Asia in general, appear to have long been regarded as a separate social group, holding lower societal status. Boat people have been referenced in Chinese writings for thousands of years, and more recently by Western accounts, however, these references are scattered, brief, mostly descriptive and of little substance (Anderson 1970). Therefore, while the presence of these people in society is well known, their status at the fringes of conventional society has led to limited understanding.

Japan and Korea

In Japan and Korea, the recorded history of oyster reefs is primarily associated with aquaculture. Oysters are stated as one of the most popular taxa to farm in Japan, and the practice is said to originate from a fisherman in Hiroshima sometime between 1624 and 1643. The observations by fishermen of an oyster attached to a rock and bamboo branches used in the construction of fish traps are believed to comprise the basis for the ancient method of oyster farming, which is substantially similar to that used in China from a comparable period. Oyster spat settling on stones and bamboo poles of fish traps was replanted in intertidal zones surrounded by bamboo fences, which excluded predators (Fujiya 1970, Ventilla 1984). Hiroshima Bay provided ideal sheltered conditions, a wide mud-free shoreline and the nearby market of Japan's new thriving merchant class in Osaka (Ventilla 1984). Similarly, Korea is concerned with oysters in relation to aquaculture but little information of relevance in English was able to be located.

India and Sri Lanka

In Bangladesh, there is increasing contemporary interest in cultivating oyster reefs, primarily as a defence against sea level rise, but detailed and well-studied reports on the production, utilisation and history of oysters are lacking. A single exception that we could find is a Department of Fisheries report from 1988 that details the annual production of shell weight of oysters (50–70 t) in 1984 (Sarker & Alam 1988, Islam 2003).

In contrast, the Indian subcontinent possessed banks (reefs) of pearl oysters. The history of this region, therefore, tends to be concerned with the harvest and trade of pearls. Pearl oysters were found along the coasts of India, Bangladesh and Pakistan (Southgate & Lucas 2008). The most notable of which, and with the most complete and detailed records, is the history of pearl oyster fisheries in Sri Lanka. The abundance of pearl oysters in Sri Lanka, and the subsequent colonialisations, resource tensions and formation of economic/trade networks, has resulted in the history of pearl oysters being well documented and studied not only from a historical perspective but also from political, economic and social perspectives (Ostroff 2016, 2020). Throughout history, the North-Western coastal belt has been known for its pearl banks (Gunawardana 2000, de Silva 2005, Ostroff 2016, 2020). Pearl beds were found in the Gulf of Mannar and were particularly dense on the Sri Lankan side (Sudarmawathie 2016).

The first records of pearls in Sri Lanka originate from Greek and Roman writers at the end of the fourth century BCE to mid-sixth century CE. Their interest in Sri Lanka was in the commercial value of its natural resources, and it was considered the only place yielding good pearls. Accounts detail how locals used to dive to collect large oysters, "which hid in covered places underwater", presumably on subtidal banks with pearl oyster habitats. A visiting Chinese Buddhist priest also noted the presence of pearls in the fifth century CE, including the economic importance of pearl fishing for local livelihoods (Sudarmawathie 2016, Legge 1886).

Locally, the first mention of pearls was found to be in the sixth century BCE, where pearls were sent as a gift to King Pandu of Madurapura in South India. This trend continued throughout the following centuries with various reports detailing pearls from Sri Lanka being gifted to kings or abroad to India and China (Sudarmawathi 2016, Katupotha 2019). How these pearls were obtained is described by Marco Polo in the thirteenth century CE; he details not only the practice of harvesting oysters by diving but also the social and economic practices involved. These include the hiring of 'fish charmers' that prevent "great fishes" from attacking the divers and the need to pay the king royalties on what they gather. He accounts that the divers were able to retrieve a vast quantity of good quality pearls that were a sought-after commodity at that time (Benedetto 1994).

These accounts provide clear evidence that pearl oyster fishing was carried out in Sri Lanka from the Anuradhapura era (377 BCE–1017 CE) to the Kotte period (1412–1597), producing pearls in large quantities. Pearls were economically and culturally important; the kings held monopoly over the pearl reefs, giving the king a large stake in the fisheries. The pearl divers had to make payments to the king as well as contribute a large part of their harvest, providing the king with substantial income (Sudarmawathie 2016).

At the start of the sixteenth century CE, the Portuguese invaded and captured parts of Sri Lanka. They were especially interested in the pearl fisheries, monopolising them and bringing in South Indian labourers to continue pearl diving in Mannar Bay and Kalipitiya. They bought in their own boats to collect oysters from the coast, developing their own system that differed from the traditional method (Sudarmawathie 2016). During this period, 50,000–60,000 persons, including divers and merchants, were involved in pearl fishing and the associated industries (Katupotha 2019). Accounts and reports from the Dutch, Portuguese and British provide further details of the pearl fisheries up to the twentieth century (Herdman 1903, Steuart 1834, Ostroff 2016, 2020, Katupotha 2019).

The Portuguese were expelled from Sri Lanka by the Dutch in 1658, who continued the development and exploitation of the oyster reefs under the auspices of the Dutch East Indian Company and, from 1746, began to rent the reefs out. In 1796, the British annexed the island from the Dutch and, like the countries before it, continued to earn considerable revenue from Sri Lanka's pearls. The pearls were considered to be the best in the world, fetching high prices in Europe, and so were a sought-after resource. According to one account, from 1828 to 1837, 227,131 pound sterling was credited as revenue into the Ceylon Treasury from pearl fisheries, and in 1891, they were the most profitable fishery of the century (Bastiampillai 1980, Katupotha 2019).

For over one and a half centuries, the colonial British attempted to regularise the harvest of pearl oysters to maximise profits (Fernando 2022). It is this misguided attempt to control biology with science that ultimately is attributed to the demise of the industry. The British government in Sri Lanka sought to increase production in the industry through privatisation, to reduce the number of people involved and the associated pollution produced. This rationalisation also involved modification of the habitats, "cleaning" the fishery banks to increase production (though the process undertaken and the rationale underlying this are unclear), resulting in the "disturbance of the oysters". The resulting rapid decline in pearl oyster populations has been interpreted as the "dearth of oysters [being] due to the premature decay of oysters", and as such, the pearl fisheries never recovered from the ecological disturbance of the 'cleaning' activities (Mahroof 1995). Consequently, large-scale pearl fishing ended in 1925, and while oysters were still sold on a smaller scale in provincial markets until 1961, the collapse of the oyster populations effectively ended the pearl oyster industry in the region (Katupotha 2019).

Discussion

There is a long, rich history of a strong socio-ecological relationship between the communities of the PRD and oysters which developed and changed over time. Here, we document this history from all of the available sources and make it accessible for the first time. Early accounts indicate the importance of oysters for tools, food and culture (e.g., Atha & Yip 2016, Peregrine 2001, Weng 2002). From the 618 CE to the 1960s, the role of oysters in the lime industry became more frequently cited (Supplementary Table 6), following the growth and decline of the industry. Concurrently, references to the aquaculture industry increase from just one in 1678 to several mentions in the 1800–1960s period, indicating the rise in prominence of this industry. What is clear is that oysters have been intertwined in the culture of southern China for millennia and that the reliance of communities on wild oyster habitats progressed from subsistence to industrial level harvesting for industry (rather than food consumption), which led to over-harvesting and almost complete removal of naturally occurring oyster habitats, driving the need for aquaculture to supply oysters for food. The history of oysters, therefore, needs to be understood on the background of societal as well as environmental change.

Socio-cultural significance and role of oysters

Over the years, oysters became associated with certain socio-economic and cultural groups. Initially, around 6500–4500 BCE, archaeological and environmental evidence indicates that the PRD region was occupied by a culturally similar group of roaming fisher-hunter-foragers. Evidence gathered from this period indicates that shellfish and fish were a prominent contribution to diets during this time period (Atha & Yip 2016, Weng 2002).

During the late Tang period (867–904), references to the 'Luting' clan increased in frequency. This clan became associated as 'only [eating] oysters' and stacked the shells to make walls. There is also a report during this period of Lutings trading oysters for liquor, demonstrating not only the importance of oysters for substance but also trade. This trade would likely have been particularly important for communities of "boat people" who spent their entire lives on boats in a nomadic lifestyle, so perhaps had less opportunity to access goods other than by trading what they harvested. Later, during the Song dynasty (960–1279 CE), two groups of boat people were documented inhabiting the region, the Lutings and the Dan (Tanka), both of which are described as living on the geographic and social edges of society. These groups are recorded as trading pearls and other products from the sea for resources such as grain. They are described as being in 'poverty', and both the Lutings and Dan groups were characterised by eating raw shellfish (Hansson 1996). Mention is again made to the 'Luting barbarians' in the fourteenth century and how they gather "fish and oysters, vines and bamboos" (Schafer 1957).

While our review did not find any further reference to 'Lutings' after the fourteenth century, the stigma and social separation of certain groups with a working relationship with oysters remained.

In the seventeenth century, we find an account that describes how those engaged in "oyster tending, lime smelting, reed gathering and offshore farming" are 'indelibly stigmatised' and kept at 'arms length' by nearby farmers. These groups working with oysters appeared to form their own social groups with long histories, so they likely stemmed from the Lutings or Dan groups (Watson & Rawski 1990). Similarly, in 1883, a news report described the oyster cultivators of the region as 'lawless and turbulent' and linked to illegal and immoral practices, such as piracy, opium smuggling and even murder (Hong Kong Daily Press 1883).

The final reference to these groups, from 1918, is much milder and states how the Dan (Tanka) and Hoklos (Hokkien) groups undertake almost the entirety of the fishing, including harvesting rock oysters by women and children. The Dan people also used large junks to trawl around the islands of Hong Kong (Schofield 1983). It is unknown if the milder language in this account towards these communities is due to the academic nature of the reference (Journal of the Hong Kong Branch of the Royal Asiatic Society), foreign nature of the author (and so unfamiliar with societal hierarchies and cultures) or if attitudes, in general, had softened and social integration was occurring.

There is, therefore, a long history and a clear socio-economic connection between certain social groups and oysters. These social groups appear to be lower socio-economic, being described as being in poverty, even as barbarians, and outcasts from society. This lack of societal integration and impoverishment may be the reason behind the lack of official government records and documentation of their work and history. This connection and association of these groups with oysters also goes some way to explaining the lack of documentation in the region of oyster habitats and reefs – these groups were not likely to be educated and so records would be oral.

The rise and fall of oyster industries in the PRD

A number of drivers have emerged as key factors in declining oyster habitats in the PRD, including extraction (over-harvesting for lime), reclamation (agricultural and "hard" coastline) and sedimentation. The historical evidence drawn together here shows that prior to the Tang dynasty (pre 618 CE), there was little impact on oyster habitats other than direct usage by mobile human communities. These communities had minimal impact and so extensive oyster habitats were abundant along the coastline of the region. As pressures from human populations increased, however, oyster habitats began to be impacted in a number of ways.

The historical documents drawn together here demonstrate how the decline of oysters in the Tang Dynasty (618–906 CE) was driven largely by increased coastal human populations and development (Weng 2002, 2007). After the available low-lying arable lands in the PRD were claimed for farming, deforestation of uphill areas to create land for food production increased river sedimentation, impacting downstream habitats. Once these uphill areas were fully utilised, successive dynasties implemented dyke-building as a major method of agricultural reclamation for food production (Weng 2007).

Dykes had a 3-fold impact on oysters. Firstly, agricultural reclamation through dykes replaced low-lying deltas and estuarine mudflats that previously would have been suitable habitats for natural oyster habitats and reefs (Liu & Deng 2007, Wang & Wall 2010). Secondly, dyke-building increased river sedimentation, changing the environmental conditions of downstream oyster habitats. Dyke building served to fasten river channels, simplifying the drainage system by silting up smaller tributaries and consolidating larger ones. As the river channels were positioned after the building of dykes, less silt was deposited into the delta, meaning more entered the sea (Weng 2007). Although new land was created from sedimentation, the delta expanded towards the ocean at an increasingly faster pace. While the new shoreline was a potential new oyster habitat, the rate of delta expansion increased from 10 m yr^{-1} pre-Tang to 20 m yr^{-1} after Song/Ming, with the West and North River deltas increasing at a rate of 33 m yr^{-1} (Weng, 2007). Vertical relief of oyster reef structures is critical for reef growth to outpace burial by sediment deposition (Colden et al. 2017, Hemraj et al. 2022).

This rapid increase in sediment deposition rate (Weng 2007) likely smothered oyster habitat as accretion occurred faster than the vertical growth capacity of reef building (Smith et al. 2005). Thirdly, dyke-building to make agricultural fields changed hydrological conditions. Flood frequency increased with each successive dynasty building dykes from once per 36 years during the Song (960–1279 CE) to once per 6 years during the Yuan (1279–1368 CE), and every 2 years during the Ming (1368–1644 CE) and Qing (1644–1912 CE) dynasties (Weng 2007). This increase in flooding was followed by more dyke-building, becoming an increasingly vicious cycle that reduced suitable availability for natural oyster populations (Weng 2007). During the Ming, Qing and later dynasties, dyke consolidation increasingly joined newly formed islands together, removing most of their shorelines altogether and destroying extensive oyster habitat (Weng 2002).

The lime industry is perhaps the most extensively documented extractive industry, with the largest direct impact on oyster habitats. Dating back at least as far as the Tang period (618–906 CE) (Atha & Yip 2016, Meacham 2009), lime making has a long history in the PRD region. Lime is thought to have been a significant export industry from the PRD between the Tang and Song period (618 CE–960 CE) (Meacham 2009), but it was also used in local construction and in making salt (Wong 2008). Similarly, the lime industry of the nineteenth century to the twentieth century was noted to have exported its product as far as Borneo and to be a prominent resource in local industries (Wong 1984). There are also accounts of synergies between the lime industry and the aquaculture industry, with the lime industry purchasing shells from the latter even though the aquaculture industry was primarily established to provide oysters for consumption (Wong 1984).

While there is some debate over the exact dates of when the lime industry began to decline, the collapses of the tenth century and the twentieth century would have been linked to over-extraction and almost complete removal of reefs from coastal areas, at least to the point of commercial collapse. During these periods, large-scale dredging of shellfish beds as a raw material for lime would have had a heavy impact on oyster beds (Figure 6). So prolific was the dredging that the local boat people vied to heap up shells in their boat, ferrying them back and forth to the lime kilns (Wong 1984). Intense extractive fisheries are likely to be the main cause of oyster reef extinction globally (Beck et al. 2011, Thurstan et al. 2013, Alleway & Connell 2015, Hernandez et al. 2018). Hard substrate, particularly oyster shells, is necessary for the formation of self-sustaining oyster reefs (Hemraj et al. 2022). Dredging removes the shells, reducing the structure, complexity and substrate provision of reefs, causing the collapse of the oyster populations (Lenihan & Peterson 2004). While returning shells and smaller oysters to the seabed following dredging can slow the pace of declines through overharvesting (Schulte 2017), the target of the lime industry is the shells, meaning that all reef structure is removed. Therefore, dredging of oyster habitats to fuel the lime industry would have rapidly driven oyster reefs to functional extinction, removing the related services (e.g., associated fisheries; Grabowski et al. 2012) and causing the crash of the lime industry itself.

In the tenth century collapse, lime making completely stopped. It is unclear whether the industry moved to other regions, or if the industry collapsed on a broader scale. One hypothesis suggests that the environmental burden of the lime-making industry had depleted local oyster reefs, and therefore, shell resources, drawing lessons from the second industry from the 1900s to the 1930s that also collapsed at least partially from overharvesting (Meacham 2009). It is also possible that continued decreased shorelines for oyster habitats due to human impacts led to a decline in oyster shell beds to harvest for lime production.

It is noteworthy that the first references to oyster aquaculture in the Asian region tend to coincide with the collapse of wild oyster fisheries or overuse of the resources to the point that the effort put into culture was less than that of searching for profitable wild oysters. While reference is made to aquaculture practices in the PRD region in 1678, ~90% of references to aquaculture are between 1870 and 1960, likely following the growth of the industry. The success of the oyster industry carried into the early 1990s, with remarks that it is another "source of wealth" and that the oysters from Canton (the provincial name at the time) were "highly valued". The account describes how oysters

Figure 6 (A) Oyster shell mound in Hong Kong being prepared for processing into lime. Note the person in the middle of the mound (on top) for scale. This mound is one of many reported such mounds and show the scale of the wild harvest of oysters. (B) Lime production company building containing lime kilns and a private dock for offloading oyster shells. (Photographers unknown. Modified from Wong (1984) with permission from the *Journal of the Hong Kong Branch of the Royal Asiatic Society*.)

are dried for export as well as how nearly "every bay and hamlet" possessed oyster beds (1912–21 Decennial Reports). There is even an account of land being rented out on the specific condition that it be used for oyster cultivation (Hong Kong Telegraph 1933), and many of these traditional oyster leases are still held by the modern-day villagers.

The discovery that oysters would collect on rocks placed in the sea has been attributed to fishermen and developed and refined over time to form a profitable industry supplying oysters to the Far East (de Thiersant 1875). Initially, oysters were cultivated on rocks placed into the sea, which had first been treated with fire to remove any harmful substances, but there is also reference made to oysters growing on bamboo sticks on the intertidal mudflats (de Thiersant 1875). This method advanced to placing rectangular granite stones, cleaned and treated with fire, at specific distances (10–15 cm) and depths (30–40 cm) in coastal areas to maximise yield. Interestingly, a recent meta-analysis of global oyster reef restoration programmes suggests that vertical relief of oyster reefs ~30 cm above the sediment surface maximises survival and growth (Hemraj et al. 2022). In addition, some farmers were recorded as throwing shells between the stones, providing more hard substrate for attachment of spat. The oysters were harvested twice a year, in April and September, and men were employed to keep the beds clear of predators. The account also demonstrates that the cultivators possessed sophisticated knowledge of the impact of environmental factors, such as rainfall, drought and salinity, on oyster yield. Despite these challenges, the industry was both profitable and politically important as it was exempt from taxes (de Thiersant 1875).

In 1883, there was a record of oyster aquaculture developing to use large old oyster shells instead of tiles or similar substrates. The oysters were harvested using 'oyster junks' flat-bottomed narrow boats, the smallest of which could carry 10 tonnes of oysters. The account gives details as to the extent and profitability of the industry. In Deep Bay and Nam Tau, the oyster grounds in four of the villages in this area measured approximately 50–60 square miles (1883 Hong Kong Daily Press), and oyster aquaculture is still seen as a part of the culture of these villages. From an ecological perspective, these benthic farming practices would have served to offset some of the loss of functions and services from the decline in natural oyster reefs. Current-day abandoned benthic farms support higher diversity, abundance and biomass of invertebrates than adjacent mudflats (Chan et al. 2022), and even sparse oyster habitat can maintain high water filtration rates (Lau et al. 2020). Farms would also have acted as a source of larvae to support recruitment to natural reefs (Lau et al. 2020) as production of triploid oysters was not in practice at the time. These functions and services would, however, be at a lower level as they have a positive relationship with the age of oysters and the reef formation itself (Brown et al. 2014, Smith et al. 2022). Therefore, while not as spatially extensive or developed (old) as the natural reefs before industrial harvesting, the broad-scale nature of benthic oyster farming would have performed many of the same functions as oyster reefs.

The oyster aquaculture industry was historically an extensive and politically influential industry in the region. Both the number of people employed and the financial accounts referenced in the sources uncovered by our review the importance of the industry. More importantly, however, the political influence of the industry is demonstrated in its exemption from taxes and the government's specification that certain land had to be used exclusively for oyster cultivation. This demonstrates the political interest in maintaining such a profitable industry, likely due to its importance as an export industry for the region. Irrespective of this social and political status, however, the production of the modern aquaculture industry in Hong Kong peaked in the 1950s and it rapidly declined in the 1960s to stabilise at <15% of this peak (Figure 3). This low-level production has continued until recently, when there has been a resurgence of interest in rejuvenating the industry and making it profitable again.

Modern shoreline reclamation: the final nail in the coffin

From the 2000s onwards, the impact of a continued wave of upstream dam and reservoir building starting in 1992 decreased sediment loads dramatically, counterbalancing continued deposits of silt into the river from deforestation-related soil erosion (Tan et al. 2016). Over 14,000 reservoirs have since been built in the Pearl River drainage basin, causing a substantial decrease in sediment transportation (Wu et al. 2016). This did not, however, indicate a reprieve for intertidal mudflats suitable for oyster reefs. New pressures for land reclamation from rapidly increasing populations and urbanisation have continued to increase in the PRD region since around the year 2000, leading to declining intertidal areas (Wu et al. 2016).

In the cases of 'hard' coastal reclamation, natural shorelines are transformed into concrete-filled built-up areas, resulting in the loss of the entire habitat (and associated species) previously occupying that area (Bulleri et al. 2020). According to an analysis of a series of British and Chinese navigation charts, since the 1850s, the PRD has gained ~1160 km^2 of new land, all of which replaced shallow coastal habitats. Most of this new land is reclaimed, derived from the disposal of dredge spoils, which not only replaces the immediate habitat but also smothers habitat within a large distance of the development (Wu et al. 2018).

Conclusions and future directions

The history of oysters in the Asian region lacks substantive academic documentation. Our review of the different and shifting oyster social-ecological systems (SES) in the Pearl River Delta (PRD) region substantially adds to knowledge regarding the relationship between oysters and humans over millennia across the continent. It is worth noting that uncovering the links between the extensive nature of oyster habitats in the region, the SES that grew around this resource, including regional development and industries, and the extensive exploitation and ultimate collapse of the systems was only possible by doing an in-depth search for documents in the regional languages; the published English literature was missing critical information in this system and future regional assessments should delve into the literature in the local languages.

While much of Asia is known for its history with aquaculture of native oyster species (Botta et al. 2020), oysters are ultimately natural organisms. Natural wild populations were first harvested and then declined or collapsed for secondary aquaculture practices to develop (Schulte 2017, Ridlon et al. 2023). Using the PRD as the model historical example, our review unearthed detailed accounts of the large geographical extent and productive nature of native oyster habitats, their decline because of overfishing and environmental changes, and the subsequent development of oyster cultivation. As ~50% of these historical documents were in languages other than English, they cover not only the evolution and development of biological practices employed but also the social, economic and political nuances involved with the industry. Interestingly, we bring to light that the people and communities employed first in oyster harvest, but then in developing aquaculture production, formed culturally distinct social groups that, despite being largely ostracised from the rest of society, came to provide an important international export for the PRD region. As far as we are aware, this could be the earliest documented international export industry from the region (Meacham 2009). The importance of oysters as an export commodity is highlighted by tax exemptions, in a similar way that the pearl industry in Sri Lanka was encouraged (De Thiersant 1875, Bastiampillai 1980, Katupotha 2019).

Our review points to an extensive trade in oyster products throughout the region and globally, in both the ancient and contemporary world (e.g., 1912–21 Decennial Reports, Wong 1984, Lam & Morton 2004, Meacham 2009). In the PRD, lime formed a regional export during the Tang-Song period (618 CE–1279 CE) and then again in the1800s during the industry's second resurgence. Oysters for consumption were exported internationally, notably from the late 1800s to the early 1900s, reportedly being sought after in the West (1912–21 Decennial Reports). These trade links and the socio-economic facets of these industries, particularly the lime industry, lend insights into the wider development of the Asian region; lime being a crucial component in the construction industry points to a human-oyster relationship centred around increasingly urbanised, regionalised and later globalised, development trajectories. Ultimately, however, it is the intertwined components of SES and the conflict between species biology and market demand that led to overharvesting and collapse of oyster reefs shortly followed by the associated industries in the region.

The significance of oyster-related industries for the regions' economies and the creation of a separate stigmatised community indicates that there are likely greater socio-political factors embedded within the governance of oyster reefs. A deeper look into these relationships would likely provide insights into the evolution of governance practices and societal dynamics, which could indicate how more culturally sensitive, acceptable and sustainable practices could be implemented.

Ultimately the success of the aquaculture and lime industries contributed to the demise of oyster reefs in the PRD. Unsustainable and ecologically destructive practices such as dredging fed the lime industry while undermining the finite (oyster) resources that it depended on; these practices combined with changing sedimentation regimes from upstream agricultural practices, and shoreline reclamation for urbanisation and infrastructure development, ultimately led to the disappearance of oysters in the region before modern times.

Beyond the PRD, the limited documentation that was sourced indicates a similarly long history of human interactions with oysters and their habitats. Accounts of pearl oysters and the political and socio-economics involved in their harvest are the most extensive of the records surveyed. These accounts provide insights into the relationship between economics and politics on the one hand and ecological systems on the other, echoing the historical case of oyster reef collapse we observed in the PRD. Demonstrating that despite efforts to create productive pearl oyster beds by modifying them, the lack of local ecological and cultural knowledge of the ecosystems ultimately ended the industry. Moving further east, records are more scarce but the focus is on consumption, particularly aquaculture and the role of boat people (e.g., referred to in the PRD as Lutings). Social and practical similarities can be found across the wider Asian region (e.g., Anderson 1970) suggesting a socio-cultural link between such practices, which warrants further exploration.

Finally, this review provides evidence of a 7000-year+ history of oyster-related SESs in the PRD region, with additional historical evidence in the broader Asian continent. Documents indicate not only the presence but also a historical abundance of oyster reefs and the associated habitat/ecosystems. The fact that oyster reefs were once prominent in the PRD provides a promising basis for oyster-related revitalisation and restoration projects. Such projects could reconstruct lost ecosystem functions and contribute ecosystem services as well as socio-economic benefits to associated communities. Our review highlights, however, that the socio-economic interactions between communities and oyster reefs drove the demise of these SESs. As such, for the successful revitalisation of oyster reefs in the PRD region, and likely for the broader Asian continent, an approach that integrates both the social and ecological systems involved is required to prevent a similar tragedy of the commons type scenario.

Acknowledgements

We thank Ruth Thurstan for extensive conversations and guidance on methods for reconstructing the historical ecology of marine ecosystems using fragmented documentary evidence.

References

AFCD (Agriculture, Fisheries and Conservation Department). 2019–2023. *Marine fish culture, pond fish culture, and oyster culture.* Online. https://www.afcd.gov.hk/english/fisheries/fish_aqu/fish_aqu_mpo/fish_aqu_mo.html

Alleway, H.K. & Connell, S.D. 2015. Loss of an ecological baseline through the eradication of oyster reefs from coastal ecosystems and human memory. *Conservation Biology* **29**(3), 795–804.

Ambrosio-Albala, M. & Delgado, M. 2008 Understanding rural areas dynamics from a complex perspective. In *An Application of Prospective Structural Analysis, European Association of Agricultural Economists. 2008 International Congress*, August 26–29. Ghent, Belgium.

Anderson, E.N. 1970. The boat people of South China. *Anthropos* **65**(1/2), 248–256.

Atha, M. & Yip, K. 2016. *Piecing Together Sha Po: Archaeological Investigations and Landscape Reconstruction.* Hong Kong: HKU Press.

Baggett, L. 2014. *Oyster Habitat Restoration Monitoring and Assessment Handbook.* Nature Conservancy.

Bastiampillai, B. 1980. The pearl fishery of Ceylon (Sri Lanka) under British administration in the 19th Century. *Vidyodaya: Journal of Arts, Science and Letters* **8**(1&2), 25–41.

Beck, M.W., Brumbaugh, R.D., Airoldi, L., Carranza, A., Coen, L.D., Crawford, C., Defeo, O., Edgar, G.J., Hancock, B., Kay, M.C. & Lenihan, H.S. 2011. Oyster reefs at risk and recommendations for conservation, restoration, and management. *Bioscience* **61**(2), 107–116.

Benedetto, L.F. 1994. *The Travels of Marco Polo*. India: Asian Educational Services.

Bennett, N.J. & Dearden, P. 2014. Why local people do not support conservation: community perceptions of marine protected area livelihood impacts, governance and management in Thailand. *Marine Policy* **44**, 107–116.

Bie, J., de Jong, M. & Derudder, B. 2015. Greater Pearl River delta: historical evolution towards a global city-region. *Journal of Urban Technology* **22**(2), 103–123, doi:10.1080/10630732.2014.971575

Botta, R., Asche, F., Borsum, J.S., Camp, E.V. 2020. A review of global oyster aquaculture production and consumption. *Marine Policy* **117**, 103952.

Brohmanonda, P., Mutarasint, K., Chongpeepien, T. & Amornjaruchit, S. 1988. Oyster culture in Thailand. In *Bivalve Mollusc Culture Research in Thailand*, E.W. McCoy & T. Chongpeepien (eds.). Eschborn, Germany: GmbH, 31–39. OCLARM Technical Reports 19. International Center for Living Aquatic Resources Management, Manila; and Deutsche Gesellschaft filr Technische Zusarnmenarbeit (GTZ).

Brown, L.A., Furlong, J.N., Brown, K.M. & La Peyre, M.K. 2014. Oyster reef restoration in the Northern Gulf of Mexico: effect of artificial substrate and age on nekton and benthic macroinvertebrate assemblage use. *Restoration Ecology* **22**, 214–222.

Bulleri, F., Batten, S., Connell, S.D., Benedetti-Cecchi, L., Gibbons, M., Nugues, M.M. & Gribben, P. 2020. Human pressures and the emergence of novel marine ecosystems. *Oceanography and Marine Biology: An Annual Review* **58**, 441–494.

Carpenter, S.R., Folke, C., Norström, A., Olsson, O., Schultz, L., Agarwal, B., Blavenera, P., Campbell, B., Castilla, J.C., Cramer, W., DeFries, R., Eyzaguirre, P., Hughes, T.P., Polasky, S., Sanusi, Z., Scholes, R. & Spierenburg, M. 2012. Program on ecosystem change and society: an international research strategy for integrated social-ecological systems. *Current Opinion in Environmental Sustainability* **4**(1), 134–138.

Carran, D., Hughes, J., Leslie, A. & Kennedy, C. 2010. A short history of the use of lime as a building material beyond Europe and North America. *International Journal of Architectural Heritage* **6**(2), 117–146, doi: 10.1080/15583058.2010.511694

Caswell, B.A., Klein, E.S., Alleway, H.K., Ball, J.E., Botero, J., Cardinale, M., Eero, M., Engelhard, G.H., Fortibuoni, T., Giraldo, A.J. & Hentati-Sundberg, J. 2020. Something old, something new: historical perspectives provide lessons for blue growth agendas. *Fish and Fisheries* **21**(4), 774–796.

Chan, S.S., Wong, H.T., Thomas, M., Alleway, H.K., Hancock, B. & Russell, B.D. 2022. Increased biodiversity associated with abandoned benthic oyster farms highlight ecosystem benefits of both oyster reefs and traditional aquaculture. *Frontiers in Marine Science* **9**, 862548.

Chen, C.T.A., Wang, S.L., Lu, X.X., Zhang, S.R., Lui, H.K., Tseng, H.C., Wang, B.J. & Huang, H.I. 2008. Hydrogeochemistry and greenhouse gases of the Pearl River, its estuary and beyond. *Quaternary International* **186**(1), 79–90, doi:10.1016/j.quaint.2007.08.024

Colden, A.M., Latour, R.J. & Lipcius, R.N. 2017. Reef height drives threshold dynamics of restored oyster reefs. *Marine Ecology Progress Series* **582**, 1–13.

Coleman, F.C. & Williams, S.L. 2002. Overexploiting marine ecosystems engineers: potential consequences for biodiversity. *Trends in Ecology and Evolution* **7**(1), 40–44, doi:10.1016/S0169-5347(01)02330-8

Crain, C.M., Halpern, B.S., Beck, M.W. & Kappel, C.V. 2009. Understanding and managing human threats to the coastal marine environment. *Annals of the New York Academy of Sciences* **1162**(1), 39–62.

Cranfield, H.J., Carbines, G., Michael, K.P., Dunn, A., Stotter, D.R. & Smith, D.J. 2001. Promising signs of regeneration of blue cod and oyster habitat changed by dredging in Foveaux Strait, southern New Zealand. *New Zealand Journal of Marine and Freshwater Resources* **35**(5), 897–908.

Decennial Reports on the Trade Navigation Industries of the Ports Open to Foreign Commerce in China and Korea, and on the Conditions and Development of the Treaty Port Provinces. 1924. Published by order of the InsPeclor general of customs. Shanghai: Published at the Statistical Department of the Inspectorate General of Customs.

Deng, J.Y. & Jin, X.S. 2000. Study on fishery biodiversity and its conservation in Laizhou Bay and Yellow River estuary. *Zoological Research* **21**(1), 76–82.

de Silva, K.M. 2005. *A History of Sri Lanka*. Colombo, Sri Lanka: Vijitha Yapa Publications.

De Thiersant, D. 1875. Ostriculture in China. *The China Review, or Notes & Queries on the Far East* **4**(1).

Du, J. 2020 Part II: regional history. In *The Shenzhen Experiment: The Story of China's Instant City*. Harvard University Press, 333.

Fang, E., Li, W. & Yu, J. 2007. Sustainable use of live oyster reef in Bohai Gulf. *Modern Fisheries Information-Changshu-* **22**, 12–14, doi:10.3969/j.issn.1004-8340.2007.11.003

Fernando, T. 2022. Seeing like the sea: a multispecies history of the Ceylon pearl fishery 1800–1925. *Past & Present*, **254**(1), 127–160.

Folke, C., Hahn, T., Olsson, P. & Norberg, J. 2005. Adaptive governance of social-ecological systems. *Annual Review of Environment and Resources* **30**, 441–473.

Fujiya, M. 1970. Oyster farming in Japan. *Helgoländer wissenschaftliche Meeresuntersuchungen* **20**(1), 464–479.

Garrett, R. 2010. *The Defences of Macau: Forts, Ships and Weapons over 450 Years*. Hong Kong: Hong Kong University Press.

Gillies, C.L., McLeod, I.M., Alleway, H.K., Cook, P., Crawford, C., Creighton, C., Diggles, B., Ford, J., Hamer, P., Heller-Wagner, G. & Lebrault, E. 2018. Australian shellfish ecosystems: past distribution, current status and future direction. *PLoS One*, **13**(2).

Global Administrative Areas. 2018. *University of California, Berkely*. [Digital geospatial data]. Online. https://www.gadm.org. (Accessed 13 Apr 2023).

Grabowski, J.H, Brumbaugh, R.D, Conrad, R.F., Keeler, A.G., Opaluch, J.J., Peterson, C.H., Piehler, M.F., Powers, S.P. & Smyth, A.R. 2012. Economic valuation of ecosystem services provided by oyster reefs. *Bioscience* **62**, 900–909.

Gunawardana, R.A.L.H. 2000. Prelude to the state: an early phase in the evolution of political institutions in ancient Sri Lanka. In *Reflection on a Heritage, History and Archaeology of Sri Lanka*, Volume I, Part I, In: *Central Cultural Fund. Ministry of Cultural and Religious Affairs*, R.A.L.H. Gunawardana, S. Pathmanathan & M. Rohanadeera (eds.), 83–122.

Guo, X., Ford, S.E. & Zhang, F. 1999. Molluscan aquaculture in China. *Journal of Shellfish Research* **18**(1), 19–31.

Hanazaki, N., Herbst, D.F., Marques, M.S. & Vandebroek, I. 2013. Evidence of the shifting baseline syndrome in ethnobotanical research. *Journal of ethnobiology and ethnomedicine* **9**, 1–11.

Hansson, A. 1996. Chinese outcasts: discrimination and emancipation in late imperial China. *Sinica Leidensia*.

Hayes, J.W. 1964. Peng Chau between 1798–1899. *Journal of the Hong Kong Branch of the Royal Asiatic Society* **4**.

Hemraj, D.A., Bishop, M.J., Hancock, B., Minuti, J.J., Thurstan, R.H., Zu Ermgassen, P.S. & Russell, B.D. 2022. Oyster reef restoration fails to recoup global historic ecosystem losses despite substantial biodiversity gain. *Science Advances* **8**(47), eabp8747.

Herdman, A.W. 1903. *Report to the Government of Ceylon on the Pearl Oyster Fisheries of the Gulf of Manaarwith Supplementary Reports upon the Marine Biology of Ceylon. Part I*. London: The Royal Society, 1–307.

Hernández, A.B., Brumbaugh, R.D., Frederick, P., Grizzle, R., Luckenbach, M.W., Peterson, C.H. & Angelini, C. 2018. Restoring the eastern oyster: how much progress has been made in 53 years? *Frontiers in Ecology and the Environment* **16**(8), 463–471.

Hingham, C.F.W. 1996. A review of archaeology in mainland Southeast Asia. *Journal of Archaeological Research* **4**(1), 3–49.

Hong Kong Daily Press. 1883. *Oyster culture in the Sun On district*. Newspaper article (1883-06-11).

Hong Kong Government Gazette (no 557). Sept 26 1901.

Hong Kong Telegraph (1933-10-06).

Huang, Z., Li, P., Zhang, Z., Li, K. & Qiao, P. 1982. The formation, development, and evolution of Zhujiang Delta. Kexue Puji Chubanshe Guangzhou Fenshe, Guangzhou (in Chinese). Cited in Weng, Q. 2007. A historical perspective of river basin management in the Pearl River Delta of China. *Journal of Environmental Management* **85**, 1048–1062.

Islam, M.S. 2003. Perspectives of the coastal and marine fisheries of the Bay of Bengal, Bangladesh. *Ocean & Coastal Management* **46**(8), 763–796.

Katupotha, J. 2019. Pearl fishery industry in Sri Lanka. *WILDLANKA Journal of the Department of Widllife Conservation of Sri Lanka* **7**(1).

Kellogg, L.M., Smyth, A.R., Luckenback, M.W., Carmichael, R.H., Brown, B.L., Cornwell, J.C., Piehler, M.F., Owens, M.S., Dalrymple, J.D. & Higgens, C.B. 2014. Use of oysters to mitigate eutrophication in coastal waters. *Estuarine, Coastal and Shelf Science* **151**, 156–168, doi:10.1016/j.ecss.2014.09.025

Laing, I., Walker, P. & Areal, F. 2006. Return of the native-is European oyster (Ostrea edulis) stock restoration in the UK feasible? *Aquatic Living Resources* **19**(3), 283–287.

Lam, K. & Morton, B. 2004. The oysters of Hong Kong (bivalvia: ostreidae and gryphaeidae). *The Raffles Bulletin of Zoology* **52**(1), 11–28.

Lau, S.C., Thomas, M., Hancock, B. & Russell, B.D. 2020. Restoration potential of Asian oysters on heavily developed coastlines. *Restoration Ecology* **28**(6), 1643–1653.

Legge J. 1886. *Fahien's Record of Buddhist Kingdom*. Oxford.

Lenihan, H.S. & Peterson, C.H. 2004. Conserving oyster reef habitat by switching from dredging and tonging to diver-harvesting. *Fisheries Bulletin* **102**, 298–305.

Lin, H., Yu, Q., Du, Z., Fan, Y., Wang, Y. & Gao, S. 2021. Geomorphology and sediment dynamics of the Liyashan oyster reefs, Jiangsu Coast, China. *Acta Oceanologica Sinica* **40**, 118–128.

Lin, H.J., Hung, J.J., Shao, K.T. & Kuo, F. 2001. Trophic functioning and nutrient flux in a highly productive tropical lagoon. *Oecologia*, 395–406.

Liu, W. & Deng, P.Y. 2007. Accumulation of cadmium, copper, lead and zinc in the pacific oyster, *Crassostrea gigas*, collected from the Pearl River Estuary, Southern China. *Bulletin of Environmental Contamination and Toxicology* **78**, 535–538.

Mahroof, M.M.M. 1995. The story of the Sri Lankan Pearl. *The Journal of Gemology* **24**(5).

McAfee, D., McLeod, I.M., Boström-Einarsson, L. & Gillies, C.L. 2020. The value and opportunity of restoring Australia's lost rock oyster reefs. *Restoration Ecology* **28**, 304–314, doi:10.1111/rec.13125

McDonnell, M.J. & MacGregor-Fors, I. 2016. The ecological futures of cities. *Science* **352**, 936–938.

Meacham. 1978 & Au et al. 1990, cited in Li, G. 2002. *Subsistence of Neolithic Pearl River Estuary area, South China*. MSc thesis, The University of British Columbia, 21.

Meacham, W. 1994. Archaeological discovery on Chek Lap Kok. Hong Kong. *Archaeological Society*.

Meacham, W. 2009. Hong Kong's first industry. In *The Archaeology of Hong Kong*. Hong Kong University Press, 149–156.

Morton, B. & Wong, P.S. 1975. THE PACIFIC OYSTER INDUSTRY IN HONG KONG (香港 的蠔業). *Journal of the Hong Kong Branch of the Royal Asiatic Society*, 139–149.

Newspaper article. 广州日报 (Guangzhou Daily). 2018-4-24. 蚝壳为墙 冬暖夏凉 (Oyster shells as walls, cool in summer warm in winter). Newspaper article. https://news.dayoo.com/guangzhou/201804/24/139995_52154453.htm

Ostroff, S. 2020. Can the oyster speak? Pearling empires and the marine environments of South India and Sri Lanka, c. 1600–1900. In *Animal Trade Histories in the Indian Ocean World*, 65–98.

Ostroff, S.M. 2016. *The Beds of Empire: Power and Profit at the Pearl Fisheries of South India and Sri Lanka, c. 1770–1840*. University of Pennsylvania.

Papworth, S.K., Rist, J., Coad, L. and Milner-Gulland, E.J. 2009. Evidence for shifting baseline syndrome in conservation. *Conservation letters*, **2**(2), 93–100.

Pauly, D. 1995. Anecdotes and the shifting baseline syndrome of fisheries. *Trends in Ecology and Evolution* **10**, 430.

Peregrine, P. N. 2001. Encyclopedia of prehistory. In *Volume 3: East Asia and Oceania*, P.N. Peregrine (ed.). Melvin Ember, 325.

Pogoda, B. 2019. Current status of European oyster decline and restoration in Germany. *Humanities* **8**(1), 9.

Quan, W., An, C., Ma, C., Huang, H., Cheng, W., Wang, Y., Shen, X. & Chen, Y. 2012. Biodiversity and community structure of benthic macroinvertebrates on the Xiaomiaohong oyster reef in Jiangsu province, China. *Oceanologia et Limnologia Sinica/Hai Yang Yu Hu Chao* **43**(5), 992–999.

Ridlon, A.D., Grosholz, E.D., Hancock, B., Miller, M.W., Bickel, A., Froehlich, H.E., Lirman, D., Pollock, F.J., Putnam, H.M., Tlusty, M.F., Waters, T.J. & Wasson, K. 2023. Culturing for conservation: the need for timely investments in reef aquaculture. *Frontiers in Marine Science* **10**, 1069494.

Robinson, N.A. 2012. Beyond sustainability: environmental management for the Anthropocene Epoch. *Journal of Public Affairs* **12**(3), 181–194, doi:10.1002/pa.1432

Rodriguez, A.B., Fodrie, F.J., Ridge, J.T., Lindquist, N.L., Theuerkauf, E.J., Coleman, S.E., Grabowski, J.H., Brodeur, M.C., Gittman, R.K., Keller, D.A. & Kenworthy, M.D. 2014. Oyster reefs can outpace sea-level rise. *Nature Climate Change* **4**(6), 493–497.

Rogers, P. R. & Widdowson, E. 1996. Midden excavation in theory and practice: a Han period midden site at Tung Wan Tsai, Ma Wan Island, Hong Kong. *Papers from the Institute of Archaeology* **7**, 31–56.
Rusher, K. 2003. The Bluff Oyster festival and regional economic development: festivals as culture commodified. *Food Tourism around the World: Development, Management and Markets* **110**(9), 192–205.
Sarker, M.N. & Alam, F. 1988. Prospects of culture of commercially important marine bivalves in Bangladesh. In *Compilation of Reports on the Marine Fishery Survey and Research*. Bangladesh: Department of Fisheries, 357.
Schafer, E.H. 1957. A fourteenth century gazetteer of Canton. Oriente Poliano: Studi e Conferenze Tenute all'Is. M.E.O. In *Occasione del VII Centenario Della Nascita di Marco Polo*, 1254–1954.
Schofield, W. 1983. The islands around Hong Kong. *Journal of the Hong Kong Branch of the Royal Asiatic Society* **23**, 91–112.
Schouten, M.A., van der Heide, M.M. & Heijman, W.J. 2009. Resilience of social-ecological systems in European rural areas: theory and prospects. In *113th EAAE Seminar "The Role of Knowledge, Innovation and Human Capital in Multifunctional Agriculture and Territorial Rural Development*", December 9–11. Belgrade, Republic of Serbia.
Schulte, D.M. 2017. History of the Virginia oyster fishery, Chesapeake Bay, USA. *Frontiers in Marine Science* **4**, 127.
Schulte, D.M. & Burke, R.P. 2014. Recruitment enhancement as an indicator of oyster restoration success in Chesapeake Bay. *Ecological Restoration* **32**(4), 434–440.
Smaal, A.C., Ferreira, J.G., Grant, J., Petersen, J.K. & Strand, Ø. 2019. *Goods and Services of Marine Bivalves*. Springer Nature.
Smith, C.S., Ito, M., Namba, M. & Nakaoka, M. 2018. Oyster aquaculture impacts Zostera marina epibiont community composition in Akkeshi-ko estuary, Japan. *PLoS One* **13**(5).
Smith, G.F., Bruce, D.G., Roach, E.B., Hansen, A., Newell, R.I.E. & McManus, A.M. 2005. Assessment of recent habitat conditions of eastern oyster Crassostrea virginica bars in mesohaline Chesapeake Bay. *North American Journal of Fisheries Management* **25**(4), 1569–1590.
Smith, R.S., Lusk, B. & Castorani, M.C.M. 2022. Restored oyster reefs match multiple functions of natural reefs within a decade. *Conservation Letters* **15**, e12883.
Southgate, P.C. & Lucas, J.S. (eds.). 2008. *The Pearl Oyster*. Netherlands: Elsevier.
Steuart, J. 1834. Account of the pearl fisheries of the North-West Coast of the Island of Ceylon. *Transactions of the Royal Asiatic Society of Great Britain and Ireland* **3**(3), 452–462.
Strickland, J. 2010. *Southern District Officer Reports: Islands and Villages in Rural Hong Kong 1910–60*. Hong Kong: Hong Kong University Press.
Sudarmawathie, J.M. 2016. Pearl industry of Sri Lanka: from ancient times to the 15th CE. *Humanities and Social Science Review* **6**(1), 101–108.
Swetnam, T.W., Allen, C.D. & Betancourt, J.L. 1999. Applied historical ecology: using the past to manage for the future. *Ecological Applications* **9**(4), 1189–1206.
Tan, C., Huang, B., Liu, F., Yang, Q., Jiang, C. & Zhang, S. 2016. Transformation of the three largest Chinese river deltas in response to the reduction of sediment discharges. *Journal of Coastal Research* **32**(6), 1402–1416.
The Directory & Chronicle for China, Japan, Corea, Indo-China, Straits Settlements, Malay States, Sian, Netherlands India, Borneo, the Philippines, &c: With Which Are Incorporated "The China Directory" and "The Hong Kong List for the Far East" 1908, Hongkong Daily Press Office.
Thurstan, R.H. 2022. The potential of historical ecology to aid understanding of human-ocean interactions throughout the Anthropocene. *Journal of Fish Biology* **101**, 351–364, doi:10.1111/jfb.15000
Thurstan, R.H., Hawkins, J.P., Raby, L. & Roberts, C.M. 2013. Oyster (Ostrea edulis) extirpation and ecosystem transformation in the Firth of Forth, Scotland. *Journal for Nature Conservation* **21**, 253–261.
Ting, V.K. & Donnelly, I.A. 1925. Things produced by the works of nature. Published 1639. *The Mariner's Mirror* **11**(3), 234–250, doi:10.1080/00253359.1925.10655321
Tittensor, D.P., Mora, C., Jetz, W., Lotze, H.K., Ricard, D., Berghe, E.V. & Worm, B. 2010. Global patterns and predictors of marine biodiversity across taxa. *Nature* **466**(7310), 1098–1101.
Translated text **1**. 420–479. Zheng Qizhi (鄭緝之). 31st chapter: Oyster (三十一. 蠣). In Yongjia jun ji (永嘉郡記). Digitised at: https://ctext.org/library.pl?if=gb&file=1000217&page=43
Translated text **2**. 867–904. Liu Xun (劉恂). Upper scroll In Ling biao lu yi ji (嶺表錄異). Digitised at: https://ctext.org/wiki.pl?if=gb&res=837558

Translated text 3. 867–904. Liu Xun (劉恂). Ling Table Records, In Ling biao lu yi ji (嶺表錄異)

Translated text 4. 1687. Dajun Qu (屈大均). Kangxi 17th year, 23rd scrol (康熙十七年, 第二十三卷介语), In Guangdong xin yu (廣東新語). Digitised at: https://ctext.org/wiki.pl?if=gb&chapter=222923&remap=gb

Translated text 5. 1596. Li Shizhen (李時珍). Chapter 53 (介之二 （蛤蚌類二十九種) In BěncǎO GāNgmÙ (本草綱目). Digitised at: https://zh.wikisource.org/wiki/%E6%9C%AC%E8%8D%89%E7%B6%B1%E7%9B%AE

Turvey, S.T., Barrett, L.A., Yujiang, H.A.O., Lei, Z., Xinqiao, Z., Xianyan, W., Yadong, H., Kaiya, Z., Hart, T.O.M. & Ding, W. 2010. Rapidly shifting baselines in Yangtze fishing communities and local memory of extinct species. *Conservation Biology* **24**(3), 778–787.

Ventilla, R.F. 1984. Recent developments in the Japanese oyster culture industry. *Advances in Marine Biology* **21**, 1–57.

Walles, B., Troost, K., van den Ende, D., Neiuwhof, S., Smaal, A.C. & Ysebaert, T. 2016. From artificial structures to self sustaining oyster reefs. *Journal of Sea Research* **108**, 1–9, doi:10.1016/j.seares.2015.11.007

Wang, F. & Wall, G. 2010. Mudflat development in Jiangsu Province, China: Practices and experiences. *Ocean & Coastal Management* **53**, 691–699.

Wang, H. & van Strydonck, M. 1997. Chronology of Holocene Cheniers and oyster reefs on the coast of Bohai Bay, China. *Quaternary Research* **47**(2), 192–205.

Watson, J.L. 1990. *Death Ritual in Late Imperial and Modern China*. University of California Press.

Wei, X., Wu, C., Ni, P. & Mo, W. 2016. Holocene delta evolution and sediment flux of the Pearl River, southern China. *Journal of Quaternary Science* **31**(5), 484–494.

Weng, Q. 2002. Human-environment interactions in agricultural land use in a South China's wetland region: a study on the Zhujiang Delta in the Holocene. *GeoJournal* **51**, 191–202.

Weng, Q. 2007. A historical perspective of river basin management in the Pearl River Delta of China. *Journal of Environmental Management* **85**, 1048–1062.

Williams, G.A., Helmuth, B., Russell, B.D., Dong, Y.W., Thiyagarajan, V. & Seuront, L. 2016. Meeting the climate change challenge: pressing issues in southern China and SE Asian coastal ecosystems. *Regional Studies in Marine Science* **8**, 373–381.

Williams, J.M., Chu, H.Y.V., Lam, W.F. & Law, W.W.Y. 2021. *Revitalising Rural Communities*. Springerbrief on Sustainability. Singapore: Springer

Wong, T-Y. 1984. Lime making on Tsing-yi. *Journal of the Royal Asiatic Society Hong Kong Branch* **24**.

Wong, W.H. 2008. Towards urbanisation: Shuen Wan and Plover Cove reservoir. In *Tai Po Book*, 234–255.

Worcester, G.R.G. 1958. Some brief notes on fishing in China. *Mariner's Mirror* **44**, 49–63.

Wu, Z., Milliman, J., Zhao, D., Cao, Z., Zhou, J. & Zhou, C. 2018. Geomorphologic changes in the lower Pearl River Delta, 1850–2015, largely due to human activity. *Geomorphology* **314** 42–54.

Wu, Z., Saito, Y., Zhao, D., Zhou, J.Q., Cao, Z.Y., Li, S.J., Shang, J.H. & Liang, Y.Y. 2016. Impact of human activities on subaqueous topographic change in Lingding Bay of the Pearl River estuary, China, during 1955–2013. *Scientific Reports* **6**, 37742, doi:10.1038/srep37742

Yang, J., Fu, Z., Yu, G., Ma, Z. & Wang, X. 2022. Combined effects of temperature and salinity on antioxidants in the immune system of the Pearl Oyster *Pinctada fucata*. *Fishes* **7**, 260.

Yeung, Y.M. 2001. Coastal mega-cities in Asia: transformation, sustainability and management. *Ocean & Coastal Management* **44**(5–6), 319–333.

Ysebaert, T., Walles, B., Haner, J. & Hancock, B. 2018. Habitat modification and coastal protection by ecosystem-engineering reef-building bivalves. In *Goods and Services of Marine Bivalves*, A.C. Smaal et al. (eds.). Cham, Switzerland: Springer, 253–273.

Zong, Y., Huang, G., Switzer, A.D. & Yim, W.W.S. 2009. An evolutionary model for the Holocene formation of the Pearl River delta, China. *The Holocene* **19**, 129–142.

Zu Ermgassen, P.S., Grabowski, J.H., Gair, J.R. & Powers, S.P. 2016. Quantifying fish and mobile invertebrate production from a threatened nursery habitat. *Journal of Applied Ecology* **53**(2), 596–606.

Oceanography and Marine Biology: An Annual Review, 2024, **62**, 192-214

Taylor & Francis

INTEGRATING BELOWGROUND INTERACTIONS INTO SEAGRASS RESTORATION STRATEGIES

RENSKE JONGEN[1], EZEQUIEL M. MARZINELLI[1], ANA B. BUGNOT[1], ANGUS FERGUSON[2], MATTHEW W. FRASER[3], TIM M. GLASBY[4], EMMA L. JACKSON[5], BELINDA C. MARTIN[3], CRAIG D.H. SHERMAN[6], ELIZABETH A. SINCLAIR[3], STACEY M. TREVATHAN-TACKETT[6], ADRIANA VERGÉS[7], MICHELLE WAYCOTT[8], JEFFREY T. WRIGHT[9], GARY A. KENDRICK[3], AND PAUL E. GRIBBEN[7]

[1]*The University of Sydney, School of Life and Environmental Sciences, Sydney, NSW 2006, Australia*
[2]*NSW Department of Planning, Industry and Environment, EES Laboratories, 480 Weeroona Road, Lidcombe, NSW 2141, Australia*
[3]*School of Biological Sciences and Oceans Institute, The University of Western Australia, Crawley, WA 6009, Australia*
[4]*New South Wales Department of Primary Industries, Port Stephens Fisheries Institute, Locked Bag 1 New South Wales, 2315 Nelson Bay, Australia*
[5]*Central Queensland University Australia, Coastal Marine Ecosystems Research Centre, Gladstone, Queensland, Australia*
[6]*Deakin Marine Research and Innovation Centre, School of Life and Environmental Sciences, Deakin University, Geelong, Victoria 3220 Australia*
[7]*Centre for Marine Science and Innovation, School of Biological Earth and Environmental Sciences, The University of New South Wales, Sydney, NSW 2052, Australia*
[8]*School of Biological Sciences, Faculty of Science, The University of Adelaide, Adelaide, SA 5005, Australia*
[9]*Institute for Marine and Antarctic Studies, University of Tasmania, Hobart, TAS 7004, Australia*

Author emails

Renske Jongen: renske.jongen@sydney.edu.au
Ezequiel M. Marzinelli: e.marzinelli@sydney.edu.au
Ana B. Bugnot: Ana.Bugnot@csiro.au
Angus Ferguson: Angus.Ferguson@environment.nsw.gov.au
Matthew W. Fraser: matthew.fraser@uwa.edu.au
Tim M. Glasby: tim.glasby@dpi.nsw.gov.au
Emma L. Jackson: emma.jackson@cqu.edu.au
Belinda C. Martin: belinda.martin@uwa.edu.au
Craig D.H. Sherman: craig.sherman@deakin.edu.au
Elizabeth A. Sinclair: Elizabeth.Sinclair@uwa.edu.au
Stacey M. Trevathan-Tackett: s.trevathantackett@deakin.edu.au
Adriana Vergés: a.verges@unsw.edu.au
Michelle Waycott: michelle.waycott@adelaide.edu.au
Jeffrey T. Wright: jeffrey.wright@utas.edu.au
Gary Kendrick: gary.kendrick@uwa.edu.au
Paul E. Gribben: p.gribben@unsw.edu.au

DOI: 10.1201/9781003477518-4

Corresponding Author:
Renske Jongen
Email address: renske.jongen@sydney.edu.au

Abstract Seagrass meadows are declining globally at alarming rates due to multiple stressors. Efforts are underway to preserve and restore these valuable ecosystems. Several frameworks for seagrass restoration exist, and some restoration attempts are being made at scale; however, poor outcomes are widespread, suggesting our understanding of the processes that drive seagrass performance is still limited. We argue that plant-sediment interactions influence almost all components of restoration, yet they are poorly incorporated in restoration design and frameworks. We synthesise the current information on how belowground interactions influence seagrass health and identify which of these are amenable to manipulation and at what scale they could be employed to improve restoration strategies and decision-making frameworks. We identified three key, interdependent areas by which belowground interactions influence seagrass health and performance. Those three areas include how (1) sediment microbial communities control nutrient and chemical cycling for seagrass, (2) the response of seagrass to sediment processes may be influenced by their life history and genetics of seagrass species targeted for restoration and (3) other species may interact with sediments to promote or inhibit seagrass survivorship and growth. Decision-making frameworks should be extended to include belowground interactions because they are of fundamental concern for determining initial site suitability and affect subsequent decisions and planning for restoration. This, in turn, will improve restoration outcomes and community engagement, critical for the successful management of these endangered ecosystems.

Keywords: Seagrass Restoration; Sediments; Microbes; Seagrass Genetics; Species Interactions; Decision-making Frameworks

Introduction

Seagrass meadows provide critical nursery grounds for fisheries (Whitfield 2017), nutrient cycling (Costanza et al. 2014), coastal protection (Boudouresque et al. 2016) and habitat for many species including those of conservation concern (Hughes et al. 2009). Seagrasses also sequester a globally significant amount of carbon, which is known as blue carbon (Macreadie et al. 2019, Serrano et al. 2019; https://www.iucn.org/resources/issues-briefs/blue-carbon). Through these services, seagrasses provide an estimated US$6.8 trillion to the global economy (Costanza et al. 2014). One of the consequences of human development and exploitation of marine coastal and estuarine ecosystems is the severe reduction (Green et al. 2021, Evans et al. 2018) and, in some instances, the complete loss of seagrasses and the ecological and socio-economic services they provide. Globally, seagrasses are disappearing at an alarming rate of $110\,km^2\,yr^{-1}$ (Waycott et al. 2009). Since the 1980s, an estimated 19%–29% of seagrass cover has been lost worldwide (Orth et al. 2006, Waycott et al. 2009, Dunic et al. 2021). As a result of the worldwide decline in seagrass habitat, 14% of the world's 72 seagrass species are now at an elevated risk of extinction (Short et al. 2011). Importantly, this loss of seagrass represents an associated loss of ecosystem services valued at US$318 million annually (based on 2011 seagrass service values of US$28,916 ha^{-1}; Costanza et al. 2014). Enhancing the resilience of seagrasses to environmental stress (i.e. the ability to resist and/or recover from environmental stress) and restoring lost seagrass habitat is critical for improving the health and function of marine coastal and estuarine ecosystems, buffering coastal ecosystems from ongoing climate change and sustaining the coastal communities and industries that depend on them.

Many factors have contributed to seagrass declines (reviewed in Statton et al. 2018) but chiefly responsible among these are massive historical and continuing land-based sediment and nutrient inputs (Orth et al. 2006, Dunic et al. 2021), as well as contemporary climate-related stressors

(Short et al. 2016). For example, following an extreme heatwave event in 2010/11, 1310 km^2 seagrass habitat was lost from the Shark Bay World Heritage Area (Strydom et al. 2020). In Tampa Bay (USA), large reductions in seagrass coverage were observed in the 1970s mainly due to dredging operations and eutrophication (Johansson & Greening 1999). To date, management actions have mainly focussed on improving light quality by mitigating sediment and nutrient loads (Orth et al. 2006). Previous reviews have highlighted the need for large-scale restoration (van Katwijk et al. 2009, 2016, Sinclair et al. 2021a), discussed how we might achieve that and subsequently developed ecological decision-making frameworks for seagrass restoration (Statton et al. 2021). Similarly, Tan et al. (2020) reviewed the various methodologies employed in seagrass restoration and the relative successes achieved by those methods. Indeed, roadmaps for coordinated landscape-scale coastal and marine ecosystem restoration have been developed in several reviews (Fortes et al. 2018, van Katwijk et al. 2021, Saunders et al. 2022). Belowground interactions that influence critical sediment processes, however, have been largely ignored in restoration and initial decision-making frameworks but are critical to seagrass health because they influence all aspects of restoration, from selecting donor and restoration sites to developing improved methods and monitoring regimes. Here, we review how belowground interactions can be incorporated into these seagrass restoration frameworks to enhance restoration success.

Managements' actions to improve water quality have led to successful restoration of seagrasses (de los Santos et al. 2019, Lefcheck et al. 2018, McLeod et al. 2018, but see Orth et al. 2012 as not all restoration depends on water quality improvements). Yet, seagrass restoration is often seen as a risky undertaking, because restoration trials often fail to deliver the desired outcomes. The perceived lack of restoration success has resulted in hesitancy from policymakers and resource managers to provide the necessary funding to develop adequate restoration methods. This has led to a 'stagnation loop' whereby a lack of funding to improve the science behind restoration leads to a lack of restoration success and community disengagement and continued low restoration success (Abelson et al. 2020). To break this stagnation loop, we need more successful restoration through a better scientific understanding of the factors influencing seagrass performance and growth (Abelson et al. 2016, 2020). Ecological knowledge of seagrasses can then be utilised to develop better tools to accelerate recovery and enhance the resilience of extant and restored seagrass ecosystems.

Fundamentally, successful restoration is contingent on appropriate site selection and removal/abatement of stressors that provide barriers to natural recovery or restoration. In general, high-energy environments (e.g., shallow sediments subject to high bed shear stress due to wind waves and/or tidal currents) may preclude the recruitment of seagrasses due to physical disturbance associated with sediment resuspension and smothering. These sites likely have coarser sediments, with low organic carbon and nutrient contents. In these high-energy environments, sulphide stress will be low; however, nutrients may be limiting, especially before a seagrass meadow is well established (Figure 1). Importantly, sediment and nutrient deposition also affect sediment grain size, nutrient and chemical cycling under microbial control, which can impact seagrass performance and natural recovery long after water quality is improved (Fortes et al. 2018, Zabarte-Maeztu et al. 2021). A particular challenge for the restoration of seagrass meadows is creating conditions that promote the establishment or 'seeding' of critical feedbacks that allow plants to survive and persist in a given site (van Katwijk et al. 2016). For example, better incorporation of sediment quality will determine site selection and whether any environmental intervention/enhancement is required (e.g., microbial or nutrient seeding) which in turn will influence the methods used for restoration.

In addition, many restoration efforts are focused on developing improved methods to stabilise the sediment for seeding and anchoring of shoots (see Tan et al. 2020). These restoration efforts are improving seagrass growth and survivorship, particularly at smaller scales (Tan et al. 2020). For some seagrass systems, these experiments are beginning to lead to larger-scale efforts. Following experimental testing of methodologies (e.g., https://www.seagrassresearch.net/restoration), seagrasses are now being restored at scale in Cockburn Sound and Shark Bay, Western Australia, as

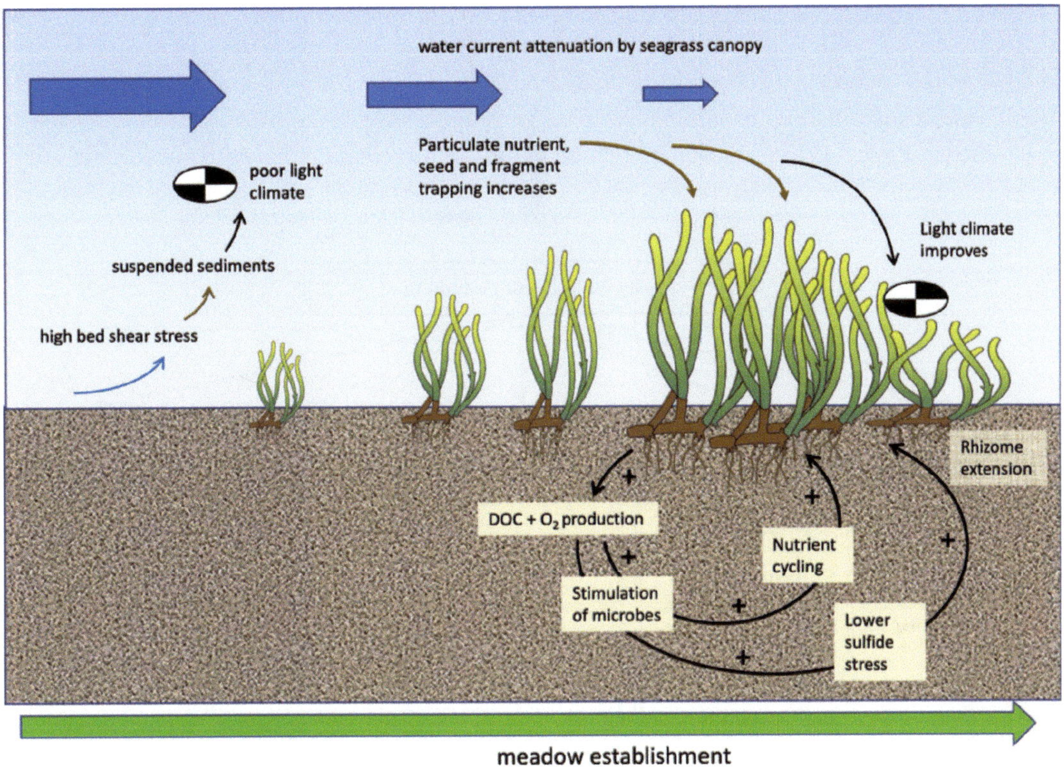

Figure 1 Seagrasses are keystone species that create feedbacks that influence hydrodynamics, particulate trapping, sediment quality, microbial processes and redox-based cycling of sediment chemistry.

well as in The Dutch Wadden Sea (Govers et al. 2022). However, current restoration methods make little provision for the inclusion of sediment processes and belowground interactions in planning restoration. Sediment processes are at the heart of these feedbacks and thus critical for improving seagrass health and restoration outcomes (Figure 2).

Ultimately, the health of seagrass meadows will be determined by their genetic architecture and their interaction with the biotic (e.g., animals and microbes) and abiotic (e.g., light availability, temperature and sediment chemistry) environments. Seagrasses create feedbacks that influence hydrodynamics, particulate trapping, sediment quality, microbial processes and redox-based cycling of sediment chemistry (Figure 1). A growing body of research demonstrates a critical role for belowground microbes in supporting seagrass health by reducing environmental stress (Crump et al. 2018, Martin et al. 2019, Fraser et al. 2023, Fuggle et al. 2023) and controlling interactions with other macrophytes (Gribben et al. 2017, 2018). Importantly, the interactions between seagrass life history, genetics and sediment conditions will likely play an important role in determining restoration success within highly modified environments (de Boer 2007, Kilminster et al. 2015, Pazzaglia et al. 2021), and more consideration needs to be given to genetic provenance when selecting material for restoration (Sinclair et al. 2014). Sediment conditions at restoration sites may have switched to an alternative stable state (McGlathery et al. 2013), requiring careful matching of seagrass species/functional types and/or genotypes to local environmental conditions at restoration sites. Genomic approaches now provide a powerful set of tools to understand adaptive and neutral genetic variation and guide the selection of species/functional types/genotypes best suited for restoration (Wood et al. 2020). In addition, seagrasses do not exist in isolation and interact with many other macro-organisms, and such interactions can influence seagrass survivorship and growth. For

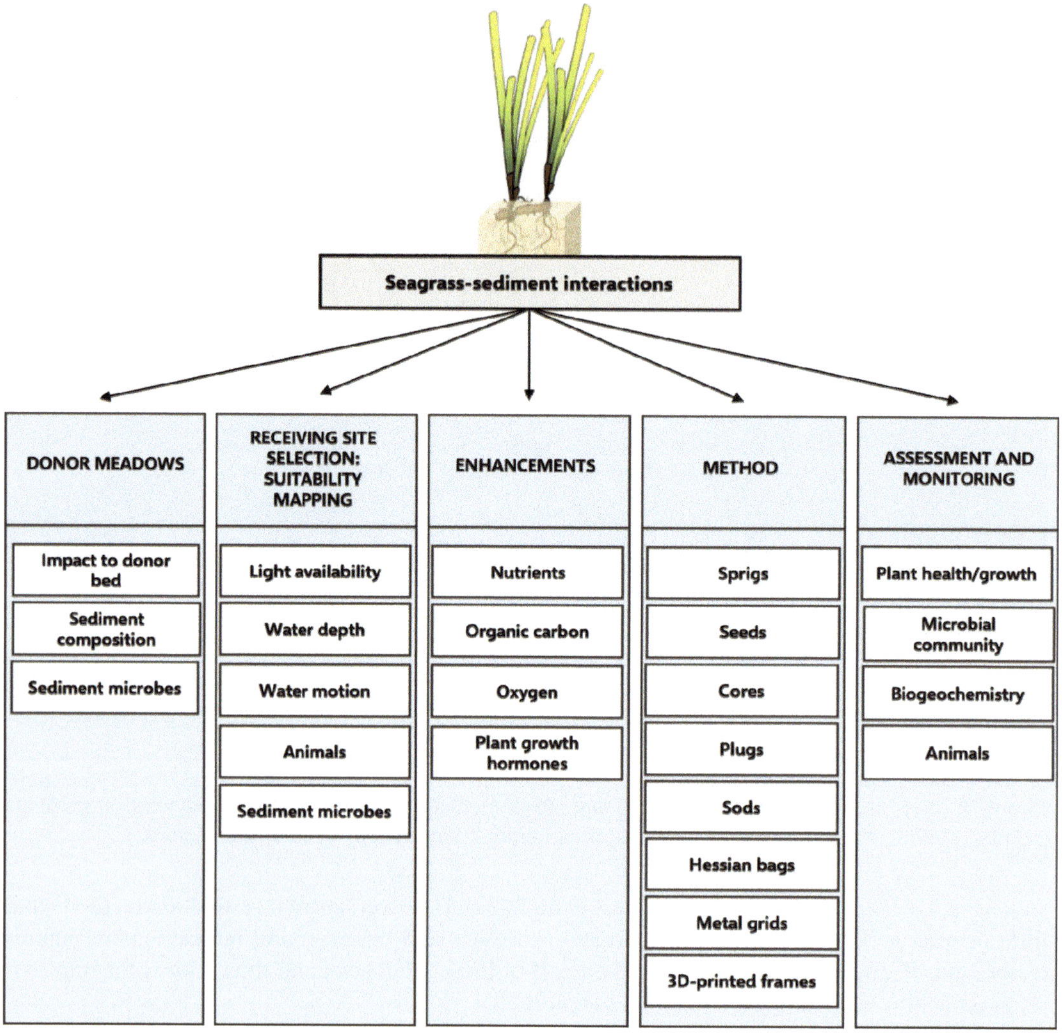

Figure 2 Seagrass-sediment interactions are at the heart of many components involved in restoration success, starting from donor and receiving site selection to methods and enhancements and ending with the assessment and monitoring of restoration success.

example, bioturbating organisms can improve sediment conditions by enhancing belowground oxygen conditions that can enhance seagrass performance (Aller 1988, Webb & Eyre 2004). However, bioturbation can also result in the dislodgement of seagrass seeds decreasing survivorship (Johnson et al. 2018). In addition, the presence of a healthy belowground microbiome can influence the outcomes of interactions with other macrophytes (e.g., Gribben et al. 2017, 2018).

In this review, we synthesise the current information on how belowground interactions influence seagrass health and identify which processes are amenable to manipulation and at what scale they could be employed to improve restoration strategies and decision-making frameworks. To address these questions, we used two approaches: (1) a workshop held in December 2021 with seagrass experts across Australia and (2) a selective literature search. Participants came from a range of backgrounds representing state government agencies charged with managing Australia's marine estates, university researchers and leading restoration practitioners. Based on the workshop, we identified three key, interdependent areas by which belowground interactions influence seagrass health and performance. Those three areas included how (1) sediment microbial communities control nutrient and chemical cycling for seagrass, (2) seagrass response to sediment processes may be influenced by their life history and genetics of seagrass species targeted for restoration and (3) interactions with other species that promote and inhibit seagrass survivorship and growth.

For each area, co-authors were divided into groups based on their area of expertise, reviewed the primary published scientific literature and provided a critical overview of the current state of knowledge. We then identified the key opportunities for utilising this knowledge to improve restoration strategies and what the scientific knowledge gaps are to realise their use in restoration. This review is not intended to contain a complete synopsis of the published literature for seagrass restoration as many aspects on frameworks, methodologies and successes have been extensively reviewed elsewhere (van Katwijk et al. 2016, Orth et al. 2020, Sinclair et al. 2021a, Pansini et al. 2022). This review focuses on how to improve restoration outcomes through the inclusion of belowground interactions and identify scientific knowledge gaps in the processes controlling seagrass performance and current knowledge that could be employed to enhance restoration success.

Current state of knowledge

Seagrass sediment microbes

In terrestrial systems, the importance of plant-soil interactions and the link between belowground processes mediated by microbes and aboveground performance is well recognised. As plants germinate and grow in the soil, they can influence the biotic and abiotic properties of the soil, e.g., through the release of chemical compounds and input of organic matter, which can result in changes of the microbial community composition and function (Van der Putten et al. 2013, Veen et al. 2019). These changes in soil properties and its microbiome can in turn affect the establishment and performance of individual plants of the same or a different species, a process known as plant-soil feedbacks (PSF) (Bever et al. 1997). PSF can be negative (resulting in a net growth-reducing effect to individuals of the same species), neutral (the net effect of all influences on plant growth is zero) or positive (growth of the same plants or other individuals of the same species is promoted) (Van der Putten et al. 2013). PSF are now widely recognised as important drivers of plant community dynamics, plant coexistence and plant responses to environmental change (Kardol et al. 2006, Heinen et al. 2020). Given the analogies between seagrasses and their terrestrial counterparts, there is huge potential for using these aboveground-belowground links to advance the field of seagrass restoration.

Our understanding of the effects of microbial communities on seagrass-sediment interactions and feedbacks has only recently started to develop (Piercey et al. 2021). Emerging studies show that seagrasses harbour diverse microbial communities on the surfaces and within their above- and belowground structures (Garcias-Bonet et al. 2012, Ugarelli et al. 2017, Tarquinio et al. 2019), as well as on their seeds (Tarquinio et al. 2021). These studies show that microbes have the potential to strongly influence above- and belowground processes related to carbon-, nitrogen- and sulphur cycling, all of which are linked to seagrass performance and the diverse ecological communities they support. For example, seagrass root-associated and rhizosphere (i.e., the region of sediment in the immediate vicinity of the roots) microbiomes are typically enriched in sulphur cycling bacteria, many of which likely play key roles in sulphide detoxification, ultimately contributing to the ability of seagrasses to persist in high sulphide environments (e.g. Martin et al. 2020, Scholz et al. 2021). Some of these bacteria also possess the genes to fix nitrogen (Petersen et al. 2016, Mohr et al. 2021) and break down organic matter, releasing nutrients that contribute to plant growth in otherwise oligotrophic oceans (Fraser et al. 2018). Epiphytic bacteria on leaf surfaces are a key component in the ability of seagrass to uptake dissolved organic nitrogen from the water column (Tarquinio et al. 2018). Some bacteria that occur on seagrass tissues can help defend against pathogens and saprophytes (Tarquinio et al. 2019). Seagrasses also host diverse fungal communities on and inside their leaves, roots and rhizomes, although their role in seagrass functioning remains poorly known (Ettinger & Eisen 2019, 2020).

There is significant variation in the composition of the microbiome between seagrass species (Garcias-Bonet et al. 2021), particularly between different functional types of seagrasses (colonising vs opportunistic vs persistent, *sensu* Kilminster et al. 2015). Significant variation can also occur among plants of the same species growing in different locations (Martin et al. 2022), which suggests that microbial composition is mediated by both the host and the environment. There is now experimental evidence that has linked alterations in seagrass microbiomes with changes in light (Martin et al. 2018), sediment organic matter (Fraser et al. 2016, 2023), nutrient enrichment and temperature (Nguyen et al. 2021). Such changes in environmental conditions can not only impact the health of seagrasses directly but also through changes to the associated microbiome.

Seagrass life history, genetics and sediment processes

The loss of seagrass from an area can lead to a rapid change in local sediment conditions that can make it difficult to re-establish the seagrass in those areas. Restoration has typically focused on technical and logistical issues relating to transplanting methods, seed collection and the physical suitability of areas for restoration (Bastyan & Cambridge 2008, Fonseca 2011, Tan et al. 2020). There is, however, a growing realisation that modification of sediment conditions may be needed for successful restoration. Key to this is understanding how seagrass life histories (e.g., Sinclair et al. 2021a,b), growth forms and physiological tolerances (e.g., Kilminster et al. 2015) influence and modify sediment conditions. For example, some seagrasses include short-lived, colonising species with fast growth rates (e.g., those from the genus *Ruppia* and *Halophila*) through to longer lived species with moderate growth rates (e.g., *Zostera*) and long-lived species with relatively slow growth rates that form persistent meadows (e.g., *Amphibolis* and *Posidonia*; Kilminster et al. 2015). We are becoming increasingly aware of how restoration success may be influenced by these different growth strategies and sediment conditions and how seagrasses modify the sediment environment as they colonise and expand within an area. Growth of transplanted *Posidonia australis* rhizomes can vary greatly within and among sites (e.g., 9.1–22.3 cm yr^{-1} in the same New South Wales (NSW) estuary (Meehan & West 2002) and 10–35 cm yr^{-1} in Western Australia (Renton et al. 2011). *P. australis* seedlings are slower growing, with rhizome extension ranging from 5 to 17 cm yr^{-1} in NSW (Meehan & West 2004). Growth rates of leaves and rhizomes can depend on sediment nutrient levels (Cambridge & Kendrick 2009), and it is likely that some genotypes perform better under some sediment conditions. Sediment properties (grain size, nutrients, microbiome) can play a role in influencing rhizome elongation and root architecture, which may in turn influence establishment success. For example, species of *Zostera* tend to produce greater belowground biomass and longer rhizome nodes in oligotrophic sediments (e.g. Ferguson et al. 2016, Song et al. 2021), while sediment type (specifically grain compactness; Statton et al. 2013) and nutrients (Hovey et al. 2011) can affect root architecture and hence anchoring of seagrass.

Similar to patterns in terrestrial plants, weedy fast-growing species (e.g., *Halophila* spp.) that are often first colonisers of disturbed seagrass beds (Smulders et al. 2017) may be less reliant on their microbial communities, and they more quickly overcome adverse sediment conditions (Bauer et al. 2018, Koziol & Bever 2015, 2016; but see Lee et al. (2018) as *Halophila ovalis* might be dependent on microbes for nitrogen acquisition because of the loss of genes encoding NADH dehydrogenase-like complex assembly). Variation in seagrass life histories will also influence the spatial and temporal patterns of genetic diversity and connectivity which may be important for sediment processes within restoration sites. Seagrass life history reflects the ability to disperse either via current-mediated transport of floating propagules along the water surface (e.g., *Posidonia*, *Thalassia*, *Enhalus* and *Zostera*) or the dispersal of negatively buoyant seeds in the water column or the sediment-water interface (e.g. *Halophila* and *Halodule*; Kendrick et al. 2012). The genetic structure of seagrasses can be heavily influenced by these modes of dispersal resulting in chaotic genetic patchiness, clonality and polyploidy (McMahon et al. 2017, Sinclair et al. 2020, Edgeloe et al. 2022).

Species interactions and their effects on seagrass-sediment interactions

Bioengineers have the capacity to affect seagrass by changing sediment conditions and influencing restoration success. Some bioturbators can rework large amounts of sediments, leading to sediment destabilisation (e.g., sea urchins and rays) negatively affecting seagrass or stabilisation that can have positive effects on seagrasses (e.g., tube building worms; Volkenborn et al. 2009, Figure 3A).

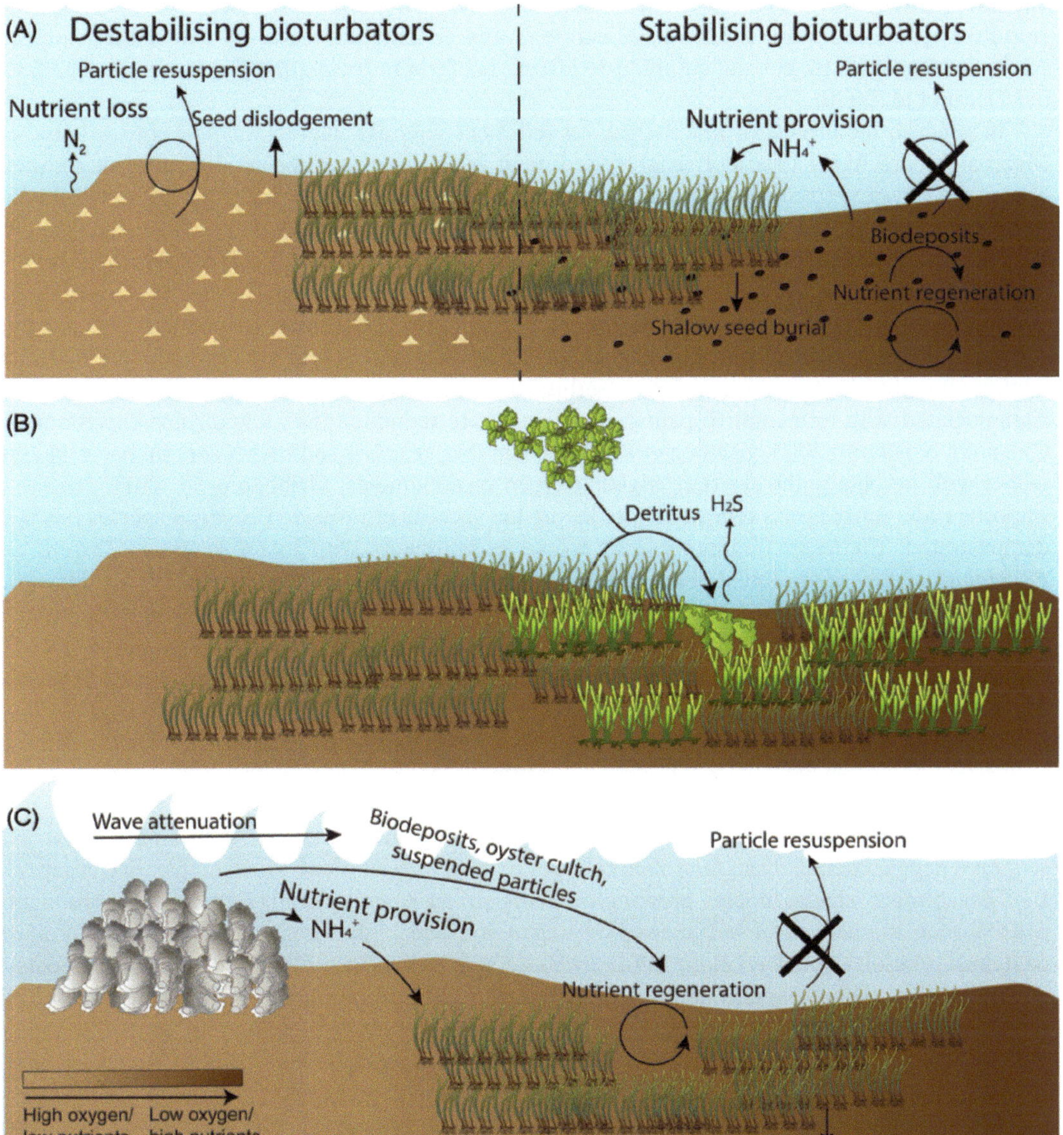

Figure 3 The effect of (A) bioturbators, (B) free-floating seaweed and *Caulerpa* and (C) shellfish reefs on seagrass-sediment interactions.

For example, Callianassid shrimps can reduce seagrass growth via sediment resuspension during burrow creation, reducing light availability and sediment stability (Suchanek 1983, Siebert & Branch 2006), while sea urchins and sand dollars can dislodge seeds (Johnson et al. 2018). For example, coir mats have been successfully used to exclude Callianassid shrimps and improve transplant success and shoot growth of *Zostera muelleri* seagrass in Australia (Wendländer et al. 2019). Studies are now underway to test the applicability of this technique at large scales.

In contrast, the activity of clams and some worms can bury seagrass seeds to a sediment depth suitable for germination within a few days after seed deposition, providing escape from predation and promoting seed retention (Blackburn & Orth 2013, Li et al. 2017, Fales et al. 2020). Alternatively, bioturbators can provide nutrients for seagrass growth in areas where nutrients are limited, such as sandy environments (de Boer 2007, Gilbertson et al. 2012, Stief 2013). Importantly, in eutrophic sediments, Lucinidae and Solemyidae bivalves can reduce sulphide stress in seagrasses by hosting sulphide-oxidising bacteria in their gills (Gagnon et al. 2020, van der Geest et al. 2020).

In general, the effects of macroalgae on sediment-seagrass interactions are not well studied. Macroalgae are likely to affect seagrass-sediment interactions via the modification of sediment chemistry, physical structure and microbial communities due to increased input of seaweed detritus (Fraser et al. 2016). *Caulerpa* species are among the few seaweeds that grow directly in soft sediments, and several species co-occur with seagrass (Ceccherelli et al. 2000, Glasby 2013). At high densities, *Caulerpa* sediments generally have higher organics (due to a high detrital input of fragments (Wright 2005), silt content, anoxia and sulphide concentration compared to seagrass sediments (Holmer et al. 2009, McKinnon et al. 2009). Moreover, there are very different microbial communities in *Caulerpa* and seagrass sediments. *Caulerpa* sediments typically contain bacteria associated with fermentative pathways and sulphate reduction (i.e., low oxygen environments; Chisholm & Moulin 2003, Gribben et al. 2017), whereas seagrass sediments contain bacteria associated with aerobic pathways (i.e., higher oxygen environments; Gribben et al. 2017, 2018). An important role for seagrass sediment in reducing the success of invasive *Caulerpa* species has been demonstrated: *Caulerpa* fragments grow worse in sediments with intact seagrass (*Zostera* and *Posidonia*) sediment microbial communities compared to disturbed seagrass sediments. However, once *Caulerpa* is established in seagrass beds, it creates reduced conditions to which it is tolerant suggesting positive feedback (self-facilitation) between *Caulerpa* sediment and propagule success. Overall, modifications to the sediment caused by *Caulerpa* are likely to influence seagrass-sediment feedbacks via modification of sediment properties and microbial communities. This more anoxic and sulphidic sediment may limit seagrass recovery in areas that may now be occupied by *Caulerpa* and/or may have legacy effects on sediment condition even if *Caulerpa* is removed that impede restoration (Gribben et al. 2017, 2018).

Similarly, high densities of seaweed attached to biogenic structures (including epiphytic on seagrass) trigger chemical and microbial changes to seagrass sediment. For example, in Chesapeake Bay, dense macroalgal canopies growing on *Zostera* had negative effects on seagrass production possibly due to changes in sediments (lowered redox and higher concentrations of ammonium) or light limitation (Hauxwell et al. 2001, 2003). Other studies have also shown negative effects of epiphytic seaweed on seagrass due to a reduction in light (Drouin et al. 2012), but in general, the effects of increased detrital input of epiphytic seaweed on seagrass-sediment interactions are not well tested. In addition, blooms of free-floating seaweed, typically the result of increased nutrient loading to a waterway, can settle on seagrass beds and change water and sediment conditions (Figure 3B, Goodman et al. 1995). Modification of sediment properties in this way, combined with flow-on effects on microbial communities, may hinder seagrass restoration in areas where free-floating seaweeds do not commonly co-exist with seagrasses. Free-floating seaweeds are not always 'bad' as they form a major component of the tropical marine seagrass meadows and coexist with seagrasses in these areas.

Shellfish also affect sediments by the addition of oyster cultch and the release of biodeposits (faeces and pseudo-faeces; Figure 3C), which increases sediment nutrient levels, sediment stabilisation and seed trapping (Meysick et al. 2019, Gagnon et al. 2020). Increases in sediment nutrients can boost growth rates of seagrass *Halodule wrightii* at intermediate oyster densities, which were related to intermediate levels of ammonia and phosphate in sediments (Booth & Heck 2009). Survival of *Zostera marina* transplants was increased by the presence of mussels in high and medium exposure sites, but not in sheltered sites, indicating an effect on sediment stabilisation and storm protection (Bos & Katwijk 2007). Even though most studies assessing interactions between seagrass and reef-building shellfish have found positive effects, negative interactions can also occur mostly due to eutrophication effects, epiphyte growth and competition for space (Gagnon et al. 2020). For example, high biodeposit loads can intensify sediment eutrophication, driving sulphide production and reducing seagrass growth (Vinther & Holmer 2008). Negative effects have mainly been identified when seagrass and shellfish occur in the same habitat, while positive effects have mainly been recorded when they are located in independent, but fringing habitats (Gagnon et al. 2020).

Science needs for increased management and restoration success

Microbes

There are an increasing number of studies that show changes in the seagrass belowground microbiome with changes in environmental conditions (Fraser et al. 2016, Martin et al. 2018, Wang et al. 2020, Nguyen et al. 2021, Fuggle et al. 2023). However, experiments that establish causal effects of microbial changes on seagrass performance are rare (but see Fuggle et al. 2023). Given the fundamental role that microorganisms have in biogeochemical cycling in seagrass ecosystems and their potential effects on plant performance, more experimental studies are needed to gain a detailed understanding of the role of belowground microbes in seagrass health and performance. Incorporating these seagrass-microbiome interactions into restoration planning could significantly improve restoration success rates now and into the future (Wood et al. 2019). In addition, seagrasses also host diverse fungal communities but their role in seagrass functioning is largely unknown (Ettinger & Eisen 2019, 2020). As evidenced from terrestrial ecosystems (Maltz & Treseder 2015, Koziol et al. 2018, Neuenkamp et al. 2019), exploring seagrass-fungal relationships and how they contribute to seagrass performance may be critical for building a holistic understanding of how belowground processes contribute to seagrass health and improving restoration strategies.

The choice of appropriate restoration strategies depends on an assessment of site-specific environmental factors that influence seagrass microbiome interactions, with a view to working within environmental constraints and promoting beneficial microbial processes. Once we have a better idea of which microbes promote or hinder seagrass growth under specific environments, knowledge of seagrass and sediment microbiomes could be used for monitoring and active intervention. Molecular tools can be used to (1) establish a baseline against which the health status of meadows can be assessed and (2) track microorganisms or functional groups linked with poor plant performance as early warning signals to trigger preventative management actions. In addition, information on plant and/or sediment microbiomes can help (3) identify 'healthy' meadows that can be used as donors for restoration, either of shoots, seeds or sediment; (4) select sites where microbiome composition is likely to increase the chances of restoration success; and (5) check that the manipulation of plants, seeds or sediments does not lead to microbial changes that may negatively affect restoration success (van Katwijk et al. 2016, Koziol et al. 2018). Using these molecular tools may in practice be challenging for managers and restoration practitioners depending on the scale of the project and because of the associated costs and expertise that are needed. The cost of DNA sequencing has, however, declined to a point where the generation of large amounts of sequence data is within the budget of most research groups. A partnership between restoration practitioners and scientists would therefore be the best way forward to deploy these methods.

Seagrass restoration practitioners may still face considerable challenges that require active manipulations of sediments or plants focused on optimising their microbiomes. Restoration of aboveground plant community properties may require steering of belowground ecosystem processes and those organisms that drive them (Wubs et al. 2016). Manipulations may thus focus on promoting the conditions that would facilitate the desired sediment microbiome, for example, by providing appropriate levels and types of nutrients or by manipulating the physical environment (e.g., grain size). An alternative approach for promoting positive seagrass-sediment interactions is to directly inoculate with desired microbiomes, e.g., through inoculations with bulk sediment from sites with healthy seagrass communities, as done in terrestrial systems to enhance the establishment of plants in degraded sites (Wubs et al. 2016, Grman et al. 2020) or through targeted inoculations of beneficial microbes such as by coating seeds or large-scale spraying of these inoculants on degraded sites prior to restoration, a commonly used method in land bioremediation (Kuiper et al. 2004).

The microbiome monitoring and manipulations outlined above rely on new and emerging techniques, such as (1) high-throughput culturing methods to isolate, screen and identify microbes conferring a beneficial trait/function to seagrass and that can be used as probiotic inoculations (e.g., mangrove bacteria; Soldan et al. 2019); (2) enriched bulk or compound-specific stable isotopes (Kaldy et al. 2006, 2013) and isotope probes viewed over highly resolved spatial scales within the rhizosphere or plant tissue (i.e., NanoSIMS; Tarquinio et al. 2018) to understand links between microbial and plant functions; and (3) bioinformatics pipelines for metagenome-assembled genomes (MAGs) and long-read technologies to improve seagrass and sediment-specific sequencing databases that allow the development of specific markers/primers (Duffin et al. 2020). The consistent use of best-practice methodology for sample collection and processing, culturing protocols, primers and bioinformatics pipelines (e.g., BioPlatforms Australia, Earth Microbiome Project; Thompson et al. 2017, Phelps et al. 2021) is critical to facilitate standardisation and to provide a place for sequence deposits and metadata. Integrating these technologies with *in situ* manipulations is key to determine causation and mechanisms that underpin plant performance and health in real-world scenarios.

Life history and genetics

The conditioning of sediments through a succession of different species could provide restoration practitioners with a valuable tool for modifying less suitable sediment conditions at a restoration site to more favourable conditions. Restoration of belowground ecosystem functions may also be achieved by using early colonising plants as tools to promote beneficial microbiomes for subsequent colonisers/desired species that can be planted after the sediment is stabilised and microbial associations are established (Van der Putten et al. 1993, Angers & Caron 1998, Kardol & Wardle 2010). In addition, genetic diversity represents the raw material that natural selection acts on and is likely to be important for seagrass to establish in a range of sediment conditions within restoration sites. There is, however, a lack of knowledge associated with links between genetic diversity in seagrass meadows and their associated microbial communities, despite health and nutrient benefits for seagrasses and adjacent ecosystems (e.g., Hughes & Stachowicz 2004, Reynolds et al. 2012). Further research is needed to understand the interaction between genotype and microbiome and how mixed species and successional approaches can be used in seagrass restoration programmes.

Some genotypes may be locally adapted to particular sediment conditions, and for successful restoration outcomes, we need to match restoration material (seeds or shoots) from environments that more closely match the conditions at the restoration site (van Katwijk et al. 2009). However, detailed range-wide knowledge of patterns of genetic diversity is limited to only several seagrass species (e.g., *Posidonia australis*, *Posidonia oceanica*, *Zostera marina*, *Zostera noltii*, *Thalassia hemprichii*, *Thalassia testudinum* and *Halodule wrightii*; Coyer et al. 2004, Olsen et al. 2004,

Arnaud-Haond et al. 2007, Dijk et al. 2009, Hernawan et al. 2017, Duffy et al. 2022, Sinclair et al. 2023, Tavares et al. 2023). The spatial scale of genetic studies varies considerably from range-wide studies providing overall patterns of genetic diversity and connectivity over hundreds to thousands of kilometres to detailed studies within meadows at the scale of metres. Finer spatial scale studies provide useful estimates within highly modified industrialised environments where active restoration is ongoing, for example, *Zostera muelleri* in Port of Gladstone (Australia; Jackson et al. 2021), *Posidonia australis* in Cockburn Sound (Australia; Sinclair et al. 2014, 2021b) and *Zostera marina* in Bodega Harbor (USA; Kollars et al. 2022). A better understanding of the genetic structure and patterns of local adaptation in relation to sediment conditions is needed both at the local and range-wide scales to guide what restoration material may be best suited for the sediment conditions at a restoration site.

From a restoration perspective, differences in life histories could be exploited by using fast-growing species (of seagrass or even algae) to rapidly colonise an area and modify and improve sediment conditions that then enable longer-lived species to establish and grow in these areas (Williams 1990). Indeed, in small-scale experiments, the growth and survivorship of slower growing species (e.g., *Thalassia testudinium* and *Posidonia sinuosa*) were enhanced when transplanted into sediments containing early colonising species (Van Keulen et al. 2003, Kenworthy et al. 2018).

The rapid development of 'omic' technologies also provides unprecedented opportunities to integrate genetic data to inform and enhance seagrass restoration. For example, the use of population genomic analysis can provide a more detailed understanding of population connectivity and patterns of genetic diversity. In addition, it can provide an understanding of the genetic basis underlying adaptive variation associated with particular environments and how a seagrass species may respond to local sediment conditions. This can assist in identifying and matching genotypes from donor meadows to environmental conditions at restoration sites, or ensuring restoration material contains genetic variants that allow for adaptation to future projected environmental conditions (Coleman et al. 2020). Transcriptomic studies of gene expression also provide significant opportunities for the use of genetic information to inform restoration as they allow for the identification of the genes underlying responses to specific environmental stressors (Mohammadi et al. 2019, Nguyen et al. 2020, Booth et al. 2022). These analyses provide important insights into the tolerance of individuals to different sediment stressors and an understanding of the resilience of restored populations to future environmental change.

Species interactions

To inform restoration, it is important to understand the full array of negative and positive seagrass-bioturbator interactions. For example, the potential beneficial effects of animal bioturbators on seagrass growth, survival and reproduction, including the provisioning of nutrients and microclimates for beneficial microbial community development and microbial function, remain largely unassessed. Moreover, in addition to providing benefits to seagrass health, bioturbators may also affect the provision of ecosystem services by seagrasses. Bioturbators can decrease carbon sequestration in saltmarshes by oxygenating sediments and boosting aerobic decomposition of organic carbon, while at the same time facilitating arbuscular mycorrhizal fungi that promote saltmarsh growth and carbon sequestration capacity (Martinetto et al. 2016). Large-scale field surveys of bioturbators co-occurring with seagrass and how they relate to seagrass health and habitat function are needed.

While studies comparing the effects of a few species of bioturbators have suggested that idiosyncratic effects play a key role in seagrass-bioturbator interactions (Johnson et al. 2018, Lacoste et al. 2018), no studies have experimentally assessed if the nature of these interactions are related to bioturbator functional groups. For example, seed burial capacity has been related to bioturbation rates (as measured by grams of sediment moved per day per individual), but only three species were assessed (Blackburn & Orth 2013). To inform these approaches, studies are needed to identify taxonomical groups or functional traits associated with positive or negative effects on seagrasses. The information produced can inform global seagrass efforts, as the focus on functional traits will make results transferable between systems, even for new sites where there is little understanding of the ecology of sediment biodiversity.

In addition, more studies are needed to understand which seagrass species benefit from associations with lucinid bivalves, and how widespread these positive interactions are. Lastly, the role of shellfish as nutrient providers for seagrasses has been mostly assessed for those shellfish species that occur in close association with seagrasses, such as the mussel *Modiolus americanus* (Peterson & Heck 1999), *Mytilus edulis* (Reusch et al. 1994) and the oyster *Crassostrea virginica* (Booth & Heck 2009). It is, however, less clear how shellfish reefs occurring in the surrounding areas affect seagrasses. Studies are needed to test how shellfish reefs affect seagrass health and understand context-specific effects to pre-empt any potential negative effects of shellfish on seagrass via eutrophication.

Given the influence bioturbators can have on seagrass restoration success, site suitability models should incorporate information on naturally occurring 'stabilising' and 'destabilising' bioturbators. A recent study in the Wadden Sea showed that the incorporation of biotic factors such as macro zoobenthos in seagrass habitat suitability models could improve the models and consequently restoration outcomes (Gräfnings et al. 2023). Moreover, in addition to designing strategies to minimise the impact of destabilising bioturbators, for example, by the use of coir mats (Wendländer et al. 2019), harmful bioturbators can be removed at the planting stage in areas where historical records show that the local presence of these species increased after seagrass loss. Strategies can also be developed to harness the beneficial effects of bioturbators to improve restoration outcomes. For example, practitioners can use stabilising bioturbators to pre-condition sediments for seagrass planting which can help exclude harmful bioturbators, increase nutrient availability in low-nutrient sediments, decrease sulphide stress in eutrophic sediments and/or inoculate with or promote the growth of beneficial microbial communities. For example, in areas with high sediment sulphide levels, co-planting seagrass with lucinid bivalves, where available, might help transplants overcome sulphide stress (Derksen-Hooijberg et al. 2018, Chin et al. 2021). The natural capacity for reproduction and dispersal of bioturbators makes this strategy ideal for projects across large scales.

High levels of some benthic and free-floating seaweed are often the result of increased nutrients and other catchment disturbances. Thus, ongoing monitoring and management of nutrient inputs is critical to reduce the biomass of these algae and their potential impacts on sediment conditions for seagrass. Other species such as *Caulerpa* are highly invasive in many areas of the world interacting with local seagrasses (Gribben et al. 2017, 2018). Manual removal of these species and sediment pre-conditioning might be necessary before seagrass restoration efforts begin.

The extensive number of shellfish reef restoration projects worldwide provide opportunities for the development of co-planting initiatives with seagrass. With a better understanding of how shellfish reefs affect seagrass performance, reefs could be used to pre-condition sediments for seagrass planting in wave-exposed areas where there are low levels of nutrients in sediments. In these areas, shellfish reefs attenuate waves promoting sediment stabilisation (Meyer et al. 1997) and contribute biodeposits that can increase nutrient content in sediments at distances up to 100 m away from reefs (Bugnot et al. 2022). Therefore, there is potential for shellfish reefs to enhance seed retention and seagrass growth and reproduction via the provision of nutrients in more exposed and nutrient-deprived areas.

Conclusion

2021–2030 is the *UN Decade on Ecosystem Restoration*, and there are good opportunities and desire from government, community groups and the scientific community to restore degraded ecosystems. However, successful restoration and building climate-resilient coastal ecosystems are contingent on understanding the key environmental processes that both inhibit and promote restoration and recovery. Despite the well-described influence of seagrass sediment interactions on seagrass performance, they are rarely considered in seagrass restoration, despite their potential to transform restoration success at scale.

This review identified significant opportunities to enhance seagrass restoration by explicitly including seagrass-sediment interactions. This includes enhanced methods for restoration, more effective monitoring and assessment of restoration success. Importantly, we suggest extending decision-making frameworks to include sediment processes as they are of primary concern in determining initial site suitability and affect subsequent decisions and planning for restoration (Figure 4, Panel 1). At the outset, we recognise that hydrodynamics play a key role in determining sediment factors that may both benefit and hinder the successful restoration of seagrasses. We suggest therefore that an assessment of hydrodynamics and its implications for sediment properties and microbial community development is an important first step in site selection, the choice of restoration strategies and the selection of suitable donor material.

Across the areas we identified, it is evident that seagrass sediment interactions are at the heart of many feedback processes influencing seagrass health. It is now more important than ever to understand the role of sediment interactions in seagrass health as climatic changes and human disturbances will increasingly impact these interactions between seagrasses and the sediment. Indeed, emerging research is showing that belowground microbes become even more important to plant performance and resilience with increasing stress (Fuggle et al. 2023). Thus, in the face of ever-increasing environmental changes, we predict that understanding these interactions will only become more important and thus critical for seagrass restoration.

Future studies on seagrasses should focus on a better understanding of (1) the functioning of the seagrass microbiome and its relationship with seagrass genetics; (2) seagrass-sediment interactions across ecological contexts; and especially (3) how the seagrass microbiome, genetics and bioengineers can increase seagrass resilience to environmental changes. These studies not only are critical to broaden our understanding of important aspects of seagrass ecology but will also be extremely

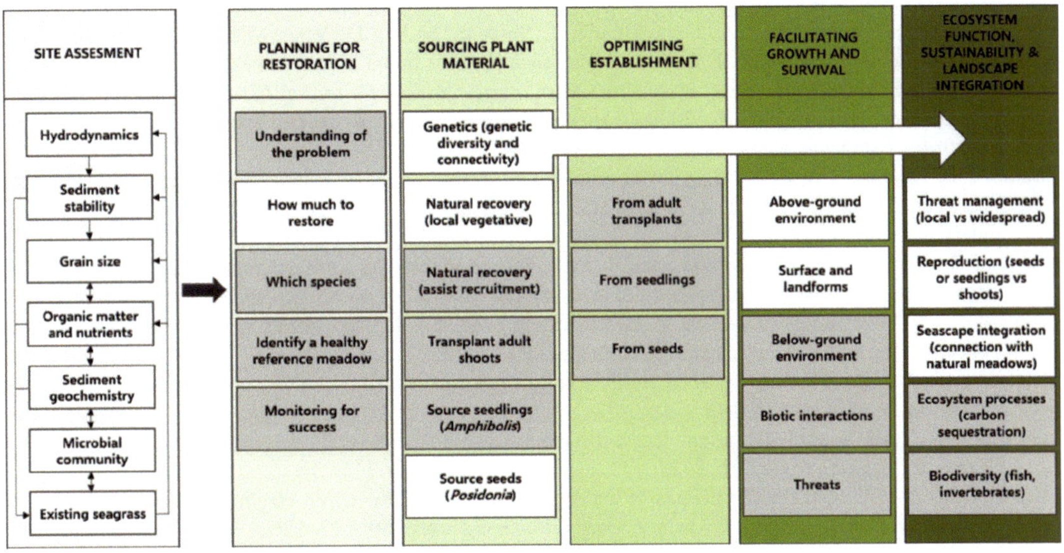

Figure 4 Extended ecological decision-making framework for seagrass restoration that explicitly includes belowground interactions. Boxes in grey indicate which actions or processes of restoration are influenced by belowground interactions. Vertical panels 2–5 to the right adapted from Miller et al. (2017) for marine restoration (Statton et al. 2021). The additional first vertical panel explicitly incorporates the sediment process as a primary consideration into the decision-making framework for seagrass restoration.

useful for seagrass restoration activities. The outcomes of these studies would help us identify how and which seagrass species or habitats will be most affected by climate change and which sediment processes promote or decrease seagrass health under stressful conditions. As a result, we could get an understanding of how to manage sediment conditions that promote the growth of seagrasses across different environments (Birnbaum & Trevathan-Tackett 2022). For example, mutualistic microbes and bioengineers could potentially be used to enhance the resilience of existing seagrass meadows or the survival of transplanted seagrasses. The right experimental strategies aimed at understanding the role of sediment processes in seagrass health will provide evidence-based insight into what management actions, at small and large scales, will improve seagrass restoration efforts now and into the future.

Acknowledgements

This work was funded under the Australian Government's National Environment Science Program (NESP).

References

Abelson, A., Halpern, B.S., Reed, D.C., Orth, R.J., Kendrick, G.A., Beck, M.W., Belmaker, J., Krause, G., Edgar, G.J., Airoldi, L., Brokovich, E., France, R., Shashar, N., de Blaeij, A., Stambler, N., Salameh, P., Shechter, M. & Nelson, P.A. 2016. Upgrading marine ecosystem restoration using ecological-social concepts. *BioScience* **66**, 156–163.

Abelson, A., Reed, D.C., Edgar, G.J., Smith, C.S., Kendrick, G.A., Orth, R.J., Airoldi, L., Silliman, B., Beck, M.W., Krause, G., Shashar, N., Stambler, N. & Nelson, P. 2020. Challenges for restoration of coastal marine ecosystems in the anthropocene. *Frontiers in Marine Science* **7**, 544105.

Aller, R.C. 1988. Benthic fauna and biogeochemical processes in marine sediments: the role of burrow structures. *Nitrogen Cycling in Coastal Marine Environments* **1988**, 301–338.

Angers, D.A. & Caron, J. 1998. Plant-induced changes in soil structure: processes and feedbacks. *Biogeochemistry* **42**, 55–72.

Arnaud-Haond, S., Migliaccio, M., Diaz-Almela, E., Teixeira, S., Van De Vliet, M.S., Alberto, F., Procaccini, G., Duarte, C.M. & Serrão, E.A. 2007. Vicariance patterns in the Mediterranean Sea: east-west cleavage and low dispersal in the endemic seagrass *Posidonia oceanica*. *Journal of Biogeography* **34**, 963–976.

Bastyan, G.R. & Cambridge, M.L. 2008. Transplantation as a method for restoring the seagrass *Posidonia australis*. *Estuarine, Coastal and Shelf Science* **79**, 289–299.

Bauer, J.T., Koziol, L. & Bever, J.D. 2018. Ecology of floristic quality assessment: testing for correlations between coefficients of conservatism, species traits and mycorrhizal responsiveness. *AoB Plants* **10**, plx073.

Bever, J.D., Westover, K.M. & Antonovics, J. 1997. Incorporating the soil community into plant population dynamics: the utility of the feedback approach. *Journal of Ecology* **85**, 561–573.

Birnbaum, C. & Trevathan-Tackett, S.M. 2022. Aiding coastal wetland restoration via the belowground soil microbiome: an overview. *Restoration Ecology* **31**, e13824.

Blackburn, N.J. & Orth, R.J. 2013. Seed burial in eelgrass *Zostera marina*: the role of infauna. *Marine Ecology Progress Series* **474**, 135–145.

de Boer, W.F. 2007. Seagrass-sediment interactions, positive feedbacks and critical thresholds for occurrence: a review. *Hydrobiologia* **591**, 5–24.

Booth, D.M. & Heck, K.L. 2009. Effects of the American oyster *Crassostrea virginica* on growth rates of the seagrass *Halodule wrightii*. *Marine Ecology Progress Series* **389**, 117–126.

Booth, M.W., Breed, M.F., Kendrick, G.A., Bayer, P.E., Severn-Ellis, A.A. & Sinclair, E.A. 2022. Tissue-specific transcriptome profiles identify functional differences key to understanding whole plant response to life in variable salinity. *Biology Open* **11**, bio059147.

Bos, A.R. & Katwijk, M.M.V. 2007. Planting density, hydrodynamic exposure and mussel beds affect survival of transplanted intertidal eelgrass. *Marine Ecology Progress Series* **336**, 121–129.

Boudouresque, C.F., Pergent, G., Pergent-Martini, C., Ruitton, S., Thibaut, T. & Verlaque, M. 2016. The necromass of the *Posidonia oceanica* seagrass meadow: fate, role, ecosystem services and vulnerability. *Hydrobiologia* **781**, 25–42.

Bugnot, A.B., Dafforn, K.A., Coleman, R.A., Ramsdale, M., Gibbeson, J.T., Erickson, K., Vila-Concejo, A., Figueira, W.F. & Gribben, P.E. 2022. Linking habitat interactions and biodiversity within seascapes. *Ecosphere* **13**, e4021.

Cambridge, M.L. & Kendrick, G.A. 2009. Contrasting responses of seagrass transplants (*Posidonia australis*) to nitrogen, phosphorus and iron addition in an estuary and a coastal embayment. *Journal of Experimental Marine Biology and Ecology* **371**, 34–41.

Ceccherelli, G., Piazzi, L. & Cinelli, F. 2000. Response of the non-indigenous *Caulerpa racemosa* (Forsskål) J. Agardh to the native seagrass *Posidonia oceanica* (L.) delile: effect of density of shoots and orientation of edges of meadows. *Journal of Experimental Marine Biology and Ecology* **243**, 227–240.

Chin, D.W., de Fouw, J., van der Heide, T., Cahill, B.V., Katcher, K., Paul, V.J., Campbell, J.E. & Peterson, B.J. 2021. Facilitation of a tropical seagrass by a chemosymbiotic bivalve increases with environmental stress. *Journal of Ecology* **109**, 204–217.

Chisholm, J.R.M. & Moulin, P. 2003. Stimulation of nitrogen fixation in refractory organic sediments by *Caulerpa taxifolia* (Chlorophyta). *Limnology and Oceanography* **48**, 787–794.

Coleman, M.A., Wood, G., Filbee-Dexter, K., Minne, A.J.P., Goold, H.D., Vergés, A., Marzinelli, E.M., Steinberg, P.D. & Wernberg, T. 2020. Restore or redefine: future trajectories for restoration. *Frontiers in Marine Science* **7**, 237.

Costanza, R., de Groot, R., Sutton, P., van der Ploeg, S., Anderson, S.J., Kubiszewski, I., Farber, S. & Turner, R.K. 2014. Changes in the global value of ecosystem services. *Global Environmental Change* **26**, 152–158.

Coyer, J.A., Diekmann, O.E., Serrão, E.A., Procaccini, G., Milchakova, N., Pearson, G.A., Stam, W.T. & Olsen, J.L. 2004. Population genetics of dwarf eelgrass *Zostera noltii* throughout its biogeographic range. *Marine Ecology Progress Series* **281**, 51–62.

Crump, B.C., Wojahn, J.M., Tomas, F. & Mueller, R.S. 2018. Metatranscriptomics and amplicon sequencing reveal mutualisms in seagrass microbiomes. *Frontiers in Microbiology* **9**, 388.

Derksen-Hooijberg, M., Angelini, C., Lamers, L.P.M., Borst, A., Smolders, A., Hoogveld, J.R.H., de Paoli, H., van de Koppel, J., Silliman, B.R. & van der Heide, T. 2018. Mutualistic interactions amplify saltmarsh restoration success. *Journal of Applied Ecology* **55**, 405–414.

Dijk, J.K.V., Tussenbroek, B.I.V., Jiménez-Durán, K., Márquez-Guzmán, G.J. & Ouborg, J. 2009. High levels of gene flow and low population genetic structure related to high dispersal potential of a tropical marine angiosperm. *Marine Ecology Progress Series* **390**, 67–77.

Drouin, A., McKindsey, C.W. & Johnson, L.E. 2012. Detecting the impacts of notorious invaders: experiments versus observations in the invasion of eelgrass meadows by the green seaweed *Codium fragile*. *Oecologia* **168**, 491–502.

Duffin, P., Martin, D.L., Pagenkopp Lohan, K.M. & Ross, C. 2020. Integrating host immune status, Labyrinthula spp. load and environmental stress in a seagrass pathosystem: assessing immune markers and scope of a new qPCR primer set. *PLoS One* **15**, e0230108.

Duffy, J.E., Stachowicz, J.J., Reynolds, P.L., Hovel, K.A., Jahnke, M., Sotka, E.E., Boström, C., Boyer, K.E., Cusson, M., Eklöf, J., Engelen, A.H., Eriksson, B.K., Fodrie, F.J., Griffin, J.N., Hereu, C.M., Hori, M., Hughes, A.R., Ivanov, M.V., Jorgensen, P., Kruschel, C., Lee, K.-S., Lefcheck, J.S., Moksnes, P.-O., Nakaoka, M., O'Connor, M.I., O'Connor, N.E., Orth, R.J., Peterson, B.J., Reiss, H., Reiss, K., Richardson, J.P., Rossi, F., Ruesink, J.L., Schultz, S.T., Thormar, J., Tomas, F., Unsworth, R., Voigt, E., Whalen, M.A., Ziegler, S.L. & Olsen, J.L. 2022. A Pleistocene legacy structures variation in modern seagrass ecosystems. *Proceedings of the National Academy of Sciences* **119**, e2121425119.

Dunic, J.C., Brown, C.J., Connolly, R.M., Turschwell, M.P. & Côté, I.M. 2021. Long-term declines and recovery of meadow area across the world's seagrass bioregions. *Global Change Biology* **27**, 4096–4109.

Edgeloe, J.M., Severn-Ellis, A.A., Bayer, P.E., Mehravi, S., Breed, M.F., Krauss, S.L., Batley, J., Kendrick, G.A. & Sinclair, E.A. 2022. Extensive polyploid clonality was a successful strategy for seagrass to expand into a newly submerged environment. *Proceedings of the Royal Society B* **289**, 20220538.

Ettinger, C.L. & Eisen, J.A. 2019. Characterization of the Mycobiome of the Seagrass, *Zostera marina*, reveals putative associations with marine chytrids. *Frontiers in Microbiology* **10**, 2476.

Ettinger, C.L. & Eisen, J.A. 2020. Fungi, bacteria and oomycota opportunistically isolated from the seagrass, *Zostera marina*. *PLoS One* **15**, e0236135.

Evans, S. M., Griffin, K. J., Blick, R. A., Poore, A. G., & Vergés, A. 2018. Seagrass on the brink: Decline of threatened seagrass Posidonia australis continues following protection. *Plos one* **13**(4), e0190370.

Fales, R.J., Boardman, F.C. & Ruesink, J.L. 2020. Reciprocal interactions between bivalve molluscs and seagrass: a review and meta-analysis. *Journal of Shellfish Research* **39**, 547–562.

Ferguson, A.J., Gruber, R.K., Orr, M. & Scanes, P. 2016. Morphological plasticity in *Zostera muelleri* across light, sediment, and nutrient gradients in Australian temperate coastal lakes. *Marine Ecology Progress Series* **556**, 91–104.

Fonseca, M.S. 2011. Addy revisited: what has changed with seagrass restoration in 64 years? *Ecological Restoration* **29**, 73–81.

Fortes, M.D., Ooi, J.L.S., Tan, Y.M., Prathep, A., Bujang, J.S. & Yaakub, S.M. 2018. Seagrass in Southeast Asia: a review of status and knowledge gaps, and a road map for conservation. *Botanica Marina* **61**, 269–288.

Fraser, M.W., Statton, J., Hovey, R.K., Laverock, B. & Kendrick, G.A. 2016. Seagrass derived organic matter influences biogeochemistry, microbial communities, and seedling biomass partitioning in seagrass sediments. *Plant and Soil* **400**, 133–146.

Fraser, M.W., Gleeson, D.B., Grierson, P.F., Laverock, B. & Kendrick, G.A. 2018. Metagenomic evidence of microbial community responsiveness to phosphorus and salinity gradients in seagrass sediments. *Frontiers in Microbiology* **9**, 1703.

Fraser, M.W., Martin, B.C., Wong, H.L., Burns, B.P. & Kendrick, G.A. 2023. Sulfide intrusion in a habitat forming seagrass can be predicted from relative abundance of sulfur cycling genes in sediments. *Science of the Total Environment* **864**, 161144.

Fuggle, R.E., Gribben, P.E. & Marzinelli, E.M. 2023. Experimental evidence root-associated microbes mediate seagrass response to environmental stress. *Journal of Ecology* **111**, 1079–1093.

Gagnon, K., Rinde, E., Bengil, E.G.T., Carugati, L., Christianen, M.J.A., Danovaro, R., Gambi, C., Govers, L.L., Kipson, S., Meysick, L., Pajusalu, L., Tüney Kızılkaya, İ., van de Koppel, J., van der Heide, T., van Katwijk, M.M. & Boström, C. 2020. Facilitating foundation species: the potential for plant-bivalve interactions to improve habitat restoration success. *Journal of Applied Ecology* **57**, 1161–1179.

Garcias-Bonet, N., Arrieta, J.M., de Santana, C.N., Duarte, C.M. & Marbà, N. 2012. Endophytic bacterial community of a Mediterranean marine angiosperm (*Posidonia oceanica*). *Frontiers in Microbiology* **3**, 342.

Garcias-Bonet, N., Eguíluz, V.M., Díaz-Rúa, R. & Duarte, C.M. 2021. Host-association as major driver of microbiome structure and composition in Red Sea seagrass ecosystems. *Environmental Microbiology* **23**, 2021–2034.

Gilbertson, W.W., Solan, M. & Prosser, J.I. 2012. Differential effects of microorganism-invertebrate interactions on benthic nitrogen cycling. *FEMS Microbiology Ecology* **82**, 11–22.

Glasby, T.M. 2013. *Caulerpa taxifolia* in seagrass meadows: killer or opportunistic weed? *Biological Invasions* **15**, 1017–1035.

Goodman, J.L., Moore, K.A. & Dennison, W.C. 1995. Photosynthetic responses of eelgrass (*Zostera marina L.*) to light and sediment sulfide in a shallow barrier island lagoon. *Aquatic Botany* **50**, 37–47.

Govers, L.L., Heusinkveld, J.H.T., Gräfnings, M.L.E., Smeele, Q. & van der Heide, T. 2022. Adaptive intertidal seed-based seagrass restoration in the Dutch Wadden Sea. *PLoS One* **17**, e0262845.

Gräfnings, M.L.E., Govers, L.L., Heusinkveld, J.H.T., Silliman, B.R., Smeele, Q., Valdez, S.R. & van der Heide, T. 2023. Macrozoobenthos as an indicator of habitat suitability for intertidal seagrass. *Ecological Indicators* **147**, 109948.

Green, A. E., Unsworth, R. K., Chadwick, M. A., & Jones, P. J. 2021. Historical analysis exposes catastrophic seagrass loss for the United Kingdom. *Frontiers in plant science* **12**, 629962.

Gribben, P.E., Nielsen, S., Seymour, J.R., Bradley, D.J., West, M.N. & Thomas, T. 2017. Microbial communities in marine sediments modify success of an invasive macrophyte. *Scientific Reports* **7**, 9845.

Gribben, P.E., Thomas, T., Pusceddu, A., Bonechi, L., Bianchelli, S., Buschi, E., Nielsen, S., Ravaglioli, C. & Bulleri, F. 2018. Below-ground processes control the success of an invasive seaweed. *Journal of Ecology* **106**, 2082–2095.

Grman, E., Allen, J., Galloway, E., McBride, J., Bauer, J.T. & Price, P.A. 2020. Inoculation with remnant prairie soils increased the growth of three native prairie legumes but not necessarily their associations with beneficial soil microbes. *Restoration Ecology* **28**, S393–S399.

Hauxwell, J., Cebrián, J., Furlong, C. & Valiela, I. 2001. Macroalgal canopies contribute to eelgrass (*Zostera marina*) decline in temperate estuarine ecosystems. *Ecology* **82**, 1007–1022.

Hauxwell, J., Cebrián, J. & Valiela, I. 2003. Eelgrass *Zostera marina* loss in temperate estuaries: relationship to land-derived nitrogen loads and effect of light limitation imposed by algae. *Marine Ecology Progress Series* **247**, 59–73.

Heinen, R., Hannula, S.E., Long, J.R.D., Huberty, M., Jongen, R., Kielak, A., Steinauer, K., Zhu, F. & Bezemer, T.M. 2020. Plant community composition steers grassland vegetation via soil legacy effects. *Ecology Letters* **23**, 973–982.

Hernawan, U.E., van Dijk, K., Kendrick, G.A., Feng, M., Biffin, E., Lavery, P.S. & McMahon, K. 2017. Historical processes and contemporary ocean currents drive genetic structure in the seagrass *Thalassia hemprichii* in the Indo-Australian Archipelago. *Molecular Ecology* **26**, 1008–1021.

Holmer, M., Marbà, N., Lamote, M. & Duarte, C.M. 2009. Deterioration of sediment quality in seagrass meadows (*Posidonia oceanica*) invaded by macroalgae (*Caulerpa sp.*). *Estuaries and Coasts* **32**, 456–466.

Hovey, R.K., Cambridge, M.L. & Kendrick, G.A. 2011. Direct measurements of root growth and productivity in the seagrasses *Posidonia australis* and *P. sinuosa*. *Limnology and Oceanography* **56**, 394–402.

Hughes, A.R. & Stachowicz, J.J. 2004. Genetic diversity enhances the resistance of a seagrass ecosystem to disturbance. *Proceedings of the National Academy of Sciences* **101**, 8998–9002.

Hughes, A.R., Williams, S.L., Duarte, C.M., Heck Jr, K.L. & Waycott, M. 2009. Associations of concern: declining seagrasses and threatened dependent species. *Frontiers in Ecology and the Environment* **7**, 242–246.

Jackson, E.L., Smith, T.M., York, P.H., Nielsen, J., Irving, A.D. & Sherman, C.D. 2021. An assessment of the seascape genetic structure and hydrodynamic connectivity for subtropical seagrass restoration. *Restoration Ecology* **29**, e13269.

Johansson, J.O.R. & Greening, H.S. 1999. Seagrass restoration in Tampa Bay: a resource-based approach to estuarine management. In *Seagrasses*, Stephen A. Bortone (ed.). Boca Raton, Florida: CRC Press, 297–312.

Johnson, A.J., Statton, J., Orth, R.J. & Kendrick, G.A. 2018. A sediment bioturbator bottleneck to seedling recruitment for the seagrass *Posidonia australis*. *Marine Ecology Progress Series* **595**, 89–103.

Kaldy, J.E., Eldridge, P.M., Cifuentes, L.A. & Jones, W.B. 2006. Utilization of DOC from seagrass rhizomes by sediment bacteria: 13C-tracer experiments and modeling. *Marine Ecology Progress Series* **317**, 41–55.

Kaldy, J.E., Brown, C.A. & Andersen, C.P. 2013. In situ 13C tracer experiments elucidate carbon translocation rates and allocation patterns in eelgrass *Zostera marina*. *Marine Ecology Progress Series* **487**, 27–39.

Kardol, P. & Wardle, D.A. 2010. How understanding aboveground-belowground linkages can assist restoration ecology. *Trends in Ecology & Evolution* **25**, 670–679.

Kardol, P., Martijn Bezemer, T. & van der Putten, W.H. 2006. Temporal variation in plant-soil feedback controls succession. *Ecology Letters* **9**, 1080–1088.

Kendrick, G.A., Waycott, M., Carruthers, T.J.B., Cambridge, M.L., Hovey, R., Krauss, S.L., Lavery, P.S., Les, D.H., Lowe, R.J., Vidal, O.M.I, Ooi, J.L.S., Orth, R.J., Rivers, D.O., Ruiz-Montoya, L., Sinclair, E.A., Statton, J., van Dijk, J.K. & Verduin, J.J. 2012. The central role of dispersal in the maintenance and persistence of seagrass populations. *BioScience* **62**, 56–65.

Kenworthy, W. J., Hall, M. O., Hammerstrom, K. K., Merello, M., & Schwartzschild, A. 2018. Restoration of tropical seagrass beds using wild bird fertilization and sediment regrading. *Ecological Engineering*, **112**, 72–81.

Kilminster, K., McMahon, K., Waycott, M., Kendrick, G.A., Scanes, P., McKenzie, L., O'Brien, K.R., Lyons, M., Ferguson, A., Maxwell, P., Glasby, T. & Udy, J. 2015. Unravelling complexity in seagrass systems for management: Australia as a microcosm. *Science of the Total Environment* **534**, 97–109.

Kollars, N.M., Abbott, J.M. & Stachowicz, J.J. 2022. Hidden biodiversity: spatial mosaics of eelgrass genotypic diversity at the centimeter to meadow scale. *Ecology* **103**, e3813.

Koziol, L. & Bever, J.D. 2015. Mycorrhizal response trades off with plant growth rate and increases with plant successional status. *Ecology* **96**, 1768–1774.

Koziol, L. & Bever, J.D. 2016. AMF, phylogeny, and succession: specificity of response to mycorrhizal fungi increases for late-successional plants. *Ecosphere* **7**, e01555.

Koziol, L., Schultz, P.A., House, G.L., Bauer, J.T., Middleton, E.L. & Bever, J.D. 2018. The plant microbiome and native plant restoration: the example of native mycorrhizal fungi. *BioScience* **68**, 996–1006.

Kuiper, I., Lagendijk, E.L., Bloemberg, G.V. & Lugtenberg, B.J. 2004. Rhizoremediation: a beneficial plant-microbe interaction. *Molecular Plant-Microbe Interactions* **17**, 6–15.

Lacoste, É., Piot, A., Archambault, P., McKindsey, C.W. & Nozais, C. 2018. Bioturbation activity of three macrofaunal species and the presence of meiofauna affect the abundance and composition of benthic bacterial communities. *Marine Environmental Research* **136**, 62–70.

Lee, H., Golicz, A.A., Bayer, P.E., Severn-Ellis, A.A., Chan, C.-K.K., Batley, J., Kendrick, G.A. & Edwards, D. 2018. Genomic comparison of two independent seagrass lineages reveals habitat-driven convergent evolution. *Journal of Experimental Botany* **69**, 3689–3702.

Lefcheck, J.S., Orth, R.J., Dennison, W.C., Wilcox, D.J., Murphy, R.R., Keisman, J., Gurbisz, C., Hannam, M., Landry, J.B., Moore, K.A., Patrick, C.J., Testa, J., Weller, D.E. & Batiuk, R.A. 2018. Long-term nutrient reductions lead to the unprecedented recovery of a temperate coastal region. *Proceedings of the National Academy of Sciences* **115**, 3658–3662.

Li, C.-J., Li, W.-T., Liu, J., Zhang, X. & Zhang, P. 2017. Zostera marina seed burial can be enhanced by Manila clam *Ruditapes philippinarum*: a microcosm study. *Ocean Science Journal* **52**, 221–229.

Macreadie, P.I., Atwood, T.B., Seymour, J.R., Fontes, M.L.S., Sanderman, J., Nielsen, D.A. & Connolly, R.M. 2019. Vulnerability of seagrass blue carbon to microbial attack following exposure to warming and oxygen. *Science of The Total Environment* **686**, 264–275.

Maltz, M.R. & Treseder, K.K. 2015. Sources of inocula influence mycorrhizal colonization of plants in restoration projects: a meta-analysis. *Restoration Ecology* **23**, 625–634.

Martin, B.C., Statton, J., Siebers, A.R., Grierson, P.F., Ryan, M.H. & Kendrick, G.A. 2018. Colonizing tropical seagrasses increase root exudation under fluctuating and continuous low light. *Limnology and Oceanography* **63**, S381–S391.

Martin, B.C., Bougoure, J., Ryan, M.H., Bennett, W.W., Colmer, T.D., Joyce, N.K., Olsen, Y.S. & Kendrick, G.A. 2019. Oxygen loss from seagrass roots coincides with colonisation of sulphide-oxidising cable bacteria and reduces sulphide stress. *The ISME Journal* **13**, 707–719.

Martin, B.C., Alarcon, M.S., Gleeson, D., Middleton, J.A., Fraser, M.W., Ryan, M.H., Holmer, M., Kendrick, G.A. & Kilminster, K. 2020. Root microbiomes as indicators of seagrass health. *FEMS Microbiology Ecology* **96**, fiz201.

Martin, B.C., Middleton, J.A., Skrzypek, G., Kendrick, G.A., Cosgrove, J. & Fraser, M.W. 2022. Composition of seagrass root associated bacterial communities are linked to nutrients and heavy metal concentrations in an anthropogenically influenced estuary. *Frontiers in Marine Science* **8**, 768864.

Martinetto, P., Montemayor, D.I., Alberti, J., Costa, C.S.B. & Iribarne, O. 2016. Crab bioturbation and herbivory may account for variability in carbon sequestration and stocks in south west Atlantic salt marshes. *Frontiers in Marine Science* **3**, 122.

McGlathery, K.J., Reidenbach, M.A., D'odorico, P., Fagherazzi, S., Pace, M.L. & Porter, J.H. 2013. Nonlinear dynamics and alternative stable states in shallow coastal systems. *Oceanography* **26**, 220–231.

McKinnon, J.G., Gribben, P.E., Davis, A.R., Jolley, D.F. & Wright, J.T. 2009. Differences in soft-sediment macrobenthic assemblages invaded by *Caulerpa taxifolia* compared to uninvaded habitats. *Marine Ecology Progress Series* **380**, 59–71.

McLeod, I.M., Boström-Einarsson, L., Johnson, C.R., Kendrick, G., Layton, C., Rogers, A.A. & Statton, J. 2018. *The Role of Restoration in Conserving Matters of National Environmental Significance in Marine and Coastal Environments*. Hobart, Tasmania: NESP Marine Biodiversity Hub.

McMahon, K.M., Evans, R.D., van Dijk, K., Hernawan, U., Kendrick, G.A., Lavery, P.S., Lowe, R., Puotinen, M. & Waycott, M. 2017. Disturbance is an important driver of clonal richness in tropical seagrasses. *Frontiers in Plant Science* **8**, 2026.

Meehan, A.J. & West, R.J. 2002. Experimental transplanting of *Posidonia australis* seagrass in Port Hacking, Australia, to assess the feasibility of restoration. *Marine Pollution Bulletin* **44**, 25–31.

Meehan, A.J. & West, R.J. 2004. Seedling development and patch formation of the seagrass *Posidonia australis* in a southeast Australian estuary. *Aquatic Botany* **79**, 1–14.

Meyer, D.L., Townsend, E.C. & Thayer, G.W. 1997. Stabilization and erosion control value of oyster cultch for intertidal marsh. *Restoration Ecology* **5**, 93–99.

Meysick, L., Infantes, E. & Boström, C. 2019. The influence of hydrodynamics and ecosystem engineers on eelgrass seed trapping. *PLoS One* **14**, e0222020.

Miller, B.P., Sinclair, E.A., Menz, M.H., Elliott, C.P., Bunn, E., Commander, L.E. & Stevens, J.C. 2017. A framework for the practical science necessary to restore sustainable, resilient, and biodiverse ecosystems. *Restoration Ecology* **25** (4), 605–617.

Mohammadi, N.S., Buapet, P., Pernice, M., Signal, B., Kahlke, T., Hardke, L. & Ralph, P.J. 2019. Transcriptome profiling analysis of the seagrass, *Zostera* muelleri under copper stress. *Marine pollution bulletin* **149**, 110556.

Mohr, W., Lehnen, N., Ahmerkamp, S., Marchant, H.K., Graf, J.S., Tschitschko, B., Yilmaz, P., Littmann, S., Gruber-Vodicka, H., Leisch, N., Weber, M., Lott, C., Schubert, C.J., Milucka, J. & Kuypers, M.M.M. 2021. Terrestrial-type nitrogen-fixing symbiosis between seagrass and a marine bacterium. *Nature* **600**, 1–5.

Neuenkamp, L., Prober, S.M., Price, J.N., Zobel, M. & Standish, R.J. 2019. Benefits of mycorrhizal inoculation to ecological restoration depend on plant functional type, restoration context and time. *Fungal Ecology* **40**, 140–149.

Nguyen, H.M., Kim, M., Ralph, P.J., Marín-Guirao, L., Pernice, M. & Procaccini, G. 2020. Stress memory in seagrasses: first insight into the effects of thermal priming and the role of epigenetic modifications. *Frontiers in Plant Science* **11**, 494.

Nguyen, H.M., Bulleri, F., Marín-Guirao, L., Pernice, M. & Procaccini, G. 2021. Photo-physiology and morphology reveal divergent warming responses in northern and southern hemisphere seagrasses. *Marine Biology* **168**, 129.

Olsen, J.L., Stam, W.T., Coyer, J.A., Reusch, T.B., Billingham, M., Boström, C., Calvert, E., Christie, H., Granger, S. & Lumiere, R.L. 2004. North Atlantic phylogeography and large-scale population differentiation of the seagrass *Zostera marina* L. *Molecular Ecology* **13**, 1923–1941.

Orth, R.J., Carruthers, T.J.B., Dennison, W.C., Duarte, C.M., Fourqurean, J.W., Heck, K.L., Hughes, A.R., Kendrick, G.A., Kenworthy, W.J., Olyarnik, S., Short, F.T., Waycott, M. & Williams, S.L. 2006. A global crisis for seagrass ecosystems. *BioScience* **56**, 987.

Orth, R.J., Moore, K.A., Marion, S.R., Wilcox, D.J. & Parrish, D.B. 2012. Seed addition facilitates eelgrass recovery in a coastal bay system. *Marine Ecology Progress Series* **448**, 177–196.

Orth, R.J., Lefcheck, J.S., McGlathery, K.S., Aoki, L., Luckenbach, M.W., Moore, K.A., Oreska, M.P.J., Snyder, R., Wilcox, D.J. & Lusk, B. 2020. Restoration of seagrass habitat leads to rapid recovery of coastal ecosystem services. *Science Advances* **6**, eabc6434.

Pansini, A., Bosch-Belmar, M., Berlino, M., Sarà, G. & Ceccherelli, G. 2022. Collating evidence on the restoration efforts of the seagrass *Posidonia oceanica*: current knowledge and gaps. *Science of The Total Environment* **851**, 158320.

Pazzaglia, J., Nguyen, H.M., Santillán-Sarmiento, A., Ruocco, M., Dattolo, E., Marín-Guirao, L. & Procaccini, G. 2021. The genetic component of seagrass restoration: what we know and the way forwards. *Water* **13**, 829.

Peterson, B.J. & Heck, K.L. 1999. The potential for suspension feeding bivalves to increase seagrass productivity. *Journal of Experimental Marine Biology and Ecology* **240**, 37–52.

Petersen, J.M., Kemper, A., Gruber-Vodicka, H., Cardini, U., Van Der Geest, M., Kleiner, M., Bulgheresi, S., Mußmann, M., Herbold, C. & Seah, B.K. 2016. Chemosynthetic symbionts of marine invertebrate animals are capable of nitrogen fixation. *Nature microbiology* **2**, 1–11.

Phelps, C.M., McMahon, K., Bissett, A., Bernasconi, R., Steinberg, P.D., Thomas, T., Marzinelli, E.M. & Huggett, M.J. 2021. The surface bacterial community of an Australian kelp shows cross-continental variation and relative stability within regions. *FEMS Microbiology Ecology* **97**, fiab089.

Piercey, R.S., Gribben, P.E., Hanley, T.C., Moles, A.T. & Hughes, A.R. 2021. Incorporating marine macrophytes in plant-soil feedbacks: emerging evidence and opportunities to advance the field. *Journal of Ecology* **109**, 614–625.

Renton, M., Airey, M., Cambridge, M.L. & Kendrick, G.A. 2011. Modelling seagrass growth and development to evaluate transplanting strategies for restoration. *Annals of Botany* **108**, 1213–1223.

Reusch, T.B.H., Chapman, A.R.O. & Gröger, J.P. 1994. Blue mussels *Mytilus edulis* do not interfere with eelgrass *Zostera marina* but fertilize shoot growth through biodeposition. *Marine Ecology Progress Series* **108**, 265–282.

Reynolds, L.K., McGlathery, K.J. & Waycott, M. 2012. Genetic diversity enhances restoration success by augmenting ecosystem services. *Plos One* **7**, e38397.

de los Santos, C.B., Krause-Jensen, D., Alcoverro, T., Marbà, N., Duarte, C.M., van Katwijk, M.M., Pérez, M., Romero, J., Sánchez-Lizaso, J.L., Roca, G., Jankowska, E., Pérez-Lloréns, J.L., Fournier, J., Montefalcone, M., Pergent, G., Ruiz, J.M., Cabaço, S., Cook, K., Wilkes, R.J., Moy, F.E., Trayter, G.M.-R., Arañó, X.S., de Jong, D.J., Fernández-Torquemada, Y., Auby, I., Vergara, J.J. & Santos, R. 2019. Recent trend reversal for declining European seagrass meadows. *Nature Communications* **10**, 3356.

Saunders, M., Waltham, N., Cannard, T., Sheppard, M., Fischer, M., Twomey, A., Bishop, M., Boody, K., Callaghan, D. & Fulton, B. 2022. A roadmap for coordinated landscape-scale coastal and marine ecosystem restoration. Report to the Reef and Rainforest Research Centre, Cairns, Queensland.

Scholz, V.V., Martin, B.C., Meyer, R., Schramm, A., Fraser, M.W., Nielsen, L.P., Kendrick, G.A., Risgaard-Petersen, N., Burdorf, L.D. & Marshall, I.P. 2021. Cable bacteria at oxygen-releasing roots of aquatic plants: a widespread and diverse plant-microbe association. *The New Phytologist* **232**, 2138–2151.

Serrano, C, Lovelock, C.E.B. Atwood, T., Macreadie, P.I., Canto, R., Phinn, S., Arias-Ortiz, A., Bai, L., Baldock, J., Bedulli, C., Carnell, P., Connolly, R.M., Donaldson, P., Esteban, A., Ewers Lewis, C.J., Eyre, B.D., Hayes, M.A., Horwitz, P., Hutley, L.B., Kavazos, C.R.J., Kelleway, J.J., Kendrick, G.A., KilmiVnster, K., Lafratta, A., Lee, S., Lavery, P.S., Maher, D.T., Marbà, N., Masque, P., Mateo, M.A., Mount, R., Ralph, P.J., Roelfsema, C., Rozaimi, M., Ruhon, R., Salinas, C., Samper-Villarreal, J., Sanderman, J., J. Sanders, C., Santos, I., Sharples, C., Steven, A.D.L., Cannard, T., Trevathan-Tackett, S.M. & Duarte, C.M. 2019. Australian vegetated coastal ecosystems as global hotspots for climate change mitigation. *Nature Communications* **10**, 4313.

Short, F.T., Polidoro, B., Livingstone, S.R., Carpenter, K.E., Bandeira, S., Bujang, J.S., Calumpong, H.P., Carruthers, T.J.B., Coles, R.G., Dennison, W.C., Erftemeijer, P.L.A., Fortes, M.D., Freeman, A.S., Jagtap, T.G., Kamal, A.H.M., Kendrick, G.A., Judson Kenworthy, W., La Nafie, Y.A., Nasution, I.M., Orth, R.J., Prathep, A., Sanciangco, J.C., Tussenbroek, B. van, Vergara, S.G., Waycott, M. & Zieman, J.C. 2011. Extinction risk assessment of the world's seagrass species. *Biological Conservation* **144**, 1961–1971.

Short, F.T., Kosten, S., Morgan, P.A., Malone, S. & Moore, G.E. 2016. Impacts of climate change on submerged and emergent wetland plants. *Aquatic Botany* **135**, 3–17.

Siebert, T. & Branch, G.M. 2006. Ecosystem engineers: interactions between eelgrass *Zostera capensis* and the sandprawn *Callianassa kraussi* and their indirect effects on the mudprawn *Upogebia africana. Journal of Experimental Marine Biology and Ecology* **338**, 253–270.

Sinclair, E.A., Krauss, S.L., Anthony, J., Hovey, R. & Kendrick, G.A. 2014. The interaction of environment and genetic diversity within meadows of the seagrass *Posidonia australis* (Posidoniaceae). *Marine Ecology Progress Series* **506**, 87–98.

Sinclair, E.A., Edgeloe, J.M., Anthony, J.M., Statton, J., Breed, M.F. & Kendrick, G.A. 2020. Variation in reproductive effort, genetic diversity and mating systems across *Posidonia australis* seagrass meadows in Western Australia. *AoB Plants* **12**, plaa038.

Sinclair, E.A., Sherman, C.D.H., Statton, J., Copeland, C., Matthews, A., Waycott, M., van Dijk, K.-J., Vergés, A., Kajlich, L., McLeod, I.M. & Kendrick, G.A. 2021a. Advances in approaches to seagrass restoration in Australia. *Ecological Management & Restoration* **22**, 10–21.

Sinclair, E.A., Sherman, C.D.H., Statton, J., Copeland, C., Matthews, A., Waycott, M., van Dijk, K.-J., Vergés, A., Kajlich, L., McLeod, I.M. & Kendrick, G.A. 2021b. Advances in approaches to seagrass restoration in Australia. *Ecological Management & Restoration* **22**, 10–21.

Sinclair, E.A., Hovey, R.K., Krauss, S.L., Anthony, J.M., Waycott, M. & Kendrick, G.A. 2023. Historic and contemporary biogeographic perspectives on range-wide spatial genetic structure in a widespread seagrass. *Ecology and Evolution* **13**, e9900.

Smulders, F.O.H., Vonk, J.A., Engel, M.S. & Christianen, M.J.A. 2017. Expansion and fragment settlement of the non-native seagrass *Halophila stipulacea* in a Caribbean bay. *Marine Biology Research* **13**, 967–974.

Soldan, R., Mapelli, F., Crotti, E., Schnell, S., Daffonchio, D., Marasco, R., Fusi, M., Borin, S. & Cardinale, M. 2019. Bacterial endophytes of mangrove propagules elicit early establishment of the natural host and promote growth of cereal crops under salt stress. *Microbiological Research* **223**, 33–43.

Song, X., Zhou, Y., Zeng, J., Shou, L., Zhang, X., Yue, S., Gao, W., Feng, W., Wang, Z. & Du, P. 2021. Distinct root system acclimation patterns of seagrass *Zostera japonica* in sediments of different trophic status: a research by X-ray computed tomography. *Journal of Oceanology and Limnology* **39**, 2267–2280.

Statton, J., Cambridge, M.L., Dixon, K.W. & Kendrick, G.A. 2013. Aquaculture of *Posidonia australis* seedlings for seagrass restoration programs: effect of sediment type and organic enrichment on growth. *Restoration Ecology* **21**, 250–259.

Statton, J., Dixon, K.W., Irving, A.D., Jackson, E.L., Kendrick, G.A., Orth, R.J. & Sinclair, E.A. 2018. Decline and restoration ecology of Australian seagrasses. In *Seagrasses of Australia: Structure, Ecology and Conservation*, A.W.D. Larkumet al. (eds.). Cham: Springer International Publishing, 665–704.

Statton, J., Sinclair, E.A., McNeair, S., Kendrick, A. & Kendrick, G.A. 2021. Assisting recovery of seagrass in Shark Bay, Gathaagudu. Final Report to the National Environmental Science Program, Marine Biodiversity Hub, 89.

Stief, P. 2013. Stimulation of microbial nitrogen cycling in aquatic ecosystems by benthic macrofauna: mechanisms and environmental implications. *Biogeosciences* **10**, 7829–7846.

Strydom, S., Murray, K., Wilson, S., Huntley, B., Rule, M., Heithaus, M., Bessey, C., Kendrick, G.A., Burkholder, D., Fraser, M.W. & Zdunic, K. 2020. Too hot to handle: unprecedented seagrass death driven by marine heatwave in a world heritage area. *Global Change Biology* **26**, 3525–3538.

Suchanek, T.H. 1983. Control of seagrass communities and sediment distribution by *Callianassa* (*Crustacea, Thalassinidea*) bioturbation. *Journal of Marine Research* **41**, 281–298.

Tan, Y.M., Dalby, O., Kendrick, G.A., Statton, J., Sinclair, E.A., Fraser, M.W., Macreadie, P.I., Gillies, C.L., Coleman, R.A., Waycott, M., van Dijk, K., Vergés, A., Ross, J.D., Campbell, M.L., Matheson, F.E., Jackson, E.L., Irving, A.D., Govers, L.L., Connolly, R.M., McLeod, I.M., Rasheed, M.A., Kirkman, H., Flindt, M.R., Lange, T., Miller, A.D. & Sherman, C.D.H. 2020. Seagrass restoration is possible: insights and lessons from Australia and New Zealand. *Frontiers in Marine Science* **7**, 617.

Tarquinio, F., Bourgoure, J., Koenders, A., Laverock, B., Säwström, C. & Hyndes, G.A. 2018. Microorganisms facilitate uptake of dissolved organic nitrogen by seagrass leaves. *The ISME Journal* **12**, 2796–2800.

Tarquinio, F., Hyndes, G.A., Laverock, B., Koenders, A. & Säwström, C. 2019. The seagrass holobiont: understanding seagrass-bacteria interactions and their role in seagrass ecosystem functioning. *FEMS Microbiology Letters* **366,** fnz057.

Tarquinio, F., Attlan, O., Vanderklift, M.A., Berry, O. & Bissett, A. 2021. Distinct endophytic bacterial communities inhabiting seagrass seeds. *Frontiers in Microbiology* **12**, 14.

Tavares, A.I., Assis, J., Larkin, P.D., Creed, J.C., Magalhães, K., Horta, P., Engelen, A., Cardoso, N., Barbosa, C., Pontes, S., Regalla, A., Almada, C., Ferreira, R., Abdoul, B.M., Ebaye, S., Bourweiss, M., dos Santos, C.V.-D., Patrício, A.R., Teodósio, A., Santos, R., Pearson, G.A. & Serrao, E.A. 2023. Long range gene flow beyond predictions from oceanographic transport in a tropical marine foundation species. *Scientific Reports* **13**, 9112.

Thompson, L.R., Sanders, J.G., McDonald, D., Amir, A., Ladau, J., Locey, K.J., Prill, R.J., Tripathi, A., Gibbons, S.M., Ackermann, G., Navas-Molina, J.A., Janssen, S., Kopylova, E., Vázquez-Baeza, Y., González, A., Morton, J.T., Mirarab, S., Zech Xu, Z., Jiang, L., Haroon, M.F., Kanbar, J., Zhu, Q., Jin Song, S., Kosciolek, T., Bokulich, N.A., Lefler, J., Brislawn, C.J., Humphrey, G., Owens, S.M., Hampton-Marcell, J., Berg-Lyons, D., McKenzie, V., Fierer, N., Fuhrman, J.A., Clauset, A., Stevens, R.L., Shade, A., Pollard, K.S., Goodwin, K.D., Jansson, J.K., Gilbert, J.A. & Knight, R. 2017. A communal catalogue reveals Earth's multiscale microbial diversity. *Nature* **551**, 457–463.

Ugarelli, K., Chakrabarti, S., Laas, P. & Stingl, U. 2017. The seagrass holobiont and its microbiome. *Microorganisms* **5**, 81.

van Katwijk, M.M., Bos, A.R., de Jonge, V.N., Hanssen, L.S.A.M., Hermus, D.C.R. & de Jong, D.J. 2009. Guidelines for seagrass restoration: importance of habitat selection and donor population, spreading of risks, and ecosystem engineering effects. *Marine Pollution Bulletin* **58**, 179–188.

van Katwijk, M.M., Thorhaug, A., Marbà, N., Orth, R.J., Duarte, C.M., Kendrick, G.A., Althuizen, I.H.J., Balestri, E., Bernard, G., Cambridge, M.L., Cunha, A., Durance, C., Giesen, W., Han, Q., Hosokawa, S., Kiswara, W., Komatsu, T., Lardicci, C., Lee, K.-S., Meinesz, A., Nakaoka, M., O'Brien, K.R., Paling, E.I., Pickerell, C., Ransijn, A.M.A. & Verduin, J.J. 2016. Global analysis of seagrass restoration: the importance of large-scale planting. *Journal of Applied Ecology* **53**, 567–578.

van Katwijk, M.M., van Tussenbroek, B.I., Hanssen, S.V., Hendriks, A.J. & Hanssen, L. 2021. Rewilding the Sea with domesticated seagrass. *BioScience* **71**, 1171–1178.

Van Keulen, M., Paling, E. I., & Walker, C. J. 2003. Effect of planting unit size and sediment stabilization on seagrass transplants in Western Australia. *Restoration Ecology* **11**(1), 50–55.

van der Geest, M., van der Heide, T., Holmer, M. & de Wit, R. 2020. First field-based evidence that the seagrass-lucinid mutualism can mitigate sulfide stress in seagrasses. *Frontiers in Marine Science* **7**, 11.

Van der Putten, W.H., Van Dijk, C. & Peters, B.A.M. 1993. Plant-specific soil-borne diseases contribute to succession in foredune vegetation. *Nature* **362**, 53–56.

Van der Putten, W.H., Bardgett, R.D., Bever, J.D., Bezemer, T.M., Casper, B.B., Fukami, T., Kardol, P., Klironomos, J.N., Kulmatiski, A., Schweitzer, J.A., Suding, K.N., Voorde, T.F.J.V. de & Wardle, D.A. 2013. Plant-soil feedbacks: the past, the present and future challenges. *Journal of Ecology* **101**, 265–276.

Veen, G.F., Fry, E.L., ten Hooven, F.C., Kardol, P., Morriën, E. & De Long, J.R. 2019. The role of plant litter in driving plant-soil feedbacks. *Frontiers in Environmental Science* **7**, 168.

Vinther, H.F. & Holmer, M. 2008. Experimental test of biodeposition and ammonium excretion from blue mussels (*Mytilus edulis*) on eelgrass (*Zostera marina*) performance. *Journal of Experimental Marine Biology and Ecology* **364**, 72–79.

Volkenborn, N., Robertson, D.M. & Reise, K. 2009. Sediment destabilizing and stabilizing bio-engineers on tidal flats: cascading effects of experimental exclusion. *Helgoland Marine Research* **63**, 27–35.

Wang, L., Tomas, F. & Mueller, R.S. 2020. Nutrient enrichment increases size of *Zostera marina* shoots and enriches for sulfur and nitrogen cycling bacteria in root-associated microbiomes. *FEMS Microbiology Ecology* **96**, fiaa129.

Waycott, M., Duarte, C.M., Carruthers, T.J.B., Orth, R.J., Dennison, W.C., Olyarnik, S., Calladine, A., Fourqurean, J.W., Heck, K.L., Hughes, A.R., Kendrick, G.A., Kenworthy, W.J., Short, F.T. & Williams, S.L. 2009. Accelerating loss of seagrasses across the globe threatens coastal ecosystems. *Proceedings of the National Academy of Sciences* **106**, 12377–12381.

Webb, A.P. & Eyre, B.D. 2004. Effect of natural populations of burrowing thalassinidean shrimp on sediment irrigation, benthic metabolism, nutrient fluxes and denitrification. *Marine Ecology Progress Series* **268**, 205–220.

Wendländer, N.S., Lange, T., Connolly, R.M., Kristensen, E., Pearson, R.M., Valdemarsen, T. & Flindt, M.R. 2019. Assessing methods for restoring seagrass (*Zostera muelleri*) in Australia's subtropical waters. *Marine and Freshwater Research* **71**, 996–1005.

Whitfield, A.K. 2017. The role of seagrass meadows, mangrove forests, salt marshes and reed beds as nursery areas and food sources for fishes in estuaries. *Reviews in Fish Biology and Fisheries* **27**, 75–110.

Williams, S.L. 1990. Experimental studies of Caribbean seagrass bed development. *Ecological Monographs* **60**, 449–469.

Wood, G., Marzinelli, E.M., Coleman, M.A., Campbell, A.H., Santini, N.S., Kajlich, L., Verdura, J., Wodak, J., Steinberg, P.D., Vergés, A., Wood, G., Marzinelli, E.M., Coleman, M.A., Campbell, A.H., Santini, N.S., Kajlich, L., Verdura, J., Wodak, J., Steinberg, P.D. & Vergés, A. 2019. Restoring subtidal marine macrophytes in the Anthropocene: trajectories and future-proofing. *Marine and Freshwater Research* **70**, 936–951.

Wood, G., Marzinelli, E.M., Vergés, A., Campbell, A.H., Steinberg, P.D. & Coleman, M.A. 2020. Using genomics to design and evaluate the performance of underwater forest restoration. *Journal of Applied Ecology* **57**, 1988–1998.

Wright, J.T. 2005. Differences between native and invasive *Caulerpa taxifolia*: a link between asexual fragmentation and abundance in invasive populations. *Marine Biology* **147**, 559–569.

Wubs, E.R.J., van der Putten, W.H., Bosch, M. & Bezemer, T.M. 2016. Soil inoculation steers restoration of terrestrial ecosystems. *Nature Plants* **2**, 1–5.

Zabarte-Maeztu, I., Matheson, F.E., Manley-Harris, M., Davies-Colley, R.J. & Hawes, I. 2021. Fine sediment effects on seagrasses: a global review, quantitative synthesis and multi-stressor model. *Marine Environmental Research* **171**, 105480.

Oceanography and Marine Biology: An Annual Review, 2024, **62**, 215-247

Taylor & Francis

DRIVERS OF CORAL REEF BENTHIC CHANGES AND IMPLICATIONS ON ECOSYSTEM FUNCTIONING AND SERVICES

MIRIAM REVERTER[1*], STEPHANIE B. HELBER[2], SVEN ROHDE[3], JASPER M. DE GOEIJ[4], AND PETER J. SCHUPP[3,5]

[1]School of Biological and Marine Sciences, University of Plymouth, Drake Circus, PL4 8AA, UK
[2]Department of Geography and Environmental Sciences, Faculty of Engineering and Environment, Northumbria University, Newcastle upon Tyne, UK
[3]Institute for Chemistry and Biology of the Marine Environment (ICBM), Carl von Ossietzky University Oldenburg, Wilhelmshaven, Germany
[4]Department of Freshwater and Marine Ecology, Institute for Biodiversity and Ecosystem Dynamics (IBED), University of Amsterdam, Amsterdam, Netherlands
[5]Helmholtz Institute for Functional Marine Biodiversity at the University of Oldenburg (HIFMB), Ammerländer Heerstrasse 231, D-26129 Oldenburg, Germany
**Correspondence should be sent to:*
Miriam Reverter, miriam.reverter@plymouth.ac.uk
Author's contributions: *M.R. conceived the initial idea. M.R. wrote the first draft. All authors (M.R., S.B.H., S.R., J.M.G. and P.J.S.) contributed to the ideas and the writing of the manuscript.*

Abstract Increasing anthropogenic stressors will likely result in many coral reefs undergoing compositional and functional benthic changes, including towards non-hard coral-dominated communities. The decline of reef-building species and the increase of alternative (i.e., non-reef-building) organisms profoundly affect the functioning of coral reefs and the services provided, but this remains poorly understood, especially for non-algal alternative communities. Here, we propose a standardised framework to study coral reef benthic changes and review the knowledge on coral reef shifts and their implications on coral reef functioning and service provisioning using this framework, which contains the five following elements: (1) contextual information, (2) drivers and triggers, (3) feedback mechanisms, (4) changes in ecosystem functions and (5) effects on ecosystem services. We highlight the importance of biotic and abiotic interactions in determining the likelihood, magnitude and direction of community changes, showing that similar drivers and triggers can lead to multiple alternative communities, depending on the environmental biotic and abiotic conditions. We stress the need to build a robust and comparable database in different ecoregions that includes both biotic information about the community and its environmental characteristics over a relevant timescale. Finally, we stress the need for understanding not only the causes (i.e., drivers and triggers) behind the changes but also the processes behind the persistence of the novel communities (i.e., feedback mechanisms), as well as their broader socio-ecological relevance (e.g., impact on ecosystem functions and services). Addressing all these knowledge shortfalls is vital to increase the prediction power of coral reef benthic changes, understand their persistence and design effective restoration measures to facilitate coral recovery.

DOI: 10.1201/9781003477518-5

Keywords: Benthic Organisms; Community Turnover; Coral Reef Shifts; Drivers and Triggers; Ecosystem Functions; Ecosystem Services

Glossary

Alternative organisms: Coral reef benthic organisms other than hard coral (e.g., macroalgae, turf algae, cyanobacteria, soft coral, sponges and ascidians). Please note that this terminology is not related to the alternate state theory and the presence/absence of hysteresis.

Alternative communities: Communities that are dominated (2-dimensional projected cover) by alternative organisms (*see definition in glossary*)

Benthic dominance: Most abundant benthic organisms in a community when analysing projected 2-dimensional cover, which is currently the most common benthic abundance variable measured.

Biological response variables: In this review, we refer to the abundance (2-dimensional projected cover) of benthic organisms.

Contextual information: Reef environmental characteristics (e.g., geological, physicochemical, biotic and socioeconomic characteristics).

Driver: Slow, generally low-impact environmental changes (e.g., eutrophication and overfishing). Drivers generally need to be sustained over time to cause a community shift.

Feedback mechanism: Processes where the interaction between biological response variables (*see definition in glossary*) or between response variables and drivers reinforce the dominance of alternative organisms. Also referred to as **response mechanisms**.

Phase or regime shift: Abrupt community changes, that are thought to occur when an ecological threshold or tipping point has been reached, in which a small cumulative increase of a stressor drives a much larger ecological change than expected from linear effects and results in a functionally different stable system. Since evidencing ecological thresholds in the field is highly challenging, we use the term **benthic/community shifts** throughout this manuscript, which are defined as changes between two communities that are dominated by different organisms (*see dominance definition in glossary*), without consideration of ecological thresholds.

Stressor: Altered environmental variables, many times as a result of anthropogenic activities, that generally have a negative effect on organisms and ecosystems (e.g., elevated temperature, eutrophication and overfishing).

Trigger: Abrupt large disturbances (e.g., heatwave, storm and disease outbreak) that have a direct and large effect on biological responses, which may lead to a fast regime shift. Often referred as pulse disturbance.

Introduction

Shallow warm-water coral reefs support about 25% of marine biodiversity (Fisher et al. 2015) and provide ecosystem services to nearly a billion people who live within 100 km of a coral reef (Sing Wong et al. 2022). However, as many tropical ecosystems, they are increasingly threatened by local and global stressors (e.g., overfishing, pollution and ocean warming) (Hughes et al. 2017a, Williams et al. 2019). A recent report showed that 14% of the world's coral was lost between 2009 and 2018 (Souter et al. 2021), with estimates predicting further coral loss as climate change effects worsen and bleaching events increase in frequency and severity (Hughes et al. 2017b, 2019).

As the cover of some coral groups decreases, the newly free space can become colonised by fast-growing organisms (i.e., macroalgae, turf algae, cyanobacteria, soft corals, sponges), causing major ecosystem restructuring (Chaves-Fonnegra et al. 2018, Tebbett et al. 2019, Reverter et al.

2020). Similarly, changing environmental conditions, especially rising seawater temperatures, eutrophication and ocean acidification, can allow more tolerant or adapted organisms to outcompete and overgrow physiologically compromised coral species (i.e., competitive replacement), also leading to community shifts (Kelly et al. 2012, Baum et al. 2016). For example, a turnover between Acroporidae, which are generally regarded as being temperature sensitive, and more robust opportunistic coral species (e.g., Poritidae and Pocilloporidae) has been reported in multiple reefs (Cannon et al. 2021, Moritz et al. 2021). Shifts towards sponge-dominated communities have also been observed in highly turbid reefs (Biggerstaff et al. 2015, 2017b). These differences in vulnerabilities can lead to altered coral community assemblages with reduced biodiversity, which may further suppress the growth and survivorship of the remaining coral species on impacted reefs (Clements & Hay 2019, 2021). In fact, it is predicted that most coral reefs will undergo compositional and functional changes in the coming years in response to local stressors (i.e., overfishing and eutrophication) and increasingly frequent thermal anomalies and storms (Hughes et al. 2017a, Williams et al. 2019, Woodhead et al. 2019, Obura et al. 2022).

In a recent systematic review and meta-analysis, up to eight different alternative communities (i.e., not dominated by hard corals) in shallow tropical coral reefs were identified: algal, soft coral, sponge, zoanthid, cyanobacteria, ascidian, anemone and corallimorpharian-dominated communities (Reverter et al. 2022) (Figure 1). Furthermore, two types of compositional changes within hard coral communities were detected: (1) a turnover from branching to massive and encrusting species and (2) a turnover between scleractinian corals and hydrocorals (Reverter et al. 2022). The decline of key reef-building species and the increase of alternative (i.e., non-reef-building) or less structurally complex organisms (e.g., encrusting corals) have ripple effects throughout the ecosystem, such as decreasing structural complexity and modifying biotic interactions (Alvarez-Filip et al. 2009, Newman et al. 2015). Although an increasing amount of scientific literature shows that coral reef benthic changes, often referred as phase or regime shifts, are widespread throughout the tropics, and may be increasing in frequency (Norström et al. 2009, Reverter et al. 2022, Tebbett et al. 2023a), their study has been far from simple. The complexity and heterogeneity of these changes, together with the use of variable methods and uneven biogeographic efforts, have led to different perspectives and potential biases in the understanding of global coral reef changes (Reverter et al. 2022, Tebbett et al. 2023b). For example, using different categories (e.g., use of different taxonomic ranks, including/excluding abiotic categories) or estimates of abundance (e.g., 2-dimensional projected cover vs. 3-dimensional estimates) has led to different results (Kornder et al. 2021, Reverter et al. 2021, Tebbett et al. 2023b). Global controversy on the definition of phase shifts (e.g., minimum threshold change, minimum persistence time of the communities involved) has resulted in the development of different terminologies, further challenging the comparative study of phase shifts (Hughes et al. 2013, Pulsford et al. 2016). The frequency of coral benthic shifts has also been long discussed. Some studies have suggested that coral reef benthic shifts are rare at the global scale, but these studies focused only on coral-macroalgae changes, overlooking other types of community changes (Bruno et al. 2009, Souter et al. 2021, Chan et al. 2023). In a recent study, Tebbett et al. (2023a) showed that changes in dominance from hard corals to macroalgae were relatively rare, while changes from hard corals to algal turfs were much more frequent (>9-fold). Data on the frequency and persistence of shifts from hard corals to other invertebrates are scarce, which could arise from their rare nature or from having been previously overlooked in the scientific literature (Reverter et al. 2022).

A number of recent systematic reviews and meta-analyses have started to highlight research gaps and provide guidelines to homogenise the study of coral reef shifts (Reverter et al. 2022, Tebbett et al. 2023a,b). For example, a re-definition of phase shifts has been recently proposed by Crisp et al. (2022) with the aim to unify the research field, defining phase shifts as a transition between the most dominant benthic groups, whose dominance persists at least 3 years before and after the shift. Producing comparative data is vital to advance the knowledge on coral reef phase

Figure 1 Examples of different coral reef benthic alternative communities. (A) Algal (*Crysocystis fragilis*)-dominated reef in Tumon Bay, Guam; (B) sponge (*Pseudoceratina* sp.)-dominated reef in Koh Phangan, Thailand; (C) soft coral (*Capnella imbricata*)-dominated reef in Bangka island, Indonesia; (D) zoanthid (*Zoanthus* sp.)-dominated reef in Kimbe island, Papua New Guinea; (E) anemone (*Heteractis magnifica*)-dominated reef in Moorea, French Polynesia; (F) corallimorpharian-dominated reef in Zanzibar, Tanzania; (G) cyanobacteria (*Lyngbya bouillionii*)-dominated reef in Cocos Lagoon, Guam; and (H) ascidian (Didemnidae)-dominated reef, Kimbe island, Papua New Guinea. (© M. Reverter and P. Schupp.)

shifts, yet this is contingent on standardising the study of coral reef shifts. In this chapter, we have used an extended version of a previously proposed framework by Bestelmeyer et al. (2011) to study ecological changes, with the aim of providing a roadmap to studying coral reef shifts. We have put particular emphasis on (1) the review of the drivers that lead to the different coral reef alternative communities and the implications of these benthic changes on (2) coral reef functioning and (3) the provision of ecosystem services, since these three elements have received little attention (Newman et al. 2015, McWilliam et al. 2020), especially in non-algal shifts. Finally, we also highlight research gaps in the study of different coral reef community shifts and identify timely research directions.

A roadmap to study coral reef shifts

Ecological changes or shifts are extremely diverse globally (e.g., different timescales and magnitudes of change, variable persistence and different alternative communities) and challenging to study (Pulsford et al. 2016, Connell et al. 2017). This has led to a large variability in the terminology and methods used to investigate coral reef community changes, making studies hard to compare (Bruno et al. 2009, Dudgeon et al. 2010, van de Leemput et al. 2016). Here, in an attempt to provide a roadmap to study coral reef shifts that allows comparison and knowledge transfer between the different types of coral reef shifts, we adapted and extended a previously proposed common framework to study ecological changes (Bestelmeyer et al. 2011). We systematically searched all peer-reviewed journals and PhD theses in Web of Science (up to 15 December 2020) that investigated coral reef community changes or shifts following the PRISMA (Preferred Reporting Items for Systematic Reviews and Meta-Analyses) guidelines (Moher et al. 2009). The following keywords were used: (phase OR alternate OR abrupt) AND (change OR shifts OR regime OR state) AND (coral reefs). This yielded 1724 articles, of which 523 were selected for further analyses (refer to Reverter et al. 2022 for details on inclusion criteria, full list provided in Supplementary Material 1). For each type of alternative communities identified, we screened the scientific literature retained from the systematic research and compiled information on five comparable elements: (1) Contextual information, which documents the characteristics of the sites studied; (2) Drivers and triggers of community change, which are the chronic (drivers) or abrupt (triggers) conditions that drive a change in the biological response (i.e., community composition); (3) Feedback mechanisms, which can happen when the biological responses interact between themselves or with the drivers/triggers and reinforce the persistence of alternative organisms; (4) Changes in ecosystem functions; and (5) Effects on ecosystem services (Figure 2).

Contextual information

Environmental physicochemical characteristics are amongst the most important forces in coral reef structuring. At the global scale, latitudinal gradients (e.g., temperature, light and aragonite saturation) determine coral reef geographic ranges (Kleypas et al. 1999), and past biogeographical history and tectonic events underpin current macroecological diversity gradients (Renema et al. 2008, Keith et al. 2013, Pellissier et al. 2014). However, fine spatial scale processes such as wave exposure, sediment composition, depth, temperature and oceanic productivity also play major roles in coral reef zonation and community composition (Williams et al. 2013, Robinson et al. 2018, Otaño-Cruz et al. 2019, McClanahan 2020). Therefore, understanding how physicochemical characteristics drive different community compositions and how they might interact with other local and global stressors is key to the study of coral reef shifts.

Temperature has been acknowledged as the most important physicochemical parameter shaping coral reef ecosystems. However, spatio-temporal thermal regimes (temperature value and variability) are often overlooked when studying coral reef community composition and dynamics. McClanahan (2020) showed that temperature displayed a strong influence on coral assembly along the East African coast, with competitive coral taxa being more abundant in thermally stable sites. Historical temperature variability has also been associated with heat and bleaching-resistant corals (Carilli et al. 2012, Safaie et al. 2018) and has been suggested to explain changes in coral mortality

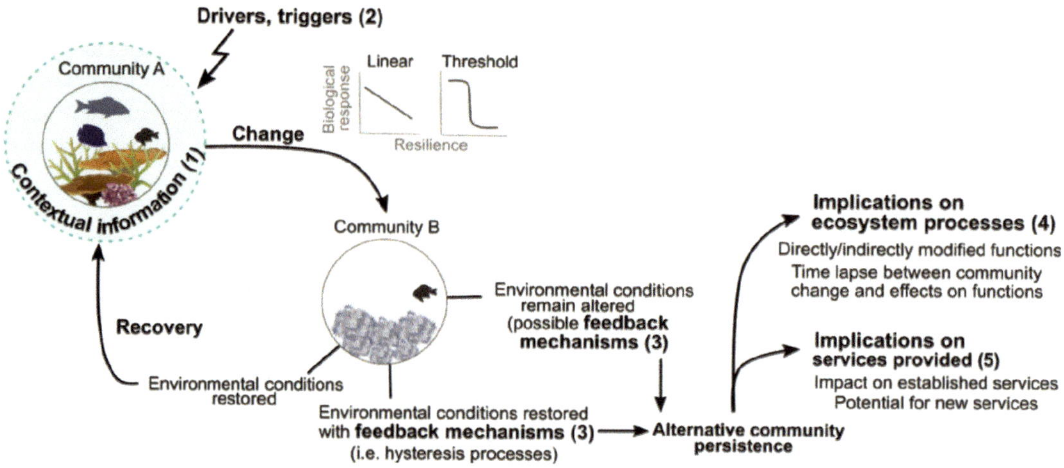

Figure 2 Schema illustrating ecological changes and the five elements (numbered from 1 to 5) that we propose as a roadmap to understand coral reef benthic changes. Briefly, when a given community (e.g., community A, which exists in a given set of environmental conditions, i.e., contextual information (1) is exposed to external drivers and/or triggers (2), the community changes its composition towards an alternative community (i.e., community B). The likelihood and direction of change will depend both on the initial community, the contextual information (1), and the drivers/triggers (2). If the environmental conditions are restored (i.e., back to pre-disturbance levels), the alternative community might recover and shift back to community A, unless feedback mechanisms (3) have been established, which will result in the persistence of the alternative community. If the alternative community persists in time, then it is important to understand what ecosystem processes (4) and services (5) will be affected.

during heatwaves (Ateweberhan & McClanahan 2010). There is evidence that some alternative organisms may be more tolerant to high temperatures than hard corals and can increase after coral bleaching events (Aronson et al. 2002, Tebbett et al. 2019). However, the effect that temperature variability and historical thermal regimes have on alternative organism physiology and abundance is poorly characterised. As thermal regimes keep changing and becoming more extreme, it is key to better understand the role that historical thermal regimes play in the likelihood of different coral reef benthic shifts. Wave exposure has been shown to increase the likelihood and magnitude of coral-algal phase shifts (Roff et al. 2015, Arif et al. 2022). Exposed reefs are generally inhabited by a higher proportion of algae than sheltered reefs, as calcifying species, especially branching morphologies such as Acroporidae corals, can suffer colony dislodgement or abrasion under high energy conditions (Madin & Connolly 2006, Williams et al. 2013, Sangil & Guzman 2016). Although under optimal conditions, many exposed reefs are still dominated by hard corals, they might be more susceptible to coral-algal community shifts. This was evidenced by Roff et al. (2015), who showed that after an acute disturbance (i.e., a cyclone), only exposed reefs in Palau suffered an immediate coral-algal shift. Wave exposure was also associated with a decrease in herbivory pressure, causing subtle community composition changes (i.e., higher turf and fleshy algae abundance in micro-crevices) that in turn lowered coral recruitment (Doropoulos et al. 2017), highlighting the importance of considering the effect of physical conditions in modulating community recovery and the establishment of feedback mechanisms.

Depth, accompanying light levels and food availability, also drives strong zonation in shallow tropical coral reefs. Generally, the abundance of heterotrophic and filter-feeding organisms such as soft corals and sponges increases with depth, whereas scleractinian corals, CCA and fleshy macroalgae are more abundant in shallow waters (Williams et al. 2013). For example, research from the Red Sea showed that deeper reefs (16 m vs 7 m and 3 m) were largely dominated by xeniid soft corals, with the most exposed deep reef presenting the largest cover in soft corals (Tilot et al. 2008). Otaño-Cruz et al. (2019) showed that even small depth differences (1–2 m vs 2–4 m) significantly explained differences in coral and macroalgal abundances. Previous studies have found that depth could help predict the likelihood of coral-macroalgal climate-induced shifts, with deeper reefs being more resilient (Graham et al. 2015, Arif et al. 2022). Biotic community attributes such as structural complexity, and fish, coral and algal densities have also been shown to affect the likelihood and magnitude of coral-algal phase shifts (Graham et al. 2015, Arif et al. 2022). Similarly, other social environmental characteristics such as distance to coast, rivers, crops and nearest population may have important effects on coral reef resilience and community composition (Darling et al. 2019). Altogether, this highlights the importance of adequately characterising a variety of contextual reef information (e.g., geological, physicochemical, biotic and socioeconomic) to better understand coral reef changes. To date, most studies investigating coral reef benthic changes have overlooked the importance of studying and reporting the site's contextual information. Furthermore, the few studies that investigated the role of different ecosystem attributes on the likelihood, magnitude and persistence of coral reef shifts, focused on coral-macroalgal shifts (Graham et al. 2015, Roff et al. 2015, Arif et al. 2022). We argue here that all coral reef community changes, and especially those that are poorly understood (e.g., sponge, soft coral, ascidians, zoanthid and turf algae-dominated reefs), need to be studied in light of their environmental context, as only with this, robust global patterns of similar changes can emerge. To facilitate spatio-temporal comparisons, a standard framework for reporting reef contextual information that lists the parameters to be recorded and provides guidance on the methods to be used should also be established (Table 1). There are some large initiatives (e.g. the Allen Coral Atlas; Allen-Coral-Atlas 2022) that have started gathering contextual information on the world's reefs, which could serve as a baseline. One of the key questions that remains controversial is the use of biotic categories to characterise coral reef benthic communities and their changes. Scientific evidence has shown that using limited categories, and in particular reporting only hard coral and algae, severely limits the study of coral reef dynamics by overlooking many types of organisms (Reverter et al. 2021, Tebbett et al. 2023a). However, visual identification of coral reef benthic organisms to species, and sometimes genera or family level, is not possible. This remains one of the big challenges in studying and comparing global reef changes, which needs to be urgently addressed by the international coral reef community.

Drivers and triggers and their effects on biological responses

Ecological changes in stable communities are caused directly or indirectly by changes in the environmental conditions (i.e., drivers and triggers). Drivers are slow, generally low-impact environmental changes (e.g., eutrophication and overfishing). Drivers can induce slow changes in the biological responses (e.g. coral cover linearly decreasing with nutrient levels) (Hughes et al. 2013). But often, they progressively decrease the system's resilience without visible changes in the biological response, until a threshold or tipping point is reached, from which an abrupt ecological change happens (i.e., phase or regime shift) (Connell et al. 2017, Cooper et al. 2020). In contrast, triggers, are abrupt pulse disturbances (e.g., heatwave, storm, disease outbreak) that have a direct and large effect on biological responses (Roff et al. 2015, Bertani et al. 2016). Identifying drivers and triggers behind community changes is crucial to understand, prevent and manage ecological changes. However, different drivers and triggers often interact, and disentangling their contribution is challenging. Here, we review the different drivers and triggers leading to different coral reef alternative communities and we classify them in two major categories depending on the mechanism underpinning the biological responses observed: triggers leading to newly vacant substrate recolonisation (1) and drivers and triggers promoting competitive replacement (2) (Figure 3).

Table 1 Summary of Proposed Contextual Information That Needs to Be Reported When Studying Coral Reef Communities and Changes with References Providing Guidance on How to Measure the Different Attributes

Macroecological Biogeographic Characteristics	Geological Properties	Physico-chemical Properties	Socioeconomic Parameters	Biotic Attributes
• Biogeographic regions and provinces (Spalding et al. 2007)	• Reef sedimentology (% abiotic substrate: rock, rubble, sand and silt) • Geomorphic zone (Kennedy et al. 2021) • Distance to coast (m) • Distance to nearest river (m)	• Wave exposure • Depth • Slope • Nutrient content • Oxygen concentrations • Speed of water currents • Turbidity (Historical) Temperature	• Human gravity, which is an index relative to distance and size of populations to reefs (Cinner et al. 2018b). • Protection level and use of resources (fisheries, etc.)	• Structural complexity (Ferrari et al. 2016) • Initial benthic cover of different organisms (hard coral, soft coral, sponges, macroalgae, turf algae, etc.) using digital photo-quadrats. • Coral bleaching (% of coral cover) • Density and biomass of different fish functional entities (FEs), defined by six traits (body size, diet, period of activity, vertical position, gregariousness and mobility) (Mouillot et al. 2014, Reverter et al. 2021)

Newly vacant substrate recolonisation

Major pulse disturbances (e.g., severe thermal anomalies, storms, earthquakes, destructive fishing practices and crown-of-thorns (COTS) outbreaks) often have immediate (i.e., in the period of days to weeks) catastrophic effects on ecological communities (Done et al. 1991, De'ath et al. 2012, Hughes et al. 2018). Although they vary in their degree of specificity (e.g., hurricanes damage species depending on their morphology, earthquakes decimate all species and COTS will primarily feed on selected scleractinian coral species), they all tend to decimate a large number of species and specimens, providing extensive space available for recolonisation (Aronson et al. 2012, Kayal et al. 2012). These newly available substrates are often colonised by fast-growing organisms (e.g., turf algae, CCA, macroalgae, ascidians, soft corals), which can also benefit from nutrient subsidies derived from coral mortality (Vaughan et al. 2021), leading in many cases to alternative communities of variable persistence (Fox et al. 2003, Roff et al. 2015, Tebbett et al. 2019) (Figure 3).

For example, a mass bleaching event in the Western Caribbean caused hard coral mass mortality and the subsequent substrate colonisation by the bioeroding sponge *Chondrilla caribensis*, whose

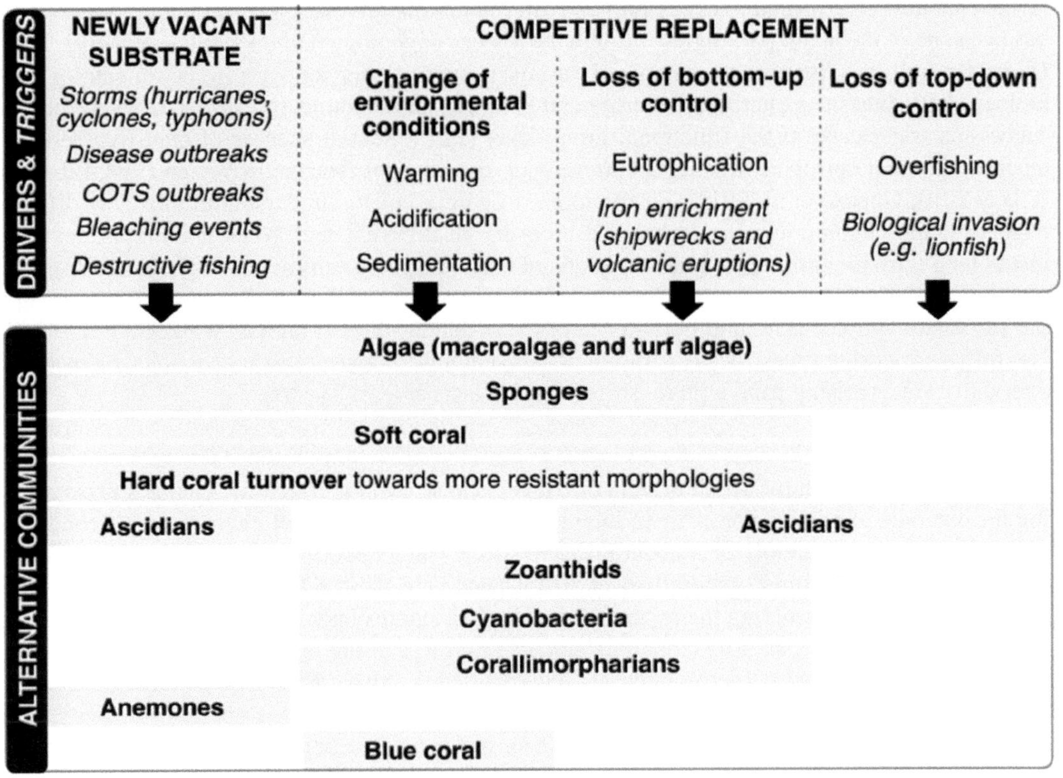

Figure 3 Schematic summary of the different drivers and triggers that have been identified in the literature to drive changes towards alternative coral reef benthic communities. The horizontal length of the alternative communities (i.e., grey bar) indicates the different types of drivers and triggers that have been shown to lead to each particular alternative community (e.g., all types of drivers and triggers have been shown to lead to coral-algal and coral-sponge shifts, but only a change in the environmental conditions has been suggested as the driver behind a shift towards blue coral-dominated communities). Different triggers under the category "Newly vacant substrate" may lead to the initial decrease of a wide range of organisms depending on their specificity (e.g., COTS will mostly affect scleractinian coral, while storms and destructive fishing such as blasting will lead to the death of many benthic organisms), but after this initial stage they lead to a change in the benthic community towards different alternative communities.

dominance lasted over 10 years (Aronson et al. 2002, 2012). Cyclones and COTS outbreaks have also led to mass coral mortalities and subsequent macroalgal proliferation in several sites of the Central and Eastern Indo-Pacific (e.g., Great Barrier Reef, Palau and Moorea in French Polynesia) (Done et al. 1991, Kayal et al. 2012, Roff et al. 2015). Blast, bomb or dynamite fishing is also one of the most destructive acute disturbances threatening coral reefs, especially in Southeast Asia and Eastern Africa (Hampton-Smith et al. 2021). Single acute blasts can kill immediately up to 70% of the live coral, creating large unstable rubble fields that can become colonised by a broad variety of fast-growing organisms such as xenid soft corals, corallimorpharians, sponges and a diversity of algae (Pet-Soede & Erdmann 1998, Fox et al. 2003, Fox & Caldwell 2006). Alternative benthic communities arising from blast fishing have been shown to be persistent over time (i.e., several decades sometimes), unless active coral restoration and rehabilitation are undertaken (Fox et al. 2019).

Although identifying triggers behind community changes may be evident due to the immediate and large damage caused, understanding the consequent community dynamics and succession and recovery patterns can be more challenging. Understanding species succession, both in newly formed habitats (i.e., primary succession) and after major disturbances (i.e., secondary succession), has been one of the major paradigms through the history of ecology (Odum 1969, Buma et al. 2017). Understanding pre-disturbance contextual variables, such as reef topography, sedimentology and biological attributes (e.g., herbivore density and biomass, larval connectivity) is vital to predict reef succession trajectories at the functional group-level (e.g., potential success of photosynthetic, filter-feeding or heterotrophic organisms) after major disturbances (Harmelin-Vivien 1994, Burkepile & Hay 2010, Gouezo et al. 2019). Furthermore, to gain a holistic understanding of how different triggers shape community reassembly and recovery in different reef types it is important to (1) install long-term monitoring programs that do not only study communities at single points (e.g., pre- and post- disturbance) and (2) to perform manipulative experiments that allow an understanding of the successional trajectories and persistence of the different states (Prach & Walker 2011). Finally, carefully considering timescales is paramount to disentangle between fast successional changes and alternative states arising from a phase shift (i.e., that lasts at least 3 years).

Competitive replacement

The second mechanism by which drivers or triggers cause coral reef benthic changes is by changing the outcome of benthic competitive interactions (i.e., the novel set of environmental conditions affects organisms differently or a novel invasive species alters ecosystem balance) (Figure 3). For example, high temperatures (generally over 30°C) negatively affect scleractinian corals by inducing oxidative stress and disrupting the relationship with their photosynthetic symbionts (i.e., bleaching) (Lesser 1997, Suggett et al. 2017), but they enhanced growth of the blue coral *Heliopora coerulea*, allowing it to overgrow scleractinian corals (Guzman et al. 2019). Similarly, elevated pCO_2 levels (≈800 µatm) decrease photosynthesis, growth, and calcification in some species of scleractinian corals (Reynaud et al. 2003, Mollica et al. 2018), but enhanced photosynthesis in the soft coral *Sarcophyton elegans,* which has been observed dominating the benthos in volcanic acidified environments (Inoue et al. 2013). Sedimentation and water turbidity arising from anthropogenic activities, such as dredging and coastal development, are also known to drive coral reef benthic changes towards communities dominated by algae (both macroalgae and low-lying algal species), sponges, or zoanthids (Segal & Castro 2011, Goatley & Bellwood 2013, Otaño-Cruz et al. 2019). High suspended sediment concentrations have detrimental effects on both coral health (i.e., by causing polyp smothering shading and tissue necrosis) and coral recruitment processes (i.e., decreased fertilisation success, larval settlement and recruit survival) (Riegl & Branch 1995, Nugues & Roberts 2003, Birrell et al. 2005, Moeller et al. 2017). Recent studies suggest that, however, some alternative species (e.g., sponges such as *Lamellodysidea herbacea*) can adapt and benefit from these conditions, giving them a competitive advantage over corals in these habitats (Biggerstaff et al. 2015, 2017a,b). Corallimorpharians are also renowned to thrive in adverse conditions (e.g., elevated temperatures, and nutrients) and have been reported to dominate highly turbid reefs close to urban locations in the Western Indo-Pacific (Muhando et al. 2002, Kuguru et al. 2004).

Two well-studied examples of competitive replacement in coral reefs are the loss of bottom-up and top-down control, which allow the proliferation of certain organisms by releasing them from demographic control (e.g., food scarcity or predation) (Smith et al. 2010a). Chronic nutrient enrichment in oligotrophic environments, such as coral reefs, induces the growth of primary producers (e.g., algae), which can lead to altered community compositions (Adam et al. 2021). In fact, eutrophication is one of the oldest and possibly best-studied drivers of coral-algal shifts (Smith et al. 1981, Hatcher & Larkum 1983). However, shifts towards communities dominated by cyanobacteria, soft corals, and zoanthids have also been reported at sites affected by nutrient enrichment (N, P) (Huang et al. 2011, Baum et al. 2016, Ford et al. 2017, Cruz et al. 2018). Several reports have also shown shifts towards communities dominated by cyanobacteria, algae, and corallimorpharians after micronutrient (e.g., Fe) enrichment derived from shipwrecks and volcanic eruptions (Work et al. 2008, Kelly et al. 2012, Schils 2012). Overfishing herbivores is the most studied example of how a decrease in algal top-down control can drive large community changes (McManus et al. 2000, Cheal et al. 2010, Dang et al. 2020). Although the effects of top-down control on other alternative organisms (i.e., non-algal) are poorly understood, few studies suggest that overfishing might also drive increases in sponge cover (Pawlik et al. 2013, Loh et al. 2015).

The introduction and proliferation of invasive species often have cascading effects on ecosystems, and as such they can also directly or indirectly promote coral benthic shifts by competitive replacement. For example, herbivore depletion in the Caribbean due to invasive lionfish has been shown to cause coral-algal shifts in Bahamian mesophotic reefs (Lesser & Slattery 2011, Kindinger & Albins 2017). More recently, an invasive xeniid soft coral native to the Indo-Pacific, *Unomia stolonifera*, steadily increased in abundance in the Caribbean, displacing native benthic species, and causing widespread benthic shifts (Ruiz Allais et al. 2014, Ruiz-Allais et al. 2021).

Disentangling species competitive replacement *in situ* from correlative abundance changes due to other factors (e.g., newly vacant substrate colonisation) is, however, extremely challenging. Complex multi-species and multi-stressor aquaria and field experiments are required to identify changes in benthic species competition. Furthermore, understanding the effect of different separate and combined abiotic and biotic variables on species' physiological performance and other phenotypic traits (e.g., production of chemical defences, microbiome and growth) is key to shed light onto the mechanisms underlying competitive replacement. During the last decade, an increasing number of *ex situ* and *in situ* experimental studies have improved our understanding on how different environmental characteristics affect coral-macroalgal interactions (Nugues & Roberts 2003, Olsen et al. 2014, Del Monaco et al. 2017). The mechanisms of competitive replacement between hard corals and other organisms (e.g., sponges, soft corals, ascidians, CCA), however, remain much less understood, highlighting the need to expand research efforts beyond coral-macroalgal interactions.

Co-occurrence and interaction of multiple drivers and triggers

Although individual drivers and triggers have been observed to cause coral reef benthic changes, they often co-occur and interact between themselves and the environmental characteristics, adding extra complexity in understanding coral reef community changes (Darling & Côté 2008, Muthukrishnan & Fong 2014, Fong et al. 2020a). This affects the capacity to determine disturbance thresholds from which certain community changes can be expected, which could facilitate the establishment of conservation or mitigation strategies. For example, a recent study suggested that increases in the abundance of the colonial ascidian *Didemnum* cf. *molle* following a bleaching event in Australia were probably related to decreased predation pressure (loss of top-down control), combined with increased space available for colonisation and increased nutrient sources (loss of bottom-up control) (Tebbett et al. 2019). Overfishing and eutrophication are in fact two ubiquitous coral-algal shift drivers that co-occur at many sites. However, despite intensive research efforts to quantify their relative importance in driving community changes, results are extremely variable, suggesting their contribution is highly context dependent (Rasher et al. 2012, Jessen et al. 2013, Adam et al. 2021). The importance of interacting local abiotic (wave exposure and water turbidity) and biotic (high local macroalgal and urchin abundances) conditions on coral mortality after a heatwave was recently demonstrated (Donovan et al. 2021). This further underlines the importance of coupling the study of drivers with environment characteristics (i.e., contextual information) when assessing conditions leading to ecological changes and the need to perform multi-stressor, multi-species experiments.

Feedback mechanisms

Feedback or response mechanisms are defined as processes where the interaction between biological response variables (e.g., coral and algal cover) or between response variables and drivers reinforces the dominance of alternative organisms (Bestelmeyer et al. 2011). Feedback mechanisms are established after the community change and are often responsible for the persistence of the alternative community, which sometimes can prevail even after the pre-disturbance environmental conditions have been restored (i.e., hysteresis concept within the alternate state theory) (Beisner et al. 2003). After reviewing all the known mechanisms by which alternative organisms reinforce their dominance in coral reefs, we classified feedback mechanisms into three categories: (1) mechanisms that have direct negative impacts on scleractinian corals, (2) mechanisms that directly positively affect alternative organisms and (3) co-reinforcement between alternative organisms (Figure 3).

Negative impacts on hard corals

The increase in abundance of alternative organisms can sometimes act as a mechanism driving further coral declines. Many alternative organisms (e.g., macroalgae, cyanobacteria, sponges, soft corals and zoanthids) possess strong chemical defences. Chemical defences can either directly damage corals upon contact and induce mortality (Sammarco et al. 1983, Morrow et al. 2011, Rasher et al. 2011, Cruz et al. 2016, Helber et al. 2018) or provide a competitive advantage to alternate organisms by protecting them against predators or pathogens (Rohde et al. 2015, Reverter et al. 2016, Helber et al. 2018). Extensive research has shown that macroalgal allelochemicals decrease coral fecundity, recruitment and settlement (Birrell et al. 2008, Evensen et al. 2019, Ritson-Williams et al. 2020), which has been suggested as a possible bottleneck limiting recovery of macroalgal-dominated reefs (Chong-Seng et al. 2014). Although coral settlement inhibition by other alternative organisms is considerably less understood, research suggests that cyanobacteria, soft corals and *Heliopora coerulea* can inhibit coral settlement (Maida et al. 1995, Atrigenio et al. 2017, Ritson-Williams et al. 2020). Brandt et al. (2019) also found that coral recruitment was higher in areas with no sponges, suggesting a potential role of sponges in coral settlement inhibition.

Contact with macroalgae and turf algae can also induce coral microbial dysbiosis both at the interaction interface and slightly beyond the contact site (Barott et al. 2012, Fong et al. 2020b). Macroalgae-associated bacteria can cause coral bleaching (Vieira et al. 2016) or trigger coral disease (Nugues et al. 2004). Recent research found that at the interaction front, turf algal chemical exudates feed coral-associated microbial communities resulting in higher microbial abundances, a turnover to heterotrophic bacteria, and decreases in oxygen concentrations, ultimately facilitating coral overgrowth by turf algae (Roach et al. 2020). The coral-killing sponge *Terpios hoshinota* also induced an increase in pathogen-like bacteria in *Porites* (Tang et al. 2011). However, whether coral dysbiosis is a mechanism enabling coral overgrowth by other alternate organisms remains largely unexplored.

Different benthic organisms, such as hard corals, soft corals and algae, also release compositionally different dissolved and particulate organic matter (DOM and POM) (Nelson et al. 2013, Nakajima et al. 2018, Wegley Kelly et al. 2022). Therefore, changes in their abundance can potentially initiate complex cascade processes that can negatively affect corals. This has been shown in algal-dominated reefs, where increases in algal biomass have resulted in changed ambient DOM profiles and bacterioplankton communities dominated by heterotrophic and coral-pathogenic bacteria, which has been referred as the "Dissolved organic carbon, Disease, Algae, Microorganisms" (DDAM) negative feedback loop (Barott & Rohwer 2012, Nelson et al. 2013, Haas et al. 2016). Recent research showed that CO_2 enrichment stimulates algal DOM release, which could further exacerbate algal dominance under near-future environmental conditions (Diaz-Pulido & Barrón 2020). Other organisms, such as zoanthids, also modify the microbial communities in the surrounding water, but the community-level impacts remain unknown (Walsh et al. 2017).

In summary, benthic organisms possess a wide array of strategies that can have direct (e.g., contact damage via allelopathy or microbial dysbiosis) or indirect (e.g., modification of environmental chemistry and microbial communities) deleterious effects on corals (Figure 4). While the direct effect of many organisms on corals (macroalgae, turf algae, sponges, soft corals) has been studied to some extent (Maida et al. 2002, Rasher et al. 2011, Helber et al. 2018, Roach et al. 2020), the indirect negative effects of non-benthic algal components remain virtually unknown. As in the

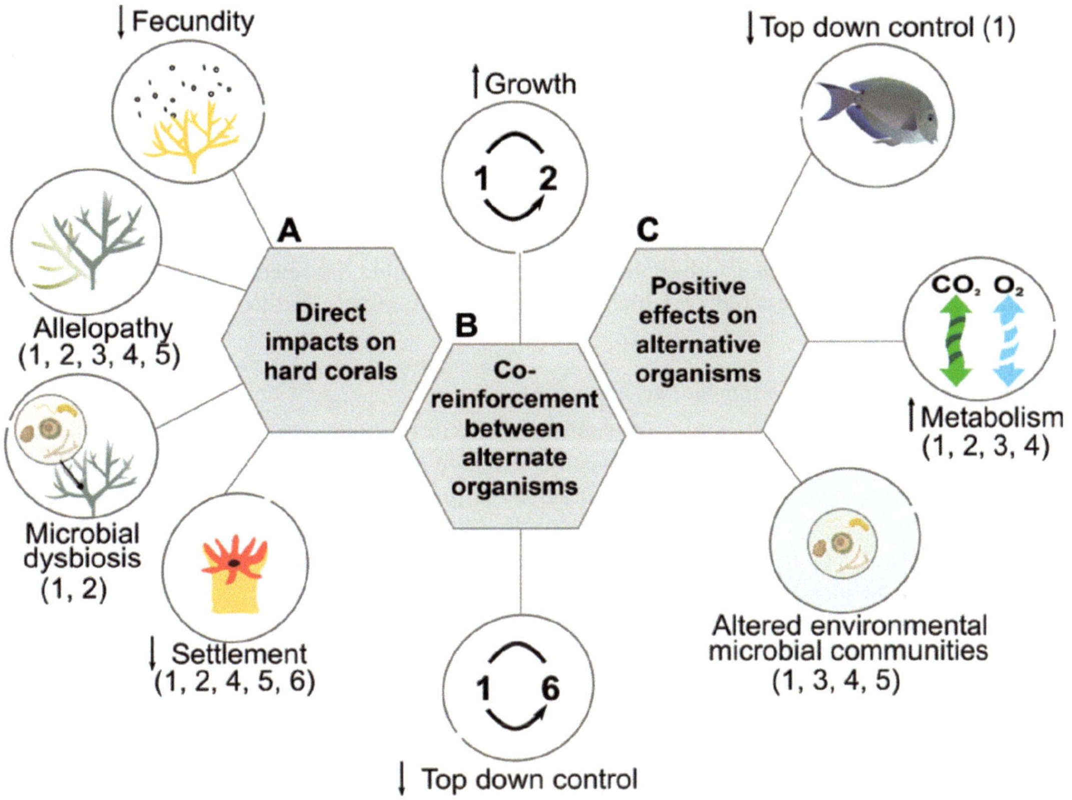

Figure 4 Schema of the three different types of feedback mechanisms that alternative benthic coral reef species may use to enhance their dominance and persistence: (A) Direct impacts on hard corals, (B) Co-reinforcement between alternate organisms and (C) Positive effects on alternative organisms. Numbers indicate the type of alternative organisms (1: Algae, 2: Sponges, 3: Zoanthids, 4: Soft corals, 5: Cyanobacteria, 6: Blue coral) for which each feedback mechanism has been described to date in the scientific literature.

evaluation of competitive replacement mechanisms, more experimental science is needed to elucidate how traits and processes from alternative species affect scleractinian corals under a wide range of environmental conditions.

Positive effects on alternative organisms

The second mechanism by which alternative organisms can reinforce their dominance and persistence is by displaying enhanced fitness under novel conditions. Increases in pCO_2 have been shown to mitigate temperature stress in phototrophic sponges (Bennett et al. 2017) and increase bioerosion rates in the boring sponge *Cliona caribbaea* (Webb et al. 2017). Del Monaco et al. (2017) also showed that high pCO_2 increased the allelopathic capacity of the algae *Canistrocarpus cervicornis*. A recent study suggested that nutrient enrichment could increase the resistance of the fast-growing soft coral *Xenia umbellata* to warming, which might facilitate and increase the persistence of soft coral-dominated reefs (Vollstedt et al. 2020). Similarly, scleractinian corals with higher heterotrophic feeding capacity may be able to withstand warmer temperatures, potentially driving community turnover between scleractinian corals (Grottoli et al. 2006, Conti-Jerpe et al. 2020). Finally, some reports indicate that increases in macroalgae also reinforce their dominance by positive feedbacks that suppress herbivory through several mechanisms, such as fish avoidance of high-density algal beds (Hoey & Bellwood 2011, Dell et al. 2016), and by providing refuge to palatable algal recruits (Davis 2018).

Although some evidence suggests that some alternative organisms may benefit from modified environmental conditions (e.g., turbid waters, high nutrient levels and warmer temperatures), this research field is still in its infancy (Figure 4). Coral reef research has historically been focused on the main ecosystem engineers of such ecosystems, scleractinian corals, with only a small fraction of research conducted on other benthic reef organisms (Lesser 2004, Burt 2013). As coral reefs restructure in the light of global change, it is increasingly urgent to expand research efforts to understand the biology and ecophysiology of other reef benthic organisms. This information will be vital for ongoing discussions on which organisms may dominate Anthropocene reefs (Lesser & Slattery 2020).

Co-reinforcement between alternative organisms

The third and most misunderstood mechanism that can reinforce the dominance of alternative organisms is the positive interaction between alternative organisms (Figure 4). For example, epiphytic cyanobacteria lowered the grazing pressure of macroalgae, ensuring its persistence after a nutrient pulse associated with la Niña (Smith et al. 2010b). Previous research has also showed the positive effects of sponge contact and abundance on algae by providing generalist fish an alternate food source (Gonzalez-Rivero et al. 2012) and by stimulating DOM recycling through the sponge loop (de Goeij et al. 2013), which could further promote both algal and sponge growth (Pawlik et al. 2016, de Goeij et al. 2017, Rix et al. 2017). This enhanced processing of DOM by sponges has been hypothesised as a "vicious circle", which could explain the lower resilience of Caribbean corals compared to Indo-Pacific reefs (Pawlik et al. 2016). Increased processing of DOM by sponges in algal-reefs may also lead to more turbid waters due to enhanced suspended detritus or changed light conditions, further inducing coral stress (Mumby & Steneck 2018). A recent study reported that algal contact may also confer a competitive advantage to the sponge *Desmapsamma anchorata*, which overgrew the coral faster when in contact with algae (Olinger et al. 2020). Although these observations remain limited to few taxa (sponges and algae mostly), further research is required to understand the positive feedback mechanisms between non-hard coral organisms that could exacerbate coral benthic community changes.

Changes in ecosystem functions

Ecosystem functions or processes, defined as the movement or storage of energy or material within an ecosystem, underpin ecosystem resilience and the provisioning of ecosystem services (Bellwood et al. 2019). As ecosystems transition to novel configurations (e.g., alternative communities), it is paramount to understand what functions will be affected and if new functions will emerge (Brandl et al. 2019, Williams & Graham 2019). Here, we provide an overview of the main ecosystem functions that have been observed to be affected by a shift towards alternative communities in tropical shallow coral reefs (Figure 4). As the functions of reefs dominated by non-algal alternative species have been very rarely studied, we use previous knowledge on coral-macroalgal shifts, which are by far the most studied community shifts on coral reefs (Done 1992, McManus & Polsenberg 2004, Reverter et al. 2022), to postulate hypotheses on reef functions that might be affected in other alternative communities.

Calcification, reef accretion and loss of structural complexity

A major consequence of changes towards communities dominated by non-calcifying alternative organisms and non-framework building corals is the decline in carbonate production rates (Perry & Alvarez-Filip 2019) (Figure 5). Positive carbon budgets are required for reef accretion, which is vital to withstand sea-level rise (Perry et al. 2018), and for maintaining structural complexity, which

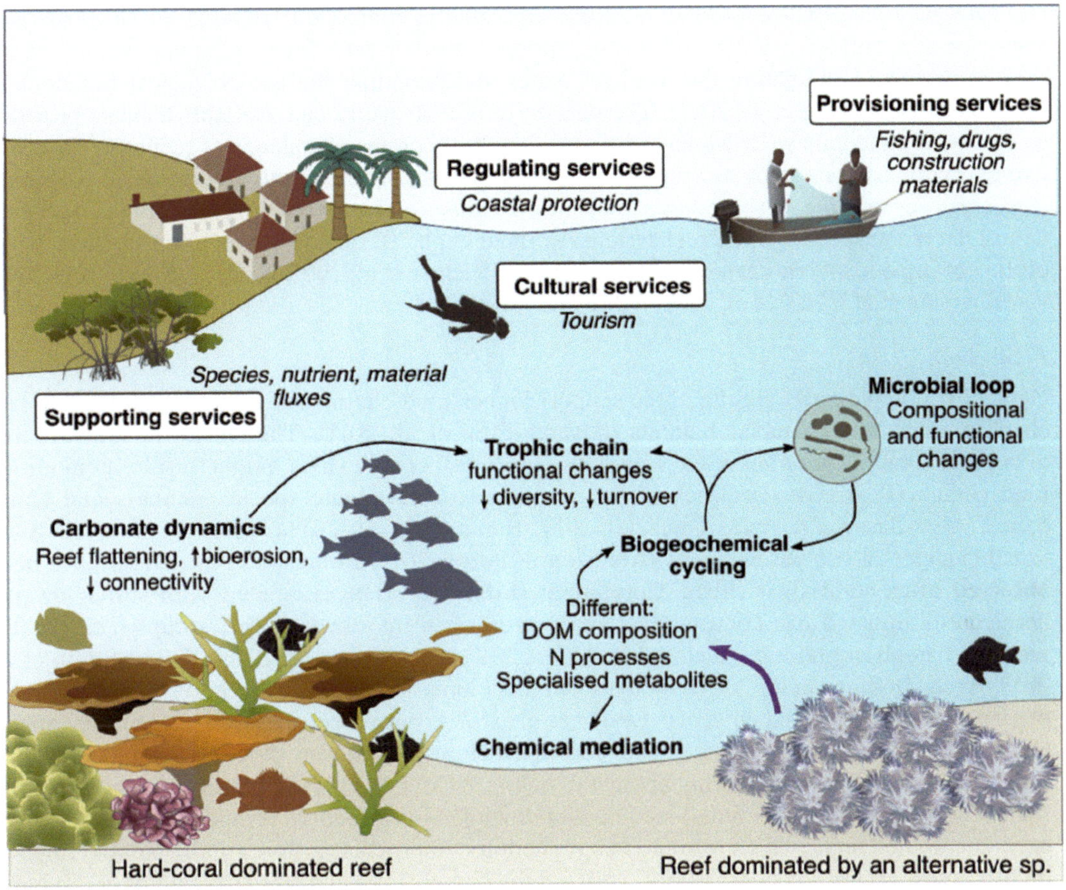

Figure 5 Main ecosystem processes (i.e., carbonate dynamics, trophic chain, biogeochemical cycling, microbial loop and chemical mediation) and ecosystem services (i.e., provisioning, regulating, cultural and supporting services) often affected by coral reef benthic changes towards alternative communities (i.e., not dominated by hard corals).

is key to sustain fisheries (Graham 2014) and protect coastlines (Harris et al. 2018). Both the decline in coral cover and changes towards scleractinian species with low calcification rates may be shifting reefs towards neutral or negative carbonate budgets (Januchowski-Hartley et al. 2017, Cornwall et al. 2021, González-Barrios et al. 2021), resulting in progressive structural flattening and substrate homogenisation (Perry & Alvarez-Filip 2019). However, the effect of benthic community changes on reef carbonate budgets is still highly misunderstood (Pisapia et al. 2019). Courtney et al. (2020) and Gischler et al. (2023) showed that community turnover between fast-growing branching corals (e.g., Acroporidae) and weedy corals (Pocilloporidae, Agariicidae) decreased reef-scale calcification and accretion. However, other studies suggest that massive and corymbose coral colonies (e.g., Poritidae, Pocilloporidae) may be able to maintain coral reef growth and positive carbon budgets in some reefs (Ryan et al. 2019, Kench et al. 2022). Recent evidence showed that complex growth forms that break easily (e.g., branching corals) disintegrate within few years, calling for a re-evaluation of the role that different corals play on reef calcification (Morais et al. 2022). Non-coral calcifiers such as crustose coralline algae (CCA) are also important reef builders. Although calcifying algae (such as *Halimeda* and *Galaxaura*) are not considered reef builders, they also contribute to coral carbonate budgets. However, both (CCA and calcifying algae) are often not included in reef carbonate budgets (Cornwall et al. 2023, McLaughlin et al. 2023). In addition, shifts towards some non-reef builders, such as bioeroding sponges, further contribute to the degradation of the reef framework, highlighting the need for better understanding the geo-ecological functions of these reefs (Schönberg et al. 2017). The decline in reef structure and available habitats provided by reef-building corals will also have tremendous effects on the abundance of reef organisms that live in cryptic habitats, such as crevices, cavities and caves. These highly understudied coelobites (e.g., crustose coralline algae, sponges, tunicates, ascidians and polychaetes) can make up more than half of the biomass of an entire reef system (Kornder et al. 2021) and have an important role in the cycling of organic and inorganic nutrients on reefs (Richter et al. 2001, de Goeij & van Duyl 2007, van Hoytema et al. 2023).

Food chain

Generally, complex reef architectures support longer food chains and higher mean trophic levels than structurally simpler habitats (Alvarez-Filip et al. 2011). Therefore, changes towards alternative benthic communities with low structural complexities might result in modified food chains (Figure 4). Declines in fish and macro-invertebrate species richness and abundance (Alva-Basurto & Arias-González 2014, Hempson et al. 2018a) as well as overall functional changes in fish assemblages (Alevizon & Porter 2014, Robinson et al. 2019a) have been observed after coral-algal shifts. Coral-algal shifts are often associated with shifts towards herbivore-dominated fish communities and decreases in intermediate fish groups (i.e., corallivores and small predators) (Kuempel & Altieri 2017, Hempson et al. 2018b, Stuart-Smith et al. 2018, Arias-Godínez et al. 2019). Increases in fish biomass (mostly due to herbivores) have been observed in algal-dominated reefs (Lester et al. 2020), but recent research showed that in the Great Barrier Reef, this was paired with decreased biomass turnover, shedding light on how benthic changes can affect underlying energetic pathways (Morais et al. 2020). Fish from algal-dominated reefs were also observed to feed further down the food chain in comparison with fish from coral-dominated reefs, which resulted in a reduction in their energy stores with potential impacts on fish fitness (Hempson et al. 2018b). Shifts in the composition and size structure of herbivorous fish assemblages following shifts from coral to algal-dominated reefs can also have profound impacts on bottom-up and top-down processes reinforcing algal dominance (Munsterman et al. 2021). Increases in small-bodied parrotfish and large-bodied detritivores in algal-dominated reefs resulted in higher removal rates of epibionts and detritus as well as in a decline of the N:P ratios due to higher P excretion rates of detritivores, which might negatively impact coral health (Allgeier et al. 2014, Munsterman et al. 2021).

A decrease in fish richness and functional changes in the fish community (i.e., increases in invertivore fishes) was also observed after coral-zoanthid shifts (Cruz et al. 2015). The effects that other benthic alternative communities (e.g., soft coral or sponge-dominated communities) have on the reef trophic chain remain largely unknown. However, similar to algal and zoanthid-dominated communities, they likely lead to significant functional fish and invertebrate changes that could affect reef resilience, stressing the urgency of addressing these knowledge gaps.

Despite numerous evidence suggesting shifts towards algal-dominated communities have deleterious effects on invertebrate and fish communities, some research indicates that reefs dominated by canopy-forming macroalgae can still provide important functions (Fulton et al. 2019). For example, algae with complex canopies such as *Sargassum* provide key habitats for a diverse community of epifauna and juvenile fishes, which in turn can support higher trophic levels (i.e., invertivore/carnivorous species) (Tano et al. 2016, 2017, Wenger et al. 2018). While these functions might not be equivalent to those furnished by highly structurally complex and diverse hard-coral reefs, it is important to identify potentially novel functions that might support, albeit differently, reef functioning in alternative communities.

Biogeochemistry, DOM- and N-cycling

In recent decades, the role of DOM has been identified as a key component of coral reef biogeochemistry and consequently of coral reef benthic changes (Kelly et al. 2022). DOM is one of the largest sources of organic matter on Earth (Hansell & Carlson 2001) and on reefs and is mainly produced by primary producers, like corals and algae (Wild et al. 2004, Haas et al. 2011). The turnover of DOM in the oceans is mostly mediated by microbes through the well-established microbial loop (Azam et al. 1983) and, on coral reefs, by sponges, through the so-called sponge loop (de Goeij et al. 2013). Through the processing of DOM, microbes and sponges provide other reef inhabitants, for which DOM is less available as food source, with relatively large amounts of new organic resources, thereby bringing life to the reef in the oligotrophic, desert-like reef waters. However, when benthic organisms' biomass changes (e.g., increasing algae), these biological processes (production and recycling of DOM) may catalyse feedback loops and alter overall ecosystem functioning (Haas et al. 2016, Pawlik et al. 2016, de Goeij et al. 2017). For example, the negative feedback loop DDAM (see section 'Negative Impacts on Hard Coral' for details) and the proliferation of DOM-producing benthic cyanobacterial mats are suggested to lead to the "microbialisation" (i.e., increases in microbial biomass, mostly copiotrophs) of coral reefs worldwide (Haas et al. 2016, Webb et al. 2021). The enhanced processing of DOM by sponges, leading to increased production of suspended detritus (Alexander et al. 2014, Hildebrand et al. 2022, Kornder et al. 2022) and inorganic nutrients, has been hypothesised to cause a "vicious circle", which could explain the lower resilience of Caribbean corals compared to Indo-Pacific reefs (Pawlik et al. 2016). Increased production of suspended detritus may also lead to more turbid waters and different light conditions, further inducing coral stress (Mumby & Steneck 2018). On the other hand, sponges, as well as other suspension- or filter-feeders, can decrease and modify ambient bacterioplankton communities, thereby potentially mitigating the effects towards the microbialisation of reefs (due to their enormous filtration capacity and very efficient feeding on bacterioplankton) (Reiswig 1971, Lesser 2006, de Goeij et al. 2017, Hildebrand et al. 2022). However, it remains generally unknown how these relatively recently discovered pathways and processes drive shifts in coral reef communities.

Nitrogen (N) cycling is also a key function in oligotrophic coral reefs, whose productivity is severely limited by the low availability of N compounds (e.g., ammonia and nitrate) in oligotrophic reef waters (O'Neil 2008, Rädecker et al. 2015). Specifically, the balance between N_2 fixation and denitrification, which provides stable quantities of available N, is vital in sustaining ecosystem functioning (Capone et al. 1992, Cardini et al. 2016). Since benthic organisms contribute differently to the uptake and release of N (Bednarz et al. 2015, Rix et al. 2015), benthic community changes might alter ecosystem N-cycling. For example, El-Khaled et al. (2021) showed that algal-dominated communities fixated more N_2, than hard-coral-dominated communities, while keeping the same denitrification capacity, resulting in higher nitrogen content, and promoting primary production. Interestingly, the authors also showed that in communities dominated by xenid soft corals, nitrogen fixation remained stable in comparison to hard-coral communities, but denitrification increased, indicating shifts towards certain soft coral species could exacerbate N limitation in coral reefs (El-Khaled et al. 2021). While the effect of other alternative organisms (e.g., sponges, zoanthids and ascidians) on N cycling remains limited, contrasting effects of algae and soft-coral-dominated communities on N bioavailability underline the importance of investigating how different benthic organisms might affect key coral reef functions.

Effects on ecosystem services

Reef ecosystem functioning is directly connected to the benefits or ecosystem services that coral reefs provide to humanity (Moberg & Folke 1999, Lau et al. 2019). Changing communities and ecosystem functions due to global threats have been forecasted to have a huge impact on people relying on coral reef goods and services (Hoegh-Guldberg et al. 2019, Woodhead et al. 2019). The capacity for coral reefs to sustain ecosystem services over time depends on their geo-ecological functions, such as maintenance of architectural complexity and reef growth potential which most likely will be negatively impacted by alternate reef communities (Perry & Alvarez-Filip 2019). However, to date, specific research on how different alternative communities impact humanity and whether new services will emerge from these communities is scarce (Daw et al. 2016). Here, we review the different ecosystem services provided by coral reefs and discuss the small accumulating evidence on how coral reef benthic changes affect known ecosystem services (Figure 5).

Provisioning services

Provisioning services are products that are extracted from ecosystems, such as food, livelihood opportunities, biomaterials for construction, or novel marine-derived pharmaceutical and agrochemical products (Albert et al. 2015). Fisheries are one of the most important services provided by coral reefs, with small-scale fisheries providing subsistence food and livelihood for hundreds of millions of people (Hicks et al. 2014, Grafeld et al. 2017). While flattening of the reefs has previously been associated with pronounced declines in fisheries (Rogers et al. 2014, 2018), recent research has suggested that novel fish assemblages in alternative benthic communities (i.e., increases in herbivores in algal-dominated communities) might sustain fisheries catch, but could decrease fisheries value and stability (Ainsworth & Mumby 2015, Robinson et al. 2019b, Yan & Bellwood 2023). Another recent study showed that fish caught in macroalgal-dominated habitats were enriched in micronutrients (e.g., Fe and Zn), suggesting fishing on reefs dominated by certain alternative organisms, such as *Sargassum* macroalgae, could remain an important micronutrient source for local human populations (Robinson et al. 2022). The effect of other (i.e., non-algal) alternative communities on fisheries remains completely unexplored. However, if alternative benthic organisms alter fish and invertebrate communities (Cruz et al. 2015), it could be expected they also impact fisheries catch and value. The harvest of other coral-based products (e.g., sand, rubble, stone and coral limestone) for construction purposes and wild animals for the curio- and aquarium trade also constitutes an important income source in many coastal rural communities (Albert et al. 2015) that might be highly disturbed by changing benthic communities. Finally, another service that might be impacted by biodiversity decreases caused by alternative benthic communities is the ability of coral reefs to provide opportunities for the discovery of high-value bioactive products (i.e., antiviral, antimicrobial, antifungal and antitumour) that are vital in this era of emerging diseases and rising antimicrobial resistance (Adey 2000, Liu et al. 2019).

Some recent work suggests, however, that macroalgal-dominated reefs could provide alternative services such as harvesting of macroalgal biomass that could be used in a wide-range of products in the food, biofuel, cosmetic, pharmaceutical and agriculture industries (Zubia et al. 2014, Fulton et al. 2019). The study of novel ecosystem services by alternative reef communities warrants further research, and it is crucial to critically evaluate their long-term viability, overall value and capacity to provide local human communities with irreplaceable services such as food subsistence and coastal protection.

Regulating services

Regulating services are defined as the benefits provided by ecosystems that regulate natural phenomena. Low-lying coastal areas are at high risk from natural hazards that are forecast to increase with global change (e.g., floods, storms, waves and rising sea level) (Barbier 2015, Vitousek et al. 2017). The coral reef limestone framework provides a first line of defence against these events, dissipating about 97% of the energy that would otherwise hit the coast (Ferrario et al. 2014), protecting 5.3 million people (Burke & Spalding 2022) and reducing expected coastal damage by nearly US$2 billion annually in the US (Reguero et al. 2021). Calcifying organisms also underpin other processes such as the generation of sand and island elevation that are vital to withstand sea-level rise (Perry et al. 2015, Masselink et al. 2020). While the speed at which changes towards alternative communities will affect regulating services in coral reefs (i.e., rate at which the consolidated reef framework will disappear) is currently unknown, emerging research on reef flattening following coral loss (Alvarez-Filip et al. 2009) suggests many coral reef regulating services could be severely compromised as communities shift towards novel benthic configurations.

Cultural services

Cultural ecosystem services are cognitive or experiential benefits gained from the interaction between people and ecosystems. This includes the aesthetic value of coral reefs, which translates into tourism activities (Pert et al. 2020). Reef tourism is one of the most significant reef services, with a global annual value of US$36 billion (Spalding et al. 2017). While highly biodiverse hard coral-dominated reefs attract most of the tourism, research has shown that the relationship between biodiversity and aesthetic value is not linear (Tribot et al. 2019). For example, in the Great Barrier Reef, reef topography and fish abundance were related to aesthetic scores, but coral cover was not (Marshall et al. 2019). Furthermore, sometimes the value of reefs is more highly associated with emblematic species than with their coral cover (Vianna et al. 2012, Shideler & Pierce 2016). However, reefs with structurally complex corals inhabited by fish communities with brighter and more diverse colouration were shown to be of the highest aesthetic value to humans (Hemingson et al. 2022). Altogether this suggests that while benefits from tourism might be lost in alternative communities with extremely low biodiversity and structural complexity, other communities (e.g., dominated by macroalgae or soft coral) could, at least partly, sustain tourism activities. To date, most of the studies evaluating coral reef cultural services have focused on coral reef degradation (in terms of coral cover loss) but have not yet evaluated the tourist perception towards distinct benthic communities. To understand the full implication of benthic community changes, research needs to specifically evaluate the tourist value and local perception of the different alternative communities.

Supporting services

Finally, supporting services are services that underpin the provision of other services. Coral reef supporting services include functional and genetic diversity and habitat provisioning (Fisher et al. 2015). These supporting services not only underlie direct provisioning services (i.e., fisheries and drug discovery) but also support reef adjoining ecosystems (such as mangroves and seagrass beds) through the exchange of nutrient fluxes, materials and species (Moberg & Folke 1999, Gillis et al. 2014). As coral reef benthic communities change, many of these reciprocal facilitation processes between ecosystems (e.g., different ecosystems providing habitat for different life stages) will likely be modified, highlighting the need to investigate the landscape effects of coral reef algal shifts. Cascading effects of coral reef shifts can result in dangerous domino effects (one-way interactions) and hidden feedbacks (two-way interactions) that can potentially impair ecosystem functioning and services in other inter-connected ecosystems (Rocha et al. 2018). To date, research investigating the connectivity between systems and how it is affected by changing communities is largely lacking.

Conclusions

Coral reef shifts towards non-hard coral-dominated communities, especially non-algal-dominated communities, remain puzzling despite several decades of scientific research. Recent research has started to highlight knowledge gaps in the study of coral reef benthic changes and redefine terms (i.e., what constitutes a phase shift) in an attempt to homogenise and guide future research (Crisp et al. 2022, Reverter et al. 2022, Tebbett et al. 2023b). Here, we build on this emerging research body by using and extending a previously proposed common framework to study ecological changes (Bestelmeyer et al. 2011) with the aim to provide a roadmap for studying coral reef shifts. Global change is threatening the stability of many ecosystems, including coral reefs, and therefore, now more than ever, scientists need to combine efforts to produce comparative data that allow research gaps to be filled and provide guidelines for coral reef conservation and recovery.

The field of coral reef shifts has been suffering from heterogeneous methodology, and therefore, the establishment of a standardised framework allows to compare and exchange knowledge acquired by different research groups. We propose to build a global database of coral reef benthic changes using the following five elements: (1) contextual information, (2) drivers and triggers, (3) feedback mechanisms, (4) functional implications and (5) implications on ecosystem services. Studying these different elements will enable us to start drawing global patterns of occurrence, persistence and socioeconomic impacts of coral reef benthic changes. Such a database should include detailed biodiversity and abundance data for algal and non-algal competitors and affected scleractinian corals (i.e., biotic attributes) by ecoregions, along with relevant physicochemical (e.g., temperature, salinity, oxygen concentrations and nutrient levels), geological (e.g., reef sedimentology and geomorphic zone) and socio-environmental data (e.g., human gravity). Furthermore, where resources allow it, long-term coral reef monitoring programmes, following the previous guidelines, should be implemented, as a comprehensive understanding of coral reef phase shifts will only be attained if robust data are collected at a relevant timescale and across different ecoregions (e.g., at least 3 years before and 3 years after a phase shift occurs; Crisp et al. 2022).

One of the most important patterns that emerged by using this framework is the importance of biotic-abiotic interactions in determining the likelihood, magnitude, and direction of community changes (Gonzalez-Rivero et al. 2012). Similar drivers and triggers can lead to multiple alternative communities depending on the reef context (Roff et al. 2015, Donovan et al. 2021). Research is starting to elucidate the role of some environmental characteristics leading to coral-algal shifts (e.g., higher wave energy increases the likelihood of disturbed reefs to shift towards algal-dominated communities; Roff et al. 2015). However, how environmental conditions affect and maybe enhance other alternative organisms remains poorly studied. Data on the physiological performance of different alternative organisms (e.g., in the field or under different experimental environmental conditions) and on the competitiveness of alternate organisms and scleractinian corals (e.g., observed interactions in the field and data arising from multi-species, multi-stressor experiments) are urgently needed to understand the mechanisms underlying the proliferation of certain alternative organisms and their persistence (e.g., macroalgae, turf algae, sponges and soft corals) at different types of reefs. Significant research gaps in feedback mechanisms in non-algal alternative communities were also identified, highlighting that experimental data (e.g., manipulative field experiments) are crucial to uncover potential feedback mechanisms and phase shift persistence. A good understanding of existing feedback mechanisms that prevent community recovery can also be extremely useful in the design of targeted restoration measures aiming to facilitate coral recovery. For example, the removal of macroalgae and turf algae as a strategy to facilitate coral recruitment and juvenile survival has been proposed based on the knowledge of algal feedback mechanisms (Cetz-Navarro et al. 2015, Ceccarelli et al. 2018).

Our review also highlights the need to shift towards understanding microscale processes and functions, as they often underpin the persistence of alternative communities. Accumulating evidence shows that unseen players such as microbes, biogeochemical processes (e.g., DOM turnover) and chemical mediation are important drivers fuelling community changes and reinforcing the persistence of alternative organisms (Haas et al. 2016, Kelly et al. 2022). However, these processes remain virtually unknown in non-algal alternative communities. Addressing these knowledge shortfalls is key to increase the prediction power of coral reef benthic changes, as well as to understand their persistence.

Most studies on coral benthic changes have so far focused on their description and the identification of different conditions (the element "drivers and triggers") leading to these changes. However, understanding their broader socio-ecological relevance is paramount to fully evaluate their importance. Here, we highlight that changes from hard coral-dominated habitats to non-calcifying species will have ripple effects on coral reef functions and people relying on their services, which could extend to neighbouring ecosystems through interrupted nutrient, material and species fluxes (Rocha et al. 2018). Therefore, there is a strong need to invest in research that investigates the magnitude and speed at which the different ecosystem functions and services will be impacted by the different community changes. Only by filling these crucial knowledge gaps can appropriate actions be undertaken to build the adaptive capacity of communities relying on coral reefs and minimise impacts on their well-being (Cinner et al. 2018a).

Acknowledgements

This research was funded by an Alexander von Humboldt post-doctoral fellowship to M.R.

References

Adam, T.C., Burkepile, D.E., Holbrook, S.J., Carpenter, R.C., Claudet, J., Loiseau, C., Thiault, L., Brooks, A.J., Washburn, L. & Schmitt, R.J. 2021. Landscape-scale patterns of nutrient enrichment in a coral reef ecosystem: implications for coral to algae phase shifts. *Ecological Applications* **31**, e2227.

Adey, W. 2000. Coral reef ecosystems and human health: biodiversity counts! *Ecosystem Health* **6**, 227–236.

Ainsworth, C.H. & Mumby, P.J. 2015. Coral-algal phase shifts alter fish communities and reduce fisheries production. *Global Change Biology* **21**, 165–172.

Albert, J.A., Olds, A.D., Albert, S., Cruz-Trinidad, A. & Schwarz, A.-M. 2015. Reaping the reef: provisioning services from coral reefs in Solomon Islands. *Marine Policy* **62** 244–251.

Alevizon, W. & Porter, J. 2014. Coral loss and fish guild stability on a Caribbean coral reef: 1974–2000. *Environmental Biology of Fishes* **98**.

Alexander, B.E., Liebrand, K., Osinga, R., van der Geest, H.G., Admiraal, W., Cleutjens, J.P.M., Schutte, B., Verheyen, F., Ribes, M., van Loon, E. & de Goeij, J.M. 2014. Cell turnover and detritus production in marine sponges from tropical and temperate benthic ecosystems. *PLoS One* **9**, e109486.

Allen-Coral-Atlas. 2022. *Imagery, maps and monitoring of the world's tropical coral reefs.*

Allgeier, J.E., Layman, C.A., Mumby, P.J. & Rosemond, A.D. 2014. Consistent nutrient storage and supply mediated by diverse fish communities in coral reef ecosystems. *Global Change Biology* **20**, 2459–2472.

Alva-Basurto, J.C. & Arias-González, J.E. 2014. Modelling the effects of climate change on a Caribbean coral reef food web. *Ecological Modelling* **289**, 1–14.

Alvarez-Filip, L., Dulvy, N.K., Gill, J.A., Côté, I.M. & Watkinson, A.R. 2009. Flattening of Caribbean coral reefs: region-wide declines in architectural complexity. *Proceedings of the Royal Society B: Biological Sciences* **276**, 3019–3025.

Alvarez-Filip, L., Gill, J.A. & Dulvy, N.K. 2011. Complex reef architecture supports more small-bodied fishes and longer food chains on Caribbean reefs. *Ecosphere* **2**, art118.

Arias-Godínez, G., Jiménez, C., Gamboa, C., Cortés, J., Espinoza, M. & Alvarado, J.J. 2019. Spatial and temporal changes in reef fish assemblages on disturbed coral reefs, north Pacific coast of Costa Rica. *Marine Ecology* **40**, e12532.

Arif, S., Graham, N.A.J., Wilson, S. & MacNeil, M.A. 2022. Causal drivers of climate-mediated coral reef regime shifts. *Ecosphere* **13**, e3956.

Aronson, R., Precht, W., Toscano, M. & Koltes, K. 2002. The 1998 bleaching event and its aftermath on a coral reef in Belize. *Marine Biology* **141**, 435–447.

Aronson, R.B., Precht, W.F., Macintyre, I.G. & Toth, L.T. 2012. Catastrophe and the life span of coral reefs. *Ecology* **93**, 303–313.

Ateweberhan, M. & McClanahan, T.R. 2010. Relationship between historical sea-surface temperature variability and climate change-induced coral mortality in the western Indian Ocean. *Marine Pollution Bulletin* **60**, 964–970.

Atrigenio, M., Aliño, P. & Conaco, C. 2017. Influence of the blue coral *Heliopora coerulea* on scleractinian coral larval recruitment. *Journal of Marine Biology* 2017.

Azam, F., et al. 1983. The ecological role of water-column microbes in the sea. *Marine Ecology Progress Series* **10**, 257–263.

Barbier, E.B. 2015. Climate change impacts on rural poverty in low-elevation coastal zones. *Estuarine, Coastal and Shelf Science* **165**, A1–A13.

Barott, K.L., Rodriguez-Mueller, B., Youle, M., Marhaver, K.L., Vermeij, M.J., Smith, J.E. & Rohwer, F.L. 2012. Microbial to reef scale interactions between the reef-building coral Montastraea annularis and benthic algae. *Proceedings of the Royal Society B: Biological Sciences* **279**, 1655–1664.

Barott, K.L. & Rohwer, F.L. 2012. Unseen players shape benthic competition on coral reefs. *Trends in Microbiology* **20**, 621–628.

Baum, G., Januar, I., Ferse, S.C., Wild, C. & Kunzmann, A. 2016. Abundance and physiology of dominant soft corals linked to water quality in Jakarta Bay, Indonesia. *PeerJ* **4**, e2625.

Bednarz, V.N., Cardini, U., van Hoytema, N., Al-Rshaidat, M.M.D. & Wild, C. 2015. Seasonal variation in dinitrogen fixation and oxygen fluxes associated with two dominant zooxanthellate soft corals from the northern Red Sea. *Marine Ecology Progress Series* **519**, 141–152.

Beisner, B.E., Haydon, D.T. & Cuddington, K. 2003. Alternative stable states in ecology. *Frontiers in Ecology and the Environment* **1**, 376–382.

Bellwood, D.R., Streit, R.P., Brandl, S.J. & Tebbett, S.B. 2019. The meaning of the term 'function' in ecology: a coral reef perspective. *Functional Ecology* **33**, 948–961.

Bennett, H.M., Altenrath, C., Woods, L., Davy, S.K., Webster, N.S. & Bell, J.J. 2017. Interactive effects of temperature and pCO(2) on sponges: from the cradle to the grave. *Global Change Biology* **23**, 2031–2046.

Bertani, I., Primicerio, R. & Rossetti, G. 2016. Extreme climatic event triggers a lake regime shift that propagates across multiple trophic levels. *Ecosystems* **19**, 16–31.

Bestelmeyer, B.T., Ellison, A.M., Fraser, W.R., Gorman, K.B., Holbrook, S.J., Laney, C.M., Ohman, M.D., Peters, D.P.C., Pillsbury, F.C., Rassweiler, A., Schmitt, R.J. & Sharma, S. 2011. Analysis of abrupt transitions in ecological systems. *Ecosphere* **2**, art129.

Biggerstaff, A., Jompa, J. & Bell, J.J. 2017a. Increasing benthic dominance of the phototrophic sponge *Lamellodysidea herbacea* on a sedimented reef within the Coral Triangle. *Marine Biology* **164**, 220.

Biggerstaff, A., Smith, D.J., Jompa, J. & Bell, J.J. 2015. Photoacclimation supports environmental tolerance of a sponge to turbid low-light conditions. *Coral Reefs* **34**, 1049–1061.

Biggerstaff, A., Smith, D.J., Jompa, J. & Bell, J.J. 2017b. Metabolic responses of a phototrophic sponge to sedimentation supports transitions to sponge-dominated reefs. *Scientific Reports* **7**, 2725.

Birrell, C.L., McCook, L.J. & Willis, B.L. 2005. Effects of algal turfs and sediment on coral settlement. *Marine Pollution Bulletin* **51**, 408–414.

Birrell, C.L., McCook, L.J., Willis, B.L. & Harrington, L. 2008. Chemical effects of macroalgae on larval settlement of the broadcast spawning coral *Acropora millepora. Marine Ecology Progress Series* **362**, 129–137.

Brandl, S.J., Rasher, D.B., Côté, I.M., Casey, J.M., Darling, E.S., Lefcheck, J.S. & Duffy, J.E. 2019. Coral reef ecosystem functioning: eight core processes and the role of biodiversity. *Frontiers in Ecology and the Environment* **17**, 445–454.

Brandt, M.E., Olinger, L.K., Chaves-Fonnegra, A., Olson, J.B. & Gochfeld, D.J. 2019. Coral recruitment is impacted by the presence of a sponge community. *Marine Biology* **166**, 49.

Bruno, J.F., Sweatman, H., Precht, W.F., Selig, E.R. & Schutte, V.G. 2009. Assessing evidence of phase shifts from coral to macroalgal dominance on coral reefs. *Ecology* **90**, 1478–1484.

Buma, B., Bisbing, S., Krapek, J. & Wright, G. 2017. A foundation of ecology rediscovered: 100 years of succession on the William S. Cooper plots in Glacier Bay, Alaska. *Ecology* **98**, 1513–1523.

Burke, L. & Spalding, M. 2022. Shoreline protection by the world's coral reefs: mapping the benefits to people, assets, and infrastructure. *Marine Policy* **146**, 105311.

Burkepile, D.E. & Hay, M.E. 2010. Impact of herbivore identity on algal succession and coral growth on a Caribbean reef. *PLoS One* **5**, e8963.

Burt, J.A. 2013. The growth of coral reef science in the Gulf: a historical perspective. *Marine Pollution Bulletin* **72**, 289–301.

Cannon, S.E., Aram, E., Beiateuea, T., Kiareti, A., Peter, M. & Donner, S.D. 2021. Coral reefs in the Gilbert Islands of Kiribati: resistance, resilience, and recovery after more than a decade of multiple stressors. *PLoS One* **16**, e0255304.

Capone, D.G., Dunham, S.E., Horrigan, S.G. & Duguay, L. 1992. Microbial nitrogen transformations in unconsolidated marine sediment. *Marine Ecology-Progress Series* **80**, 75–88.

Cardini, U., Bednarz, V., van Hoytema, N., Rovere, A., Naumann, M., Al-Rshaidat, M. & Wild, C. 2016. Budget of primary production and dinitrogen fixation in a highly seasonal Red Sea coral reef. *Ecosystems* **19**.

Carilli, J., Donner, S.D. & Hartmann, A.C. 2012. Historical temperature variability affects coral response to heat stress. *PLoS One* **7**, e34418.

Ceccarelli, D.M., Loffler, Z., Bourne, D.G., Al Moajil-Cole, G.S., Boström-Einarsson, L., Evans-Illidge, E., Fabricius, K., Glasl, B., Marshall, P., McLeod, I., Read, M., Schaffelke, B., Smith, A.K., Jorda, G.T., Williamson, D.H. & Bay, L. 2018. Rehabilitation of coral reefs through removal of macroalgae: state of knowledge and considerations for management and implementation. *Restoration Ecology* **26**, 827–838.

Cetz-Navarro, N.P., Carpizo-Ituarte, E.J., Espinoza-Avalos, J. & Chee-Barragán, G. 2015. The effect of filamentous turf algal removal on the development of gametes of the coral Orbicella annularis. *PLoS One* **10**, e0117936.

Chan, Y.K.S., Affendi, Y.A., Ang, P.O., Baria-Rodriguez, M.V., Chen, C.A., Chui, A.P.Y., Giyanto, Glue, M., Huang, H., Kuo, C.Y., Kim, S.W., Lam, V.Y.Y., Lane, D.J.W., Lian, J.S., Lin, S.M.N.N., Lunn, Z., Nañola, C.L., Nguyen, V.L., Park, H.S., Suharsono, Sutthacheep, M., Vo, S.T., Vibol, O., Waheed, Z., Yamano, H., Yeemin, T., Yong, E., Kimura, T., Tun, K., Chou, L.M. & Huang, D. 2023. Decadal stability in coral cover could mask hidden changes on reefs in the East Asian Seas. *Communications Biology* **6**, 630.

Chaves-Fonnegra, A., Riegl, B., Zea, S., Lopez, J.V., Smith, T., Brandt, M. & Gilliam, D.S. 2018. Bleaching events regulate shifts from corals to excavating sponges in algae-dominated reefs. *Global Change Biology* **24**, 773–785.

Cheal, A.J., MacNeil, M.A., Cripps, E., Emslie, M.J., Jonker, M., Schaffelke, B. & Sweatman, H. 2010. Coral-macroalgal phase shifts or reef resilience: links with diversity and functional roles of herbivorous fishes on the Great Barrier Reef. *Coral Reefs* **29**, 1005–1015.

Chong-Seng, K., Graham, N. & Pratchett, M. 2014. Bottlenecks to coral recovery in Seychelles. *Coral Reefs* **33**, 449–461.

Cinner, J.E., Adger, W.N., Allison, E.H., Barnes, M.L., Brown, K., Cohen, P.J., Gelcich, S., Hicks, C.C., Hughes, T.P., Lau, J., Marshall, N.A. & Morrison, T.H. 2018a. Building adaptive capacity to climate change in tropical coastal communities. *Nature Climate Change* **8**, 117–123.

Cinner, J.E., Maire, E., Huchery, C., MacNeil, M.A., Graham, N.A.J., Mora, C., McClanahan, T.R., Barnes, M.L., Kittinger, J.N., Hicks, C.C., D'Agata, S., Hoey, A.S., Gurney, G.G., Feary, D.A., Williams, I.D., Kulbicki, M., Vigliola, L., Wantiez, L., Edgar, G.J., Stuart-Smith, R.D., Sandin, S.A., Green, A., Hardt, M.J., Beger, M., Friedlander, A.M., Wilson, S.K., Brokovich, E., Brooks, A.J., Cruz-Motta, J.J., Booth, D.J., Chabanet, P., Gough, C., Tupper, M., Ferse, S.C.A., Sumaila, U.R., Pardede, S. & Mouillot, D. 2018b. Gravity of human impacts mediates coral reef conservation gains. *Proceedings of the National Academy of Sciences* **115**, E6116–E6125.

Clements, C.S. & Hay, M.E. 2019. Biodiversity enhances coral growth, tissue survivorship and suppression of macroalgae. *Nature Ecology & Evolution* **3**, 178–182.

Clements, C.S. & Hay, M.E. 2021. Biodiversity has a positive but saturating effect on imperiled coral reefs. *Science Advances* **7**, eabi8592.

Connell, S.D., Fernandes, M., Burnell, O.W., Doubleday, Z.A., Griffin, K.J., Irving, A.D., Leung, J.Y.S., Owen, S., Russell, B.D. & Falkenberg, L.J. 2017. Testing for thresholds of ecosystem collapse in seagrass meadows. *Conservation Biology* **31**, 1196–1201.

Conti-Jerpe, I.E., Thompson, P.D., Wong, C.W.M., Oliveira, N.L., Duprey, N.N., Moynihan, M.A. & Baker, D.M. 2020. Trophic strategy and bleaching resistance in reef-building corals. *Science Advances* **6**, eaaz5443.

Cooper, G.S., Willcock, S. & Dearing, J.A. 2020. Regime shifts occur disproportionately faster in larger ecosystems. *Nature Communications* **11**, 1175.

Cornwall, C.E., Carlot, J., Branson, O., Courtney, T.A., Harvey, B.P., Perry, C.T., Andersson, A.J., Diaz-Pulido, G., Johnson, M.D., Kennedy, E., Krieger, E.C., Mallela, J., McCoy, S.J., Nugues, M.M., Quinter, E., Ross, C.L., Ryan, E., Saderne, V. & Comeau, S. 2023. Crustose coralline algae can contribute more than corals to coral reef carbonate production. *Communications Earth & Environment* **4**, 105.

Cornwall, C.E., Comeau, S., Kornder, N.A., Perry, C.T., van Hooidonk, R., DeCarlo, T.M., Pratchett, M.S., Anderson, K.D., Browne, N., Carpenter, R., Diaz-Pulido, G., D'Olivo, J.P., Doo, S.S., Figueiredo, J., Fortunato, S.A.V., Kennedy, E., Lantz, C.A., McCulloch, M.T., González-Rivero, M., Schoepf, V., Smithers, S.G. & Lowe, R.J. 2021. Global declines in coral reef calcium carbonate production under ocean acidification and warming. *Proceedings of the National Academy of Science USA* **118**.

Courtney, T., Barnes, B., Chollett, I., Elahi, R., Gross, K., Guest, J., Kuffner, I., Lenz, E., Nelson, H., Rogers, C., Toth, L. & Andersson, A. 2020. Disturbances drive changes in coral community assemblages and coral calcification capacity. *Ecosphere* **11**.

Crisp, S.K., Tebbett, S.B. & Bellwood, D.R. 2022. A critical evaluation of benthic phase shift studies on coral reefs. *Marine Environmental Research* **178**, 105667.

Cruz, I.C.S., Loiola, M., Albuquerque, T., Reis, R., de Anchieta C.C. Nunes, J., Reimer, J.D., Mizuyama, M., Kikuchi, R.K.P. & Creed, J.C. 2015. Effect of phase shift from corals to zoantharia on reef fish assemblages. *PLoS One* **10**, e0116944.

Cruz, I.C.S., Meira, V.H., de Kikuchi, R.K.P. & Creed, J.C. 2016. The role of competition in the phase shift to dominance of the zoanthid *Palythoa* cf. *variabilis* on coral reefs. *Marine Environmental Research* **115**, 28–35.

Cruz, I.C.S., Waters, L.G., Kikuchi, R.K.P., Leão, Z.M.A.N. & Turra, A. 2018. Marginal coral reefs show high susceptibility to phase shift. *Marine Pollution Bulletin* **135**, 551–561.

Dang, V.D.H., Cheung, P.-Y., Fong, C.-L., Mulla, A.J., Shiu, J.-H., Lin, C.-H. & Nozawa, Y. 2020. Sea urchins play an increasingly important role for coral resilience across reefs in Taiwan. *Frontiers in Marine Science* **7**.

Darling, E.S. & Côté, I.M. 2008. Quantifying the evidence for ecological synergies. *Ecology Letters* **11**, 1278–1286.

Darling, E.S., McClanahan, T.R., Maina, J., Gurney, G.G., Graham, N.A.J., Januchowski-Hartley, F., Cinner, J.E., Mora, C., Hicks, C.C., Maire, E., Puotinen, M., Skirving, W.J., Adjeroud, M., Ahmadia, G., Arthur, R., Bauman, A.G., Beger, M., Berumen, M.L., Bigot, L., Bouwmeester, J., Brenier, A., Bridge, T.C.L., Brown, E., Campbell, S.J., Cannon, S., Cauvin, B., Chen, C.A., Claudet, J., Denis, V., Donner, S., Estradivari, Fadli, N., Feary, D.A., Fenner, D., Fox, H., Franklin, E.C., Friedlander, A., Gilmour, J., Goiran, C., Guest, J., Hobbs, J.-P.A., Hoey, A.S., Houk, P., Johnson, S., Jupiter, S.D., Kayal, M., Kuo, C.-y., Lamb, J., Lee, M.A.C., Low, J., Muthiga, N., Muttaqin, E., Nand, Y., Nash, K.L., Nedlic, O., Pandolfi, J.M., Pardede, S., Patankar, V., Penin, L., Ribas-Deulofeu, L., Richards, Z., Roberts, T.E., Rodgers, K.u.S., Safuan, C.D.M., Sala, E., Shedrawi, G., Sin, T.M., Smallhorn-West, P., Smith, J.E., Sommer, B., Steinberg, P.D., Sutthacheep, M., Tan, C.H.J., Williams, G.J., Wilson, S., Yeemin, T., Bruno, J.F., Fortin, M.-J., Krkosek, M. & Mouillot, D. 2019. Social-environmental drivers inform strategic management of coral reefs in the Anthropocene. *Nature Ecology & Evolution* **3**, 1341–1350.

Davis, S. 2018. Associational refuge facilitates phase shifts to macroalgae in a coral reef ecosystem. *Ecosphere* **9**, e02272.

Daw, T.M., Hicks, C.C., Brown, K., Chaigneau, T., Januchowski-Hartley, F.A., Cheung, W.W.L., Rosendo, S., Crona, B., Coulthard, S., Sandbrook, C., Perry, C., Bandeira, S., Muthiga, N.A., Schulte-Herbrüggen, B., Bosire, J. & McClanahan, T.R. 2016. Elasticity in ecosystem services: exploring the variable relationship between ecosystems and human well-being. *Ecology and Society* **21**.

De'ath, G., Fabricius, K.E., Sweatman, H. & Puotinen, M. 2012. The 27-year decline of coral cover on the Great Barrier Reef and its causes. Proceedings of the National Academy of Sciences USA **109**, 17995–17999.

de Goeij, J.M., Lesser, M.P. & Pawlik, J.R. 2017. Nutrient fluxes and ecological functions of coral reef sponges in a changing ocean. In *Climate Change, Ocean Acidification and Sponges: Impacts across Multiple Levels of Organization*, J.L. Carballo & J.J. Bell (eds.). Cham: Springer International Publishing, 373–410.

de Goeij, J.M., Oevelen, D.v., Vermeij, M.J.A., Osinga, R., Middelburg, J.J., Goeij, A.F.P.M.d. & Admiraal, W. 2013. Surviving in a marine desert: the sponge loop retains resources within coral reefs. *Science* **342**, 108–110.

de Goeij, J.M. & van Duyl, F.C. 2007. Coral cavities are sinks of dissolved organic carbon (DOC). *Limnology and Oceanography* **52**, 2608–2617.

Dell, C.L.A., Longo, G.O. & Hay, M.E. 2016. Positive feedbacks enhance macroalgal resilience on degraded coral reefs. *PLoS One* **11**, e0155049.

Del Monaco, C., Hay, M.E., Gartrell, P., Mumby, P.J. & Diaz-Pulido, G. 2017. Effects of ocean acidification on the potency of macroalgal allelopathy to a common coral. *Scientific Reports* **7**, 41053.

Diaz-Pulido, G. & Barrón, C. 2020. CO2 enrichment stimulates dissolved organic carbon release in coral reef macroalgae. *Journal of Phycology* **56**, 1039–1052.

Done, T.J. 1992. Phase shifts in coral reef communities and their ecological significance. *Hydrobiologia* **247**, 121–132.

Done, T.J., Dayton, P.K., Dayton, A.E. & Steger, R. 1991. Regional and local variability in recovery of shallow coral communities: Moorea, French Polynesia and central Great Barrier Reef. *Coral Reefs* **9**, 183–192.

Donovan, M.K., Burkepile, D.E., Kratochwill, C., Shlesinger, T., Sully, S., Oliver, T.A., Hodgson, G., Freiwald, J. & van Woesik, R. 2021. Local conditions magnify coral loss after marine heatwaves. *Science* **372**, 977–980.

Doropoulos, C., Roff, G., Visser, M.-S. & Mumby, P.J. 2017. Sensitivity of coral recruitment to subtle shifts in early community succession. *Ecology* **98**, 304–314.

Dudgeon, S.R., Aronson, R.B., Bruno, J.F. & Precht, W.F. 2010. Phase shifts and stable states on coral reefs. *Marine Ecology Progress Series* **413**, 201–216.

El-Khaled, Y.C., Roth, F., Rädecker, N., Tilstra, A., Karcher, D.B., Kürten, B., Jones, B.H., Voolstra, C.R. & Wild, C. 2021. Nitrogen fixation and denitrification activity differ between coral- and algae-dominated Red Sea reefs. *Scientific Reports* **11**, 11820.

Evensen, N.R., Doropoulos, C., Wong, K.J. & Mumby, P.J. 2019. Stage-specific effects of Lobophora on the recruitment success of a reef-building coral. *Coral Reefs* **38**, 489–498.

Ferrari, R., Bryson, M., Bridge, T., Hustache, J., Williams, S.B., Byrne, M. & Figueira, W. 2016. Quantifying the response of structural complexity and community composition to environmental change in marine communities. *Global Change Biology* **22**, 1965–1975.

Ferrario, F., Beck, M.W., Storlazzi, C.D., Micheli, F., Shepard, C.C. & Airoldi, L. 2014. The effectiveness of coral reefs for coastal hazard risk reduction and adaptation. *Nature Communications* **5**, 3794.

Fisher, R., O'Leary, R.A., Low-Choy, S., Mengersen, K., Knowlton, N., Brainard, Russell E. & Caley, M.J. 2015. Species richness on coral reefs and the pursuit of convergent global estimates. *Current Biology* **25**, 500–505.

Fong, C.R., Gaynus, C.J. & Carpenter, R.C. 2020a. Complex interactions among stressors evolve over time to drive shifts from short turfs to macroalgae on tropical reefs. *Ecosphere* **11**, e03130.

Fong, J., Deignan, L.K., Bauman, A.G., Steinberg, P.D., McDougald, D. & Todd, P.A. 2020b. Contact- and water-mediated effects of macroalgae on the physiology and microbiome of three Indo-Pacific coral species. *Frontiers in Marine Science* **6**, 831.

Ford, A., van Hoytema, N., Moore, B., Pandihau, L., Wild, C. & Ferse, S. 2017. High sedimentary oxygen consumption indicates that sewage input from small islands drives benthic community shifts on overfished reefs. *Environmental Conservation* **44**, 1–7.

Fox, H.E. & Caldwell, R.L. 2006. Recovery from blast fishing on coral reefs: a tale of two scales. *Ecological Applications* **16**, 1631–1635.

Fox, H.E., Harris, J.L., Darling, E.S., Ahmadia, G.N., Estradivari & Razak, T.B. 2019. Rebuilding coral reefs: success (and failure) 16 years after low-cost, low-tech restoration. *Restoration Ecology* **27**, 862–869.

Fox, H.E., Pet, J.S., Dahuri, R. & Caldwell, R.L. 2003. Recovery in rubble fields: long-term impacts of blast fishing. *Marine Pollution Bulletin* **46**, 1024–1031.

Fulton, C.J., Abesamis, R.A., Berkström, C., Depczynski, M., Graham, N.A.J., Holmes, T.H., Kulbicki, M., Noble, M.M., Radford, B.T., Tano, S., Tinkler, P., Wernberg, T. & Wilson, S.K. 2019. Form and function of tropical macroalgal reefs in the Anthropocene. *Functional Ecology* **33**, 989–999.

Gillis, L., Bouma, T., Jones, C., Katwijk, M., Nagelkerken, I., Cjl, J., Herman, P. & Ziegler, A. 2014. Potential for landscape-scale positive interactions among tropical marine ecosystems: a review. *Marine Ecology Progress Series* **503**, 289–303.

Gischler, E., Hudson, J.H., Eisenhauer, A., Parang, S. & Deveaux, M. 2023. 9000 years of change in coral community structure and accretion in Belize reefs, western Atlantic. *Scientific Reports* **13**, 11349.

Goatley, C.H.R. & Bellwood, D.R. 2013. Ecological consequences of sediment on high-energy coral reefs. *PLoS One* **8**, e77737–e77737.

González-Barrios, F.J., Cabral-Tena, R.A. & Alvarez-Filip, L. 2021. Recovery disparity between coral cover and the physical functionality of reefs with impaired coral assemblages. *Global Change Biology* **27**, 640–651.

Gonzalez-Rivero, M., Ferrari Legorreta, R., Schoenberg, C. & Mumby, P. 2012. Impacts of macroalgal competition and parrotfish predation on the growth of a common bioeroding sponge. *Marine Ecology Progress Series* **444**, 133–142.

Gouezo, M., Golbuu, Y., Fabricius, K., Olsudong, D., Mereb, G., Nestor, V., Wolanski, E., Harrison, P. & Doropoulos, C. 2019. Drivers of recovery and reassembly of coral reef communities. *Proceedings of the Royal Society B: Biological Sciences* **286**, 20182908.

Grafeld, S., Oleson, K.L.L., Teneva, L. & Kittinger, J.N. 2017. Follow that fish: uncovering the hidden blue economy in coral reef fisheries. *PLoS One* **12**, e0182104.

Graham, N.A.J. 2014. Habitat complexity: coral structural loss leads to fisheries declines. *Current Biology* **24**, R359–R361.
Graham, N.A.J., Jennings, S., MacNeil, M.A., Mouillot, D. & Wilson, S.K. 2015 Predicting climate-driven regime shifts versus rebound potential in coral reefs. *Nature* **518**, 94–97.
Grottoli, A., Rodrigues, L. & Palardy, J. 2006. Heterotrophic plasticity and resilience in bleached corals. *Nature* **440**, 1186–1189.
Guzman, C., Atrigenio, M., Shinzato, C., Aliño, P. & Conaco, C. 2019. Warm seawater temperature promotes substrate colonization by the blue coral, *Heliopora coerulea. PeerJ* **7**, e7785.
Haas, A.F., Fairoz, M.F.M., Kelly, L.W., Nelson, C.E., Dinsdale, E.A., Edwards, R.A., Giles, S., Hatay, M., Hisakawa, N., Knowles, B., Lim, Y.W., Maughan, H., Pantos, O., Roach, T.N.F., Sanchez, S.E., Silveira, C.B., Sandin, S., Smith, J.E. & Rohwer, F. 2016. Global microbialization of coral reefs. *Nature Microbiology* **1**, 16042.
Haas, A.F., Nelson, C.E., Wegley Kelly, L., Carlson, C.A., Rohwer, F., Leichter, J.J., Wyatt, A. & Smith, J.E. 2011. Effects of coral reef benthic primary producers on dissolved organic carbon and microbial activity. *PLoS One* **6**, e27973.
Hampton-Smith, M., Bower, D.S. & Mika, S. 2021. A review of the current global status of blast fishing: causes, implications and solutions. *Biological Conservation* **262**, 109307.
Hansell, D. & Carlson, C.A. 2001. Marine dissolved organic matter and the carbon cycle. *Oceanography* **14**, 41–49.
Harmelin-Vivien, M.L. 1994. The effects of storms and cyclones on coral reefs: a review. *Journal of Coastal Research*, 211–231.
Harris, D.L., Rovere, A., Casella, E., Power, H., Canavesio, R., Collin, A., Pomeroy, A., Webster Jody, M. & Parravicini, V. 2018. Coral reef structural complexity provides important coastal protection from waves under rising sea levels. *Science Advances* **4**, eaao4350.
Hatcher, B.G. & Larkum, A.W.D. 1983. An experimental analysis of factors controlling the standing crop of the epilithic algal community on a coral reef. *Journal of Experimental Marine Biology and Ecology* **69**, 61–84.
Helber, S.B., Hoeijmakers, D.J.J., Muhando, C.A., Rohde, S. & Schupp, P.J. 2018. Sponge chemical defenses are a possible mechanism for increasing sponge abundance on reefs in Zanzibar. *PLoS One* **13**, e0197617.
Hemingson, C.R., Mihalitsis, M. & Bellwood, D.R. 2022. Are fish communities on coral reefs becoming less colourful? *Global Change Biology* **28**, 3321–3332.
Hempson, T.N., Graham, N.A.J., MacNeil, M.A., Bodin, N. & Wilson, S.K. 2018a. Regime shifts shorten food chains for mesopredators with potential sublethal effects. *Functional Ecology* **32**, 820–830.
Hempson, T.N., Graham, N.A.J., MacNeil, M.A., Hoey, A.S. & Wilson, S.K. 2018b. Ecosystem regime shifts disrupt trophic structure. *Ecological Applications* **28**, 191–200.
Hicks, C., Stoeckl, N., Cinner, J. & Robinson, J. 2014. Fishery benefits and stakeholder priorities associated with a coral reef fishery and their implications for management. *Environmental Science & Policy* **44**.
Hildebrand, T., Osterholz, H., Bunse, H., Grotheer, H., Dittmar, T., Schupp, P.J. 2022. Transformation of dissolved organic matter by two Indo-Pacific sponges. *Limnology and Oceanography* **67**, 2483–2496.
Hoegh-Guldberg, O., Pendleton, L. & Kaup, A. 2019. People and the changing nature of coral reefs. *Regional Studies in Marine Science* **30**, 100699.
Hoey, A.S. & Bellwood, D.R. 2011. Suppression of herbivory by macroalgal density: a critical feedback on coral reefs? *Ecology Letters* **14**, 267–273.
Huang, Y.C.A., Hsieh, H.J., Huang, S.C., Meng, P.J., Chen, Y.S., Keshavmurthy, S., Nozawa, Y. & Chen, C.A. 2011. Nutrient enrichment caused by marine cage culture and its influence on subtropical coral communities in turbid waters. *Marine Ecology Progress Series* **423**, 83–93.
Hughes, T.P., Barnes, M.L., Bellwood, D.R., Cinner, J.E., Cumming, G.S., Jackson, J.B.C., Kleypas, J., van de Leemput, I.A., Lough, J.M., Morrison, T.H., Palumbi, S.R., van Nes, E.H. & Scheffer, M. 2017a. Coral reefs in the Anthropocene. *Nature* **546**, 82–90.
Hughes, T.P., Kerry, J.T., Álvarez-Noriega, M., Álvarez-Romero, J.G., Anderson, K.D., Baird, A.H., Babcock, R.C., Beger, M., Bellwood, D.R., Berkelmans, R., Bridge, T.C., Butler, I.R., Byrne, M., Cantin, N.E., Comeau, S., Connolly, S.R., Cumming, G.S., Dalton, S.J., Diaz-Pulido, G., Eakin, C.M., Figueira, W.F., Gilmour, J.P., Harrison, H.B., Heron, S.F., Hoey, A.S., Hobbs, J.-P.A., Hoogenboom, M.O., Kennedy, E.V., Kuo, C.-y., Lough, J.M., Lowe, R.J., Liu, G., McCulloch, M.T., Malcolm, H.A., McWilliam, M.J.,

Pandolfi, J.M., Pears, R.J., Pratchett, M.S., Schoepf, V., Simpson, T., Skirving, W.J., Sommer, B., Torda, G., Wachenfeld, D.R., Willis, B.L. & Wilson, S.K. 2017b. Global warming and recurrent mass bleaching of corals. *Nature* **543**, 373–377.

Hughes, T.P., Kerry, J.T., Baird, A.H., Connolly, S.R., Chase, T.J., Dietzel, A., Hill, T., Hoey, A.S., Hoogenboom, M.O., Jacobson, M., Kerswell, A., Madin, J.S., Mieog, A., Paley, A.S., Pratchett, M.S., Torda, G. & Woods, R.M. 2019. Global warming impairs stock-recruitment dynamics of corals. *Nature* **568**, 387–390.

Hughes, T.P., Kerry, J.T., Baird, A.H., Connolly, S.R., Dietzel, A., Eakin, C.M., Heron, S.F., Hoey, A.S., Hoogenboom, M.O., Liu, G., McWilliam, M.J., Pears, R.J., Pratchett, M.S., Skirving, W.J., Stella, J.S. & Torda, G. 2018. Global warming transforms coral reef assemblages. *Nature* **556**, 492–496.

Hughes, T.P., Linares, C., Dakos, V., van de Leemput, I.A. & van Nes, E.H. 2013. Living dangerously on borrowed time during slow, unrecognized regime shifts. *Trends in Ecology and Evolution* **28**, 149–155.

Inoue, S., Kayanne, H., Yamamoto, S. & Kurihara, H. 2013. Spatial community shift from hard to soft corals in acidified water. *Nature Climate Change* **3**, 683–687.

Januchowski-Hartley, F.A., Graham, N.A.J., Wilson, S.K., Jennings, S. & Perry, C.T. 2017. Drivers and predictions of coral reef carbonate budget trajectories. *Proceedings of the Royal Society B: Biological Sciences* **284**, 20162533.

Jessen, C., Roder, C., Villa Lizcano, J.F., Voolstra, C.R. & Wild, C. 2013. In-Situ effects of simulated overfishing and eutrophication on benthic coral reef algae growth, succession, and composition in the Central Red Sea. *PLoS One* **8**, e66992.

Kayal, M., Vercelloni, J., Lison de Loma, T., Bosserelle, P., Chancerelle, Y., Geoffroy, S., Stievenart, C., Michonneau, F., Penin, L., Planes, S. & Adjeroud, M. 2012. Predator crown-of-thorns starfish (Acanthaster planci) outbreak, mass mortality of corals, and cascading effects on reef fish and benthic communities. *PLoS One* **7**, e47363.

Keith, S.A., Baird, A.H., Hughes, T.P., Madin, J.S. & Connolly, S.R. 2013. Faunal breaks and species composition of Indo-Pacific corals: the role of plate tectonics, environment and habitat distribution. *Proceedings of the Royal Society B: Biological Sciences* **280**, 20130818.

Kelly, L.W., Barott, K.L., Dinsdale, E., Friedlander, A.M., Nosrat, B., Obura, D., Sala, E., Sandin, S.A., Smith, J.E., Vermeij, M.J.A., Williams, G.J., Willner, D. & Rohwer, F. 2012. Black reefs: iron-induced phase shifts on coral reefs. *The ISME Journal* **6**, 638–649.

Kelly, L.W., Nelson, C.E., Petras, D., Koester, I., Quinlan, Z.A., Arts, M.G.I., Nothias, L.-F., Comstock, J., White, B.M., Hopmans, E.C., Duyl, F.C.v., Carlson, C.A., Aluwihare, L.I., Dorrestein, P.C. & Haas, A.F. 2022. Distinguishing the molecular diversity, nutrient content, and energetic potential of exometabolomes produced by macroalgae and reef-building corals . *Proceedings of the National Academy of Sciences USA* **119**, e2110283119.

Kench, P.S., Beetham, E.P., Turner, T., Morgan, K.M., Owen, S.D. & McLean, R.F. 2022. Sustained coral reef growth in the critical wave dissipation zone of a Maldivian atoll. *Communications Earth & Environment* **3**, 9.

Kennedy, E.V., Roelfsema, C.M., Lyons, M.B., Kovacs, E.M., Borrego-Acevedo, R., Roe, M., Phinn, S.R., Larsen, K., Murray, N.J., Yuwono, D., Wolff, J. & Tudman, P. 2021. Reef Cover, a coral reef classification for global habitat mapping from remote sensing. *Scientific Data* **8**, 196.

Kindinger, T.L. & Albins, M.A. 2017. Consumptive and non-consumptive effects of an invasive marine predator on native coral-reef herbivores. *Biological Invasions* **19**, 131–146.

Kleypas, J.A., McManus, J.W. & Meñez, L.A.B. 1999. Environmental limits to coral reef development: where do we draw the line? *American Zoologist* **39**, 146–159.

Kornder, N.A., Cappelletto, J., Mueller, B., Zalm, M.J.L., Martinez, S.J., Vermeij, M.J.A., Huisman, J. & de Goeij, J.M. 2021. Implications of 2D versus 3D surveys to measure the abundance and composition of benthic coral reef communities. *Coral Reefs* **40**, 1137–1153.

Kornder, N.A., Esser, Y., Stoupin, D., Leys, S.P., Mueller, B., Vermeij, M.J.A., Huisman, J. & de Goeij, J.M. 2022. Sponges sneeze mucus to shed particulate waste from their seawater inlet pores. *Current Biology* **32**, 3855–3861.e3853.

Kuempel, C.D. & Altieri, A.H. 2017. The emergent role of small-bodied herbivores in pre-empting phase shifts on degraded coral reefs. *Scientific Reports* **7**, 39670.

Kuguru, B., Mgaya, Y., Öhman, M. & Wagner, G. 2004. The reef environment and competitive success in the Corallimorpharia. *Marine Biology* **145**, 875–884.

Lau, J.D., Hicks, C.C., Gurney, G.G. & Cinner, J.E. 2019. What matters to whom and why? Understanding the importance of coastal ecosystem services in developing coastal communities. *Ecosystem Services* **35**, 219–230.

Lesser, M.P. 1997. Oxidative stress causes coral bleaching during exposure to elevated temperatures. *Coral Reefs* **16**, 187–192.

Lesser, M.P. 2004. Experimental biology of coral reef ecosystems. *Journal of Experimental Marine Biology and Ecology* **300**, 217–252.

Lesser, M.P. 2006. Benthic-pelagic coupling on coral reefs: feeding and growth of Caribbean sponges. *Journal of Experimental Marine Biology and Ecology* **328**, 277–288.

Lesser, M.P. & Slattery, M. 2011. Phase shift to algal dominated communities at mesophotic depths associated with lionfish (*Pterois volitans*) invasion on a Bahamian coral reef. *Biological Invasions* **13**, 1855–1868.

Lesser, M.P. & Slattery, M. 2020. Will coral reef sponges be winners in the Anthropocene? *Global Change Biology* **26**, 3202–3211.

Lester, S.E., Rassweiler, A., McCoy, S.J., Dubel, A.K., Donovan, M.K., Miller, M.W., Miller, S.D., Ruttenberg, B.I., Samhouri, J.F. & Hay, M.E. 2020. Caribbean reefs of the Anthropocene: variance in ecosystem metrics indicates bright spots on coral depauperate reefs. *Global Change Biology* **26**, 4785–4799.

Liu, M., El-Hossary, E.M., Oelschlaeger, T.A., Donia, M.S., Quinn, R.J. & Abdelmohsen, U.R. 2019. Potential of marine natural products against drug-resistant bacterial infections. *The Lancet Infectious Diseases* **19**, e237–e245.

Loh, T.-L., McMurray, S.E., Henkel, T.P., Vicente, J. & Pawlik, J.R. 2015. Indirect effects of overfishing on Caribbean reefs: sponges overgrow reef-building corals. *PeerJ* **3**, e901–e901.

Madin, J.S. & Connolly, S.R. 2006. Ecological consequences of major hydrodynamic disturbances on coral reefs. *Nature* **444**, 477–480.

Maida, M., Sammarco, P. & Coll, J.C. 1995. Effects of soft corals on scleractinian coral recruitment. I: directional allelopathy and inhibition of settlement. *Marine Ecology Progress Series* **121**, 191–202.

Maida, M., Sammarco, P. & Coll, J.C. 2002. Effects of soft corals on scleractinian coral recruitment. II: allelopathy, spat survivorship and reef community structure. *Marine Ecology* **22**, 397–414.

Marshall, N., Marshall, P., Curnock, M., Pert, P., Smith, A. & Visperas, B. 2019. Identifying indicators of aesthetics in the Great Barrier Reef for the purposes of management. *PLoS One* **14**, e0210196.

Masselink, G., Beetham, E. & Kench, P. 2020. Coral reef islands can accrete vertically in response to sea level rise. *Science Advances* **6**, eaay3656.

McClanahan, T.R. 2020. Chapter Eleven – Coral community life histories and population dynamics driven by seascape bathymetry and temperature variability. In *Advances in Marine Biology*, B.M. Riegl (ed.). Academic Press, 291–330.

McLaughlin, M.J., Bessey, C., Kendrick, G.A., Keesing, J. & Olsen, Y.S. 2023. Production and accumulation of reef framework by calcifying corals and macroalgae on a remote Indian Ocean cay. *Biogeosciences* **20**, 1011–1026.

McManus, J.W., Meñez, L.A.B., Kesner-Reyes, K.N., Vergara, S.G. & Ablan, M.C. 2000. Coral reef fishing and coral-algal phase shifts: implications for global reef status. *ICES Journal of Marine Science* **57**, 572–578.

McManus, J.W. & Polsenberg, J.F. 2004. Coral-algal phase shifts on coral reefs: ecological and environmental aspects. *Progress in Oceanography* **60**, 263–279.

McWilliam, M., Pratchett, M.S., Hoogenboom, M.O. & Hughes, T.P. 2020. Deficits in functional trait diversity following recovery on coral reefs. *Proceedings of the Royal Society B: Biological Sciences* **287**, 20192628.

Moberg, F. & Folke, C. 1999. Ecological goods and services of coral reef ecosystems. *Ecological Economics* **29**, 215–233.

Moeller, M., Nietzer, S., Schils, T. & Schupp, P.J. 2017. Low sediment loads affect survival of coral recruits: the first weeks are crucial. *Coral Reefs* **36**, 39–49.

Moher, D., Liberati, A., Tetzlaff, J. & Altman, D.G. 2009. Preferred reporting items for systematic reviews and meta-analyses: the PRISMA statement. *BMJ* **339**, b2535.

Mollica, N.R., Guo, W., Cohen, A.L., Huang, K.-F., Foster, G.L., Donald, H.K. & Solow, A.R. 2018. Ocean acidification affects coral growth by reducing skeletal density. Proceedings of the National Academy of Sciences USA **115**, 1754–1759.

Morais, J., Morais, R., Tebbett, S.B. & Bellwood, D.R. 2022. On the fate of dead coral colonies. *Functional Ecology* **36**, 3148–3160.

Morais, R.A., Depczynski, M., Fulton, C., Marnane, M., Narvaez, P., Huertas, V., Brandl, S.J. & Bellwood, D.R. 2020. Severe coral loss shifts energetic dynamics on a coral reef. *Functional Ecology* **34**, 1507–1518.

Moritz, C., Brandl, S.J., Rouzé, H., Vii, J., Pérez-Rosales, G., Bosserelle, P., Chancerelle, Y., Galzin, R., Liao, V., Siu, G., Taiarui, M., Nugues, M.M. & Hédouin, L. 2021. Long-term monitoring of benthic communities reveals spatial determinants of disturbance and recovery dynamics on coral reefs. *Marine Ecology Progress Series* **672**, 141–152.

Morrow, K., Paul, V., Liles, M. & Chadwick, N. 2011. Allelochemicals produced by Caribbean macroalgae and cyanobacteria have species-specific effects on reef coral microorganisms. *Coral Reefs* **30**, 309–320.

Mouillot, D., Villéger, S., Parravicini, V., Kulbicki, M., Arias-González, J.E., Bender, M., Chabanet, P., Floeter, S.R., Friedlander, A., Vigliola, L. & Bellwood, D.R. 2014. Functional over-redundancy and high functional vulnerability in global fish faunas on tropical reefs. Proceedings of the National Academy of Sciences USA **111**, 13757–13762.

Muhando, C.A., Kuguru, B.L., Wagner, G.M., Mbije, N.E. & Ohman, M.C. 2002. Environmental effects on the distribution of corallimorpharians in Tanzania. *Ambio* **31**, 558–561.

Mumby, P.J. & Steneck, R.S. 2018. Paradigm lost: dynamic nutrients and missing detritus on coral reefs. *BioScience* **68**, 487–495.

Munsterman, K.S., Allgeier, J.E., Peters, J.R. & Burkepile, D.E. 2021. A view from both ends: shifts in herbivore assemblages impact top-down and bottom-up processes on coral reefs. *Ecosystems* **24**, 1702–1715.

Muthukrishnan, R. & Fong, P. 2014. Multiple anthropogenic stressors exert complex, interactive effects on a coral reef community. *Coral Reefs* **33**, 911–921.

Nakajima, R., Haas, A.F., Silveira, C.B., Kelly, E.L.A., Smith, J.E., Sandin, S., Kelly, L.W., Rohwer, F., Nakatomi, N. & Kurihara, H. 2018. Release of dissolved and particulate organic matter by the soft coral Lobophytum and subsequent microbial degradation. *Journal of Experimental Marine Biology and Ecology* **504**, 53–60.

Nelson, C.E., Goldberg, S.J., Wegley Kelly, L., Haas, A.F., Smith, J.E., Rohwer, F. & Carlson, C.A. 2013. Coral and macroalgal exudates vary in neutral sugar composition and differentially enrich reef bacterioplankton lineages. *The ISME Journal* **7**, 962–979.

Newman, S.P., Meesters, E.H., Dryden, C.S., Williams, S.M., Sanchez, C., Mumby, P.J. & Polunin, N.V.C. 2015. Reef flattening effects on total richness and species responses in the Caribbean. *Journal of Animal Ecology* **84**, 1678–1689.

Norström, A.V., Nyström, M., Lokrantz, J. & Folke, C. 2009. Alternative states on coral reefs: beyond coral-macroalgal phase shifts. *Marine Ecology Progress Series* **376**, 295–306.

Nugues, M.M. & Roberts, C.M. 2003. Coral mortality and interaction with algae in relation to sedimentation. *Coral Reefs* **22**, 507–516.

Nugues, M.M., Smith, G.W., van Hooidonk, R.J., Seabra, M.I. & Bak, R.P.M. 2004. Algal contact as a trigger for coral disease. *Ecology Letters* **7**, 919–923.

Obura, D., Gudka, M., Samoilys, M., Osuka, K., Mbugua, J., Keith, D.A., Porter, S., Roche, R., van Hooidonk, R., Ahamada, S., Araman, A., Karisa, J., Komakoma, J., Madi, M., Ravinia, I., Razafindrainibe, H., Yahya, S. & Zivane, F. 2022. Vulnerability to collapse of coral reef ecosystems in the Western Indian Ocean. *Nature Sustainability* **5**, 104–113.

Odum, E.P. 1969. The strategy of ecosystem development. *Science* **164**, 262–270.

Olinger, L., Chaves-Fonnegra, A., Enochs, I. & Brandt, M. 2020. Three competitors in three dimensions: photogrammetry reveals rapid overgrowth of coral during multispecies competition with sponges and algae. *Marine Ecology Progress Series* **657**.

Olsen, K., Ritson-Williams, R., Paul, V.J. & Ross, C. 2014. Combined effects of macroalgal presence and elevated temperature on the early life-history stages of a common Caribbean coral. *Marine Ecology Progress Series* **509**, 181–191.

O'Neil, J. 2008. Nitrogen Cycling in Coral Reef Environments, 949–989.

Otaño-Cruz, A., Montañez-Acuña, A.A., García-Rodríguez, N.M., Díaz-Morales, D.M., Benson, E., Cuevas, E., Ortiz-Zayas, J. & Hernández-Delgado, E.A. 2019. Caribbean near-shore coral reef benthic community response to changes on sedimentation dynamics and environmental conditions. *Frontiers in* Marine *Science* **6**.

Pawlik, J., Burkepile, D. & Vega Thurber, R. 2016. A vicious circle? Altered carbon and nutrient cycling may explain the low resilience of Caribbean coral reefs. *BioScience* **66**, biw047.

Pawlik, J.R., Loh, T.-L., McMurray, S.E. & Finelli, C.M. 2013. Sponge communities on Caribbean coral reefs are structured by factors that are top-down, not bottom-up. *PLoS One* **8**, e62573.

Pellissier, L., Leprieur, F., Parravicini, V., Cowman, P.F., Kulbicki, M., Litsios, G., Olsen, S.M., Wisz, M.S., Bellwood, D.R. & Mouillot, D. 2014. Quaternary coral reef refugia preserved fish diversity. *Science* **344**, 1016–1019.

Perry, C.T. & Alvarez-Filip, L. 2019. Changing geo-ecological functions of coral reefs in the Anthropocene. *Functional Ecology* **33**, 976–988.

Perry, C.T., Alvarez-Filip, L., Graham, N.A.J., Mumby, P.J., Wilson, S.K., Kench, P.S., Manzello, D.P., Morgan, K.M., Slangen, A.B.A., Thomson, D.P., Januchowski-Hartley, F., Smithers, S.G., Steneck, R.S., Carlton, R., Edinger, E.N., Enochs, I.C., Estrada-Saldívar, N., Haywood, M.D.E., Kolodziej, G., Murphy, G.N., Pérez-Cervantes, E., Suchley, A., Valentino, L., Boenish, R., Wilson, M. & Macdonald, C. 2018. Loss of coral reef growth capacity to track future increases in sea level. *Nature* **558**, 396–400.

Perry, C.T., Kench, P.S., O'Leary, M.J., Morgan, K.M. & Januchowski-Hartley, F. 2015. Linking reef ecology to island building: parrotfish identified as major producers of island-building sediment in the Maldives. *Geology* **43**, 503–506.

Pert, P.L., Thiault, L., Curnock, M.I., Becken, S. & Claudet, J. 2020. Beauty and the reef: evaluating the use of non-expert ratings for monitoring aesthetic values of coral reefs. *Science of the Total Environment* **730**, 139156.

Pet-Soede, L. & Erdmann, M.V. 1998. Blast fishing in Southwest Sulawesi, Indonesia.

Pisapia, C., Hochberg, E.J. & Carpenter, R. 2019. Multi-decadal change in reef-scale production and calcification associated with recent disturbances on a Lizard Island reef flat. *Frontiers in Marine Science* **6**.

Prach, K. & Walker, L.R. 2011. Four opportunities for studies of ecological succession. *Trends in Ecology & Evolution* **26**, 119–123.

Pulsford, S.A., Lindenmayer, D.B. & Driscoll, D.A. 2016. A succession of theories: purging redundancy from disturbance theory. *Biological Reviews* **91**, 148–167.

Rädecker, N., Pogoreutz, C., Voolstra, C.R., Wiedenmann, J. & Wild, C. 2015. Nitrogen cycling in corals: the key to understanding holobiont functioning? *Trends in Microbiology* **23**, 490–497.

Rasher, D.B., Engel, S., Bonito, V., Fraser, G.J., Montoya, J.P. & Hay, M.E. 2012. Effects of herbivory, nutrients, and reef protection on algal proliferation and coral growth on a tropical reef. *Oecologia* **169**, 187–198.

Rasher, D.B., Stout, E.P., Engel, S., Kubanek, J. & Hay, M.E. 2011. Macroalgal terpenes function as allelopathic agents against reef corals. *Proceedings of the National Academy of Science USA* **108**, 17726–17731.

Reguero, B.G., Storlazzi, C.D., Gibbs, A.E., Shope, J.B., Cole, A.D., Cumming, K.A. & Beck, M.W. 2021. The value of US coral reefs for flood risk reduction. *Nature Sustainability* **4**, 688–698.

Reiswig, H.M. 1971. Particle feeding in natural populations of three marine demosponges. *Biological Bulletin* **141**, 568–591.

Renema, W., Bellwood, D.R., Braga, J.C., Bromfield, K., Hall, R., Johnson, K.G., Lunt, P., Meyer, C.P., McMonagle, L.B., Morley, R.J., O'Dea, A., Todd, J.A., Wesselingh, F.P., Wilson, M.E. & Pandolfi, J.M. 2008. Hopping hotspots: global shifts in marine biodiversity. *Science* **321**, 654–657.

Reverter, M., Helber, S.B., Rohde, S., de Goeij, J.M. & Schupp, P.J. 2022. Coral reef benthic community changes in the Anthropocene: biogeographic heterogeneity, overlooked configurations, and methodology. *Global Change Biology* **28**, 1956–1971.

Reverter, M., Jackson, M., Daraghmeh, N., von Mach, C. & Milton, N. 2020. 11-yr of coral community dynamics in reefs around Dahab (Gulf of Aqaba, Red Sea): the collapse of urchins and rise of macroalgae and cyanobacterial mats. *Coral Reefs* **39**, 1605–1618.

Reverter, M., Jackson, M., Rohde, S., Moeller, M., Bara, R., Lasut, M.T., Segre Reinach, M. & Schupp, P.J. 2021. High taxonomic resolution surveys and trait-based analyses reveal multiple benthic regimes in North Sulawesi (Indonesia). *Scientific Reports* **11**, 16554.

Reverter, M., Perez, T., Ereskovsky, A.V. & Banaigs, B. 2016. Secondary metabolome variability and inducible chemical defenses in the Mediterranean sponge *Aplysina cavernicola*. *Journal of Chemical Ecology* **42**, 60–70.

Reynaud, S., Leclercq, N., Romaine-Lioud, S., Ferrier-Pagés, C., Jaubert, J. & Gattuso, J.-P. 2003. Interacting effects of CO2 partial pressure and temperature on photosynthesis and calcification in a scleractinian coral. *Global Change Biology* **9**, 1660–1668.

Richter, C., Wunsch, M., Rasheed, M., Kötter, I. & Badran, M.I. 2001. Endoscopic exploration of Red Sea coral reefs reveals dense populations of cavity-dwelling sponges. *Nature* **413**, 726–730.

Riegl, B. & Branch, G.M. 1995. Effects of sediment on the energy budgets of four scleractinian (Bourne 1900) and five alcyonacean (Lamouroux 1816) corals. *Journal of Experimental Marine Biology and Ecology* **186**, 259–275.

Ritson-Williams, R., Arnold, S.N. & Paul, V.J. 2020. The impact of macroalgae and cyanobacteria on larval survival and settlement of the scleractinian corals *Acropora palmata, A. cervicornis* and *Pseudodiploria strigosa*. *Marine Biology* **167**, 31.

Rix, L., Bednarz, V.N., Cardini, U., van Hoytema, N., Al-Horani, F.A., Wild, C. & Naumann, M.S. 2015. Seasonality in dinitrogen fixation and primary productivity by coral reef framework substrates from the northern Red Sea. *Marine Ecology Progress Series* **533**, 79–92.

Rix, L., de Goeij, J.M., van Oevelen, D., Struck, U., Al-Horani, F.A., Wild, C. & Naumann, M.S. 2017. Differential recycling of coral and algal dissolved organic matter via the sponge loop. *Functional Ecology* **31**, 778–789.

Roach, T.N.F., Little, M., Arts Milou, G.I., Huckeba, J., Haas Andreas, F., George Emma, E., Quinn Robert, A., Cobián-Güemes Ana, G., Naliboff Douglas, S., Silveira Cynthia, B., Vermeij Mark, J.A., Kelly Linda, W., Dorrestein Pieter, C. & Rohwer, F. 2020. A multiomic analysis of in situ coral-turf algal interactions. Proceedings of the National Academy of Sciences USA **117**, 13588–13595.

Robinson, J., Wilson, S., Jennings, S. & Graham, N. 2019a. Thermal stress induces persistently altered coral reef fish assemblages. *Global Change Biology* **25**.

Robinson, J.P.W., Maire, E., Bodin, N., Hempson, T.N., Graham, N.A.J., Wilson, S.K., MacNeil, M.A. & Hicks, C.C. 2022. Climate-induced increases in micronutrient availability for coral reef fisheries. *One Earth* **5**, 98–108.

Robinson, J.P.W., Williams, I.D., Yeager, L.A., McPherson, J.M., Clark, J., Oliver, T.A. & Baum, J.K. 2018. Environmental conditions and herbivore biomass determine coral reef benthic community composition: implications for quantitative baselines. *Coral Reefs* **37**, 1157–1168.

Robinson, J.P.W., Wilson, S.K., Robinson, J., Gerry, C., Lucas, J., Assan, C., Govinden, R., Jennings, S. & Graham, N.A.J. 2019b. Productive instability of coral reef fisheries after climate-driven regime shifts. *Nature Ecology & Evolution* **3**, 183–190.

Rocha, J.C., Peterson, G., Bodin, Ö. & Levin, S. 2018. Cascading regime shifts within and across scales. *Science* **362**, 1379–1383.

Roff, G., Chollett, I., Doropoulos, C., Golbuu, Y., Steneck, R.S., Isechal, A.L., van Woesik, R. & Mumby, P.J. 2015. Exposure-driven macroalgal phase shift following catastrophic disturbance on coral reefs. *Coral Reefs* **34**, 715–725.

Rogers, A., Blanchard, J.L. & Mumby, P.J. 2014. Vulnerability of coral reef fisheries to a loss of structural complexity. *Current Biology* **24**, 1000–1005.

Rogers, A., Blanchard, J.L. & Mumby, P.J. 2018. Fisheries productivity under progressive coral reef degradation. *Journal of Applied Ecology* **55**, 1041–1049.

Rohde, S., Nietzer, S. & Schupp, P.J. 2015. Prevalence and mechanisms of dynamic chemical defenses in tropical sponges. *PLoS One* **10**, e0132236.

Ruiz-Allais, J., Benayahu, Y. & Lasso-Alcalá, O. 2021. The invasive octocoral Unomia stolonifera (Alcyonacea, Xeniidae) is dominating the benthos in the Southeastern Caribbean Sea. *Memoria de la Fundación La Salle de Ciencias Naturales* **79**, 63–80.

Ruiz Allais, J.P., Amaro, M.E., McFadden, C.S., Halász, A. & Benayahu, Y. 2014. The first incidence of an alien soft coral of the family Xeniidae in the Caribbean, an invasion in eastern Venezuelan coral communities. *Coral Reefs* **33**, 287–287.

Ryan, E.J., Hanmer, K. & Kench, P.S. 2019. Massive corals maintain a positive carbonate budget of a Maldivian upper reef platform despite major bleaching event. *Scientific Reports* **9**, 6515.

Safaie, A., Silbiger, N.J., McClanahan, T.R., Pawlak, G., Barshis, D.J., Hench, J.L., Rogers, J.S., Williams, G.J. & Davis, K.A. 2018. High frequency temperature variability reduces the risk of coral bleaching. *Nature Communications* **9**, 1671.

Sammarco, P.W., Coll, J.C., La Barre, S. & Willis, B. 1983. Competitive strategies of soft corals (Coelenterata: Octocorallia): allelopathic effects on selected scleractinian corals. *Coral Reefs* **1**, 173–178.

Sangil, C. & Guzman, H.M. 2016. Macroalgal communities on multi-stressed coral reefs in the Caribbean: long-term changes, spatial variations, and relationships with environmental variables. *Journal of Sea Research* **117**, 7–19.

Schils, T. 2012. Episodic eruptions of volcanic ash trigger a reversible cascade of nuisance species outbreaks in pristine coral habitats. *PLoS One* **7**, e46639.

Schönberg, C.H.L., Fang, J.K.-H. & Carballo, J.L. 2017. Bioeroding sponges and the future of coral reefs. In *Climate Change, Ocean Acidification and Sponges: Impacts Across Multiple Levels of Organization,* J.L. Carballo & J.J. Bell (eds.). Cham: Springer International Publishing, 179–372.

Segal, B. & Castro, C. 2011. Coral community structure and sedimentation at different distances from the coast of the Abrolhos Bank, Brazil. *Brazilian Journal of Oceanography* **59**, 119–129.

Shideler, G.S. & Pierce, B. 2016. Recreational diver willingness to pay for goliath grouper encounters during the months of their spawning aggregation off eastern Florida, USA. *Ocean & Coastal Management* **129**, 36–43.

Sing Wong, A., Vrontos, S. & Taylor, M.L. 2022. An assessment of people living by coral reefs over space and time. *Global Change Biology* **28**, 7139–7153.

Smith, J.E., Hunter, C.L. & Smith, C.M. 2010a. The effects of top-down versus bottom-up control on benthic coral reef community structure. *Oecologia* **163**, 497–507.

Smith, S.V., Kimmerer, W.J., Laws, E.A., Brock, R.E. & Walsh, T.W. 1981. Kaneohe Bay sewage diversion experiment – Perspectives on ecosystem responses to nutritional perturbation. *Pacific Science* **35**, 279–402.

Smith, T.B., Fong, P., Kennison, R. & Smith, J. 2010b. Spatial refuges and associational defenses promote harmful blooms of the alga Caulerpa sertularioides onto coral reefs. *Oecologia* **164**, 1039–1048.

Souter, D., Planes, S., Wicquart, J., Logan, M., Obura, D., Staub, F. 2021. *Status of Coral Reefs of the World: 2020 Report.* Global Coral Reef Monitoring Network (GCRMN) and International Coral Reef Initiative (ICRI), doi:10.59387/WOTJ9184

Spalding, M., Burke, L., Wood, S.A., Ashpole, J., Hutchison, J. & zu Ermgassen, P. 2017. Mapping the global value and distribution of coral reef tourism. *Marine Policy* **82**, 104–113.

Spalding, M.D., Fox, H.E., Allen, G.R., Davidson, N., Ferdaña, Z.A., Finlayson, M., Halpern, B.S., Jorge, M.A., Lombana, A., Lourie, S.A., Martin, K.D., McManus, E., Molnar, J., Recchia, C.A. & Robertson, J. 2007. Marine ecoregions of the world: a bioregionalization of coastal and shelf areas. *BioScience* **57**, 573–583.

Stuart-Smith, R.D., Brown, C.J., Ceccarelli, D.M. & Edgar, G.J. 2018. Ecosystem restructuring along the Great Barrier Reef following mass coral bleaching. *Nature* **560**, 92–96.

Suggett, D.J., Warner, M.E. & Leggat, W. 2017. Symbiotic dinoflagellate functional diversity mediates coral survival under ecological crisis. *Trends in Ecology & Evolution* **32**, 735–745.

Tang, S.L., Hong, M.J., Liao, M.H., Jane, W.N., Chiang, P.W., Chen, C.B. & Chen, C.A. 2011. Bacteria associated with an encrusting sponge (*Terpios hoshinota*) and the corals partially covered by the sponge. *Environmental Microbiology* **13**, 1179–1191.

Tano, S., Eggertsen, M., Wikström, S.A., Berkström, C., Buriyo, A.S. & Halling, C. 2016. Tropical seaweed beds are important habitats for mobile invertebrate epifauna. *Estuarine, Coastal and Shelf Science* **183**, 1–12.

Tano, S.A., Eggertsen, M., Wikström, S.A., Berkström, C., Buriyo, A.S. & Halling, C. 2017. Tropical seaweed beds as important habitats for juvenile fish. *Marine and Freshwater Research* **68**, 1921–1934.

Tebbett, S.B., Connolly, S.R. & Bellwood, D.R. 2023a. Benthic composition changes on coral reefs at global scales. *Nature Ecology & Evolution* **7**, 71–81.

Tebbett, S.B., Crisp, S.K., Evans, R.D., Fulton, C.J., Pessarrodona, A., Wernberg, T., Wilson, S.K. & Bellwood, D.R. 2023b. On the challenges of identifying benthic dominance on Anthropocene coral reefs. *BioScience* **73**, 220–228.

Tebbett, S.B., Streit, R.P. & Bellwood, D.R. 2019. Expansion of a colonial ascidian following consecutive mass coral bleaching at Lizard Island, Australia. *Marine Environmental Research* **144**, 125–129.

Tilot, V., Leujak, W., Ormond, R.F.G., Ashworth, J.A. & Mabrouk, A. 2008. Monitoring of South Sinai coral reefs: influence of natural and anthropogenic factors. *Aquatic Conservation: Marine and Freshwater Ecosystems* **18**, 1109–1126.

Tribot, A.-S., Deter, J., Claverie, T., Guillhaumon, F., Villéger, S. & Mouquet, N. 2019. Species diversity and composition drive the aesthetic value of coral reef fish assemblages. *Biology Letters* **15**, 20190703.

van de Leemput, I.A., Hughes, T.P., van Nes, E.H. & Scheffer, M. 2016. Multiple feedbacks and the prevalence of alternate stable states on coral reefs. *Coral Reefs* **35**, 857–865.

van Hoytema, N., de Goeij, J.M., Kornder, N.A., El-Khaled, Y., van Oevelen, D., Rix, L., Cardini, U., Bednarz, V.N., Naumann, M.S., Al-Horani, F.A. & Wild, C. 2023. A carbon cycling model shows strong control of seasonality and importance of sponges on the functioning of a northern Red Sea coral reef. *Coral Reefs* **42**, 367–381.

Vaughan, E.J., Wilson, S.K., Howlett, S.J., Parravicini, V., Williams, G.J. & Graham, N.A.J. 2021. Nitrogen enrichment in macroalgae following mass coral mortality. *Coral Reefs* **40**, 767–776.

Vianna, G.M.S., Meekan, M.G., Pannell, D.J., Marsh, S.P. & Meeuwig, J.J. 2012. Socio-economic value and community benefits from shark-diving tourism in Palau: a sustainable use of reef shark populations. *Biological Conservation* **145**, 267–277.

Vieira, C., Engelen, A.H., Guentas, L., Aires, T., Houlbreque, F., Gaubert, J., Serrão, E.A., De Clerck, O. & Payri, C.E. 2016. Species specificity of bacteria associated to the brown seaweeds Lobophora (Dictyotales, Phaeophyceae) and their potential for induction of rapid coral bleaching in *Acropora muricata*. *Frontiers in Microbiology* **7**, 316.

Vitousek, S., Barnard, P.L., Fletcher, C.H., Frazer, N., Erikson, L. & Storlazzi, C.D. 2017. Doubling of coastal flooding frequency within decades due to sea-level rise. *Scientific Reports* **7**, 1399.

Vollstedt, S., Xiang, N., Simancas-Giraldo, S.M. & Wild, C. 2020. Organic eutrophication increases resistance of the pulsating soft coral Xenia umbellata to warming. *PeerJ* **8**, e9182.

Walsh, K., Haggerty, J.M., Doane, M.P., Hansen, J.J., Morris, M.M., Moreira, A.P.B., de Oliveira, L., Leomil, L., Garcia, G.D., Thompson, F. & Dinsdale, E.A. 2017. Aura-biomes are present in the water layer above coral reef benthic macro-organisms. *PeerJ* **5**, e3666.

Webb, A.E., de Bakker, D.M., Soetaert, K., da Costa, T., van Heuven, S.M.A.C., van Duyl, F.C., Reichart, G.J. & de Nooijer, L.J. 2021. Quantifying functional consequences of habitat degradation on a Caribbean coral reef. *Biogeosciences* **18**, 6501–6516.

Webb, A.E., van Heuven, S.M.A.C., de Bakker, D.M., van Duyl, F.C., Reichart, G.-J. & de Nooijer, L.J. 2017. Combined effects of experimental acidification and eutrophication on reef sponge bioerosion rates. *Frontiers in* Marine *Science* **4**.

Wegley Kelly, L., Nelson Craig, E., Petras, D., Koester, I., Quinlan Zachary, A., Arts Milou, G.I., Nothias, L.-F., Comstock, J., White Brandie, M., Hopmans Ellen, C., van Duyl Fleur, C., Carlson Craig, A., Aluwihare Lihini, I., Dorrestein Pieter, C. & Haas Andreas, F. 2022. Distinguishing the molecular diversity, nutrient content, and energetic potential of exometabolomes produced by macroalgae and reef-building corals. Proceedings of the National Academy of Sciences USA **119**, e2110283119.

Wenger, L.N., van Lier, J.R. & Fulton, C.J. 2018. Microhabitat selectivity shapes the seascape ecology of a carnivorous macroalgae-associated tropical fish. *Marine Ecology Progress Series* **590**, 187–200.

Wild, C., Huettel, M., Klueter, A., Kremb, S.G., Rasheed, M.Y.M. & Jørgensen, B.B. 2004. Coral mucus functions as an energy carrier and particle trap in the reef ecosystem. *Nature* **428**, 66–70.

Williams, G.J. & Graham, N.A.J. 2019. Rethinking coral reef functional futures. *Functional Ecology* **33**, 942–947.

Williams, G.J., Graham, N.A.J., Jouffray, J.-B., Norström, A.V., Nyström, M., Gove, J.M., Heenan, A. & Wedding, L.M. 2019. Coral reef ecology in the Anthropocene. *Functional Ecology* **33**, 1014–1022.

Williams, G.J., Smith, J.E., Conklin, E.J., Gove, J.M., Sala, E. & Sandin, S.A. 2013. Benthic communities at two remote Pacific coral reefs: effects of reef habitat, depth, and wave energy gradients on spatial patterns. *PeerJ* **1**, e81.

Woodhead, A.J., Hicks, C.C., Norström, A.V., Williams, G.J. & Graham, N.A.J. 2019. Coral reef ecosystem services in the Anthropocene. *Functional Ecology* **33**, 1023–1034.

Work, T.M., Aeby, G.S. & Maragos, J.E. 2008. Phase shift from a coral to a corallimorph-dominated reef associated with a shipwreck on Palmyra atoll. *PLoS One* **3**, e2989.

Yan, H.F. & Bellwood, D.R. 2023. Multi-decadal stability of fish productivity despite increasing coral reef degradation. *Functional Ecology* **37**, 1245–1255.

Zubia, M., Andréfouët, S. & Payri, C. 2014. Distribution and biomass evaluation of drifting brown algae from Moorea lagoon (French Polynesia) for eco-friendly agricultural use. *Journal of Applied Phycology* **27**.

Supplementary Materials are provided online at: https://www.routledge.com/9781032761961

Oceanography and Marine Biology: An Annual Review, 2024, **62**, 248-266

Taylor & Francis

CLONAL GROWTH PATTERNS IN COLONIAL ANTHOZOAN CORALS

ELEONORA RE[1], SEBASTIAN SCHMIDT-ROACH[1], EVA LLABRES[2,3], TOMÁS SINTES[2], AND CARLOS M. DUARTE[1]

[1]*Marine Science Program, Biological and Environmental Science and Engineering Division, King Abdullah University of Science and Technology (KAUST), Thuwal 23955-6900, Kingdom of Saudi Arabia*
[2]*Institute for Cross-Disciplinary Physics and Complex Systems, IFISC (CSIC-UIB), Universitat de les Illes Balears, E-07122 Palma de Mallorca, Spain*
[3]*Hawai'i Institute of Marine Biology, University of Hawai'i at M anoa, Kaneohe, Hawaii, USA*
Corresponding Author:
Eleonora Re
Email address: eleonora.re@kaust.edu.sa

Abstract Corals are modular organisms that use asexual reproduction (i.e., budding) to increase the size of the colony by adding new polyps. Clonal modularity offers several advantages, such as the ability to sustain partial mortality, redistribute resources internally, replace or repair single modules, and delay senescence, potentially supporting millenary growth. Despite the global importance of coral reefs, little is known about clonal growth rules in anthozoan corals. Coral clonal growth was reviewed to synthetise current understandings and identify gaps driving future research efforts. Despite corals present high plasticity, their growth is dictated by strong intrinsic regularities at different life stages and modularity levels. For example, a six-polyps crown with fixed distances among polyps is typically formed in the early development stages of *Stylophora pistillata* larvae. Similarly, specific developmental regularities are observed in the budding of azooxanthellate Dendrophylliidae polyps, which are consistently maintained across generations and species. In Octocorallia, colony shape is preserved by maintaining a constant ratio between the total number of branches and mother branches. Concurrently, environmental factors (i.e., light and hydrodynamics) play a fundamental role in shaping the final morphology of the colony, driving the architectural design at different levels of modularity. Some species revealed higher plasticity at the branching level in contrast with the predetermined shape assumed by the colony. Several models have been proposed to describe the environmental modulation of coral growth, mostly in branching forms. However, a holistic, universal model applicable to a broader range of coral taxa is still lacking. Understanding the fundamental rules underpinning coral clonal growth is essential to improving predictions of coral reef recovery, inferring stress on coral colonies and guiding restoration efforts.

Keywords: Clonal Growth; Budding; Modularity; Asexual Reproduction; Growth Model; Coral Morphology; Growth Pattern

 DOI: 10.1201/9781003477518-6

Introduction

Coral reefs are the largest structures of biological origin in the biosphere, engineered by tiny clonal organisms. Coral reef frameworks develop by the growth of coral colonies, supplemented by contributions from other calcifiers. In addition, coral clonal growth is fundamental in responding to disturbances. For instance, the remarkably rapid recovery of coral cover from recent bleaching events in the Great Barrier Reef (AIMS 2022) has been supported largely by clonal growth. Yet, coral clonal growth, as well as other aspects of coral development (i.e., developmental biology, Ball et al. 2004) has received limited attention, particularly compared to the attention received by coral spawning and sexual reproduction, which has been studied on more than 400 coral species (Harrison 2011). A comprehension of coral clonal growth processes is required to understand coral reef formation and recovery from disturbance and to design effective restoration projects. This is particularly important considering that the global coverage of living corals is estimated to have declined by 50% since the 1950s (Eddy et al. 2021).

In contrast to the paucity of studies on coral clonal growth (Supplementary Figure 1), research on clonal growth rules in terrestrial and marine plants has delivered important insights into their growth rules and their application to predict space occupation by clonal plants (Routledge 1990, Marbà & Duarte 1998, Sintes et al. 2005, 2006), the extent and density of plant populations (Bell & Tomlinson 1980, Duarte et al. 1994, Sintes et al. 2005), resource acquisition (Cain et al. 1996), shoot dynamics (Bell & Tomlinson 1980, Marbà & Duarte 1998, 2003) and interactions between neighbouring ramets (Kenkel 1993, Marbà & Duarte 1998, 2003) and species within the community (Llabrés et al. 2022, 2023).

Corals can develop in a solitary growth form, a discrete and physiologically independent organism, or in a colonial form, where a prolonged tissue continuity is maintained across the individual modules, allowing for colony integration. The growth of coral colonies involves two processes, tissue growth (horizontal) and skeleton deposition (vertical), tightly linked to each other (Figure 1).

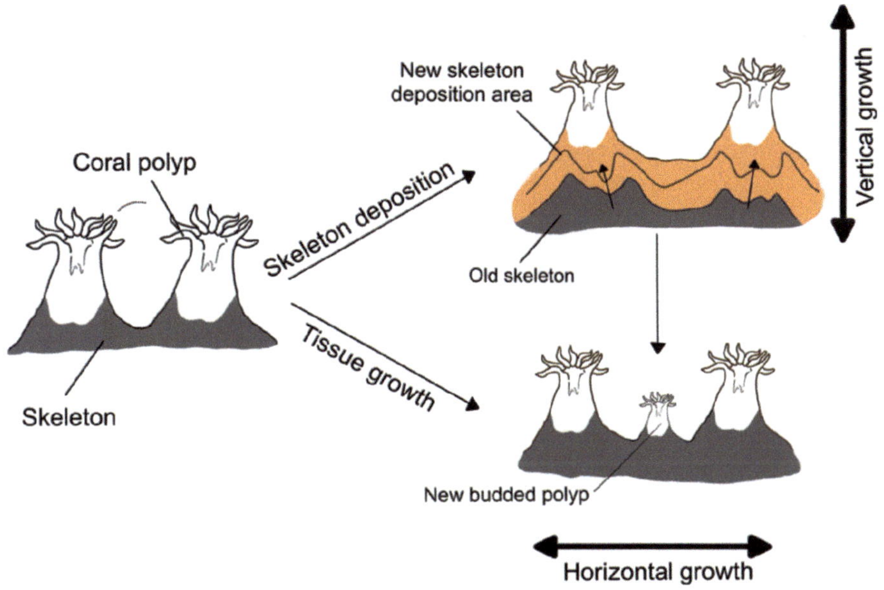

Figure 1 A schematic view of skeletal deposition and budding processes in colonial corals. Skeletal deposition is facilitated by the polyps' calcium carbonate deposition, which is responsible for both the vertical growth of the colony and the peripheral expansion that enables the formation of new polyps and horizontal tissue growth (Medellín-Maldonado et al. 2022).

Clonal growth is initiated by the iterative replication and addition of modules (Figure 2) from an initial founder polyp (Boardman et al. 1973). The term 'module', initially coined to define plant growth (Harper 1977), describes the basic unit that is replicated to support clonal growth. Modularity brings several advantages, such as the ability to sustain partial mortality and increase fitness by enhancing the total reproductive output of the colonial genotype. Resources are distributed internally from areas of abundance to sites where they are scarce (Hester et al. 1994, Kemball & Marshall 1995, Hutchings & Wijesinghe 1997, Oren et al. 2001). Moreover, modularity implies the capacity to replace or repair single modules, delaying senescence and leading, potentially, to indefinite growth (Harrington & Hughes 1990, Winston 2010). The basic module supporting coral clonal growth is the polyp, defined as the conical organism composed of a tentacle crown, and the exoskeleton hosting the polyp (Figure 2).

Coral colonies extend vertically by accumulating new skeleton. Modification, i.e., growth of the skeleton, is fundamental to create available space for new polyps to arise (Medellín-Maldonado et al. 2022). Despite being closely coupled, tissue growth and skeleton deposition are limited by

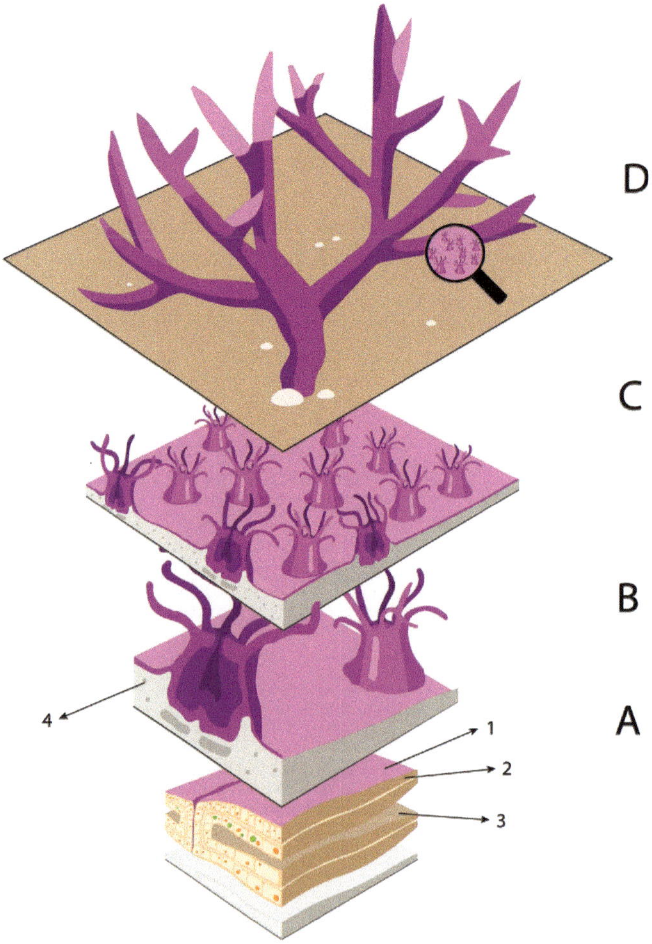

Figure 2 Modularity in coral colonies and coral reefs. Colonial corals are modular organisms, which consist of up to three levels of organisation, individual polyps (B), individual branches (in branching growth forms) made up by the repetition of single polyps (C) and the whole colony (D). In (A), the outer layer of the coral colony, also known as ectoderm (1), covers the mesoglea (2) and the gastrovascular cavity (3). The skeletal corallite underneath the polyp is visible in section (4).

different factors (Barnes & Lough 1992, Darke & Barnes 1993). Light availability, saturation state of calcium carbonate (Marubini et al. 2001) and dissolved inorganic nutrients concentration (Lesser et al. 1994) have been shown to play a role in skeletal growth, whereas organic nutrients availability appears to determine tissue growth rate (Barnes 1973, Graus & Macintyre 1982, Darke & Barnes 1993). Corals have evolved a repertoire of colony morphologies to maximise space occupation and resource acquisition.

Coral polyps multiply clonally through a process known as budding, defined as a gemmation process, where the daughter polyp, or bud, develops its own set of organs, borrowing nothing but an undifferentiated body wall from the parent, which otherwise remains complete (Hill 1935). Coral clonal growth is governed by a combination of a genetically mediated developmental programme and the plastic response to the surrounding environment. In each species, polyps stop growing when they reach a determined size; however, the colony may potentially grow indeterminately, achieving enormous sizes (Hughes et al. 1992), with the largest coral clones reported to reach 13 m across at 500 years of age.

In cnidarians, research has focussed on understanding the variation of colony forms in response to environmental factors (i.e., phenotypic plasticity; Graus & Macintyre 1976, 1982, Bruno & Edmunds 1997, Muko et al. 2000, Kaandorp & Kübler 2001, Todd et al. 2004, Rodolfo-Metalpa et al. 2008, Todd 2008, Filatov et al. 2010, Chindapol et al. 2013). In contrast, little attention has been paid to the more challenging task of defining and elucidating the set of rules that control clonal growth and, therefore, colony shape (Rinkevich 2002, Gateno & Rinkevich 2003, Lasker et al. 2003, Ohno et al. 2015, Guerrini et al. 2021). Different aspects of coral clonal growth have been reviewed in the past by Matthai (1926), Buddemeier and Kinzie (1976), Rosen (1986), Hughes et al. (1992), Lasker and Sànchez (2002), Sànchez (2004), Hughes (2005) and Todd (2008).

Growth models applied to describe coral growth were often borrowed from plant-based models (e.g., "tree architectural model", Dauget 1991, 1994), given their similar ecology and analogous shape (Connell 1978). In addition, branching colonial organisations have been analysed by mimicking the rules of colony formation through mathematical models based on the accretive growth of the skeleton subjected to the diffusion of resources (e.g., Kaandorp & Sloot 2001). Despite the apparent relative simplicity of each modular unit, corals and colonial marine invertebrates can generate complex architectures (Kaandorp & Kübler 2001, Sánchez & Lasker 2003, Shaish et al. 2007), involving a highly coordinated growth leading to emerging, non-linear properties, as reported for clonal seagrass (Sintes et al. 2005).

Here we review the current understanding of anthozoan coral clonal growth, as a basis to understand and predict patterns of space occupation, resource acquisition, and the competitive and recovery potential of different coral species. Our review also focuses on synthesising past research efforts in describing coral colony growth through mathematical models. Finally, we identify knowledge gaps and key future research directions that need to be addressed to expand our capacity to model and predict coral clonal growth.

Growth patterns in Hexacorallia

Massive Faviidiae corals

Gateño and Rinkevich (2003) analysed the rules governing budding in the massive coral *Favia favus*, to report that polyp budding was initiated once the polyp reached a threshold of two morphometric fields, consisting of the perimeter of the polyp and its area. They report that a ratio of the two radii (r_p and r_a) greater or equal to the average value of 1.14 can trigger intra-tentacular budding in the species, thereby proposing the first allometric threshold ratio for bud formation described in cnidarians (Gateño & Rinkevich 2003).

Kai and Sakai (2008) found that budding in single polyp and small (5–8 polyps) colonies of two massive faviids species (i.e., *Goniastrea aspera* and *Favites chinensis*) was initiated when space was available around the polyps, showing that the availability of open area around the polyp can affect budding rate and mode. Budding mode (intra- versus extra-tentacular budding) varied depending on the position of the polyp in the colony. Polyps in the colony margins usually show extra-tentacular budding, whereas polyps surrounded by other polyps usually exhibit intra-tentacular budding. The budding rate was found to be inversely proportional to colony size, as the perimeter-to-area ratio decreases with colony size, resulting in less marginal polyps and space available to expand in larger colonies. The budding response to open space was proposed to be a mechanism to increase colony fitness (Kai & Sakai 2008), as the mortality rate is higher when colonies are young and smaller (Connell 1973, Hughes & Jackson 1980). As a consequence, small colonies, with a low number of polyps and high perimeter-to-area ratios, tend to grow faster, allowing restoration practitioners to use this strategy in a technique known as micro-fragmentation (Page & Vaughan 2014, Page et al. 2018).

Branching Pocilloporidae family

The clonal growth process of several Pocilloporidae coral species has been extensively studied and modelled over time (Supplementary Table 1) and is believed to be genetically controlled for *Stylophora pistillata* (Rinkevich 2002, Shaish et al. 2006). Fifty-three to 121 days after *S. pistillata* larval metamorphosis, a 'six-polyp crown' or 'six-polyp circlet' is established by budding around the primary founder polyp (FP), maintaining fixed distances (1.59±0.29 mm) between polyps (Figure 3A). This basic morphometric landscape had already been observed in another Pocilloporidae species, *Pocillopora damicornis* (*bulbosa*) by Stephenson (1931), and reflects a strong genetic clonal growth programme (Guerrini et al. 2021). The authors hypothesised that the sites of budding are governed by a morphogen ('isomone' *sensu* Rinkevich & Loya 1985a) frontier region, with high and low concentration areas, establishing fixed and regular distances among polyps. However, the nature of the hypothetical hormone-regulating spacing has not been elucidated yet, giving rise to questions about the plausibility of a hormone being released in a highly hydrodynamic environment.

An early observation of the branching process in *P. damicornis* came from Stephenson (1931), who described, through longitudinal growth (i.e., branching process), the lateral expansion of the polyps, until reaching a branching stage. After almost a century, Guerrini et al. (2021) described the branching process of another Pocilloporidae species, *S. pistillata*. The authors found that when the distances between neighbouring polyps of the six-polyp crown are shorter (in the range of 1.08±0.28 mm) than the average fixed distance, there is no possibility of expanding the space by lateral retreat growth. Therefore, branching is the solution to resume growth while conforming with the allometric rules (Guerrini et al. 2021).

Different *S. pistillata* genotypes exhibited variable and different architectural designs during growth, each characterised by specific morphometric parameters (Shaish et al. 2006). However, a more comprehensive analysis revealed that clonal growth is characterised by a continuum of architectural patterns of several distinct stages. In particular, three phases can be distinguished. In the beginning, colonies display fast vertical growth and the initiation of some side branches; vertical growth then declines to favour lateral extension, leading to an increase in the ecological volume and bifurcation of branches. Finally, in the last phase, lateral extension is reduced, ecological volume is increased and older branches start to bifurcate. Moreover, Shaish et al. (2007) investigated plasticity at the branch and colony levels. The branching process is also characterised by more plastic morphometries, with individual branch modules exhibiting diverse architectures and developmental trajectories, whereas clonal growth traits are more conserved at the colony level. Conserved colony

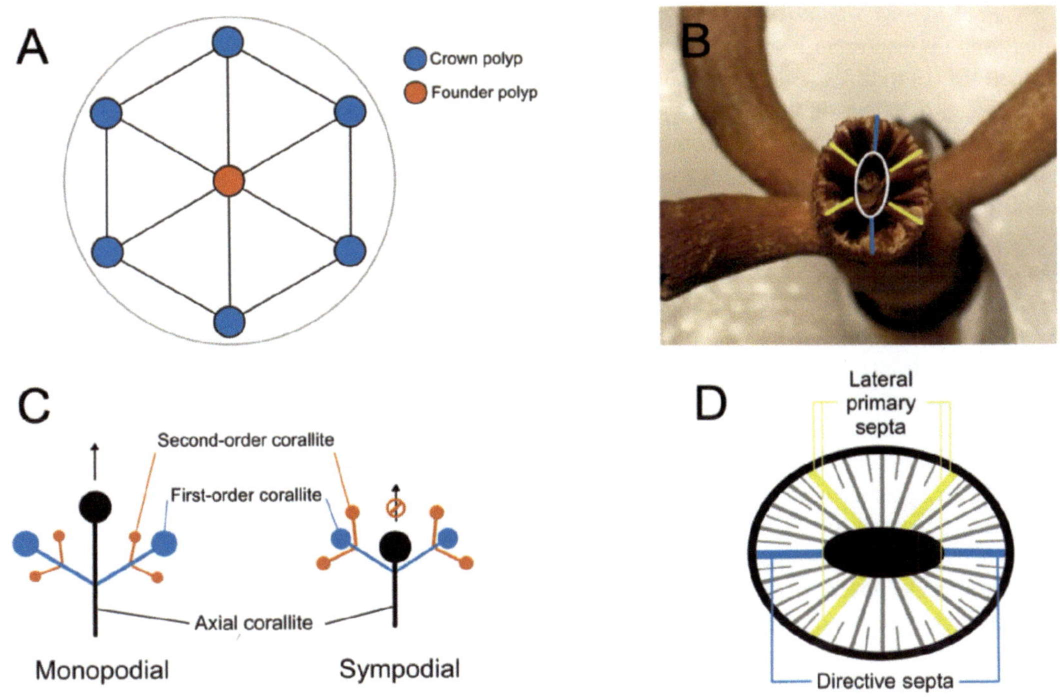

Figure 3 (A) A schematic view of the six-polyp circlet in *Stylophora pistillata* early stages after metamorphosis. Around the founder polyp (in red), a crown of 6 polyps (in blue) arises at a fixed distance of 1.59±0.29 mm (six equilateral triangles). Modified from Guerrini et al. (2021). (B) Corallite of a dendrophylliids species from the Eastern Red Sea. In blue are the directive septa; in yellow are the four lateral primary septa. (C) Monopodial growth: The axial corallite (in black) continues to grow upwards giving rise to lateral corallites. Sympodial growth: the axial corallite stops growing; instead, it produces lateral corallites. Modified from Sentoku et al. (2013). (D) A schematic view of the dendrophylliids corallite Portuales plan. Two opposite directive septa are in blue, and four lateral primary septa are in yellow. Modified from Sentoku and Ezaki (2012b).

traits, with plasticity at the branching level, result in variability within species-specific fixed colonial shapes (Loya 1976).

Pocillopora species has a widespread distribution across the Red Sea, Indian Ocean and Pacific Ocean to the Eastern Tropical Pacific (Veron 2000) and shows remarkable variation in colony morphology, making identification a challenging task. Several authors focused on modelling the effects of environmental factors in shaping the whole coral colony architecture (Kaandorp 1999, Chindapol et al. 2013). Paz-García et al. (2015) found that under high-flow rates, the morphology of different branching *Pocillopora* species (*P. damicornis*, *Pocillopora verrucosa* and *Pocillopora meandrina*) was more similar among morphospecies compared to low-flow conditions. In particular, branch thickness was significantly higher under high-flow rates, seemingly a coping mechanism of the colony (Paz-García et al. 2015). Branch modularity (i.e., the likelihood of different sets of branch traits to vary in a coordinated way) varied under different flow rates, with branch density, branch diameter and distance among branches showing a stronger connection at the high-flow rate site (Paz-García et al. 2015). Similarly, *S. pistillata* shows morphological plasticity under different light and water flow regimes. Colonies subjected to high water flow and light resulted in more densely branching colonies, whereas thinner and more open branches arose under low light intensity and low flow (Shwartsberg et al. 2012). Under mesophotic conditions, *S. pistillata* has been shown to develop a more plate-like morphology in order to capture more light (Mass et al. 2010).

Pocillopora damicornis is one of the most extensively studied coral species, yet its growth and budding pattern remain poorly unknown. Li et al. (2020) used high-resolution computed tomography (HRCT) to investigate the 3D skeletal structure of *P. damicornis* to understand budding, growth and polyp distribution during the growth process. Results revealed that the average diameter of the oral and aboral surface of *P. damicornis* is generally between 0.5 mm and 1 mm, and the average polyps height is between 0.1 mm and 0.6 mm, allowing to reconstruct the average polyp volume, 3600 cm^2. In *P. damicornis*, although corallites are not related to each other at the micro level, coral growth is a holistic and synchronised process, where the living polyps extend to the surface of the coral from their initial growth point, with no dead polyp present in the internal part of the coral (Li et al. 2020). Whereas the latter could be considered as a species-specific feature, it may not apply to other species (i.e., *Porites*, Darke & Barnes 1993). When a polyp secretes a new calix, it enters the newly created chamber, horizontally closing the previous one in a process known as dissepiment. Each polyp grows along an axis, which points in the direction of the branch. Polyps deviating from the axis commonly stop after four or five inter-septa spaces, thickening the side branches of the colony, which follows a dichotomous branching rule during the colony growth (Li et al. 2020).

Plasticity of growth processes to environmental conditions found for the genus *Pocillopora* was also reported for *P. damicornis*, where light can affect the final colony shape (Li et al. 2020). In particular, calix volume and inter-septal space exhibit larger average volume and inter-septal space when exposed to higher light levels, with growth slowing gradually towards shaded areas. Transplant experiments with the extensively studied and modelled Pocilloporidae species *Madracis mirabilis* revealed that both genotype and environment affect the morphometric traits, in particular, corallite architecture and density, branch diameter and spacing (Bruno & Edmunds 1997).

Growth regularities in the Dendrophylliidae family

Intrinsic regularities in the budding of azooxanthellate corals Dendrophylliidae have been extensively studied (Sentoku & Ezaki 2012a–d, 2013, Ohno et al. 2015, Sentoku et al. 2015a,b). Colonial growth of corals lacking symbiotic algae is less influenced by light compared to the zooxanthellate counterpart (Ohno et al. 2015). However, competition for space is severe among corallites, especially in massive and encrusting forms compared to branching forms. In Dendrophylliidae corals, the axial corallite is produced sexually, whereas the first-order and subsequent lateral corallites arise by budding from the preceding corallite. Moreover, a single polyp is present at the individual corallite tip. Corallites are oval with a large calicular diameter (*sensu* Cairns 1994) located in the region containing two directive septa. Four primary septa, also known as lateral primary septa, arise around the mouth (Figure 3B and D).

Colonial azooxanthellate dendrophylliid bud according to specific developmental regularities despite the diversity of growth forms (Sentoku & Ezaki 2012a). Results from a diverse array of species highlighted the fact that different growth forms (bushy, massive, dendroid) can be explained with subtle differences in one or more of the clonal growth parameters, such as budding sites, orientation of directive septa, inclination of budding and the interval of budding (Supplementary Table 2, Figure 4).

Regularities in growth patterns are maintained across all generations, suggesting the presence of strict developmental constraints on asexual reproduction. Lateral corallites bud in various directions to maximise the use of available space and avoid direct contact between polyps. In branching forms (e.g., *Dendrophyllia arbuscula*, *Tubastrea coccinea*, *Dendrophyllia ehrenbergiana*), the resulting colonies can be classified as monopodial, when the axial corallite prolongs continuously upwards, giving rise to several lateral buds, and sympodial, when the axial corallite ceases to grow upwards and through growing and budding, colony extension occurs in one or two lateral corallites (Figure 3C). In monopodial forms, budding sites are located at four lateral primary septa (Sentoku & Ezaki 2012a,c) or on one side of the two primary septa (Sentoku & Ezaki 2012c).

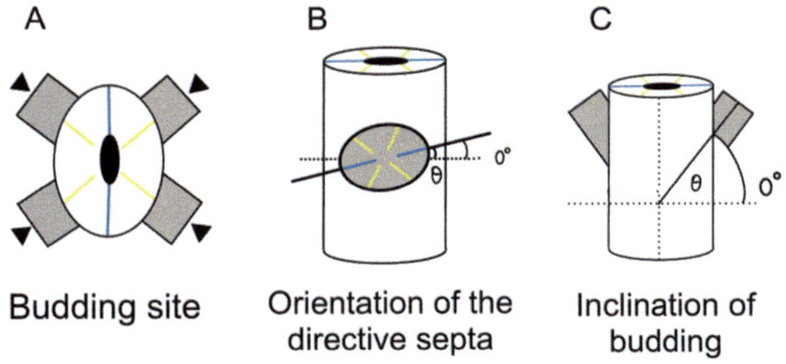

Figure 4 Schematic diagram of the morphometric measurements conducted. (A) Budding sites. The two directive septa are indicated in blue, and the four lateral primary septa are indicated in yellow. Lateral primary corallites are in light grey. (B) Orientation of the directive septa, which is defined as the angle (θ) between a line connecting the two directive septa (solid line) and a horizontal line (dotted). (C) Inclination of budding, which is defined as the angle between the direction of the lateral corallites (solid line) and a horizontal line (dotted line) oriented perpendicular to the growth direction of the parental corallite (vertical dotted line). Modified from Sentoku and Ezaki (2012c).

Conversely, in sympodial forms, budding sites are restricted to a single primary septum (Sentoku & Ezaki 2012b, 2013). In both sympodial and monopodial species (*D. ehrenbergiana*, *Dendrophyllia boschmai*, *Dendrophyllia cribrosa*, and *Turbinaria peltata*), a distinct polarity at individual corallites has been found to affect the budding site and, eventually, the coral morphology (Sentoku & Ezaki 2012d, 2013). This striking characteristic implies the presence of radio-bilateral symmetrical constraints on the asexual reproduction of a Scleractinia group (Sentoku & Ezaki 2012d). Finally, the same regularities in budding have been found for the zooxanthellate species *T. peltata*, suggesting a strong influence of those underlying regularities on the final growth form in both species whether they contain or lack zooxanthellae (Sentoku et al. 2015a).

Azooxanthellate cold-water corals

Morphological diversity in cold-water corals is still poorly understood due to the limited accessibility of the deep ocean they inhabit. Sanna and Freiwald (2021) investigated the intraspecific morphological variation in the azooxanthellate caryophyllid *Lophelia pertusa* (recently synonymised with *Desmophyllum pertusum*; Addamo et al. 2016), showing remarkable morphological diversities in both corallites and colonies. However, corallite morphology was mostly decoupled from colony morphology. Three major branching patterns were identified, asymmetrical (uniplanar growth along numerous branches), bushy (simultaneous multidirectional growth) and columnar (longitudinal growth along few branches, often in a zig-zag mode). The number of buds per corallite varied in response to the growth pattern, with higher values in bushy forms. These variations may reflect plastic strategies in response to environmental factors, such as water movement, which implies mechanical stress and food supply. Asymmetric and bushy forms minimise the hydraulic stress in high-energy settings, orientating the branches or reducing the branch length towards the current flow. Conversely, columnar forms are more suitable for low-energy settings, associated with scarcer food availability.

Growth patterns in Octocorallia

Gorgonian octocorals present a highly complex colonial organisation, spanning from the distribution of single polyps to the branching pattern. At the branching level, Sánchez et al. (2004)

showed that the common assumption of branching as a bifurcation process does not apply to modular marine organisms, which conform to a sub-apical process, where mother branches give rise to daughter branches at a fixed internode length. The highly organised gorgonian tree-like shape is preserved by maintaining a constant ratio between mother and daughter branches (Sanchez 2004).

Branching process in gorgonian has been extensively studied in different species (Sánchez & Lasker 2003, Sánchez et al. 2004, Cadena et al. 2010). Cadena et al. (2010) analysed the association between the branching process and stem canals present at the base and at the terminal segment of the branches of the colony. In fact, an internal tree-like structure is present in gorgonian octocorals, equally diametric-sized hollow vessels, running all along the colony, connected with gastrovascular canal (solenia; Bayer 1973) to exchange nutrients (Gateño et al. 1998). Results revealed that a species-specific constant ratio (*c*) of total and mother branches was maintained in *Eunicea mammosa*, as previously found in Sánchez et al. (2004) (in Caribbean corals, *c* ranges between 2 and 13). Moreover, a linear relationship between branches and stem canals was found, suggesting an important role in colony growth. In particular, the number of stems in the holdfast was correlated with the total length of the colony, independently from the number of mother and daughter branches (Cadena et al. 2010).

Growth models

Morphogenesis is a complex biological phenomenon, whose fundamental principles are still not completely understood. Hence, mathematical modelling is considered an important tool to test hypotheses and help to answer specific questions about developmental processes (Tomlin & Axelrod 2007).

Coral growth models have been mostly focused on the role of environmental factors (i.e., light intensity and water flow) in shaping coral morphology (Kaandorp & Sloot 2001, Chindapol et al. 2013). The ultimate goal was to resolve the interplay between environmental and genetic constraints in defining growth forms and coral shapes. Results supported the hypothesis of very plastic species or levels of modularity (e.g., branching versus whole colony level, see Loya 1976, Shaish et al. 2007, as *Pocillopora damicornis* and Pocilloporidae in general, Kaandorp et al. 2005), and less plastic groups or modules, where tight genetic constraints preclude morphological plasticity (Dauget 1991, Shaish et al. 2007) (Table 1).

Table 1 Summary of the Principal Models Applied to Simulate Coral Morphogenesis

Model	References
Theoretical model for coral growth	Mistr and Bercovici (2003)
Model for Dendrophyllidae corals modified from Niklas and Kercher (1984)	Ohno et al. (2015)
Tree Architectural model	Dauget (1991)
Hydrodynamically influenced radiate accretive model (HIRAG)	Kaandorp et al. (2001)
Hydrodynamically influenced radiate accretive model (HIRAG) modified from Kaandorp and Sloot (2001)	Merks et al. (2003)
Polyp-oriented radiate accretive growth model (PORAG)	Merks et al. (2004)
PORAG model coupled with diffusion model in *Madracis mirabilis*	Kaandorp et al. (2005)
PORAG model in *Madracis mirabilis*	Filatov et al. (2010)
Flow-driven accretive growth model from PORAG Merks et al. (2004) in *Pocillopora verrucosa*	Chindapol et al. (2013)
Branching and self-organisation model in gorgonians	Sánchez et al. (2004)
Stochastic model for *Corallium rubrum* growth	Kahramanoğulları et al. (2019)

Theoretical model of coral growth

Among the earliest models developed to describe coral growth, Mistr and Bercovici (2003) adopted a simplified mathematical model to describe growth in a coral subjected to unidirectional flow. Linear stability analysis of the system of equations predicts that the growth of solid (coral) structures will be aligned perpendicular to flow, propagating against flow direction (Mistr & Bercovici 2003).

Plant-based models

The analogy between plants and corals has been extensively reviewed by Connell (1978), based on a long tradition of ecologists making direct comparisons between plants and corals (Crossland 1913, Squires 1964). Ohno et al. (2015) proposed a voxel-based programme used in addition to a vector model based on the plant modular model of Niklas and Kerchner (1984) to explain colonial dendrophylliids morphogenesis. The model was originally developed for terrestrial plants to predict architectural solutions to maximise plant photosynthetic efficiency, reduce mechanical stress attending vertical growth and optimise the translocation of metabolites. The model aims at predicting architectural solutions that are more efficient than others in plants (Niklas & Kerchner 1984). In this study, the Niklas and Kerchner (1984) model has been further developed by adding a local reference system, similar to the growing tube model of Okamoto (1988). Three parameters were adopted to simulate a range of colonial growth forms, budding orientation, branching angle and interval of budding (Figure 4, Ohno et al. 2015). The regulation of two of the parameters produced two distinct morphologies, branching and massive, represented in nature by two distinct species with different ecological and physical traits, *Tubastrea coccinea* and *Dendrophyllia arbuscula*. However, the model could not generate all coral growth forms available in nature. Thus, Ohno et al. (2015) suggested that other controlling factors (e.g., density of nutrients, water flow and sedimentary conditions), not included in the model, might also affect the growth and morphogenesis of corallites.

Another application of the plant model in coral morphogenesis is Dauget (1991), where architectural methods used for tropical trees were applied to explain constant traits in coral colony forms. Each coral growth form was described using four simple architectural models (Corners's, Schoute's, Attims' and a newly named Wood-Jones' model) and their iteration over time. In particular, the author identified species-specific architecture based on graphic methods (Figure 5). The fundamental element is the axis, which could develop in at least three different ways, by a unique apical polyp (e.g., *Lobophyllia corymbosa*), by multiple apical polyps (e.g., *Acropora palifera*) or by the apical zone of the coenosarc (e.g., *Montipora ramosa*). Moreover, the ramification type affects the general architecture of the colony. The Corners's model (Figure 5A) represents solitary-polyp non-ramified colonies (e.g., *Fungia fungites*), with a single mouth. The Schoute's model (Figure 5B) describes the orthotropic growth with apical ramification, typical of *Acropora* or *L. corymbosa* colonies. Orthotropic growth with lateral and diffuse branching is represented by the Attims' model (Figure 5C), displayed by colonies of *Acropora formosa*. Lastly, the modified Wood Jones' model (Figure 5D) depicts the growth of tabular or plate-like colonies, where growth is plagiotropic in the basal portion and orthotropic in the apical with lateral ramification, resulting in the table forms of *Acropora hyacinthus* colonies.

Despite being able to describe some of the typical growth forms available in nature, those models represent just a visual description of the general coral architecture.

Hydrodynamically influenced radiate accretive growth model (HIRAG)

The hydrodynamically influenced radiate accretive growth model (Kaandorp & Sloot 2001) represents an important milestone, aiming at understanding which aspects of coral growth can be

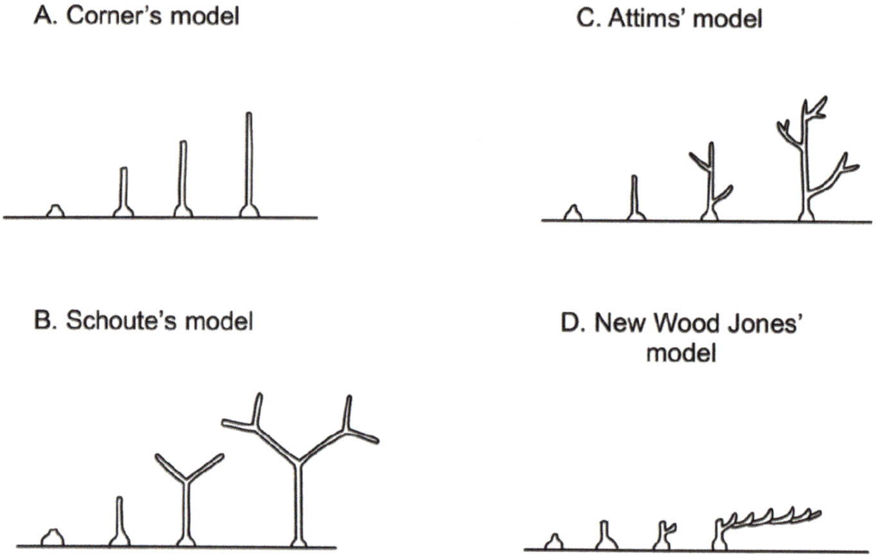

Figure 5 Graphic visualisation of the four main architectural models adapted for corals. Modified from Dauget (1991).

explained from environmental and biological constraints, such as genetic regulation and differential tissue and skeletal growth. In their model, growth proceeds by the iterative accretion of layers whose thickness depends on the resource fluxes over the coral transported by advection and diffusion, where the coral surface is represented by a triangular mesh and where vertices and inter-vertex links define the curved coral surface (Kaandorp & Sloot 2001).

However, in this preliminary HIRAG model, two unresolved issues remained. Firstly, branch splitting didn't occur unless a curvature rule was enforced in the model, allowing to move from lobated shapes to branching objects (Kaandorp & Sloot 2001). Further progresses in the model allowed to resolve the branch splitting mechanism and address the reason for morphological variety under the model (Merks et al. 2003).

Secondly, in the early work by Kaandorp and Sloot (2001), a compactification mechanism emerged under high water flow conditions (HIRAG model, Kaandorp & Sloot 2001; Diffusion Limited Aggregation model, Kaandorp et al. 1996). Despite being in line with previous findings in sponges, scleractinian corals and hydrozoans (Kaandorp 1999), the mechanisms behind this phenomenon were not understood. The positive correlation between the object's compactness and Péclet number was not supported by subsequent simulations, suggesting a lack of computational evidence that hydrodynamics influences the compactness of corals (Merks et al. 2003).

Polyp-oriented radiate accretive growth model (PORAG) and its applications

Merks et al. (2004) developed a polyp-oriented model, PORAG, in which coral growth is described as the "collective results of a growth process taking place in the polyps", as an extension of the HIRAG model proposed by Kaandorp and Sloot (2001). Polyps are modelled individually, as separate entities, which exploit resources, deposit skeleton, bud off new polyps and die. In the PORAG model, the coral surface is represented as a mesh, containing vertices (representing the centre of the coral polyp) and inter-vertex links (representing the coenosarc of the coral), compared to the HIRAG model, where vertices and inter-vertex links do not represent any biological entity. The model assumes that new polyps arise as space becomes available, to maintain a fixed "inter-polyp" distance, which has a dominant effect on the growth form, especially on the thickness of the branches and the compactness of the colony. Morphogenesis is, hence, a natural consequence of the way in which individual polyps are deployed at the surface of the coral colony. In summary, this model concludes that genetics sets the properties of the individual polyps, whereas environment influences the morphology of the individual polyp and colony (Merks et al. 2004).

Subsequent applications of the PORAG model implemented the effects of environmental factors, such as water flow and inorganic nutrients diffusion, to determine the degree of phenotypic plasticity among colonies of the same species. Computer simulations were supported by 3D images of actual colonies to compare the results. Quantitative comparison of complex real colonies with model outputs is a fundamental prerequisite to verify the models. Complex branching organisms need 3D accurate reconstruction that could include internal structures of the colony. The use of 3D scanning techniques allowed to simplify the objects into a network of lines (skeletonization algorithm; Kruszyński et al. 2007), or using a voxel thinning algorithm (Ramírez-Portilla et al. 2022), where a set of morphometric variables can be extracted.

The PORAG model has been coupled with the diffusion model and applied to simulate morphology processes in *Madracis mirabilis*, assuming a crucial role of dissolved inorganic carbon (DIC) as a limiting factor (Kaandorp et al. 2005). Polyp height and distance were the only species-specific information included in the model. Results revealed that gradients of DIC around coral colonies are responsible for branching colony morphology, highlighting the importance of including advection-diffusion model (Kaandorp et al. 2003) in combination with a morphogenetic model of coral, as done by Chindapol et al. (2013). Light intensity, nutrient availability and the degree of diffusion of the nutrients across the object surface were further implemented in the PORAG model by Filatov et al. (2010) to show how those factors affect growth forms of species belonging to the genus *Madracis*. Under high light intensity and surface diffusion, a transition from thin-branched morphologies to more compact growth forms was observed (Filatov et al. 2010). However, the computational model was not able to generate more irregular shapes (e.g., *Madacis formosa*) (Filatov et al. 2010).

Chindapol et al. (2013) applied the radial accretive growth model (PORAG) coupled with an advection-diffusion equation to study the impact of hydrodynamics on the growth of the scleractinian coral *P. verrucosa*. Simulated morphologies and computed tomography (CT) scans on real colonies exposed to various flow rates confirmed a strong influence of flow conditions on coral morphologies. In particular, results showed an asymmetrical growth of colonies subjected to unidirectional flow conditions, with branching developing towards the incoming currents (Chindapol et al. 2013). However, the model could not resolve finer details (i.e., small-scale roughness of the bumps typical of the species) nor the process of anastomosis – the fusion of branches – which might play an important role in coral morphogenesis.

Moreover, Kücken et al. (2011) argued that the Kaandorp-Merks' models were not able to describe the morphogenesis of *Stylophora pistillata*, as significant skeletal growth takes place mostly at the branch tips, even though all the parts of the branch have equal access to resources. This contrasts with the Kaandorp-Merks model, where local growth is related to the amount of available resources. A more detailed representation of *S. pistillata* morphogenesis has been developed (Kücken et al. 2011). The mathematical model includes resources and photosynthate transport across the colony, which provides positional information for coral morphogenesis; growth is uniform and takes place only at the tips; and branch splitting is a stochastic event regulated by the presence of resources (Kücken et al. 2011).

Branching and self-organisation model in Gorgonians

Sánchez et al. (2004) presented a model of the dynamical behaviour of branching through colony development based on empirical observations from the octocorals *Pseudopterogorgia bipinnata* and *Plexaura flexuosa*, where colonies branch subapically, generating hierarchical mother-daughter relationships among branches, rather than a simple bifurcation with two "sister" branches. The model included both the intrinsic subapical branching and an ecological-physiological limit to growth, a maximum number of mother branches.

Stochastic model to describe Corallium rubrum *growth*

The red coral *Corallium rubrum* is a Mediterranean octocoral, an overexploited, slow-growing species with a calcium carbonate skeleton deposited by the polyps, which can be found between 10 m and 800 m depth. The model developed by Kahramanoğulları et al. (2019) consisted of three rules parameterised with stochastic rates, describing the growth and branching dynamics of *C. rubrum* to capture the phenotypic variability of coral structures in the field. The influence of the environment on individual colonies was well represented, and the model suggested that the morphology of a *C. rubrum* colony can result from two competing processes, branching and growth, the priority of which is regulated by environmental factors (Kahramanoğulları et al. 2019).

Knowledge gaps

Colony morphology is the result of a single developmental plan, characterised by a continuum of architectural rules, each characterised by specific morphometric parameters. Yet, the body of research that has been focused on elucidating growth form and defining growth rules in colonial anthozoans is rather limited, comprising only 0.05% of the research efforts on coral reefs, despite the importance of clonal growth in the recovery of coral reefs from disturbance (AIMS 2022). This is in contrast to the attention on clonal growth for other habitat-forming marine clonal species, specifically seagrasses (Hemminga & Duarte 2000). As discussed in Marbà and Duarte (2003), different seagrass species display a diverse repertoire of growth programmes, which can be modelled, across species, considering rhizome extension rates, branching rates and angles, shoot mortality rates and the per capita space required by the shoots (Sintes et al. 2005).

Several studies across taxa showed that the body plans of modular organisms are constrained by developmental programmes and environments. In corals, the interest in defining a set of rules describing clonal growth has focused on a limited set of species (Supplementary Table 1) compared to the high species richness of corals (>1000 species of Hexacorallia and >3000 species of Octocorallia). The polyp-oriented model (Merks et al. 2004) emphasised the importance of understanding how single modules bud with respect to the mother polyps, and if a distinct morphogenetic landscape, as described in *S. pistillata* by Guerrini et al. (2021), takes place in different coral species. Models developed to explain coral growth have been mostly focussed on the effect of environmental factors (i.e., light and water flow), often limited to single species and growth forms. Hence, a parsimonious, universal model, able to reproduce coral growth across coral species with a simple set of parameters, as available for seagrasses (Sintes et al. 2005), is yet missing; however, the complexity of coral forms, their high plasticity (Todd 2008) and the gaps in our understanding of the mechanisms that underlie coral development and gene regulation pose a significant challenge.

Another important gap in coral clonal growth research is colony integration. The existence and transmission of cues, molecules and energy resources between remote parts of a colonial structure are prerequisites for colonial holism, fundamental to distinguish "true" colonies from mere aggregates (Jackson 1979, Silén 1981, Rinkevich 2000). However, previous models (i.e., HIRAG and PORAG models) assumed the independence of single polyps, as it seemed a fundamental precondition for the hypothesised coral growth mechanism (Merks 2003). Noteworthy, from a model point of view, in the coral morphogenesis process such as the branches avoidance mechanism, the occurrence of an extracellular hormone (i.e., isomone) would be indistinguishable from a nutrient depletion situation (Merks 2003), both resulting in branches avoidance behaviour. In seagrass, as in clonal plants in general, resources are transported along rhizomes, and apical dominance, exerted through hormonal control, regulates branching processes (Schwarzschild & Zieman 2008, Terrados et al. 1997). However, the existence and drivers of mechanisms for clonal integration within coral colonies remain, surprisingly, to be explored, despite clues that clonal integration is likely present in some species. Within the same colony, polyps have integrated physiology, and the shape of the colony is coordinated. For instance, Loya (1976) discovered a tendency of *S. pistillata* colonies to regain the initial symmetry of a sphere when damaged, which requires some degree of clonal integration (Loya 1976). Also, changes in growth directionality were detected in *S. pistillata* branches when confronting isogenic branches (Rinkevich & Loya 1985b) and in *Dendrophyllia cribrosa* branches (Sentoku et al. 2015a,b). These findings reveal the existence of feedback mechanisms able to 'consider' the current shape and future growth under a pre-planned coral pattern or the presence of conserved 'morphogenetic codes' (Hogan 1999), species-specific rules repeatedly expressed to develop the colony. Sentoku et al. (2015b) proposed the role of individual corallites in sensing and maintaining the distance between branches through sensory recognition of tentacles. Rinkevich and Loya (1985a) proposed the presence of a hypothetical coral 'isomone', yet to be identified, with a role as a chemical cue for signalling. Although this hypothesis has never been questioned before, the energetic requirements for releasing a pheromone in a highly hydrodynamic environment raise some doubts about the existence of a chemical compound released into the flow involved in colony integration. Indeed, the underline mechanism to explain existing evidence of communication and coordination remains unknown. A recent experiment by Bouderlique et al. (2022) used particle tracking and mathematical modelling to reveal polyps' connection and integration, either on the surface or inside the colony. Results showed a strong species-specific integration among polyps via complex liquid dynamics highly dependent on the cilia orientation typical of every species, suggesting a potential mechanism of colony coordination (Bouderlique et al. 2022).

Conclusions

The quest to develop a more holistic approach towards coral growth patterns, aiming to be informative and meaningful at the same time, requires an effort to extract a parsimonious and yet universal set of clonal growth rules that can be used to understand and model clonal growth across colonial anthozoans. These models can be used for heuristic purposes, to better understand coral clonal growth and the resulting growth forms and are also essential to guide restoration and conservation efforts attending to light availability and the response to hydrodynamic disturbances (Cresswell et al. 2020), which are accelerating with increasing global coral loss.

Acknowledgements

E.L. and T.S. acknowledge the financial support from the following projects: PID2021-12372 3OB-C22 and CEX2021-001164-M funded by MICIU/AEI/10.13039/501100011033, and by ERDF, a way of making Europe, and the Vicenç Mut post-doctoral fellowship funded by ERDF, a way of making Europe, and the Conselleria de Fons Europeus, Universitat i Cultura (CAIB). ELL has received funding from the European Union's Horizon 2021 research and innovation programme under the Marie Skłodowska-Curie grant agreement No 101063295 (CoralMath).

This work was supported by King Abdullah University of Science and Technology (KAUST) baseline fundings BAS/1/1071-01-01.

All the authors acknowledge the participants of the "Coral clonal growth" workshops held in Majorca and KAUST, Professors Manuel Matías, Prof. Damià Gomila, Prof. Nuria Marbà, Dr. Shannon Klein, Dr. Alexandra Steckbauer, Dr. Lutfi Rosli, Alex Gimenez, Miguel Alvarez, Pablo Moreno and Nayra Pluma for their constructive and proficient feedbacks.

Contribution

This work was conceptualised by C.M.D. E.R. performed the data analysis and produced the first draft with inputs from C.M.D., S.S.R., E.L. and T.S. All authors contributed critically to the drafts.

References

Addamo, A.M., Vertino, A., Stolarski, J., García-Jiménez, R., Taviani, M. & Machordom, A. 2016. Merging scleractinian genera: the overwhelming genetic similarity between solitary *Desmophyllum* and colonial *Lophelia*. *BMC Evolutionary Biology* **16**(1), 1–17.

AIMS. 2022. *Annual Summary Report of Coral Reef Condition* 2021/22. Australian Institute of Marine Science Online. https://www.aims.gov.au/monitoring-great-barrier-reef/gbr-condition-summary-2021-22 (Accessed 2 May 2023).

Ball, E.E., Hayward, D.C., Saint, R. & Miller, D.J. 2004. A simple plan-cnidarians and the origins of developmental mechanisms. *Nature Reviews Genetics* **5**(8), 567–577.

Barnes, D.J. 1973. Coral reef project-papers in memory of Dr. Thomas F. Goreau. 12. Growth in colonial scleractinians. *Bulletin of Marine Science* **23**(2), 280–298.

Barnes, D.J. & Lough, J.M. 1992. Systematic variations in the depth of skeleton occupied by coral tissue in massive colonies of porites from the Great Barrier Reef. *Journal of Experimental Marine Biology and Ecology* **159**(1), 113–128, doi:10.1016/0022-0981(92)90261-8

Bayer, F.M. 1973. Colonial organization in octocorals. *Animal Colonies* **82**, 69–93

Bell, A.D. & Tomlinson, P.B. 1980. Adaptive architecture in rhizomatous plants. *Botanical Journal of the Linnean Society* **80**(2), 125–160, doi:10.1111/j.1095-8339.1980.tb01662.x

Boardman, R.S., Cheetham, A.H. & Oliver, W.A. 1973. Degrees of colony dominance in stenolaemate and gymnolaemate Bryozoa. *Animal Colonies: Development and Function through Time* **603**, 121–220.

Bouderlique, T., Petersen, J., Faure, L., Abed-Navandi, D., Bouchnita, A., Mueller, B., Nazarov, M., Englmaier, L., Tesarova, M., Frade, P.R., Zikmund, T., Koehne, T., Kaiser, J., Fried, K., Wild, C., Pantos, O., Hellander, A., Bythell, J. & Adameyko, I. 2022. Surface flow for colonial integration in reef-building corals. *Current Biology* **32**(12), 2596–2609, doi:10.1016/J.CUB.2022.04.054

Bruno, J.F. & Edmunds, P.J. 1997. Clonal variation for phenotypic plasticity in the coral *Madracis mirabilis*. *Ecology* **78**(7), 2177–2190.

Buddemeier, R.W. & Kinzie, R.A. 1976. Coral growth. *Oceanography and Marine Biology: An Annual Review* **14**, 183–225.

Cadena, N.J., Rey, C., Hernández-Hoyos, M., Sánchez, J.D., Teillaud, S., Ardila, N. & Sánchez, J.A. 2010. Linking local to global properties in branching modular networks: gorgonian coral colonies. *Marine Biology* **157**, 1003–1010.

Cain, M.L., Dudle, D.A. & Evans, J.P. 1996. Spatial models of foraging in clonal plant species; spatial models of foraging in clonal plant species. *American Journal of Botany* **83**, 76–85, doi:10.1002/j.1537-2197.1996.tb13877.x

Cairns, S.D. 1994. Scleractinia of the temperate North Pacific. *Smithsonian Contributions to Zoology* **557**, 150.

Chindapol, N., Kaandorp, J.A., Cronemberger, C., Mass, T. & Genin, A. 2013. Modelling growth and form of the scleractinian coral *Pocillopora verrucosa* and the influence of hydrodynamics. *PLoS Computational Biology* **9**(1), 1002849.

Connell, J.H. 1973. Population ecology of reef-building corals. *Biology and Geology of Coral Reefs* **2**, 205–245.

Connell, J.H. 1978. Diversity in tropical rain forests and coral reefs: high diversity of trees and corals is maintained only in a nonequilibrium state. *Science* **199**(4335), 1302–1310.

Cresswell, A.K., Thomson, D.P., Haywood, M.D.E. & Renton, M. 2020. Frequent hydrodynamic disturbances decrease the morphological diversity and structural complexity of 3D simulated coral communities. *Coral Reefs* **39**(4), 1147–1161.

Crossland, C. 1913. *Desert and Water Gardens of the Red Sea: Being an Account of the Natives and the Shore Formations of the Coast*. Cary, North Carolina: University Press.

Darke, W.M. & Barnes, D.J. 1993. Growth trajectories of corallites and ages of polyps in massive colonies of reef-building corals of the genus Porites. *Marine Biology* **117**, 321–326.

Dauget, J.M. 1991. Application of tree architectural models to reef-coral growth forms. *Marine Biology* **111**, 157–165.

Dauget, J.M. 1994. Essai de comparaison entre l'architecture des coraux et celle des végétaux. *Bulletin du Muséum national d'histoire Naturelle. Section A, Zoologie, Biologie et* écologie *Animales* **16**(1), 39–53.

Duarte, C.M., Marba, N., Agawin, N., Cebrian, J., Enriquez, S., Fortes, M.D., Gallegos, M.E., Merino, M., Olesen, B., Sand-Jensen, K., Uri, J. & Vermaat, J. 1994. Reconstruction of seagrass dynamics - age determinations and associated tools for the seagrass ecologist. *Marine Ecology Progress Series* **107**(1–2), 195–209, doi:10.3354/meps107195

Eddy, T.D., Lam, V.W., Reygondeau, G., Cisneros-Montemayor, A.M., Greer, K., Palomares, M.L.D., Bruno, J.F., Ota, Y. & Cheung, W.W. 2021. Global decline in capacity of coral reefs to provide ecosystem services. *One Earth* **4**(9), 1278–1285.

Filatov, M.V., Kaandorp, J.A., Postma, M., Van Liere, R., Kruszyński, K.J., Vermeij, M.J., Streekstra, G.J. & Bak, R.P. 2010. A comparison between coral colonies of the genus *Madracis* and simulated forms. *Proceedings of the Royal Society B: Biological Sciences* **277**(1700), 3555–3561.

Gateno, D. & Rinkevich, B. 2003. Coral polyp budding is probably promoted by a canalized ratio of two morphometric fields. *Marine Biology* **142**, 971–973, doi:10.1007/s00227-003-1009-8

Gateno, D., Israel, A., Barki, Y. & Rinkevich, B. 1998. Gastrovascular circulation in an octocoral: evidence of significant transport of coral and symbiont cells. *The Biological Bulletin* **194**(2), 178–186.

Graus, R.R. & Macintyre, I.G. 1976. Light control of growth form in colonial reef corals: computer simulation. *Science* **193**(4256), 895–897.

Graus, R.R. & Macintyre, I.G. 1982. Variation in growth forms of the reef coral *Montastrea annularis* (Ellis and Solander): a quantitative evaluation of growth response to light distribution using computer simulation. *Smithsonian Contributions to the Marine Sciences* **12**, 441–464.

Guerrini, G., Shefy, D., Shashar, N., Shafir, S. & Rinkevich, B. 2021. Morphometric and allometric rules of polyp's landscape in regular and chimeric coral colonies of the branching species *Stylophora pistillata. Developmental Dynamics* **250**(5), 652–668.

Harper, J.L. 1977. *Population Biology of Plants*. London: Academic Press.

Harrington, R. & Hughes, R.N. 1990. A functional biology of clonal animals. *The Journal of Animal Ecology* **59**(2), 6, doi:10.2307/4898

Harrison, P.L. 2011. Sexual reproduction of scleractinian corals. In *Coral Reefs: An Ecosystem in Transition*, Z. Dubinsky & N. Stambler (eds.). Dordrecht: Springer, doi:10.1007/978-94-007-0114-4_6

Hemminga, M.A. & Duarte, C.M. 2000. *Seagrass Ecology*. Cambridge: Cambridge University Press.

Hester, M.W., McKee, K.L., Burdick, D.M., Koch, M.S., Flynn, K.M., Patterson, S. & Mendelssohn, I.A. 1994. Clonal integration in Spartina patens across a nitrogen and salinity gradient. *Canadian Journal of Botany* **72**(6), 767–770.

Hill, D. 1935. British terminology for rugose corals. *Geological Magazine* **72**(11), 481–519, doi:10.1017/S0016756800094577

Hogan, B.L. 1999. Morphogenesis. *Cell* **96**(2), 225–233.

Hughes, R.N. 2005. Lessons in modularity: the evolutionary ecology of colonial invertebrates. *Scientia Marina* **69**(S1), 169–179.

Hughes, T.P. & Jackson, J.B.C. 1980. Do corals lie about their age? Some demographic consequences of partial mortality, fission, and fusion. *Science* **209**(4457), 713–715.

Hughes, T.P., Ayre, D. & Connell, J.H. 1992. The evolutionary ecology of corals. *Trends in Ecology & Evolution* **7**(9), 292–295.

Hutchings, M.J. & Wijesinghe, D.K. 1997. Patchy habitats, division of labour and growth dividends in clonal plants. *Trends in Ecology & Evolution* **12**(10), 390–394.

Jackson, J.B. 1979. Morphological strategies of sessile animals. *Biology and Systematic of Colonial Organisms* **767**, 499–555.

Kaandorp, J.A. 1999. Morphological analysis of growth forms of branching marine sessile organisms along environmental gradients. *Marine Biology* **134**, 295–306.

Kaandorp, J.A. & Kübler, J.E. 2001. *The Algorithmic Beauty of Seaweeds, Sponges, and Corals*. Heidelberg, Germany: Springer Science & Business Media.

Kaandorp, J.A. & Sloot, P.M. 2001. Morphological models of radiate accretive growth and the influence of hydrodynamics. *Journal of Theoretical Biology* **209**(3), 257–274.

Kaandorp, J.A., Lowe, C.P., Frenkel, D. & Sloot, P.M. 1996. Effect of nutrient diffusion and flow on coral morphology. *Physical Review Letters* **77**(11), 2328.

Kaandorp, J.A., Koopman, E.A., Sloot, P.M., Bak, R.P., Vermeij, M.J. & Lampmann, L.E. 2003. Simulation and analysis of flow patterns around the scleractinian coral *Madracis mirabilis* (Duchassaing and Michelotti). *Philosophical Transactions of the Royal Society of London. Series B: Biological Sciences* **358**(1437), 1551–1557.

Kaandorp, J.A., Sloot, P.M., Roeland, A., Merks, R.M.H., Bak, R.P.M., Vermeij, M.J.A. & Maier, C. 2005. Morphogenesis of the branching reef coral *Madracis mirabilis. Proceedings of the Royal Society B* **272**, 127–133, doi:10.1098/rspb.2004.2934

Kahramanoğulları, O., Bramanti, L. & Benedetti, M.C. 2019. Stochastic mechanisms of growth and branching in Mediterranean coral colonies. In *Theory and Practice of Natural Computing: 8th International Conference, TPNC 2019, Kingston, ON, Canada, December 9-11, 2019, Proceedings 8*. Heidelberg, Germany: Springer International Publishing, 57–69.

Kai, S. & Sakai, K. 2008. Effect of colony size and age on resource allocation between growth and reproduction in the corals *Goniastrea aspera* and *Favites chinensis. Marine Ecology Progress Series* **354**, 133–139.

Kemball, W.D. & Marshall, C. 1995. Clonal integration between parent and branch stolons in white clover: a developmental study. *New Phytologist* **129**(3), 513–521.

Kenkel, N.C. 1993. Modeling markovian dependence in populations of aralia nudicaulis. *Ecology* **74**(6), 1700–1706, doi:10.2307/1939928

Kruszyński, K.J., Kaandorp, J.A. & Van Liere, R. 2007. A computational method for quantifying morphological variation in scleractinian corals. *Coral Reefs* **26**, 831–840.

Kücken, M., Rinkevich, B., Shaish, L. & Deutsch, A. 2011. Nutritional resources as positional information for morphogenesis in the stony coral *Stylophora pistillata. Journal of Theoretical Biology* **275**(1), 70–77.

Lasker, H.R. & Sanchez, J.A. 2002. Astogeny and allometry of modular colonial organisms. *Reproductive Biology of Invertebrates* **11**, 207–253.

Lasker, H.R., Boller, M.L., Castanaro, J. & Sanchez, J.A. 2003. Determinate growth and modularity in a gorgonian octocoral. *The Biological Bulletin* **205**(3), 319–330.

Lesser, M.P., Weis, V.M., Patterson, M.R. & Jokiel, P.L. 1994. Effects of morphology and water motion on carbon delivery and productivity in the reef coral, *Pocillopora damicornis* (Linnaeus): diffusion barriers, inorganic carbon limitation, and biochemical plasticity. *Journal of Experimental Marine Biology and Ecology* **178**(2), 153–179.

Li, Y., Han, T., Bi, K., Liang, K., Chen, J., Lu, J., He, C. & Lu, Z. 2020. The 3D reconstruction of *Pocillopora* colony sheds light on the growth pattern of this reef-building coral. *Iscience* **23**(6), 101069.

Llabrés, E., Mayol, E., Marbà, N. & Sintes, T. 2022. A mathematical model for inter-specific interactions in seagrasses. *Oikos* **2022**, e09296, doi:10.1111/oik.09296

Llabrés, E., Blanco-Magadán, A., Sales, M. & Sintes, T. 2023. Effect of global warming on western mediterranean seagrasses: a preliminary agent-based modelling approach. *Marine Ecology Progress Series* **710**, 43–56, doi:10.3354/meps14298

Loya, Y. 1976. Skeletal regeneration in a Red Sea scleractinian coral population. *Nature* **261**, 490–491.

Marbà, N. & Duarte C.M. 1998. Rhizome elongation and seagrass clonal growth. *Marine Ecology Progress Series* **174**, 269–280, doi:10.3354/meps174269

Marbà, N. & Duarte, C.M. 2003. Scaling of ramet size and spacing in seagrasses: implications for stand development. *Aquatic Botany* **77**(2), 87–98, doi:10.1016/S0304-3770(03)00079-2

Marubini, F., Barnett, H., Langdon, C. & Atkinson, M.J. 2001. Dependence of calcification on light and carbonate ion concentration for the hermatypic coral *Porites compressa. Marine Ecology Progress Series* **220**, 153–162.

Mass, T., Genin, A., Shavit, U., Grinstein, M. & Tchernov, D. 2010. Flow enhances photosynthesis in marine benthic autotrophs by increasing the efflux of oxygen from the organism to the water. *Proceedings of the National Academy of Sciences* **107**(6), 2527–2531.

Matthai, G. 1926. VIII. Colony-formation in astrœid corals. *Philosophical Transactions of the Royal Society of London. Series B, Containing Papers of a Biological Character* **214**(411–420), 313–367.

Medellín-Maldonado, F., López-Pérez, A., Ruiz-Huerta, L. & Carricart-Ganivet, J.P. 2022. Understanding corallite demography to comprehend potential bias in sclerochronology: analysis of coral modular growth by micro-computed tomography. *Limnology and Oceanography* **67**, 2665–2676.

Merks, R.M.H. 2003. *Branching growth in stony corals: a modelling approach*. PhD thesis, University of Amsterdam, Netherlands.

Merks, R., Hoekstra, A., Kaandorp, J. & Sloot, P. 2003. Models of coral growth: spontaneous branching, compactification and the Laplacian growth assumption. *Journal of Theoretical Biology* **224**(2), 153–166.

Merks, R.M., Hoekstra, A.G., Kaandorp, J.A. & Sloot, P.M. 2004. Polyp oriented modelling of coral growth. *Journal of Theoretical Biology* **228**(4), 559–576.

Mistr, S. & Bercovici, D. 2003. A theoretical model of pattern formation in coral reefs. *Ecosystems* **6**, 0061–0074.

Muko, S., Kawasaki, K., Sakai, K., Takasu, F. & Shigesada, N. 2000. Morphological plasticity in the coral *Porites sillimaniani* and its adaptive significance. *Bulletin of Marine Science* **66**(1), 225–239.

Niklas, K.J. & Kerchner, V. 1984. Mechanical and photosynthetic constraints on the evolution of plant shape. *Paleobiology* **10**(1), 79–101.

Ohno, R., Sentoku, A., Masumoto, S. & Ezaki, Y. 2015. Morphological variability in azooxanthellate scleractinian dendrophylliids governed by regular modes of asexual reproduction: a computer simulation approach. *Paleontological Research* **19**(3), 195–203.

Okamoto, T. 1988. Analysis of heteromorph ammonoids by differential geometry. *Palaeontology* **31**, 35–52.

Oren, U., Benayahu, Y., Lubinevsky, H. & Loya, Y. 2001. Colony integration during regeneration in the stony coral *Favia favus*. *Ecology* **82**(3), 802–813.

Page, C.A. & Vaughan, D. 2014. The cultivation of massive corals using 'micro-fragmentaion' for the 'reskinning' of degraded coral reefs. *Benthic Ecology Meeting*, Jacksonville, Florida: University of North Florida.

Page, C.A., Muller, E.M. & Vaughan, D.E. 2018. Microfragmenting for the successful restoration of slow growing massive corals. *Ecological Engineering*, **123**, 86–94.

Paz-García, D.A., Aldana-Moreno, A., Cabral-Tena, R.A., García-De-León, F.J., Hellberg, M.E. & Balart, E.F. 2015. Morphological variation and different branch modularity across contrasting flow conditions in dominant *Pocillopora* reef-building corals. *Oecologia* **178**, 207–218.

Ramírez-Portilla, C., Bieger, I.M., Belleman, R.G., Wilke, T., Flot, J.F., Baird, A.H., Harii, S., Sinniger, F. & Kaandorp, J.A. 2022. Quantitative three-dimensional morphological analysis supports species discrimination in complex-shaped and taxonomically challenging corals. *Frontiers in Marine Science* **9**, 955582.

Rinkevich, B. 2000. A critical approach to the definition of Darwinian units of selection. *The Biological Bulletin* **199**(3), 231–240.

Rinkevich, B. 2002. The branching coral *Stylophora pistillata*: contribution of genetics in shaping colony landscape. *Israel Journal of Zoology* **48**(1), 71–82.

Rinkevich, B. & Loya, Y. 1985a. Coral isomone: a proposed chemical signal controlling intraclonal growth patterns in a branching coral. *Bulletin of Marine Science* **36**(2), 319–324.

Rinkevich, B. & Loya, Y. 1985b. Intraspecific competition in a reef coral: effects on growth and reproduction. *Oecologia* **66**, 100–105.

Rodolfo-Metalpa, R., Peirano, A., Houlbrèque, F., Abbate, M. & Ferrier-Pagès, C. 2008. Effects of temperature, light and heterotrophy on the growth rate and budding of the temperate coral *Cladocora caespitosa*. *Coral Reefs* **27**, 17–25.

Rosen, B.R. 1986. Modular growth and form of corals: a matter of metamers? *Philosophical Transactions of the Royal Society of London. B, Biological Sciences* **313**(1159), 115–142.

Routledge, R.D. 1990. Spatial patterns arising from plant dispersal as modelled by a correlated random walk. *Journal of Applied Probability* **27**(1), 1–13, doi:10.2307/3214591

Sanchez, J.A. 2004. Evolution and dynamics of branching colonial form in marine modular cnidarians: gorgonian octocorals. *Hydrobiologia* **530**, 283–290.

Sanchez, J.A. & Lasker, H.R. 2003. Patterns of morphological integration in marine modular organisms: supra-module organization in branching octocoral colonies. *Proceedings of the Royal Society of London. Series B: Biological Sciences* **270**(1528), 2039–2044.

Sánchez, J.A., Lasker, H.R., Nepomuceno, E.G., Sánchez, J.D. & Woldenberg, M.J. 2004. Branching and self-organization in marine modular colonial organisms: a model. *The American Naturalist* **163**(3), E24–E39.

Sanna, G. & Freiwald, A. 2021. Deciphering the composite morphological diversity of *Lophelia pertusa*, a cosmopolitan deep-water ecosystem engineer. *Ecosphere* **12**(11), e03802.

Schwarzschild, A.C. & Zieman, J.C. 2008. Apical dominance and the importance of clonal integration to apical growth in the seagrass *Syringodium filiforme*. *Marine Ecology Progress Series* **360**, 37–46.

Sentoku, A. & Ezaki, Y. 2012a. Constraints on the formation of colonies of the extant azooxanthellate scleractinian coral *Dendrophyllia arbuscula*. *Lethaia* **45**(1), 62–70.

Sentoku, A. & Ezaki, Y. 2012b. Regularity in budding mode and resultant growth morphology of the azooxanthellate colonial scleractinian *Cyathelia axillaris*: effective and adaptive ways of utilizing habitat resources. *Paleontological Research* **16**(3), 252–259.

Sentoku, A. & Ezaki, Y. 2012c. Regularity in budding mode and resultant growth morphology of the azooxanthellate colonial scleractinian *Tubastraea coccinea*. *Coral Reefs* **31**, 67–74.

Sentoku, A. & Ezaki, Y. 2012d. Regularity and polarity in budding of the colonial scleractinian *Dendrophyllia ehrenbergiana*: consequences of radio-bilateral symmetry of the scleractinian body plan. *Lethaia* **45**(4), 586–593.

Sentoku, A. & Ezaki, Y. 2013. Intrinsic constraints on sympodial growth morphologies of azooxanthellate scleractinian coral *Dendrophyllia*. *Plos One* **8**(5), e63790.

Sentoku, A., Ishibashi, M., Masumoto, S., Ohno, R., Tomiyama, T., Machiyama, H., Tadai, O. & Ezaki, Y. 2015a. Regular budding modes in a zooxanthellate dendrophylliid *Turbinaria peltata* (Scleractinia) revealed by X-ray CT imaging and three-dimensional reconstruction. *Journal of Morphology* **276**(9), 1100-1108.

Sentoku, A., Morisaki, H., Masumoto, S., Ohno, R., Tomiyama, T. & Ezaki, Y. 2015b. Internal skeletal analysis of the colonial azooxanthellate scleractinian *Dendrophyllia cribrosa* using microfocus X-ray CT images: underlying basis for its rigid and highly adaptive colony structure. *Journal of Structural Biology* **189**(1), 37–43.

Shaish, L., Abelson, A. & Rinkevich, B. 2006. Branch to colony trajectory in a modular organism: pattern formation in the Indo-Pacific coral *Stylophora pistillata*. *Developmental Dynamics* **235**(8), 2111–2121.

Shaish, L., Abelson, A. & Rinkevich, B. 2007. How plastic can phenotypic plasticity be? The branching coral *Stylophora pistillata* as a model system. *PloS One* **2**(7), e644.

Shwartsberg, M., Kizner, Z., Dubinsky, Z. & Bachar, A. 2012. Morphological growth response of *Stylophora pistillata* TO *in-situ* manipulations of light intensity and water flow regime. *Israel Journal of Ecology and Evolution* **58**(1), 69–85.

Silén, L. 1981. Colony structure in Flustra foliacea (Linnaeus) (Bryozoa, Cheilostomata). *Acta Zoologica* **62**(4), 219–232.

Sintes, T., Marbà, N., Duarte, C.M. & Kendrick, G.A. 2005. Nonlinear processes in seagrass colonisation explained by simple clonal growth rules. *Oikos* **108**(1), 165–175, doi:10.1111/J.0030-1299.2005.13331.X

Sintes, T., Marbà, N. & Duarte, C.M. 2006. Modeling nonlinear seagrass clonal growth: assessing the efficiency of space occupation across the seagrass flora. *Estuaries and Coasts* **29**(1), 72–80, doi:10.1007/BF02784700

Squires, D.F. 1964. Fossil coral thickets in Wairarapa, New Zealand. *Journal of Paleontology* **38**, 904–915.

Stephenson, T.A. 1931. Development and the formation of colonies in *Pocillopora* and *Porites*. *British Museum* **47**, 55–67.

Terrados, J., Duarte, C.M. & Kenworthy, W.J. 1997. Is the apical growth of *Cymodocea nodosa* dependent on clonal integration?. *Marine Ecology Progress Series* **158**, 103–110.

Todd, P.A. 2008. Morphological plasticity in scleractinian corals. *Biological Reviews* **83**(3), 315–337.

Todd, P.A., Sidle, R.C. & Lewin-Koh, N.J.I. 2004. An aquarium experiment for identifying the physical factors inducing morphological change in two massive scleractinian corals. *Journal of Experimental Marine Biology and Ecology* **299**(1), 97–113.

Tomlin, C.J. & Axelrod, J.D. 2007. Biology by numbers: mathematical modelling in developmental biology. *Nature Reviews Genetics* **8**(5), 331–340.

Veron, J.E.N. 2000. Corals of the world. *Australian Institute of Marine Science* **1**, 382.

Winston, J.E. 2010. Life in the colonies: learning the alien ways of colonial organisms. *Integrative and Comparative Biology* **50**(6), 919–933, doi:10.1093/ICB/ICQ146

Supplementary Materials are provided online at: https://www.routledge.com/9781032761961

Oceanography and Marine Biology: An Annual Review, 2024, **62**, 267-302

Taylor & Francis

THE CHANGING FOCUS OF CORAL-MACROPHYTE INTERACTIONS: A REVIEW OF FOUR DECADES OF RESEARCH

MARTINA BURGO[1,2] AND ANDREW S. HOEY[1*]

[1]*College of Science and Engineering, James Cook University, Townsville Queensland 4810, Australia*
[2]*Australian Institute of Marine Science, Townsville Queensland 4810, Australia*
**Corresponding author*
Email: Andrew.hoey1@jcu.edu.au
Ph: +61 (0)7 4781 5979

Abstract Coral reefs are one of the world's most biodiverse ecosystems, yet they are also one of the most threatened. The effects of local anthropogenic stressors are being greatly compounded by those of climate change, leading to widespread declines in coral cover over the past five decades. Declines in coral cover are inevitably accompanied by increases in other benthic taxa, including algae and cyanobacteria, that rapidly colonise the dead coral skeletons. Any recovery of coral assemblages, and the associated communities they support, therefore must generally occur against a backdrop of elevated macrophyte (i.e., algae and cyanobacteria) cover, and in some cases in areas dominated by large fleshy macroalgae. Understanding the mechanisms and outcomes of coral-macrophyte interactions, and how they vary among coral and macrophyte taxa, and among coral life stages is crucial for forecasting future reef configurations. We conducted a systematic review of the evidence on coral-macrophyte interactions to investigate taxonomic, functional and life stage biases in terms of both corals and macrophytes, and how these influence the outcomes of coral-macrophyte interactions. Across 169 studies and >1300 coral-macrophyte interactions, the majority of available data (73%) was based on adult corals, with relatively few studies investigating the potential effects of macrophytes on early life stages of corals (juvenile corals: 10%; larvae and recently settled corals: 17%). This bias towards adult corals is somewhat surprising given the importance of early-life stages to the replenishment of coral assemblages. Indeed, the negative effects of macrophyte interactions on coral health (largely survival and growth) were greatest for early-life stages (i.e., coral larvae, 45%; and juveniles, 38%) and considerably reduced for adult corals (3%). There were also regional differences in the taxonomic and functional identity of the coral and macrophyte taxa investigated, with coral-macrophyte interactions across the tropical Atlantic focusing on *Porites*, *Orbicella* and other massive and submassive corals, while those in the Indo-Pacific focused on *Acropora* and other branching corals. While macrophyte communities on coral reefs are typically diverse, and spatially and temporally variable, the vast majority of studies investigated the effect of a single macrophyte taxon in direct contact with a coral over a relatively short temporal scale. Very few studies considered the potential effects of multiple macrophyte taxa or broader habitat-scale effects of macrophyte assemblages on corals. Given the ongoing effects of climate change on coral reefs, and concerns that reefs may increasingly be overgrown by macroalgae, there is a pressing need to better understand the nature and outcomes of coral-macrophyte interactions to predict the recovery potential of coral assemblages.

DOI: 10.1201/9781003477518-7

Keywords: Competition; Macroalgae; Coral Reefs; Inhibition; Facilitation; Abrasion; Overgrowth; Allelopathy

Introduction

Coral reefs are one of the world's most biodiverse ecosystems, yet they are also one of the most threatened. The combined effect of climate change and local anthropogenic stressors have led to global declines in coral cover over the past five decades (Souter et al. 2021). Shallow water coral cover on the Great Barrier Reef (GBR) declined by ~50% from the 1960s to the early 2000s, and by ~80% on Caribbean reefs between 1977 and 2001 (Gardner et al. 2003, Bellwood et al. 2004, De'Ath et al. 2012). Following coral mortality, the dead coral skeletons are rapidly colonised by a diversity of other benthic taxa, in particular algae (Diaz-Pulido & McCook 2002), and in some cases, they may become the dominant benthic taxa. Indeed, reefs in the Atlantic (e.g., Jamaica: Hughes 1994), Indian (e.g. Seychelles: Graham et al. 2015) and Pacific Oceans (e.g., Fiji: Rasher et al. 2013, inshore GBR: Diaz-Pulido et al. 2009, Johns et al. 2018, Ceccarelli et al. 2020) have been overgrown by macroalgae. Once established, these alternate benthic configurations can be difficult to reverse due to a series of positive feedbacks that reinforce the new communities and inhibit the recovery of coral populations (Hoey & Bellwood 2011, Dell et al. 2016, Van De Leemput et al. 2016, Hughes et al. 2017, Clements et al. 2018). These changes in the composition and structure of benthic communities invariably affect the associated fish and invertebrate communities (Pratchett et al. 2011, Stella et al. 2011, Richardson et al. 2018), and ecosystems goods and services they provide (Graham & Nash 2013, Woodhead et al. 2019). Predicted increases in the frequency, intensity and diversity of disturbances that coral reefs will be exposed under ongoing and future climate change will undoubtedly lead to further declines in coral cover (Hughes et al. 2018a) and have led to concerns that reefs may be increasingly overgrown by macroalgae (Bellwood et al. 2019).

While the prevalence of shifts to macroalgal-dominance is debated (Bruno et al. 2009), the maintenance and recovery of coral populations will depend on the ability of corals to settle, grow and persist in areas of high macrophyte cover (Diaz-Pulido & McCook 2002, Diaz-Pulido et al. 2009, Ritson-Williams et al. 2009) (Figure 1). Corals, algae and cyanobacteria (together with other benthic organisms) compete for space on the benthos through a diverse range of mechanisms including shading, abrasion, allelopathy and disease. Abrasion and shading from algae can reduce the growth and fecundity of corals and cause local tissue loss (River & Edmunds 2001, Hughes et al. 2007, Brown & Carpenter 2015, Brown et al. 2019, Thinesh et al. 2019, Clements et al. 2020). Some algae and cyanobacteria have been shown to produce chemical compounds that can induce bleaching and subsequent tissue mortality in corals following direct contact (Rasher & Hay 2010, Bonaldo & Hay 2014). Close or direct contact between corals and macrophytes (i.e., algae and cyanobacteria) can also alter microbiome composition, increasing the abundance of coral pathogens (Morrow et al. 2017, Briggs et al. 2021). Moreover, algae can release dissolved organic matter (DOM) that stimulates microbial growth and thus the incidence of coral diseases (Roach et al. 2017). Co-occurrence of algae, DOM, pathogens and coral diseases has been referred to as the DDAM (DOM, diseases, algae and microbes) model that promotes coral mortality, thus facilitating further algal colonisation and expansion (Barott & Rohwer 2012).

This body of research on coral-macrophyte interactions has led to a widely accepted view that macroalgae *sensu lato* are a sign of reef degradation and inhibit the recovery of coral populations. However, much of the research to date has focused on a limited number of coral and macrophyte species, and/or specific life stages, and as such, the ubiquity of the effects of macroalgae is largely unknown. There is considerable variation in morphology, chemical and microbial composition among algal and cyanobacteria taxa (Vieira 2020), and likely their effect on corals. However, evidence on whether certain coral morphologies and life stages are more susceptible to interactions

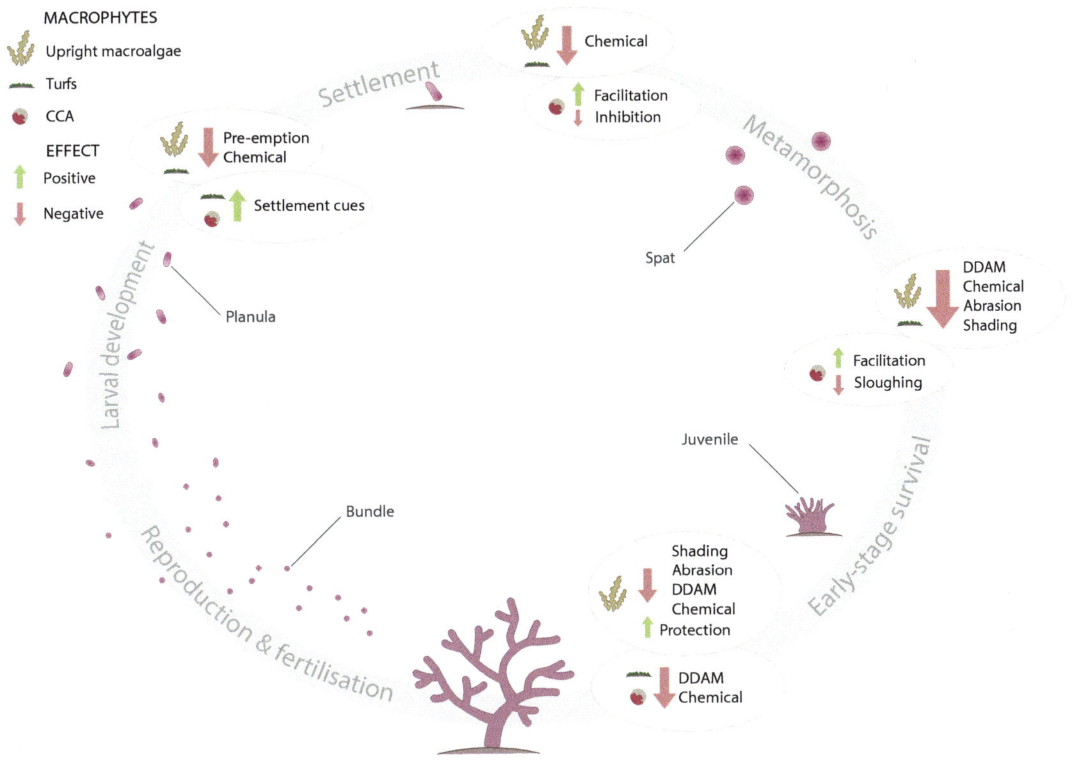

Figure 1 Diagram of the life cycle of corals showing the potential positive and negative effects of macrophytes on different coral life stages. White bubbles show the potential mechanisms by which each macrophyte category (i.e., upright macroalgae, turfs and CCA) has been shown to facilitate or inhibit corals. The arrows indicate the positive (green) and negative (pink) effects of macrophytes. 'DDAM' refers to a model of coral-macrophyte interactions that promotes coral mortality through co-occurrence of macrophytes, dissolved organic matter, pathogens and coral diseases (Barott & Rohwer 2012).

with algae or cyanobacteria or whether some macrophyte functional groups are better competitors is scarce. Moreover, the likelihood of encounter and the proximity between corals and macrophytes under natural settings are likely to be important in predicting competitive outcomes, yet there is limited evidence on spatial thresholds of coral-algal competition (Clements et al. 2018). Further, the vast majority of studies of coral-macrophyte interactions have focused on testing hypotheses of negative effects of macrophytes on corals, and consequently, there is an ascertainment bias against detecting positive effects of algae on corals or any effects of corals on macrophytes. Algae and/or cyanobacteria may have positive (e.g., Harrington et al. 2004, Ritson-Williams et al. 2010), negative (e.g., Rasher & Hay 2010, Arnold et al. 2010) or neutral (e.g., Olsen et al. 2016, Thinesh et al. 2019) on corals, while some corals may be able to outcompete algae (e.g., Diaz-Pulido et al. 2009).

We conducted a systematic literature review to synthesise current knowledge on (1) the effects of macrophytes throughout coral life stages, how these effects vary among macrophyte taxa, the spatial scale at which macrophytes can affect corals and the importance of physical versus chemical mechanisms. We then (2) evaluate the use of a trait-based approach to coral-macrophyte interactions and (3) examine the effect of corals on macrophytes. We restricted our analyses to macroscopic algae (namely, filamentous or turf algae, crustose coralline algae and macroalgae) and cyanobacteria and excluded microalgae. While cyanobacteria are not algae, they share many characteristics of algae, including the ability to rapidly colonise available benthic space and thereby compete with corals. By evaluating the effect of macrophytes on corals throughout their life cycle, we aim to provide a perspective of what is currently known on the role of macrophytes in reef recovery and identify future research directions. In a context of global loss of coral cover due to the ongoing effects of climate change and concerns that reefs may increasingly be overgrown by macroalgae, there is a pressing need to better understand the nature and outcomes of coral-macrophyte interactions to predict the recovery potential of coral assemblages and support reef management practices.

Materials and methods

Data collection

A systematic literature search was conducted to identify original research articles on coral-macrophyte interactions published between January 1980 and May 2022 using Scopus. To identify publications that had investigated coral-macrophyte competition, the search terms 'coral* AND *alga* OR seaweed* AND competition OR interaction' were used. The search targeted publications that included the search terms in their Title, Abstract and/or Keywords. The Scopus search returned 1504 potential publications (updated May 2022). To be included in the subsequent analyses the following selection criteria were applied: (1) interactions were between one or more species of hard (scleractinian) coral and macrophyte (interactions with seagrass were excluded); and (2) the outcome of interaction was explicitly quantified. Studies reporting the frequency of coral-algal interactions but not the outcome were excluded. Citations within all publications that met our selection were also examined to identify additional studies not captured within our Scopus search.

From each of the filtered publications, we extracted the following information: (1) publication metadata; (2) type (e.g., field experiment, lab experiment, survey), location and duration of the study; (3) taxonomic information and functional groups of both corals and macrophytes; (4) parameters measured (e.g., growth) and the interaction outcome (i.e., 'Positive', 'Neutral' or 'Negative'); (5) distance between interacting macrophytes and corals; (6) other abiotic factors (e.g., temperature); and (7) potential mechanisms of competition (e.g., physical through shading) and if/how they were tested (e.g., use of algal mimics and addition of antibiotics). If a study included multiple interactions (e.g., *Acropora* sp. vs *Lobophora* sp. and *Acropora* sp. vs *Dictyota* sp.), then a separate entry was recorded for each interaction. To allow comparisons among publications, interaction outcomes were classified as 'positive' when the presence of macrophytes had a significant positive effect on the survival and/or health of corals (e.g., protection from predation), 'negative' when macrophytes had a significant negative effect on corals (e.g., slower growth) or 'neutral' when the study did not find a significant difference between the presence or absence of macrophytes. We did not use proportional or absolute change in the measured parameters due to variation in the methods, parameter/s quantified and temporal scales among the filtered publications.

To investigate whether sensitivity to macrophyte competition changes with coral life stage, coral growth form and macrophyte functional group, organisms were classified into broader categories. Macrophytes were divided into three broad categories (turfs, crustose coralline algae: CCA, and upright macroalgae) and, where possible, into six of the morphological groups proposed by Steneck and Dethier (1994) (i.e., crustose algae, filamentous algae, leathery macrophytes, corticated macrophytes, articulated calcareous algae, corticated foliose algae and foliose algae). Corals were assigned to three life stages: larvae, juveniles and adults. Larvae include the free-swimming planulae and newly settled corals (or spat) up to 5 days post-settlement. Juvenile coral colonies were those settled corals >5 days old and up to 5 cm in maximum diameter (Rylaarsdam 1983), and adults were all colonies >5 cm in diameter. Adult corals were divided into branching, massive, tabular, plating, encrusting, free-living, solitary, corymbose and digitate growth forms.

Taxonomic classification was based on the information provided in each publication and was updated according to the Word Register of Marine Species (WoRMS) (WoRMS Editorial Board 2021) for corals, and AlgaeBase (Guiry & Guiry 2021) and WoRMS for macrophytes. Study sites were divided into regions according to the Sixth Status of Corals of the World 2020 Report (Souter et al. 2021, Australia, Brazil, Caribbean, East Asia, Eastern Tropical Pacific, Pacific, Red Sea and Gulf of Aden, South Asia, West Indian Ocean), with the addition of the Mediterranean region to include studies from that area. The duration of the study was also recorded, and where the study duration was reported as a range (e.g., 10–12 days), the average duration was used in the analyses.

To enable a comparison of the number of publications on coral-macrophyte interactions published each year relative to the total number of coral reef studies, a search was conducted on Scopus using the terms 'coral AND reef*' for studies published between January 1980 (earliest coral-algal interactions study) and May 2022.

Statistical analysis

To test whether the outcome of coral-macrophyte interactions could be explained by macrophyte functional groups, coral life stage and proximity between corals and macrophytes, a series of Cumulative Link Mixed Models (CLMMs) were used. CLMMs were selected as they allow an ordinal response variable (i.e., negative vs neutral vs positive effect) and the inclusion of random effects (i.e., macrophyte broad categories and coral health parameter). CLMMs were built using the package 'ordinal' (Christensen 2019). The package 'emmeans' was used to interrogate all model outputs (Lenth 2023).

A CLMM was used to investigate differences in the effect of macrophytes on the survival of coral among coral life stages. The model was standardised across broad macrophyte categories (i.e., upright macroalgae, turfs and CCA). To investigate the relative effect of broad macrophyte categories across coral life stages, a CLMM was built and standardised across different study types (i.e., laboratory experiments, field experiments and field surveys).

To investigate whether coral-macrophyte interaction outcomes (i.e., positive vs. neutral vs. negative) could be predicted by using macrophyte functional groups, another CLMM was built using coral health parameter as random effect. Due to the paucity of data, it was not possible to identify clear trends and further compare whether macrophyte morphological groups were better predictors than broad macrophyte categories.

Finally, a CLMM was built to investigate the effect of distance on the outcomes of coral-macrophyte interactions. However, further analysis was not possible (e.g., to identify threshold distances) due to the paucity of available data reporting the distance between organisms.

All analyses were performed in R (4.3).

Results

Summary of available data

Spatial and temporal distribution of publications

Of the 1,504 publications returned by Scopus, 118 publications satisfied our selection criteria. A further 51 papers were added outside of the Scopus systematic search (total of 169 publications). Of these, the greatest number of studies was conducted in the Caribbean (66 of 169 publications; 39%), followed by the Pacific (n=40; 24%) and Australia (n=32; 19%) regions (Figure 2A). Of the 32 studies conducted in Australia, the vast majority (n=30 out of 32) were from the GBR. Similarly, of the 1317 coral-macrophyte interactions identified across the 169 studies, the Caribbean, Pacific and Australia accounted for 38.0%, 30.8% and 12.3% of interactions, respectively. In contrast, only

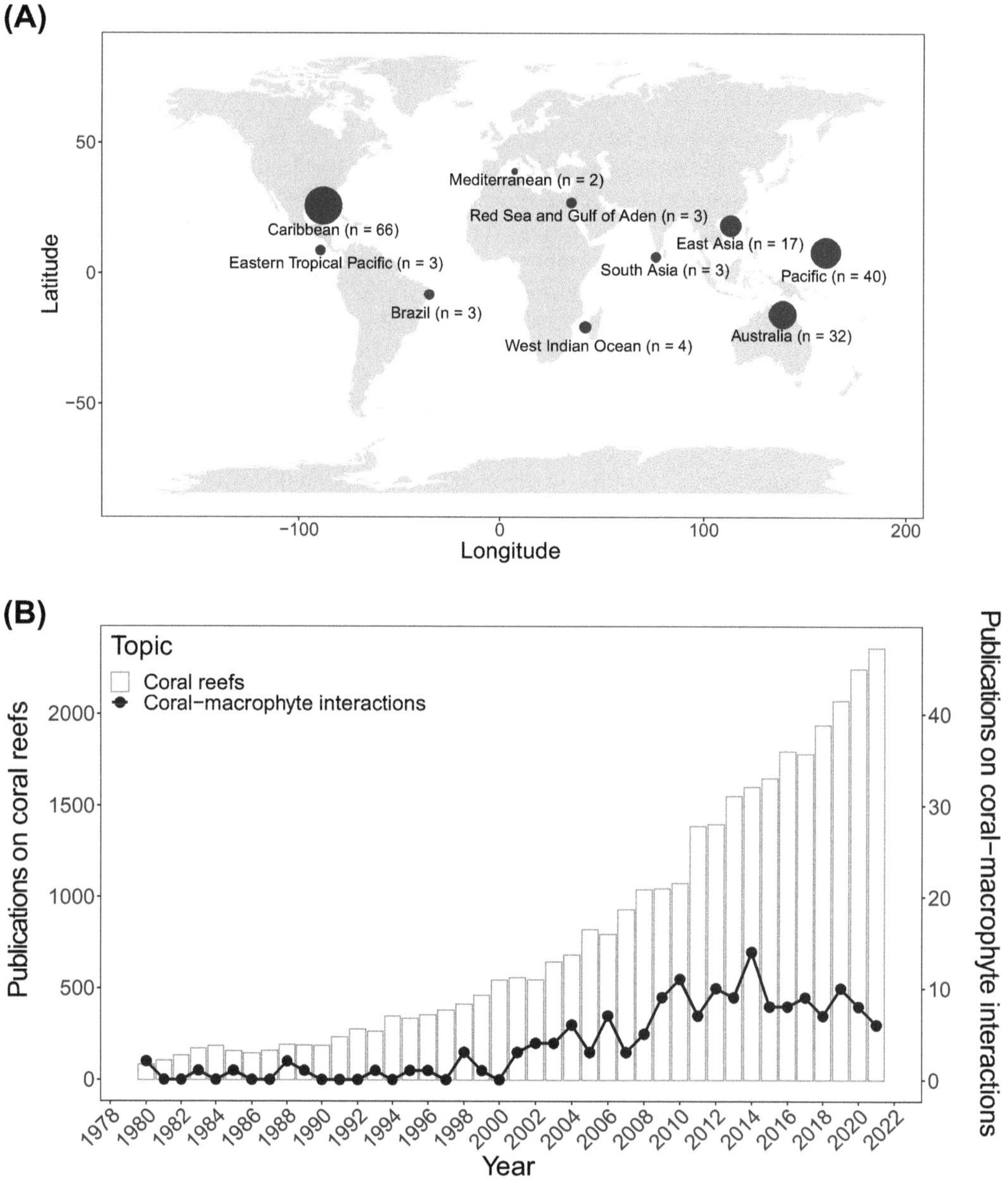

Figure 2 Spatial and temporal variation of published studies on coral-macrophyte interactions. (A) Spatial distribution of studies that have examined coral-macrophyte interactions among eleven regions globally; (B) Number of studies published each year on coral reefs (white bars; left *Y*-axis) and on coral-macrophyte interactions (black line; right *Y*-axis). Data for 'coral reefs' are based on a systematic search in Scopus (1980–2022) using search terms 'coral AND reef*'.

11.9% of interactions were from studies in the East Asia region, and no interactions or studies were reported from the Arabian plate region (Figure 2A).

Our search found that while the annual number of publications on "coral reefs" has increased at an increasing rate between 1980 and 2022, the annual number of publications on coral-macrophyte interactions has remained low and has not kept pace with the increases in publications on coral

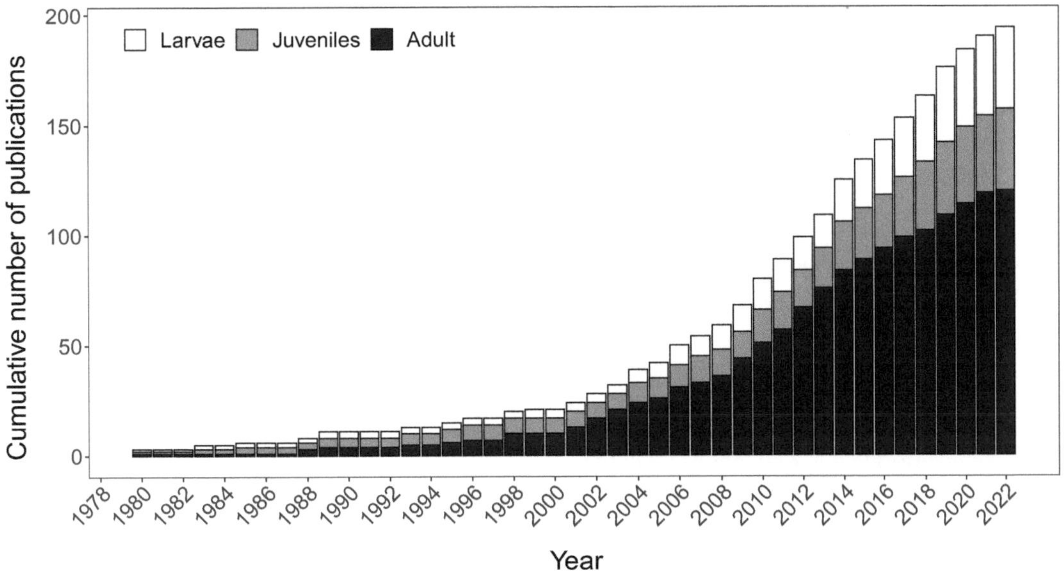

Figure 3 Cumulative number of publications on coral-macrophyte interactions from 1978 to 2022. The life stage of corals is indicated by the different shades; adults: *black*; juveniles: *grey*; larvae: *white*. See Methods for coral life stage classification.

reefs (Figure 1B). Consequently, the number of publications on coral-macrophyte competition relative to the total number of coral reefs has decreased from an average of 0.7% in 1980–1990 to 0.5% in 2011–2020 (Figure 2B). Initially, the number of publications on coral-macrophyte interactions was relatively consistent (<4 publications per year) from 1980 to 2000, increased gradually to ~5–8 publications per year in 2010 and stabilised at ~ 8 publications from 2010 to 2022.

There was considerable variation in the coral life stages examined, with the vast majority of all coral-macrophyte interactions being based on adult corals (72.6%), with relatively few focused on coral larvae (17.1%), and juvenile corals (9.9%). A small percent of studies (0.5%) did not report the life stage or provide any information on the size of the coral colonies. Interestingly, the number of studies on coral-macrophyte interactions was relatively evenly distributed among life stages from 1980 to 2000 (<1 study year^{-1}); however, from 2000, there was a disproportionate increase in studies focused on adults (~5 studies year^{-1}) as opposed to early-life stage corals (~1 study year^{-1}; Figure 3). These trends were largely consistent among regions (Supplementary Figure 1; Supplementary Table 1).

Taxonomic diversity

The selected 169 publications recorded a total of 1317 interactions between 71 coral species from 34 genera and 15 families, and 115 macrophyte species from 68 genera. For both corals and macrophytes, a few taxa were disproportionately represented (Supplementary Figure 2). Of the 34 coral genera investigated, *Porites* was the most common coral (24% interactions), followed by *Acropora* (21.4%), *Orbicella* (9.9%) and *Pocillopora* (8.4%). The most common families were Acroporidae (28.2% interactions), Poritidae (24.6%) and Merulinidae (13.6%). In terms of growth forms, branching and massive colonies were the most common, being studied in 41.8% and 40.2% of coral-macrophyte interactions, respectively (Supplementary Figure 3, Supplementary Table 2). Coral taxa were not reported in 6.6% of interactions. Many of these were field studies of coral settlement (44 of 87 interactions) where identifying recently settled corals to genus level is difficult (Babcock et al. 2003), or field surveys (35 of 87 interactions) quantifying overgrowth of corals by macrophytes.

The number of coral genera included in studies of coral-macrophyte interactions was relatively similar between the tropical Atlantic and Indo-Pacific (18 and 16 genera, respectively); however, the identity of the coral taxa differed between regions (Figure 4). The most common coral genera included in studies from the tropical Atlantic were *Porites* and *Orbicella* (each 25.5% of interactions), and *Acropora* (11.9%). In contrast, *Acropora* (27.9%), *Porites* (23.9%) and *Pocillopora* (13.7%) were the most common coral taxa investigated in the Indo-Pacific.

The most common coral morphology was massive (71.4%) in the tropical Atlantic, while data from the Indo-Pacific are more diverse, with branching (59.2%) being the most common coral growth form followed by massive (21.8%) (Supplementary Figure 3, Supplementary Table 2). Such differences likely reflect the limited number of branching coral taxa in the tropical Atlantic (Souter et al. 2021).

The taxonomic identity of the macrophyte/s was not reported in 19.4% of interactions, with only broad classifications (e.g., CCA, turf and upright macroalgae) provided, with no information beyond 'algae' or 'macroalgae' in 1.4% of interactions (Supplementary Table 3). Of the interactions with reported taxonomic information, *Lobophora* accounted for 16.9% of interactions, followed by *Halimeda* (7.5%), *Dictyota* (7.4%) and *Sargassum* (4.6%). In terms of broad macrophyte categories, most interactions involved upright macroalgae (67.7%), followed by turfs (14.9%) and crustose algae (16.0%).

For publications that did report the genus or some morphological information, macrophytes were classified into functional groups. Corticated foliose algae (e.g., *Lobophora* and *Padina*) accounted for 27.7% of interactions, followed by crustose algae (16.0%; e.g., *Porolithon* and *Peyssonnelia*) and articulated calcareous algae (12.2%; e.g., *Galaxaura* and *Halimeda*). Approximately one in every five coral-macrophyte interactions (i.e., 18.5%) did not report any information that could allow further classification.

As for corals, there were marked regional differences in the macrophytes considered in studies of coral-macrophyte interactions. Most interactions in the tropical Atlantic focused on three genera (*Lobophora*, 16.0%; *Dictyota*, 15.2%; *Halimeda*, 15.0%) with other genera accounting for <2% each (total number of genera=33). Interestingly, *Sargassum* was included in <1% of interactions in the tropical Atlantic, despite reports of this genera overgrowing large parts of reefs in the Caribbean (Wilkinson & Souter 2008). *Lobophora* (18.5%) was also the most common macrophyte included in Indo-Pacific studies; however, the remaining interactions were distributed among a greater number of genera ($n=50$), with *Sargassum*, *Chlorodesmis* and *Galaxaura* accounting for 7.2%, 5.8% and 5.2% of interactions, respectively.

In contrast to coral morphologies, the relative proportion of macrophyte morphological groups used in coral-macrophyte interactions was broadly consistent between regions. Corticated foliose algae and crustose algae are the most common groups in both the tropical Atlantic and Indo-Pacific regions, followed by articulated calcareous algae and leathery macrophytes, respectively (Supplementary Table 4).

Experimental approach and duration

Of the 169 publications examined, 138 publications conducted manipulative experiments, either in the field ($n=95$) or in the lab ($n=52$), while 38 were based on field surveys/observations. While the rank order of the experimental approach was consistent between the tropical Atlantic and Indo-Pacific, there were marked differences in the relative proportion of studies between regions. Studies in the tropical Atlantic were biased towards field experiments (62.7% of publications) with the remaining publications spread relatively evenly between lab experiments (22.7%) and field surveys (14.7%). In contrast, studies from the Indo-Pacific were more evenly spread among field experiments (44.2% of publications), lab experiments (30.8%) and field surveys (25.0%). Relatively few studies have incorporated multiple approaches ($n=14$). Six studies used a combination of lab and field experiments, and six studies combined lab or field experiments with surveys in situ. Only two publications used a combination of all three study types.

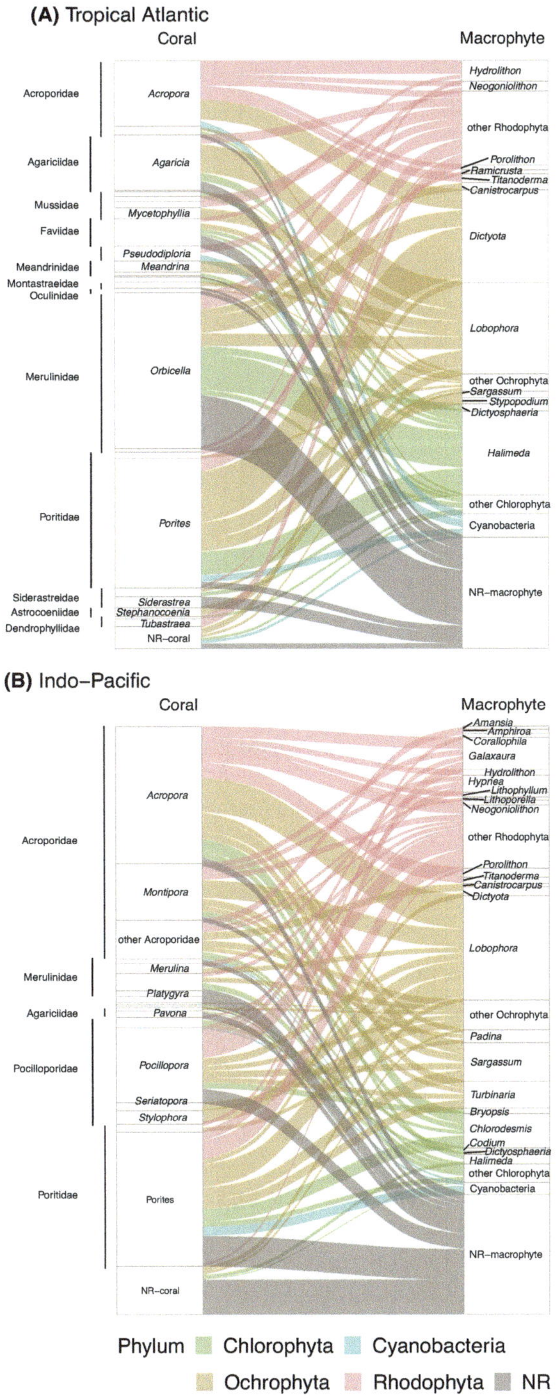

Figure 4 Differences in the coral and macrophyte genera used in studies of coral-macrophyte interactions between the (A) tropical Atlantic and (B) Indo-Pacific regions. Relative frequency of coral-macrophyte interactions between coral (left) and macrophyte genera (right). Coral and macrophyte genera with a relative frequency < 0.005 were grouped into 'other' phyla and family, respectively. Colours indicate macrophyte phyla. 'NR-coral' and 'NR-macrophyte' refer to interactions for which the taxonomic information was not reported in the original publication.

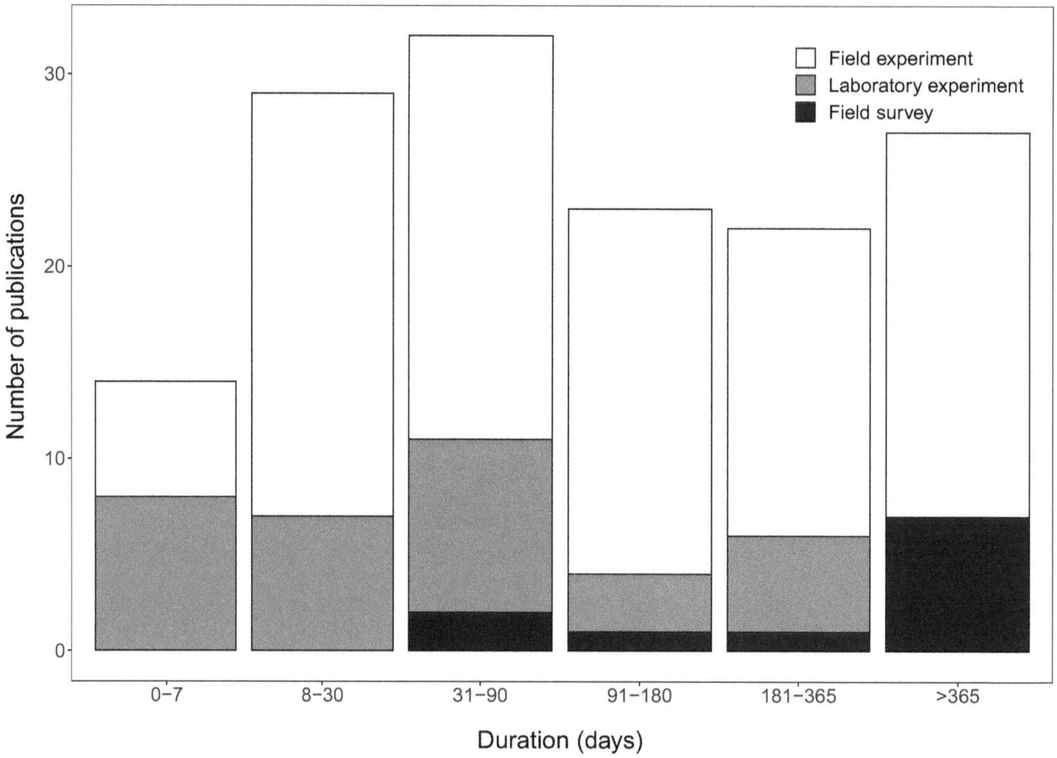

Figure 5 The frequency distribution of the duration of studies of coral-macrophyte interactions. Studies that did not directly investigate outcomes of coral-macrophyte interactions through time (e.g., one-off surveys) were excluded. Shades indicate the type of study; field survey: *black*; laboratory experiment: *grey*; field experiment: *white*.

The median duration of experiments was 74 days (range: 2–1460 days; Figure 5), with laboratory experiments generally being the shortest duration (median: 33 days, range: 2–240 days), followed by field experiments (median: 90 days, range: 2–1095 days) and field-based surveys (median: 545 days, range: 60–1460 days). Out of the 11 field surveys (i.e., studies recording the natural occurrence of coral-macrophyte interactions) that included a temporal aspect of coral-macrophyte interactions (e.g., seasonality of interactions), one study did not follow the same colonies over time but rather sampled random colonies for each trip.

Studies based on field experiments and field surveys were generally of longer duration in the tropical Atlantic (field experiments: median = 82 days, field surveys: 1200 days), but lab experiments were shorter (median: 4 days) than those in the Indo-Pacific (field experiments: 90 days, field surveys: 378 days; lab experiments: 56 days).

Of the 169 publications, almost all ($n = 164$) were focused on the effect of macrophytes on corals, with most coral responses being recorded as survival (57 publications), macrophyte overgrowth ($n = 43$), settlement ($n = 39$), growth/calcification ($n = 30$), bleaching ($n = 30$), and partial mortality/tissue loss ($n = 29$). Interestingly, only five publications considered the effect of macrophytes on the fecundity of corals, despite reproduction being an important component of coral replenishment following disturbances (Supplementary Tables 5 and 6).

There were considerable differences in both the coral and macrophyte taxa used among the three coral life stages across both field and laboratory experiments (Figure 6). Most studies involving coral larvae focused on the potential negative effects of macrophytes (24 studies out of 50) or

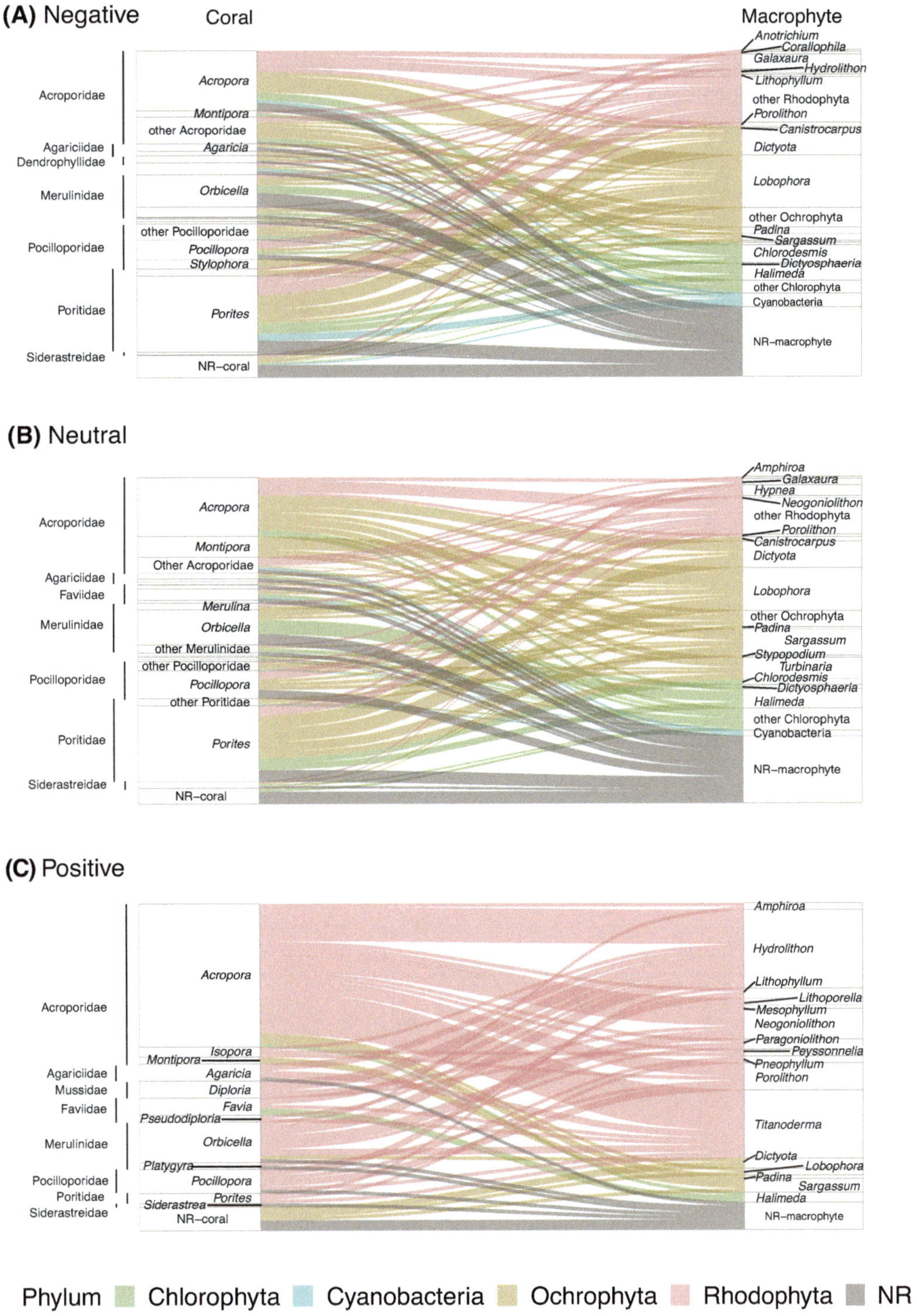

Figure 6 Relative frequency of coral-macrophyte interactions by observed outcome (A) Negative, (B) Neutral and (C) Positive. The relative frequency of coral-macrophyte interactions between various coral (left) and macrophyte (right) genera. Corals and macrophytes with frequency < 0.005 were grouped into 'other' phyla and family, respectively. Colours indicate macrophyte phyla. 'NR-coral' and 'NR-macrophyte' refer to interactions for which taxonomic information was not reported in the original publication.

the positive effects of CCA in inducing settlement (n=15), while very few publications (n=10) considered the effects of turfs. One study did not provide any detail of the macrophytes involved. Over a third of experiments on the effect of macrophytes on coral larvae used *Acropora* (34.7%), followed by *Porites* (20.4%) and *Pocillopora* (10.2%). Similarly, experiments on juvenile corals were largely based on the effect of macroalgae (26 studies out of 41), while few interactions considered the effects of CCA (n=9) and turfs (n=6), and three studies provided no information on the identity of the macrophytes. The juvenile coral taxa used were evenly distributed between *Acropora* and *Porites* (17.4% each), while 26.1% of studies did not report the coral genus. For adult corals, 88 out of 154 studies were based on interactions with macroalgae, 46 on turfs, 16 on CCA and 4 provided no information on the identity or composition of the macrophytes. The most common genus used in adult coral experiments was *Porites* (26.7%), followed by *Acropora* (12.6%) and *Orbicella* (11.2%).

Although most interactions among corals and macrophytes on reefs will involve multiple macrophyte taxa and/or thalli, most studies (n=92 out of 169) investigated interactions between single coral-macrophyte pairs (i.e., *Dictyota* vs *Acropora*), 47 examined the effect of the overall macrophyte assemblage (e.g., using macrophyte removal) and a further 30 studies investigated both. Of the total 77 studies that quantified the effect of macrophyte assemblages on corals, only 16 provided any detail of the composition of the macrophyte assemblage. Most studies only reported the dominant taxa.

Coral replenishment in algal habitats

The likelihood of a positive outcome of coral-macrophyte interactions differed depending on both the coral life stage and the algal broad category (Figure 7). For coral larvae, CCA had a higher probability of having a positive effect on corals (mainly through settlement facilitation), while turfs

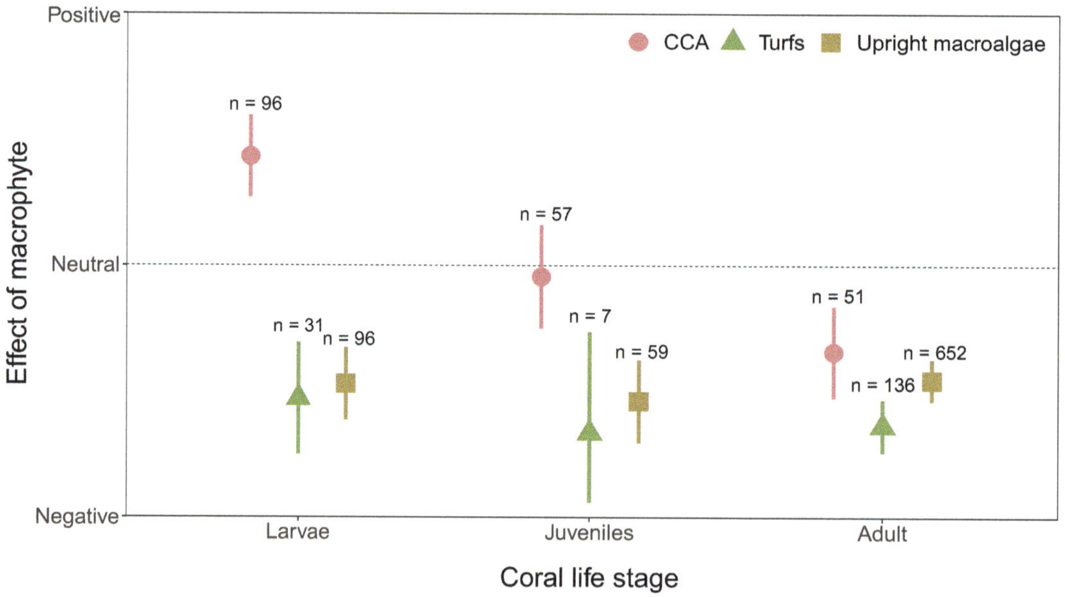

Figure 7 Relative effect of broad macrophyte categories across coral life stages. The likelihood of a positive effect of macrophyte on corals is shown. The bars indicate the confidence interval. Colours and shapes indicate different macrophyte broad categories; CCA: pink circles; turfs: green triangles; upright macroalgae: brown squares. The numbers above each bar indicate the number of coral-macrophyte interactions used in the analysis. The data used for this analysis include both lethal (i.e., mortality) and sub-lethal (e.g., growth and tissue loss) effects of macrophytes on corals.

and upright algae showed a higher probability of a negative effect (e.g., reduced survival or inhibiting settlement). While the likelihood of upright macroalgae having a negative effect on corals was relatively consistent across coral life stages, the probability of CCA negatively affecting corals increased with ontogeny. Turfs followed a similar pattern to CCA, but the change was less pronounced among life stages. Due to limited data available for each macrophyte morphological group, it was not possible to investigate the relative effect of these on corals.

Across all life stages, existing studies indicate that macrophytes can negatively affect coral health with 51.4% of interactions resulting in reduced growth and/or survival of the corals, 39.7% reported no effect, while 8.9% reported positive effects of macrophytes on corals. Comparing the outcomes of coral-macrophyte interactions across coral life stages revealed that any effects on adult corals were more likely to be sub-lethal (i.e., reduced growth or partial mortality) with few studies reporting colony mortality of adult colonies (2.5% of interactions lead to mortality; Figure 8). In contrast, more than half of macrophyte interactions with larval or recently settled corals and a third of interactions with juvenile corals lead to mortality (coral larvae: 44.9%; juvenile corals: 37.5%; Figure 8).

Together with mortality, interactions with macrophytes were reported to cause bleaching (8.7% of interactions), reduced photosynthesis (6.8%), tissue loss (4.3%) and reduced growth (4.2%) in corals (Supplementary Tables 5 and 6). These effects were not, however, ubiquitous with almost equal numbers of interactions failing to detect any effect of macrophytes on bleaching (7% of interactions), photosynthesis (5.2%), growth (3.1%) or tissue loss (2.5%) of corals. Some coral-macrophyte interactions (8.8% of all interactions) were positive, but almost all came from studies looking at the potential role of CCA as a facilitator of coral settlement (5.9% of all interactions). However, 7% of coral-algal interactions examined still resulted in a negative effect of algae on coral settlement

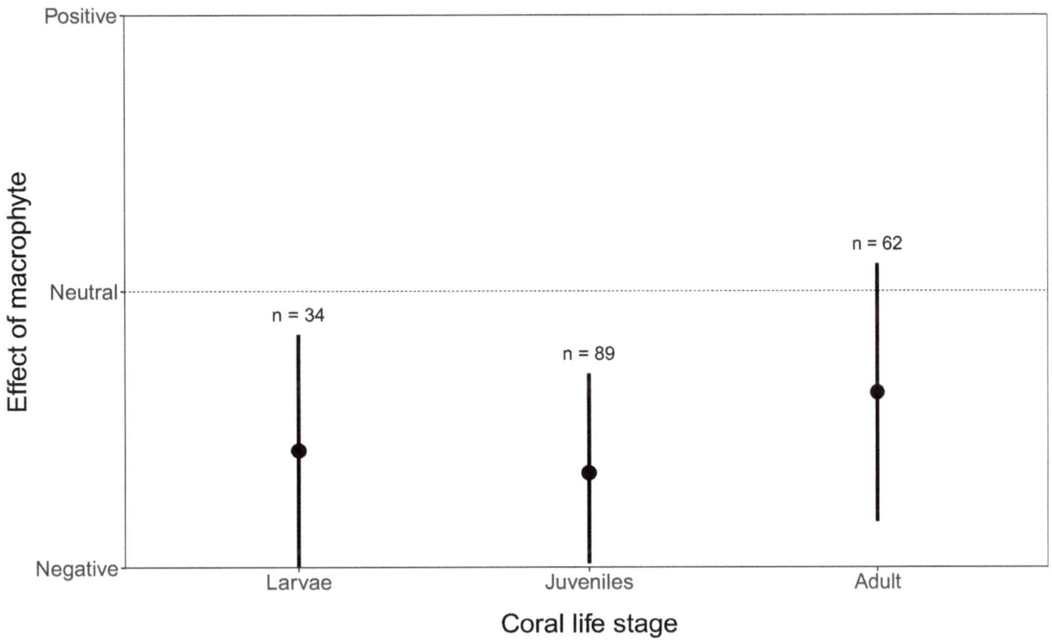

Figure 8 Variation in the effect of macrophytes on coral survival among coral life stages. The likelihood of a positive effect of macrophytes on coral survival is shown. The bars indicate the confidence interval. The number above each bar indicates the number of coral-macrophyte interactions used for this analysis. While there was no difference in the likelihood of survival between coral larvae and juveniles, adult corals were more likely to survive in the presence of macrophytes compared to early life stages (p-value = 0.05).

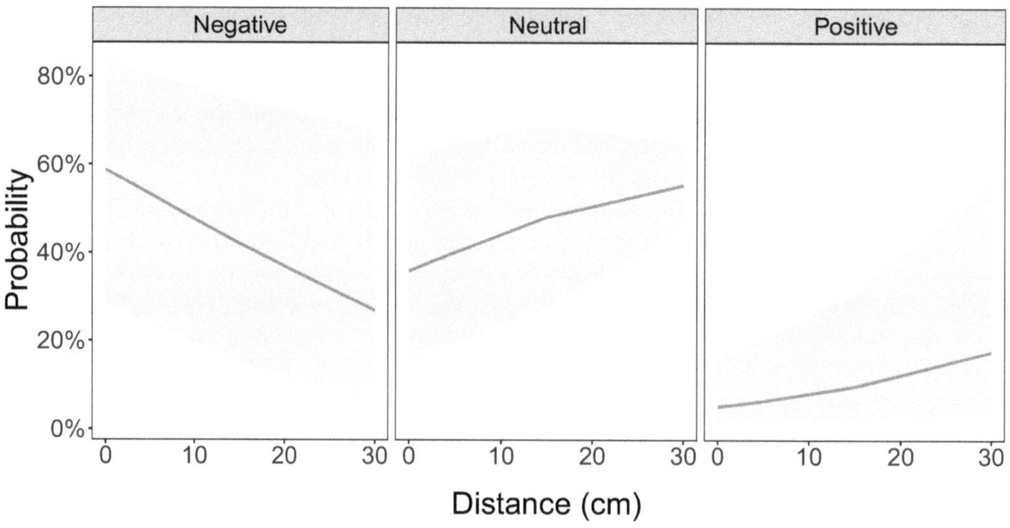

Figure 9 The effect of proximity between coral and macrophytes on the likelihood of a negative, neutral or positive effect on corals. The increasing confidence intervals with distance are due to the limited number of studies that included distances > 5cm between interacting coral and macrophyte taxa.

(i.e., settlement inhibition). Very few interactions (<1% of interactions for each parameter) resulted in a positive effect of macrophytes on coral health (e.g., protection from bleaching). Few studies are available on the effect of macrophytes on coral diseases (n=6) and fecundity (n=5).

Of the 169 studies considered, 118 studies reported the proximity of the corals and macrophytes, and the remaining 51 studies provided no information on the distance between corals and macrophytes. Of all interactions that reported the distance between corals and macrophytes, 83.4% were in direct contact and 16.6% were nearby but not touching (range: 0.5–30cm; median: 2cm). The likelihood of having a negative effect of macrophyte on corals decreased with distance between organisms, while the likelihood of having a neutral or positive effect increased (p-value <0.05; Figure 9), although the nature and strength of these relationships may be influenced by the few studies at which macrophytes were 20–30cm from the focal coral/s.

Do corals affect macrophytes?

Of the 169 studies, 138 investigated the effects of macrophytes on corals, five examined the effects of corals on macrophytes and 26 examined both. Of the 31 studies on the effect of corals on macrophytes, most are from the Indo-Pacific (n=20), ten from the tropical Atlantic and one from the Mediterranean. Upright macroalgae (n=24) were the most studied group, while fewer publications focused on turfs (n=10) and CCA (n=5). Most studies (n=13) recorded the effect of corals on macrophytes as 'overgrowth' but did not quantify changes in macrophyte health (e.g., photosynthesis).

Overall, most studies recorded negative (n=22) or neutral (n=14) outcomes of corals on macrophytes (Supplementary Tables 7 and 8). Only two studies reported a positive effect of corals on macrophytes (Figure 10). The effects of corals on CCA were always negative, while the effects of corals on turfs and macroalgae were neutral or positive suggesting that these macrophyte groups have more effective defence mechanisms against corals. The limited number of studies that quantified the effects of corals on macrophytes precluded any meaningful statistical comparisons.

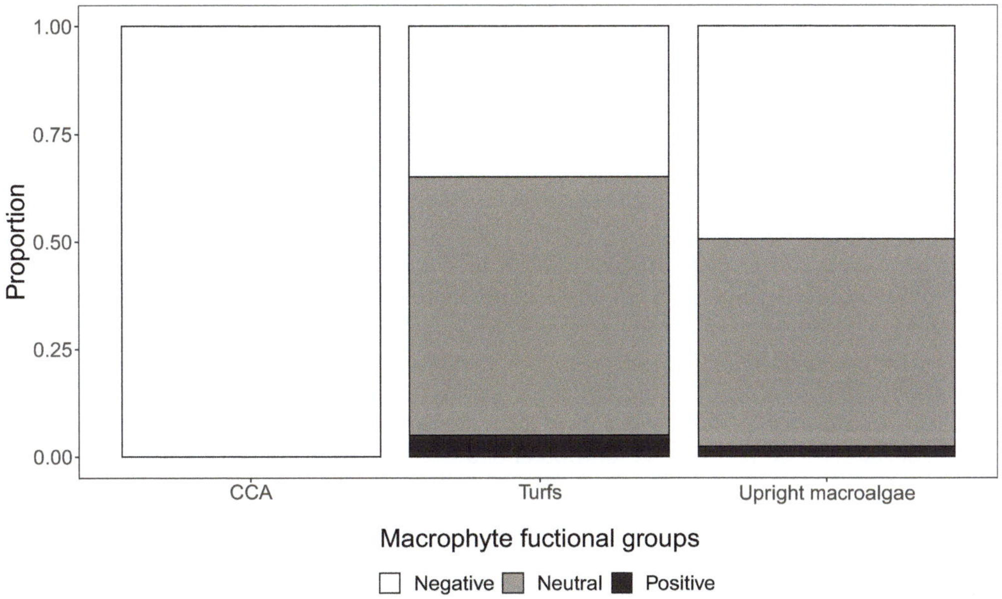

Figure 10 Differences in the effect of corals on crustose coralline algae, turfs, and upright macroalgae. The shades indicate whether macrophytes had a positive (black), neutral (grey) or negative (white) effect on the macrophyte.

Mechanisms of coral-algal interactions

Fifty-six studies attempted to identify the mechanism driving the interactions between corals and macrophytes (37.5% of all interactions, and all concerned with the effects of macrophytes on corals) using algal mimics, algal extracts and microbial sampling. Most studies only considered a single mechanism, so we were not able to use statistical analysis to identify trends. For example, studies that investigated the potential for macrophyte chemicals to influence coral health did not consider the potential importance of microbial or physical mechanisms. As such, the number of studies that identified chemical (n=48) and physical (n=15) mechanisms likely reflects the interests of specific researchers rather than the relative prevalence of mechanisms among macrophyte taxa. Of the 38 macrophyte genera that were reported to affect corals through chemical mechanisms, *Lobophora* (n=16) and *Dictyota* (n=13) were the most common genera. *Dictyota* was also found to affect corals through physical mechanisms (n=4). Twelve genera were found to physically affect corals, with *Sargassum* (n=4) and *Dictyota* being the most common. The most common physical mechanism identified was abrasion (n=6), although studies did not rule out the potential effect of shading.

Discussion

Interactions between corals and benthic macrophytes are widespread on coral reefs and have the potential to shape the future configuration of coral reefs, and the goods and services they provide. Over the past several decades, regional declines in coral cover have been accompanied by increases in macrophytes, as well as other benthic organisms (Souter et al. 2021). The persistence and recovery of coral populations will depend, at least in part, on the outcomes of diverse interactions between coral and macrophytes under changing environmental conditions. Comparison among 169 studies and >1300 individual coral-macrophyte interactions revealed that the majority of studies and hence our understanding of coral-macrophyte interactions come from two regions: the Caribbean and Australia's Great Barrier Reef. Importantly, most studies focused on a limited suite of both coral and macrophyte taxa, with interactions in the tropical Atlantic being dominated by *Porites* and *Orbicella* with *Dictyota, Lobophora* and *Halimeda*, while in the Indo-Pacific were dominated by *Acropora*, *Pocillopora* and *Porites* with *Lobophora*. Despite the importance of the early life stages of corals (i.e., larvae, recently settled spat and juveniles) in the replenishment and recovery of populations following disturbance, and the generally greater effect of macrophytes on early life stages, the vast majority of studies investigating coral-macrophyte interactions (73%) have focused on adult life stages of coral. While macrophyte communities on coral reefs are typically diverse, and spatially and temporally variable, most studies focused on the effect of a single macrophyte species when in direct contact with a coral over a relatively short temporal scale, with few studies considering broader habitat-scale effects of macrophyte assemblages on corals.

The spatial distribution of available data on coral-macrophyte interactions was biased towards the tropical Atlantic region, with over a third of all studies being conducted in this region. Although this area hosts ~11% of the world's reef (Souter et al. 2021), it has a long history of human use and exploitation (Pandolfi 2003, Jackson et al. 2014) and provides the seminal example of a phase shift from coral- to macroalgal-dominance (Hughes 1994, Edmunds & Carpenter 2001). The persistence of this shift to macroalgal dominance and the generally high macrophyte cover on many Caribbean reefs (Williams & Polunin 2001, Jackson et al. 2014, Souter et al. 2021) likely contributed to the elevated interest in coral-macrophyte interactions in the region. In contrast, East Asian reefs account for ~30% of global shallow water coral reefs, yet only *ca.*12% of publications on coral-algal interactions were conducted in this region. Coastal regions throughout East Asia support dense human populations that are largely dependent on coral reefs as a source of food and/or income (Sing Wong et al. 2022). Given these local human stressors, coupled with ongoing and future effects of climate change, future declines in coral and increases in macrophytes on these reefs are likely.

Together with the regional variation, our analyses revealed temporal variation in studies of coral-macrophyte interactions. Prior to 2000, the number of studies on coral-macrophyte interactions was relatively low and consistent among years (~1 study per year); however, this increased to *ca.* 5–8 studies per year (2000–2022). This increase in studies of coral-algal interactions broadly coincided with the 1998 global coral bleaching event (Berkelmans & Oliver 1999, Goreau et al. 2000, Glynn et al. 2001) and likely reflected a growing concern of the influence macrophytes could have on the future structure of coral reefs. Despite this initial increase, and the growing number and diversity of threats to which coral reefs are being exposed, studies of coral-macrophyte interactions have not kept pace with the increasing number of studies on coral reefs more broadly.

Bottlenecks to coral replenishment

Across all regions, we found a strong bias in the life history stages of corals examined, with *ca.* 75% of coral-macrophyte interactions being based on adult (>5 cm diameter) corals. While adult corals are easier to manipulate and track through time, early life stages (larvae, recently-settled spat and juveniles) are critical for the replenishment of coral populations and recovery of populations following disturbance (e.g., Gilmour et al. 2013). Our comparisons show that the early life stages of corals were the most adversely affected by contact with, or the presence of macrophytes, with these interactions resulting in mortality of coral larvae and juveniles in 45% and 38% of interactions, respectively, compared to 3% for adult corals.

There was also a taxonomic bias in studies of the effects of macrophytes on the early life stages of corals with almost half (45%) of the interactions between macrophytes and coral larvae being based on a single genus, *Acropora*. While branching and plating *Acropora* corals can be key in the recovery of coral cover, especially on Indo-Pacific reefs, due to their fast growth rates (Pratchett et al. 2015, 2020, Ortiz et al. 2021), they are also one of the most sensitive taxa to both thermal and mechanical stress (Loya et al. 2001, Madin & Connolly 2006) and are often the preferred prey of coral predators (Baird et al. 2013). However, the increasing intensity of climate change stressors is altering the differential bleaching susceptibility among taxa (Hughes et al. 2018a, Burn et al. 2023), with *Acropora* being lost, or markedly reduced, from some Indo-Pacific regions following disturbance (Berumen & Pratchett 2006, Bento et al. 2016). Moreover, *Acropora* is only represented by two species (*Acropora palmata* and *Acropora cervicornis*) in the tropical Atlantic, both of which are listed as critically endangered by the International Union for Conservation of Nature (IUCN). Together with *Acropora,* a considerable number of studies on interactions between macrophytes and coral larvae focused on *Porites* (20%) and *Pocillopora* (10%). Collectively, these three coral genera accounted for 65% of all studies between macrophyte and coral larvae, with relatively little or no investigation of the remaining *ca.* 230 genera (Kitahara et al. 2016). Greater focus is needed on the effects of macrophytes on the early life stages of corals and across a broad range of taxa, morphologies and life histories. Moreover, there is a need to investigate the interactions between early life stages of both macrophytes and corals (e.g., algal propagules vs coral larvae) as in many regions the timing of reproduction and settlement coincide, and the outcomes of such interactions may have considerable influence on the recruitment to juvenile populations.

As well as the direct effects of macrophytes on coral larvae, macrophytes have been shown to influence the microbiome and behaviour of coral larvae. For example, Beatty et al. (2018) found that macroalgal assemblages altered the microbiome of coral larvae and speculated that could affect settlement success and post-settlement survival. Macrophytes have also been shown to alter the behaviour of coral larvae, with larvae either avoiding cues from diverse macrophyte assemblages and delaying initial settlement in the presence of the green alga *Chlorodesmis fastigiata*, although these larvae did eventually settle with final settlement rates being comparable to those under control (i.e., no algae) conditions (Birrell et al. 2008a). The latent effect of macrophyte-driven delayed settlement of corals is unknown but it may be energetically costly and ultimately affect the growth and/or survival of recently settled corals (Birrell et al. 2008b, Ritson-Williams et al. 2009, Beatty et al. 2018).

The fecundity of corals, and hence the supply of coral larvae, is also the key aspects of the replenishment and recovery of coral populations. Studies of the effects of macrophytes on adult corals have, however, focused on the growth, bleaching/discolouration and survival of corals, with only *ca.* 2% of the interactions ($n=5$ studies) considering changes in fecundity. This is despite previous calls for a greater focus on the effects of algae on coral reproduction (Birrell et al. 2008a). Contact with fleshy macroalgae has been shown to reduce egg size, number of eggs per gonad and number of gonads per polyp in *Montipora digitata*, *Echinopora lamellosa*, *Merulina ampliata* and *Orbicella annularis* (Hughes et al. 2007, Foster et al. 2008, Monteil et al. 2020). Direct contact with CCA was also found to reduce oocyte size in *O. annularis* (Cetz-Navarro et al. 2015), although to a lesser extent than filamentous turfs. However, the effects of algae on coral fecundity appear dependent on coral-algal contact during gametogenesis as opposed to any longer-term effects of prior contact, with fecundity increasing significantly following the removal of algae immediately prior gametogenesis (Foster et al. 2008, Cetz-Navarro et al. 2015). These results suggest that corals quickly reallocate energy towards gamete production when competition with algae is reduced, prioritising reproduction over other functions (Leuzinger et al. 2012). As coral taxa vary in their reproductive modes (i.e., broadcast vs brooding), output and energy allocation, and these differences may be amplified when stressed (e.g., due to interactions with macrophytes), greater research is needed to assess the impact of macrophytes on coral fecundity (Leuzinger et al. 2012, Álvarez-Noriega et al. 2016).

While the prevailing view is that macroalgae *sensu lato* negatively affects corals (Nugues & Bak 2006, Hughes et al. 2007, Rasher & Hay 2010, Bonaldo & Hay 2014, Vieira et al. 2016), our analyses question the ubiquity of this relationship given taxonomic biases (in both coral and macrophytes), and the variation in responses among and within macrophyte taxa, and among coral life stages. For the early life stages of corals, several field studies have shown that the recruitment of corals is reduced in areas with high cover of fleshy macroalgae and/or turf assemblages (Hughes et al. 2007, Arnold et al. 2010, Steneck et al. 2014), yet these studies did not differentiate among the effects of individual macrophyte taxa. Studies investigating the effect of individual macrophyte taxa, however, have reported varied responses, with some macrophytes having no detectable effect on coral settlement (Paul et al. 2011, Olsen et al. 2016, Ritson-Williams et al. 2020), while other macrophytes elicited a strong negative effect on the survival and/or settlement of coral larvae (Paul et al. 2011, Ritson-Williams et al. 2020). This variation was evident both among and within macrophyte genera. For example, the brown macroalgae *Dictyota pinnatifida* reduced the survival and recruitment of larvae of *Porites astreoides*, whereas the congener *Dictyota pulchella* had no detectable effect (Paul et al. 2011). The effects of macrophytes on the early life stage of corals have also been shown to vary among coral taxa and years. For example, the brown macroalga *D. pulchella* reduced the survival of *A. palmata* larvae in one of the two years tested, whereas *Dictyota bartayresiana* had no effect on survival of *A. palmata* larvae in either year, and neither *D. pulchella* or *D. bartayresiana* affected the survival of *A. cervicornis* or *Pseudodiploria strigosa* larvae (Ritson-Williams et al. 2020). The cyanobacterium *Caldora pencillata* also reduced the survival, settlement and metamorphosis of *A. palmata* larvae, yet had no detectable effect on the larvae of *P. strigosa* (Ritson-Williams et al. 2020). Similarly, while turfs are generally considered to have a negative effect on corals (Kuffner & Paul 2004, Kuffner et al. 2006, Birrell et al. 2008a, Ritson-Williams et al. 2016a), Diaz-Pulido et al. (2010) found that mixed filamentous turfs composed of *Hincksia mitchelliae*, *Polysiphonia* spp. and *Sphacelaria* spp. enhanced the settlement of *Platygyra daedalea* relative to algal mimics in an aquaria study.

Perhaps the greatest variation in responses to coral-macrophyte interactions is seen among CCA taxa (Ritson-Williams et al. 2014, Abdul Wahab et al. 2023). CCAs are often lumped into a single taxonomic unit in ecological studies (Wismer et al. 2009), yet there are marked differences in the effect of CCA species on the settlement of coral larvae. Some CCA taxa are largely considered enhancers or facilitators of coral settlement and are often used to induce the settlement of coral larvae in experimental settings (Harrington et al. 2004), while other CCA taxa may inhibit the settlement of coral larvae (Ritson-Williams et al. 2014). Moreover, the effect of different CCA taxa can vary among coral species. Ritson-Williams et al. (2016a) found that two species of CCA (*Titanoderma prototypum* and *Hydrolithon boergesenii*) enhanced the settlement of larvae of broadcast spawning corals (*P. strigosa* and *Orbicella faveolata*), but had no effect on the settlement of the majority of the brooding coral species examined (*Agaricia agaricites, Favia fragum* and *P. astreoides*). Rather the larvae of brooding corals settled on substrata with biofilms. Similarly, the CCA *Porolithon onkodes* induced settlement of *P. daedalea* larvae on the GBR (Diaz-Pulido et al. 2010), but had a negative impact on the settlement of *Acropora*, *Pocillopora* and *Porites* species in French Polynesia (Price 2010, Jorissen et al. 2021). The health of CCAs can also influence the outcomes of interactions with coral larvae, with the disease on *H. boergesenii* reducing the settlement and/or survival of *O. faveolata* and *Diploria labyrinthiformis* (Quéré & Nugues 2015).

Patterns in coral-algal interactions

Bias in studies of coral-macrophyte interactions and the prevalence of positive interactions

The widely held view that macroalgae are a sign of reef degradation and inhibit the recovery of coral populations has led to the majority of studies of coral-macrophyte interactions testing hypotheses of negative interactions. In doing so, there has been an ascertainment bias against the detection of positive interactions (except the role of some CCA species as facilitators of coral settlement described above). Despite this bias, 32 of the 169 studies we interrogated reported that macrophytes can positively affect adult corals. While shading from fleshy macroalgae is often reported to reduce coral growth (Box & Mumby 2007), it can also protect corals from predation by Crown-of-Thorns Starfish (*Acanthaster* spp.) and parrotfish (Venera-Ponton et al. 2011, Clements & Hay 2015). Further, Jompa and McCook (1998) suggested the *Sargassum* canopy protected corals from bleaching by reducing the incident UV light intensity, although a recent field study showed the *Sargassum* canopy offered limited protection to corals from bleaching during a period of thermal stress (Smith et al. 2022). The effects of macrophytes on corals appear to be more nuanced and context specific and emphasise the need to reconsider the currently negative view of algae and cyanobacteria on coral reefs.

The role of proximity and colony size in predicting outcomes

Distance between macrophytes and corals was important in determining the outcome of interactions. Coral-macrophyte interactions were less likely to have a negative effect on corals as distance to the macrophyte increased, with significant shifts in interaction outcome being realised at relatively small scales (De Ruyter Van Steveninck et al. 1988). For example, *Galaxaura rugosa* and *Sargassum polycystum* in direct contact with corals were most likely to cause a negative effect, while algae just 1.5 cm away from a colony had no effect on coral growth and photosynthesis (Clements et al. 2020). Other experiments have demonstrated that there is no effect of algae on coral fitness from 5 cm (Pratte et al. 2018, Fong et al. 2020) or 15 cm (Clements et al. 2018). Similarly, the growth of *Acropora millepora* and *Porites cylindrica* was negatively correlated with the density of *Sargassum* fronds that were held adjacent to, but not in direct contact, the corals (Clements et al. 2018). Higher density patches of macrophytes, especially for canopy-forming algae such as *Sargassum*, likely limit the light available to the corals and increase the likelihood of macrophytes being in proximity with corals. Spatial variation in the composition of macrophyte assemblages and the proximity to different macrophytes may contribute to the within-habitat variability in coral settlement, for example, reported on an inshore reef of the GBR (Johns et al. 2018), and highlights the need to consider broader habitat-scale effects of macrophyte assemblages on corals.

Understanding the importance of proximity in coral-macrophyte interactions, as well as broader habitat-scale effects of macrophyte assemblages, is particularly important when considering that under natural settings corals are likely to be interacting with multiple macrophyte species simultaneously. However, most studies of coral-macrophyte interactions are based on single species interactions, or 'macroalgae' assemblages where the composition of those assemblages is not reported. Only 16 publications reported the composition of the macrophyte assemblage interacting with corals, and none reported the proximity of the different taxa to the corals. Titlyanov et al. (2006) found that coral growth was not affected by mixed macrophyte assemblages, even though the macrophyte assemblage had species known to have a negative effect on corals (e.g., *Lyngbya*, Titlyanov et al. 2005; and *Anotrichium*, Jompa and McCook 2003a). Similarly, a field study on the central GBR showed that a monospecific *Anotrichium tenue* algal turf killed coral tissue by overgrowth, while a mixed-species turf assemblage had no effect, although it was not reported whether *A. tenue* was present in the mixed-species assemblage (Jompa & McCook 2003a).

Colony size has also been reported to affect the outcomes of coral-macrophyte interactions; however, we were unable to compare the effect of colony size among studies as most did not report the size of the colony used. Interestingly, the studies that have considered the effect of colony size have reported contrasting results. For example, smaller colonies (<10 cm diameter) of various genera (mainly *Porites*) were more resistant to a diverse macrophyte assemblage (dominant taxa=*Halimeda*) than larger colonies in a field study in the Maldives (Brown et al. 2017). Conversely, larger colonies (~50 cm) of *Colpophyllia*, *Porites* and *Agaricia* were reported to be better competitors against *Lobophora* and *Halimeda* in a field study in the Caribbean (Ferrari et al. 2012). Similarly, Swierts and Vermeij (2016) also observed that larger colonies (>80 cm) of various massive, plating, encrusting and solitary corals experienced less competitive interactions with macrophytes than smaller conspecifics, and suggested that colony growth form is a better predictor of competition outcome. Yet, other studies have found coral colony size didn't influence the outcomes of coral-algal interactions (Bonaldo & Hay 2014, Fong & Todd 2021). Although some mechanisms for colony size-related differences in the outcomes of coral-macrophyte interactions have been proposed (e.g., smaller corals actively fight algae by investing energy towards competition; Barott et al. 2012), the taxonomic identity of both corals and macrophytes are likely to be important considerations.

Functional groups and morphologies

While the mechanisms and outcomes of coral-macrophyte interactions are complex, identifying algae and cyanobacteria to species in the field is difficult, and as such, macrophytes are often reported in broad groupings (e.g., upright macroalgae, turfs and CCA). While our analyses show that these broad categories can predict interaction outcomes to some extent, with CCAs having a net neutral effect and turfs being the most damaging to corals (Barott et al. 2012, Brown et al. 2020), such an approach overlooks the complexities of these interactions. When using the six macrophyte functional groups proposed by Steneck & Dethier (1994), we were unable to identify distinct patterns. While the lack of detectable effects may be due to small sample size, the macrophyte groups are based largely on morphology and do not capture differences in life history strategies, chemical compounds produced or their microbiome. New morphological groups have been proposed that better capture macrophyte life strategies, but these have been only tested in temperate areas, and the application to tropical reefs and coral-macrophyte interactions is unknown (Balata et al. 2011). As trait-based approaches are being increasingly used for other taxa (e.g., corals: Madin et al. 2016; reef fish: McLean et al. 2021), more research is needed to investigate the use of new functional groups for tropical macrophytes and their application for investigating coral-macrophyte interactions.

Functional group approaches offer considerable promise in detecting and describing generalities of the outcomes of coral-macrophyte interactions; however, like all reductionist approaches overlook variation among taxa within each group. This is important when considering coral-macrophyte interactions as some closely related and morphologically similar macrophyte species have been shown to have significantly different effects on corals. For example, some species of CCA are facilitators of coral settlement, while others inhibit or have a negligible effect on the settlement of coral larvae (Harrington et al. 2004, Doropoulos et al. 2012a, Ritson-Williams et al. 2014, Ritson-Williams et al. 2016a). Even congeneric species have been shown to have markedly different effects on corals, with *D. pinnatifida* reducing the survival and settlement of *P. astreoides* larvae, whereas the congener *D. pulchella* had no detectable effect (Paul et al. 2011). Similarly, the capacity to overgrow corals varies greatly among species of *Lobophora* (Vieira et al. 2015).

Coral-macrophyte interactions in space and time

Seasonality of coral-macrophyte interactions

Duration of studies on coral-macrophyte interactions showed a large difference between laboratory (or aquaria) experiments (median = 33 days) and field-based surveys (median = 545 days). While such differences are likely due to the difficulty of keeping macrophytes and/or coral in laboratory conditions for extended periods of time, the extent to which they represent longer-term processes is unknown, especially given the temporal variations in environmental conditions and the highly seasonal nature of algal communities. Light, water temperature and nutrients influence macrophyte growth and reproduction, and thus, macrophyte assemblages typically present seasonal changes in their species composition and abundance (Benedetti-Cecchi & Cinelli 1993, Venera-Ponton et al. 2011, Castellani et al. 2017, Hutchings et al. 2019, Jauffrais et al. 2022). Peaks in abundance and reproductive activity vary between genera, location and reef zone. For example, on inshore reefs of the GBR, *Sargassum* biomass generally increases from winter to spring, reaching a peak during the summer months, before releasing propagules and senescing (Martin-Smith 1993, Vuki & Price 1994, Schaffelke & Klumpp 1997, Lefèvre & Bellwood 2010, Loffler & Hoey, 2018). The timing of the peak in biomass is, however, species specific (Martin-Smith 1993, Vuki & Price 1994, Schaffelke & Klumpp 1997, Lefèvre & Bellwood 2010), and differs in other regions globally (e.g., Singapore: Low et al. 2019). Similar seasonal variation is evident for other macrophyte taxa (Schaffelke & Klumpp 1997, Clifton & Clifton 1999, Diaz-Pulido & Garzón-Ferreira 2002, Burgess 2006).

Due to macrophyte seasonality and the dynamic nature of macrophyte assemblages, the occurrence, frequency and strength of coral-macrophyte interactions are temporally variable (Haas et al. 2010, Zaneveld et al. 2016, Brown et al. 2020). For example, seasonal variation in coral growth in the Caribbean was attributed to seasonal shifts in algal assemblages (Mejía Niño & Garzón Ferreira 2003, Jiménez-López et al. 2020). Reductions in the cover of canopy-forming algae can also release competitive pressure on other understory algae (Ateweberhan et al. 2006). Indeed, on Heron Island on the southern GBR, CCAs were reported to be three times more competitive with corals when fleshy macroalgal cover was at its lowest (Brown et al. 2020). Similar dynamics were observed in the southern Red Sea, Eritrea (Ateweberhan et al. 2006). Further, the effect of specific macrophytes on coral has been shown to vary seasonally. In a laboratory study, *Halimeda* enhanced photosynthesis in *Acropora* corals in summer, but had the opposite effect in winter conditions (Brown et al. 2019).

The seasonality of macrophytes may also allow corals to recover from any negative effects short-term (i.e., seasonal) effects of interactions following the senescence of the interacting macrophytes (Lirman & Biber 2000, Bonaldo & Hay 2014). For example, algal overgrowth on damaged adult corals was found to be only temporary, with several examples of corals recovering following algal senescence in the field (e.g., Caribbean: Bak & Steward-Van Es 1980; Japan: Van Woesik 1998; GBR: Diaz-Pulido et al. 2009). Similarly (and as described above), the effects of macrophytes on coral fecundity are lost following the removal of the interacting macrophyte (Foster et al. 2008). Consequently, short-term studies on coral-macrophyte interactions likely overestimate the effect of macrophytes on corals, the exception being where there is whole colony mortality. Moreover, while field surveys record the effect of macrophytes on corals, often the longer-term fate of corals is not reported, and whether corals recover is unknown (Burgess 2006, Smith et al. 2008). For example, *Dictyota* is sensitive to storm and physical damage, and its biomass changes dramatically throughout the year (Mumby et al. 2005). As a result, the findings of short-term studies that have reported *Dictyota* to cause localised tissue mortality and/or bleaching of corals (e.g., 20 days: Rasher & Hay 2010; 15 days: Wolf et al. 2012) may not be representative of longer-term outcomes.

In contrast, some perennial macrophytes may present a greater risk to corals due to their year-round presence. For example, turfs damaged corals in more than 95% of interactions and showed little seasonality in cover, although their composition was not quantified (Jompa & McCook 2003b, Brown et al. 2018, 2020). Moreover, the perennial species *Chlorodesmis fastigiata* and *Galaxaura filamentosa* had greater negative effects on several common coral species, than the seasonal alga *Dictyota* (Bonaldo & Hay 2014). Although the effect of ephemeral algae on corals may be overestimated in the literature, the timing of their blooms or peaks in biomass relative to key events in the coral life cycle is likely to be important. For example, algal blooms prior to coral spawning can have catastrophic consequences on reefs and can cause complete failure of coral settlement (Doropoulos et al. 2014).

Spatial dynamics of coral-macrophyte interactions

The outcomes of coral-macrophyte interactions may be influenced by numerous extrinsic, or external, factors including herbivore abundance and activity, nutrients, sediment loads and disturbance history. Herbivores, through the mechanical removal of macrophyte material when feeding, have been hypothesised to aid reefs in avoiding macrophyte overgrowth, by reducing interactions between corals and fleshy macroalgae, and clearing space for the settlement of corals (Rogers & Miller 2006, Hughes et al. 2007, Norström et al. 2009). The relationship between herbivory and macrophytes is complex, with some studies reporting negative correlations between the abundance and grazing rates of herbivorous fish and urchins with turf height, sediment load and cover of macroalgae (Cheal et al. 2013, Suchley & Alvarez-Filip 2017, Jiménez-López et al. 2020), while others have found no such relationship (Russ et al. 2015). Further, herbivorous fishes have been shown to have a limited effect on the survival of *Sargassum* holdfasts and hence the regrowth of *Sargassum* biomass (Loffler et al. 2018, Loffler & Hoey 2018). Grazing by some nominally herbivorous taxa has also been reported to be a source of mortality for recently settled corals (Doropoulos et al. 2012b, Trapon et al. 2013).

Poor water quality can influence the prevalence and outcomes of coral-macrophyte competition by enhancing macrophyte growth and lowering coral health and limiting replenishment. High nutrient and sediment loads inhibit coral fecundity, gamete fertilisation and settlement (Koop et al. 2001, Fabricius, 2005). Nutrients may also increase coral disease outbreaks, facilitating macrophyte overgrowth on weakened corals (Hayes et al. 2001, Voss & Richardson, 2006, Bruno et al. 2007). At the same time, eutrophication has been generally associated with elevated macrophyte growth, although the effects vary among macrophyte groups and taxa (Littler & Littler 1985, Jompa & McCook 2002, Diaz-Pulido & McCook 2005, Titlyanov et al. 2009, Ceccarelli et al. 2020, López-Jiménez et al. 2021, Ramseyer et al. 2021). Elevated levels of suspended and/or deposited sediments also tend to favour macrophytes over corals. Elevated levels of suspended sediments reduce the light available for corals (Storlazzi et al. 2015, Bessell-Browne et al. 2017) and suspended sediments are energetically costly for corals to remove from their surface (Lasker 1980), while elevated sediments can enhance the inhibitory effect of turfs on coral settlement and increase turf recruitment (Birrell et al. 2005, Clausing et al. 2016, López-Jiménez et al. 2021, Roth et al. 2021).

Coral-macrophyte interactions in the Anthropocene

The effect of climate change on corals and macrophytes

As oceans are predicted to become warmer and more acidic, understanding the dynamics of coral-macrophyte interactions under different environmental conditions is critical (Hoey et al. 2016, Bellwood et al. 2019, Masson-Delmotte et al. 2021). A large body of research has shown that exposure to elevated seawater temperatures can push corals beyond their thermal limits, leading to partial or whole colony bleaching, and ultimately mortality if exposure to elevated temperatures is prolonged (Baird & Marshall 2002, McWilliams et al. 2005, Lesser 2011, Van Oppen & Lough 2018, Cziesielski et al. 2019, Hughes et al. 2019, Sully et al. 2019). It is also well established that sensitivity to high temperatures varies significantly among coral taxa and bleaching affects some colonies more than others (Marshall & Baird 2000, Van Woesik et al. 2011, Grottoli et al. 2014, Hughes et al. 2018b). While many corals may die following extreme heat stress, other corals will experience sub-lethal effects (Anthony et al. 2009, Leinbach et al. 2021) that may reduce their competitive abilities with macrophytes. Thus, widespread mass mortality events caused by marine heatwaves will change the structure and composition of coral assemblages, and potentially the strength of interactions with macrophytes.

Less is known on the effect of climate change on algae and cyanobacteria, especially tropical macrophyte taxa (Diaz-Pulido et al. 2007, Pires 2017), with most research focused on commercially useful or temperate algae (Reed et al. 2016, Graba-Landry et al. 2018). Many temperate and sub-tropical algae occur across a broad range of temperatures, and as such, it has been suggested that predicted increases in temperatures may not exceed their physiological limits (Singh &

Singh 2015). The few studies that have been conducted on tropical algae suggest that the canopy-forming macroalgae *Sargassum* spp. may be particularly sensitive to heat stress (Fulton et al. 2019, Graba-Landry et al. 2020), while the calcification rates of CCAs are negatively impacted by elevated seawater temperatures (Johnson & Carpenter 2012, Cornwall et al. 2019). Moreover, increasing temperatures have been suggested to alter phenological cycles (Benedetti-Cecchi & Cinelli 1993, Venera-Ponton et al. 2011, Castellani et al. 2017, Hutchings et al. 2019, Jauffrais et al. 2022) and physiology of algae (Benedetti-Cecchi & Cinelli 1993, Venera-Ponton et al. 2011, Castellani et al. 2017, Hutchings et al. 2019, Jauffrais et al. 2022). Overall, it appears that increased temperatures are unlikely to drive widespread mortality, but changes in the composition of macrophyte assemblages are to be expected (Diaz-Pulido et al. 2007, Fulton et al. 2019).

How will climate change affect coral-macrophyte interactions?

Research to date suggests that climate change, and in particular elevated seawater temperature, is likely to shift the outcomes of coral-macrophyte interactions in favour of macrophytes. The effect of temperature has been associated with the higher competitive ability of macrophytes against corals, although the relative effect varies among life stages and metrics of coral health. For example, elevated temperature (+3.5°C) had no detectable impact on the larval survival and settlement, or post-settlement survival of *Porites astreoides* exposed to *Dictyota menstrualis*, however, resulted in a synergistic effect on oxidative damage in *P. astreoides* larvae (Olsen et al. 2014). Similarly, Ritson-Williams et al. (2016b) found that elevated temperature (+3°C–5°C) had no effect on the survival of *P. astreoides* larvae, however, exposure to elevated temperature, and microcolin A (a natural product from the common cyanobacterium *Okeania erythroflocculosa*) led to a synergistic reduction in survival and increase in oxidative stress of the larvae. In contrast, elevated temperature and contact with *Halimeda heteromorpha* were found to act independently on the survival and calcification of *Acropora intermedia* fragments (i.e., elevated seawater temperature didn't influence the outcome of the coral-algal interaction; Brown et al. 2019). Moreover, heat stress has also been linked with an increasing incidence of coral disease (Selig et al. 2006, Bruno et al. 2007) and an increase in virulence of the coral pathogen *Vibrio* (Rosenberg & Ben-Haim 2002). These differences among studies may be related to differences in the coral and/or macrophyte species, the coral life stage, environmental conditions, microbial communities or the temperatures used and should be the focus of future studies.

Together with the effects of ocean warming, ocean acidification is predicted to reduce calcification rates in corals, coralline algae and other calcified taxa (Anthony et al. 2008, Campbell et al. 2014, Comeau et al. 2018), enhance bacterial activity due to increased algal dissolved organic carbon release (Diaz-Pulido & Barrón 2020) and thereby influence the outcomes of coral-macrophyte interactions. Indeed, the settlement and/or survival of coral larvae (*Acropora millepora* and *Pocillopora acuta*) has been shown to be reduced under elevated CO_2 conditions (Doropoulos et al. 2012a, Putnam et al. 2020), with reductions in settlement of *A. millepora* being linked to reductions in the recruitment and cover of CCAs, in particular the settlement inducing genera *Titanode rma* and *Hydrolithon* (Doropoulos et al. 2012a). Further, elevated CO_2 conditions exacerbated the effect of the macroalga *Stypopodium zonale* on the survival and settlement of *P. astreoides* larvae (Campbell et al. 2017). Ocean acidification has also been shown to adversely affect the outcomes of interactions between adult corals and macroalgae. For example, tissue mortality of *A. intermedia* branches in contact with *Lobophora papenfussii, Canistrocarpus* (previously *Dictyota*) *cervicornis, Chlorodesmis fastigiata* or *Amansia glomerata* increased when held under elevated CO_2 conditions (Diaz-Pulido et al. 2011, Del Monaco et al. 2017). Further, lipid-soluble surface extracts from *C. cervicornis* (but not *C. fastigiata* or *A. glomerata*) thalli cultured under elevated CO_2 conditions were more damaging to *A. intermedia* in the field, than thalli cultured under ambient conditions (Del Monaco et al. 2017).

In conclusion, although some research suggests that ongoing climate change will lead to high macrophyte abundance and low coral cover on reefs, the paucity of studies limits our ability to predict future reef conditions and the composition of benthic communities. Most research to date has focused on the effect of increased temperatures and high CO_2 concentrations on corals (McLachlan et al. 2020), while relatively little is known on the effect of ocean warming and/or ocean acidification on tropical macrophytes, or the outcomes of coral-macrophyte interactions. In particular, our understanding of the effect of temperature on the health and phenology of algae is very limited (Diaz-Pulido et al. 2007). While some studies suggest that the competitive ability of macrophytes against corals may be enhanced under ocean warming and/or acidification, the abundance, biomass, composition and health of macrophyte assemblages may also be negatively influenced by climate change.

The effect of water quality on coral-macrophyte interactions

Human pressure on coastlines has been associated with higher nutrient and sediment loads (Wooldridge et al. 2006). Extreme weather events, such as floods, can further reduce water quality, especially on reefs adjacent to estuaries (Lam et al. 2018). Eutrophication and elevated input of sediments have been shown to reduce coral growth, skeletal density, reproductive output, settlement and/or survivorship and increase the incidence of coral disease (Koop et al. 2001, Nugues & Roberts 2003, Fabricius 2005, Vega Thurber et al. 2014). Moreover, high nutrient concentrations can exacerbate the negative effects of heat stress on corals (Donovan et al. 2020). The effect of terrestrial runoff is more complex for macrophytes (Miller et al. 1999). Increased sediment loads and reductions in light attenuation have a general negative effect on most algae, excluding turfs (McClanahan et al. 2002). In contrast, increased nutrients can enhance the growth of some fleshy macroalgae, have a limited effect on other fleshy macroalgae and have a negative effect on CCA (Jompa & McCook 2002, Fabricius 2005, López-Jiménez et al. 2021, Sura et al. 2021). Fecundity of the common brown alga *Sargassum siliquosum* was reduced in high nutrient conditions in a laboratory study, while the growth and reproduction of turfs may be enhanced in high sediment conditions (Diaz-Pulido & McCook 2005, Tebbett & Bellwood 2019, López-Jiménez et al. 2021).

As water quality affects both the abundance and physiology of corals and macrophytes, the frequency and outcomes of coral-macrophyte interactions may vary along gradients of human pressures in coastal environments (e.g., coastal development, agriculture and land use; Vermeij et al. 2010, Barott et al. 2012). A 3-year field experiment study conducted in the Florida Keys showed that nutrient addition increased macroalgal and turf cover and shifted coral microbiomes, increasing the incidence of coral diseases, with the effects being exacerbated under high temperatures (Zaneveld et al. 2016). In contrast, Jompa & McCook (2002) found limited to no effect of nutrient additions on competition between *Porites cylindrica* and *Lobophora variegata* on the GBR. Similarly, Nugues & Roberts (2003) found no evidence that interactions with algae (e.g., algal overgrowth) were responsible for the partial mortality of corals with increasing proximity to river mouths, rather attributing the mortality to the direct effects of sedimentation. Sediment-laden turfs have also been shown to markedly reduce the settlement of corals (*Acropora palmata* and *Orbicella faveolata*), and while this reduction was not related to the turfs per se (Speare et al. 2019), it is the ability of turfs to trap and retain sediments. The effects of sediments and nutrients on coral-macrophyte interactions appear variable, although the limited number of studies to date precludes any generalisations.

Conclusions

Given sustained declines in the cover of hard corals in many major regions and commensurate increases in the cover or abundance of algae on reefs (Souter et al. 2021), understanding the nature and outcomes of coral-macrophyte interactions is crucial to assess the current, and predict the future, state of reefs. The vast majority of coral-macrophyte interactions we interrogated resulted in either a negative (52.5%) or neutral (40.7%) effect on corals, with only 6.8% of interactions having a positive effect on corals. The lack of positive interactions may be an artefact of the experimental designs focused on testing hypotheses of negative interactions. Future studies need to address this ascertainment bias against the detection of positive interactions.

Existing knowledge of coral-macrophyte interactions largely comes from studies of a few select coral and macrophyte taxa and typically single coral-macrophyte pairs. While physical mechanisms of competition are maintained within functional groups and may be inferred from morphologically similar species, allelopathic and microbial properties can vary among and within species, making it difficult to infer outcomes across the wide diversity of corals and algae. The inherent focus on a limited suite of common coral and macrophyte taxa has revealed that the effects of macrophytes on corals can be varied and highlights the need for future studies to consider interactions among a greater diversity of both coral and macrophyte taxa. Further, the focus on single coral-macrophyte pairs overlooks the potential effects of broader macrophyte assemblages and their associated microbial and environmental conditions on coral health. Moreover, most studies to date have focused on the health of adult corals, while key measures for assessing the recovery potential of coral populations, such as larval settlement, post-settlement survival and fecundity, are often overlooked. There is a clear need to consider the effects of macrophytes on early life history stages and colony fecundity, in particular, across a diversity of coral species.

Coral reefs are subject to spatially and temporally variable environmental conditions (i.e., temperature, light, herbivory, nutrients, sediments, and water flow), and together with the seasonal nature of many macrophyte taxa, they may strongly influence the outcomes of these interactions. Caution needs to be applied when extrapolating findings from short-term field and laboratory studies of single coral-macrophyte pairs to natural reef environments. Moreover, long-term studies are scarce limiting our ability to detect delayed and/or legacy effects of interactions with algae on corals.

Local anthropogenic stressors and climate change are predicted to enhance the competitive ability of macrophytes with corals. While there is a large body of research investigating the response of corals to climate change stressors (namely, ocean warming and acidification), less is known on the effect of these stressors on tropical macrophytes or the interactions between corals and macrophytes. Research is urgently needed to understand the relative effect of anthropogenic change (namely, ocean warming, ocean acidification, eutrophication and sedimentation) on the frequency, intensity and outcomes of coral-algal interactions. In particular, there is a need to consider how multiple stressors may interact to effect the outcomes of coral-macrophyte interactions. This will require scaling up of single stressor experiments to simultaneously consider multiple stressors and ideally multiple levels of each stressor.

To date, relatively few studies have attempted to identify the causes, or mechanisms, underlying the outcomes of coral-macrophyte interactions. Further, most of these studies have only considered a single mechanism. It is possible, and indeed quite likely, that macrophytes interact with corals through multiple mechanisms. Studies investigating a single mechanism may, therefore, report a neutral interaction when other mechanisms may have yielded a negative or positive outcome. Designing and implementing experiments to test all possible outcomes (positive, neutral, and negative) of coral-macrophyte interactions, while simultaneously investigating the importance of multiple mechanisms, and how these vary with environmental conditions will greatly advance our understanding of coral-macrophyte interactions.

Acknowledgements

This contribution was supported by the Australian Research Council Centre of Excellence for Coral Reef Studies, the Australian Institute of Marine Science and the EcoRRAP subprogram of the

Reef Restoration and Adaptation Program. We further acknowledge AIMS@JCU Postgraduate Research Scholarship (James Cook University) support for M.B. We are grateful to M.S. Pratchett, K. Fabricius and R. Ritson-Williams who provided feedback on an earlier draft of this manuscript.

References

Abdul Wahab, M.A., Ferguson, S., Snekkevik, V.K., McCutchan, G., Jeong, S., Severati, A., Randall, C.J., Negri, A.P. & Diaz-Pulido, G. 2023. Hierarchical settlement behaviours of coral larvae to common coralline algae. *Scientific Reports* **13**, 5795, doi:10.1038/s41598-023-32676-4

Álvarez-Noriega, M., Baird, A.H., Dornelas, M., Madin, J.S., Cumbo, V.R. & Connolly, S.R. 2016. Fecundity and the demographic strategies of coral morphologies. *Ecology* **97**, 3485–3493.

Anthony, K.R.N., Kline, D.I., Diaz-Pulido, G., Dove, S. & Hoegh-Guldberg, O. 2008. Ocean acidification causes bleaching and productivity loss in coral reef builders. *Proceedings of the National Academy of Sciences of the United States of America* **105**, 17442–17446.

Anthony, K.R.N., Hoogenboom, M.O., Maynard, J.A., Grottoli, A.G. & Middlebrook, R. 2009. Energetics approach to predicting mortality risk from environmental stress: a case study of coral bleaching. *Functional Ecology* **23**, 539–550.

Arnold, S., Steneck, R. & Mumby, P. 2010. Running the gauntlet: inhibitory effects of algal turfs on the processes of coral recruitment. *Marine Ecology Progress Series* **414**, 91–105.

Ateweberhan, M., Bruggemann, J.H. & Breeman, A.M. 2006. Effects of extreme seasonality on community structure and functional group dynamics of coral reef algae in the southern Red Sea (Eritrea). *Coral Reefs* **25**, 391–406.

Babcock, R., Baird, A.H., Piromvaragorn, S., Thomson, D.P. & Willis, B.L. 2003. Identification of scleractinian coral recruits from Indo-Pacific reefs. *Zoological Studies* **42**, 211–226.

Baird, A. & Marshall, P. 2002. Mortality, growth and reproduction in scleractinian corals following bleaching on the Great Barrier Reef. *Marine Ecology Progress Series* **237**, 133–141.

Baird, A.H., Pratchett, M.S., Hoey, A.S., Herdiana, Y. & Campbell, S.J. 2013. *Acanthaster planci* is a major cause of coral mortality in Indonesia. *Coral Reefs* **32**, 803–812.

Bak, R.P. & Steward-Van Es, Y. 1980. Regeneration of superficial damage in the scleractinian corals *Agaricia agaricites f. purpurea* and *Porites astreoides*. *Bulletin of Marine Science* **30**, 883–887.

Balata, D., Piazzi, L. & Rindi, F. 2011. Testing a new classification of morphological functional groups of marine macroalgae for the detection of responses to stress. *Marine Biology* **158**, 2459–2469. Barott, K.L. & Rohwer, F.L. 2012. Unseen players shape benthic competition on coral reefs. *Trends in Microbiology* **20**, 621–628.

Barott, K., Williams, G., Vermeij, M., Harris, J., Smith, J., Rohwer, F. & Sandin, S. 2012. Natural history of coral–algae competition across a gradient of human activity in the Line Islands. *Marine Ecology Progress Series* **460**, 1–12.

Beatty, D.S., Clements, C.S., Stewart, F.J. & Hay, M.E. 2018. Intergenerational effects of macroalgae on a reef coral: major declines in larval survival but subtle changes in microbiomes. *Marine Ecology Progress Series* **589**, 97–114.

Bellwood, D.R., Hughes, T.P., Folke, C. & Nyström, M. 2004. Confronting the coral reef crisis. *Nature* **429**, 827–833.

Bellwood, D.R., Pratchett, M.S., Morrison, T.H., Gurney, G.G., Hughes, T.P., Álvarez-Romero, J.G., Day, J.C., Grantham, R., Grech, A., Hoey, A.S., Jones, G.P., Pandolfi, J.M., Tebbett, S.B., Techera, E., Weeks, R. & Cumming, G.S. 2019. Coral reef conservation in the anthropocene: confronting spatial mismatches and prioritizing functions. *Biological Conservation* **236**, 604–615.

Benedetti-Cecchi, L. & Cinelli, F. 1993. Seasonality and reproductive phenology of algae inhabiting littoral pools in the western mediterranean. *Marine Ecology* **14**, 147–157.

Bento, R., Hoey, A.S., Bauman, A.G., Feary, D.A. & Burt, J.A. 2016. The implications of recurrent disturbances within the world's hottest coral reef. *Marine Pollution Bulletin* **105**, 466–472.

Berkelmans, R. & Oliver, J.K. 1999. Large-scale bleaching of corals on the Great Barrier Reef. *Coral Reef* **18**, 55–60.

Berumen, M.L. & Pratchett, M.S. 2006. Recovery without resilience: persistent disturbance and long-term shifts in the structure of fish and coral communities at Tiahura Reef, Moorea. *Coral Reefs* **25**, 647–653.

Bessell-Browne, P., Negri, A.P., Fisher, R., Clode, P.L., Duckworth, A. & Jones, R. 2017. Impacts of turbidity on corals: the relative importance of light limitation and suspended sediments. *Marine Pollution Bulletin* **117**, 161–170.

Birrell, C.L., McCook, L.J. & Willis, B.L. 2005. Effects of algal turfs and sediment on coral settlement. *Marine Pollution Bulletin* **51**, 408–414.

Birrell, C., McCook, L., Willis, B. & Harrington, L. 2008a. Chemical effects of macroalgae on larval settlement of the broadcast spawning coral *Acropora millepora*. *Marine Ecology Progress Series* **362**, 129–137.

Birrell, C.L., McCook, L.J., Willis, B.L. & Diaz-Pulido, G.A. 2008b. Effects of benthic algae on the replenishment of corals and the implications for the resilience of coral reefs. *Oceanography and Marine Biology: An Annual Review* **46**, 25–63.

Bonaldo, R.M. & Hay, M.E. 2014. Seaweed-coral interactions: variance in seaweed allelopathy, coral susceptibility, and potential effects on coral resilience. *PLoS One* **9**, e85786, doi:10.1371/journal.pone.0085786

Box, S. & Mumby, P. 2007. Effect of macroalgal competition on growth and survival of juvenile Caribbean corals. *Marine Ecology Progress Series* **342**, 139–149.

Briggs, A.A., Brown, A.L. & Osenberg, C.W. 2021. Local versus site-level effects of algae on coral microbial communities. *Royal Society Open Science* **8**, 210035, doi:10.1098/rsos.210035

Brown, A.L. & Carpenter, R.C. 2015. Water flow influences the mechanisms and outcomes of interactions between massive *Porites* and coral reef algae. *Marine Biology* **162**, 459–468.

Brown, K.T., Bender-Champ, D., Bryant, D.E.P., Dove, S. & Hoegh-Guldberg, O. 2017. Human activities influence benthic community structure and the composition of the coral-algal interactions in the central Maldives. *Journal of Experimental Marine Biology and Ecology* **497**, 33–40.

Brown, K.T., Bender-Champ, D., Kubicek, A., van der Zande, R., Achlatis, M., Hoegh-Guldberg, O. & Dove, S.G. 2018. The dynamics of coral-algal interactions in space and time on the southern Great Barrier Reef. *Frontiers in Marine Science* **5**, 13, doi:10.3389/fmars.2018.00181

Brown, K.T., Bender-Champ, D., Kenyon, T.M., Rémond, C., Hoegh-Guldberg, O. & Dove, S. 2019. Temporal effects of ocean warming and acidification on coral-algal competition. *Coral Reefs* **38**, 297–309.

Brown, K.T., Bender-Champ, D., Hoegh-Guldberg, O. & Dove, S. 2020. Seasonal shifts in the competitive ability of macroalgae influence the outcomes of coral-algal competition. *Royal Society Open Science* **7**, 8, doi:10.1098/rsos.201797

Bruno, J.F., Selig, E.R., Casey, K.S., Page, C.A., Willis, B.L., Harvell, C.D., Sweatman, H. & Melendy, A.M. 2007. Thermal stress and coral cover as drivers of coral disease outbreaks. *PLoS Biology* **5**, e124, doi:10.1371/journal.pbio.0050124

Bruno, J.F., Sweatman, H., Precht, W.F., Selig, E.R. & Schutte, V.G.W. 2009. Assessing evidence of phase shifts from coral to macroalgal dominance on coral reefs. *Ecology* **90**, 1478–1484.

Burgess, S.C. 2006. Algal blooms on coral reefs with low anthropogenic impact in the Great Barrier Reef. *Coral Reefs* **25**, 390–390.

Burn, D., Hoey, A.S., Matthews, S., Harrison, H.B. & Pratchett, M.S. 2023. Differential bleaching susceptibility among coral taxa and colony sizes, relative to bleaching severity across Australia's Great Barrier Reef and Coral Sea Marine Parks. *Marine Pollution Bulletin* **191**, 114907.

Campbell, J., Craft, J., Muehllehner, N., Langdon, C. & Paul, V. 2014. Responses of calcifying algae (Halimeda spp.) to ocean acidification: implications for herbivores. *Marine Ecology Progress Series* **514**, 43–56.

Campbell, J., Sneed, J., Johnston, L. & Paul, V. 2017. Effects of ocean acidification and contact with the brown alga *Stypopodium zonale* on the settlement and early survival of the coral *Porites astreoides*. *Marine Ecology Progress Series* **577**, 67–77.

Castellani, G., Losch, M., Lange, B.A. & Flores, H. 2017. Modelling Arctic sea-ice algae: physical drivers of spatial distribution and algae phenology. *Journal of Geophysical Research: Oceans* **122**, 7466–7487.

Ceccarelli, D.M., Evans, R.D., Logan, M., Mantel, P., Puotinen, M., Petus, C., Russ, G.R. & Williamson, D.H. (2020). Long-term dynamics and drivers of coral and macroalgal cover on inshore reefs of the Great Barrier Reef Marine Park. *Ecological Applications* **30**, e02008, doi:10.1002/eap.2008

Cetz-Navarro, N.P., Carpizo-Ituarte, E.J., Espinoza-Avalos, J. & Chee-Barragán, G. 2015. The effect of filamentous turf algal removal on the development of gametes of the coral *Orbicella annularis*. *PLoS One* **10**, e0117936, doi:10.1371/journal.pone.0117936

Cheal, A.J., Emslie, M., Macneil, M.A., Miller, I. & Sweatman, H. 2013. Spatial variation in the functional characteristics of herbivorous fish communities and the resilience of coral reefs. *Ecological Applications* **23**, 174–188.

Christensen, R.H.B. 2019. A tutorial on fitting cumulative link mixed models with clmm2 from the ordinal Package. https://cran.r-project.org/web/packages/ordinal/Accessed.

Clausing, R.J., Bittick, S.J., Fong, C.R. & Fong, P. 2016. Sediments influence accumulation of two macroalgal species through novel but differing interactions with nutrients and herbivory. *Coral Reefs* **35**, 1297–1309.

Clements, C.S. & Hay, M.E. 2015. Competitors as accomplices: seaweed competitors hide corals from predatory sea stars. *Proceedings of the Royal Society B: Biological Sciences* **282**, 20150714, doi:10.1098/rspb.2015.0714

Clements, C., Rasher, D., Hoey, A., Bonito, V. & Hay, M. 2018. Spatial and temporal limits of coral-macroalgal competition: the negative impacts of macroalgal density, proximity, and history of contact. *Marine Ecology Progress Series* **586**, 11–20.

Clements, C.S., Burns, A.S., Stewart, F.J. & Hay, M.E. 2020. Seaweed-coral competition in the field: effects on coral growth, photosynthesis and microbiomes require direct contact. *Proceedings of the Royal Society B: Biological Sciences* **287**, 20200366, doi:10.1098/rspb.2020.0366

Clifton, K.E. & Clifton, L.M. 1999. The phenology of sexual reproduction by green algae (Bryopsidales) on Caribbean coral reefs. *Journal of Phycology* **35**, 24–34.

Comeau, S., Cornwall, C.E., Decarlo, T.M., Krieger, E. & McCulloch, M.T. 2018. Similar controls on calcification under ocean acidification across unrelated coral reef taxa. *Global Change Biology* **24**, 4857–4868.

Cornwall, C.E., Diaz-Pulido, G. & Comeau, S. 2019. Impacts of ocean warming on coralline algal calcification: meta-analysis, knowledge gaps, and key recommendations for future research. *Frontiers in Marine Science* **6**, 186, doi:10.3389/fmars.2019.00186

Cziesielski, M.J., Schmidt-Roach, S. & Aranda, M. 2019. The past, present, and future of coral heat stress studies. *Ecology and Evolution* **9**, 10055–10066.

De'Ath, G., Fabricius, K.E., Sweatman, H. & Puotinen, M. 2012. The 27-year decline of coral cover on the Great Barrier Reef and its causes. *Proceedings of the National Academy of Sciences of the United States of America* **109**, 17995–17999.

De Ruyter Van Steveninck, E.D., Van Mulekom, L.L. & Breeman, A.M. 1988. Growth inhibition of *Lobophora variegata* (Lamouroux) Womersley by scleractinian corals. *Journal of Experimental Marine Biology and Ecology* **115**, 169–178.

Del Monaco, C., Hay, M.E., Gartrell, P., Mumby, P.J. & Diaz-Pulido, G. 2017. Effects of ocean acidification on the potency of macroalgal allelopathy to a common coral. *Scientific Reports* **7**, 41053, doi:10.1038/srep41053

Dell, C.L.A., Longo, G.O. & Hay, M.E. 2016. Positive feedbacks enhance macroalgal resilience on degraded coral reefs. *PLoS One* **11**, e0155049, doi:10.1371/journal.pone.0155049

Diaz-Pulido, G. & Garzón-Ferreira, J. 2002. Seasonality in algal assemblages on upwelling-influenced coral reefs in the Colombian Caribbean. *Botanica Marina* **45**, 284–292.

Diaz-Pulido, G. & McCook, L.J. 2002. The fate of bleached corals: patterns and dynamics of algal recruitment. *Marine Ecology Progress Series* **232**, 115–128, doi:10.3354/meps232115

Diaz-Pulido, G. & McCook, L.J. 2005. Effects of nutrient enhancement on the fecundity of a coral reef macroalga. *Journal of Experimental Marine Biology and Ecology* **317**, 13–24.

Diaz-Pulido, G. & Barrón, C. 2020. CO2 enrichment stimulates dissolved organic carbon release in coral reef macroalgae. *Journal of Phycology* **56**, 1039–1052.

Diaz-Pulido, G., McCook, L.J., Larkum, A.W.D., Lotze, H.K., Raven, J.A., Schaffelke, B., Smith, J.E. & Steneck, R.S. (2007). Vulnerability of macroalgae of the Great Barrier Reef to climate change. In *Climate Change and the Great Barrier Reef: A Vulnerability Assessment*, J.E. Johnson & P.A. Marshall (eds.). Townsville, Queensland, Australia: The Great Barrier Reef Marine Park Authority, 153–192.

Diaz-Pulido, G., McCook, L.J., Dove, S., Berkelmans, R., Roff, G., Kline, D.I., Weeks, S., Evans, R.D., Williamson, D.H. & Hoegh-Guldberg, O. 2009. Doom and boom on a resilient reef: climate change, algal overgrowth and coral recovery. *PLoS One* **4**, e5239.

Diaz-Pulido, G., Harii, S., McCook, L.J. & Hoegh-Guldberg, O. 2010. The impact of benthic algae on the settlement of a reef-building coral. *Coral Reefs* **29**, 203–208.

Diaz-Pulido, G., Gouezo, M., Tilbrook, B., Dove, S. & Anthony, K.R.N. 2011. High CO_2 enhances the competitive strength of seaweeds over corals. *Ecology Letters* **14**, 156–162.

Donovan, M.K., Adam, T.C., Shantz, A.A., Speare, K.E., Munsterman, K.S., Rice, M.M., Schmitt, R.J., Holbrook, S.J. & Burkepile, D.E. 2020. Nitrogen pollution interacts with heat stress to increase coral bleaching across the seascape. *Proceedings of the National Academy of Sciences of the United States of America* **117**, 5351–5357.

Doropoulos, C., Ward, S., Diaz-Pulido, G., Hoegh-Guldberg, O. & Mumby, P.J. 2012a. Ocean acidification reduces coral recruitment by disrupting intimate larval-algal settlement interactions. *Ecology Letters* **15**, 338–346.

Doropoulos, C., Ward, S., Marshell, A., Diaz-Pulido, G. & Mumby, P.J. 2012b. Interactions among chronic and acute impacts on coral recruits: the importance of size-escape thresholds. *Ecology* **93**, 2131–2138.

Doropoulos, C., Roff, G., Zupan, M., Nestor, V., Isechal, A.L. & Mumby, P.J. 2014. Reef-scale failure of coral settlement following typhoon disturbance and macroalgal bloom in Palau, Western Pacific. *Coral Reefs* **33**, 613–623.

Downs, C.A., McDougall, K.E., Woodley, C.M., Fauth, J.E., Richmond, R.H., Kushmaro, A., Gibb, S.W., Loya, Y., Ostrander, G.K. & Kramarsky-Winter, E. 2013. Heat-stress and light-stress induce different cellular pathologies in the symbiotic dinoflagellate during coral bleaching. *PLoS One* **8**, e77173, doi:10.1371/journal.pone.0077173

Edmunds, P.J. & Carpenter, R.C. 2001. Recovery of *Diadema antillarum* reduces macroalgal cover and increases abundance of juvenile corals on a Caribbean reef. *Proceedings of the National Academy of Sciences of the United States of America* **98**, 5067–5071.

Fabricius, K.E. 2005. Effects of terrestrial runoff on the ecology of corals and coral reefs: review and synthesis. *Marine Pollution Bulletin* **50**, 125–146.

Ferrari, R., Gonzalez-Rivero, M. & Mumby, P. 2012. Size matters in competition between corals and macroalgae. *Marine Ecology Progress Series* **467**, 77–88.

Fong, J. & Todd, P.A. 2021. Spatio-temporal dynamics of coral-macroalgal interactions and their impacts on coral growth on urbanised reefs. *Marine Pollution Bulletin* **172**, 112849, doi:10.1016/j.marpolbul.2021.112849

Fong, J., Deignan, L.K., Bauman, A.G., Steinberg, P.D., McDougald, D. & Todd, P.A. 2020. Contact- and water-mediated effects of macroalgae on the physiology and microbiome of three Indo-Pacific coral species. *Frontiers in Marine Science* **6**, doi:10.3389/fmars.2019.00831

Foster, N., Box, S. & Mumby, P. 2008. Competitive effects of macroalgae on the fecundity of the reef-building coral *Montastraea annularis*. *Marine Ecology Progress Series* **367**, 143–152.

Fulton, C.J., Abesamis, R.A., Berkström, C., Depczynski, M., Graham, N.A.J., Holmes, T.H., Kulbicki, M., Noble, M.M., Radford, B.T., Tano, S., Tinkler, P., Wernberg, T. & Wilson, S.K. (2019). Form and function of tropical macroalgal reefs in the Anthropocene. *Functional Ecology* **33**, 989–999.

Gardner, T.A., Côté, I.M., Gill, J.A., Grant, A. & Watkinson, A.R. 2003. Long-term region-wide declines in Caribbean corals. *Science* **301**, 958–960.

Glynn, P.W., Maté, J.L., Baker, A.C. & Calderón, M.O. 2001. Coral bleaching and mortality in Panama and Ecuador during the 1997–1998 El Niño-Southern Oscillation event: spatial/temporal patterns and comparisons with the 1982–1983 event. *Bulletin of Marine Science* **69**, 79–109.

Goreau, T., McClanahan, T., Hayes, R. & Strong, A. 2000. Conservation of coral reefs after the 1998 global bleaching event. *Conservation Biology* **14**, 5–15.

Graba-Landry, A., Hoey, A., Matley, J., Sheppard-Brennand, H., Poore, A., Byrne, M. & Dworjanyn, S. 2018. Ocean warming has greater and more consistent negative effects than ocean acidification on the growth and health of subtropical macroalgae. *Marine Ecology Progress Series* **595**, 55–69.

Graba-Landry, A.C., Loffler, Z., McClure, E.C., Pratchett, M.S. & Hoey, A.S. 2020. Impaired growth and survival of tropical macroalgae (*Sargassum* spp.) at elevated temperatures. *Coral Reefs* **39**, 475–486.

Graham, N.A.J. & Nash, K.L. 2013. The importance of structural complexity in coral reef ecosystems. *Coral Reefs* **32**, 315–326.

Graham, N.A.J., Jennings, S., Macneil, M.A., Mouillot, D. & Wilson, S.K. (2015). Predicting climate-driven regime shifts versus rebound potential in coral reefs. *Nature* **518**(7537), 94–97, doi:10.1038/nature14140

Grottoli, A.G., Warner, M.E., Levas, S.J., Aschaffenburg, M.D., Schoepf, V., McGinley, M., Bauman, J. & Matsui, Y. 2014. The cumulative impact of annual coral bleaching can turn some coral species winners into losers. *Global Change Biology* **20**, 3823–3833.

Guiry, M.D. & Guiry, G.M. 2021. AlgaeBase. https://www.algaebase.org.

Haas, A., El-Zibdah, M. & Wild, C. 2010. Seasonal monitoring of coral-algae interactions in fringing reefs of the Gulf of Aqaba, Northern Red Sea. *Coral Reefs* **29**, 93–103.

Harrington, L., Fabricius, K., De'Ath, G. & Negri, A. 2004. Recognition and selection of settlement substrata determine post-settlement survival in corals. *Ecology* **85**, 3428–3437.

Hayes, M.L., Bonaventura, J., Mitchell, T.P., Prospero, J.M., Shinn, E.A., Van Dolah, F. & Barber, R.T. 2001. How are climate and marine biological outbreaks functionally linked? *Hydrobiologia* **460**, 213–220.

Hoey, A.S. & Bellwood, D.R. 2011. Suppression of herbivore by macroalgal density: a critical feedback on coral reefs? *Ecology Letters* **14**, 267–273.

Hoey, A.S., Howells, E., Johansen, J.L., Hobbs, J.P.A., Messmer, V., McCowan, D.M., Wilson, S.K. & Pratchett, M.S., 2016. Recent advances in understanding the effects of climate change on coral reefs. *Diversity* **8**, doi:10.3390/d8020012

Hughes, T.P. 1994. Catastrophes, phase shifts, and large-scale degradation of a Caribbean coral reef. *Science* **265**, 1547–1551.

Hughes, T.P., Rodrigues, M.J., Bellwood, D.R., Ceccarelli, D., Hoegh-Guldberg, O., McCook, L., Moltschaniwskyj, N., Prathett, M.S., Steneck, R.S. & Willis, B. 2007. Phase shifts, herbivory, and the resilience of coral reefs to climate change. *Current Biology* **17**, 360–365.

Hughes, T.P., Kerry, J.T., Álvarez-Noriega, M., Álvarez-Romero, J.G., Anderson, K.D., Baird, A.H., Babcock, R.C., Beger, M., Bellwood, D.R., Berkelmans, R., Bridge, T.C., Butler, I.R., Byrne, M., Cantin, N.E., Comeau, S. Connolly, S.R., Cumming, G.S., Dalton, S.J., Diaz-Pulido, G., Eakin, E.M., Figueira, W.F., Gilmour, J.P., Harrison, H.B., Heron, S.F., Hoey, A.S., Hobbs, J.-P.A., Hoogenboom, M.O., Kennedy, E.M., Kuo, C.-Y., Lough, J.M., Lowe, R.J., Liu, G., McCulloch, M.T., Malcolm, H.A., McWilliam, M.J., Pandolfi, J.M., Pears, R.J., Pratchett, M.S., Schoepf, V., Simpson, T., Skirving, W.J., Sommer, B., Torda, G., Wachenfeld, D.R., Willis, B.L. & Wilson, S.K. 2017. Global warming and recurrent mass bleaching of corals. *Nature* **543**, 373–377.

Hughes, T.P., Anderson, K.D., Connolly, S.R., Heron, S.F., Kerry, J.T., Lough, J.M., Baird, A.H., Baum, J.K., Berumen, M.L., Bridge, T.C., Claar, D.C., Eakin, C.M., Gilmour, J.P., Graham, N.A.J., Harrison, H., Hobbs, J.-P.A., Hoey, A.S., Hoogenboom, M.O., Lowe, R.J., McCulloch, M.T., Pandolfi, J.M., Prathcett, M.S., Schoepf, V., Torda, G. & Wilson, S.K. 2018a. Spatial and temporal patterns of mass bleaching of corals in the Anthropocene. *Science* **359**, 80–83.

Hughes, T.P., Kerry, J.T., Baird, A.H., Connolly, S.R., Dietzel, A., Eakin, C.M., Heron, S.F., Hoey, A.S., Hoogenboom, M.O., Liu, G., McWilliam, M.J., Pears, R.J., Pratchett, M.S., Skirving, W.J., Stella, J.S. & Torda, G. (2018b). Global warming transforms coral reef assemblages. *Nature* **556**, 492–496.

Hughes, T.P., Kerry, J.T., Connolly, S.R., Baird, A.H., Eakin, C.M., Heron, S.F., Hoey, A.S., Hoogenboom, M.O., Jacobson, M., Liu, G., Pratchett, M.S., Skirving, W.J. & Torda, G. (2019). Ecological memory modifies the cumulative impact of recurrent climate extremes. *Nature Climate Change* **9**, 40–43.

Hutchings, P., Kingsford, M. & Hoegh-Guldberg, O. 2019. *The Great Barrier Reef: Biology, Environment and Management.* Clayton South: CSIRO publishing.

Jackson, J., Donovan, M., Cramer, K. & Lam, V. 2014. *Status and Trends of Caribbean Coral Reefs: 1970-2012.* Gland, Switzerland: Global Coral Reef Monitoring Network, IUCN.

Jauffrais, T., Brisset, M., Lagourgue, L., Payri, C.E., Gobin, S., Le Gendre, R. & van Wynsberge, S. 2022. Seasonal changes in the photophysiology of *Ulva batuffolosa* in a coastal barrier reef. *Aquatic Botany* **179**, 103515, doi:10.1016/j.aquabot.2022.103515

Jiménez-López, I.T., Flórez-Leiva, L. & Quan-Young, L.I. 2020. Influencia de la herbivoría sobre la interacción alga-coral en un arrecife coralino de bahía Capurganá, Caribe colombiano. *Revista de Biología Tropical* **68**, 729–742.

Johns, K.A., Emslie, M.J., Hoey, A.S., Osborne, K., Jonker, M.J. & Cheal, A.J. 2018. Macroalgal feedbacks and substrate properties maintain a coral reef regime shift. *Ecosphere* **9**, 15, doi:10.1002/ecs2.2349

Johnson, M.D. & Carpenter, R.C. 2012. Ocean acidification and warming decrease calcification in the crustose coralline alga *Hydrolithon* onkodes and increase susceptibility to grazing. *Journal of Experimental Marine Biology and Ecology* **434**, 94–101.

Jompa, J. & McCook, L. 1998. Seaweeds save the reef?: *Sargassum* canopy decreases coral bleaching on inshore reefs. *Reef Research* **8**, 5. www.aims.gov.au/docs/projectnet/seaweeds.html

Jompa, J. & McCook, L.J. 2002. The effects of nutrients and herbivory on competition between a hard coral (*Porites cylindrica*) and a brown alga (*Lobophora variegata*). *Limnology and Oceanography* **47**, 527–534.

Jompa, J. & McCook, L.J. 2003a. Contrasting effects of turf algae on corals: massive Porites spp. are unaffected by mixed-species turfs, but killed by the red alga *Anotrichium tenue*. *Marine Ecology Progress Series* **258**, 79–86.

Jompa, J. & McCook, L.J. 2003b. Coral-algal competition: macroalgae with different properties have different effects on corals. *Marine Ecology Progress Series* **258**, 87–95.

Jorissen, H., Galand, P.E., Bonnard, I., Meiling, S., Raviglione, D., Meistertzheim, A.-L., Hedouin, L., Banaigs, B., Payri, C. & Nugues, M.M. 2021. Coral larval settlement preferences linked to crustose coralline algae with distinct chemical and microbial signatures. *Scientific Reports* **11**, 14610, doi:10.1038/s41598-021-94096-6

Kitahara, M.V., Fukami, H., Benzoni, F. & Huang, D. 2016. The new systematics of Scleractinia: integrating molecular and morphological evidence. In *The Cnidaria, Past, Present and Future: The World of Medusa and Her Sisters*, S. Goffredo & Z. Dubinsky (eds.). Cham, Switzerland: Springer International Publishing, 41–59.

Koop, K., Booth, D., Broadbent, A., Brodie, J., Bucher, D., Capone, D., Coll, J., Dennison, W., Erdmann, M., Harrison, P., Hoegh-Guldberg, O., Hutchings, P., Jones, G.B., Larkhum, A.W.D., O'Neil, J., Steven, A., Tentori, E., Ward, S., Williamson, J. & Yellowlees, D. 2001. ENCORE: the effect of nutrient enrichment on coral reefs. Synthesis of results and conclusions. *Marine Pollution Bulletin* **42**, 91–120.

Kuffner, I., Walters, L., Becerro, M., Paul, V., Ritson-Williams, R. & Beach, K. 2006. Inhibition of coral recruitment by macroalgae and cyanobacteria. *Marine Ecology Progress Series* **323**, 107–117.

Kuffner, I.B. & Paul, V.J. 2004. Effects of the benthic cyanobacterium *Lyngbya majuscula* on larval recruitment of the reef corals *Acropora surculosa* and *Pocillopora damicornis*. *Coral Reefs* **23**, 455–458.

Lam, V.Y.Y., Chaloupka, M., Thompson, A., Doropoulos, C. & Mumby, P.J. 2018. Acute drivers influence recent inshore Great Barrier Reef dynamics. *Proceedings of the Royal Society B-Biological Sciences* **285**, 9, doi:10.1098/rspb.2018.2063

Lasker, H.R. 1980. Sediment rejection by reef corals: the roles of behavior and morphology in *Montastrea cavernosa* (Linnaeus). *Journal of Experimental Marine Biology and Ecology* **47**, 77–87.

Lefèvre, C.D. & Bellwood, D.R. 2010. Seasonality and dynamics in coral reef macroalgae: variation in condition and susceptibility to herbivory. *Marine Biology* **157**, 955–965.

Leinbach, S.E., Speare, K.E., Rossin, A.M., Holstein, D.M. & Strader, M.E. 2021. Energetic and reproductive costs of coral recovery in divergent bleaching responses. *Scientific Reports* **11**, doi:10.1038/s41598-021-02807-w

Lenth, R.V. 2023. Emmeans: estimated marginal means, aka least-squares means (Version 1.8.6). https://CRAN.R-project.org/package=emmeans

Lesser, M.P. 2011. Coral bleaching: causes and mechanisms. In *Coral Reefs: An Ecosystem in Transition*, Z. Dubinsky & N. Stamblet (eds.). Dordrecht, Netherlands: Springer Dordrecht, 405–419.

Leuzinger, S., Willis, B.L. & Anthony, K.R.N. 2012. Energy allocation in a reef coral under varying resource availability. *Marine Biology* **159**, 177–186.

Lirman, D. & Biber, P. 2000. Seasonal dynamics of macroalgal communities of the northern Florida Reef Tract. *Botanica Marina* **43**, 305–314.

Littler, M.M. & Littler, D.S. 1985. Factors controlling relative dominance of primary producers on biotic reefs. *Proceedings of the Fifth International Coral Reef Congress, Tahiti* **4**, 35–39.

Loffler, Z. & Hoey, A.S. 2018. Canopy-forming macroalgal beds (*Sargassum*) on coral reefs are resilient to physical disturbance. *Journal of Ecology* **106**, 1156–1164.

Loffler, Z., Graba-Landry, A., Kidgell, J.T., McClure, E.C., Pratchett, M.S. & Hoey, A.S. 2018. Holdfasts of *Sargassum swartzii* are resistant to herbivory and resilient to damage. *Coral Reefs* **37**, 1075-1084.

López-Jiménez, I.T., Quan-Young, L.I. & Florez-Leiva, L. 2021. Effect of terrigenous sediments on macroalgae functional-form groups of coral reefs in Capurganá, Colombian Caribbean. *Scientia Marina* **85**, 125–135.

Low, J.K.Y., Fong, J., Todd, P.A., Chou, L.M. & Bauman, A.G. 2019. Seasonal variation of *Sargassum ilicifolium* (Phaeophyceae) growth on equatorial coral reefs. *Journal of Phycology* **55**, 289–296.

Loya, Y., Sakai, K., Yamazato, K., Nakano, Y., Sambali, H. & Van Woesik, R. 2001. Coral bleaching: the winners and the losers. *Ecology Letters* **4**, 122–131.

Madin, J.S. & Connolly, S.R. 2006. Ecological consequences of major hydrodynamic disturbances on coral reefs. *Nature* **444**, 477–480.

Madin, J.S., Anderson, K.D., Andreasen, M.H., Bridge, T.C.L., Cairns, S.D., Connolly, S.R., Darling, E.S., Diaz, M., Falster, D.S., Franklin, E.C., Gates, R.D., Harmer, A.M.T., Hoogenboom, M.O., Huang, D., Keith, S.A., Kosnik, M.A., Kuo, C.-Y., Lough, J.M., Lovelock, C.E., Luiz, O., Martinelli, J., Mizerek, T., Pandolfi, J.M., Pochon, X., Pratchett, M.S., Putman, H., Roberts, T.E., Stat, M., Wallace, C.C., Widman, E. & Baird, A.H. 2016. The Coral Trait Database, a curated database of trait information for coral species from the global oceans. *Scientific Data* **3**, 160017, doi:10.1038/sdata.2016.17

Marshall, P.A. & Baird, A.H. 2000. Bleaching of corals on the Great Barrier Reef: differential susceptibilities among taxa. *Coral Reefs* **19**, 155–163.

Martin-Smith, K.M. 1993. The phenology of four species of *Sargassum* at Magnetic Island, Australia. *Botanica Marina* **36**, 327–334.

Masson-Delmotte, V., Zhai, P., Pirani, A., Connors, S.L., Pean, C., Chen, Y., Goldfarb, L., Gomis, M.I., Matthews, J.B.R., Berger, S., Huang, M., Yelekci, O., Yu, R. & Zhou, B. (eds.) *Climate Change 2021: The Physical Science Basis. Contribution of Working Group I to the Sixth Assessment Report of the Intergovernmental Panel on Climate Change*. Cambridge, UK: Cambridge University Press.

McClanahan, T.R., Cokos, B.A. & Sala, E. 2002. Algal growth and species composition under experimental control of herbivory, phosphorus and coral abundance in Glovers Reef, Belize. *Marine Pollution Bulletin* **44**, 441–451.

McLachlan, R.H., Price, J.T., Solomon, S.L. & Grottoli, A.G. 2020. Thirty years of coral heat-stress experiments: a review of methods. *Coral Reefs* **39**, 885–902.

McLean, M., Stuart-Smith, R.D., Villéger, S., Auber, A., Edgar, G.J., Macneil, M.A., Loiseau, N., Leprieur, F. & Mouillot, D. 2021. Trait similarity in reef fish faunas across the world's oceans. *Proceedings of the National Academy of Sciences of the United States of America* **118**, e2012318118, doi:10.1073/pnas.2012318118

McWilliams, J.P., Côté, I.M., Gill, J.A., Sutherland, W.J. & Watkinson, A.R. 2005. Accelerating impacts of temperature-induced coral bleaching in the Caribbean. *Ecology* **86**, 2055–2060.

Mejía Niño, N. & Garzón Ferreira, J. 2003. Dinámica de las interacciones algal-coral en dos Bahías de la región de Santa Marta (caribe colombiano) con distinto grado de influencia antropogénica. *Boletín de Investigaciones Marinas y Costeras-INVEMAR* **32**, 243–261.

Miller, M.W., Hay, M.E., Miller, S.L., Malone, D., Sotka, E.E. & Szmant, A.M. 1999. Effects of nutrients versus herbivores on reef algae: a new method for manipulating nutrients on coral reefs. *Limnology and Oceanography* **44**, 1847–1861.

Monteil, Y., Teo, A., Fong, J., Bauman, A.G. & Todd, P.A. 2020. Effects of macroalgae on coral fecundity in a degraded coral reef system. *Marine Pollution Bulletin* **151**, 110890, doi:10.1016/j.marpolbul.2020.110890

Morrow, K.M., Bromhall, K., Motti, C.A., Munn, C.B. & Bourne, D.G. 2017. Allelochemicals produced by brown macroalgae of the *Lobophora* genus are active against coral larvae and associated bacteria, supporting pathogenic shifts to vibrio dominance. *Applied and Environmental Microbiology* **83**, e02391-16, doi:10.1128/aem.02391-16

Mumby, P.J., Foster, N.L. & Fahy, E.A.G. 2005. Patch dynamics of coral reef macroalgae under chronic and acute disturbance. *Coral Reefs* **24**, 681–692.

Naumann, M.S., Bednarz, V.N., Ferse, S.C.A., Niggl, W. & Wild, C. 2015. Monitoring of coastal coral reefs near Dahab (Gulf of Aqaba, Red Sea) indicates local eutrophication as potential cause for change in benthic communities. *Environmental Monitoring and Assessment* **187**, doi:10.1007/s10661-014-4257-9

Norström, A., Nyström, M., Lokrantz, J. & Folke, C. 2009. Alternative states on coral reefs: beyond coral-macroalgal phase shifts. *Marine Ecology Progress Series* **376**, 295–306.

Nugues, M. & Bak, R. 2006. Differential competitive abilities between Caribbean coral species and a brown alga: a year of experiments and a long-term perspective. *Marine Ecology Progress Series* **315**, 75–86.

Nugues, M.M. & Roberts, C.M. 2003. Coral mortality and interaction with algae in relation to sedimentation. *Coral Reefs* **22**, 507–516.

Olsen, K., Paul, V.J. & Ross, C. 2015. Direct effects of elevated temperature, reduced pH, and the presence of macroalgae (Dictyota spp.) on larvae of the Caribbean coral *Porites astreoides*. *Bulletin of Marine Science* **91**, 255–270.

Olsen, K., Ritson-Williams, R., Paul, V. & Ross, C. 2014. Combined effects of macroalgal presence and elevated temperature on the early life-history stages of a common Caribbean coral. *Marine Ecology Progress Series* **509**, 181–191.

Olsen, K., Sneed, J.M. & Paul, V.J. 2016. Differential larval settlement responses of *Porites astreoides* and *Acropora palmata* in the presence of the green alga *Halimeda opuntia*. *Coral Reefs* **35**, 521–525.

Ortiz, J.C., Pears, R.J., Beeden, R., Dryden, J., Wolff, N.H., Gomez Cabrera, M.D.C. & Mumby, P.J. 2021. Important ecosystem function, low redundancy and high vulnerability: the trifecta argument for protecting the Great Barrier Reef's tabular *Acropora*. *Conservation Letters* **14**, doi:10.1111/conl.12817

Pandolfi, J.M. 2003. Global trajectories of the long-term decline of coral reef ecosystems. *Science* **301**, 955–958.

Paul, V., Kuffner, I., Walters, L., Ritson-Williams, R., Beach, K. & Becerro, M. 2011. Chemically mediated interactions between macroalgae *Dictyota spp*. and multiple life-history stages of the coral *Porites astreoides*. *Marine Ecology Progress Series* **426**, 161–170.

Pires, J.C.M. 2017. COP21: the algae opportunity? *Renewable and Sustainable Energy Reviews* **79**, 867–877.

Pratchett, M.S., Hoey, A.S., Wilson, S.K., Messmer, V. & Graham, N.A.J. 2011. Changes in biodiversity and functioning of reef fish assemblages following coral bleaching and coral loss. *Diversity* **3**, 424–452.

Pratchett, M.S., Anderson, K.D., Hoogenboom, M.O., Widman, E., Baird, A.H., Pandolfi, J.M., Edmunds, P.J. & Lough, J.M. 2015. Spatial, temporal and taxonomic variation in coral growth-implications for the structure and function of coral reef ecosystems. *Oceanography and Marine Biology: An Annual Review* **53**, 215–295.

Pratchett, M.S., McWilliam, M.J. & Riegl, B. 2020. Contrasting shifts in coral assemblages with increasing disturbances. *Coral Reefs* **39**, 783–793.

Pratte, Z.A., Longo, G.O., Burns, A.S., Hay, M.E. & Stewart, F.J. 2018. Contact with turf algae alters the coral microbiome: contact versus systemic impacts. *Coral Reefs* **37**, 1–13.

Price, N. 2010. Habitat selection, facilitation, and biotic settlement cues affect distribution and performance of coral recruits in French Polynesia. *Oecologia* **163**, 747–758.

Putnam, H.M., Ritson-Williams, R., Cruz, J.A., Davidson, J.M. & Gates, R.D. 2020. Environmentally-induced parental or developmental conditioning influences coral offspring ecological performance. *Scientific Reports* **10**, 1, doi:10.1038/s41598-020-70605-x

Quéré, G. & Nugues, M.M. 2015. Coralline algae disease reduces survival and settlement success of coral planulae in laboratory experiments. *Coral Reefs* **34**, 863–870.

Ramseyer, T., Tronholm, A., Turner, T., Brandt, M. & Smith, T. 2021. Elevated nutrients and herbivory negatively affect *Dictyota* growth dynamics. *Marine Ecology Progress Series* **671**, 81–95.

Rasher, D.B. & Hay, M.E. 2010. Chemically rich seaweeds poison corals when not controlled by herbivores. *Proceedings of the National Academy of Sciences of the United States of America* **107**, 9683–9688.

Rasher, D.B., Hoey, A.S. & Hay, M.E. 2013. Consumer diversity interacts with prey defenses to drive ecosystem function. *Ecology* **94**, 1347–1358.

Reed, D., Washburn, L., Rassweiler, A., Miller, R., Bell, T. & Harrer, S. 2016. Extreme warming challenges sentinel status of kelp forests as indicators of climate change. *Nature Communications* **7**, 13757, doi:10.1038/ncomms13757

Richardson, L.E., Graham, N.A.J., Pratchett, M.S., Eurich, J.G. & Hoey, A.S. 2018. Mass coral bleaching causes biotic homogenization of reef fish assemblages. *Global Change Biology* **24**, 3117–3129.

Ritson-Williams, R., Arnold, S., Fogarty, N., Steneck, R.S., Vermeij, M. & Paul, V.J. 2009. New perspectives on ecological mechanisms affecting coral recruitment on reefs. *Smithsonian Contributions to the Marine Sciences* **38**, 437–457.

Ritson-Williams, R., Paul, V.J., Arnold, S.N. & Steneck, R.S. 2010. Larval settlement preferences and post-settlement survival of the threatened Caribbean corals *Acropora palmata* and *A. cervicornis*. *Coral Reefs* **29**, 71–81.

Ritson-Williams, R., Arnold, S.N., Paul, V.J. & Steneck, R.S. 2014. Larval settlement preferences of *Acropora palmata* and *Montastraea faveolata* in response to diverse red algae. *Coral Reefs* **33**, 59–66.

Ritson-Williams, R., Arnold, S. & Paul, V. 2016a. Patterns of larval settlement preferences and post-settlement survival for seven Caribbean corals. *Marine Ecology Progress Series* **548**, 127–138.

Ritson-Williams, R., Ross, C. & Paul, V.J. 2016b. Elevated temperature and allelopathy impact coral recruitment. *PLoS One* **11**, e0166581, doi:10.1371/journal.pone.0166581

Ritson-Williams, R., Arnold, S.N. & Paul, V.J. 2020. The impact of macroalgae and cyanobacteria on larval survival and settlement of the scleractinian corals *Acropora palmata*, *A. cervicornis* and *Pseudodiploria strigosa*. *Marine Biology* **167**, doi:10.1007/s00227-019-3639-5

River, G.F. & Edmunds, P.J. 2001. Mechanisms of interaction between macroalgae and scleractinians on a coral reef in Jamaica. *Journal of Experimental Marine Biology and Ecology* **261**, 159–172.

Roach, T.N.F., Abieri, M.L., George, E.E., Knowles, B., Naliboff, D.S., Smurthwaite, C.A., Kelly, L.W., Haas, A.F. & Rohwer, F.L. 2017. Microbial bioenergetics of coral-algal interactions. *PeerJ* **5**, e3423, doi:10.7717/peerj.3423

Rogers, C. & Miller, J. 2006. Permanent phase shifts or reversible declines in coral cover? Lack of recovery of two coral reefs in St. John, US Virgin Islands. *Marine Ecology Progress Series* **306**, 103–114.

Rosenberg, E. & Ben-Haim, Y. 2002. Microbial diseases of corals and global warming. *Environmental Microbiology* **4**, 318–326.

Roth, F., El-Khaled, Y.C., Karcher, D.B., Rädecker, N., Carvalho, S., Duarte, C.M., Silva, L., Calleja, M.L., Moran, X.A.G., Jones, B.H., Voolstra, C.R. & Wild, C. 2021. Nutrient pollution enhances productivity and framework dissolution in algae- but not in coral-dominated reef communities. *Marine Pollution Bulletin* **168**, 112444, doi:10.1016/j.marpolbul.2021.112444

Russ, G.R., Questel, S.-L.A., Rizzari, J.R. & Alcala, A.C. 2015. The parrotfish-coral relationship: refuting the ubiquity of a prevailing paradigm. *Marine Biology* **162**, 2029-2045.

Rylaarsdam, K.W. 1983. Life histories and abundance patterns of colonial corals on Jamaican reefs. *Marine Ecology Progress Series* **13**, 249–260.

Schaffelke, B. & Klumpp, D.W. 1997. Biomass and productivity of tropical macroalgae on three nearshore fringing reefs in the central Great Barrier Reef, Australia. *Botanica Marina* **40**, 373–384.

Selig, E.R., Drew Harvell, C., Bruno, J.F., Willis, B.L., Page, C.A., Casey, K.S. & Sweatman, H. 2006. Analyzing the relationship between ocean temperature anomalies and coral disease outbreaks at broad spatial scales. In *Coral Reefs and Climate Change: Science and Management*, J.T. Phinney et al. (eds.). Washington, District of Columbia: American Geophysical Union, 111–128.

Sing Wong, A., Vrontos, S. & Taylor, M.L. 2022. An assessment of people living by coral reefs over space and time. *Global Change Biology* **28**, 7139–7153.

Singh, S.P. & Singh, P. 2015. Effect of temperature and light on the growth of algae species: a review. *Renewable and Sustainable Energy Reviews* **50**, 431–444.

Smith, J.E., Kuwabara, J., Coney, J., Flanagan, K., Beets, J., Brown, D., Stanton, F., Takabayashi, M., du Plessis, S., Griesemer, B.K., Barnes, S. & Turner, J. 2008. An unusual cyanobacterial bloom in Hawai`i. *Coral Reefs* **27**, 851–851.

Smith, H.A., Prenzlau, T., Whitman, T., Fulton, S.E., Borghi, S., Logan, M., Heron, S.F. & Bourne, D.G. 2022. Macroalgal canopies provide corals limited protection from bleaching and impede post-bleaching recovery. *Journal of Experimental Marine Biology and Ecology* **553**, 151762, doi:10.1016/j.jembe.2022.151762

Souter, D., Planes, S., Wicquart, J., Logan, M., Obura, D. & Staub, F. (eds.) 2021. *Status of Coral Reefs of the World: 2020*. Townsville, Australia: Global Coral Reef Monitoring Network.

Speare, K.E., Duran, A., Miller, M.W. & Burkepile, D.E. 2019. Sediment associated with algal turfs inhibits the settlement of two endangered coral species. *Marine Pollution Bulletin* **144**, 189–195.

Stella, J.S., Pratchett, M.S., Hutchings, P. & Jones, G.P. 2011. Coral-associated invertebrates: diversity, ecological importance and vulnerability to disturbance. *Oceanography and Marine Biology: An Annual Review* **49**, 43–104.

Steneck, R.S. & Dethier, M.N. 1994. A functional group approach to the structure of algal-dominated communities. *Oikos* **69**, 476–498.

Steneck, R., Arnold, S. & Mumby, P. 2014. Experiment mimics fishing on parrotfish: insights on coral reef recovery and alternative attractors. *Marine Ecology Progress Series* **506**, 115–127.

Storlazzi, C.D., Norris, B.K. & Rosenberger, K.J. 2015. The influence of grain size, grain color, and suspended-sediment concentration on light attenuation: why fine-grained terrestrial sediment is bad for coral reef ecosystems. *Coral Reefs* **34**, 967–975.

Suchley, A. & Alvarez-Filip, L. 2017. Herbivory facilitates growth of a key reef-building Caribbean coral. *Ecology and Evolution* **7**, 11246–11256.

Sully, S., Burkepile, D.E., Donovan, M.K., Hodgson, G. & van Woesik, R. 2019. A global analysis of coral bleaching over the past two decades. *Nature Communications* **10**, 1264, doi:10.1038/s41467-019-09238-2

Sura, S.A., Bell, A., Kunes, K.L., Turba, R., Songer, R. & Fong, P. 2021. Responses of two common coral reef macroalgae to nutrient addition, sediment addition, and mechanical damage. *Journal of Experimental Marine Biology and Ecology* **536**, 151512, doi:10.1016/j.jembe.2021.151512

Swierts, T. & Vermeij, M.J. 2016. Competitive interactions between corals and turf algae depend on coral colony form. *PeerJ* **4**, e1984, doi:10.7717/peerj.1984

Tebbett, S.B. & Bellwood, D.R. 2019. Algal turf sediments on coral reefs: what's known and what's next. *Marine Pollution Bulletin* **149**, 110542, doi:10.1016/j.marpolbul.2019.110542

Thinesh, T., Jose, P.A., Ramasamy, P., Meenatchi, R., Selvan, K.M. & Selvin, J. 2019. Differential coral response to algae contact: *Porites* tissue loss, praise for *Halimeda* interaction at southeast coast of India. *Environmental Science and Pollution Research* **26**, 17845–17852.

Titlyanov, E.A., Titlyanova, T.V., Yakovleva, I.M., Nakano, Y. & Bhagooli, R 2005. Regeneration of artificial injuries on scleractinian corals and coral/algal competition for newly formed substrate. *Journal of Experimental Marine Biology and Ecology* **323**, 27–42.

Titlyanov, E.A., Titlyanova, T.V., Yakovleva, I.M. & Sergeeva, O.S. 2006. Influence of winter and spring/summer algal communities on the growth and physiology of adjacent scleractinian corals. *Botanica Marina* **49**, 200–207.

Titlyanov, E.A., Titlyanova, T.V. & Arvedlund, M. 2009. Finding the winners in competition for substratum between coral polyps and epilithic algae on damaged colonies of the coral *Porites lutea*. *Marine Biodiversity Records* **2**, e85, doi:10.1017/S175526720900089X

Trapon, M.L., Pratchett, M.S., Hoey, A.S. & Baird, A.H. 2013. Influence of fish grazing and sedimentation on the early post-settlement survival of the tabular coral *Acropora cytherea*. *Coral Reefs* **32**, 1051–1059.

Tsai, C.-C., Chang, J.-S., Sheu, F., Shyu, Y.-T., Yu, A.Y.-C., Wong, S.-L., Dai, C.-F. & Lee, T.-M. 2005. Seasonal growth dynamics of *Laurencia papillosa* and *Gracilaria coronopifolia* from a highly eutrophic reef in southern Taiwan: temperature limitation and nutrient availability. *Journal of Experimental Marine Biology and Ecology* **315**, 49–69.

Van De Leemput, I.A., Hughes, T.P., Van Nes, E.H. & Scheffer, M. 2016. Multiple feedbacks and the prevalence of alternate stable states on coral reefs. *Coral Reefs* **35**, 857–865.

Van Oppen, M.J.H. & Lough, J.M. 2018. Synthesis: coral bleaching: patterns, processes, causes and consequences. In *Coral Bleaching: Patterns, Processes, Causes and Consequences*, M.J.H. van Oppen & J.M. Lough (eds.). Cham, Switzerland: Springer International Publishing, 343–348.

Van Woesik, R. 1998. Lesion healing on massive *Porites* spp. corals. *Marine Ecology Progress Series* **164**, 213–220.

Van Woesik, R., Sakai, K., Ganase, A. & Loya, Y. 2011. Revisiting the winners and the losers a decade after coral bleaching. *Marine Ecology Progress Series* **434**, 67–76.

Vega Thurber, R.L., Burkepile, D.E., Fuchs, C., Shantz, A.A., McMinds, R. & Zaneveld, J.R. 2014. Chronic nutrient enrichment increases prevalence and severity of coral disease and bleaching. *Global Change Biology* **20**, 544–554.

Venera-Ponton, D., Diaz-Pulido, G., McCook, L. & Rangel-Campo, A. 2011. Macroalgae reduce growth of juvenile corals but protect them from parrotfish damage. *Marine Ecology Progress Series* **421**, 109–115.

Vermeij, M.J.A., Van Moorselaar, I., Engelhard, S., Hörnlein, C., Vonk, S.M. & Visser, P.M. 2010. The effects of nutrient enrichment and herbivore abundance on the ability of turf algae to overgrow coral in the Caribbean. *PLoS One* **5**, e14312, doi:10.1371/journal.pone.0014312

Vieira, C. 2020. *Lobophora*-coral interactions and phase shifts: summary of current knowledge and future directions. *Aquatic Ecology* **54**, 1–20.

Vieira, C., Payri, C. & De Clerck, O. 2015. Overgrowth and killing of corals by the brown alga *Lobophora hederacea* (Dictyotales, Phaeophyceae) on healthy reefs in New Caledonia: a new case of the epizoism syndrome. *Phycological Research* **63**, 152–153.

Vieira, C., Thomas, O.P., Culioli, G., Genta-Jouve, G., Houlbreque, F., Gaubert, J., de Clerck, O. & Payri, C.E. 2016. Allelopathic interactions between the brown algal genus *Lobophora* (Dictyotales, Phaeophyceae) and scleractinian corals. *Scientific Reports* **6**, 18637, doi:10.1038/srep18637

Voss, J.D. & Richardson, L.L. (2006). Nutrient enrichment enhances black band disease progression in corals. *Coral Reefs* **25**, 569–576.

Vuki, V.C. & Price, I.R. 1994. Seasonal changes in the *Sargassum* populations on a fringing coral reef, Magnetic Island, Great Barrier Reef region, Australia. *Aquatic Botany* **48**, 153–166.

Wilkinson, C.R. & Souter, D. (eds.) 2008. *Status of Caribbean Coral Reefs After Bleaching and Hurricanes in 2005*. Townsville, Australia: Global Coral Reef Monitoring Network.

Williams, I. & Polunin, N. 2001. Large-scale associations between macroalgal cover and grazer biomass on mid-depth reefs in the Caribbean. *Coral Reefs* **19**, 358–366.

Wismer, S., Hoey, A.S. & Bellwood, D.R. 2009. Cross-shelf benthic community structure on the Great Barrier Reef: relationships between macroalgal cover and herbivore biomass. *Marine Ecology Progress Series* **376**, 45–54.

Wolf, A.T., Wild, C. & Nugues, M. 2012. Contact with macroalgae causes variable coral mortality in *Montastraea faveolata*. *Proceedings of the 12th International Coral Reef Symposium*. Australia: Cairns.

Woodhead, A.J., Hicks, C.C., Norström, A.V., Williams, G.J. & Graham, N.A.J. 2019. Coral reef ecosystem services in the Anthropocene. *Functional Ecology* **33**, 1023–1034.

Wooldridge, S., Brodie, J. & Furnas, M. 2006. Exposure of inner-shelf reefs to nutrient enriched runoff entering the Great Barrier Reef Lagoon: post-European changes and the design of water quality targets. *Marine Pollution Bulletin* **52**, 1467–1479.

WoRMS Editorial Board 2021. World Register of Marine Species (WoRMS). https://www.marinespecies.org.

Zaneveld, J.R., Burkepile, D.E., Shantz, A.A., Pritchard, C.E., McMinds, R., Payet, J.P., Welsh, R., Correa, A.M.S., Lemoine, N.P., Rosales, S., Fuchs, C., Maynard, J. & Thurber, R.V. 2016. Overfishing and nutrient pollution interact with temperature to disrupt coral reefs down to microbial scales. *Nature Communications* **7**, 11833, doi:10.1038/ncomms11833

Supplementary Materials are provided online at: https://www.routledge.com/9781032761961

Oceanography and Marine Biology: An Annual Review, 2024, **62**, 303-349
© P. A. Todd, B. D. Russell, A. J. Lemasson, A. L. Allcock, M. Byrne, L. B. Firth, C. H. Lucas, E. M. Marzinelli, P. J. Mumby, J. Sharples, I. P. Smith, S. E. Swearer, and M. Thiel, Editors
Taylor & Francis

IMPORTANCE OF SUBLETHAL PREDATION IN NON-MIGRATORY MEGAFAUNAL AND MACROFAUNAL ASSEMBLAGES IN SOFT SEDIMENTS

DENNIS CODY LEES*

**Littoral Ecological & Environmental Services, Leucadia, California, USA*

Abstract Sublethal predation on non-migratory megafaunal and macrofaunal assemblages at the organismic level has been described in a wide range of autecological studies for six higher taxa over the past 90 years. However, these important data have been virtually absent from synecological discussions of infaunal communities or analyses. This is especially important when considering secondary production, where sublethal predation likely plays a highly significant role.

In this study, I examine 77 species for which investigators have described sublethal predation on non-migratory megafaunal (larger longer-lived but less common) and macrofaunal (small shorter-lived and far more abundant) infaunal invertebrates living in or on soft sediments. The review focuses on the importance of sublethal predation in infaunal assemblages, its effects on the target species and its contributions to higher trophic levels. Most research has been conducted on non-migratory megafauna but a considerable body of data for macrofauna is available. The review presents examples based on (1) observed or documented sublethal predation or regeneration of structures in prey species; (2) analysis based on gut contents of predators; and (3) the wide variety of effects observed in these studies. I also provide (4) examples that provide insight into the substantial importance of sublethal predation on non-migratory megafauna to secondary production. The volume of research conducted varies widely among these topics. The discussion makes clear that, while largely overlooked in previous discussions of ecology and ecosystem dynamics for infaunal ecosystems, sublethal predation is an activity of major importance in the energy dynamics of these ecosystems, especially for non-migratory megafauna and the higher trophic levels that prey on them.

Keywords: Predation; Sublethal; Nonlethal; Behaviour; Regeneration; Infauna; Megafauna; Macrofauna; Secondary Production

Introduction

In most discussions of benthic ecosystem dynamics, predation generally has been considered a lethal interaction in which prey animals are killed, consumed and removed from prey standing stocks as the consumed energy is transferred to higher trophic levels (Lawrence & Vasquez 1996). Thus, trophic studies for infaunal assemblages, which generally have focused on macrofaunal assemblages, assumed that lethal predation (1) predominates in energy transfer from lower to higher trophic levels (Virnstein 1977) and (2) is the major component of secondary production on soft sediments (e.g., Zahorczak et al. 2000). Nevertheless, a wide range of autecological and feeding studies document sublethal predation, an interaction analogous to herbivory on plants in terrestrial ecosystems. Most

DOI: 10.1201/9781003477518-8

of these studies focus on non-migratory megafauna [the much larger longer-lived but less common 'trees' *sensu* Lees (2021)] as opposed to macrofauna (the much smaller, shorter-lived and relatively far more abundant weeds). Harris (1989) pointed out that nonlethal injury, i.e., sublethal predation, also may play an important role in the regulatory biology of population size in many species.

Non-migratory megafaunal organisms as defined herein are large, relatively long-lived basically non-migratory infaunal organisms that live on or in the sediments, i.e., they live out their lives within several metres of where their recruits settle out of the plankton. They do not exhibit seasonal or other types of migration like many crab or shrimp species (e.g., Boddeke 1976) or sea stars. Bivalves like *Macoma balthica* are included even though juveniles often relocate several metres up or down a beach as they mature. Sand dollars are included even though they may move tens of metres onshore or offshore to accommodate for variations in seasonal swell activity (Merrill & Hobson 1970, Morin et al. 1988). Although they are motile, epibenthic snails and brittle stars are included because their range of movements is limited to tens of metres.

In terms of longevity, typically the life span considered herein for non-migratory megafauna averages over 18 years and ranges from 3 to 4 years to over 150 years (Orensanz et al. 2004); in contrast, the life span for macrofauna is typically less than 3 years. In this framework, macrofaunal species are considered opportunistic, i.e., they recruit to disturbed areas, whereas non-migratory megafauna are considered relatively stable perennials. Generally, macrofaunal organisms live within a few centimetres of the sediment/water interface, whereas most of the non-migratory megafauna bury up to metres into the sediment, often creating complex burrow systems (Dworschak 2001). Thus, macrofauna are easy to sample, whereas non-migratory megafaunal organisms are somewhat more difficult to census. In terms of biomass, standing stocks for macrofauna are generally in tens of grams wet weight m^{-2}, whereas standing stocks for non-migratory megafauna often exceed a kg m^{-2} (Lees et al. 1980; unpublished results).

De Vlas (1981) noted that many infaunal animals must expose parts of their bodies to predators to feed, access oxygen, defecate, or relocate. Moreover, many burrowing predators intercept their prey beneath the sediment/water interface and nip off buried parts. These interactions can lead to appreciable sublethal energy transfer from prey species to higher trophic levels (Lindsay 2010).

The number of studies for non-migratory megafauna and macrofaunal polychaetes and the variety and importance of the organisms studied indicate that sublethal predation is a major pathway for energy transfer from these assemblages to higher trophic levels and is a normal and widespread occurrence in infaunal assemblages. Ansell (1995) used underwater television to document the cues that some clams and polychaetes provide to predators that graze on exposed parts of the prey and how some predators use those cues for grazing. Nevertheless, the interaction has been ignored in traditional models and in estimates of energy flow or secondary production (Pape-Lindstrom et al. 1997).

Three reviews on this topic have been published previously. Lawrence and Vasquez (1996) described sublethal predation for epifaunal and infaunal echinoderms. They noted ca. 85 taxa among the five classes comprising Echinodermata. Ophiuroids, with 32 taxa listed, were by far the most well-researched echinoderm. They discussed the effects of sublethal predation on growth, reproduction, acquisition and distribution of nutrient reserves, movement, survivorship, tissue production, regeneration of body parts and relationships between the effects of sublethal predation and life-history strategies. They noted that sublethal predation has been only occasionally considered in discussions of marine communities, which have focused on lethal predation. They concluded that sublethal predation exerts effects on behaviour, morphology and physiology of individuals of a prey species but also reaches far beyond those effects on an individual to both population and community levels.

Lindsay (2010) covered injury for a much wider range of epifaunal and infaunal invertebrate taxa and approached the topic from different perspectives. A listing of >130 species was provided in which sublethal predation has been reported. It included cnidarians, polychaetes, brachiopods, bivalves, gastropods, crustaceans, crinoids, ophiuroids, echinoids and sea stars. Focusing on frequency and types of injury and regeneration, she provided a broad summary of studies describing types of injury, their frequency and the source of the injuries. She discussed how effects can modify ecological patterns and processes. She concluded that a large percentage of the injuries resulted from sublethal predation, but that frequency of injury appears to vary substantially seasonally, indicating that single samplings do not suffice, especially when related to estimating energy flow.

She stated [in the supplementary table], "If the minimum and maximum incidences of injury are averaged across all taxa, types of injury, and records (i.e., each reported estimate of the incidence of injury for a given species), the proportion of injured individuals in marine benthic invertebrate populations ranged from 33% to 47%". She pointed out many ways this interaction is important in the ecology of many non-migratory megafaunal and macrofaunal polychaete taxa living in unconsolidated sediments throughout the world and to the epibenthic browsers such as demersal fish species or invertebrate predators associated with these assemblages. In an earlier paper, Lindsay and Woodin (1992) opined that sublethal predation changes both the condition of the individual and the cost or risk associated with further predation.

Most recently, Rennolds and Bely (2023) provided a comprehensive review of injury in animals in general. They covered terrestrial, aquatic and marine organisms, including molecular, cellular, behavioural and ecological effects of injury; they included some of the examples provided below in this review. One of their final conclusions was that greater taxonomic coverage of injury needs to be conducted to expand our ability to predict how injury influences animal biology, and how injury acts as a variable in ecosystems and influences evolution.

The concept of sublethal predation likely applies to a broad variety of other non-migratory megafaunal and macrofaunal organisms, at least sea pens, nemerteans, sipunculids, echiurids and enteropneusts. However, as Pape-Lindstrom et al. (1997) pointed out, the process of sublethal predation by epibenthic or infaunal browsers is rarely mentioned or woven into the fabric of infaunal community dynamics. For example, the text by Gray and Elliott (2009, p. 101) provided a single unreferenced mention of "sublethal browsers", following epibenthic and infaunal predators.

The intent of the review is to consolidate a substantial amount of the literature available for injury experienced by infaunal organisms. The major objectives are to describe (1) the importance of sublethal predation to ecosystem dynamics for non-migratory megafauna and macrofauna living in unconsolidated sediments; (2) the wide variety of predators that depend on these unevenly studied infaunal sub-assemblages; (3) the wide variety of effects occurring in prey organisms and populations resulting from sublethal predation; and (4) the value of the resulting energy transfer to higher trophic levels. The discussion focuses on observed injury, regeneration, structures of infaunal organisms reported in gut contents of predators (mainly fishes) and effects of this injury on the prey species. In addition, I have included a short discussion of the contributions from sublethal predation to secondary production.

It is likely that sublethal predation is a major component of secondary production, perhaps even more important than lethal predation in some ecosystems. Its importance will be described in separate chapters by providing examples of sublethal predation based on (1) documented observed injury or regeneration/recovery for injured megafauna; (2) gut contents of predators; (3) effects of sublethal predation on the prey activities and associated sediments; and (4) secondary production of the prey. Most of these examples refer to a broad range of megafauna but some examples are cited for macrofaunal polychaetes and a gammarid amphipod. While the assessment provides a wide variety of examples, it should not be considered complete. Many other examples likely will be discovered.

Species in the appendices are listed alphabetically, by higher taxon (phylum or class) or family or a functional trait. Higher taxa are listed for organisational purposes. Subordinate taxa are listed separately as an aid to understanding taxonomic diversity within higher taxa. Reports in the text are listed chronologically within species to demonstrate evolution in thinking by investigators based on earlier studies. However, reports are listed alphabetically by species within subordinate taxa (mainly family) and then chronologically in the appendices for ease of finding reports. Indicators of variation indicated in the text, tables and appendices for species are standard deviations but where summarising for major taxa, they represent standard errors.

Initial references for this review were part of a discussion in Lees (2021) of differences between what I refer to herein as non-migratory megafauna and macrofauna. That discussion included a broad range of ecological and behavioural papers on specific infaunal organisms that included reports of sublethal predation. I also included references from the Lawrence and Vasquez (1996) and Lindsay (2010) reviews. Further references were obtained by intensive searches on the internet using the terms "sublethal" and "nonlethal" combined with "predation" and "injury", "regeneration", "recovery" and by examining reference sections of papers that were found in those searches for relevant papers.

Sublethal predation based on observed injury or regeneration

This section covers reports of observed injuries or regenerating structures (recovery) from nipping or cropping of (1) feeding structures of cnidarians; (2) anterior or posterior ends of polychaetes; (3) clam siphons and feet; (4) claws or legs of crabs, (5) arms or discs of ophiuroids; or (6) cropping the ambiti of sand dollars. It also includes many lab studies initiated by investigators to study regeneration.

Regeneration is a process that is of paramount importance for infaunal organisms that experience sublethal predation. Reports on regeneration associated with sublethal predation appear to be limited to polychaetes, bivalves, gammarid amphipods and ophiuroids. In many cases, investigators reported on field observations of injury and later followed up with lab studies to observe regeneration.

In most cases, injured individuals retreat into the sediment or their tubes or burrows, where the wound heals and, over time, the cropped structure regenerates. Following a few to several weeks of recovery, they return to their normal mode of life to feed and be fed upon again. In many cases, individuals may be preyed upon numerous times during their lives. Data for observed losses or regeneration from sublethal predation on the various taxa are summarised in Table 1 and detailed in Supplementary Appendix 1.

Non-migratory Megafauna

Cnidaria: Pennatulacea

In a study of the effects of man-made structures on adjacent benthic assemblages, Davis et al. (1982) reported that fish species inhabiting a newly placed artificial reef grazed on the leaves bearing the feeding polyps of nearshore sea pens (*Stylatula elongata*), progressing from the tip down the rachis of a sea pen until eventually no feeding polyps remained. The sea pens within 30 m of the reef survived for months but eventually disappeared, presumably succumbing to injury and starvation.

Polychaeta

In polychaetes, external scars are not visible after lost tissue has been regenerated. Therefore, researchers have estimated the incidence of injury by counting the number of individuals with visibly regenerating segments and appendages or counting and weighing segments. In many cases, these methods almost certainly underestimate the actual frequency of injury sustained by worms as a result of rapid recovery.

Arenicolidae: Arenicolid worms are deposit feeders that burrow head-down in sediments to feed; they extend their posteriors out of the sediment to defecate and spawn (Fauchald & Jumars 1979). De Vlas (1979a,b) reported sublethal predation for posterior segments of *Arenicola marina*

Table 1 Species of Non-migratory Megafauna or Macrofauna in Which Sublethal Predation or Regeneration of Injured Structures Was Observed and Numbers of Predators Reported

Higher Taxon	Subordinate Taxon	Number of Prey Species	Type of Structure Affected	Percent of Structures Lost or Population Affected	Number of Regenerating Structure Types	% of Population Regenerating	Time Required for Recovery	Number of Predators	Number of Reports
Non-migratory Megafauna									
Cnidaria	Pennatulacea								
	Virgulariidae	1	Feeding leaves	100% of population ultimately lost	0	0		Several fishes	1
Polychaeta		**26**						**>23**	**34**
		15						**16**	**23**
	Arenicolidae	1	Posterior segments	20% of total weight; 40%–50% of available tail segments/year	1	0	No regeneration but tail segments swell to compensate	3	9
	Maldanidae	6	Anterior & posterior segments	Anterior: 8.8 ± 8.2 Posterior: 35.0 ± 23.2	2	Anterior: 11.0 ± 14.1 Posterior: 38.4 ± 23.4	15–45 days	3	8
	Onuphidae	8	Antennae, heads, anterior & posterior segments	Antennae: 38.5 ± 18.0 Head & anterior segments: 17.5 ± 21.5 Posterior segments: 3.9 ± 3.5	4	36.5% ± 33.7%	50–70 days	9	6
Macrofauna		**11**						**6**	**11**
Polychaeta									
	Capitellidae	1	Posterior segments	100% within 3 days due to natural predation	1			1	1
	Spionidae	10	Palps & posterior segments	Palps: 23.3% ± 15.2 Anterior segments: 7% Posterior segments: 14.9	2			5	10

(Continued)

Table 1 (*Continued*) Species of Non-migratory Megafauna or Macrofauna in Which Sublethal Predation or Regeneration of Injured Structures Was Observed and Numbers of Predators Reported

Higher Taxon	Subordinate Taxon	Number of Prey Species	Type of Structure Affected	Percent of Structures Lost or Population Affected	Number of Regenerating Structure Types	% of Population Regenerating	Time Required for Recovery	Number of Predators	Number of Reports
Crustacea		**1**							
	Ampeliscidae	1	Feeding antennae & urosomal appendages	Antennae: 84 Urosomal appendages: 12	2		1 or more moults	>4	1
Bivalvia		**19**						**29**	**28**
	Pectinidae	1	Unid. Tissue					2	1
	Tellinoidea	10	Siphons & feet	Siphons: 37.9 ± 43.5 Feet: 15.0 ± 9.8	2	25.7 ± 29.6	Prob. within 10 days	13	16
	Mesodesmatidae	1							1
	Veneridae	5	Siphons & feet	Siphons: 10%–37% Feet: 14%–34%		21.5 ± 17.2		7	7
	Myidae	1	Siphons			0		1	2
	Hiatellidae	1	Siphons					3	1
Arthropoda: Crustacea/ Decapoda		**2**						**2**	**2**
	Ocypodidae	1	Feeding claws	38				1	1
	Portunidae	1	Chelipeds & legs	28.0 ± 15.6				1	1
Ophiuroida		**22**						**9**	**21**

(Continued)

Table 1 (*Continued*) Species of Non-migratory Megafauna or Macrofauna in Which Sublethal Predation or Regeneration of Injured Structures Was Observed and Numbers of Predators Reported

Higher Taxon	Subordinate Taxon	Number of Prey Species	Type of Structure Affected	Percent of Structures Lost or Population Affected	Number of Regenerating Structure Types	% of Population Regenerating	Time Required for Recovery	Number of Predators	Number of Reports
	Burrowing	8			**4**			**5**	**16**
	Amphiuridae	8	Arms, disks, & body weight	Arms: 72.0 ± 32.0 Disks: 16.2 ± 27.3 Body weight: 44.3 ± 17.2		Arms: 70.0 ± 21.0 Disk: 1.8% + 1.4 % of arm population: 54.0% ± 5.6 % Body weight: 38.0 ± 45.3	Est. during growth season, 100–120 days for tissue. 135 days for skeletal regeneration, 160 days for complete replacement	5	16
	Epibenthic	14			**2**			**4**	**6**
	Ophiocanthidae	1	Arms	<0.1					1
	Ophiocomidae	4	Arms	37.2±14.4	1		521 days for 1 arm; 1158 days for 3 arms	1	3
	Ophiodermatidae	3	Arms & disks	Arms: 47.7±21.0 Disks: 1.1±0.6	1			2	2
	Ophiolepididae	1	Arms	28	1				1
	Ophiomyxidae	1	Arms	49	1				1
	Ophionereidae	1	Arms	83.3±12.6	1	92.2; ca. 3.52 arms/ individual		1	2
	Ophiothricidae	1	Arms	46	46				1
	Ophiuridae	2	Arms	20.0±9.9	1				1
Echinoidea		**4**						**4**	**5**
	Clypeasteridae	1	Ambitus	21				1	2
	Dendrasteridae	1	Ambitus	<5				4	3
	Mellitidae	2	Ambitus	76.5 ± 16.3				1	2

Note: Range of variation is standard deviation.

in tidal-flat populations in the Dutch Wadden Sea. He reported the worm contributed about 120 mg ash-free dry weight (AFDW) per lugworm annually or about 20% of total weight and 40%–50% of available tail segments to a variety of juvenile flatfishes. He estimated that the number of tail-nippings by juvenile plaice (*Pleuronectes platessa*) and flounder (*Platichthys flesus*) ranged from 2.2 to 3.8 per month per individual worm in the Balgzand tidal flat and that flatfish removed only one or a few tail segments per attack. From March to December, he estimated consumption of lugworm tail segments by plaice and flounder at two sites accounted for most of the estimated loss of tail segments in field populations. Biomass of cropped tail-tips almost equaled the standing stock of lugworms in the area. At one site, the loss rate of *Arenicola* tail tips was about twice the flatfish consumption rate and de Vlas suggested that a large predatory polychaete, *Nereis* sp., was the likely predator. De Vlas (1979b) reported that posterior segments of the lugworm, *A. marina*, nipped by flatfish in the Dutch Wadden Sea, were not replaced with new segments, i.e., regeneration of nipped structures did not occur. Instead, lugworms restored the weight of their tails by lengthening the remaining tail segments and keeping their weight almost constant. Therefore, the number of tail segments declined from about 100 in adult worms to "some tens or even less" over a worm's life. He found this kept tail size in balance with trunk size.

Bergman et al. (1988) conducted lab studies into the effects of tail amputation on *A. marina*. Across a wide range of nipping frequencies, loss of body weight or reduced reproduction due to amputation was compensated for by increased growth rather than regeneration until the frequency of amputation rose to once per week. Beyond that point, mortality rates increased, and reproductive effort decreased.

Maldanidae: Maldanids typically are tubicolous deposit-feeders that live head down in tubes they construct (Fauchald & Jumars 1979) and, as in arenicolids, they expose their posterior segments to defecate and spawn.

Sayles conducted two studies of external features in *Clymenella torquata* at Woods Hole, Massachusetts, USA. In the first, Sayles (1932) reported that 8.8%–78.6% of the populations surveyed exhibited loss of posterior segments but only 0.5%–5.5% had lost anterior segments. In the second, Sayles (1934) reported that 25.6% were regenerating posterior segments whereas only 1.4% were regenerating anterior segments. These patterns fit well with expected structure exposure related to the head-down feeding mode of maldanid polychaetes.

Based on lab studies at Woods Hole, MA, USA, Sayles (1932) reported that *C. torquata* regenerated from the loss of 5 anterior segments in about 15 days. In a subsequent study, he observed similar patterns after the removal of 1–6 anterior segments. About 85% of the injured worms survived 10 days or longer.

Mangum (1964) reported sublethal predation in five species of maldanid worms in Beaufort Harbor, North Carolina, USA. She reported minimal [*Branchioasychis americana* (0%–8%) and *Petaloproctus socialis* (12%)] to substantial [*Clymenella mucosa* (47%–50%), *C. torquata* (23%–56%), *C. zonalis* (48%–54%)] losses of posterior segments in the sampled populations. These differences suggest behavioural differences among the genera.

In a study of the regulation of species density of seagrass-associated macrobenthos in the Indian River estuary, Florida, USA, Young and Young (1978) observed that 46% of a population of *C. mucosa* was regenerating tails. They regenerated up to the normal 22 posterior segments. Reported predators were blue crabs (*Callinectes sapidus*) and pinfish (*Lagodon rhomboides*).

In a study of community structure on sedimentary reefs constructed by *P. socialis* near Cape Lookout, North Carolina, USA, Wilson (1979) reported that 11.8% of the worms were regenerating heads and 5.2% were regenerating tails. Considering that maldanids typically reside head-down in their tubes and significantly more worms had lost their heads, he concluded that infaunal predators, possibly a predatory worm (*Glycera* sp.), were more important than epibenthic predators or shorebirds.

Clavier (1984) reported populations of *Euclymene oerstedi* were regenerating posterior (41%) or anterior (22%) ends of their bodies. Losses ranged from 3 to 18 segments. Based on caging experiments, it appeared that, because of normal head-down behaviour of maldanid worms in their tubes, exposure of posterior segments was a greater risk. He postulated that posterior loss was due to epibenthic predation whereas anterior loss likely was due to infaunal predators. He reported that 22% were regenerating anterior ends of their bodies and 41%, posterior ends, i.e., infaunal predators were more important in this population. Mean recovery times were 1.5 and 1 month, respectively.

These data suggest substantial differences among genera but relative similarity between losses of anterior and posterior structures in maldanids. Mean losses for *Clymenella* were 3.0%±3.5% for anteriors and 45.7%±19.9% for posteriors. For the five species of maldanids overall, the mean loss of posteriors was 38.4%±23.4%.

Onuphidae: Most onuphids live in self-constructed parchment tubes, sometimes extending more than a metre into the sediment (Emerson 1975, Fauchald & Jumars 1979). They have varied feeding strategies but generally feed with the head extending out the top of the tube and with antennae (palpi) waving in the water column and exposed to potential predators.

Sublethal predation has been reported for eight species by six investigators. Head and antenna nipping were commonly reported but tail nipping was uncommon. Emerson (1975) reported that 2%–4% of the population of *Diopatra ornata* at Big Fisherman's Cove, Santa Catalina Island, CA, USA, was undergoing anterior regeneration at any time during his 18 monthly surveys. He noted that some animals in marked tubes would not be seen for ≈60 days. He assumed they disappeared after losing their heads. He surmised the losses were due to predatory fishes, especially California sheepshead (*Pimelometopon pulchrum*) and senorita (*Oxyjulis californica*), which often picked up and masticated tube segments discarded by divers. Complete field-collected specimens showed no evidence of posterior regeneration although he noted in the lab that posterior regeneration can occur. He conducted lab and field experiments on *Diopatra ornata* to determine the time required to regenerate the anterior and posterior ends of the worm. In the lab, he found that worms with the head and several anterior segments clipped off completely regenerated the head and segments within 5 or 6 weeks, but feeding was not observed until the seventh week. He also found regeneration in specimens that lost posterior segments during collection but did not indicate the time required.

In their description of *Diopatra dexiospira* in the nearshore sand flats on the south side of Oahu, Hawaii, USA, Paxton and Bailey-Brock (1986) indicated the worms were subject to sublethal predation. They reported that a large number of worms were damaged and regenerating anterior and posterior segments.

When describing a new species, *Diopatra tuberculantennata*, in Belize, Budaeva and Fauchald (2008) noted that 18.5% of the specimens examined were regenerating anterior ends of the body. They described five stages of regeneration, indicating that loss had been recurrent in the sampled population.

In describing the ecology of mound-forming populations of *Diopatra leuckarti* in Oahu, Hawaii, USA, Bailey-Brock (2008) reported that 11.8% of the worms were regenerating heads and 7.2% were regenerating tails. She tentatively attributed these losses to sublethal grazing by crustaceans, fishes, or birds, but indicated they could result from bait digging by fishermen.

Nipped heads were observed in 12.8%±8.5% of *Diopatra cuprea* populations in a summer month in Florida and Virginia (Berke et al. 2009). About 40%–60% of these populations (mean 38.1%±19.0%) experienced antenna nipping. Possible sublethal predators include several fishes [Atlantic croaker (*Micropogonias undulatus*), windowpane flounder (*Scophthalmus aquosus*) and spot (*Leiostomus xanthurus*)]. Estimated loss in nipped individuals averaged 30 mg dry weight (≈420 mg wet weight, based on a conversion factor from Ricciardi & Bourget (1998)). In lab experiments, 80% of specimens of *D. cuprea* for which 5–10 anterior segments had been experimentally ablated showed significant recovery of the head within 7 days; 80% recovered within 14 days at 28°C.

They also reported sublethal predation for the sympatric onuphid, *Americonuphis magna*, in Florida, USA. The species experienced more antennal loss (ca. 80%) than sympatric *D. cuprea*; 40% regenerated antennae within 14 days at 28°C. However, *A magna* was incapable of regenerating heads and died.

In a study to assess regeneration from sublethal predation or bait digging on *Diopatra neopolitana* and *Diopatra marocensis* in Portugal, Pires et al. (2012) found tissue loss in field-collected specimens of both species. For *D. neopolitana*, they reported that 77 of 1246 specimens (6.2%) were regenerating the anterior end and 4.2% the posterior end. For *D. marocensis*, 13.6% were regenerating the anterior and 0.3% the posterior end. These species were regenerating, on average, 9.0±2.5 and 7.5±1.9 chaetigers, respectively. The proportion of the population regenerating was substantially higher in summer than in winter for both species.

In lab experiments of regeneration for *Diopatra neopolitana*, Pires et al. (2012) found that when posterior portions were amputated at the 15th chaetiger, which is in the brachial region, to emulate predation by fish, etc., those segments were able to regenerate the anterior of the worm, but the anterior part of the worm did not survive. With amputation at chaetiger 20 (middle of the brachial region), neither piece was able to regenerate and did not survive. For specimens amputated at or beyond chaetiger 25, the anterior portion regenerated the posterior segments, but the posterior part was unable to regenerate the anterior and succumbed. Full recovery required 60–70 days. This is relatively consistent with Emerson's (1975) findings on *Diopatra ornata*.

In summary for Polychaeta, sublethal predation has been reported for macrofaunal and non-migratory megafaunal species of mostly tubicolous or burrowing polychaetes (Table 1). Most data for non-migratory megafauna are available for maldanids and onuphids, both tubicolous. For *Diopatra* spp., on average, 11.8%±7.3% (3%–23%) exhibited loss of heads, 2.9%±1.7% (0%–7.2%) loss of tails, and time required to recover heads averaged 51.0±12.5 (14–70 days). Reports of sublethal predation on macrofaunal polychaetes are limited to one capitellid and eight spionid species. Again, most are tubicolous.

Lost structures and regeneration varied considerably by family. Loss patterns depended upon which structure was most exposed to predators. *Arenicola* lost 40%–50% of tail segments and up to 20% of total weight annually. Tail segments were not regenerated. For maldanid populations, which live head-down in constructed tubes, loss of posterior segments was distinctly greater than anterior loss (38.4%±23.4% vs. 3.0%±3.5%); frequency of observed regeneration of lost structures was comparable. Time required for maldanids to regenerate lost structures ranged from 15 to 45 days. Onuphids exhibited 42.5%±21.5% loss of antennae, 11.8%±7.3% of populations lost heads and anterior segments, and only 3.9%±3.5% lost tail segments. Regeneration was observed in 36.5%±33.7% of the populations. Full regeneration of the head and return to feeding required 50–70 day. Time required for regeneration varied from no regeneration for *Arenicola* and *Americonuphis* to 50–70 days for onuphids. A wide variety of predators was reported but fish dominated.

Bivalvia

Sublethal predation and regeneration have been reported for a wide variety of bivalve taxa, especially Tellinoidea and Veneridae. Most reported lost structures are siphon or siphon tips and feet.

Pectinidae: Members of this family typically live at the sediment/water interface. Only their very short siphons and mantle are accessible to sublethal predators in the water column. Data are available for only one species, bay scallops (*Argopecten irradians concentricus*). Irlandi and Mehlich (1996) reported that scallops caged with browsing fishes in the fall resulted in lighter siphons, an injury, and less shell growth, an effect, but no differences in summer. When caged with seven browsers per cage, juvenile scallops grew about 25% faster than those with ≈16 fish per cage.

Tellinoidea[1]: Tellinoid clams typically have long separate incurrent and excurrent siphons. Depending on size, they generally burrow as far below the sediment/water interface as their siphon

[1] This order designation includes families Donacidae, Psammobiidae, Semelidae, and Tellinidae, which are characterized by exposure of long thin siphons that extend above the surface of the sediments to facilitate feeding, respiration, reproduction, and defecation (Coan et al. 2000).

length and feeding mode allow. Feeding modes include suspension- and surficial deposit-feeding. Eleven investigators reported sublethal predation for nine species in this order.

During studies of growth rates of 0-year class plaice, Edwards et al. (1970) observed sublethal predation on siphons of the shallow-burrowing tellinoid clam, *Tellina tenuis*, stocked in outdoor tanks. Reporting further on that study, Trevallion (1971) observed the longer inhalant siphons of the clam were preyed upon by 0-year class plaice. Rate of predation varied by year partly due to changes in relative density of predator and prey and the abundance of other prey such as polychaetes. She measured a rate of regeneration of the inhalant siphon of ca. 0.3 mg dry tissue/week/individual. Frequency of siphon predation and regeneration varied widely by season, weather, abundance of alternate prey and abundance of the plaice.

Hodgson (1982a) reported from lab experiments that wound healing of siphon tips of the inhalant siphons of the bean clam, *Donax serra*, which lives on wave-swept beaches in South Africa is complete within 48 hours of amputation. After the wound is sealed to prevent blood loss, regeneration of the siphon tentacles commences. The siphon tentacles grow to their original size within 5 days. Alacrity is necessary because the clam is unable to respire or feed until it can reopen the inhalant siphon.

Based on studies at West Sands beach, St. Andrews, Scotland, Ansell et al. (1999) reported that between 2.5% and 18% of *Donax vittatus* showed damage to inhalant and/or exhalant siphons as a result of predation by 0- and 1-year class flatfish such as plaice. Those removed from sand by sublethal predation rapidly reburied. He further reported wound healing and regeneration of siphonal tentacles following experimental amputation of siphonal tips of *D. vittatus* on the beach at West Sands, St. Andrews, Scotland. Siphonal tips include tentacle crowns formed by siphonal tentacles. Newly formed tentacles appeared almost normal after 10 days. Siphonal tentacles included the full complement of ciliated sensory organs.

Luzzatto and Penchaszadeh (2001) conducted *in vivo* studies to examine the process of regeneration of the inhalant siphon of an Argentinian bean clam (*Donax hanleyanus*), which live in the surf zone on exposed beaches. They noted that the siphon is a complex organ with a tip bearing branched tentacles (Figure 1A). Following amputation of the tip, rudiments of the primary tentacles were observed within 24 hours. Subsequently, a very intensive process of regeneration of the primary, secondary and tertiary tentacles occurred over a period of 5 days. They observed the ramification process for 10 days after amputation, but the siphons were still less complex than those of control clams.

Complexity of siphon tips varies substantially, as is demonstrated by Figures 1A, 1B and C. It varies likely according to response to exposure to suspended particles and detrital material to which a species is exposed. Genera like *Donax* (Figure 1A) live in or near the swash zone, is which large quantities of sand are suspended by wave action. Genera like *Tresus* (Figure 1B) generally inhabit habitats with lower quantities of suspended sediment and detrital material but still require protection. Finally, genera like *Panopea* (Figure 1C) live in habitats where suspended particles are rare and therefore require no protection. However, all of these taxa likely experience sublethal predation.

Salas et al. (2001) studied sublethal foot predation in four donacid clams in Malaga, Spain and Ré Island in the French Atlantic. In Malaga, an average of 18.3% *Donax trunculus* individuals were nipped with up to 48% in September. On Ré Island, the average was 27%. Overall, nipping was observed in 10.6% of *D. venustus*, 8.4% of *D. vittatus* and 4.2% in *D. semistriatus*. Likely predators were carcinid and cancroid crabs, viz. *Portumnus latipes, Liocarcinus vernalis* and *Atelecyclus undecimdentatus*).

To study the rate of regeneration in the Venezuelan *Donax denticulatus* under laboratory conditions, Miloslavich et al. (2004) amputated their inhalant siphons and placed them into aquaria. They then sacrificed two individuals twice a day for microscopic examination of regeneration over a 9-day period. Initially, regeneration was slow, with only rudiments of the primary tentacle showing. However, between the second and third day, the secondary and tertiary tentacles became

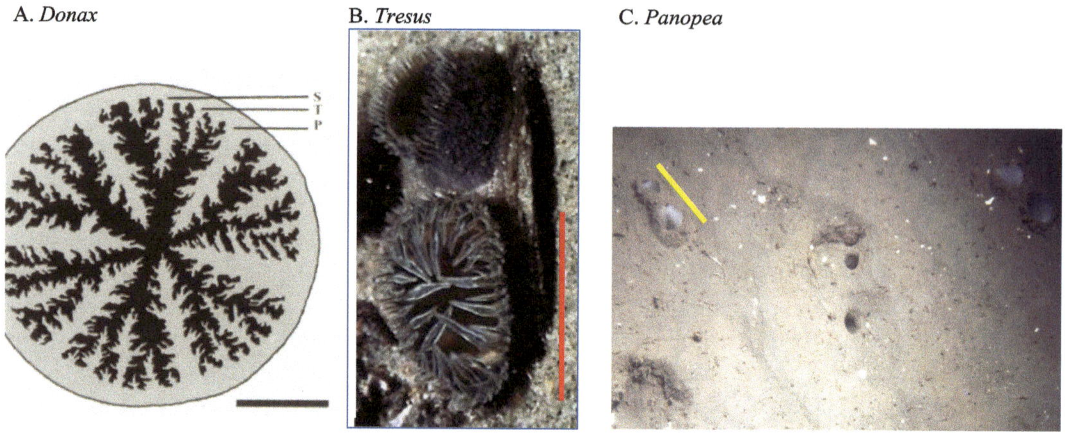

Figure 1 Variation in siphonal tentacles that exists in bivalve siphons, likely in response to differences in exposure to suspended sand particles. (A) Highly developed tentacles in a bean clam (*Donax hanleyanus*) in the surf zone; black scale bar=0.5 mm; P, S and T=primary, secondary and tertiary tentacles. (B) Moderate developed tentacles in a gaper clam (*Tresus nuttallii*) inhabiting mud flats or shallow subtidal areas; red scale bar=2.5 cm. (C) Absence of tentacles in the geoduck (*Panopea abrupta*) inhabiting deep water or mud flats; yellow scale bar=8 cm. *Sources*: A, Luzzatto and Penchaszadeh (2001); B, Lovell and Libby Langstroth © California Academy of Sciences; C, personal photo.

conspicuous. The siphons were fully regenerated on the fifth day after amputation, showing no difference a month later. They concluded that development of primary and secondary tentacles must be very important to basic functionality of the inhalant siphon.

When Pekkarinen (1984) amputated one-third of the inhalant and exhalant siphons of *Macoma balthica*, muscles at the cut end of the siphons were observed to contract to close the wound. Traces of tentacles were observed forming, usually within 48 hours. Tentacles were fully reformed by a week later. Siphons had regenerated to normal size and configuration under laboratory conditions within 3 months. When half the siphons were removed twice within 3 months, regeneration was slower; siphons were smaller than normal 6 months later.

De Vlas (1985) reported that regeneration of the siphons of *M. balthica* was rapid under simulated tidal flat conditions, hardly depended on temperature and season, and was faster in smaller (lighter) siphons.

In a study to determine the secondary production of *M. balthica* in the Wadden Sea, de Vlas (1985) found that only about 13% of the of siphon tips lost were consumed by flatfish. He estimated that individual clams lost unrestored tips of siphons on average several times per day. He estimated the amount of *Macoma* siphon tissue lost during 1975 at four tidal-flat transects varied from 0.31 to 1.66 g AFDW m^{-2} yr^{-1} (2.0–10.4 g shell-free wet weight), based on a conversion factor in Ricciardi and Bourget (1998). Other epibenthic predators like gobies, shrimps and shore crabs cropped the other 87%.

Peterson and Skilleter (1994) reported on siphon cropping in *M. balthica* by juvenile demersal fishes in the Neuse River estuary in North Carolina. Effects of croppers were more substantial in early to midsummer when juvenile fish were abundant. As part of the same study, Skilleter and Peterson (1994) reported that the presence of cropping fish resulted in ≈35% loss of ash-free dry weight for *M. balthica* siphons at a density of 110 clams m^{-2} and ≈14% at 370 clams m^{-2}.

Bonsdorff et al. (1995) observed siphon cropping of *M. balthica*, the dominant infaunal animal in the north Baltic Sea, by large predatory isopods (*Saduria entomon*) and brown shrimp (*Crangon crangon*). The regression between *Macoma* shell length and dry weight was significantly stronger in control treatments than in *Saduria* and *Crangon* treatments, indicating greater tissue loss in the siphon cropping experiments.

De Goeij et al. (2001) conducted lab enclosure experiments with *M. balthica.* Some clams were exposed to siphon nipping by plaice and others were not. Those exposed to nipping were buried less deeply, which the investigators concluded made them more susceptible to lethal predation by predators such as shorebirds.

Although considerable effort has been invested in studies of sublethal predation in *M. balthica*, most examined recovery rates in lab studies. Few studies have quantified siphon loss rates in nature.

Tomiyama and Ito (2006) studied the rate of siphon regeneration in *Nuttallia olivacea* with shell lengths >30 mm in laboratory and field experiments in the Natori River estuary, Japan. Treatment clams had ca. 15% of their siphons amputated. In the lab, amputated clams were sampled periodically for a month. At a temperature of 15°C or 25°C, estimated siphon production was quite high within a few days of amputation but decreased greatly thereafter and was incomplete after a month. Considerable regeneration was observed in 3 months in field caging experiments but siphon weight was significantly less in the amputated clams than in the controls, indicating that regeneration was still incomplete.

Zwarts (1986) reported that Eurasian Curlews (*Numenius arquata*) graze siphons of *Scrobicularia plana* at a rate of 1–10 per minute and indicated that Eurasian Oystercatchers (*Haematopus ostralegus*) are another major predator.

Similar to his findings with the tellinid *D. serra* in South Africa, Hodgson (1982b) found that after siphon tips of the semelid, *Scrobicularia plana*, were amputated in the lab, they sealed up within 48 hours and epithelial tissue began to grow. Subsequent regeneration of the siphons, complete with siphon tentacles, was accomplished within 5 days. Regeneration rate is ≈6 mg week^{-1}, or about 20% of the siphon weight, until the original weight of the siphon is attained.

Mesodesmatidae: Noting bivalve siphons as a renewable resource, Nuñez et al. (2010) investigated the seasonal regeneration response of inhalant siphonal tips of *Mesodesma mactroides* on a beach in Punta Mogotes, Argentina. Siphons had been clipped in the laboratory at 1 and 5 cm from the distal ends. Generally, siphons recovered fastest in the spring. Length of primary siphonal tentacles on nipped siphons grew significantly faster in winter than in summer until after 12 days after nipping, after which they grew at the same rate. Length of primary siphonal tentacles on nipped siphons with 1-cm cuts consistently recovered faster for 2 weeks than those with 5-cm cuts but the rate of regeneration between 1-cm and 5-cm cuts was not significantly different.

Veneridae: Siphon length in venerid clams varies from relatively short to moderately long and the depth to which they burrow varies accordingly. Four investigators have reported observing sublethal predation on four species of venerid clams.

Based on field caging experiments, Irlandi and Mehlich (1996) examined the effects of tissue cropping and disturbance on the shallow-burrowing suspension feeding *Mercenaria mercenaria.* During fall experiments, hard clams caged with potential siphon nipping fishes had lighter siphon weight and less shell growth than clams caged without siphon-nipping fishes. During summer experiments, clams in cages exhibited no difference in siphon weight but added 3.0 mm of new shell, whether or not siphon-nipping pinfish (*Lagodon rhomboides*) and filefish (*Monacanthus hispidus*) were present. This indicates the absence of siphon nipping in summer.

Whitlatch et al. (1997) and Mouritsen and Poulin (2003) reported that 11%–37% and 14%–34% (24.2% ± 8.9%) of *Austrovenus* populations suffered partial loss of feet, respectively. Whitlatch et al. (1997) found that the density of nipped *Austrovenus* ranged from 11.5 to 26.9 individuals m^{-2} (17%–31%), depending on the area surveyed (0.25–9.0 m^2) but that mean nipping rate (24.1% ± 6.2%) did not vary significantly with major differences in density (0–1200 clam/1.0 m^2 treatment plot. Mouritsen and Poulin (2005) later reported that the frequency of cropped cockles declined significantly at lower elevations, from 52% to 4% because high levels of echinosome trematode parasitism reduce the ability of the clam to bury. Predators included a variety of siphon-nipping fishes such as soles (*Rhombosolea leporina*, *Rhombosolea plebeia*) and a wrasse (*Notolabrus celidotus*).

Meyer and Byers (2005) observed siphons of the venerid clams, *Protothaca* (now *Leukoma) staminea* and *Venerupis philippinarum*, were cropped by two fish species (sculpins *Leptocottus armatus* and *Myoxocephalus polyacanthocephalus*) around San Juan Islands, WA, USA. They found siphons in 25% and 31% of the stomachs of the respective fishes. In a field experiment, they excised 40% of *P. staminea* and *V. philippinarum* siphons, retaining one group of each species in the lab as controls and placing another group in appropriate sediment in the field. After a month, both treatments were regenerating siphon tissue, exhibited orange coloration, but not the typical black tips seen in uncropped specimens. All had regenerated up to 65% of the lost siphon tissue in that period.

To investigate the effect of temperature on siphon regeneration, Nuñez et al. (2013) examined growth rates of tentacle crowns for ca. 1-cm long inhalant siphons of *Amiantis purpurata*, from which 0.5 cm had been snipped. Clams were exposed to seawater temperatures of 12°C, 15°C, 20°C and 22°C in the laboratory. Regeneration was most rapid at 22°C. Regeneration experiments for the tentacle crowns ran until long siphonal tentacles were between 0.6 mm and 0.7 mm long. The siphons were all still growing at an appreciable rate at the end of the respective experiments, which ranged from 18 days at 22°C to 36 days at 12°C. They stated, "Regeneration was slower than in any other species previously reported in the literature and decreasing temperature slowed down the process."

Tomiyama (2016) conducted field experiments in Matsukawaura Lagoon, Japan, with caged siphon-clipped Manila clams (*Ruditapes philippinarum*), which have short siphons. Based on his observations on the clipped clams, he concluded that clams with short siphons regenerated their siphons at a slower rate than rates in literature for venerid clams with longer siphons. He suggested that regeneration may be less important for clams with short siphons.

Myidae: Softshell clams have long joined siphons and tend to bury as deeply as the siphon length permits in relatively consolidated sediments. The tubular chamber formed in the sediment by the siphon usually persists when the siphon is retracted.

Welch and Martin-Bergmann (1990) reported that, based on stomach contents, walrus (*Odobenus rosmarus*) feed on siphons of truncate soft-shell clams (*Mya truncata*) in the western Arctic. To assess the regeneration of lost siphons, they excised various lengths of the siphon from 27 clams and placed them in plastic cylinders which they backfilled with sand and mud. They placed other similarly arranged uninjured clams alternately in a tray and placed the tray in the ocean at a depth of 10 m for a year. After a year, all excised clams were dead and 31 of the 33 unexcised clams remained alive. They concluded that *M. truncata* is not capable of regenerating from loss of the siphon due to walrus predation. However, based on design issues with this experiment, their conclusion cannot be considered definitive for *M. truncata*.

Smith et al. (1999) conducted a field study of the foraging behaviour of red rock crabs (*Cancer productus*) on soft-shell clams (*Mya arenaria*), a commercially valuable species. During these studies, they observed a crab digging out, pinching off and eating a soft-shell clam siphon but presented no data on survival or regeneration. Consequently, no studies have confirmed that *Mya* spp. experience regeneration from predation, i.e., sublethal predation. This comprises a major data gap.

Hiatellidae: Geoducks (*Panopea abrupta*), one of the largest clams, have the largest siphon of all burying clams and bury deeply in sand or muddy sand on the west coast of the USA. The species is commercially important. Andersen (1971) reported that live geoducks are occasionally collected in Puget Sound, WA, USA, with pieces of their siphons missing but studies have not demonstrated regeneration. However, despite the absence of reports of regeneration, in view of the reported number of injured individuals, it seems safe to assume injured clams have survived and are recovering. Gut studies indicate several large natural predators that might cause partial siphon removal. Species include spiny dogfish (*Squalus acanthias*), Pacific halibut (*Hippoglossus stenolepis*), great sculpin

(*M. polyacanthocephalus*) and sea otters (*Enhydra lutris*). It is doubtful that the geoduck fishery would result in siphon removal.

In summary for Bivalvia, siphon or foot nipping and regeneration have been observed in many types of bivalve. Injuries have been observed in the families Pectinidae, Veneridae, Myidae, Hiatellidae and a broad range of the superfamily Tellinoidea, all of which expose a pair of long siphons above the sediment/water interface. Most studies have been conducted on siphon loss in tellinids and venerids but reports of foot loss are also common, especially in tellinoids, many of which are often awash in the swash zone.

Lost structures varied considerably by higher taxon. For Tellinoidea, predation is reported for siphon tips, siphons and feet for *Donax* spp. Siphons were lost in 37.9%±43.5% of the Tellinoidea and feet were lost in 15.0%±9.8%. Siphons were lost in 30.6%±46.4% (4.2%–48%) of the *Donax* populations sampled and 28.2%±14.0% had injured feet. Loss of siphon biomass in *Macoma* populations ranged from 4% to 25% after 10-day exposure in enclosures with predatory species. Lindsay (2010) commented on the repeated nature of siphon loss in tellinoid bivalves and suggested these clams contribute substantially to local food webs. Venerid populations lost 10% of siphons and feet were injured in 25.3%±18.5%. The wide variety of predators includes large isopods (*Saduria*), shrimp (*Crangon*), a variety of crabs, gobies and flatfish. Greatest observed loss occurred in summer when juvenile flatfish were occupying bays or tidal flats. However, it is likely that similar or greater losses occur in subtidal sedimentary ecosystems.

Regeneration was studied for 13 species in ten bivalve genera, including Tellinoidea, Mesodesmatidae, Veneridae and one myid. Regenerated structures included siphons and feet. Regeneration of siphonal tips (tentacle crowns), a critical structure in respiration, feeding and defecation, is a complex process. Percent of populations regenerating siphons or feet ranged from 25.6%±28.1% (Tellinoidea; $n=11$) to 21.5%±17.2% (Veneridae; $n=8$). Some research indicated that regeneration rates vary directly with water temperature, among and within the families, and between clam species with short and long siphons. Rates can range from 5 days to over 10 weeks. However, while *M. truncata* siphons were observed in walrus gut contents, it is unclear whether they can regenerate; experiments to investigate regeneration were inconclusive.

Arthropoda: Crustacea: Decapoda

Although capabilities for regeneration of lost appendages by crustaceans are well known, reports on sublethal predation in these animals are uncommon. Only two reports were found for two megafaunal crustaceans.

In a study of autotomy in blue crab in Chesapeake Bay, Smith and Hines (1991) reported that 28.0%±15.6% (17%–39%) of the crabs in various populations were missing or regenerating one or more limbs. Most frequently, the crabs were missing a cheliped but seldom a swimming leg. Loss of more than a single limb was uncommon. They surmised that intraspecific aggression and cannibalism were major causes of limb loss.

Oliveira et al. (2000) found that ≈38% of the large feeding claw for adult fiddler crabs in Ria Formosa, southern Portugal, were missing or had regenerating claws. The predators were fishermen that take the large claws for consumption as a delicacy. Male fiddler crabs in an unexploited population at a remote site were not missing large claws.

Echinodermata

Sublethal predation in echinoderms has been reported for a wide variety of ophiuroids, but only a few echinoids (Table 1; Supplementary Appendix 1). Losses in ophiuroids were predominantly arms.

Ophiuroida: I have separated ophiuroids into two categories, burrowing and epibenthic, on the basis of life-style to facilitate comparison. Family Amphiuridae has been categorised as burrowing. The remaining ophiuroid families studied, all somewhat motile, are categorised as epibenthic.

Missing arms and disks in specimens observed in the field and structures in gut contents are the basis for enumeration.

Burrowing Ophiuroids: Amphiuridae: Several investigators have reported that 70-97% of individuals in populations of these species exhibited arm regeneration. Martin (1968) reported that 80% of the viviparous *Axiognathus squamata* collected in pooled intertidal areas in the vicinity of Auckland, New Zealand, were partially regenerating up to four arms, with an average of 1.6 regenerating arms per individual. The substrate comprised coarse shell fragments with loose stones and the calcareous red alga, *Corallina*. Only three specimens were found that had lost the complete aboral part of the disk, the stomach and gonads. The role of sublethal predation was mentioned but not quantified.

Several investigators have reported that up to 96% of individuals in populations of *Amphiura chiajei* and *A. filiformis* exhibited arm regeneration. Bowmer and Keegan (1983) reported consistent signs of recent recovery from arm damage in *Amphiura filiformis* in Galway Bay, Ireland; 96% of the population exhibited regenerating arms. Roughly 3% of the population was regenerating the disk cap or covering. Based on the frequency of scars on arms and biomass assay over 2 years, an average of ≈25%, and up to 50%, of total body weight may be regenerated tissue. Indications of arm breakage (scars) were more frequent in the proximal or mid-arm areas than in the distal ends of the arms. They reported that plaice and dabs were among the principal sublethal predators.

The proportion of total production allocated to regeneration was estimated to be 16% in Galway Bay (O'Connor et al. 1986) whereas respiration accounted for 77.4% and gonad output accounted for only 6.6% of energy flow. They reported that for a stable, high-density population of *A. filiformis* in Galway Bay, somatic growth is almost entirely restricted to regeneration.

Reporting on *A. filiformis* sampled on the Oyster Ground in the southern North Sea, Duineveld and Van Noort (1986) reported that 96% of individuals had lost arms; 55% had one or more recent losses, which equated to 20% of all arms. They reported that common dab (*Limanda limanda*) was an important predator for this brittle star.

Based on SCUBA studies in Killary Harbour, Ireland, Munday (1993) found that 99.1% of the dense (≈700 individuals m^{-2}) population of burrowing *A. chiajei* was regenerating arms but only 0.5% was regenerating disks. Over 21 months, the overall estimated mean length of regenerated arms was 33.5±6.5 mm per individual. He estimated 4.2 arms per individual and 57.9% of total body weight were lost. Arm loss appeared to be seasonal and due to fishes, especially pleuronectid flatfish, but bottom trawling may have been a factor.

For *A. filiformis* in the Skagerrak, west Sweden, 13% (0.34 g AFDW m^{-2} yr^{-1}) was allocated to regeneration, primarily of arms, indicating that this ophiuran is an important food source (Skold et al. 1994). Mean regenerated biomass ranged from 12% to 30% of total biomass and the annual P/B ratio was 0.46 yr^{-1}.

Sköld and Rosenberg (1996) observed that burrowing deposit feeders such as *A. filiformis* and *A. chiajei* had significantly more scars per arm and greater tissue loss than more exposed epibenthic suspension feeders, carnivores, or deposit feeders. Based on their other finding that arm tissue for *Amphura* spp. had higher ash-free dry weight and nitrogen content, they proposed that demersal fishes were cropping selectively on more nutritious prey.

In laboratory studies, Nilsson and Sköld (1996) and Nilsson (2000) assessed the effects of differences in normoxia (>80% O_2) and hypoxia and low and moderate flow rates on regeneration rates for suspension-feeding ophiuran, *A. filiformis*, after amputation of one or three arms or the disk. Nilsson and Sköld observed reduced arm growth in hypoxic conditions (19% and 29% O_2 saturation). Nilsson (2000) found that arm regeneration rates were higher in normoxic than in hypoxic conditions (18% saturation) at both flow velocities and that mean arm regeneration rates were significantly higher at moderate flow rates. He further reported that in normoxia, arm regeneration rates did not exhibit a response to differences in flow velocity, indicating that *A. filiformis* is able to adequately ventilate its burrow at low flow velocities. The number of arms amputated did not affect mean arm regeneration rates. A significant interaction in regeneration rates for both arms and disks was observed between oxygen concentration and disk amputation, indicating that the disk is the major organ for gas exchange in this species.

Noting substantial variation in regeneration rates of grazed arms for *A. filiformis* in Sweden, Dupont and Thorndyke (2006) assessed how the species allocates energy between increasing arm length (growth) and developing function for feeding in regenerating arms. They commented: "There is a trade-off between regeneration in length and functional recovery for feeding", reporting that the amount of tissue lost determines whether priority will go to regeneration or differentiation of feeding structures.

Reporting on *Amphioplus coniortodes* in the Caribbean Sea, Singletary and Moore (1974) and Singletary (1980) reported observing regeneration of arms in 87% of the population and regeneration of disks in 3%.

In a study of the ophiuroid *Acrocnida brachiata* in the Bay of Douarnenez, France, Bourgoin and Guillou (1994) reported that nearly 70% of the total arm population (i.e., number of ophiuroids times five) exhibited arm regeneration in intertidal and subtidal sampling areas. Breakages occurred in the mid-arm area intertidally and on the distal third subtidally. Actively feeding arms appeared to be the least damaged, suggesting that nipped arms were held underground while healing. Biomass of regenerating tissue accounted for 11.1% and 6.4% of total biomass in intertidal and subtidal areas, respectively. Subtidal production of regenerating tissue was estimated to be 33 g dry weight (19 g AFDW) m^{-2} yr^{-1}.

Between 52% and 94% of *Ophiophragmus filograneus* burrowing in natural and planted seagrass beds in Florida exhibited arm regeneration, sometimes on two or more arms (Clements et al. 1994). Approximately one-third of the total individuals collected were regenerating only one arm with smaller numbers regenerating more arms. Rates of soft tissue regeneration did not differ between planted and natural beds. However, rate of skeletal regeneration was three times greater in planted than in natural beds, suggesting that burrowing was less successful in sediments that were disturbed by replanting activities. Rate of regeneration was greater overall for specimens that had lost more tissue but the rate per arm was greater in specimens that had lost one arm than in specimens that had lost three arms. Regeneration rates were lower than has been reported for other burrowing ophiuroids.

In natural populations of *Microphiopholis gracillima*, Singletary and Moore (1974) found that 77.0% of the individuals had lost arms. Subsequently, Stancyk et al. (1994) reported that arms removed in lab experiments did not exhibit regeneration during cooler months of the year. Only 15.1% of the individuals collected in 13 monthly surveys from subtidal mud flats in North Inlet, South Carolina, exhibited regeneration, mainly in spring and in summer. Disk loss was substantial in 1985 and 1986 (40%–70%) but substantially lower in 1990 (20%–40%). A smaller proportion of the population showed disk regeneration in the summer. They estimated skeletal regeneration would take 135 days in the field and nearly 200 days in the lab; complete replacement of arm tissue would take 160 days under May conditions. Between 70.9% and 91.8% of the 2405 arms observed in monthly collections over 13 months were regenerating at least one arm. Greater than 50% had one scar; some arms had up to 4 scars, indicating at least four attacks. Between 20% and 70% of the individuals examined were regenerating disks. Disk predation appeared more frequent in warmer months but no such pattern was observed for arms. The species appears capable of regenerating at least 50% of its total body weight during a single growing season. In a subsequent study, Pape-Lindstrom et al. (1997) estimated the species was losing ca. 4% of total arm tissue/day but regenerating up to 3%/day. They stated the population, "can probably regenerate arm tissue almost as rapidly as it is lost to sublethal predation".

Ophiuroida: Epibenthic Ophiuroids

Sides (1987) reported that arm loss indicating sublethal predation varied significantly among eight species of epibenthic ophiuroids, i.e., *Ophiomyxa flaccida, Ophionereis reticulata, Ophiocoma echinata, Ophiocoma pumila, Ophiocoma wendtii, Ophioderma appressum, Ophioderma cinereum* and *Ophiolepis impressa* in back-reef populations in Jamaica. Rates, ranging from 74% for *O. reticulata* to 28% for *O. impressa*, varied significantly among species. Injury rates did not appear to be simply relatable to palatability or behaviour. He also compared regeneration rates for a single arm loss for these eight species. Regeneration rates ranged from 55% replacement of its mean maximum arm length over a 3-month period for *Ophiocoma wendtii* to 23% for *Ophiolepis impressa*. Highest levels of injury were observed for *Ophiocoma pumila* and *Ophionereis reticulata*; *O. impressa* suffered the least injury. Injury rates did not appear related to either palatability or behaviour.

In a study of *Ophiocoma echinata* in the Florida Keys, Pomory and Lawrence (2001) observed that 18% and 47% of two populations exhibited regeneration. They reported that regeneration was estimated to require 0.07 kJ $m^{-2}day^{-1}$ on average. Based on this energy budget calculation, they estimated it took, on average, ca. 720 days for individuals to completely regenerate three arms (521 days at summer rates and 1158 days at winter rates).

Aronson (1991) compared sublethal arm damage in lagoonal populations of the epibenthic ophiuroids, *Ophioderma brevispinum* in Belize and back-reef populations of *Ophioderma appressum* over 3 years in Belize and Jamaica to museum specimens of fossils of two ophiodermatid species living during the Lower Jurassic period. Arm injuries were significantly more common in the two living populations of *O. appressum* in Belize and Jamaica (30%–74%) and *O. brevispinum* in Belize (46%–57%) than in the two fossil species (0%). The differences were not significant. Disk injuries were uncommon in populations of both ophiodermatids (0%–6%) but absent in the two fossil species. Since predation is the primary source of arm damage in living *Ophioderma* spp., Aronson hypothesised that the evolution of durophagous fishes and decapod crustaceans in the late Mesozoic is the cause of current arm damage in these species. He also examined arm and disk injury in *O. appressum* and *O. brevispinum*, in Jamaica and Belize in four surveys over 3 years. Based on the difference in regenerating arms between extant and fossil ophiodermatids, he concluded that "predation on ophiuroids increased when durophagous teleostean fishes and decapod crustaceans diversified in the late Mesozoic", where 'durophagous' refers to animals capable of feeding on and crushing hard structures in prey such as clams, snails and brittle stars. He noted that injured disks were rare in all populations.

To investigate the effects of different feeding strategies, Sköld and Rosenberg (1996) observed the effects of arm nipping on six ophiuran species in the northern Kattegat-eastern Skagerrak and three species in the northern Adriatic Sea. These included burrowing deposit feeders (*Amphiura filiformis* and *A. chiajei*), which had significantly more scars per arm than epibenthic suspension feeders (*Ophiothrix fragilis* and *Ophiocomina nigra*), epibenthic carnivores and deposit feeders (*Ophiura ophiura* and *Ophiura albida*). Greatest tissue loss was in *Amphiura* spp. and the least in *Ophiura* spp. Based on their finding that arm tissue of burrowing suspension- and deposit-feeding *Amphiura* spp. had the highest arm loss, highest AFDW and nitrogen content in their arms, they proposed that demersal fishes were cropping selectively more nutritious prey.

Based on ROV surveys, Metaxas and Giffin (2004) described dense aggregations of *Ophiacantha abyssicola* at depths between 430m and 498 m in Northeast Channel on the continental shelf of Nova Scotia, Canada. Such dense beds of ophiuroids are rare in the bathyal and abyssal environments. They found only two individuals with evidence of arm regeneration out of 50 examined per dive in more than six dives. They concluded predation intensity is low in these deep exposed brittle star beds, suggesting that abyssal fishes have not adapted to feeding on brittle stars.

Yokoyama and Amaral (2010) compared the frequency of arm regeneration in intertidal epibenthic populations of *Ophionereis reticulata* between two beaches in Southeast Brazil. The beaches were mixed boulder and sand with different wave exposure. Arm regeneration, observed in 92.2%

of the 1170 individuals sampled, was slightly higher on the more exposed beach. The number of regenerating arms, ca. 3.5 arms per individual, was similar between sexes. They surmised the main cause of arm loss was sublethal predation but suggested hydrodynamic factors might contribute.

In summary, many reports are available describing sublethal predation for a wide variety of ophiuroids. Specimens that are regenerating arms or disks potentially provide indirect evidence that sublethal predation has occurred. However, physical disturbances can also cause arm or disk loss that does not represent sublethal predation, especially in epibenthic brittle stars that are more exposed to physical disturbance. The different causes of arm loss generally cannot be identified.

Arm loss in burrowing ophiuroids (Amphiuridae) is far more common than in epibenthic species. Loss of arms was observed in 83.7%±11.8% of the populations of burrowing ophiuroids. Observed loss of disks in populations where reported was 16.1%±27.4%. Average body weight reported by three investigators in observed populations declined 37.5%±19.6%. Relative to the five-arm arm population, 51.3%±39.5% of arms were regenerating. Regeneration of arms was observed in 70.0%±21.0% of the populations but disk regeneration was minimal (1.8%±1.4%), suggesting that loss of disk is often fatal. Estimated time required for full recovery was 160 days, depending on the number of arms regenerating. Burrowing ophiuroids generally keep arms that are regenerating in a cavity in the sediment. Predators for burrowing ophiuroids included penaeid shrimp, blue crabs, Norwegian lobster, dab, plaice, sand gobies and haddock.

Loss of arms observed in epibenthic ophiuroids was 39.0%±18.2, excluding *Ophiocantha abyssicola*, which apparently lives below depths where ophiuroid grazers thrive. For that species, <0.1% of the observed population suffered arm loss. Regeneration for the 5-arm population averaged 40.4%±19.2, quite comparable to arm loss. Time required for full recovery, reported only once, was estimated to be 523 days for one arm and 1158 days for three arms. The few reports of predators for epibenthic ophiuroids included "durophagous" fishes such as labrids and gobies and decapod crustaceans.

Arm loss in epibenthic ophiuroids was significantly lower than in burrowing species (Figure 2; 44%±21% vs. 82%±13%; $p \ll 0.0001$. one-way resampling comparison of means). A likely reason is selective feeding by predators. Investigators have concluded that burrowing species, with higher AFDW and nitrogen content, are more nutritious and attractive to predators, even though arms are less accessible than for epibenthic species, especially the highly mobile types.

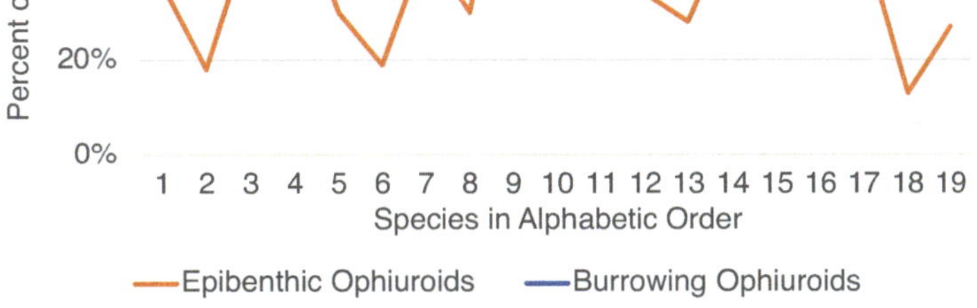

Figure 2 Comparison of percentage of brittle star arms lost by burrowing and epibenthic ophiuroids in sublethal predation reports.

Echinoidea

Because of their globular structure and open coelomic cavity, regular echinoids do not survive predatory attacks. Therefore, all reports of sublethal predation are for irregular discoid species of echinoids with a flat or thickened ambitus.

Birkeland and Chia (1971) reported <5% of the populations of the sand dollar, *Dendraster excentricus*, at Alki Point, Seattle, WA. USA, had a broken and healed ambitus, possibly due to crab grazing. I also observed grazing on the ambitus of this sand dollar in Mission Bay, San Diego, CA, USA, by white sea urchins (*Lytechinus pictus*; unpublished results, 1970).

Tennakoon et al. (2019) and Negron et al. (2020) reported on indications of sublethal predation on recent and fossil sand dollars by triggerfish. They observed cuspate-shaped marks along the ambitus of the tests of notched sand dollars (*Encope michelini*), keyhole sand dollars (*Mellita tenuis*) and sea biscuit (*Clypeaster subdepressus*) in the northeastern Gulf of Mexico. These marks extended only as deep as the lunules. They were observed in 88% of *E. michelini*, 65% of *M. tenuis* and 21% of *C. subdepressus*. Injuries in the tropical species were observed in 58%±34% of the populations.

In addition, they observed similar healed traces of nipping in fossil sand dollars from the Pliocene portion of the Tamiami Formation in specimens of *Encope tamiamiensis, Mellita aclinensis* and *Clypeaster sunnilandensis*. Durophagous fish and crustaceans evolved during this period. Negron et al. (2020) noted that traces were more frequent on flatter sand dollars (e.g., *Encope* spp.) than on the biscuit urchins (*Clypeaster*). These were attributed to gray triggerfish (*Ballistes capriscus*). Injuries to *Dendraster*, observed in <5% of the populations, were attributed to large majid or cancrid crabs and a sea urchin (*Lytechinus pictus*).

Asterozoa

I considered including some sea stars in this review but concluded they are too mobile to truly be considered non-migratory megafauna. Sublethal predation is common among members of this taxon both among and within species. As described above, Lawrence and Vasquez (1996) provided a good review of this topic and several papers have been published since then.

Macrofauna

Although macrofaunal assemblages include many taxa besides polychaetes and amphipods, all are quite small and fragile. Specimens are often damaged by sampling and sieving. Moreover, they are quickly digested as gut contents following consumption. Consequently, little observational research has been conducted on sublethal predation for macrofaunal organisms in the lab or field except on polychaetes. Regeneration has been reported for four species of spionid polychaete and one species of gammarid amphipod by four investigators.

Polychaeta

Capitellidae: Bonsdorff and Pearson (1997) conducted laboratory experiments on *Capitella capitata* to examine its response to (1) sublethal predation of the worms' posterior by shrimp (*Crangon crangon*) and (2) mechanical disturbance of the sediments in which the worm tubes were constructed. Tubes at the sediment/water interface disappeared immediately when the sediment was disturbed mechanically but reappeared within 3 days in the shrimp browsing treatment (Supplementary Appendix 1). Recovery from sublethal predation took longer than from disturbance, likely because regeneration of the grazed posterior structures required more time and energy than reconstructing tubes.

Spionidae: Many spionids are tubicolous whereas others, like *Scolelepis*, are relatively free-living but with some constructing temporary tubes and remaining stationary for a period (Fauchald and Jumars 1979). They have varied feeding strategies but expose their palpi to potential predators to feed by sweeping the sediment/water interface or waving them in the water column. Regeneration of feeding palps has been observed for 11 species by several investigators (Table 1).

Woodin (1982) reported that regenerating individuals were common in all populations of the tubicolous spionid worms that she had observed on the northwest coast of the U.S. and in the U.K. These included *Pygospio elegans, Rhyncospio glutaeus, Malacoceros fuliginosus, Spiophanes bombyx* and *Pseudopolydora kempi*. Lindsay and Woodin (1992) reported that, in field populations, twice as many *Rhyncospio* were found regenerating one palp as *Pseudopolydora* (17% vs 8%) but the percentage observed missing two palps was the same (10%).

Zajac (1995) found that the percentage of adult *Polydora cornuta* regenerating feeding palps or posterior segments in southeast Connecticut, USA, varied seasonally from May to November. On average, 14.9% were regenerating posterior segments and 7% were regenerating palps. Worms lost, on average, 19.1% of their posterior segments. Larger worms lost more segments. The incidence of regeneration increased directly with the density of adult worms in a population. Part of the loss was due to predatory activities by a phyllodocid polychaete (*Eteone heteropoda*). Demographic modelling suggested that up to 25% of the population could be subjected to sublethal predation before population growth would fall below population maintenance levels.

In areas in which populations of the errant spionid, *Scolelepis squamata*, and the predatory phyllodocid, *Eteone longa*, overlap, the latter preyed on the former (Michaelis and Vennemann 2005). *Eteone* does not ingest whole prey individuals but only feeds on parts of the body that *S. squamata* can regenerate. Predatory behaviour follows a distinct pattern that leaves distinctive tracks in the sand.

From lab studies, Hentschel and Harper (2006) reported that *Polydora cornuta* can complete regeneration of feeding palps in 3 days, and the palps of recovered individuals were larger than their initial size. Worms for which both palps were excised grew significantly more slowly than those that lost one palp or none, but those that lost only one palp, or none, grew at the same rate. In another experiment, the investigators removed ca. 18% of body volume from the posterior of the worms. At the end of the 3-day flume experiment, these worms were six times larger than at the start of the experiment. They commented that since regeneration in this species is so rapid, rates of sublethal predation are likely underestimated.

Lindsay et al. (2007) studied regeneration following ablation of various lengths for anterior structures or the anterior end of the body of two spionids (*Dipolydora quadrilobata* and *Pygospio elegans*). In various laboratory treatments, they amputated palps only, anterior tissue through the first setiger, the fifth setiger, half of the gill-bearing setigers and through the last gill-bearing setiger. Treated specimens of both species regenerated lost tissues over time regardless of how much tissue was lost. Following amputation of the first setiger, palp growth did not commence for 3 days following amputation. First, the wounds healed. This was followed by the formation of the prostomium and the peristomium, i.e., the head, within about 6 days. Worms appeared to have grown ‘normal’ but smaller heads and palps within 9–12 days. *Pygospio* added segments at a similar rate regardless of whether 50% or 70% of the original segments were amputated. In contrast, the rate of segment addition in *Dipolydora* was directly related to the degree of tissue loss.

Whitford and Williams (2016) reported that 7% of specimens of the deep-burrowing tubicolous *Marenzelleria viridis* collected in Long Island Sound, New York, USA, were regenerating their anterior ends. Likely predators include fishes, birds and crabs (https://invasions.si.edu/nemesis/species_summary/-47). To examine regenerative capacity, they documented regeneration in worms for which from the 10th to the 50th chaetigers had been ablated. Anterior morphogenesis was similar to that observed in other spionids and was completed within 14 days. Up to ten segments were replaced normally but increasingly fewer segments were replaced than the number ablated when greater numbers of segments were removed. Survival and growth decreased when more chaetigers were removed.

In summary, observed examples of sublethal predation involving macrofaunal polychaetes have been conducted on a deposit-feeding capitellid and eight deposit- or suspension-feeding spionids. Lost structures varied considerably between the two families of macrofaunal polychaete and among the polychaete genera. Deposit-feeding capitellids live head down in tubes in the sediment and extend their posteriors out into the water column for respiration, reproduction, or defecation. In contrast, spionids live in tubes with the heads and tentacles extending out to feed but expose their posterior ends to defecate or reproduce. Consequently, capitellid lost only posterior segments. These segments were restored within 3 days. For the spionids, which expose both heads and posterior segments, 27.8% ± 36.5 of the populations lost one or both palps, 7% lost anterior segments (single observation) and 31.2% ± 29.8 lost posterior segments. Recovery occurred within 3–14 days. However, because macrofaunal polychaetes recover so rapidly, it is likely that rates of loss are underestimated (Hentschel & Harper 2005). Predators include phyllodocid polychaetes (e.g., *Eteone* spp.), crustaceans, fishes and birds.

Arthropoda: Crustacea; Amphipoda

Only one report was found for sublethal predation on macrofaunal crustaceans. Sheader (1998) conducted a comprehensive study of the effects of grazing predation on the tubicolous gammarid amphipod, *Ampelisca tenuicornis*, off the east coast of the Isle of Wight, England, U.K. When immature, this amphipod spends most of its time in a tube partially buried in the sediment. In this situation, its antennae are waving in the water column collecting food and are available to predators. In his study, 84% of the benthic populations of *A. tenuicornis* lost antennae. After maturity, individuals spend an appreciable amount of time in the water column, where their urosomal appendages are subject to predation; 12% lost urosomal appendages. Regeneration occurs over one or more moult cycles. Potential predators accounting for antennal loss during the benthic stage include small epibenthic predators such as peracarid crustaceans, polychaetes, post-larval fish and juvenile decapods. During the pelagic phase, small nektonic predators account for urosomal losses.

Summary for observed injury and regeneration

Earliest reports of sublethal predation based on observation were by Sayles (1932) for the maldanid polychaete, *Clymenella torquata* (Supplementary Appendix 1). Injured structures were those most frequently exposed and accessible to predators. Burrowing ophiuroids, which have arms extending from subterranean burrows onto the sediment/water interface or into the water column, exhibited the highest percentages of observed evidence of sublethal predation within populations (82%—13%); average for epibenthic ophiuroids, which live on the surface of the sediment, was 44%±21%. Sköld and Rosenberg (1996) reported that amphiurid brittle stars have higher AFDW and nitrogen content in their arms than epibenthic species, possibly leading to selective predation.

Observed evidence of predation was considerably less among polychaetes, bivalves and crustaceans than in burrowing ophiuroids. Most observations have been made on non-migratory megafaunal and macrofaunal polychaete worms (32 species; Supplementary Appendix 1). Non-migratory megafaunal polychaetes, especially maldanids and onuphids, dominated the reports for worms but macrofaunal polychaetes were well represented. Ophiuroids (22 species) and bivalves (20 species) were also well represented.

Among the bivalves, sublethal predation was most frequently reported for Tellinoidea and venerids. A wide range of predators has been reported (Supplementary Appendix 1). These range from polychaetes through sea otters and humans. Fish, especially flatfish, dominated. Crabs and other crustaceans also were common predators. The greatest number of predators has been reported for bivalves and non-migratory megafaunal polychaetes.

Sublethal predation based on predator gut contents

This section discusses sublethal predation on non-migratory megafaunal polychaetes, bivalves, decapods and ophiuroids, and macrofaunal polychaetes based on prey structures observed in the gut contents of captured predators. The types of structures identified are the same as those identified in the previous section. Data for sublethal predation among the various taxa based on stomach contents are summarised in Table 2 and detailed in Supplementary Appendix 2.

Non-migratory Megafauna

Polychaeta

Arenicolidae: As indicated above, lugworms live buried heads down in sediment and extend their posterior segments out to defecate and spawn. De Vlas (1979a,b) and Beyst et al. (1999) reported that appreciable quantities of posterior segments of *Arenicola marina* were observed in gut contents of dabs, flounder, plaice and sole in the Dutch Wadden Sea and Belgium.

Magelonidae: In their study, Beyst et al. (1999) also observed that palps of *Magelona papillicornis*, a sessile burrower (Encyclopedia of Life) or motile surface deposit-feeder (Fauchald &Jumars 1979), were common (52.9% by count, 13.5% by weight) in the stomachs of juvenile plaice.

Bivalvia

In samples from the surf zone on sandy beaches on the Belgian coast, Beyst et al. (1999) reported finding bivalve siphons in 38% of stomachs of juvenile plaice (15.9% by count, 2.1% by weight) and 7.4% of sole.

Cardiidae: Members of this family typically live buried near the surface of sediments and mainly only their short siphons are apparent to nektonic predators. In a study of annual consumption

Table 2 Summary of Structures for Non-Migratory Megafaunal Species Observed in Gut Contents of Predators as Evidence of Sublethal Predation

Higher Taxon	Subordinate Taxa	Number of Prey Species	Number of Predator Species	Number of Structures Injured	Number of Reports
Non-migratory Megafauna		**20**	**24**	**6**	**25**
Polychaeta		**2**	**5**		**8**
	Arenicolidae	1	2	1	3
	Magelonidae	1	1	1	1
Bivalvia		**15**	**19**	**2**	**23**
	Cardiidae	1	2	2	1
	Pharidae	1	1	1	1
	Tellinoidea	7	8	1	13
	Veneridae	3	6	2	4
	Myidae	2	3	1	3
	Hiatellidae	1	4	1	3
Ophiuroida					
	Burrowing	**3**	**6**	**2**	**4**
	Amphiuridae	3	6	2	4
Macrofauna		**5**	**7**	3	**6**
Polychaeta					
	Capitellidae	2	3	1	2
	Spionidae	3	4	3	5

of benthic prey by plaice and flounder in a tidal flat in the Dutch Wadden Sea, de Vlas (1979a) reported that, in addition to tails of lugworms, these flatfish fed on siphons or foot tips of the cockle, *Cerastoderma edule*. He observed that about one-third of the total food intake of these fish consisted of siphon tips of bivalves and tentacles, tail ends and heads of polychaetes.

Pharidae: Pharid razor clams have long conjoined siphons, can burrow deeply into the sediment and move rapidly up and down semi-permanent burrows in relatively consolidated sediments. Depth of burrowing depends on the size of the clam. When feeding, their siphons are subject to nipping by predators, especially flatfish. Data on sublethal predation are available for only one species, California tagelus (*Tagelus californianus*). The predator is the diamond turbot (*Hypsopsetta guttulata*). In Anaheim Bay, California, USA, Lane (1975) estimated 3500 turbot in the middle arm of the bay consumed nearly 4.4 kg wet weight of clam siphons day^{-1}, or ≈1600 kg yr^{-1}. This comprised 10%–35% of the diet of the turbot. *Tagelus* siphons comprised 9.1% of total food weight for turbot weighing <25 g whole weight and 24.2% for those ≥25 g. Contribution of siphons from other clams was 7.35% for all turbot sizes.

Tellinoidea: In describing the ecology of 0-year class plaice and common dabs in Loch Ewe, Scotland, Edwards and Steele (1968) reported the food of these species overlapped but differed quantitatively. From 1964 to 1966, plaice focused on siphons of *Tellina* (probably *Tellina tenuis*) and dabs focused on polychaete feeding palps. However, in 1967, plaice began feeding mainly on polychaete palps. As the species grew larger, it began feeding on whole polychaetes, amphipods and cumaceans, i.e., lethal predation.

De Vlas (1979a) reported that plaice and flounder fed on siphons tips of *Macoma balthica* and the semelid *Scrobicularia plana*. Total annual consumption for plaice averaged 5.6 g shell-free wet tissue m^{-2}. About 36% of this comprised regenerable parts of megafaunal species. Annual consumption for flounder averaged 0.9 g wet tissue m^{-2}, 12% comprising regenerable parts, including tail tips of *Arenicola marina* and *Heteromastus filiformis*, heads and feeding palps of *Pygospio elegans* and other spionids, siphons and foot tips of *C. edule*, siphons of *Mya arenaria* and *S. plana*, and siphon tips of *M. balthica*.

Based on gut studies of fishes obtained in frequent seining surveys in Mugu Lagoon, CA, USA, Peterson and Quammen (1982) reported siphons of *Macoma nasuta* and *Macoma secta* were abundant in gut contents of several fish species, especially in muddy sand habitats. Important predatory fishes included a sculpin (*Leptocottus armatus*) and two flatfish (*Hypsopsetta guttulata* and *Paralichthys californicus*). Among these fishes, *Macoma* spp. siphons were observed in more *Hypsopsetta* stomachs (86%) but *Leptocottus* contained more pieces of *Macoma* spp. in their stomachs (55.9% of all pieces).

Poxton et al. (1983) reported that in early summer, young plaice and common dab in the Clyde Sea area fed on *Tellina* siphons, eating progressively larger siphons as the plaice grew. Growth rates of the plaice in the various areas surveyed appeared to be correlated to the benthic productivity of the specific areas.

In a study to determine the secondary production of *M. balthica* in the Wadden Sea, de Vlas (1985) found that only about 13% of siphon tips lost were consumed by flatfish. Other epibenthic predators included shrimp, shore crabs and gobies, which cropped the other 87%.

Based on studies on St. Andrew's Beach, Scotland, Ansell and Gibson (1990) reported finding parts of exhalant and inhalant siphons and feet of *Donax vittatus* in the guts of 0-year class flatfish such as plaice. This provided evidence of attacks while the clams were swash-ridng (where the clams leave the sand to "ride" the swash to a higher or lower level on the beach, depending on the tide stage (Ellers 1995) or secondary attacks as they attempted to rebury after initial attacks on siphons removed them from the sand. They typically were able to quickly rebury.

In laboratory studies by Kamermans and Huitema (1994), after *M. balthica* had been exposed to shrimp (*Crangon*), siphon tissue of the clam was identified immunologically in shrimp stomachs, confirming the shrimp had preyed upon siphon tips.

Moreira (1995) reported that Black-headed Gulls (*Larus ridibundus*) fed largely on siphons of *S. plana* during winter in Tagus estuary in Portugal. Clam siphons represented >80% of the total ingested biomass. Average rate of siphon consumption ranged from 7.9 to 17.0 siphons per minute. Estimated length of ingested siphons ranged from 6.5 to 32 mm.

In a series of papers, Tomiyama et al. reported that juvenile stone flounder (*Platichthys bicoloratus*) was an important sublethal predator in several estuarine nurseries in Sendai Bay, Japan. Tomiyama et al. (2007) observed that the maximum width of siphon tips of the deep-burrowing psammobiid clam (*Nuttallia olivacea*) was smaller in 0-year class fish than in 1-year class fish. They concluded that the size of fish rather than clam size availability determined the intensity of predation on clam size classes. While the shell length of grazed clams ranged from 5 mm to 30 mm, highest predation intensity was in the range of 10–25 mm, even though that size class comprised only 25% of the population. Tomiyama et al. (2005) showed that juvenile stone flounder in several estuarine nurseries fed selectively on palps of the spionid polychaete, *P. kempi* in March, but switched to siphon tips of *N. olivacea* in April through June. They concluded that sublethal predation on highly abundant infaunal prey is important in achieving the stone flounders' high growth rate. They indicated that sublethal tissue cropping of benthic invertebrates is thought to contribute to the high growth rates in the flounder in these nursery areas. Tomiyama et al. (2005) suggested that morphological development of the predator and prey vulnerability might induce the prey shift.

In another study of *N. olivacea*, Sasaki et al. (2008) described a detailed study of siphon loss to juvenile stone flounder in the Natori River estuary, Japan. They reported that clams at one station were cropped, on average, 25.8 times over four months ("22.1 mg [shell-free] wet weight) in a [≈120-day] season by fishes"). The cropped pieces are equal to approximately 1/3 of the total siphon tip weight. The interval between croppings, ranging from 2.36 to 60.1 days, averaged 17.3 ± 14.4 days and depended on location. Stone flounders accounted for 75% of cropping. The authors indicated that individual stone flounder juveniles preyed on more than 56 siphon tips per day from March to early June. This approximated the total siphon tip weight for 370 clams. They further indicated that, besides the juvenile flounder, three goby species that inhabit the estuary year-round likely accounted for 2.6 mg of the total loss of siphon tips.

Veneridae

Substantial research has been reported on venerid clams. Based on gut studies of fishes obtained in frequent seining surveys in Mugu Lagoon, CA, USA, siphons of *Protothaca staminea* were abundant in gut contents of several important predatory fish species (e.g., a sculpin (*Leptocottus armatus*) and two flatfish (*Hypsopsetta guttulata* and *Paralichthys californicus*), especially in sandy and muddy habitats (Peterson & Quammen 1982). *Hypsopsetta* were the most important grazers on *Protothaca* in terms of percent of stomachs containing *Protothaca* siphons (25.1% of stomachs) but *Leptocottus* stomachs contained more siphons than flatfish (27.8% of total items). They concluded that cropping per individual was 30%–92% lower in muddy habitats, possibly because predator attention changed to the more accessible siphons of *Macoma* spp. in that habitat.

Arrighetti et al. (2005) reported that the electric ray (*Discopyge tschudii*) fed heavily on the purple clam (*Amiantis purpurata*) on sand plains at 10–18 m depths off Argentina from June to November. Siphon tips, observed in 90% of the stomachs containing items, were the main food item. A total of 5921 siphon tips weighing 85.5 g wet weight was collected from 62 rays, i.e., 95.5 siphons per ray. Small rays ingested small siphons and large rays ingested larger siphons. While describing siphon regeneration for this species, Nuñez et al. (2013) reported these rays derive considerable nutrition from the siphons of this clam from specimens living in beaches in this region.

Meyer and Byers (2005) observed siphons of two venerid clams [*P. staminea* and *Venerupis* (now *Ruditapes*) *philippinarum*] were cropped by two sculpin species (*L. armatus* and *Myoxocephalus polyacanthocephalus*) around San Juan Islands, WA, USA. They found siphons in 25% and 31% of stomachs of the respective fishes.

Myidae: De Vlas (1979a) reported that plaice and flounder in the Dutch Wadden Sea fed on *Mya arenaria* siphons as well as siphons of other bivalve and several polychaete species. Parts of various prey, all assumed by de Vlas to be regenerating, made up about 36% of the stomach contents of these flatfish. However, regeneration was not confirmed.

In a study to determine if the truncate soft-shell clams (*Mya truncata*) can regenerate lost siphons, Welch and Martin-Bergmann (1990) cited several reports that walruses (*Odobenus rosmarus*) fed on these clams in the western Arctic, Baffin Bay and eastern Canadian Arctic. They noted that walrus stomachs in the eastern Arctic contain both *Mya* siphons and characteristic chitinous siphon sheaths but not shells or viscera, suggesting possible sublethal predation. To assess this possibility, siphons of live specimens, which constitute 45% of its soft body dry weight, were partially excised in the lab. The excised clams were placed in containers. These containers were then filled with mud from the area and returned to the seafloor. When examined 1 year later, none of the clams had regenerated or survived so it appears the observed partial predation was lethal. However, because of flaws in the design, this experiment cannot be considered definitive.

Stehlik and Meise (2012) reported on stomach contents of winter flounder (*Pseudopleuronectes americanus*) in New Jersey, USA. Siphons were absent from stomach contents of individuals smaller than 50 mm total length. In contrast, fish between 50 and 299 mm in total length contained *M. arenaria* siphons. Again however, since regeneration has not been confirmed, it is not clear that this predation is sublethal.

Hiatellidae: In his thesis on geoduck clams (*Panopea abrupta*), Andersen (1971) cited reports of sublethal predation on this large clam in Puget Sound, WA, USA, by large fish species. He reported that stomachs of spiny dogfish (*Squalus acanthias*) contained as many as seven geoduck siphons. One cabezon (*Scorpaenichthys marmoratus*) contained 14 siphon tips. Goodwin and Pease (1989) noted that fishermen have reported finding siphons in stomachs of Pacific halibut (*Hippoglossus stenolepis*). Geoduck siphons have also been observed in the stomach contents of sea otters (*Enhydris lutris*) in southeastern Alaska (B. Weitzman, USGS, unpublished results). As with *Mya*, no research has been conducted to demonstrate regeneration of nipped siphons. However, considering that: (1) geoducks experience high rates of siphon nipping; (2) are sufficiently abundant to be commercially harvestable; and (3) are reported to live >160 years (Orensanz et al. 2004), it seems reasonable to assume that loss of siphon tips is sublethal and siphons are regenerated. Meat (i.e., siphon) weights for geoducks comprise 42.5% of total weight. Thus, considering the high individual tissue weight of these clams, often in excess of 636 g wet tissue (Andersen 1971), it is clear they contribute considerable energy to higher trophic levels in areas where they are common and preyed on.

In summary, a variety of studies on sublethal predation involving gut contents of predators has been conducted on bivalves. Tellinoidea clams are most represented; fish species were the dominant predators. The most numerous structure was siphon tips, but siphons, especially from *Mya* and *Panopea*, likely comprise the highest biomass. Commenting on the repeated nature of siphon loss in tellinoid bivalves, Lindsay (2010) suggested these clams likely contribute substantially to local food webs. Quantities of structures are substantial, ranging up to 4.4 kg or ≈1600 kg yr^{-1} wet weight of clam siphons/day in a small bay, and comprising from 10% up to 35% of the diet for some fish species. Siphon tips comprise >80% of the stomach contents of a gull in an estuary in Portugal. Studies suggesting sublethal predation have been conducted on Pectinidae, Cardiidae, Tellinoidea, Pharidae, Myidae and Hiatellidae.

Echinodermata

Ophiuroida: Burrowing Ophiuroids: Amphiuridae: Just three species of burrowing ophiuroids, all amphiurids, were reported in gut contents. Bowmer and Keegan (1983) reported that arms of *Amphiura filiformis* occurred at high densities within Galway Bay, Ireland, and were found in the gut contents of young plaice and dabs. Later, Duineveld and Van Noort (1986) reported that *A. filiformis* formed an important component of the diet of dab at a site in the Oyster Ground, southern

North Sea. Only 4% of the individual ophiuroids sampled had fully intact arms and 55% of the individuals (20% of the arms) had one or more distal scars from injury. Annual consumption was estimated to be 0.84 g wet weight m^{-2}, or 6% of the arm population. They hypothesised this had little effect on population structure.

Munday (1993) reported observing arms of the abundant brittle star, *Amphiura chiajei*, in the stomach contents of fish species, especially pleuronectid flatfish, in Killary Harbour, western Ireland, especially in summer. He reported that such grazing is the likely cause of the high rate of arm regeneration observed (99.1%).

Trophic transfer resulting from sublethal predation on another amphiurid, *Microphiopholis gracillima* by fish, crabs and penaeid shrimp ranged from ≈13 to 39 g wet weight m^{-2} yr^{-1} (Pape-Lindstrom et al. 1997). Gut contents of 16 potential predators were tested by immunochemical analyses. Most important were macerating (durophagous) predators, viz., shrimp [*Penaeus setifera* (88%) and *P. aztecus* (39%)], and blue crab (70%).

Macrofauna

Polychaeta

Capitellidae: Capitellid tail tips have been reported in gut contents by two investigators. De Vlas (1979a) found tail tips of *Heteromastus filiformis* in gut contents of plaice and flounder in the Dutch Wadden Sea. In the laboratory, Bonsdorff and Pearson (1997) compared impacts of predation by *Crangon* to physical disturbance of sediments on density of *Capitella capitata* tails appearing at the surface of the sediment. They reported that physical disturbance of the sediment caused worm tails to disappear immediately whereas with predation by *Crangon*, they disappeared gradually within 3 days. Examination of the gut contents of the shrimp indicated they fed mainly on worm tails. The effect of physical disturbance was to destroy the tubes; recovery from that treatment, i.e., reappearance of tubes and tails, occurred within 48 hours. In contrast, predation removed the tails and required regeneration. Consequently, reappearance at the surface of tubes containing tails following shrimp grazing required 11 days.

Spionidae: Appearance of spionids in gut contents of fishes has been reported for at least three species. Edwards and Steele (1968) observed feeding palps for an unidentified spionid in gut contents from 0-year class plaice and common dabs in Loch Ewe, Scotland. Similarly, Poxton et al. (1983) found palps of unidentified spionids in the gut contents of 0-year class plaice and flounder in the Clyde Sea area in Scotland. De Vlas (1979a) reported finding heads and palps of *Pygospio elegans* in gut contents of flounder and plaice in the Dutch Wadden Sea.

In a feeding study of juvenile stone flounder in the Natori River estuary and Sendai Bay, northern Japan, Tomiyama et al. (2005, 2007) and Tomiyama (2012) reported that juveniles frequently fed on palps of *Pseudopolydora kempi* in March but, as they grew, they switched to the siphons of a clam (*Nuttallia olivacea*) in April through June. They postulated that sublethal tissue cropping may contribute to the high growth rates for the flounder in these estuarine nursery areas.

Summary for gut contents

Earliest reports of sublethal predation based on gut contents were by Edwards and Steele (1968) for siphons of the tellinid clam, *Tellina tenuis*, and unidentified spionid polychaetes consumed by 0-year class plaice and common dabs (Supplementary Appendix 2). The types of structures observed in gut contents were limited for each major taxon. These included palps and posterior segments for polychaetes, siphon tips, siphons and foot tips for bivalves and arms for brittle stars. Several studies indicate that sublethal predation provides large percentages of the diets of juvenile as well as adult species of predators. Juvenile flatfish especially depend on feeding on clam siphons and palps of macrofaunal polychaetes to support their growth and survival. Most observations have been made on bivalves, followed by macrofaunal polychaetes (Table 2). It is notable that amphiurid ophiuroids are common in gut contents of predators but no reports for epibenthic ophiuroids in gut contents were found. The variety of predators ranged from walrus, sea otters and birds to a variety of fish and shrimp. Fish (70%), especially flatfish, were the dominant predator.

Effects of sublethal predation on non-migratory megafauna and macrofauna

This section deals with the direct effects of injury on non-migratory megafauna and macrofauna. The effects are quite diverse, including changes in individual and population growth rates, physiology, behaviour, reproduction and survival. Effects tend to be negative, including reduced feeding, slower growth, reduced mating success and lower fecundity and mortality. Effects can also alter sediment conditions. Only one effect, escape response, is considered positive. Data for the effects of sublethal predation among the various taxa are summarised in Table 3 and detailed in Supplementary Appendix 3.

Non-migratory Megafauna

Polychaeta

Arenicolidae and *Maldanidae*: Several polychaete species need to expose their tail tips to defecate. Woodin (1984) amputated the tail tips of the arenicolid, *Abarenicola pacifica*, and the tube-building

Table 3 Summary for Frequency of Prey Species, Types of Structure Injured, and Types of Effects of Sublethal Predation on Non-Migratory Megafaunal and Macrofaunal Species

Higher Taxon	Subordinate Taxon	Number of Prey Species	Type of Structures Injured	Number of Types of Effects	Number of Reports
Non-migratory Megafauna		**22**		**10**	**35**
Polychaeta		**5**	**4**	**10**	**6**
	Arenicolidae	2	1	8	3
	Maldanidae	1	1	3	1
	Onuphidae	2	2	5	2
Bivalvia		**10**	**2**	**11**	**16**
	Tellinoidea	5	2	6	9
	Veneridae	5	2	5	5
Arthropoda					
	Crustacea/Decapoda				
	Ocypodidae	**1**	**1**	**4**	**1**
Ophiuroida		**5**	**2**	**7**	**7**
	Burrowing				
	Amphiuridae	4	2	7	6
	Epibenthic				
	Ophiocomidae	1	1	3	2
Macrofauna		**6**			**7**
Polychaeta					
	Spionidae	5	4	10	6
Arthropoda: Crustacea: Amphipoda					
	Gammaridae	1	2	Minimal	1

maldanid, *Axiothella rubrocincta*, in the Pacific Northwest to observe the effects on defecation, burrowing, or tube building. Defecation rates were significantly reduced for both species. This also indicates that the feeding rate was reduced. Neither frequency of defecation nor weight of fecal piles had recovered after 24 days for *Abarenicola,* but amputation did not affect burrowing. Percent of specimens defecating returned to control levels by Day 8 for *Axiothella,* but tube-building was not reduced. These experiments indicate that the loss of small amounts of tissue can result in changes in feeding and defecation rates, burrowing and tube building. She concluded these changes can result in important sediment effects.

Bergman et al. (1988) studied the impacts of sublethal predation on the growth and reproduction of the lugworm, *Arenicola marina*, in the Wadden Sea by amputating tail tips in the laboratory at various rates. Although they saw differences in growth between control and treatment experiments of one or two amputations per month, the differences were not significant and the worms continued to grow. When they amputated weekly, however, differences were significant; growth became negative, mortality rates were significantly higher and sulphide concentrations in sediments increased, likely because of reduced sediment mixing (bioturbation) due to reduced feeding and burrowing.

In a modelling study to examine the interactions among browsing predators, infaunal adult activity and recruitment of the lugworm, *Abarenicola pacifica*, Lindsay et al. (1996) predicted that mortality of juvenile worms would be highest at intermediate rates of nipping. Modelling results indicated that when adult densities of *Abarenicola* were low, browsing did not affect recruitment success. However, at higher adult densities, higher grazing rates resulted in increased recruitment success.

Onuphidae: Emerson (1975) observed several effects of regeneration resulting from loss of anterior or posterior tissue in *Diopatra ornata* in lab experiments on Santa Catalina Island, CA, USA. He examined the effect of regeneration on oogenesis of mature worms and found that specimens undergoing anterior regeneration contained all sizes of oocytes and appeared to be developing in a manner similar to non-regenerating specimens over the 4–6 weeks of regeneration. Specimens with mature oocytes regenerated posterior tissue more rapidly, apparently using nutrients from the oocytes for somatic growth. Spawning of reproductively mature individuals could be delayed by both anterior and posterior regeneration, possibly desynchronising spawning within a population.

Berke et al. (2009) found that when heads of *Americonuphis magna* in Fort Pierce, Florida, USA, were lost, the worms died. When posterior segments were ablated, they were incapable of reburrowing or regenerating segments and the worms succumbed.

Bivalvia

Pectinidae: Only the very short siphons and mantle of scallops are accessible to predators in the water column. Data are available for only bay scallops (*Argopecten irradians concentricus*). Irlandi and Mehlich (1996) reported that caging scallops with browsing fishes in the fall resulted in lighter siphons (injury) and less shell growth (effect); however, they observed no differences in summer. When caged with seven browsers per cage, juvenile scallops grew about 25% faster than those with ≈16 fish per cage. In the lab, their observations indicated that scallops in tanks in contact with browsing fish spent more time with their valves closed, presumably because of siphon or mantle nipping.

Tellinoidea: Based on a study of the share of cropped *Macoma balthica* siphon tips in the secondary production in the Wadden Sea, de Vlas (1985) reported that individuals lost several siphon tips per day. He stated: "The loss and subsequent regeneration of siphon tips must be an important [negative] factor in the ecology of *Macoma*, affecting feeding behaviour, burying depth and body weight."

Brey (1991) reported that where siphons were smaller, the potential feeding area that could be grazed by *M. balthica* was reduced. Cropping of siphons by shrimp (*Crangon*) resulted in reduced growth rates in clam populations (Kamermans & Huitema 1994). During a 4-week lab experiment in which replicated buckets each contained 10 clams, whole ash-free dry weight of the clams was reduced by ca. 85% in replicates containing 2 shrimp. Moreover, shorter siphons led to a change from deposit feeding to suspension feeding as deposit feeding was less 'profitable' with shorter siphons. Feeding areas were smaller. They found that clams with shorter siphons buried less deeply in the sediment, increasing the risk of predation, but that exposure to browsing caused clams with same size siphons to bury more deeply.

Bonsdorff et al. (1995) demonstrated that siphon cropping by large isopods (*Saduria*) and shrimp (*Crangon*) caused changes in length/weight relationships and general condition index in *M. balthica*, which was the dominant infaunal prey in the area in terms of abundance and biomass. Siphon nipping by *Crangon* caused significantly more clams to move nearer the sediment/water interface than *Saduria* or than was observed in control plots. Cropping by both species reduced the condition index of clams significantly below that observed in controls and changed the feeding mode from deposit feeding to suspension feeding, and because siphons became more visible to predators, it increased the risk of siphon predation.

Similarly, Peterson and Skilleter (1994) reported that cropping by juvenile demersal fish on siphons of *M. balthica*, a facultative forager, reduced the length of siphons needed for effective deposit feeding. This caused the clams to change to suspension feeding. This was most prevalent in early to midsummer when juvenile fish were most abundant.

De Goeij et al. (2001) conducted lab enclosure experiments with *M. balthica* in which some were exposed to siphon nipping by plaice and others were not. They reported siphon nipping caused significant reductions in length-weight relationships and condition index for 7–17-mm long clams in addition to greater susceptibility to lethal predation.

Sandberg-Kilpi et al. (2008) studied the effects of cropping by shrimp (*Crangon*) on morphometric, somatic and biochemical condition of *M. balthica* under normoxic and moderate hypoxic conditions. They reported a significant reduction in the condition of clams under both normoxic and hypoxic conditions when exposed to cropping by *Crangon* for 3 weeks. However, exposure to hypoxia did not increase siphon cropping.

Zwarts (1986) reported that the semelid *Scrobicularia plana* reduced its burying depth when part of its siphon was experimentally amputated. With the shorter incurrent siphon, the semelid needed to burrow shallower in order to feed. The degree to which cropped *S. plana* approached the surface appeared to depend on body condition. Thus, only specimens with underweight siphons that were also in poor condition move closer to the surface and put themselves at greater risk of further predation. He then concluded that burying depth "is the outcome of conflicting demands", i.e., need to avoid predation vs. need to enhance feeding success.

Salas et al. (2001) reported that foot nipping in two species of bean clam (*Donax trunculus* and *D. vittatus*) by three species of crabs resulted in a greater than 20% loss of dry tissue weight annually. The loss of biomass accounted for up to 37% of tissue in larger clams.

Tomiyama et al. (2010) found that siphon-cropping pressure by juvenile stone flounder on *Nuttallia olivacea*, the dominant bivalve in the Natori River estuary, exerted only a minor negative influence on somatic condition, accounting for only 1.2% of the variation. In contrast, environmental variables accounted for 30.2%. They concluded that "sublethal predation is only a potential factor affecting bivalve condition".

Veneridae: With regard to growth, Peterson and Quammen (1982) reported that littleneck clams (*Protothaca staminea*) in a clean-sand habit protected from siphon cropping by fish in control plots for 2 years exhibited 2.2 times greater average linear growth and 2.5 times greater volumetric growth than those in unprotected plots. Similar growth was not observed concurrently in muddy sand habitat, likely because co-occurring *Macoma nasuta*, which extend their siphons for deposit-feeding, were more easily available for siphon nipping to predatory fishes. Cropping of *Protothaca* siphons in the mud habitat was 30%–92% lower than in the clean-sand habitat.

Meyer and Byers (2005) amputated the top 40% of siphons of *P. staminea* and *Venerupis phillipinarum*. *Venerupis* naturally has shorter siphons than *Protothaca* and so buries significantly less deeply. In the laboratory, clipped *Venerupis* burrowed 55% shallower than unclipped specimens whereas clipped *Protothaca* burrowed 42% shallower. In field experiments, mortality rate for *Venerupis* was 37% greater than in the laboratory but did not increase for *Protothaca*.

In a comparison of experimentally excised and unexcised specimens of shallow-burrowing *Mercenaria mercenaria* in sandy unvegetated areas and areas vegetated by a subtropical seagrass, Coen and Heck (1991) reported that shell length of unexcised clams grew significantly faster than excised clams in sandy habitats (17.3% vs. 12.0%, respectively). However, growth for excised clams, 6.9% faster in sand, was not significantly different between the two habitats. They also observed that shell growth was most rapid in unexcised individuals in sand, less rapid in unexcised individuals in a seagrass bed, and least in excised individuals in grass beds. They observed no change in burrowing depth following nipping. They hypothesised that regeneration costs and reduced feeding efficiency together may explain the significant differences observed between excised and unexcised treatments where they occurred.

Mouritsen and Poulin (2003) reported on widespread sublethal predation of the foot of intertidal populations of the New Zealand cockle (*Austrovenus stutchburyi*) in Otago Harbour, New Zealand. Although the injury was sublethal, it reduced the ability of the cockle to rebury and likely resulted in lethal predation or death from environmental exposure effects.

Cledón and Nuñez (2010) studied the effects of siphon nipping on subsequent sublethal predation by snipping off the top 6.6%–30% of the siphons of the clam, *Mesodesma mactroides*. This resulted in clams burrowing 25%–75% shallower than unsnipped control clams. Predation rates on artificially snipped clams were twice those of unsnipped clams.

Arthropoda: Crustacea: Decapoda

Juanes and Smith (1995) stated, "Given the prevalence of injury in decapod crustacean populations, the costs involved, and the ecological importance of many crustacean species, nonlethal injury has the potential to affect population dynamics and community processes. Convincing evidence of autotomy's effects beyond the level of the individual, however, is, at present, lacking." However, the study by Oliviera et al. (2000) below provides convincing evidence of the accuracy of their statement.

Ocypodidae: Sublethal predation by humans on adult fiddler crabs (*Uca tangeri*) at Ria Formosa, southern Portugal, had major effects on population structure of the exploited population (Oliveira et al. 2000). They found that 37% of male fiddlers were missing or were regenerating large feeding claws whereas none were missing large claws in an unexploited population at a remote site. Lack of the large feeding claws precluded breeding displays and reduced feeding efficiency. The differences between the populations at the two sites were highly significant, with the unexploited site having 15% larger carapace size, 40% higher burrow density and 40% greater operational sex ratios. Nevertheless, the ratio of males to females was the same.

Echinodermata

Ophiuroida: Burrowing Ophiuroids: Amphiuridae Bowmer and Keeegan (1983) suggested that *Amphiura filiformis* is a long-lived species that is regularly subjected to arm cropping by predators. The continual need to replace cropped tissues must be a major drain on energy that reduces growth and reproductive output. In Nilsson's and Sköld's (1996) study of the effects of hypoxia on *A. filiformis* in Gullmarsfjord, Sweden, they reported disturbance of spawning in addition to reduced arm regeneration rates. They also indicated that arm loss may result in an increase in sulphide concentrations in sediment under moderate and severe hypoxia.

In flume experiments, Rosenberg and Selander (2000) demonstrated that excision of arms of *A. filiformis* caused a nearly simultaneous alarm response in downstream *A. filiformis*. When dabs nipped an arm of an *A. filiformis* upstream in the flume, the downstream specimens with upstretched arms bent them down close to the sediment/water interface or retracted them into burrows. When *Amphiura chiajei* was nipped, the response by downstream *A. filiformis* was similar but weaker. Homogenates of the two species introduced into the flume elicited similar response.

Fielman et al. (1991) studied *Microphiopholis gracillima* with different amounts of disks and arm tissue removed in North Inlet Estuary near Georgetown, South Carolina, USA. They found these brittle stars suffered frequent sublethal tissue loss in nature and have regeneration patterns that result from a complex interplay between time, quantity and quality of tissue lost, available nutrients, and risk of future damage or mortality. Depending on which and how much tissue was lost, the specimens allocated stored material to either the disk or arms until they could construct feeding and respiration burrows, i.e., either a gut or three whole arms were completed.

Based on a study of arm regeneration in a population of *Acrocnida brachiata* in a high-energy environment in Little Killary on the west coast of Ireland, Makra and Keegan (1999) concluded that, in addition to sublethal predation, physical stresses, probably due to wave action, caused considerable amount of arm loss. They found that injured arms were withdrawn inside the burrow and replaced at the sediment/water interface with less injured arms for feeding.

Ophiuroida: Epibenthic Ophiuroids: Ophiocomidae In two studies of effects of sublethal predation for epibenthic *Ophiocoma echinata*, Pomory and Lawrence (1999) conducted laboratory studies to assess the effects of arm regeneration. They subjected one group with all arms intact and another with three arms amputated to two different food regimes for 2 months. One food regime was adequate for maintenance and the other was below the estimated maintenance level. In the adequate food regime, both treatments stored lipids, as normal, in the stomach, but the regenerating animal stored less. In the group with an inadequate ration, both treatments lost tissue. Later, Pomory and Lawrence (2001) reported that regeneration of three arms reduced storage material in the stomach of *O. echinata* of both males and females and significantly reduced gonad production by females.

Macrofauna

Polychaeta

Spionidae: In a study of predator browsing effects on activity, Woodin (1984) reported that removal of one or two feeding palps for tubicolous *Spiophanes bombyx* resulted in a reduction of defecation frequency for 7 days and 6%–35% reduction of weight of fecal pellets. Tube-building was also significantly reduced.

Zajac (1985) studied the effects of amputating two types of structures on reproduction for *Polydora ligni*. Loss of feeding palps led to mixed results in terms of changes in numbers of gametogenic segments, capsules, eggs per capsule and fecundity, with either no change or reductions. Loss of non-gametogenic posterior segments reduced fecundity significantly and increased brood development time by 100%. Nevertheless, females continued to reproduce in both cases. Cost of regeneration in terms of lost fecundity was estimated to range between 10% and 29% from palp loss and between 49% and 80% for posterior regeneration.

In a laboratory study to assess the differences in effects of palp removal on two spionid polychaetes with different feeding mechanisms, Lindsay and Woodin (1992) excised 0, 1, or 2 palps from worms. Both *Rhyncospio glutaeus* and *Pseudopolydora kempi japonica* feed at the sedment surface. During 2-hour periods of observation after palp removal, the frequency of segment exposure (2.28–7.72 exposures), maximum number of segments exposed (1.73–3.25), and frequency of palp exposure (14.27–19.17) increased significantly for *Pseudopolydora* with increasing palp loss. Also, the maximum number of segments exposed per emergence increased from 1.21 to 1.71. For *Rhyncospio*, mean number of segments exposed in 2 hours (3.70–5.32) and mean and maximum number of segments per emergence increased significantly (3.12–4.84 and 3.72–6.52, respectively) with an increasing number of palps removed. They concluded that the loss of one or both palps significantly increased the amount and frequency of tissue exposure, but not the duration of exposure, and that the patterns were consistent with the differences in foraging behaviour.

Subsequently, Lindsay and Woodin (1995) reported on how ablation of 1 or 2 palps affects feeding mode of *R. glutaeus* and *P. kempi japonica*. Loss of both feeding palps caused both species to switch to alternative feeding modes involving mouth-feeding but did not change where the worms fed. Based on fecal output and time spent feeding, the feeding mode was effective for *Rhyncospio* but not for *Pseudopolydora*.

In further studies of effects of loss of feeding palps on disturbance of feeding areas, Lindsay and Woodin (1996) reported disturbed areas, and fecal production was reduced 30 to nearly 100% for up to a week in plots with greater numbers of regenerating *P. kempi*. Fecal production for *R. glutaeus* was reduced only when all worms were regenerating both palps. Neither species exhibited changes in rates of tube-building.

Demographic modelling suggested that sublethal predation can reduce the population growth rate of *Polydora cornuta*, but the reduction in growth rate is less than if the added predation pressure were solely lethal (Zajac 1995). It was estimated that up to 25% of the population could be subjected to sublethal predation before the potential for population growth fell below population maintenance levels.

Henschel and Harper (2006) reported that removal of feeding palps reduced growth rates significantly for *P. cornuta*. Damaged palps fully regenerated within 3 days. In contrast, removal of the posterior end of the worms (ca. 18% of body volume) did not reduce growth rates significantly. In fact, after 3 days, the damaged worms were six times larger than at the start of the experiment.

Arthropoda: Crustacea: Amphipoda

Ampeliscidae: Sheader (1998) reported that because regeneration is rapid in *Ampelisca tenuicornis* and it engages in compensatory feeding with the secondary antennae when the first antennae are lost, the effect of sublethal grazing on feeding and reproduction of this amphipod is minimal.

Summary for Effects of Sublethal Predation

The earliest report of effects of sublethal predation found in this review were for delayed and desynchronised spawning and more rapid regeneration of cropped structures in specimens with more mature oocytes for the onuphid polychaete, *Diopatra ornata* (Emerson 1975; Supplementary Appendix 3). Overall, 36 reports showed a wide variety of effects of sublethal predation for 25 species in four higher taxa (Table 3). Most observations have been made on bivalves and polychaetes. Non-migratory megafauna were represented by 23 species whereas macrofaunal polychaetes were represented by only five species. Fewer ophiuroids were studied. For brittle stars, Fielman et al. (1991) remarked about the "complex interplay between time, quantity and quality of tissue lost, available nutrients and risk of future damage or mortality" for many of these species following loss of arms. This statement applies equally to the interplay among many other effects of sublethal predation on prey species.

Overall, at least 25 types of effects have been reported. The most important effects were reduced depth of burial, which resulted in increased susceptibility to predation, and growth. The more common effects reported were for fecundity, defecation, growth and increased susceptibility to mortality. Other effects included changes in feeding mode, area, rates, regeneration rates, energy allocation and burrowing rates. Changes in defecation rates imply reduced feeding and that effect extends to altered conditions of surficial sediments and subsurface sediment chemistry. Changes in feeding behaviour were particularly important for tellinoid clams following siphon nipping.

Only one effect is positive, where nipping of an upstream individual causes an alarm response and potential escape behaviour in downstream neighbors or congeners. The remaining effects all have negative implications for the prey populations. Loss of palps or posterior segments resulted in reduced production of fecal pellets, so in cases where deposit feeders are common, sublethal predation resulted in changes to sediment quality (Rhoads & Young 1974). Loss of brittle star arms may result in increased sulphides in the sediment during moderate or severe hypoxic conditions.

Khater et al. (2016) discussed ways sublethal predation can play "crucial roles in shaping population and community dynamics". The study of the effects of human harvesting of the feeding claw of the fiddler crab, *Uca tangeri*, demonstrated the complexity of effects of interactions that can result from sublethal predation, where removal of the male feeding claw changed operational sex ratios, male growth rates, burrow defense and colony density (Oliveira et al. 2000). This is a dramatic example of the tradeoffs in population dynamics.

Effects of sublethal predation on secondary production in sedimentary assemblages

Few investigators have provided details that facilitate estimation of mean annual rates for secondary production for non-migratory megafauna (Table 4). Data for macrofauna appear to be lacking.

Non-migratory Megafauna

Polychaeta

Arenicolidae: De Vlas (1979b) observed annual losses of about 120 mg ash-free dry weight (AFDW) per lugworm (*Arenicola marina*) or about 20% of its total weight but that loss was replaced by elongation of segments, resulting in maintenance of biomass. He estimated that number of tail-nippings per month per individual worm by juvenile plaice and flounder in two areas ranged from 2.2 to 3.8 in the Balgzand tidal flat in the Wadden Sea. From March to December, he estimated consumption of lugworm tail segments by plaice, flounder and *Nereis* sp., a predatory polychaete, at a third site almost equaled the standing stock of lugworms in the area, i.e., was likely >90% (Table 4). Thus, for lugworms in this table, the annual replacement rate due to sublethal predation averaged ≈55%. They undoubtedly constitute a major renewable energy source for higher trophic levels in the Dutch Wadden Sea.

Bivalvia

Tellinoidea: De Vlas (1985) reported that individual *Macoma balthica* lost an average of several incompletely regenerated siphon tips per day. The estimated amounts of *Macoma* siphon tissue lost during 1975 at four tidal-flat transects varied from 0.31 to 1.66 g AFDW m^{-2} yr^{-1} (Table 4), i.e., was "nearly half of the annual mean biomass of soft parts and hardly different from elimination in the form of whole *Macoma*". He concluded that regeneration of siphon tips and other cropped body parts of *Macoma* must contribute significantly to benthic secondary production.

Veneridae: Based on the comparison of changes in size between caged and uncaged control clams over 2 years, Peterson and Quammen (1982) indicated that *Protothaca staminea* in Mugu Lagoon, CA, USA, grew more than two times faster in sandy plots protected from siphon nipping than in control plots. In sand habitat, they observed that caged clams increased 218% in length and 250% in volume more than clams exposed to grazing. An increase in volume is roughly equivalent to an increase in biomass. In mud habitat, they observed no significant differences, likely because *Macoma* spp. provided an alternative prey to *Protothaca*. Based on these figures, it appears that siphon nipping may result in ca. 58.5% annual loss of growth (basically biomass) in the sandy and muddy environments as a result of sublethal predation in these two habitats. Thus, for bivalves in

Table 4 Percentage of Population Biomass Lost Annually to Sublethal Predation by Three Higher Taxa of Dominant Non-Migratory Megafaunal Organisms

Species	Higher Taxon	Annual Biomass Lost or Replaced (%)	Estimated Annual Tissue Loss	Structure Lost	Reference
Arenicola marina Intertidal	Polychaete	>20[a]		Tail segment	de Vlas (1979b)
Arenicola marina Intertidal	Polychaete	>90[b]		Tail segment	de Vlas (1979b)
Macoma balthica Intertidal	Bivalve	≈50	0.31–1.66 g AFDW m^{-2} yr^{-1}	Siphon	de Vlas (1985)
Protothaca staminea Intertidal	Bivalve	≈58.5		Siphon	Peterson and Quammen (1982)
Acrocnida brachiata Subtidal	Ophiuroid	11.1		Arm	Bourgoin and Guillon (1994)
Acrocnida brachiata Subtidal	Ophiuroid	6.4	19 g AFDW (33 g DW) m^{-2} yr^{-1}	Arm	Bourgoin and Guillon (1994)
Amphiura chiajei Subtidal	Ophiuroid	33		Arm	Munday (1993)
Amphiura filiformis Subtidal	Ophiuroid	25		Arm	Bowmer and Keegan (1983)
Amphiura filiformis Subtidal	Ophiuroid	22	0.34 g AFDW m^2 yr^{-1}	Arm	Sköld et al. (1994)
Amphiura filiformis Subtidal	Ophiuroid	22.6		Arm & gonad production	O'Connor et al. (1986)
Microphiopholis gracillima Subtidal	Ophiuroid	50		Arm	Stancyk et al. (1994)
Mean±SD		35.3±24.6			

[a] Based on predation by juvenile plaice and flounder.
[b] Based on predation by juvenile plaice and flounder and a polychaete, *Nereis* sp.
DW, dry weight; AFDW, ash-free dry weight.

this table, the annual replacement rate due to sublethal predation averaged ca. 54%. They undoubtedly constitute a major renewable energy source for higher trophic levels.

Echinodermata

Ophiuroida: Burrowing Ophiurans: Amphiuridae: For *Acrocnida brachiata* in Brittany, France, Bourgoin and Guillou (1994) reported that nearly 70% of arms in intertidal and subtidal populations were regenerating. Biomass composed of regenerating arm tissue differed significantly annually between intertidal (11.1%±9.5%) and subtidal habitats (6.4%±8.4%). Estimated annual production for *Acrocnida* arm tissue lost subtidally was 33 g dry weight·(19 g AFDW m^{-2} yr^{-1}). O'Connor et al. (1986) estimated that *Amphiura filiformis* in Galway Bay, Ireland, contributed 16% of its biomass annually to higher trophic levels through loss of arms. They estimated that an additional 6.6% was contributed through gonad production, for a total of 22.6% annually. Sköld et al. (1994) reported that *A. filiformis* in the Skagerrak, Sweden, contributed 22% of its total population biomass (0.34 g AFDW m^{-2} yr^{-1}) to higher trophic levels through sublethal predation on arms. Trophic transfer resulting from sublethal predation on another amphiurid, *Microphiopholis gracillima*, by fish, crabs, and penaeid shrimp ranged from ≈13 to 39 g wet weight m^{-2} yr^{-1} (Pape-Lindstrom et al. 1997). Based on their data, that team subsequently estimated 50% production for that species (Stancyk et al. 1994). With an estimated average of 24.3%±14.4% annual replacement of biomass for these four ophiuroid species, combined with high biomass, burrowing brittle stars probably constitute a significant renewable energy source for higher trophic levels in areas where they occur in dense populations.

Summary for secondary production for non-migratory megafauna

Overall, the biomass lost or replaced annually by these seven species ranged from 6.4% to >90% and averaged 35.3%±24.6%. This value suggests that non-migratory megafauna contribute substantial energy to higher trophic levels, especially in areas where species are abundant, and biomass is high. This tissue or energy contribution to higher trophic levels must be added as an important component to secondary production calculations for infaunal assemblages as defined by Crisp (1971). Considering that non-migratory megafaunal biomass likely is much higher than macrofaunal biomass in most sedimentary habitats (Lees et al. 1980, Lees 2021), it is highly likely this sub-assemblage contributes far more energy to higher trophic levels than macrofauna.

Studies providing adequate data on the percent of population biomass lost due to sublethal predation (i.e, contributed to higher trophic levels) are sparse. Nevertheless, despite the paucity of data, it seems reasonable, when estimating secondary production for infaunal assemblages, to estimate that sublethal predation on non-migratory megafaunal organisms contributes conservatively ca. 25% biomass annually. The average for polychaetes in Table 4 is 55%, for bivalves, 54.25% and for ophiuroids, 28.4%, for the overall average cited above. Obviously, however, these are very speculative estimates for those taxa. This 25% estimate is basically a conservative place holder based on a very sparse data set. It is meant to provide some insight into a potential effect of sublethal predation on secondary production. In view of regeneration, that contribution has, on average over time, a negligible effect on standing stocks. Undoubtedly, the actual figure will vary substantially according to ecosystem and which taxa dominate, as well as addition of more data sets on production.

Unfortunately, few estimates of annual tissue loss were located. The two included from intertidal populations in this table are consistent. However, it is notable that the subtidal estimate for tissue loss (i.e., secondary production) is orders of magnitude higher. It seems reasonable to assume that this pattern is valid since subtidal populations of many prey species are denser and individuals are larger. They have more time to feed and are exposed to less stress. However, they also have more time to expose nippable structures for predators to browse on them.

Since sublethal predation has been demonstrated for numerous macrofaunal species, is seems reasonable to adopt that practice for macrofauna as well although it is likely that lethal predation will be a more important component for these much smaller organisms. Adding to the conservative nature of the 25% placeholder, estimates of secondary production or yield could be expanded to include consumable reproductive and excretory products such as eggs, sperm and mucus, which likely are substantial.

Discussion and conclusions

Eighty-eight species from six major taxa reported to have been subjected to sublethal predation or regeneration are discussed in this review. Dominant among these were polychaetes, bivalves and ophiuroids. Other higher taxa included sea pens, amphipod and decapod crustaceans, and irregular echinoids. The numbers of species observed in each of the five topics examined in this review

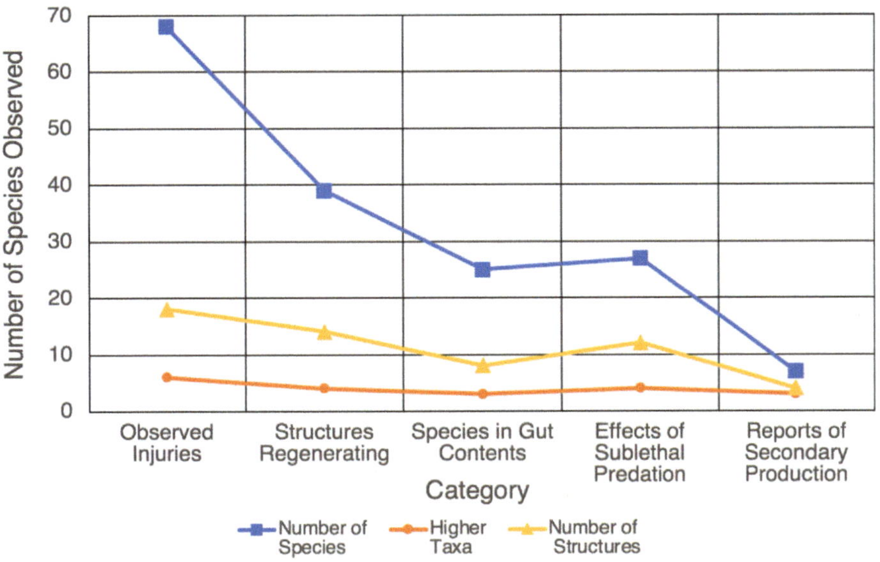

Figure 3 Distribution of research effort among the five categories of focus in this review.

vary widely by topic (Figure 3). Substantially more species were examined for observed injuries than for other categories. The least, with only about a tenth of those for observed injury, were observed for secondary production. The number observed regenerating was slightly more than half those observed with injuries. This suggests that considerably more research needs to be conducted on regeneration. Also, the number of species observed in gut contents was relatively small, suggesting that more effort is needed in studies of gut contents. The number of structures observed, generally small, declined progressively by category with a slight bump in the effects category. It is interesting that only about a third of the number of observed injured structures appeared in gut contents. Diversity of higher taxa was generally low, suggesting that a greater range of taxa should be examined.

Studies providing adequate information to determine secondary production are sparse even though that is a process of major importance for describing the value of ecosystems. Considering the importance of this topic to infaunal ecology, much more research is necessary.

Observed injuries of specific structures or their absence or observed regeneration were the basis for the initial discussion. These structures, generally exposed and accessible to predators, were reported for six higher taxa comprising 26 polychaete, 18 bivalves and 23 ophiuroids. The latter dominated studies for regeneration. The types of structure accessible to sublethal predation are unique and specific to each major taxon. The injuries were caused by at least 50 predators. Demersal fishes, from juveniles to mature adults, were the dominant predators but polychaetes, crabs, shrimp and other crustaceans, shorebirds and extending on to sea otters and humans, were also commonly reported.

Gut studies were performed on at least 24 predator species including fish, crabs, shrimp, a gull, sea otters and a walrus. Most studies of gut contents were on fish. Structures of megafaunal species suggesting sublethal predation based on examination of predator gut contents are reported here for 33 prey species from three higher taxa. Dominants among these were bivalves (17 species) and polychaetes (10 species). Macrofaunal species are likely greatly under-represented because structures resulting from sublethal predation are small, fragile and probably quickly digested in the predator.

Effects described in this review related to 27 species from four higher taxa. Most important among these are bivalves (11 species) and polychaetes (10 species), followed by ophiuroids

(6 species). Effects were complex, wide-ranging, important and mostly negative. They ranged from several individual behaviours to reproduction effects and population dynamics, 41 kinds in all. Only one of these effects, an escape response observed in downstream brittle stars, when an upstream individual is attacked, can be considered a positive effect.

The distribution of research effort varied substantially among higher taxa in terms of number of species studied and reporting effort (Figure 4). A broader range of species was studied for Polychaeta and least on Ophiuroida. Most effort on polychaetes was expended on *Arenicola marina* and *Diopatra cuprea*. Substantially more research effort was expended on Bivalvia; most studied were *Macoma balthica* and *Donax* spp. Most studied among the ophiuroids was *A. filiformis*, a burrowing species.

The number of species was evenly split among non-migratory megafaunal and macrofaunal polychaetes (16 each). *A. marina* and *D. cuprea* were the most studied megafaunal species, and *Pseudopolydora kempi japonica* was the most studied macrofauna.

Maginnis (2006) discussed how both structure loss and regeneration can result in important effects on various aspects of fitness, e.g., locomotion, foraging, survivorship, somatic growth and reproduction. In the reviewed studies, the most commonly studied of the ten general effects categories identified by Maginnis related to reproduction, somatic growth and feeding. Less frequently mentioned were effects on feeding, energy, respiration, mortality and effects on sediment quality due to changes in fecal production and foraging patterns. Because this review focuses on relatively non-migratory invertebrates, locomotion was not a factor. It is clear that the range of effects from sublethal predation is broad and complex; more research is needed.

Nine reports (polychaete-1, bivalves-2 and ophiuroids-6) provided suitable data to allow preliminary estimation of the contribution by sublethal predation to production. Types of tissue included tail segments for lugworms, siphons for bivalves and arms for ophiuroids. The reported amount of biomass lost or replaced in populations of these species ranged from 6.4% to >90% and averaged 35.3%±24.6. Obviously, considerably more research is needed to improve our knowledge for this important process.

Based on her research on spionid worms, Woodin (1982) suggested that sublethal predation (browsing) by visual predators has been important in marine sedimentary environments "in the

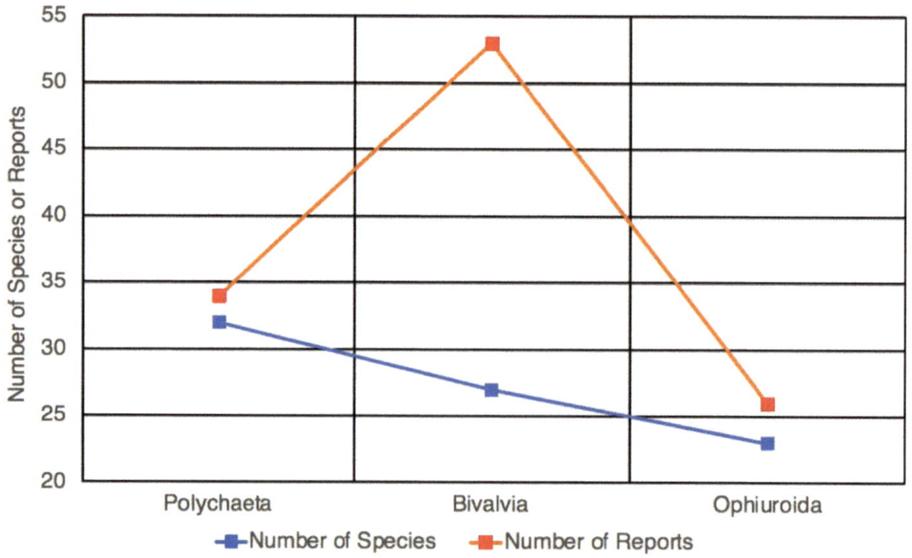

Figure 4 Distribution of research effort among the major taxonomic groups reviewed in this assessment.

evolutionary past". She reported that many spionids that expose their anterior ends to feed or defecate have evolved cryptic anteriors, likely in response to sublethal predation. In contrast, spionids not exposing their anteriors have not developed cryptic coloring. However, the importance of sublethal predation on the evolution of the affected species has received little attention.

The importance of the "evolutionary past" is also indicated by studies on fossil ophiuroids and echinoids. Aronson (1992) reported that evidence of browsing was absent in fossil ophiuroids from the Jurassic period (201–145 mya). He hypothesised that such browsing on ophiuroids did not arise until the evolution of durophagous fishes and decapod crustaceans, i.e., animals that are able to crush and digest hard skeletal parts, diversified during the late Mesozoic era (145–66 mya). Echinoids from the Pliocene epoch (5.33–2.58 mya) were also unbrowsed (Negron et al. 2020). It appears durophagous predators on sea urchins evolved much later in geologic history.

Knowledge gaps: suggested future research

Lees (2021) provided a broader description for infaunal assemblages than has been presented in more conventional research. This broader scope suggests that in order to obtain better understanding of infaunal systems, future research should focus on non-migratory megafauna and associated sublethal predation. This will require adding sampling methods that cover larger sampling areas and penetrate more deeply into the sediment (see Peterson 1977).

Two major data gaps in knowledge are suggested by these analyses. The first gap, knowledge of sublethal injury to a broad range of infaunal invertebrates, can be resolved by (1) actively documenting frequency of lost or regenerating structures in collected species, especially those previously shown to suffer partial predation (e.g., *Mya* spp. and *Panopea*) and (2) studying gut contents of more and a wider variety of predators. Resolving the second gap, a paucity of information regarding sublethal predation in a broad range of infaunal species, requires more careful assessment of sublethal predation within the many major taxa for which regeneration is possible but has not been investigated.

Major gaps for bivalves are studies demonstrating whether soft-shell clams (*Mya* spp.) or geoducks (*Panopea* spp.) are capable of regenerating partially or fully lost siphons. These clams are major inhabitants of intertidal and subtidal mudflats and nearshore sediments. Some are valuable commercial species and important sources of nutrition for several charismatic predators. Their siphons appear as major items of stomach contents of several major predators. However, studies demonstrating lost or regenerating siphons in living individuals appear to be lacking. This is an important data gap. As reported above, Welch and Martin-Bergmann (1990) reported that siphons and chitinous siphon sheaths of truncate soft-shell clams (*Mya truncata*) are common in gut contents of walrus (*Odobenus rosmarus*) in the western Arctic, Baffin Bay and eastern Canadian Arctic. Unfortunately, the design of their year-long study to assess the regeneration of experimentally excised siphons was flawed and all excised clams succumbed. *In situ* studies need to be conducted for deep-burrowing species like *Mya* spp. and *Panopea* spp., which are preyed upon by several large fish and sea otters. These studies should be conducted over a period sufficient to assess recovery and regeneration. Because of the importance of soft-shell clams as food items, they should be conducted on at least two species of *Mya*.

Taxa lacking documented sublethal predation

The species reviewed in this study are far from a complete listing of species potentially subjected to sublethal predation. For example, 27.5% of the marine genera listed in a review of segment regeneration for annelids (Bely 2006) are not included among the genera discussed in this current review. These include members of important families such as Amphinomidae, Chaetopteridae, Eunicidae, Nereididae, Oweniidae, Polynoidae, Sabellidae and Syllidae, all of which are reported by Bely to exhibit regenerative capabilities. Many of these taxa likely are subject to sublethal predation, are important members of infaunal assemblages and thus contribute to secondary production.

This review shows clearly that little information is available for a broad range of infaunal taxa that often are important in non-migratory megafaunal assemblages in sedimentary habitats. Many of these soft-bodied invertebrates have strong regeneration capabilities. However, because of their lack of hard parts on nippable organs (e.g., setae and jaws for polychaetes or arms for ophiuroids), they are not identifiable in gut contents. Nevertheless, it is likely they are also subject to sublethal predation. Zattara et al. (2019) summarised the widespread regeneration of heads in Nemertea, especially *Lineus sanguinea*. Davis et al. (1982) reported that repeated predation on sea pens by fishes, initially sublethal, was ultimately lethal. Malecha and Stone (2009) indicate that predation of pennatulaceans by nudibranchs is possibly sublethal. Rice (1970) described regeneration in sipunculids. Walls (1982) reported on regeneration for echiurids. It is surprising that sublethal predation is not commonly reported for sabellid (feather duster) worms since regeneration is widely reported for these worms, some of which form large mound-like colonies (Bely 2006, Kolbasova et al. 2013). McKinney et al. (2003) and Jørgensen et al. (2015) reported that bryozoans, often common to abundant on mid to outer shelf sediments, can be subjected to intense predation by a broad range of predators ranging from flatworms and pycnogonids to fish. This predation is likely often sublethal in some because of their colonial fleshy growth form (e.g., *Flustrellidra* spp.) Eisapour et al. (2021) described regeneration for burrowing holothurians. Francour (1997) stated that holothurians are usually consumed whole but cited papers describing predation of viscera (Bakus 1968) and tentacle crowns (Mortensen 1927) from which the prey likely survived and regenerated the lost structures. Holothurians are known to eviscerate all internal organs in response to disturbance by potential predators (Byrne 1985); this should be considered sublethal predation. Rychel and Swalla (2009) and Arimoto and Tagawa (2018) reported regeneration in enteropneusts, a poorly studied but often important non-migratory megafaunal invertebrate.

The importance of many of these infaunal assemblages indicates that substantial increases in research on various aspects of sublethal predation and regeneration, as well as more comprehensive studies of the highly productive non-migratory megafauna, are called for. Non-migratory megafaunal organisms have routinely been undersampled or actively avoided by conventional infaunal sampling methods (Lees 2021), leading to a relatively poor understanding of the value of these ecosystems worldwide. Expanding assessment of non-migratory megafaunal assemblages will greatly increase our understanding of the importance and value of infaunal ecosystems and will increase the opportunity to inspect these organisms for lost or regenerating structures. This will lead to a better understanding of the magnitude of the contribution of these assemblages to higher trophic levels, i.e., secondary production, by sublethal predation.

Importance of sublethal predation

In view of its virtual absence in discussions of community dynamics, one can ask, "How important is sublethal predation in infaunal ecology?" The answer to that question is apparent in this review, based on the importance of the types of infaunal animals that are its focus, especially non-migratory megafaunal polychaetes, bivalves and ophiuroids, and considering the proportion of community biomass these taxa generally comprise in specific ecosystems. These three higher taxa often are among the dominants in sedimentary ecosystems in terms of biomass and ecological engineering. However, other higher taxa like echiurids, burrowing shrimp, crabs or holothurians also can be important. Their combined biomass is generally orders-of-magnitude greater than that of macrofauna (e.g., Peterson 1978, Lees et al. 1980, Lees 2021). Moreover, as is well demonstrated above, they are major targets for sublethal predation and contribute substantial amounts of energy to their predators as well as other trophic levels. Consequently, their contribution to secondary production far exceeds that of macrofauna. Since secondary production is a major factor in discussions of infaunal ecosystem dynamics, the importance of sublethal predation is therefore quite clear. However, few studies have addressed the topic, particularly from the perspective of non-migratory megafauna.

Notably, sublethal predation has never been considered previously in calculations of infaunal secondary production. For future implementation, in view of the paucity of applicable data, I propose establishing a value of 25% as a conservative placeholder for use in estimating contributions by non-migratory megafauna to secondary production for infaunal ecosystems. All the taxa listed for the loss analysis above upon which that conservative estimate is based (Table 4) are non-migratory megafauna; such data appear to be lacking for macrofauna. In view of the variation in percentage contribution among those taxa, it is clear that contribution estimates will vary substantially within species and among assemblages depending on dominant species. Optimally, more data will be acquired, allowing refinement of that figure or providing specific values for particular studied ecosystems. Clearly, then, considerably more research into sublethal predation, especially on non-migratory megafauna, will be required to improve estimates for contributions by sublethal predation and provide more accurate and defensible estimates of the magnitude of the secondary production for these rich productive systems. But it seems quite clear that inclusion of the contributions of sublethal predation in these calculations will increase estimates of secondary production for infaunal assemblages by a substantial amount. This will demonstrate that these systems have been greatly undervalued in the past.

Acknowledgements

I extend my sincere thanks to Nynke Fortuin, Jonathan Houghton, Jeremy Lees, Sandra Lindstrom and Howard Teas for their thoughtful editorial reviews. Thanks to the anonymous reviewer of my earlier "weeds and trees" manuscript for the valuable suggestion to remove much of the material on sublethal predation from that manuscript and build it into a separate review paper. Furthermore, I thank the two anonymous reviewers of this manuscript for their valuable suggestions and encouragement. Finally, I wish to thank my wife, Kathleen, for tolerating my obsession and always encouraging me to keep working on my projects.

References

Andersen, A. 1971. *Spawning, growth and spatial distribution of the geoduck clam,* Panope generosa *Gould, in Hood Canal, Washington*. PhD thesis. University of Washington, USA.

Ansell, A.D. 1995. Surface activity of some benthic invertebrate prey in relation to the foraging activity of juvenile flatfishes. In *Biology and Ecology of Shallow Coastal Waters. Proceedings 28th European Mari ne Biology Symposium 1993*, A. Eleftheriou et al. (eds.). Fredensborg: Olsen & Olsen, 245–252.

Ansell, A.D. & Gibson, R.N. 1990. Patterns of feeding and movement of juvenile flatfishes on an open sandy beach. In *Trophic Relationships in the Marine Environment. 24th European Marine Biology Symposium, Oban, 1989*, M. Barnes & R.N. Gibson (eds.). Oban: Aberdeen University Press, 191–207.

Ansell, A., Harvey, R. & Günther, C.-P. 1999. Recovery from siphon damage in *Donax vittatus* (Da Costa) (Bivalvia: Donacidae). *Journal of Molluscan Studies* **65**, 223–232.

Arimoto, A. & Tagawa, K. 2018. Regeneration in the enteropneust hemichordate, *Ptychodera flava*, and its evolutionary implications. *Development, Growth & Differentiation* **60**, 400–408.

Aronson, R.B. 1991. Predation, physical disturbance, and sublethal arm damage in ophiuroids: a Jurassic-Recent comparison. *Marine Ecology-Progress Series* **74**, 91–97.

Aronson, R.B. 1992. Biology of a scale-independent predator-prey interaction. *Marine Ecology Progress Series* **89**, 1–13.

Arrighetti, F., Livore, J. & Penchaszadeh, P. 2005. Siphon nipping of the bivalve *Amiantis purpurata* by the electric ray *Discopyge tschudii* in Mar del Plata, Argentina. *Journal of the Marine Biological Association of the United Kingdom* **85**, 1151–1154.

Bailey-Brock, J. 2008. Ecology of the tube-building polychaete *Diopatra leuckarti* Kinberg, 1865 (Onuphidae) in Hawaii: community structure, and sediment stabilizing properties. *Zoological Journal of the Linnean Society* **80**, 191–199.

Bakus, G.J. 1968. Defensive mechanisms and ecology of some tropical holothurians. *Marine Biology* **2**, 23–32.

Bely, A. 2006. Distribution of segment regeneration ability in Annelida. *Integrative and Comparative Biology* **46**, 508–518.

Bergman, M.J.N., van der Veer, H. & Karczmarski, L. 1988. Impact of tail nipping on mortality, growth and reproduction of *Arenicola marina*. *Netherlands Journal of Sea Research* **22**, 83–90.

Berke, S., Cruz, V. & Osman, R.W. 2009. Sublethal predation and regeneration in two onuphid polychaetes: patterns and implications. *Biological Bulletin* **217**, 242–252.

Beyst, B., Cattrijsse, A. & Mees, J. 1999. Feeding ecology of juvenile flatfish of the surf zone of a sandy beach. *Journal of Fish Biology* **55**, 1171–1186.

Birkeland, C. & Chia, F.-S. 1971. Recruitment risk, growth, age and predation in two populations of sand dollars, *Dendraster excentricus* (Eschscholtz). *Journal of Experimental Marine Biology and Ecology* **6**, 265–278.

Boddeke, R. 1976. The seasonal migration of the brown shrimp *Crangon crangon*. *Netherlands Journal of Sea Research* **10**, 103–130.

Bonsdorff, E. & Pearson, T. 1997. The relative impact of physical disturbance and predation by *Crangon crangon* on population density in *Capitella capitata*: an experimental study. *Ophelia* **46**, 1–10.

Bonsdorff, E., Norkko, A. & Sandberg-Kilpi, E. 1995. Structuring zoobenthos: the importance of predation, siphon cropping and physical disturbance. *Journal of Experimental Marine Biology and Ecology* **192**, 125–144.

Bourgoin, A. & Guillou, M. 1994. Arm regeneration in two populations of *Acrocnida brachiata* (Montagu) (Echinodermata: Ophiuroidea) in Douarnenez Bay, (Brittany: France): an ecological significance. *Journal of Experimental Marine Biology and Ecology* **184**, 123–139.

Bowmer, T. & Keegan, B. 1983. Field survey of the occurrence and significance of regeneration in *Amphiura filiformis* (Echinodermata: Ophiuroidea) from Galway Bay, west coast of Ireland. *Marine Biology* **74**, 65–71.

Brey, T. 1991. Interactions in soft bottom benthic communities - Quantitative aspects of behaviour in the surface deposit feeders *Pygospio elegans* (Polychaeta) and *Macoma balthica* (Bivalvia). *Helgolander Meeresuntersuchungen* **45**, 301–316.

Budaeva, N. & Fauchald, K. 2008. *Diopatra tuberculantennata*, a new species of Onuphidae (Polychaeta) from Belize with a key to onuphids from the Caribbean Sea. *Zootaxa* **1795**, 29–45.

Byrne, M. 1985. Evisceration behaviour and the seasonal incidence of evisceration in the holothurian *Eupentacta quinquesemita* (Selenka). *Ophelia* **24**, 75–90.

Clavier, J. 1984. Production due to regeneration by *Euclymene oerstedi* (Claparède) (Polychaeta: Maldanidae) in the maritime basin of the Rance (Northern Britanny). *Journal of Experimental Marine Biology and Ecology* **75**, 97–106.

Cledón, M. & Nuñez, J. 2010. Siphon nipping facilitates lethal predation in the clam *Mesodesma mactroides* (Reeve, 1854) (Mollusca: Bivalva). *Marine Biology* **157**, 737–745.

Clements, L.A., Bell, S. & Kurdziel, J. 1994. Abundance and arm loss of the infaunal brittlestar *Ophiophragmus filograneus* (Echinodermata: Ophiuroidea), with an experimental determination of regeneration rates in natural and planted seagrass beds. *Marine Biology* **121**, 97–104.

Coan, E., Valentich-Scott, P. & Bernard, F. 2000. *Bivalve Seashells of Western North America. Marine Mollusks from Arctic Alaska to Baja California*. Santa Barbara, California: Santa Barbara Museum of Natural History.

Coen, L. & Heck, K. 1991. The interacting effects of siphon nipping and habitat on bivalve (*Mercenaria mercenaria* (L.)) growth in a subtropical seagrass (*Halodule wrightii Aschers*) meadow. *Journal of Experimental Marine Biology and Ecology* **145**, 1–13.

Crisp, D.J. 1971. Energy flow measurements. In *Methods for the Study of Marine Benthos*, A. Holme & A.D. McIntyre (eds.). Oxford: U.K. Blackwell Scientific Publications, 197–297.

Davis, N., vanBlaricom, G. & Dayton, P. 1982. Man-made structures on marine sediments: effects on adjacent benthic communities. *Marine Biology* **70**, 295–303.

De Goeij, P., Luttikhuizen, P., van der Meer, J. & Piersma, T. 2001. Facilitation on an intertidal mudflat: the effect of siphon nipping by flatfish on burying depth of the bivalve *Macoma balthica*. *Oecologia* **126**, 500–506.

De Vlas, J. 1979a. Annual food intake by plaice and flounder in a tidal flat area in the Dutch Wadden Sea, with special reference to consumption of regenerating parts of macrobenthic. *Netherlands Journal of Sea Research* **13**, 117–153.

De Vlas, J. 1979b. Secondary production by tail regeneration in a tidal population of lugworms (*Arenicola marina*) cropped by flatfish. *Netherlands Journal of Sea Research* **13**, 362–393.

De Vlas, J. 1981. On cropping and being cropped: the regeneration of body parts by benthic organisms. In *Feeding and Survival Strategies of Estuarine Organisms*, N.V. Jones & W.J. Wolff (eds.). Boston, Massachusetts: Springer, 173–177.

De Vlas, J. 1985. Secondary production by siphon regeneration in a tidal flat population of *Macoma balthica*. *Netherlands Journal of Sea Research* **19**, 147–164.

Duineveld, G. & Van Noort, G.J. 1986. Observations on the population dynamics of *Amphiura filiformis* (Ophiuroidea: Echinodermata) in the southern North Sea and its exploitation by the dab, *Limanda limanda*. *Netherlands Journal of Sea Research* **20**, 85–94.

Dupont, S. & Thorndyke, M. 2006. Growth or differentiation? Adaptive regeneration in the brittlestar *Amphiura filiformis*. *Journal of Experimental Biology* **209**, 3873–3881.

Dworschak, P.C. (2001). The burrows of *Callianassa tyrrhena* (Petagna 1792) (Decapoda: Thalassinidea). *Marine Ecology* **22(1–2)**, 155–166.

Edwards, R. & Steele, J.H. 1968. The ecology of 0-group plaice and common dabs at Loch Ewe I. Population and food. *Journal of Experimental Marine Biology and Ecology* **2**, 215–238.

Edwards, R.R.C., Steele, J.H. & Trevallion, A. 1970. The ecology of 0-group plaice and common dabs in Loch Ewe. III. Prey-predator experiments with plaice. *Journal of Experimental Marine Biology and Ecology* **4**, 156–173.

Eisapour, M., Salamat, N., Salari-Aliabadi, D.M.A., Bahabadi, M. & Salati, A. 2021. Digestive tract regeneration in the posteriorly eviscerating sea cucumber *Holothuria parva* (Holothuroidea, Echinodermata). *Zoomorphology* **140**, 1–15.

Ellers, O. 1995. Behavioral control of swash-riding in the clam *Donax variabilis*. *Biological Bulletin* **189**, 120–127.

Emerson, R.R. 1975. *The biology of* Diopatra ornata *at Santa Catalina Island, California*. PhD thesis, University of Southern California, USA.

Encyclopedia of Life. Online. https://eol.org. (Accessed 15 July 2021)

Fauchald, K. & Jumars, P.A. 1979. The diet of worms: a study of polychaete feeding guilds. *Oceanography and Marine Biology-An Annual Review* **17**, 193–284.

Fielman, K.T., Stancyk, S.E., Dobson, W.E & Jerome Clements, L.A. 1991. Effects of disc and arm loss on regeneration by *Microphiopholis gracillima* (Echinodermata: Ophiuroidea) in nutrient-free seawater. *Marine Biology* **111**, 121–127.

Francour, P. 1997. Predation on holothurians: a literature review. *Invertebrate Biology* **116**, 52–60.

Goodwin, C. & Pease, B. 1989. Species profiles: life histories and environmental requirements of coastal fishes and invertebrates (Pacific Northwest)-Pacific geoduck clam. *U.S. Fish and Wildlife Service Biological Report* **82(11)**, 22.

Gray, J.S. & Elliott, M. 2009. *The Ecology of Marine Sediments: From Science to Management*. London: Oxford University Press, 2nd edition.

Harris, R. 1989. Nonlethal injury to organisms as a mechanism of population regulation. *American Naturalist* **134**, 835–847.

Hentschel, B. & Harper, N. 2006. Effects of simulated sublethal predation on the growth and regeneration rates of a spionid polychaete in laboratory flumes. *Marine Biology* **149**, 1175–1183.

Hodgson, A.N. 1982a. Studies on wound healing and regeneration of the siphon of the bivalve *Donax serra* (Roding). *Transactions of the Royal Society of South Africa* **44**, 489–498.

Hodgson, A.N. 1982b. Studies on wound healing, and an estimation of the rate of regeneration, of the siphon of *Scrobicularia plana* (da Costa). *Journal of Experimental Marine Biology and Ecology* **62**, 117–128.

Irlandi, E.A. & Mehlich, M.E. 1996. The effect of tissue cropping and disturbance by browsing fishes on growth of two species of suspension-feeding bivalves. *Journal of Experimental Marine Biology and Ecology* **197**, 279–293.

Jørgensen, L., Benjamin, P., Thangstad, T. & Certain, G. 2015. Vulnerability of megabenthic species to trawling in the Barents Sea. *ICES Journal of Mar Science* **73**, i84–i97.

Juanes, F. & Smith, L. 1995. The ecological consequences of limb damage and loss in decapod crustaceans: a review and prospectus. *Journal of Experimental Marine Biology and Ecology* **193**, 197–223.

Kamermans, P. & Huitema H. 1994. Shrimp (*Crangon crangon* L.) browsing upon siphon tips inhibits feeding and growth in the bivalve *Macoma balthica* (L.). *Journal of Experimental Marine Biology and Ecology* **175**, 59–75.

Khater, M., Murariu, D. & Gras, R. 2016. Predation risk tradeoffs in prey: effects on energy and behaviour. *Theoretical Ecology* **9**, 18.

Kolbasova, G., Tzetlin, A. & Kupriyanova, E. 2013. Biology of *Pseudopotamilla reniformis* (Müller 1771) in the White Sea, with description of asexual reproduction. *Invertebrate Reproduction & Development* **57**, 264–275.

Lane, E.D. 1975. Quantitative aspects of the life history of the diamond turbot, *Hypsopsetta guttulata* (Girard). *Fisheries Bulletin* **165**, 153–173.

Lawrence, J.M. & Vasquez, J. 1996. The effect of sublethal predation on the biology of echinoderms. *Oceanologica Acta* **19**, 431–440.

Lees, D.C. 2021. Sampling nearshore infaunal 'weeds' rather than 'trees': Does this orthodoxy undervalue importance of sedimentary biomes? *Journal of Sea Research* **171**, 102021.

Lees, D.C., Houghton, J.P., Erikson, D.E., Driskell, W.B. & Boettcher, D.E. 1980. Ecological studies of intertidal and shallow subtidal habitats in lower Cook Inlet, Alaska. Final Report. National Oceanic & Atmospheric Administration, Outer Continental Shelf Environmental Assessment Program, U. S. Dept. of Commerce **44**, 1–436. https://www.osti.gov/biblio/6158236

Lindsay, S. 2010. Frequency of injury and the ecology of regeneration in marine benthic invertebrates. *Integrative and Comparative Biology* **50**, 479–493.

Lindsay, S. & Woodin, S. 1992. The effect of palp loss on feeding behavior of two spionid polychaetes: changes in exposure. *Biological Bulletin* **183**, 440–447.

Lindsay, S. & Woodin, S. 1995. Tissue loss induces switching of feeding mode in spionid polychaetes. *Marine Ecology Progress Series* **125**, 159–169.

Lindsay, S. &Woodin, S. 1996. Quantifying sediment disturbance by browsed spionid polychaetes: implications for competitive and adult-larval interactions. *Journal of Experimental Marine Biology and Ecology* **196**, 97–112.

Lindsay, S., Wethey, D. & Woodin, S. 1996. Modeling interactions of browsing predation, infaunal activity, and recruitment in marine soft-sediment habitats. *American Naturalist* **148**, 684–699.

Lindsay, S., Jackson, J. & He, S. 2007. Anterior regeneration in the spionid polychaetes *Dipolydora quadrilobata* and *Pygospio elegans*. *Marine Biology* **150**, 1161–1172.

Luzzatto, D. & Penchaszadeh, P. 2001. Regeneration of the inhalant siphon in *Donax hanleyanus* (Bivalvia; Donacidae) from Argentina. *Journal of Shellfish Research* **20**, 149–153.

Maginnis, T. 2006. The costs of autotomy and regeneration in animals: a review and framework for future research. *Behavioral Ecology* **17**, 857–872.

Makra, A. & Keegan, B.F. 1999. Arm regeneration in *Acrocnida brachiata* (Ophiuroidea) at Little Killary, West Coast of Ireland. *Biology and Environment: Proceedings of the Royal Irish Academy* **99B**, 95–102.

Malecha, P.W. & Stone, R.P. 2009. Response of the sea whip *Halipteris willemoesi* to simulated trawl disturbance and its vulnerability to subsequent predation. *Marine Ecology Progress Series* **388**, 197–206.

Mangum, C. 1964. Activity patterns in metabolism and ecology of polychaetes. *Comparative Biochemistry and Physiology* **11**, 239–256.

Martin, R.B. 1968. Aspects of the ecology and behaviour of *Axiognathus squamata* (Echinodermata, Ophiuroidea). *Tane* **14**, 65–81.

McKinney, F., Taylor, P. & Lidgard, S. 2003. Predation on bryozoans and its reflection in the fossil record. In *Topics in Geobiology Series*, P.H. Kelley, M. Kowalewski &T.A. Hansen (eds.). New York: Kluwer Academic/Plenum Publishers, 239–261.

Merrill, R. & Hobson, E. 1970. Field observations of *Dendraster excentricus*, a sand dollar of Western North America. *American Midland Naturalist* **83**, 595.

Metaxas, A. & Giffin, B. 2004. Dense beds of the ophiuroid *Ophiacantha abyssicola* on the continental slope off Nova Scotia, Canada. *Deep-Sea Research. Part I-Oceanographic Research Papers* **51**:1307–1317.

Meyer, J. & Byers, J. 2005. As good as dead? Sublethal predation facilitates lethal predation on an intertidal clam. *Ecological Letters* **8**, 160–166.

Michaelis, H. &Vennemann, L. 2005. The "piece-by-piece predation" of *Eteone longa on Scolelepis squamata* (Polychaetes)-traces on the sediment documenting chase, defence and mutilation. *Marine Biology* **147**, 719–724.

Miloslavich, P., Penchaszadeh, P., Carbonini, A.K. & Schapira, D. 2004. Regeneration time and morphology of the inhalant siphon of *Donax denticulatus* Linnaeus, 1758 (Bivalvia, Donacidae) after amputation. *Journal of Shellfish Research* **23**, 447–450.

Moreira, F. 1995. Diet of black-headed gulls *Larus ridibundus* on emerged intertidal areas in the Tagus Estuary (Portugal): predation or grazing? *Journal of Avian Biology* **26**, 277–282.

Morin, J.G., Kastendiek, J.E.A. Harrington, A. & Davis, N. 1988. Organisms of a subtidal sand community in southern California. *Bulletin of the Southern California Academy of Sciences* **87**(1), 1–11.

Mortensen, T. 1927. *Handbook of the Echinoderms of the British Isles*. London: Oxford University Press.

Mouritsen, K. & Poulin, R. 2003. The risk of being at the top: Foot-cropping in the New Zealand cockle *Austrovenus stutchburyi*. *Journal of the Marine Biological Association of the United Kingdom* **83**, 497–498.

Mouritsen, K. & Poulin, R. 2005. Parasitism can influence the intertidal zonation of non-host organisms. *Marine Biology* **148**, 1–11.

Munday, B. 1993. Field survey of the occurrence and significance of regeneration in *Amphiura chiajei* (Echinodermata: Ophiuroidea) from Killary Habrour, west coast of Ireland. *Marine Biology* **115**, 661–668.

Negron, K., Tennakoon, S., Grun, T.B., Portell, R.W., Kowaleski, M., Petsios, E. & Tyler, C.L. 2020. *Comparison of non-lethal predatory traces in recent and fossil sand dollars*. Joint 69th Annual Southeastern/5th Annual Northeastern GSA Section Meeting GSA2020. Phoenix, Arizona: Geological Society of America. doi:10.1130/abs/2020SE-345130

Nilsson, H. 2000. Interaction between water flow and oxygen deficiency on growth in the infaunal brittle star *Amphiura filiformis* (Echinodermata: Ophiuroidea). *Journal of Sea Research* **44**, 233–241.

Nilsson, H. & Sköld, M. 1996. Arm regeneration and spawning in the brittle star *Amphiura filiformis* (O.F. Müller) during hypoxia. *Journal of Experimental Marine Biology and Ecology* **199**, 193–206.

Nuñez, J., Ocampo, E., Chiaradi, N., Morsan, E. & Cledón, M. 2013. The effect of temperature on the inhalant siphon regeneration of *Amiantis purpurata* (Lamarck, 1818) (Bivalvia; Veneridae). *Marine Biology Research* **9**, 189–197.

Nuñez, J., Scelzo, M. & Cledón, M. 2010. Regeneration of the inhalant siphon of *Mesodesma mactroides* (Deshayes, 1854) (Mollusca: Bivalvia). *Malacologia* **52**, 175–179.

O'Connor, B., Bowmer, T., McGrath, D., Raine & R. 1986. Energy flow through an *Amphiura filiformis* (Ophiuroidea: Echinodermata) population in Galway Bay, west coast of Ireland: a preliminary investigation. *Ophelia* **26**, 351–357.

Oliveira, R.F., Machado, J.L., Jordao, J.M., Burford, F.L., Latruffe, C. & McGregor, P.K. 2000. Human exploitation of male fiddler crab claws: behavioural consequences and implications for conservation. *Animal Conservation* **3**, 1–5.

Orensanz, L., Hand, C., Parma, A., Valero, J. & Hilborn, R. 2004. Precaution in the harvest of Methuselah's clams-the difficulty of getting timely feedback from slow-paced dynamics. *Canadian Journal of Fisheries and Aquatic Science* **61**, 1355–1372.

Pape-Lindstrom, P., Feller, R.J., Stancyk, S.E. & Woodin, S. 1997. Sublethal predation: field measurements of arm tissue loss from the ophiuroid *Microphiopholis gracillima* and immunochemical identification of its predators in North Inlet, South Carolina, USA. *Marine Ecology Progress Series* **156**, 131–140.

Paxton, H. &, Bailey-Brock, J. 1986. *Diopatra dexiognatha*, a new species of Onuphidae (Polychaeta) from Oahu, Hawaiian Islands. *Pacific Science* **40**, 1–6.

Pekkarinen, M. 1984. Regeneration of the inhalant siphon and siphonal sense organs of brackish-water (Baltic Sea) *Macoma balthica* (Lamellibranchiata, Tellinacea). *Annales Zoologici Fennici* **21**, 29–40.

Pemberton, G.S., Risk, M.J. & Buckley, D.E. 1976. Supershrimp: deep bioturbation in the Strait of Canso, Nova Scotia. *Science* **192**, 790–791.

Peterson, C.H. & Quammen, M.L. 1982. Siphon nipping its importance to small fishes and its impact on growth of the bivalve *Protothaca staminea* Conrad. *Journal of Experimental Marine Biology and Ecology* **63**, 249–268.

Peterson, C. & Skilleter, G. 1994. Control of foraging behavior of individuals within an ecosystem context: the clam *Macoma balthica*, flow environment, and siphon-cropping fishes. *Oecologia* **100**, 256–267.

Pires, A., Freitas, R., Quintino, V. & Rodrigues, A. 2012. Can *Diopatra neapolitana* (Annelida: Onuphidae) regenerate body damage caused by bait digging or predation? *Estuarine, Coastal and Shelf Science* **110**, 36–42.

Pomory, C. & Lawrence, J. 1999. Effect of arm regeneration on energy storage and gonad production in *Ophiocoma echinata* (Echinodermata: Ophiuroidea). *Marine Biology* **135**, 57–63.

Pomory, C. & Lawrence, J. 2001. Arm regeneration in the field in *Ophiocoma echinata* (Echinodermata: Ophiuroidea): effects on body composition and its potential role in a reef food web. *Marine Biology* **139**, 661–670.

Poxton, M.G., Eleftheriou, A. & McIntyre, A.D. 1983. The food and growth of 0-group flatfish on nursery grounds in the Clyde Sea area. *Estuarine, Coastal and Shelf Science* **17**, 319–337.

Rennolds, C.W. & Bely, A.E. 2023. Integrative biology of injury in animals. *Biological Reviews* **98**, 34–62.

Ricciardi, A. & Bourget, E. 1998. Weight-to-weight conversion factors for marine benthic macroinvertebrates. *Marine Ecology Progress Series* **163**, 245–251.

Rice, M. 1970. Asexual reproduction in a sipunculan worm. *Science* **167**, 1618–1620.

Rosenberg, R. & Selander, E. 2000. Alarm signal response in the brittle star *Amphiura filiformis. Marine Biology* **136**, 43–48.

Rychel, A. & Swalla, B. 2009. Regeneration in hemichordates and echinoderms. In *Stem Cells in Marine Organisms*, B. Rinkevich & V. Matranga (eds.). London: Springer, 245–265.

Salas, C., Tirado, C. & Manjon-Cabeza, M.E. 2001. Sublethal foot- predation on Donacidae (Mollusca: Bivalvia). *Journal of Sea Research* **46**, 43–56.

Sandberg-Kilpi, E., Tallqvist, M. & Bonsdorff, E. 2008. The effects of reduced oxygen content on predation and siphon cropping by the brown shrimp, *Crangon crangon. Marine Ecology* **17**, 411–423.

Sasaki, K., Kudo, M., Tomiyama, T., Ito, K. & Omori, M. 2008. Predation pressure on the siphons of the bivalve *Nuttallia olivacea* by the juvenile stone flounder Platichthys bicoloratus in the Natori River estuary, north-eastern Japan. *Fisheries Science* **68**, 104–116.

Sayles, L.P. 1932. External features of regeneration in *Clymenella torquata. Journal of Experimental Biology* **62**, 237–257.

Sayles, L.P. 1934. Regeneration in the polychaete *Clymenella torquata*. II. Effect of level of cut on type of new structures in posterior tegeneration. *Physiology and Zoology* **7**, 1–16.

Sheader, M. 1998. Grazing predation on a population of *Ampelisca tenuicornis* (Gammaridae:Amphipoda) off the south coast of England. *Marine Ecology Progress Series* **164**, 253–262.

Sides, E.M. 1987. An experimental study of the use of arm regeneration in estimating rates of sublethal injury on brittle-stars. *Journal of Experimental Marine Biology and Ecology* **106**, 1–16.

Singletary, R. 1980. The biology and ecology of *Amphioplus coniortodes*, *Ophionepthys limicola* and *Micropholis gracillima* (Ophiuroidea: Amphiuridae). *Caribbean Journal of Science* **16**, 39–55.

Singletary, R. & Moore, H. 1974. A redescription of the *Amphioplus coniortodes-Ophionepthys limicola* community of Biscayne Bay, Florida. *Bulletin of Marine Science* **24**, 690–699.

Skilleter, G. & Peterson, C. 1994. Control of foraging behavior of individuals within an ecosystem context: the clam *Macoma balthica* and interactions between competition and siphon cropping. *Oecologia* **100**, 268–278.

Skold, M., Loo, L.O. & Rosenberg, R. 1994. Production, dynamics and demography of an *Amphiura filiformis* population. *Marine Ecology Progress Series* **103**, 81–90.

Sköld, M. & Rosenberg, R. 1996. Arm regeneration frequency in eight species of Ophiuroidea (Echinodermata) from European sea areas. *Journal of Sea Research* **35**, 353–362.

Smith, L. & Hines, A. 1991. Autotomy in blue crab (*Callinectes sapidus* Rathbun) populations: geographic, temporal, and ontogenetic variation. *Biological Bulletin* **180**, 416–431.

Smith, T., Ydenberg, R. & Elner, R. 1999. Foraging behaviour of an excavating predator, the red rock crab (*Cancer productus* Randall) on soft-shell clam (*Mya arenaria* L.). *Journal of Experimental Marine Biology and Ecology* **238**, 185–197.

Stancyk, S., Golde, H., Pape-Lindstrom, P. & Dobson, W. 1994. Born to lose. I. Measures of tissue loss and regeneration by the brittlestar *Microphiopholis gracillima* (Echinodermata: Ophiuroidea). *Marine Biology* **118**, 451–462.

Stehlik, L. & Meise, C. 2012. Diet of winter flounder in a New Jersey estuary: ontogenetic change and spatial variation. *Estuaries* **23**, 381–391.

Tennakoon, S., Jamal, F., Grun, T.B., Portell, R.W., Kowaleski, M. & Petsios, E. 2019. *Non-lethal traces of predation on tests of recent and fossil echinoids*. GSA Annual Meeting, GSA2019. Phoenix, Arizona: Geological Society of America. doi:10.1130/abs/2019AM-338144

Tomiyama, T. 2012. Seasonal and ontogenetic diet shift in juvenile stone flounder *Platichthys bicoloratus*. *Journal of Fish Biology* **81**, 1430–1435.

Tomiyama, T. 2016. Quantitative regeneration in bivalve siphons: difference between short- and long-siphoned species. *Marine Biology* **163**, 1–8.

Tomiyama, T. &, Ito, K. 2006. Regeneration of lost siphon tissues in the tellinacean bivalve *Nuttallia olivacea*. *Journal of Experimental Marine Biology and Ecology* **335**, 104–113.

Tomiyama, T., Katayama, S., Omori, M. & Honda, H. 2005. Importance of feeding on regenerable parts of prey for juvenile stone flounder *Platichthys bicoloratus* in estuarine habitats. *Journal of Sea Research* **53**, 297–308.

Tomiyama, T., Komizunai, N., Ito, K. & Omori, M. 2010. Spatial variation in the abundance and condition of the bivalve *Nuttallia olivacea* in relation to environmental factors and sublethal predation. *Marine Ecology Progress Series* **406**, 185–196.

Tomiyama, T., Omori, M. & Minami, T. 2007. Feeding and growth of juvenile stone flounder in estuaries: generality and the importance of sublethal tissue cropping of benthic invertebrates. *Marine Biology* **151**, 365–376.

Trevallion, A. 1971. Studies on *Tellina tenuis* Da Costa III. Aspects of general biology and energy flow. *Journal of Experimental Marine Biology and Ecology* **7**, 95–122.

Virnstein, R.W. 1977. The importance of predation by crabs and fishes on benthic infauna in Chesapeake Bay. *Ecology* **58**, 1199–1217.

Walls, J.G. (ed.). 1982. *Encyclopedia of Marine Invertebrates*. Neptune, New Jersey: T.F.H. Publications.

Welch, H. & Martin-Bergmann, K. 1990. Does the clam *Mya truncata* regenerate its siphon after predation by walrus? An experimental approach. *Arctic* **43**, 157–158.

Whitford, T. & Williams, J. 2016. Anterior regeneration in the polychaete *Marenzelleria viridis* (Annelida: Spionidae). *Invertebrate Biology* **135**, 357–369.

Whitlatch, R., Hines, A., Thrush, S., Hewitt, J. & Cummings, V. 1997. Benthic faunal responses to variations in patch density and patch size of a suspension- feeding bivalve inhabiting a New Zealand intertidal sand-flat. *Journal of Experimental Marine Biology and Ecology* **216**, 171–189.

Wilson, W. 1979. Community structure and species diversity of the sedimentary reefs constructed by *Petaloproctus socialis* (Polychaeta, Maldanidae). *Journal of Marine Research* **37**, 623–641.

Woodin, S. 1982. Browsing: important in marine sedimentary environments? Spionid polychaete examples. *Journal of Experimental Marine Biology and Ecology* **60**, 35–45.

Woodin, S. 1984. Effects of browsing predators: activity changes in infauna following tissue loss. *Biological Bulletin* **166**:558–573.

Yokoyama, L. & Amaral, A. 2010. Arm regeneration in two populations of *Ophionereis reticulata* (Echinodermata, Ophiuroidea). *Beringia Serie Zoologia* **100**, 123–127.

Young, D. & Young, M. 1978. Regulation of species densities of seagrass-associated macrobenthos: evidence from field experiments in the Indian River estuary, Florida. *Journal of Marine Research* **36**, 569–593.

Zahorczak, P., Silvano, R. & Sazima, I. 2000. Feeding biology of a guild of benthivorous fishes in a sandy shore on south-eastern Brazilian coast. *Revista Brasileira de Biologia* **60**, 511–518.

Zajac, R. 1985. The effects of sublethal predation on reproduction in the spionid polychaete *Polydora ligni* Webster. *Journal of Experimental Marine Biology and Ecology* **88**, 1–19.

Zajac, R. 1995. Sublethal predation on *Polydora cornuta* (Polychaeta: Spionidae): Patterns of tissue loss in a field population, predator functional response and potential demographic impacts. *Marine Biology* **123**, 531–541.

Zattara, E., Fernández-Álvarez, F.Á., Hiebert, T., Bely, A. & Norenburg, J. 2019. A phylum-wide survey reveals multiple independent gains of head regeneration in Nemertea. *Proceedings of the Royal Society B: Biological Sciences* **286**, 9.

Zwarts, L. 1986. Burying depth of the benthic bivalve *Scrobularia plana* (Da Costa) in relation to siphon-cropping. *Journal of Experimental Marine Biology and Ecology* **101**, 25–39.

Supplementary Materials are provided online at: https://www.routledge.com/9781032761961

Oceanography and Marine Biology: An Annual Review, 2024, **62**, 350-375

Taylor & Francis

AN INTEGRATED PHYLOGENOMIC APPROACH FOR POTENTIAL HOST-ASSOCIATED EVOLUTION OF MONSTRILLOID COPEPODS

DONGGU JEON[1], CHI-UNE SONG[1], HYEON GYEONG JEONG[2], SUSUMU OHTSUKA[3], WONCHOEL LEE[4], HO YOUNG SOH[5], AND SEONG-IL EYUN[1*]

[1]*Department of Life Science, Chung-Ang University, Seoul 06974, Korea*
[2]*Department of Taxonomy and Systematics, National Marine Biodiversity Institute of Korea, Seocheon 33662, Korea*
[3]*Blue Innovation Division, Setouchi Carbon Neutral Research Center, Hiroshima University, Hiroshima 725-0024, Japan*
[4]*Department of Life Science, Hanyang University, Seoul 04763, Korea*
[5]*Department of Marine Integrated Science, Chonnam National University, Yeosu 59626, Korea*
**Corresponding author: Seong-il Eyun*
Email: eyun@cau.ac.kr

Abstract Copepods, small crustaceans, exist in almost every aquatic environment on Earth, from freshwater to hypersaline and deep-sea bottom to high mountains. They play an important role in linking primary producers to higher consumers and thus contribute to the material and energy circulation in the ecosystems. The subclass Copepoda is recognised as ten orders, and each group shows highly optimised morphology and lifestyle to diverse habitat conditions. The Monstrilloida Sars, 1901 is one of the most obscure copepod groups with an unusually atrophied morphology and unique semi-parasitic lifestyle of endoparasitic juvenile and free-living, planktonic adult phases. The lack of common diagnostic features from their morphological peculiarity and little information about the endoparasitic juveniles have caused many uncertainties and ambiguities in their taxonomy and phylogeny. To elucidate phylogenetic relationships and evolutionary significance of these organisms, we first generated two genomes and three transcriptomes from four monstrilloid species (*Monstrilla grandis*, *Caromiobenella ohtsukai*, *Monstrillopsis longilobata* and *Maemonstrilla* sp.) and analysed the 25 nuclear protein-coding genes from 40 arthropod species. The molecular phylogenomic results supported the monophyly of Monstrilloida within Podoplea. Our analysis revealed that the Monstrilloida was more closely related to the Harpacticoida (Oligoarthra) than the Siphonostomatoida. These phylogenomic relationships for the Copepoda were confirmed by statistical tree topology tests, and previously known phylogenies were rejected. Our arthropod phylogeny identified a long branch leading to the Monstrilloida. Given the new phylogeny, we tested hypotheses about monstrilloid evolution by integrating the known morphological and ecological traits of four monstrilloid genera.

Keywords: Copepod Evolution; Monstrilloida; Crustacea; Phylogenomics; Divergence Time

DOI: 10.1201/9781003477518-9

Introduction

Historical overview of the order Monstrilloida

Krøyer reported the first monstrilloid *Thaumatoessa typica* in 1842 by providing illustrations without descriptive text. Seven years later, the author gave an extended description and diagnosis for the species accompanied by additional figures based on the same specimen but simultaneously changed the species name to *Thaumaleus typicus* without any emendatory note (Krøyer 1849). Grygier (1994a) later re-examined the holotype of *T. typicus* and found that the specimen is probably a congener belonging to *Monstrilla* Dana, 1849, which had been more widely recognised at that time. Therefore, the generic name *Thaumatoessa* Krøyer is a senior objective synonym of *Thaumaleus* Krøyer and a senior subjective synonym of *Monstrilla* Dana. Grygier (1995b) subsequently proposed the conservation of the latter two junior names and the suppression of *Thaumatoessa* in consideration of universal awareness to the public: *Thaumaleus* had been frequently used rather than *Thaumatoessa* in about 50 publications of the last 100 years of taxonomic works, and *Monstrilla* had also been widely recognised and used for over 140 years with more than 50 nominal species assigned into this genus. On the other hand, *Thaumatoessa* has only been used as a valid name in Hesse (1868).

Another nomenclatural issue between two different names *Thaumaleus* and *Monstrilla*, both regarded as congeneric, then arose. *Monstrilla* with the type species *M. viridis* was published on 4 September 1849, while the precise publication date for *Thaumaleus* and its type species *T. typicus* in 1849 was not given (Grygier 1995a,b). Thus, the date for the latter publication was automatically taken to be the last day of the year (i.e., 31 December 1849) according to Article 21.3 (the Determination of Date) of the International Code of Zoological Nomenclature (International Commission on Zoological Nomenclature, ICZN) (Grygier 1994a, 1995b). As a result, *Monstrilla* has priority over *Thaumaleus* according to Article 23 (the Principle of Priority) of the ICZN code and has eventually been the type genus of the family Monstrillidae Dana, 1849 and of the order Monstrilloida Sars, 1901.

The generic names *Thaumaleus* and *Monstrilla* continued to be used for some time before the nomenclatural revision of Grygier (1995b). During this time, *Thaumaleus* was considered a synonym of another genus *Cymbasoma* Thompson, 1888 rather than *Monstrilla* (Giesbrecht 1892a,b, Davis 1949, Isaac 1975). The incorrect synonymy resulted in *Thaumaleus* containing another form of monstrilloids. This conflicts with the diagnoses of *Thaumaleus* and *T. typicus*. Sars (1921) was clearly aware of the issue and proposed that most *Thaumaleus* species, except for *T. typicus* Krøyer, should be considered as the species of *Cymbasoma* (also see Rose 1933, Grygier 1994a). Sars suggested an additional genus *Monstrillopsis* for *Monstrilla dubia* Scott, 1904 (= *Monstrillopsis dubia*). This differs from the diagnoses of either *Monstrilla* Dana (= *Thaumaleus* Krøyer) or *Cymbasoma* Thompson. There were several other generic names, such as *Haemocera* Malaquin, 1896, *Thaumatohessia* Giard, 1900, and *Strilloma* Isaac, 1974, but these are now considered invalid or uncertain (Suárez-Morales & Gasca 2004, Grygier & Ohtsuka 2008). In the 2000s, there four additional genera were named, *Maemonstrilla* Grygier and Ohtsuka (2008), *Australomonstrillopsis* Suárez-Morales and McKinnon (2014), *Caromiobenella* Jeon, Lee and Soh (2018), and *Spinomonstrilla* Suárez-Morales, 2019 (Grygier & Ohtsuka 2008, Suárez-Morales & McKinnon 2014, Jeon et al. 2018a, Suárez-Morales 2019). Recent advances in the taxonomic methodology have provided detailed and definitive morphological characteristics for the monstrilloid taxonomy, but doubts about the validity of some genera remain (Jeon et al. 2018a,b).

General morphology of adult monstrilloids

The male body consists of nine somites including a cephalothorax fully incorporated with the first pedigerous somite (= first thoracic somite), free pedigerous somites 1–3 (= second to fourth thoracic somites), first urosomal somite, genital somite, postgenital somite, penultimate somite and anal somite (Figure 1). The females have a genital compound somite (sensu Grygier & Ohtsuka 2008) which is formed by a fusion of the genital and succeeding postgenital somites, and thus, they present one less body somite than their corresponding males. The species of *Cymbasoma* typically have one more less urosomal somite in each sex, and thus, eight somites in males and seven in females (Thompson 1888, Suárez-Morales & McKinnon 2016). The large cephalothorax lacks common mouth appendages from mandibles to maxillipeds including antennae. Several pairs of scars that remained after discarding feeding tubes used for their endoparasitic juvenile stages are present on the ventral surface of the cephalothorax instead (Huys & Boxshall 1991, Suárez-Morales 2011) (Figure 1). The scars vary in number, size, shape and location depending on the genus and species; three pairs of the scars are common in *Monstrilla*, *Maemonstrilla* and *Caromiobenella*, and one or two pairs in *Monstrillopsis* and *Cymbasoma*. The maximum number of five scar pairs was reportedly known from *Spinomonstrilla spinosa* (Suárez-Morales 2019). A cone-shaped oral papilla is located in the ventral medial axis of the cephalothorax, but its fore-and-aft position varies by the genera or species; it is often located in the midlength of the cephalothorax in *Monstrilla*, whereas

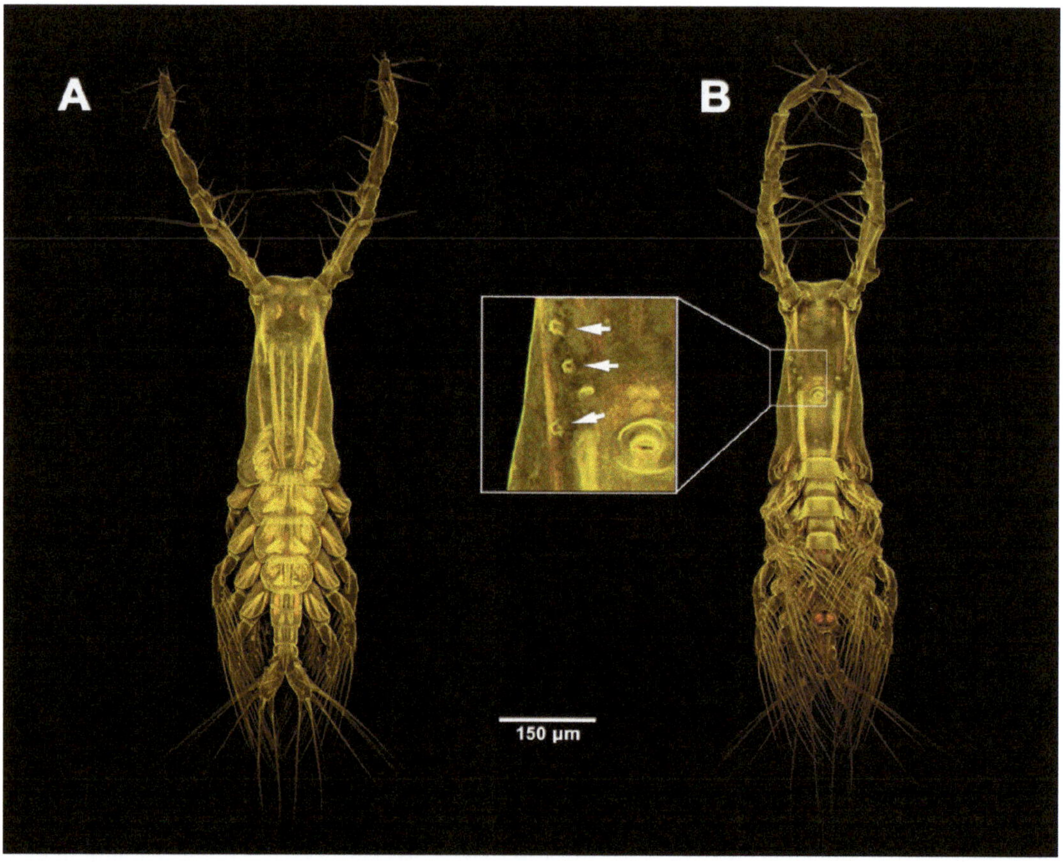

Figure 1 Confocal laser scanning microscope image of adult male *Monstrilla grandis*. (A) habitus, dorsal view. (B) habitus, ventral view. The inset illustrates three scars (arrows) on the right ventral side of cephalothorax. Monstrilloid specimen preserved in 4% borax buffered formalin was stained by dissolved Congo Red for 24 hours at room temperature. The stained specimen was washed in distilled water until no solute staining dye was present and then observed on a confocal laser scanning microscope Leica TCS5 (Leica, Germany) equipped with an optical microscope Leica DM5000 (Leica) using three visible lasers (Argon, DPSS, and HeNe). Scale bar in a micrometer. Microphotographs were provided by Dr. Seunghan Lee (Marine Act Co., Korea).

that of the other genera is more anteriorly. The developmental degree of the oral papilla also shows some differences by the genera; females of *Maemonstrilla* typically have a very large, prominent one (Grygier & Ohtsuka 2008, Suárez-Morales & McKinnon 2014), but males of *Caromiobenella* have a low, inconspicuous one (Jeon et al. 2018a).

All monstrilloids have a pair of anteriorly straight antennules each with a fundamental 5-segmented structure. Although, in females, the segmentation is often indistinct due to fusions between the segments, a quite conservative setation pattern appearing in both sexes provides a ground for assuming the putative segmentation plan for the female antennules. The last antennular segment (= fifth segment) of the males represents at least three different types of modifications, and this characteristic is regarded as one of the most evident features distinguishing the monstrilloid genera (Huys & Boxshall 1991, Suárez-Morales 2011). The first type (= type I) is a segment showing no specific modification, and this is present in the species of *Monstrilla* and *Cymbasoma*. The second type (= type II) is a segment with a hyaline bump on the inner proximal half margin and a gradually tapered distal half with a slight inner curvature. This type is currently exclusive for the males of *Monstrillopsis*. The third type (= type III) is a segment with five transverse serrate ridges on the inner distal margin and is only known from the males of *Caromiobenella*. In addition to these major types, two other modifications are also known. The male of *Cymbasoma longispinosum* presents similar ridges of the third type antennular modification, but these are much reduced (= type IV sensu Huys & Boxshall 1991). Suárez-Morales and McKinnon (2014) found another modification in which the last segment bears two inner rounded expansions along the inner margin (putatively, type V). The antennules exhibit four kinds of armature elements: aesthetascs, spines, simple setae and branched setae. The branched setae are present only in the last antennular segment, but all of these elements are replaced with simple ones in the species of *Caromiobenella* (Jeon et al. 2018a).

Monstrilloids have four well-developed pairs of swimming legs that have a similar structure (Figure 2). The protopod of each leg consists of a large coxa and a relatively small basis. The border

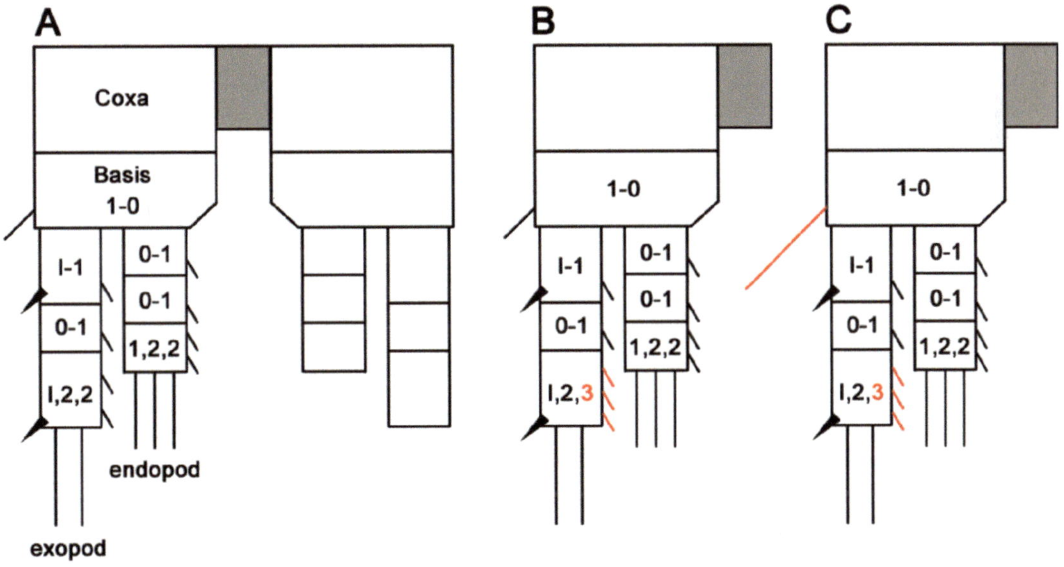

Figure 2 Schematic diagrams of swimming legs 1–4 with setation patterns. (A) pair of leg 1, each conjoined by a intercoxal sclerite (grey box). (B) legs 2 and 4. (C) leg 3. Arrowheads and lines indicate spinous and setal elements, respectively. Red-coloured numbers and elements represent the difference from the leg 1. Lines for setae, except for the basal setae, do not represent their actual length or ratio. Nomenclatural terms for describing the setation were adopted from Sewell (1949): Roman and Arabic numerals indicate the number of spines and setae, respectively.

between the coxa and basis is clearly defined from the posterior side by a clear diagonal articulation, while the anterior border is indistinctly defined by a fine seam. The basis bears two rami on its distal margin, each is typically developed into a tri-articulated endopod or exopod. Only *Cymbasoma rafaelmartinezi* is known to have a 2-segmented exopod for the swimming leg 4 (Suárez-Morales & McKinnon 2016). In addition, a similar setation pattern for the swimming legs is shared through the species. The basis of the legs 1, 2 and 4 presents a short, thin outer seta (Figure 2A and B), but this seta in leg 3 is much longer and thicker (Figure 2C). Another difference appears through the third exopodal segments of the legs. The third exopodal segments of legs 2–4 have the setation pattern of "I, 2, 3". This pattern describes one spine (in Roman numeral) on the outer distal corner, and two apical and three inner setae (in Arabic numerals) following the nomenclature of Sewell (1949). In this aspect, the third exopodal segment of leg 1 shows one less setal element on the inner margin, thus "I, 2, 2" pattern (Figure 2A). Further reduction of the setal elements occurs in the females of the *Maemonstrilla hyottoko* species group, but not for the *Maemonstrilla turgida* species group, due to the absence of inner setae from both first exopodal and endopodal segments (Grygier & Ohtsuka 2008, Suárez-Morales & McKinnon 2014).

A pair of fifth legs is another relevant morphological character for the monstrilloid taxonomy. The fifth legs are mainly present in females. The most ancestral form is biramous with unarticulate inner and outer lobes each armed with two and three setae, respectively. The most simplified structure of an elongate rod-shaped leg with two apical setae was found in the *M. hyottoko* species group (Grygier & Ohtsuka 2008). In addition, there are various types of fifth legs in the female that differ in structure and setation pattern (see section "Congruence between morphological and molecular phylogenies for Monstrilloida"). On the contrary, most males do not have a fifth pair of legs, except for some *Monstrilla* species, such as *Monstrilla grandis*, *Monstrilla longicornis* and *Monstrilla conjunctiva*. The fifth pair of legs of these latter species each are represented by a small knob-like protuberance with a long, prominent apical seta.

Males have a well-developed genital apparatus ventrally in their genital somite. The apparatus is represented by a robust, protrusive basal shaft and two lappets diverging from the distal posterior end of the shaft. The shaft can be short and broad, or elongate and cylindrical. The lappets are also various in shape including a digitiform, lamella-form or subtriangular form. The combinations of such structural variety (see Suárez-Morales 2011) result in one of the most utilisable features for identifying and distinguishing males. The genital organ of females is developed into a pair of two long spines (= ovigerous spines) and is directed posteriorly, but the females of *Maemonstrilla* have anteriorly directed spines. In *Cymbasoma*, there are two structural types of the ovigerous spines known: (1) each spine separately arises from the base as in most female monstrilloids or (2) arises as a single thread, running for a short distance, and then separates into two (especially in *C. longispinosum* species group) (Grygier 1994b, Suárez-Morales & McKinnon 2014, Üstün et al. 2014, Suárez-Morales et al. 2020).

The caudal rami present a different number of setae by the genera or species. Five or six caudal setae on each ramus are frequently found in *Monstrilla*, *Caromiobenella*, *Maemonstrilla*, *Australomonstrillopsis* and *Spinomonstrilla*. According to Suárez-Morales et al. (2006), both sexes of *Monstrillopsis* are distinct with four caudal setae. Further reduction of the number of the caudal setae is seen in *Cymbasoma*, with females always having three setae, while some males have four.

Biology and ecology

Monstrilloids have a protelean life cycle that consists of an endoparasitic juvenile phase and a free-living, planktonic adult phase (Malaquin 1901, Suárez-Morales et al. 2014). After hatching, the early lecithotrophic nauplii are initially planktonic, but soon infect various marine invertebrate hosts including polychaetes, gastropods, bivalves and sponges (Malaquin 1901, Caullery & Mesnil 1914, Pelseneer 1914, Gallien 1934, Huys & Boxshall 1991, Suárez-Morales et al. 2014). After infection, development continues in the host, fuelled by absorbing the host's body fluids using pairs of tailored feeding tubes (Malaquin 1901, Raibaut 1985, Suárez-Morales et al. 2014). These tubes may be altered antennae, adjacent mouth appendages or both. The immature adults, presumably at their copepodid V stage, emerge from the host and undergo a final moult to the mature form which leads to a planktonic life. During emergence, maturing adults discard the feeding tubes leaving specific scars on the ventral surface of the cephalothorax. The adults are non-feeding with a lack of mouthparts and are regarded as solely reproductive (Suárez-Morales 2011).

Monstrilloid parasites can inflict negative effects on the host causing physical damage, structural alteration, castration and inducing strong inflammatory response (Malaquin 1901, Suárez-Morales et al. 2010, 2014). However, it is not clear whether those effects eventually lead the hosts to mortality. In some cases, the host is able to recover quickly from the damage. The damage itself does not appear to be very detrimental to the host (Malaquin 1901, Suárez-Morales et al. 2010). Conversely, in another instance, the damage may cause mortality (Suárez-Morales et al. 2014). In regarding to this, Suárez-Morales (2018) mentioned that the relative size of the host and the intensity of infection are also important factors determining the fate of the hosts.

Adult monstrilloids are, in general, scarce in the marine water columns but are known to occur aggregated in reef-related environments (Sale et al. 1976, Sekiguchi 1982, Suárez-Morales 2001, Grygier & Ohtsuka 2008). Suárez-Morales (2001) also found that the adult monstrilloids appear in a higher density in the water column at twilight suggesting that they remain near the bottom during the day and swim up towards the surface at night.

Diversity and geographical distribution

The review of Suárez-Morales (2011) recognised about 120 species of monstrilloids in four genera (*Monstrilla*, *Cymbasoma*, *Monstrillopsis* and *Maemonstrilla*) and noted that 73% of those were known from the North Atlantic (45% from the Northeast Atlantic including the European waters; 17% from the Northwest Atlantic; 11% from the Mediterranean Sea), whereas less than 10% were from the Asian (8%) and the Australian (~3%) waters. Since then, over 50 species have been described along with the proposal of three new genera in the past decade. In particular, Suárez-Morales and McKinnon (2014, 2016) reported 36 species in four genera (more than 20% of the species known so far), including 33 then-undescribed species and an additional genus, only from the Australian waters. As such, these studies revealed a much higher regional monstrilloid diversity in a less studied area and simultaneously suggested a promising number of new species in other unexplored regions as well (Suárez-Morales 2018).

The order Monstrilloida currently contains over 170 nominal species in a single family Monstrillidae: 51 species of *Monstrilla*, 77 of *Cymbasoma*, 21 of *Monstrillopsis*, 11 of *Maemonstrilla*, ten of *Caromiobenella*, one of *Australomonstrillopsis* and one of *Spinomonstrilla* (Suárez-Morales 2000, Razouls et al. 2005–2022, Walter & Boxshall 2022, WoRMS Editorial Board 2022). With respect to regions, Australian waters have the highest species richness (36 species) followed by the Northwest Atlantic including the Caribbean Sea and the Gulf of Mexico (34), the Northeast Atlantic including the European seas (32), the Mediterranean-Black Sea region (26), the Indonesia-Malaysia-Philippines region including southern China and Japan (i.e., Okinawa) (25), the waters around northern Japan and Korea (19), and the Brazil-Argentine region (15) (Suárez-Morales & Grygier 2021). Africa and the East Pacific regions have very low monstrilloid diversity with only three species known from African waters and 13 from the East Pacific region (Suárez-Morales 2018).

Among 172 nominal species, only 38 of them (22% of total species) are known from both sexes, whereas 134 species (78%) are described based on material for one sex; 49 of from males and 85 from females. Incomplete, sex-biased species determinations have prompted nomenclatural problems: both sexes of the same species were recorded under different names (see Grygier 1994a, Suárez-Morales et al. 2008, Rosa et al. 2021), or, conversely, each sex of different species was identified as one species (see Suárez-Morales et al. 2006). As a result, there are still many uncertain species with taxonomic and nomenclatural problems; thus, the true diversity of the monstrilloids is likely to differ from what we know at present.

Phylogeny

Early phylogenetic investigations into the Copepoda including Monstrilloida were carried out mainly based on their morphological characteristics (Ho 1990, 1994, Huys & Boxshall 1991, Ho et al. 2003, Dahms 2004). These studies generally indicated that the monophyletic Copepoda is divided into two infraclasses, the Progymnoplea (order Platycopioida) and the Neocopepoda. The latter encompasses two superorders the Gymnoplea (order Calanoida) and the Podoplea. In the podoplean phylogeny, Monstrilloida had frequently shown a close relationship with another parasitic group, the Siphonostomatoida (Huys & Boxshall 1991, Ho 1994, Ho et al. 2003). As Huys et al. (2007) pointed out, the previous Siphonostomatoida-Monstrilloida clade (Huys & Boxshall 1991) was formed based on convergent features. Thus, the phylogenetic relationships between the two orders may not indicate a recent common ancestry. In addition to the issue of morphological convergence, inevitable errors arising from the process of the identification of structural homology, determination of character states (e.g., plesiomorphic or apomorphic) and application of different weighting criteria for morphological data have also been potential pitfalls leading to faulty phylogenetic inference (Ho 1994, Eyun 2017).

Most molecular research supports two main Neocopepoda clades (i.e., the superorders Gymnoplea and Podoplea) (Oakley et al. 2013, Eyun 2017, Schwentner et al. 2018, Lozano-Fernandez et al. 2019, Song et al. 2021) and is congruent with results from the previous morphology-based studies (Huys & Boxshall 1991, Ho 1994, Ho et al. 2003). However, some molecular-based studies have produced somewhat different copepod phylogenies (Braga et al. 1999, Schwentner et al. 2017) with respect to the possible subgroups and subsequent separations such as for the Poecilostomatoida (Tung et al. 2014) and the Harpacticoida (Schizas et al. 2015). Therefore, the relationships within the podoplean lineage remain highly controversial. Under these circumstances, the Monstrilloida has frequently been overlooked and its molecular phylogenetic position therefore remains unknown. Huys et al. (2007) suggested that the Monstrilloida appears to be nested within the Siphonostomatoida. Their molecular analyses based on 18S ribosomal RNA sequences (18S rRNA, a total aligned length of 1941 base pairs) revealed a specific molecular affinity to caligiform-family siphonostomatoids and implied the consequent demise of the order Monstrilloida. However, a similar molecular study using the same 18S rRNA data with the inclusion of the Harpacticoida taxa (a total aligned length of 1878 base pairs) indicated an unresolved polytomy among the three orders: the Harpacticoida, Siphonostomatoida and Monstrilloida (Ki et al. 2009). Thus, there is no agreed phylogenetic position for monstrilloids with either morphological or molecular data.

The main purpose of our study was to clarify the unknown phylogenetic position of the Monstrilloida among copepod orders. For this, 25 nuclear protein-coding genes were obtained from the draft genomes and transcriptomes of four monstrilloid species (*Monstrilla grandis* Giesbrecht 1891, *Caromiobenella ohtsukai* Jeon, Lee & Soh 2019, *Monstrillopsis longilobata* Lee, Kim & Chang 2016, and *Maemonstrilla* sp.) (four genera among seven monstrilloid genera). The orthologous sequences from the multiple species of Arthropoda (Hexapoda, Branchiopoda, Cephalocarida, Remipedia, Decapoda, and Thecostraca) including the four major copepod orders (Calanoida, Cyclopoida, Harpacticoida and Siphonostomatoida) were also obtained to reconstruct the phylogenomic tree. The inclusion of 21 copepod species from five groups greatly enhances our ability to make inferences regarding copepod evolution. Our results reaffirmed the previously known monophyly of Copepoda within Arthropoda and revealed the detailed phylogenetic position of the Monstrilloida within the Copepoda. The phylogenomic results detected a particularly long branch leading to the Monstrilloida. The present phylogenies and the sequence divergences also suggested high gene evolutionary rates for this copepod group. Given all present molecular results, we discussed the hypothesis of the evolution of the Monstrilloida by integrating their morphological and ecological characteristics.

Methods

DNA/RNA sample preparation for next-generation sequencing

Genomic DNAs were extracted from *Monstrillopsis longilobata* (83 adults) and *Maemonstrilla* sp. (101) using MagAttract HMW DNA kit (Qiagen, Germany) by following the manufacturer's instructions, and the DNA libraries were constructed using TruSeq DNA Nano kit (insert size of 550 bp; Illumina, USA). Total RNA was extracted from *M. longilobata* (ten adults), *Monstrilla grandis* (15) and *Caromiobenella ohtsukai* (26) using mirVana kit (Thermofisher, USA). Total RNA libraries were constructed using TruSeq Stranded Total RNA LT sample prep kit (insert size of 300 bp; Illumina, USA). The libraries were sequenced using either 101 or 151 bp read lengths on NovaSeq6000 and HiSeq2500 (Illumina, USA) according to the manufacturer's recommended protocol (Supplementary Table 1).

Genomic/transcriptomic data and de novo assembly

Publicly available genome and transcriptome assemblies for 11 species were downloaded from the National Center for Biotechnology Information (NCBI) including two copepod genomes (*Eurytemora affinis* (Eyun et al. 2017) and *Tigriopus californicus* (Barreto et al. 2018)) (Supplementary Table 2). Additionally, next-generation sequencing data for 24 arthropods (Supplementary Table 3) were also obtained from the NCBI Sequence Read Archive (SRA, https://www.ncbi.nlm.nih.gov/sra). The assembly pipeline in this work is discussed in our previous papers (Jung et al. 2020a,b). The quality control and trimming of the SRA data and the newly obtained sequencing reads were performed using `fastp` (ver. 0.20.1) (Chen et al. 2018). After read filtering process with the stringent criteria (with PHRED score Q28 or more and minimum length requirement of 75 bp for transcriptomic reads and of 125 bp for genomic reads), *de novo* assemblies were generated using Trinity (ver. 2.10.0) (Grabherr et al. 2011) and SPAdes (ver. 3.15.2) (Bushmanova et al. 2019): the transcriptomic assemblies for three monstrilloids, *Monstrilla grandis*, *Caromiobenella ohtsukai* and *Monstrillopsis longilobata*, were conducted using Trinity with the default option, and genome assemblies for *M. longilobata* and *Maemonstrilla* sp. were performed using SPAdes with the option "`--isolate`". Quality of the assemblies was estimated using BUSCO (ver. 5.3.2) (Manni et al. 2021) with the Arthropoda lineage dataset (`arthropoda _ odb10` as of September 10, 2020) containing a total of 1013 marker genes.

Gene mining, identification of orthologous genes and data preparation

Twenty-five nuclear protein-coding genes were initially obtained from the GenBank protein sequence datasets for 11 arthropods in the previous studies (Regier et al. 2008, Eyun 2017). With the acquired gene sequences as the queries, the orthologous sequences were searched using `tblastn` (Basic Local Alignment Search Tool; BLAST, ver. 2.10.0+) (Camacho et al. 2009) against the newly assembled genomes and transcriptomes for 30 arthropod species. The coding regions were predicted using `GeneWise` (ver. 2.2) (Birney et al. 2004) to determine the open reading frames and the intron/exon boundaries. The translated amino acid sequences were inspected using SMART (Letunic et al. 2021). All 25 protein-coding gene sequences from 40 arthropod species (Supplementary Table 4) were collected and verified again by `blastp` against the NCBI non-redundant (NR) protein database.

Three datasets were arranged by different taxon sampling. The arthropod dataset (AD) contained 40 arthropod species including two non-pancrustacean taxa, the blacklegged tick *Ixodes scapularis* (Chelicerata) and centipede *Strigamia maritima* (Myriapoda). The second and third sub-datasets consisted of only the sequences of copepods with the malacostracan and thecostracan taxa as outgroups. The difference between the datasets was an inclusion or exclusion of Harpacticoida sequences; Copepoda dataset with Harpacticoida (CD) and Copepoda dataset without Harpacticoida (CDwoH). Thus, the number of taxa is 27 in CD and 24 in CDwoH. The purpose of the sub-datasets is to reconstruct the copepod phylogenies as similar to the previous molecular results in the aspect of the number of orders. We examined the robustness of our phylogeny under various taxon-sampling conditions and further compared it more directly with previous molecular phylogenies (Ho 1990, 1994, Huys & Boxshall 1991, Ho et al. 2003, Huys et al. 2007, Ki et al. 2009), regardless of taxon bias. The amino acid sequences used in this study are available in Additional File 1.

Phylogenetic analysis

Multiple sequence alignments for the 25 nuclear protein-coding genes in three different datasets (AD, CD, and CDwoH) were individually generated using MAFFT (ver. 7.475) (Katoh & Standley 2013) with the L-INS-i algorithm. The poorly aligned or many gap regions of the alignments were removed using `trimAl` (ver. 1.4.rev15) (Capella-Gutiérrez et al. 2009) with the option "`--automated1`" for further phylogenetic tree reconstruction. Best-fit phylogenetic model searches for each gene set were performed using IQ-TREE2 (ver. 2.1.2) (Minh et al. 2020) using Bayesian Information Criterion (BIC). The target models were trained only to the nuclear protein substitution models using the option "`--msub nuclear`" because all present genes have nuclear origins. The aligned protein sequences in each dataset were then concatenated into a single alignment using a custom-written Perl script (Eyun 2017) (Additional File 2). The maximum-likelihood (ML) and Bayesian inference (BI) analyses were performed by applying the best substitution models to the gene partitions in each dataset (Supplementary Tables 5–7). The ML tree was reconstructed using RAxML-NG (ver. 1.0.1) (Kozlov et al. 2019). The best tree with the highest log-likelihood score was selected among the initial 30 random and 30 parsimony trees. Bootstrap support values ($BP_{RAxML\text{-}NG}$, in percentage) were calculated with 3000 bootstrap replicates and provided with the transfer bootstrap expectation values (Lemoine et al. 2018). BI analysis was run for 3×10^6 generations, sampling every 100 generations using MrBayes (ver. 3.2.7a) (Ronquist et al. 2012). The model parameters across the partitions were unlinked to reflect probable different evolutionary rates by genes. The first 25% of the generations were discarded for the final tree reconstruction and the nodal support values ($PP_{MrBayes}$) were provided with the Bayesian posterior probability. An additional ML phylogenetic tree based on the AD was inferred by employing the protein mixture models (C20-profile mixture model) (Le et al. 2008). The best-fit model selection and subsequent best partitioning scheme determination (Supplementary Table 8) were performed in IQ-TREE2. The phylogenetic tree was reconstructed using the ultrafast bootstrapping algorithm implemented in IQ-TREE2, and bootstrap support values ($BP_{IQ\text{-}TREE2}$, in percentage) were calculated with 3000 bootstrap replicates. Graphical presentation of the phylogenies was performed using FigTree (ver. 1.4.4) (http://tree.bio.ed.ac.uk/software/figtree).

Alternative tree topology test

Statistical tree evaluations and confidence tests for the present phylogenies and those from studies by Ho (1990), Huys and Boxshall (1991) and Huys et al. (2007) were performed. An additional tree topology from Khodami et al. (2017) was also tested for support for one of our phylogenetic hypotheses, despite recent controversy (Mikhailov & Ivanenko 2021). The tree topologies were manually edited (Additional File 3) and categorised into two groups by the number of copepod taxa involved; three topologies consisting of five copepod orders (i.e., Calanoida, Cyclopoida, Harpacticoida, Siphonostomatoida and Monstrilloid) taken from other studies (Ho 1990, Huys & Boxshall 1991, Khodami et al. 2017) and one topology (Huys et al. 2007) consisting of only four copepod orders lacking Harpacticoida. Each of these groups was compared to the two phylogenies based on CD and CDwoH of each given our 25 gene alignment. For the statistical assessment of the log-likelihood of each tree, the Kishino-Hasegawa (KH) (Kishino & Hasegawa 1989), Shimodaira-Hasegawa (SH) (Shimodaira & Hasegawa 1999) and Approximately Unbiased (AU) (Shimodaira 2002) tests were performed with 10,000 re-samplings using the RELL method in IQ-TREE2.

Sequence divergence calculation and divergence time estimation

Within-group sequence divergences for copepod orders were calculated based on the present protein and 18S rRNA sequences to detect possible high genetic mutation rates. The calculations were performed under various model parameters. For 18S rRNA sequences (see Supplemental analysis), the *p*-distance method and Kimura 2-parameter (Kimura 1980) model were used with 3000 bootstrap replicates. Similarly, the Dayhoff (Dayhoff et al. 1978) and Jones-Taylor-Thornton (JTT) (Jones et al. 1992) models including the *p*-distance method were applied for analysing protein sequence divergence. Additional attempts calculating overall mean divergence and within-group divergences at various taxonomic ranks were performed to explore the possibly different substitution rates among the arthropod groups involved in this study. This was done based only on the protein sequences under the JTT model. Divergence of each group was compared to overall mean divergence, and estimated rates were applied to the following divergence time estimation.

The divergence time of lineages was estimated using LSD2 (ver. 1.10) (To et al. 2016) with the least-squares method. The phylogenetic tree generated through ML analysis was provided as the user-defined topology. According to Wolfe et al. (2016), the minimum and maximum ages for four calibration points were applied; 497–636.1 Ma (million-year ago, in order of the minimum and maximum age boundaries) for the Branchiopoda-Hexapoda split, 405–521 Ma for the Anostraca-Cladocera split, 358.5–521 Ma for the Amphipoda-Decapoda split, and 313.7–411 Ma for the Diptera-Coleoptera split. The minimum constraint for the crown Pancrustacea was set to 525 Ma based on the fossil record of *Wujicaris muelleri* (Zhang et al. 2010). Two non-pancrustacean taxa were used as outgroups and removed from the final divergence time tree.

Results and Discussion

Genome and transcriptome assembly quality assessment

Three transcriptomes (*Monstrilla grandis*, *Caromiobenella ohtsukai* and *Monstrillopsis longilobata*) and two genomes (*Maemonstrilla* sp. and *M. longilobata*) were obtained using Illumina paired-end sequencing technology. The detailed information and statistical measurements for the sequencing and assembly results are summarised in Supplementary Table 1. The percentage of complete BUSCO gene coverage of the transcriptome assemblies ranged from 57.1% for *C. ohtsukai* to 84.4% for *M. grandis*, and those in the genome assemblies showed 45.6% for *M. longilobata* to 58.4% for *Maemonstrilla* sp. (Supplementary Figure 1). Although these assemblies generated did not contain the entire gene contents of the monstrilloids, we found many marker genes (Supplementary Table 4). We continued downstream phylogenetic analyses for the sake of providing any evidence for revealing the phylogenetic relationships of the Monstrilloida within the Copepoda.

Copepoda phylogeny within Arthropoda

Twenty-five nuclear protein-coding genes were obtained from 40 arthropod species including 21 copepods in five orders. Three concatenated sequence data (AD, CD and CDwoH) were generated, and each dataset consisted of 16,976, 17,436 and 17,478 amino acids. The ML and BI trees reconstructed based on AD showed four major splits for the pancrustacean group (Figure 3);

Hexapoda-Branchiopoda clade (HB clade) having 100/1.00/52 nodal support values (in order of $BP_{RAxML\text{-}NG}/PP_{MrBayes}/BP_{IQ\text{-}TREE2}$ analyses, see Materials for the abbreviations), Xenocarida (Remipedia-Cephalocarida clade) (Regier et al. 2010) with 65.5/1.00/100, Communostraca (Malacostraca-Thecostraca clade) (Regier et al. 2010) and Copepoda with 76.7/1.00/54. Our phylogenetic analyses at the Arthropoda level resulted in different tree topologies (RAxML-NG, MrBayes vs. IQ-TREE2 analyses). The HB clade took the most basal phylogenetic position among the given pancrustaceans in the RAxML-NG and MrBayes trees (Figure 3A), while the Xenocarida appeared as the most basal group in the IQ-TREE2 tree (the HB clade appeared as the second most basal group by following the Xenocarida; Figure 3B). The validity of the Allotriocarida clade (Oakley et al. 2013) has gained attention in recent studies (von Reumont et al. 2012, Oakley et al. 2013, Eyun 2017, Schwentner et al. 2017, 2018, Lozano-Fernandez et al. 2019). Our phylogenies failed to recover this clade. Therefore, questions on the validity of the Xenocarida or Allotriocarida remain unanswered. Nonetheless, our phylogenies still provided stable and consistent relationships within the class-level phylogeny with higher nodal support and without any unexpected inclusion of alien species into another taxon group. As the resolution of detailed relationships within the Arthropoda is beyond the scope of this study, thus the discussion should be focused on the phylogeny of the Copepoda.

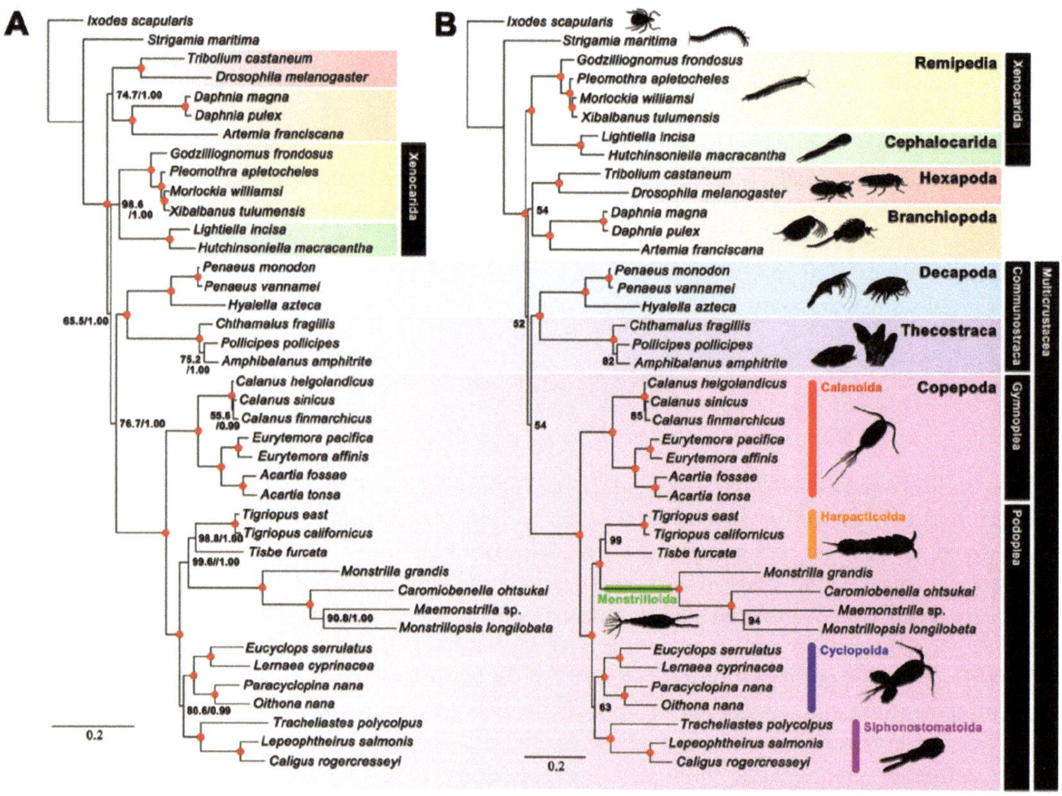

Figure 3 Phylogenies of 21 copepod and 19 other arthropod species inferred from 25 nuclear protein-coding genes. (A) Tree inferred using RAxML-NG and MrBayes. The numbers near the branching points indicate the maximum-likelihood bootstrap support values (BS, in percentage) and Bayesian posterior probabilities (PP, in probability) in order of BS/PP. (B) Tree inferred using IQ-TREE2 with the protein mixture model, C20. The numbers near the branching points indicate BS values. The red dots at the nodes indicate that BS and PP are 100% and 1.00, respectively. Non-pancrustacean species (*Ixodes scapularis*) is used as the outgroup. The clade names (Xenocarida, Communostraca, and Multicrustacea) are adopted from Regier et al. (2010).

Table 1 Statistical Comparisons among the Best Maximum-Likelihood Trees and Alternative Phylogenetic Hypotheses within the Copepoda Orders

Tree topology[a]	logL	ΔL[b]	p-KH[c]	p-SH[d]	p-AU[e]	References
(CAL,((HAR,MON),(CYC,SIP)))	−258924.0581	Best	1	1	1	This study
(CAL,(HAR,(MON,(CYC,SIP))))	−259028.8959	104.84	0	<0.001	<0.001	Ho (1990)
(CAL,(CYC,(HAR,(SIP,MON))))	−259059.0093	134.95	0	0	<0.001	Huys and Boxshall (1991)
(CAL,((HAR,CYC),(MON,SIP)))	−259039.0141	114.96	<0.001	<0.001	<0.001	Khodami et al. (2017)
(CAL,(MON,(CYC,SIP)))	−243919.5916	Best	1	1	1	This study
(CAL,(CYC,(nc-SIP,(MON,c-SIP))))	−244220.1269	300.54	0	0	<0.001	Huys et al. (2007)

[a] CAL = Calanoida, CYC = Cyclopoida, HAR = Harpacticoida, SIP = Siphonostomatoida, nc-SIP = non-caligiform siphonostomatoid (*Tracheliastes polycolpus*), c-SIP = caligiform siphonostomatoids (*Lepeophtheirus salmonis* and *Caligus rogercresseyi*), MON = Monstrilloida.

[b] ΔL: log L difference from the maximal log L in the set.

[c] *p*-KH: *p*-value of the one-sided Kishino-Hasegawa test (Kishino & Hasegawa 1989).

[d] *p*-SH: *p*-value of the Shimodaira-Hasegawa test (Shimodaira & Hasegawa 1999).

[e] *p*-AU: *p*-value of the approximately unbiased (AU) test (Shimodaira 2002).

Within the Copepoda, two superorder divisions (Gymnoplea and Podoplea) were recovered ($BS_{RAxML\text{-}NG}=100\%$, $PP_{MrBayes}=1.00$, $BS_{IQ\text{-}TREE2}=100\%$) in congruence with previous morphological and molecular studies (Huys & Boxshall 1991, Ho 1994, Ho et al. 2003, Huys et al. 2006, 2007, Oakley et al. 2013, Tung et al. 2014, Schizas et al. 2015, Eyun 2017, Schwentner et al. 2018, Lozano-Fernandez et al. 2019). However, the interrelationships within the Podoplea showed unusual topologies involving groupings of Cyclopoida-Siphonostomatoida and Harpacticoida-Monstrilloida with strong nodal support ($BS_{RAxML\text{-}NG}=100\%$, $PP_{MrBayes}=1.00$, $BS_{IQ\text{-}TREE2}=100\%$). The multicrustacean sub-phylogenies based on CD and CDwoH also showed the same subsets of the present arthropod phylogeny (Supplementary Figures 2 and 3). Notably, none of these results provide evidence for the inclusion of the Monstrilloida within the Siphonostomatoida or a close affinity between them as in other studies (Huys & Boxshall 1991, Ho 1994, Ho et al. 2003, Huys et al. 2007). The present results were inferred based on a limited number of the five copepod orders with the inclusion of three species from the Siphonostomatoida. Given the limited availability of the genomic and transcriptomic data from copepods, we also conducted a larger phylogenetic analysis using the 18S rRNA sequences from the previous molecular phylogenetic studies (Huys et al. 2006, 2007, 2009, Khodami et al. 2017, Vakati et al. 2019) to test for other possible relationships (see Supplemental analysis). We used 129 species in our analysis from nine copepod orders and 98 species from five orders (Supplementary Figures 4 and 5). Our 18S rRNA-based analyses yielded unstable phylogenetic relationships among the copepods with a general lack of support ($BS_{RAxML\text{-}NG}>43\%$, $PP_{MrBayes}>0.13$) (Supplementary Figure 4). All the 18S rRNA-based results including those from the previous studies have shown a different tree topology to each other. Furthermore, the statistical assessments on the alternative tree topology tests including the present phylogeny and the others from the previous studies (Ho 1990, Huys & Boxshall 1991, Huys et al. 2007, Khodami et al. 2017) showed that our copepod phylogenetic hypothesis as the most likely one, while rejecting the other hypotheses at the 0.001 significance level (Table 1). The choice of suitable sequence data with sufficient phylogenetic signal and securing taxon-sampling from wider taxa are probably most important for successful phylogenetic inference (Rosenberg & Kumar 2001, 2003, Blanco-Bercial et al. 2011, Eyun 2017, Mikhailov & Ivanenko 2021). In this respect, the analyses based on partial or short single gene sequences dealing with a wide range of taxa might have a limited power to reveal true relationships.

Our phylogenomic results suggest that the Monstrilloida is a monophyletic group with a closer affinity to the Harpacticoida than the Siphonostomatoida. This contrasts with morphology-based copepod phylogenies (Ho 1990, 1994, Huys & Boxshall 1991, Ho et al. 2003). In general, previous morphology-based phylogenies have formed an extended parasitic (or parasitic taxon-rich) clade involving not only the Siphonostomatoida and the Monstrilloida but also the Poecilostomatoida and the Thaumatopsylloida. However, the latter two orders were found to be the subgroups of Cyclopoida (Boxshall & Halsey 2004, Huys et al. 2006, 2012, Khodami et al. 2019). Thus, the parasitic copepod clustering in preceding morphology-based phylogenies does not reflect their true evolutionary relationships. Convergent evolution rules the appearance of similar phenotypic or analogous structures among phylogenetically distant species due to adaptive responses to similar environments or ecological niches (Huang et al. 2015, Speed & Arbuckle 2016, Castiglione et al. 2019). Thus, the previous morphology-based phylogenies which include convergent features (Ho 1994, Huys et al. 2007, Eyun 2017) show other relationships. Our phylogenetic results based on genetic data provide a different perhaps more objective view and suggest that the Monstrilloida has a different stem lineage from the Siphonostomatoida and a closer affinity to the Harpacticoida.

The relationship between the Cyclopoida and the Siphonostomatoida needs more research attention. It may be worth noting that the aesthetasc arising from the ancestral antennular segment XXI is typically stable within the Siphonostomatoida, even in the highly transformed genus *Spongiocnizon* (Huys & Boxshall 1991). This aesthetasc is also present in the Misophrioida and the Cyclopoida (including Poecilostomatoida) but not in the Harpacticoida and the Monstrilloida (Huys & Boxshall 1991). The loss of the key features in the Monstrilloida may be another piece of evidence to reject the close phylogenetic relationship with the Siphonostomatoida, supporting formations of two clades, the Cyclopoida-Siphonostomatoida and the Harpacticoida-Monstrilloida, in congruence with our molecular results.

Congruence between morphological and molecular phylogenies for Monstrilloida

This study provides the phylogenomic evidence for the relationships among four monstrilloid genera (*Monstrilla*, *Monstrillopsis*, *Maemonstrilla* and *Caromiobenella*) among seven valid genera (additionally, *Cymbasoma*, *Australomonstrillopsis* and *Spinomonstrilla*) and suggests that the last common ancestor (based on the phylogenetic tree) is *Monstrilla* and *Monstrillopsis* has emerged recently in evolution. Then, as a follow-up question, can the present molecular relationships explain the morphological evolutionary pattern or is there a mutual agreement between the morphological and molecular phylogenies? It is known that the copepod evolution is primarily proceeded by structural oligomerisation (i.e., a loss or fusion in body parts) and reappearance of ancestral (complex) character state from once derived (simplified) state in descendants is an extremely rare case (Huys & Boxshall 1991). These findings imply that a species or group retaining more intricate plesiomorphic features is morphologically closer to their last common ancestor.

The genus *Monstrilla*, particularly *Monstrilla grandis* involved in this study, exhibits the most complex morphological character combination including the largest number of body somites (eight in the female and nine in the male) and caudal setae (six in both sexes), a pair of a bilobed female fifth leg with a maximum of five setae (three on the outer and two on the inner lobes; i.e., 3-2 setal pattern; Figures 4 and 5A) and the presence of a pair of fifth legs (each a small knob-like protuberance with a long seta) even in males. In general, the species of *Monstrilla* have five or six caudal setae according to the species or sexes and the males usually do not have fifth legs (Suárez-Morales 2011). Most females also share the same bilobed fifth pair of legs, but only have a single seta on the inner lobe (3-1; Figure 5B). In this respect, *M. grandis* is one of the most plesiomorphic character-rich species either among the known congeners or also within the entire family Monstrillidae. On the other hand, *Monstrillopsis longilobata* has reduced structures of these characters: a presence of four caudal setae in both sexes, a fifth pair of legs in females with the inner lobe reduced and unarmed (3-0; Figure 5C), and an absence of the fifth pair of legs in males (Suárez-Morales et al.

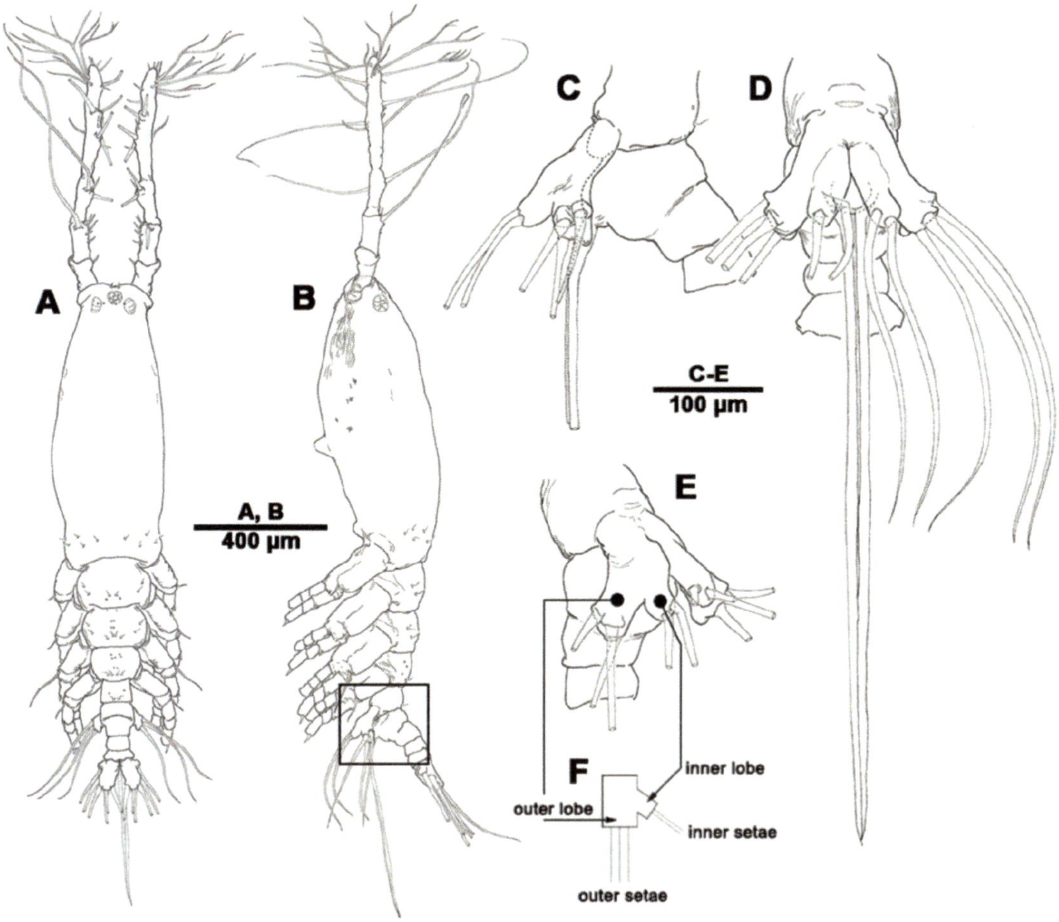

Figure 4 *Monstrilla grandis*, adult female. (A) habitus, dorsal view. (B) habitus, lateral view. Urosomal part bearing the fifth leg highlighted in the box. (C) Detail of urosomal part in B, lateral view. (D) Urosome showing a pair of fifth legs, ventral view. (E) First urosomal somite bearing a pair of fifth legs, lateroventral view. Ovigerous spines not presented. (F) Diagrammatic representation of fifth leg (cf. Figure 5).

2006, Suárez-Morales 2011). Thus, the genus *Monstrillopsis* may be a later appeared group than *Monstrilla*.

The genus *Maemonstrilla*, currently consists of two distinct subgroups, the *Maemonstrilla turgida* and *Maemonstrilla hyottoko* species groups, has only been known from the females (Grygier & Ohtsuka 2008, Suárez-Morales & McKinnon 2014). They show morphological similarities to females *Monstrillopsis* in that they have four urosomites, well-developed eyes and an anteriorly located oral papilla (Grygier & Ohtsuka 2008). The *M. turgida* species group have similarly bilobed fifth legs of *Monstrillopsis*, but simultaneously show an increase of setal elements with an additional seta on the inner lobe (3-1; Figure 5B). By contrast, the *M. hyottoko* species group which includes the present *Maemonstrilla* sp. has a derived state of the fifth leg that is reduced to an elongate rod-shape with two apical setae (Figure 5D). In addition, this group can be distinguished from the *M. turgida* group by the absence of an inner seta on each first exopodal and endopodal segment of swimming legs 1–4 (Grygier & Ohtsuka 2008). Considering the oligomerisation rule, these simplified structures could suggest that *M. hyottoko* species group might be considered to be the most recently diverged group within the Monstrilloida. However, both species

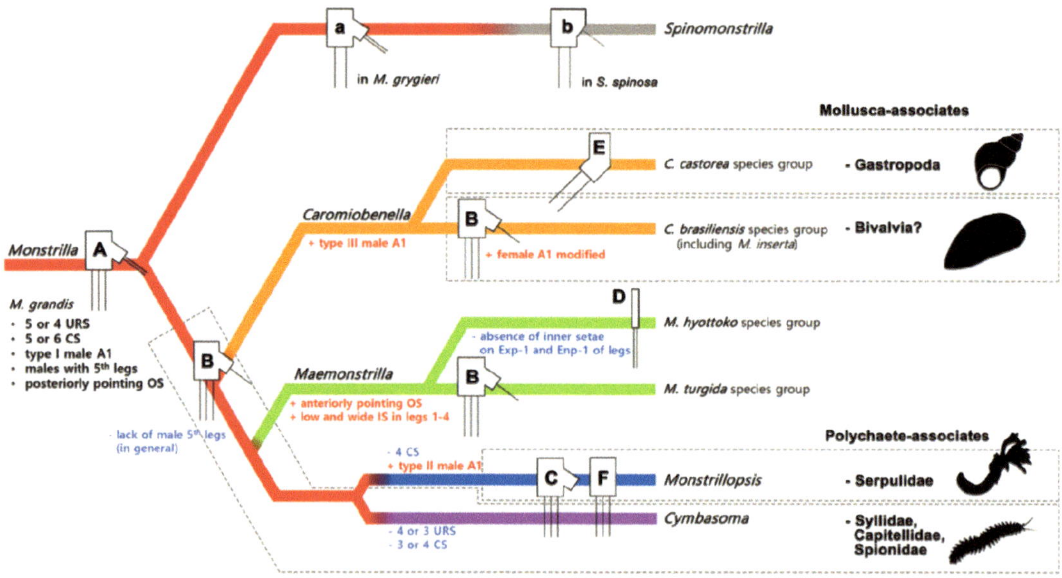

Figure 5 Schematic diagram of morphological divergence of the female fifth leg. The acquisition of new features (in red with '+' marks) and morphological oligomerisation (in blue with '-' marks) are shown along the branches. The most plesiomorphic characters from *Monstrilla grandis* were provided (in black). The host groups utilised by the monstrilloid genera were presented in dashed boxes mapping on the appropriate branches. (A) bilobed leg with 3 outer and 2 inner setae in *M. grandis*; (B) bilobed leg with 3 outer and 1 inner seta, in the majority of *Monstrilla*, *Maemonstrilla turgida* species-group and *Caromiobenella brasiliensis* species-group; (C) bilobed leg with 3 outer setae, inner lobe unarmed, in the majority of *Cymbasoma* and *Monstrillopsis*; (D) a long, rod-shaped leg with 2 apical setae, in the *Maemonstrilla hyottoko* species-group; (E), unilobed leg bent outward at half, in *Caromiobenella* species; (F) unilobed leg with 3 setae, inner lobe almost rudimentary or absent, in some *Cymbasoma* (e.g., *C. bowmani*) (Suárez-Morales & Gasca 1998) and *Monstrillopsis* (e.g., *M. planifrons*) (Delaforge et al. 2017); (a) bilobed leg with 2 outer and 2 inner setae in *Monstrilla grygieri* (Suárez-Morales 2000); (b), unilobed leg with 2 apical and 1 inner seta in *Spinomonstrilla spinosa* (Park 1967, Suárez-Morales 2019). Females of *Australomonstrillopsis* have not been reported. Branch colours of red, orange, green, blue, purple, and grey indicate the lineage of the genus *Monstrilla*, *Caromiobenella*, *Maemonstrilla*, *Monstrillopsis*, *Cymbasoma*, and *Spinomonstrilla*, respectively. Abbreviations used: URS, urosomal somites; CS, caudal setae; A1, antennules; OS, ovigerous spines; IS, intercoxal sclerites; Exp-1, first exopodal segment of leg; Enp-1, first endopodal segment of leg.

groups retain the ancestral plesiomorphic character in the caudal rami with up to six caudal setae (Grygier & Ohtsuka 2008, Suárez-Morales & McKinnon 2014). Thus, we conclude that *Maemonstrilla* is likely to be a more plesiomorphic group than *Monstrillopsis*. If this holds true, the most simplified structures inherited by the *M. hyottoko* species group are perhaps the apomorphic characters that appeared after the divergence of two *Maemonstrilla* species groups. *M. longilobata* females have mixed *Monstrillopsis* and *Maemonstrilla* morphological characteristics (Jeon et al. 2018b). This morphologically intermediate and transitional form contributes to the particularly close relationship between *Maemonstrilla* and *Monstrillopsis*.

The genus *Caromiobenella* had long been regarded as a subgroup within *Monstrilla* (Sars 1921, Huys & Boxshall 1991, Grygier & Ohtsuka 1995, Suárez-Morales et al. 2008, Jeon et al. 2018a). They are morphologically similar by sharing several ancestral plesiomorphic conditions in the aspect of the number of urosomal somites and caudal setae. Thus, it could also be considered one of the earlier clades of the Monstrilloida but might have appeared later than *M. grandis* in terms of morphological complexity. However, a distinct apomorphic character combination, each showing the structural polarity of the simpleness (e.g., a single lamella-form female fifth leg with two apical setae (Figure 5E) and the replacement of the antennular branched setae into unbranched simple ones) and complexity (e.g., the most complicate male antennular modification with its last segments bearing an additional five serrate ridges on the inner distal margins; type III antennular modification) (Jeon et al. 2018a) had raised questions about their phylogenetic position. Recently, Rosa et al. (2021) reassigned *Monstrilla brasiliensis* into the genus *Caromiobenella* based on the morphological and molecular evidence (thus now *C. brasiliensis*). This taxonomic revision consequently expands the morphological diversity of the female *Caromiobenella* to include the presence of two prominent sensilla-like structures between the antennular bases, a peculiar round protuberance on the putative third antennular segment and a bilobed fifth leg with 3-1 setal pattern (Figure 5B). Based on the morphology of the fifth pair of legs in females in *C. brasiliensis* which is expressed in a more plesiomorphic state than those of *Monstrillopsis*, *Caromiobenella* is more primitive than *Monstrillopsis*. Given our conclusions that *Caromiobenella* and *Maemonstrilla,* respectively, have a closer affinity with *Monstrilla* (primitive character-rich taxon) and *Monstrillopsis* (derived character-rich taxon), the ancestor of *Caromiobenella* might have diverged earlier than that of *Maemonstrilla*. In conclusion, the morphological evidence depicts the relationships of (*Monstrilla*, (*Caromiobenella*, (*Maemonstrilla*, *Monstrillopsis*))), which is the same topology as our phylogenomic inferences.

Another monstrilloid genus, *Cymbasoma* is characterised by the most reduced body structures such as the least number of body somites (one less than the others in each sex) and of caudal setae (exclusively three in females, whereas males have three or four setae according to species but the maximum number of caudal seta never exceeds those of *Monstrillopsis*) (Suárez-Morales & McKinnon 2016). Moreover, the similar morphological repertoires of the fifth pair of legs in females are shared with *Monstrillopsis* (Figure 5C and F) supports their closest relationship. We did not evaluate molecular data for *Cymbasoma*, and so their molecular phylogenetic position within the Monstrilloida remains unknown. However, morphological evidence showing a general agreement with the present molecular results indicates that *Cymbasoma* is the most recent genus in the Monstrilloida with the phylogenetic relationships of (*Monstrilla*, (*Caromiobenella*, (*Maemonstrilla*, (*Monstrillopsis*, *Cymbasoma*)))).

Divergence time estimation in copepod lineages and insight into monstrilloid evolution

The divergence of Copepoda lineage (Figure 6) is estimated to have occurred 494.2 Ma (476.4–500.2 Ma; 95% confidence intervals) and further divergence into two Neocopepoda groups (Gymnoplea and Podoplea) around 433.1 Ma (372.3–454.3 Ma). This suggests that the most recent common ancestor of Copepoda might have originated between the late Cambrian and the middle Silurian. However, the minimum age for the most recent common ancestor may be earlier when considering the existence of the most primitive copepod taxon, Platycopioida (not present in this study). The podoplean lineage was diverged into several ancestral groups each leading to the present Cyclopoida, Siphonostomatoida, Harpacticoida, and Monstrilloida during the Devonian (390.3–407.1 Ma).

The potential of copepod fossilisation is considerably low because of their tiny body size and unsclerotised cuticle, and thus, few body fossil records are available from limited copepod orders (i.e., Harpacticoida, Cyclopoida, and Siphonostomatoida) (Huys et al. 2016). The first fossil records were harpacticoids (Canthocamptidae: *Cletocamptus* sp.) and unidentified cyclopoids found from the Miocene Barstow Formation (13.4–19.3 Ma) in Southern California, USA (Palmer 1960). A fossil of the siphonostomatoid (Dichelesthiidae: *Kabatarina patersoni*) was collected from the gills of a fossilised fish in the Santana Formation (110–120 Ma), Brazil (Cressey & Patterson 1973, Cressey & Boxshall 1989). The fossil of a freshwater copepod, presumably assignable to the extant harpacticoid family Canthocamptidae, was found in carboniferous bitumen (Selden et al. 2010). This latter

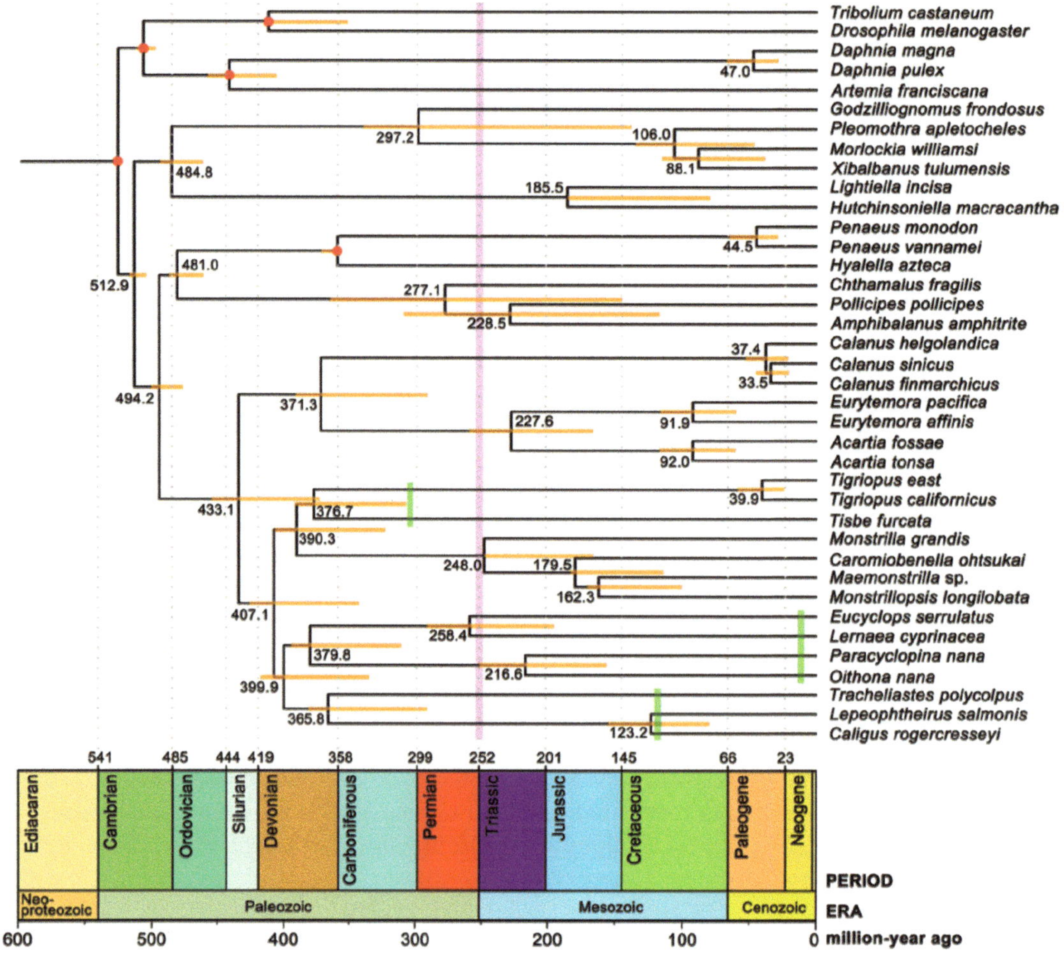

Figure 6 Estimated divergence times for arthropod species inferred using LSD2 (ver. 1.10) (To et al. 2016) with five non-copepod calibration points indicated by red dots. Orange horizontal bars near nodes indicate 95% confidence intervals and green vertical bars indicate the fossil records of Copepoda. Purple vertical bar indicates the Permian-Triassic extinction. The geologic time scale is prepared according to the International Chronostratigraphic Chart (http://www.stratigraphy.org).

find significantly extends the history of Harpacticoida back to the late Carboniferous (ca. 303 Ma). Although our divergence times were estimated without copepod fossil calibrations, the known fossil record so far fit well with our divergence time estimations. In addition, previously suggested divergence times for the copepod orders (Selden et al. 2010, Eyun 2017) are also in general agreement with our results. Boxshall and Jaume (2000) estimated that the cyclopoids probably invaded freshwaters before the breakup of Pangaea (ca. 200 Ma) (Dietz & Holden 1970, Yoshida & Hamano 2015) based on their patterns of current distribution through all continents. The suggested divergence time of 258.4 Ma for a freshwater cyclopoid *Eucyclops serrulatus* in this study also supports this hypothesis.

According to our divergence time estimates, the within-order diversifications for the Calanoida, Cyclopoida, Siphonostomatoida, and the Harpacticoida occurred in the Devonian (365.8–379.8 Ma), but it did not happen for the Monstrilloida until 248.0 Ma (165.9–248.0 Ma). The divergence time suggested for the Monstrilloida indicates that their diversification proceeded after the

Permian-Triassic mass extinction (252 Ma), which is recognised as the greatest natural disaster in magnitude and impacts (Chen & Benton 2012, Burgess et al. 2014, Stanley 2016) (Figure 6). Unfortunately, our current knowledge is insufficient to understand the relevance of the extinction event for monstrilloid evolution. We can only suggest an evolutionary hypothesis based on limited clues. Among the seven valid monstrilloid genera, at least three have been known to utilise the polychaetes as their hosts (Malaquin 1901, Caullery & Mesnil 1914, Hartman 1961, Huys & Boxshall 1991, Suárez-Morales et al. 2014). These monstrilloid genera include not only the most basal, primitive group (i.e., *Monstrilla*) but also the most recently evolved groups (*i.e.*, *Monstrillopsis* and *Cymbasoma*). Thus, extant monstrilloids may have initially stemmed from ancestors that parasitised polychaetes (Figure 5). Some marine animals, particularly polychaetes, managed to survive the mass extinction, while other marine invertebrates such as corals, brachiopods, gastropods and bivalves were severely impacted (Sepkoski 1981, Knoll et al. 2007). Monstrilloid survival relied on available parasitic hosts. Therefore, mass extinction potentially eradicated several ancestral monstrilloids that relied on parasitising other marine invertebrates rather than polychaetes due to drastic environmental changes and rapid host resource depletion. This may explain the long, seemingly trimmed branch pattern for the monstrilloid clade. The present phylogenomic results showed that the genus *Caromiobenella* evolved from the same ancestor that was associated with polychaetes. As this genus also parasitises molluscs (Pelseneer 1914, Gallien 1934), they may have undergone an independent host switch during their evolution (Figure 5). Overall bivalve and gastropod diversity prominently increased in the Mesozoic (Valentine 1969, Sepkoski 1984, Benton 1995). This pattern may also explain the later divergence of the genus *Caromiobenella* within the Monstrilloida. Coincidently, the suggested divergence time of 195.08 Ma (Lee et al. 2019) for the bivalve genus *Perna* (which is considered a potential host for *C. brasiliensis*) (Rosa et al. 2021) is congruent with the subsequent appearance of *Caromiobenella* around 179.5 Ma (113.9–182.8 Ma).

Although we confirmed monstrilloid phylogeny and parasitism pattern correlations, our comprehension is still limited by insufficient knowledge of monstrilloid hosts. Further research on endoparasitic juveniles is necessary to enhance our monstrilloid evolution understanding.

Knowledge gaps and future research priorities

In recent decades, the records of monstrilloid species have increased rapidly (Suárez-Morales 2011). On the other hand, their taxonomy based on a limited number of morphological features from adult specimens seems unable to provide sufficient criteria for their identification and classification (Suárez-Morales et al. 2017). Vague generic and specific diagnoses have caused taxonomic uncertainties along with the increase of new species. Therefore, alternative research methods in addition to traditional morphology are needed. In this context, the use of molecular tools is suggested as one of the most efficient and definitive methods for revealing their true diversity, taxonomy and phylogeny (Suárez-Morales 2011, Jeon et al. 2018b). Unfortunately, the application of molecular methods in monstrilloid research is rare.

Monstrilloida's morphological information is limited to planktonic adults, and their highly modified body structures make a direct comparative analysis with ordinary copepods for taxonomic and phylogenetic inferences almost impossible. However, their planktonic juveniles at their earliest developmental stage (i.e., nauplius stage I), which present the fundamental body plan and structures of the Copepoda (Dahms 1990, 2004, Grygier & Ohtsuka 1995, 2008), could provide additional information useful for resolving more detailed phylogenetic relationships within Monstrilloida but also among Copepoda.

As the present study shows, it would be much better to consider multilateral data together rather than a piecemeal approach to understand the monstrilloids. Particularly, research is needed to determine host diversity and the interaction between the hosts and parasites (e.g., host specificity), which are poorly known.

Table 2 Sequence Divergence within Orders Calculated Using 18S rRNA and 25 Nuclear Protein Sequences Using the *p*-Distance Method and Various Models

	Sequence Divergence				
	18S rRNA		25 Nuclear Proteins		
Copepoda Order	*p*-Distance	K2P[a]	*p*-Distance	Dayhoff	JTT[b]
Calanoida	0.084	0.093	0.104	0.114	0.117
Cyclopoida	0.080	0.087	0.134	0.148	0.152
Siphonostomatoida	0.070	0.076	0.143	0.161	0.165
Harpacticoida	0.074	0.080	0.104	0.115	0.117
Monstrilloida	0.095	0.105	0.223	0.271	0.275

[a] Kimura 2-parameter model.
[b] Jones-Taylor-Thornton model.

There are several studies into the host-parasite coevolution demonstrating that parasites have an accelerated gene mutation rate compared with their hosts (Hafner et al. 1994, Page et al. 1998, Paterson & Banks 2001, Nieberding et al. 2004, Huyse et al. 2005, Levin & Parker 2013, Booth et al. 2015). In this respect, the unusually higher sequence divergences within the Monstrilloida (Table 2) would deserve more attention for better understanding their evolutionary significance. Our phylogenetic results alluded that the monstrilloid genera appear to separate by the usage of the different hosts. Therefore, further molecular research targeting more specific genes involved in immune response/evasion mechanisms and neuro-signalling pathways could probably provide the primary ground for understanding the particular adaptation to various environmental conditions and testing the fast gene mutation phenomena for Monstrilloida.

The weakest point of this study is probably that the phylogenetic evaluation of Monstrilloida was considered only between the so-called major copepod taxa based on relatively small genetic information compared to other bioinformatic analyses. Accordingly, our present results show a partial resolution for the Copepoda phylogeny. Our first genome-scale phylogenetic analyses into Monstrilloida provide more detailed copepod relationships to date. It is clear that the amount of genetic information available is important, especially in resolving the deep nodes. Five copepod orders were not included in this study. Their phylogenetic relationships remain to be explored. As such, the continued accumulation and application of more genetic data for future phylogenetic research are needed for a better understanding of the phylogenetic relationships among broad taxa. These works will eventually provide further insights into the phylogeny of the Copepoda and, by extension, into the Arthropoda evolution.

Concluding remarks

The present molecular study based on the 25 orthologous nuclear protein-coding genes reaffirmed the monophyletic status of the Copepoda within the Arthropoda and further revealed the molecular phylogenetic position of the Monstrilloida within the Copepoda. The Monstrilloida appeared to have a closer affinity to the benthic copepod group, Harpacticoida, rather than the previously suggested affinity to another parasitic group, the Siphonostomatoida. Further investigations applying morphological evidence to the molecular phylogeny suggested a certain evolutionary hypothesis that the Monstrilloida have evolved in the sequence *Monstrilla*, *Caromiobenella*, *Maemonstrilla*, *Monstrillopsis* and *Cymbasoma* from their last common ancestor.

Most interestingly, the relationships between the monstrilloid genera showed two distinct patterns of their morphological evolutionary trend depending on the host type of each genus: (1) a gradual morphological reduction pattern (e.g., a decrease in the number of body somites and caudal setae, and the structural oligomerisation in the female fifth legs) mainly occur along the polychaete-associated lineages, *Monstrilla*, *Monstrillopsis* and *Cymbasoma*, while (2) an appearance of the new feature (e.g., type III antennular modification in males of *Caromiobenella*; low and wide intercoxal sclerites and anteriorly directed ovigerous spines in females of *Maemonstrilla*) is prominent in the groups utilising non-polychaete hosts (at least in *Caromiobenella*). The adoption of the different evolutionary strategies may represent their host-associated adaptation to simultaneously thrive in challenging environments for survival. As such, this study serves as a critical starting point for explaining hypotheses on which copepod species might have evolved to adapt to diverse ecological niches. Furthermore, their intriguing characteristics in all three aspects of morphology, molecular biology and ecology will continuously give opportunities and insights to shed more light on the copepod evolution.

Data availability

All raw sequence data are deposited in the NCBI Sequence Read Archive (SRA) under the BioProject accession PRJNA766505. The multiple sequence alignment used for the analyses is available for download at http://eyunlab.cau.ac.kr/monstrilloid_phylogenomics.

Acknowledgements

The authors thank the laboratory members for their insightful discussion on the manuscript. This work was supported by the National Research Foundation of Korea Grant (2022R1A2C4002058) and the Korea Institute of Ocean Science and Technology (PE99984 and 20220532) to SE and the Basic Science Research Program by the Ministry of Education (RS-2023-00243296) to DJ. Additionally, this research was a part of the project titled "Study on the fate of microplastics as a tracer of environmental change in Northwest Pacific Ocean (PE99653)" funded by the Ministry of Oceans and Fisheries, Korea.

References

Barreto, F.S., Watson, E.T., Lima, T.G., Willett, C.S., Edmands, S., Li, W. & Burton, R.S. 2018. Genomic signatures of mitonuclear coevolution across populations of *Tigriopus californicus*. *Nature Ecology & Evolution* **2**, 1250–1257.

Benton, M.J. 1995. Diversification and extinction in the history of life. *Science* **268**, 52–58.

Birney, E., Clamp, M. & Durbin, R. 2004. GeneWise and genomewise. *Genome Research* **14**, 988–995.

Blanco-Bercial, L., Bradford-Grieve, J. & Bucklin, A. 2011. Molecular phylogeny of the Calanoida (Crustacea: Copepoda). *Molecular Phylogenetics and Evolution* **59**, 103–113.

Booth, W., Balvín, O., Vargo, E.L., Vilímová, J. & Schal, C. 2015. Host association drives genetic divergence in the bed bug, *Cimex lectularius*. *Molecular Ecology* **24**, 980–992.

Boxshall, G. & Halsey, S.H. 2004. *An Introduction to Copepod Diversity*. London: The Ray Society.

Boxshall, G. & Jaume, D. 2000. Making waves: the repeated colonization of fresh water by copepod crustaceans. *Advances in Ecological Research* **31**, 61–79.

Braga, E., Zardoya, R., Meyer, A. & Yen, J. 1999. Mitochondrial and nuclear rRNA based copepod phylogeny with emphasis on the Euchaetidae (Calanoida). *Marine Biology* **133**, 79–90.

Burgess, S.D., Bowring, S. & Shen, S.-Z. 2014. High-precision timeline for Earth's most severe extinction. *Proceedings of the National Academy of Sciences of the United States of America* **111**, 3316–3321.

Bushmanova, E., Antipov, D., Lapidus, A. & Prjibelski, A.D. 2019. rnaSPAdes: a *de novo* transcriptome assembler and its application to RNA-Seq data. *GigaScience* **8**, giz100.

Camacho, C., Coulouris, G., Avagyan, V., Ma, N., Papadopoulos, J., Bealer, K. & Madden, T.L. 2009. BLAST+: architecture and applications. *BMC Bioinformatics* **10**, 421.

Capella-Gutiérrez, S., Silla-Martínez, J.M. & Gabaldón, T. 2009. trimAl: a tool for automated alignment trimming in large-scale phylogenetic analyses. *Bioinformatics* **25**, 1972–1973.

Castiglione, S., Serio, C., Tamagnini, D., Melchionna, M., Mondanaro, A., Di Febbraro, M., Profico, A., Piras, P., Barattolo, F. & Raia, P. 2019. A new, fast method to search for morphological convergence with shape data. *PLoS ONE* **16**, e0226949.

Caullery, M. & Mesnil, F. 1914. Sur deux Monstrillides parasites d'Annélides (*Polydora giardi Mesn. et Syllis gracilis Gr.*). *Bulletin Scientifique de la France et de la Belgique* **48**, 15–29.

Chen, Z.-Q. & Benton, M.J. 2012. The timing and pattern of biotic recovery following the end-Permian mass extinction. *Nature Geoscience* **5**, 375–383.

Chen, S., Zhou, Y., Chen, Y. & Gu, J. 2018. fastp: an ultra-fast all-in-one FASTQ preprocessor. *Bioinformatics* **34**, i884–i890.

Cressey, R. & Boxshall, G. 1989. *Kabatarinia pattersoni*, a fossil parasitic copepod (Dichelesthiidae) from a Lower Cretaceous fish. *Micropaleontology* **35**, 150–167.

Cressey, R. & Patterson, C. 1973. Fossil parasitic copepods from a Lower Cretaceous fish. *Science* **180**, 1283–1285.

Dahms, H.-U. 1990. Naupliar development of Harpacticoida (Crustacea, Copepoda) and its significance for phylogenetic systematics. *Microfauna Marina* **6**, 169–272.

Dahms, H.-U. 2004. Postembryonic apomorphies proving the monophyletic staus of the Copepoda. *Zoological Studies* **43**, 446–453.

Davis, C.C. 1949. A preliminary revision of the Monstrilloida, with descriptions of two new species. *Transactions of the American Microscopical Society* **68**, 245–255.

Dayhoff, M.O., Schwarz R.M. & Orcutt B.C. 1978. A model of evolutionary change in proteins. In *Atlas of Protein Sequence and Structure*, M.O. Dayhoff (ed.). Washington, District of Columbia: National Biomedical Research Foundation, 345–352.

Delaforge, A., Suárez-Morales, E., Walkusz, W., Campbell, K. & Mundy, C.J. 2017. A new species of *Monstrillopsis* (Crustacea, Copepoda, Monstrilloida) from the lower Northwest Passage of the Canadian Arctic. *ZooKeys* **709**, 1–16.

Dietz, R.S. & Holden, J.C. 1970. Reconstruction of Pangaea: breakup and dispersion of continents, Permian to present. *Journal of Geophysical Research* **75**, 4939–4956.

Eyun, S. 2017. Phylogenomic analysis of Copepoda (Arthropda, Crustacea) reveals unexpected similarities with earlier proposed morphological phylogenies. *BMC Evolutionary Biology* **17**, 23.

Eyun, S., Soh, H.Y., Posavi, M., Mounro, J.B., Hughes, D.S.T., Murali, S.C., Qu, J., Dugan, S., Lee, S.L., Chao, H., Dinh H., Han Y., Doddapaneni, H.V., Worley, K.C., Muzny, D.M., Park, E.-O., Silva, J.C., Gibbs, R.A., Richards, S. & Lee, C.E. 2017. Evolutionary history of chemosensory-related gene families across the Arthropoda. *Molecular Biology and Evolution* **34**, 1838–1862.

Gallien, L. 1934. Description du mâle de *Monstrilla helgolandica Claus*. Synonymie de Monstrilla serricornis G. O. Sars et de Monstrilla helgolandica Claus. *Bulletin de la Société Zoologique de France* **59**, 377–382.

Giesbrecht, W. 1892. *Systematik Und Faunistik Der Pelagischen Copepoden Des Golfes Von Neapel Und Der Angrenzenden Meeres-Abschnitte. Atlas von 54 Tafeln Nebst Den Tsfelerklärungen. Fauna Und Flora Des Golfes Von Neapel Und Der Angrenzenden Meeres-Abschnitte Herausgegeben Von Der Zoologischen Station Zu Neapel. Xix. Monographie*. Berlin: Verlag von R. Friedländer & Sohn.

Grabherr, M.G., Haas, B.J., Yassour, M., Levin, J.Z., Thompson, D.A., Amit, I., Adiconis, X., Fan, L., Raychowdhury, R., Zeng, Q., Chen, Z., Mauceli, E., Hacohen, N., Gnirke, A., Rhind, N., di Palma, F., Birren, B.W., Nusbaum, C., Lindblad-Toh, K., Friedman, N. & Regev, A. 2011. Full-length transcriptome assembly from RNA-Seq data without a reference genome. *Nature Biotechnology* **29**, 644–652.

Grygier, M.J. 1994a. Identity of *Thaumatoessa* (= *Thaumaleus*) typica Krøyer, the first described monstrilloid copepod. *Sarsia* **78**, 235–242.

Grygier, M.J. 1994b. Nomenclature, redescription, and new record from Okinawa of *Cymbasoma morii* Sekiguchi, 1982 (Monstrilloida). *Hydrobiologia* **292/293**, 23–29.

Grygier, M.J. 1995a. Annotated chronological bibliography of Monstrilloida (Crustacea: Copepoda). *Galaxea* **12**, 1–82.

Grygier, M.J. 1995b. Case 2894 *Monstrilla Dana*, 1849 and *Thaumaleus Krøyer*, 1849 (Crustacea, Copepoda): proposed conservation. *Bulletin of Zoological Nomenclature* **52**, 245–249.

Grygier, M.J. & Ohtsuka, S. 1995. SEM observation of the nauplius of *Monstrilla hamatapex*, new species, from Japan and an example of upgraded descriptive standards for monstrilloid copepods. *Journal of Crustacean Biology* **15**, 703–719.

Grygier, M.J. & Ohtsuka, S. 2008. A new genus of monstrilloid copepods (Crustacea) with anteriorly pointing ovigerous spines and related adaptations for subthoracic brooding. *Zoological Journal of the Linnean Society* **152**, 459–506.

Hafner, M.S., Sudman, P.D., Villablanca, F.X., Spradling, T.A., Demastes, J.W. & Nadler, S.A. 1994. Disparate rates of molecular evolution in cospeciating hosts and parasites. *Science* **265**, 1087–1090.

Hartman, O. 1961. A new monstrilloid copepod parasitic in capitellid polychaetes in Southern California. *Zoologischer Anzeiger* **167**, 325–334.

Hesse, M. 1868. Observations sur des Crustacés rares ou nouveaux des cotes de France (Seizième article). *Annales des Sciences Naturelles Cinquième Série, Zoologie et Paléontologie* **10**, 347–371.

Ho, J.-S. 1990. Phylogenetic analysis of copepod orders. *Journal of Crustacean Biology* **10**, 528–536.

Ho, J.-S. 1994. Copepod phylogeny: a reconsideration of Huys & Boxshall's 'parsimony versus homology' *Hydrobiologia* **292/293**, 31–39.

Ho, J.-S., Dojiri, M., Hendler, G. & Deets, G.B. 2003. A new species of Copepoda (Thaumatopsyllidae) symbiotic with a brittle star from California, U.S.A., and designation of a new order Thaumatopsylloida. *Journal of Crustacean Biology* **23**, 582–594.

Huang, S., Roy, K., Valentine, J.W. & Jablonski, D. 2015. Convergence, divergence, and parallelism in marine biodiversity trends: integrating present-day and fossil data. *Proceedings of the National Academy of Sciences of the United States of America* **112**, 4903–4908.

Huys, R. & Boxshall, G. 1991. *Copepod Evolution*. London: The Ray Society.

Huys, R., Llewellyn-Hughes, J., Olson, P.D. & Nagasawa, K. 2006. Small subunit rDNA and Bayesian inference reveal *Pectenophilus ornatus* (Copepoda *incertae sedis*) as highly transformed Mytilicolidae, and support assignment of Chondracanthidae and Xarifiidae to Lichomolgoidea (Cyclopoida). *Biological Journal of the Linnean Society* **87**, 403–425.

Huys, R., Llewellyn-Hughes, J., Conroy-Dalton, S., Olson, P.D., Spinks, J.N. & Johnston, D.A. 2007. Extraordinary host switching in siphonostomatoid copepods and the demise of the Monstrilloida: integrating molecular data, ontogeny and antennulary morphology. *Molecular Phylogenetics and Evolution* **43**, 368–378.

Huys, R., Mackenzie-Dodds, J. & Llewellyn-Hughes, J. 2009. Cancrincolidae (Copepoda, Harpacticoida) associated with land crabs: a semiterrestrial leaf of the ameirid tree. *Molecular Phylogenetics and Evolution* **51**, 143–156.

Huys, R., Fatih, F., Ohtsuka, S. & Llewellyn-Hughes, J. 2012. Evolution of the bomolochiform superfamily complex (Copepoda: Cyclopoida): new insights from ssrDNA and morphology, and origin of umazuracolids from polychaete-infesting ancestors rejected. *International Journal for Parasitology* **42**, 71–92.

Huys, R., Suárez-Morales, E., de Lourdes Serrano-Sánchez, M., Centeno-García, E. & Vega, F.J. 2016. Early Miocene amber inclusions from Mexico reveal antiquity of mangrove-associated copepods. *Scientific Reports* **6**, 34872.

Huyse, T., Poulin, R. & Théron, A. 2005. Speciation in parasites: a population genetics approach. *Trends in Parasitology* **21**, 469–475.

Isaac, M.J. 1975. Copepoda, Sub-order: Monstrilloida. *Fiches d'Identification du Zooplancton* **144/145**, 1–10.

Jeon, D., Lee, W. & Soh, H.Y. 2018a. A new genus and two new species of monstrilloid copepods (Copepoda: Monstrillidae): integrating morphological, molecular phylogenetic, and ecological evidence. *Journal of Crustacean Biology* **38**, 45–65.

Jeon, D., Lim, D., Lee, W. & Soh, H.Y. 2018b. First use of molecular evidence to match sexes in the Monstrilloida (Crustacea: Copepoda), and taxonomic implications of the newly recognized and described, partly *Maemonstrilla-like females of Monstrillopsis longilobata* Lee, Kim & Chang, 2016. *PeerJ* **6**, e4938.

Jones, D.T., Taylor, W.R. & Thornton, J.M. 1992. The rapid generation of mutation data matrices from protein sequences. *Bioinformatics* **8**, 275–282.

Jung, H., Jeon, M.-S., Hodgett, M., Waterhouse, P. & Eyun, S. 2020a. Comparative evaluation of genome assemblers from long-read sequencing for plants and crops. *Journal of Agricultural and Food Chemistry* **68**, 7670–7677.

Jung, H., Ventura, T., Chung, J.S., Kim, W.-J., Nam, B.-H., Kong, H.J., Kim, Y.-O., Jeon, M.-S. & Eyun, S.-I. 2020b. Twelve quick steps for genome assembly and annotation in the classroom. *PLoS Computational Biology* **16**, e1008325.

Katoh, K. & Standley, D.M. 2013. MAFFT multiple sequence alignment software version 7: improvements in performance and usability. *Molecular Biology and Evolution* **30**, 772–780.

Khodami, S., McArthur, J.V., Blanco-Bercial, L. & Martinez Arbizu, P. 2017. Molecular phylogeny and revision of copepod orders (Crustacea: Copepoda) (RETRACTED: Retraction published 2020, *Sci. Rep.*, 10, 17602). *Scientific Reports* **7**, 9164.

Khodami, S., Mercado-Salas, N.F., Tang, D. & Martinez Arbizu, P. 2019. Molecular evidence for the retention of the Thaumatopsyllidae in the order Cyclopoida (Copepoda) and establishment of four suborders and two families within the Cyclopoida. *Molecular Phylogenetics and Evolution* **138**, 43–52.

Ki, J.-S., Lee, K.-W., Park, H.G., Chullasorn, S., Dahms, H.-U. & Lee, J.-S. 2009. Phylogeography of the copepod *Tigriopus japonicus* along the Northwest Pacific rim. *Journal of Plankton Research* **31**, 209–221.

Kimura, M. 1980. A simple method for estimating evolutionary rates of base substitutions through comparative studies of nucleotide sequences. *Journal of Molecular Evolution* **16**, 111–120.

Kishino, H. & Hasegawa, M. 1989. Evaluation of the maximum likelihood estimate of the evolutionary tree topologies from DNA sequence data, and the branching order in hominoidea. *Journal of Molecular Evolution* **29**, 170–179.

Knoll, A.H., Bambach, R.K., Payne, J.L., Pruss, S. & Fischer, W.W. 2007. Paleophysiology and end-Permian mass extinction. *Earth and Planetary Science Letters* **256**, 295–313.

Kozlov, A.M., Darriba, D., Flouri, T., Morel, B. & Stamatakis, A. 2019. RAxML-NG: a fast, scalable and user-friendly tool for maximum likelihood phylogenetic inference. *Bioinformatics* **35**, 4453–4455.

Krøyer, H. 1849. Karcinologiske bidrag (Fortsættelse). *Naturhistorisk Tidsskrift* **2**, 561–609.

Le, S.Q., Lartillot, N. & Gascuel, O. 2008. Phylogenetic mixture models for proteins. *Philosophical Transactions of the Royal Society B* **363**, 3965–3976.

Lee, Y., Kwak, H., Shin, J., Kim, S.-C., Kim, T. & Park, J.-K. 2019. A mitochondrial genome phylogeny of Mytilidae (Bivalvia: Mytilida). *Molecular Phylogenetics and Evolution* **139**, 106533.

Lemoine, F., Domelevo Entfellner, J.-B., Wilkinson, E., Correia, D., Dávila Felipe, M., De Oliveira, T. & Gascuel, O. 2018. Renewing Felsenstein's phylogenetic bootstrap in the era of big data. *Nature* **556**, 452–456.

Letunic, I., Khedkar, S. & Bork, P. 2021. SMART: recent updates, new developments and status in 2020. *Nucleic Acids Research* **49**, D458–D460.

Levin, I.I. & Parker, P.G. 2013. Comparative host-parasite population genetic structures: obligate fly ectoparasites on Galapagos seabirds. *Parasitology* **140**, 1061–1069.

Lozano-Fernandez, J., Giacomelli, M., Fleming, J.F., Chen, A., Vinther, J., Thomsen, P.F., Glenner, H., Palero, F., Legg, D.A., Iliffe, T.M., Pisani, D. & Olesen, J. 2019. Pancrustacean evolution illuminated by taxon-rich genomic-scale data sets with an expanded remipede sampling. *Genome Biology and Evolution* **11**, 2055–2070.

Malaquin, A. 1901. Le parasitisme évolutif des Monstrillides (Crustacés Copépodes). *Archives de Zoologie Expérimentale et Générale* **9**, 81–232.

Manni, M., Berkeley, M.R., Seppey, M., Simão, F.A. & Zdobnov, E.M. 2021. BUSCO update: novel and streamlined workflows along with broader and deeper phylogenetic coverage for scoring of eukaryotic, prokaryotic, and viral genomes. *Molecular Biology and Evolution* **38**, 4647–4654.

Mikhailov, K.V. & Ivanenko, V.N. 2021. Low support values and lack of reproducibility of molecular phylogenetic analysis of Copepoda orders. *Arthropoda Selecta* **30**, 39–42.

Minh, B.Q., Schmidt, H.A., Chernomor, O., Schrempf, D., Woodhams, M.D., von Haeseler, A. & Lanfear, R. 2020. IQ-TREE 2: new models and efficient methods for phylogenetic inference in the genomic era. *Molecular Biology and Evolution* **37**, 1530–1534.

Nieberding, C., Morand, S., Libois, R. & Michaux, J.R. 2004. A parasite reveals cryptic phylogeographic history of its host. *Proceedings of the Royal Society B* **271**, 2559–2568.

Oakley, T.H., Wolfe, J.M., Lindgren, A.R. & Zaharoff, A.K. 2013. Phylotranscriptomics to bring the understudied into the fold: monophyletic Ostracoda, fossil placement, and pancrustacean phylogeny. *Molecular Biology and Evolution* **30**, 215–233.

Page, R.D.M., Lee, P.L.M., Becher, S.A., Griffiths, R. & Clayton, D.H. 1998. A different tempo of mitochondrial DNA evolution in birds and their parasitic lice. *Molecular Phylogenetics and Evolution* **9**, 276–293.

Palmer, A.R. 1960. Miocene copepods from the Mojave Desert, California. *Journal of Paleontology* **34**, 447–452.

Park, T.S. 1967. Two unreported species and one new species of *Monstrilla* (Copepoda: Monstrilloida) from the Strait of Georgia. *Transactions of the American Microscopical Society* **86**, 144–152.

Paterson, A.M. & Banks, J. 2001. Analytical approaches to measuring cospeciation of host and parasites: through a glass, darkly. *International Journal for Parasitology* **31**, 1012–1022.

Pelseneer, P. 1914. Éthologie de quelques *Odostomia* et d'un Monstrillide parasite de l'un d'eux. *Bulletin Scientifique de la France et de la Belgique* **48**, 1–14.

Raibaut, A. 1985. Les cycles évolutifs des Copépodes parasites et les modalités de l'infestation. *Anneé Biologique* **24**, 233–274.

Razouls, C., Desreumaux, N., Kouwenberg, J. & de Bovée, F. 2005-2022. Biodiversity of marine planktonic copepods (morphology, geographical distribution and biological data). Available at https://copepodes.obs-banyuls.fr/en

Regier, J.C., Shultz, J.W., Ganley, A.R., Hussey, A., Shi, D., Ball, B., Zwick, A., Stajich, J.E., Cummings, M.P., Martin, J.W. & Cunningham, C.W. 2008. Resolving arthropod phylogeny: exploring phylogenetic signal within 41 kb of protein-coding nuclear gene sequence. *Systematic Biology* **57**, 920–938.

Regier, J.C., Shultz, J.W., Zwick, A., Hussey, A., Ball, B., Wetzer, R., Martin, J.W. & Cunningham, C.W. 2010. Arthropod relationships revealed by phylogenomic analysis of nuclear protein-coding sequences. *Nature* **463**, 1079–1083.

Ronquist, F., Teslenko, M., Van Der Mark, P., Ayres, D.L., Darling, A., Höhna, S., Larget, B., Liu, L., Suchard, M.A. & Huelsenbeck, J.P. 2012. MrBayes 3.2: efficient Bayesian phylogenetic inference and model choice across a large model space. *Systematic Biology* **61**, 539–542.

Rosa, J.C.L., Dias, C.D.O., Suárez-Morales, E., Weber, L.I. & Fischer, L.G. 2021. Record of *Caromiobenella* (Copepoda, Monstrilloida) in Brazil and discovery of the male of *C. brasiliensis*: morphological and molecular evidence. *Diversity* **13**, 241.

Rose, M. 1933. Copépodes pélagiques. *Fauna de France* **26**, 1–374.

Rosenberg, M.S. & Kumar, S. 2001. Incomplete taxon sampling is not a problem for phylogenetic inference. *Proceedings of the National Academy of Sciences of the United States of America* **98**, 10751–10756.

Rosenberg, M.S. & Kumar, S. 2003. Taxon sampling, bioinformatics, and phylogenomics. *Systematic Biology* **52**, 119–124.

Sale, P.F., McWilliam, P.S. & Anderson, D.T. 1976. Composition of the near-reef zooplankton at Heron Reef, Great Barrier Reef. *Marine Biology* **34**, 59–66.

Sars, G.O. 1921. *An Account of the Crustacea of Norway with Short Descriptions and Figures of All The Species. Vol. III. Copepoda Monstrilloida & Notodelphyoida.* Bergen: The Bergen Museum.

Schizas, N.V., Dahms, H.-U., Kangtia, P., Corgosinho, P.H.C. & Galindo Estronza, A.M. 2015. A new species of *Longipedia* Claus, 1863 (Copepoda: Harpacticoida: Longipediidae) from Caribbean mesophotic reefs with remarks on the phylogenetic affinities of Polyarthra. *Marine Biology Research* **11**, 789–803.

Schwentner, M., Combosch, D.J., Nelson, J.P. & Giribet, G. 2017. A phylogenomic solution to the origin of insects by resolving crustacean-hexapod relationships. *Current Biology* **27**, 1818–1824.

Schwentner, M., Richter, S., Rogers, D.C. & Giribet, G. 2018. Tetraconatan phylogeny with special focus on Malacostraca and Branchiopoda: highlighting the strength of taxon-specific matrices in phylogenomics. *Proceedings of the Royal Society B* **285**, 20181524.

Sekiguchi, H. 1982. Monstrilloid copepods from Ago Bay, Central Japan. *Proceedings of the Japanese Society of Systematic Zoology* **22**, 24–34.

Selden, P.A., Huys, R., Stephenson, M.H., Heward, A.P. & Taylor, P.N. 2010. Crustaceans from bitumen clast in Carboniferous glacial diamictite extend fossil record of copepods. *Nature Communications* **1**, 50.

Sepkoski, J.J. 1981. A factor analytic description of the Phanerozoic marine fossil record. *Paleontology* **7**, 36–53.

Sepkoski, J.J. 1984. A kinentic model of Phanerozoic taxonomic diversity. III. Post-Paleozoic families and mass extinctions. *Paleontology* **10**, 246–267.

Sewell, R.B.S. 1949. The lottral and semi-parasitic Cyclopoida, the Monstrilloida and Notodelphyoida. *The John Murray Expedition 1933-34 Scientific Reports* **9**, 17–199.

Shimodaira, H. 2002. An approximately unbiased test of phylogenetic tree selection. *Systematic Biology* **51**, 492–508.

Shimodaira, H. & Hasegawa, M. 1999. Multiple comparisons of log-Likelihoods with applications to phylogenetic inference. *Molecular Biology and Evolution* **16**, 1114–1116.

Song, C.-U., Choi, H., Jeon, M.-S., Kim, E.-J., Jeong, H.G., Kim, S., Kim, C.-G., Hwang, H., Purnaningtyas, D.W., Lee, S., Eyun, S. & Lee, Y.-H. 2021. Zooplankton diversity monitoring strategy for the urban coastal region using metabarcoding analysis. *Scientific Reports* **11**, 24339.

Speed, M.P. & Arbuckle, K. 2016. Quantification provides a conceptual basis for convergent evolution. *Biological Reviews* **92**, 815–829.

Stanley, S.M. 2016. Estimates of the magnitudes of major marine mass extinctions in earth history. *Proceedings of the National Academy of Sciences of the United States of America* **113**, E6325–E6334.

Suárez-Morales, E. 2000. A new species and new geographic records of *Monstrilla* (Copepoda: Monstrilloida) from the Philippines. *Journal of Crustacean Biology* **20**, 680–686.

Suárez-Morales, E. 2001. An aggregation of monstrilloid copepods in a western Caribbean reef area: ecological and conceptual implications. *Crustaceana* **74**, 689–696.

Suárez-Morales, E. 2011. Diversity of the Monstrilloida (Crustacea: Copepoda). *PLoS One* **6**, e22915.

Suárez-Morales, E. 2018. Monstrilloid copepods: the best of three worlds. *Bulletin of the Southern California Academy of Sciences* **117**, 92–103.

Suárez-Morales, E. 2019. A new genus of the Monstrilloida (Copepoda) with large rostral process and metasomal spines, and redescription of *Monstrilla spinosa* Park, 1967. *Crustaceana* **92**, 1099–1112.

Suárez-Morales, E. & Gasca, R. 1998. *Cymbasoma bowmani* sp. nov., a new monstrilloid (Copepoda: Monstrilloida) from a Caribbean reff, with notes on species variation. *Journal of Marine Systems* **15**, 433–439.

Suárez-Morales, E. & Gasca, R. 2004. On the invalidity of Strilloma Isaac (Copepoda: Monstrilloida): observations from the type species. *Zoological Studies* **43**, 292–299.

Suárez-Morales, E., Bello-Smith, A. & Palma, S. 2006. A revision of the genus *Monstrillopsis* Sars (Crustacea: Copepoda: Monstrilloida) with description of a new species from Chile. *Zoologischer Anzeiger* **245**, 95–107.

Suárez-Morales, E. & McKinnon, A.D. 2014. The Australian Monstrilloida (Crustacea: Copepoda) I. *Monstrillopsis* Sars, *Maemonstrilla* Grygier & Ohtsuka, and *Australomonstrillopsis* gen. nov. *Zootaxa* **3779**, 301–340.

Suárez-Morales, E. & McKinnon, A.D. 2016. The Australian Monstrilloida (Crustacea: Copepoda) II. *Cymbasoma* Thompson, 1888. *Zootaxa* **4102**, 1–129.

Suárez-Morales, E. & Grygier, M.J. 2021. Mediterranean and Balck Sea monstrilloid copepods (Copepoda: Monstrilloida): rediscovering the diversity of transient zooplankters. *Water* **13**, 1036.

Suárez-Morales, E., Ramírez, F.C. & Derisio, C. 2008. Monstrilloida (Crustacea: Copepoda) from the Beagle Channel, South America. *Contributions to Zoology* **77**, 217–226.

Suárez-Morales, E., Paiva Scardua, M. & Da Silva, P.M. 2010. Occurrence and histopathological effects of *Monstrilla* sp. (Copepoda: Monstrilloida) and other parasites in the brown mussel *Perna perna* from Brazil. *Journal of the Marine Biological Association of the United Kingdom* **90**, 953–958.

Suárez-Morales, E., Harris, L.H., Ferrari, F.D. & Gasca, R. 2014. Late postnaupliar development of *Monstrilla* sp. (Copepoda: Monstrilloida), a protelean endoparasite of benthic polychaetes. *Invertebrate Reproduction & Development* **58**, 60–73.

Suárez-Morales, E., Goruppi, A., de Olazabal, A. & Tirelli, V. 2017. Monstrilloids (Crustacea: Copepoda) from the Mediterranean Sea (Northern Adriatic Sea), with a description of six new species. *Journal of Natural History* **51**, 1795–1834.

Suárez-Morales, E., Vásquez-Yeomans, L. & Santoya, L. 2020. A new species of the *Cymbasoma longispinosum* species-group (Copepoda: Monstrilloida) from Belize, wetern Caribbean. *Revista Mexicana de Biodiversidad* **91**, e913176.

Thompson, I.C. 1888. Copepoda of Medeira and the Canary Islands, with descriptions of new genera and species. *Journal of the Linnean Society of London Zoology* **20**, 145–156.

To, T.-H., Jung, M., Lycett, S. & Gascuel, O. 2016. Fast dating using Least-Squares criteria and algorithms. *Systematic Biology* **65**, 82–97.

Tung, C.-H., Cheng, Y.-R., Lin, C.-Y., Ho, J.-S., Kuo, C.-H., Yu, J.-K. & Su, Y.-H. 2014. A new copepod with transformed body plan and unique phylogenetic position parasitic in the acorn worm *Ptychodera flava*. *Biological Bulletin* **226**, 69–80.

Üstün, F., Terbiyik Kurt, T. & Suárez-Morales, E. 2014. A new species of *Cymbasoma* (Copepoda, Monstrilloida) from the northern coast of Turkey (Black Sea) with comments on the *C. longispinosum* species–group. *Crustaceana* **87**, 1393–1410.

Vakati, V., Eyun, S. & Lee, W. 2019. Unraveling the intricate biodiversity of the benthic harpacticoid genus *Nannopus* (Copepoda, Harpacticoida, Nannopodidae) in Korean waters. *Molecular Phylogenetics and Evolution* **130**, 366–379.

Valentine, J.W. 1969. Patterns of taxonomic and ecological structure of the shelf benthos during Phanerozoic time. *Paleontology* **12**, 684–709.

von Reumont, B.M., Jenner, R.A., Wills, M.A., Dell'Ampio, E., Pass, G., Ebersberger, I., Meyer, B., Koenemann, S., Iliffe, T.M., Stamatakis, A., Niehuis, O., Meusemann, K. & Misof, B. 2012. Pancrustacean phylogeny in the light of new phylogenomic data: support for Remipedia as the possible sister group of Hexapoda. *Molecular Biology and Evolution* **29**, 1031–1045.

Walter, T.C. & Boxshall, G. 2022. World of Copepods database. Available at https://www.marinespecies.org/copepoda

Wolfe, J.M., Daley, A.C., Legg, D.A. & Edgecombe, G.D. 2016. Fossil calibrations for the arthropod tree of life. *Earth-Science Reviews* **160**, 43–110.

WoRMS Editorial Board. 2022. World Register of Marine Species. Available at https://www.marinespecies.org

Yoshida, M. & Hamano, Y. 2015. Pangea breakup and northward drift of the Indian subcontinent reproduced by a numerical model of mantle convection. *Scientific Reports* **5**, 8407.

Zhang, X.-G., Maas, A., Haug, J., Siveter, D.J. & Waloszek, D. 2010. A eucrustacean metanauplius from the Lower Cambrian. *Current Biology* **20**, 1075–1079.

Supplementary Materials are provided online at: https://www.routledge.com/9781032761961

Oceanography and Marine Biology: An Annual Review, 2024, **62**, 376-403

Taylor & Francis

POTENTIAL RISK OF FLOATING MARINE LITTER TO CETACEANS AND SEA TURTLES: A REVIEW OF SPATIAL RISK EXPOSURE ASSESSMENTS

ARCANGELI A.[1], GREGORIETTI M.[2,3,7], PASANISI E.[1,4], SANTINI E.[1], CAMPANA I.[2,5,6], AND SARÀ G.[3,7]

[1]ISPRA Bio Dep., via Brancati 60, 00144 Roma, Italy
[2]Accademia del Leviatano, Via dell'Ospedaletto 53/55, 00054 Maccarese (RM), Italy
[3]Laboratorio di Ecologia, Dipartimento di Scienze della Terra e del Mare, Viale delle Scienze Ed. 16, 90128 Palermo, Italy
[4]Department of Environmental Biology, Sapienza University of Rome, 00185 Rome, Italy
[5]Tuscia University, Dep. of Ecological and Biological Sciences, Ichthyogenic Experimental Marine Center (CISMAR), Borgo Le Saline, 01016 Tarquinia (VT), Italy
[6]CoNISMa, National Inter-University Consortium for Marine Sciences, Piazzale Flaminio 9, 00196 Roma, Italy
[7]NBFC, National Biodiversity Future Center, Piazza Marina 61, 90133 Palermo, Italy
Antonella Arcangeli: antonella.arcangeli@isprambiente.it - Contact author
Martina Gregorietti: martina.gregorietti@unipa.it
Eugenia Pasanisi: pasanisieugenia@gmail.com
Elena Santini: elena.santini@isprambiente.it
Ilaria Campana: ilariacampana@unitus.it
Gianluca Sarà: gianluca.sara@unipa.it
Corresponding Author:
Arcangeli A.
Email address: antonella.arcangeli@isprambiente.it

Abstract Marine litter is a main threat for marine life, although the assessment of the associated risks has not yet been fully incorporated into area-based management tools. Floating litter is detrimental to cetaceans and sea turtles, and thus, these organisms are considered an effective indicator of areas where litter accumulates. Increasing our ability to predict high-exposure risk locations, i.e., where and when marine megafauna is exposed to the potential negative impacts of litter, is important for prioritising smart-conservation planning and is an essential first step in characterising the risk of real injury/damage. However, Risk Exposure Assessment (REA) is still underrepresented as a standardised procedure. Here, a literature review framed the state-of-the-art of REA approaches for cetaceans and sea turtles from floating litter supporting the standardisation of metrics and procedures. Of the 415 papers resulting from the literature search, the 23 selected (2011–2022) showed that 57% of the studies were conducted in the Western-Mediterranean Sea, evidencing inconsistent geographical applications. While a variety of REA methodological approaches revealed high informational heterogeneity, main limits and future recommendations were identified regarding raw data availability, information bias, geographical gaps, target species selection and lack of standard protocol needed to assess trends to evaluate the effectiveness of mitigation measures. Ultimately, the study showed that a spatial-contextual approach (possibly functional trait-based) is needed to effectively support long-term year-round monitoring programmes, especially in still un-surveyed regions.

 DOI: 10.1201/9781003477518-10

Introduction

Marine litter and main impact on species

Marine litter (ML) is a major form of pollution in the marine environment and is defined as any persistent, manufactured or processed solid material discarded, disposed of or abandoned in the marine and coastal environment (UNEP 2009). ML, and in particular the accumulation of plastic debris, has been identified as a global problem alongside other key issues of our time, including climate change, ocean acidification and biodiversity loss (Sutherland et al. 2007, Moore 2015).

ML itself contributes to biodiversity loss affecting many free-swimming and sessile animals (Moore & Barco 2013). The complex path of ML, from the source to dispersal, fragmentation and accumulation, determines interactions with marine life at various levels, causing mostly negative effects (Coe & Rogers 1997). Its impacts on species vary, according to its type and size and depending on the organisms that encounter it (Vegter et al. 2014, Poeta et al. 2017, Bucci et al. 2020).

Species can be exposed to ML litter via a variety of pathways, including entanglement, ingestion, chemical and biological transfer. Entanglement can cause direct harm (physical injuries) or death to individuals, as well as restrict natural movements and affect their ability to catch food, escape from predators and reproduce (Gregory 2009, Woods et al. 2019). Ingestion of ML by individuals can cause direct impacts such as intestinal blockage, malnutrition and poisoning or alter the ability to sense hunger, escape from predators and reproduce, potentially leading to severe suffering or death (Laist 1997, Derraik 2002, CBD 2012, Baulch & Perry 2014, Marn et al. 2020, López-Martínez et al. 2023). Ingestion of ML, especially microplastics, can also facilitate the transport of harmful chemicals into the organism potentially bioaccumulating along the trophic chain and increasing the risk to apex predators or filter-feeding organisms (Cole et al. 2011, Davison & Ash 2011, Fossi et al. 2012, Wright et al. 2013, Ory et al. 2017, Berlino et al. 2021, Salerno et al. 2021, López-Martínez et al. 2023). Furthermore, ML can also act as a transport vector that facilitates the opportunistic dispersal of organisms, including alien species (Wilcox et al. 2013, Kühn et al. 2015, Werner et al. 2016, Claro et al. 2019).

Direct impacts of ML have been documented for more than 1400 species, including marine megafauna such as fishes, birds, sea turtles and mammals, with the primary causes being entanglement and ingestion (Wilcox et al. 2013, Kühn et al. 2015, Claro et al. 2019). In 2012, the Convention of Biological Diversity (CBD) revealed that all known species of sea turtles, about half of all species of marine mammals, and one-fifth of all species of seabirds were affected by entanglement or ingestion of ML, with about 15% of the total number of species affected being included in the IUCN Red List (CBD 2012). In 2016, a new assessment increased this figure to 23% (CBD 2016). As for entanglement, certain categories of ML may be much more prone to cause entanglement due to their shape, size and material: in 2016, the Joint Research Centre (JRC) report ascertained that 44 out of the 217 ML categories defined within the Marine Strategy Framework Directive (MSFD, 2008/56/EC) pose an elevated risk for entanglement, particularly fishing-related items, such as nets, traps and ropes (Werner et al. 2016) and that over 80% of the impacts have been associated with plastic items (CBD 2012). Among these, net fragments, ropes and lines (e.g., gill and trawl nets, lost or discarded line for pots and traps), monofilament lines, packaging bands, plastic circular rings and packaging such as multipack can rings were identified as the items most frequently associated with entanglement (Butterworth et al. 2012).

Floating ML (hereafter FML) is a major cause of entanglement and ingestion, especially macro FML (>2.5 cm), which has previously been reported to negatively impact nearly 400 marine species (Wilcox et al. 2013, Gall & Thompson 2015, Kühn et al. 2015, Claro et al. 2019). FML is of particular concern for animals that are obligate air breathers such as cetaceans and sea turtles, which are bound to encounter these materials as they surface for air (Abreo et al. 2023). Moreover, according to the findings from the CBD, the global distribution of the FML can be an effective indicator of areas where litter accumulates, including seasonal trends, and where it can mostly affect vagrant pelagic megafauna such as cetaceans and sea turtles (CBD 2016, Poeta et al. 2017). The main EU marine policy Directive MSFD, in Annex I identifies ML as one of the eleven mandatory qualitative descriptors for assessing the Good Environmental Status of EU marine waters (GES, MSFD 2017, i.e., a state in which the seas remain healthy and productive and are used sustainably) and sets as overall provision for the GES that "composition, amount and distribution of litter and micro-litter on the coastline, in the surface layer of the water column (i.e., FML), and on the seabed, are at levels that do not cause harm to the coastal and marine environment" (MSFD primary criteria D10C1 and D10C2). According to these criteria, EU Member States should establish monitoring programmes and assess the extent of the pollution problem. Despite these, only a few member states have included FML in their monitoring strategies (Ruiz-Orejón et al. 2021, Vighi et al. 2022), and filling the data gaps in FML was recognised as a priority to improve our understanding of litter dynamics and distribution (Molina et al. 2019). This is particularly important in high-biodiversity areas, such as the Mediterranean Sea, which has been also recognised as one of the most polluted areas in the world (Coll et al. 2012, Mansui et al. 2020).

The risk assessment

In line with the legislative requirements, the MSFD task group on ML underlines the need to identify, quantify and prioritise the risks deriving from ML to ensure that there are no significant impacts on, or risks to, marine biodiversity (Werner et al. 2016). Although the traditional risk assessment is standardised to human health, it could provide a useful framework to analyse the potential harm of human activities on wildlife populations and ecosystems, and support decision-making processes that are faced with considerable uncertainty (Suter 2016). For the MSFD, cetaceans and sea turtles are considered relevant indicators to assess the risk of ML pressure on marine ecosystems, being representative components of marine ecosystems with key functional roles, sensitive to ML pressure, and present in large-scale marine areas (i.e., representative, relevant, present. European Commission 2017, Palialexis et al. 2022). The identification of high-risk areas, where marine fauna have an increased exposure to litter, is the first step for prioritising conservation measures to address the highest-risk settings (UNEP-MAP 2016). Unlike the 'traditional' risk assessment, where exposure is mostly given as ascertained, a fundamental part of the assessment of risk for highly mobile marine fauna from ML is focused on the identification of the pathways by which exposure might occur and the organisms that are more likely to be impacted by the threat (e.g., Darmon et al. 2017, Matiddi et al. 2017, Compa et al. 2019, Soto-Navarro et al. 2021, Clark et al. 2023). The focus on spatial exposure risk is also in line with the precautionary principle and is therefore considered to be adequate for wildlife conservation purposes. A spatially explicit contextualisation of the exposure risk is an essential component of the Maritime Spatial Planning Directive (MSPD, 2014/89/EU) where it has also been used to assess the risk from threats other than litter (e.g., maritime traffic, Pennino et al. 2017) and is one of the key Principles of the Integrated Monitoring and Assessment Programme (principle 5, UNEP-MAP 2016).

Despite the growing concern of the adverse effects of ML on different marine species and the potential effects on marine ecosystems, the 'Risk Assessment' topic is still underrepresented within the European research effort (Maes et al. 2019). According to Maes et al. (2019), North Sea and Mediterranean Sea regions contributed more than other countries to ML research and received more funds, while large gaps still exist in other areas and research-related topics (e.g., 'Risk Assessment' and 'Assessment Tools'). Most of the available information on ML and the risks it poses to the species is scattered widely across the literature (Soto-Navarro et al. 2021), and there is a pressing need to synthesise the information around concepts such as risk and vulnerability of species to ML.

Considering the relevance of FML and marine megafauna (i.e., cetaceans and sea turtles) as indicators to assess the risk deriving from ML pressure on marine ecosystems, this study was specifically focused on reviewing the current scientific literature on spatial Risk Exposure Assessment (hereafter REA) applied to cetaceans and sea turtles and related to the floating micro and macro ML in order to (1) identify, at global level, the main geographic areas where risk exposure studies were conducted, (2) describe the typology of datasets currently available and the methodologies used for data collection on species and threats, (3) investigate the approaches applied to carry out REA and (4) highlight the main research findings and areas of high exposure risk to prioritise mitigation measures in the Mediterranean Sea. Prior to these, a preliminary review on risk assessment terminology and framework approach was performed to contextualise the level and types of assessment methodologies reviewed (in 'Approach used'). The main objective of this review was to identify key information gaps on REA, in order to highlight areas and topics that require further research.

Methodology

Approach used

Risk is defined as "(exposure to) the possibility of loss, injury, or other adverse or unwelcome circumstance, a chance or situation involving such a possibility" (Oxford English Dictionary, OED 2005). Risk assessment is considered a technical supportive tool for decision-making when faced with uncertainty (Suter 2016) and is mostly used in a variety of human-related sectors, such as engineering, medicine, wildfire management and environmental regulation (Guzik et al. 2020, Yu et al. 2020, Sakellariou et al. 2022). In the context of the environment, the terms 'environmental risk' and 'ecological risk' are often confused in the literature, the first being commonly used to describe risk to humans due to contaminants in the environment, but being also both generally used to refer to risks to nonhuman organisms, populations, and ecosystems (Suter 2016). In wildlife management, the conventional risk framework could provide a structured process to analyse the potential risks to wild species from human activities (Gormley et al. 2011). In the case of ML, for example, risk assessments can help identify priority mitigation actions by considering the complex sources, pathways and consequences of ML on vulnerable species (i.e., cetaceans and sea turtles).

The process could be schematised into four main phases (Figure 1). The first phase is necessary to understand the source of the pressure, the pathways by which exposure might occur and the receptors (vulnerable species). Considering that environmental threats can be spatially and temporally limited, the information about where (the spatial axis) and when (the temporal axis) vulnerable species may be exposed to the threat (Werner et al. 2016) must be well understood, identifying areas/time of exposure risk (Phase 2). Once a potential risk of exposure has been identified, the risk analysis focuses on the impact (how), investigating the probability of effects and the severity (nature and magnitude) of the impact from an action or a threat (Suter 2016) (Phase 3).

To evaluate the degree of impact, information is needed on how many individuals are affected (i.e., Specificity or "Potentially Affected Fraction" in Woods et al. 2019, Hoiberg et al. 2022) and on how severely these organisms are affected (i.e., Severity). Thus, in general, the risk of being impacted represents a combination of events, starting with the vulnerable organisms being exposed to the threat, and then eventually being affected by the threat at different levels, from movement restriction, ingestion, to injury or death (Gregory 2009). Exposure to threats does not always imply impact as the latter depends on the individual behaviour, the type of litter (e.g., Duncan et al. 2019, Bucci et al. 2020) and the nature of interactions between the two, so that only a fraction of all individuals potentially exposed to the threat may be affected.

The spatial risk exposure assessment focuses on the initial phase of potential risk (Phase 2) and does not require detailed information on sensitivity and specificity (Phase 3) that are difficult to collect at sea. The lack of these latter data would delay the assessment process hampering the

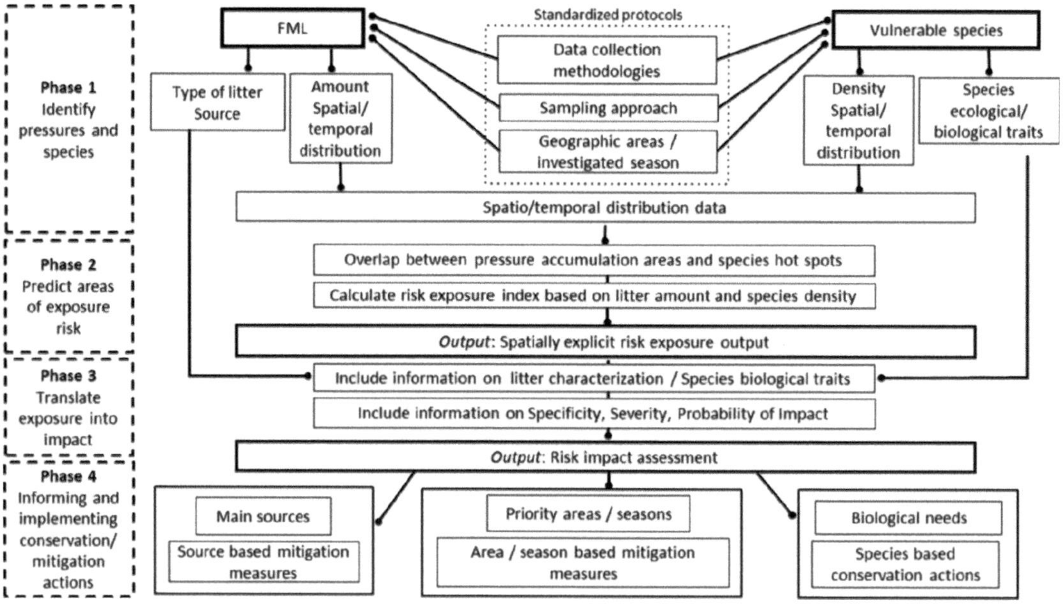

Figure 1 Risk assessment framework to analyse FML threat to marine vulnerable species and support decision-making processes.

Table 1 Eligibility Criteria in Relation to Question Key Elements following the PEC(O) Framework

Question Key Elements	Eligibility Criteria
Population (P): • Cetacean and sea turtle occurrences	*Included*: All cetacean and sea turtle species occurring at the global scale and their at-sea spatial occurrences/abundances. *Excluded*: occurrences of nesting sites, bycatch and stranding events.
Exposure (E) • Floating Marine Litter (FML)	*Included*: FML, irrespective of the object size, and their spatial occurrences/ abundances. *Excluded*: beach and seafloor litter, ingested litter.
Comparators (C) • Risk Exposure Assessment (REA) approaches	*Included*: spatially-explicit REA. *Excluded*: REA not based on geospatial analysis, or based on toxicological risk.

possibility to intervene as soon as possible to mitigate the potential impact (Phase 4). Moreover, a standardised protocol for investigating the spatial risk exposure to marine wildlife from ML would provide a key technical support for risk assessment in line with the precautionary management approach. Consequently, the articles' search was focused on spatially explicit risk exposure assessment (i.e., REA), excluding those studies not based on geospatial analysis or based on toxicological risk (Table 1).

Literature review method

Review questions, criteria for inclusion

To formulate the search and to ensure an unbiased quality of the evidence synthesis, the following research questions and associated sub-questions were defined using the standard approach for research questions definition (i.e., Population, Exposure, Comparator, Output – PEC(O) framework and CEE 2022; Table 1):

- Considering studies conducted worldwide, which is the current state of the art for REA analyses evaluating the potential risk of exposure to FML for cetaceans or sea turtles at the global scale? In particular: (1) Which geographic regions have been studied so far? (2) What datasets on species and FML occurrences are currently available, and what are the methodologies used for data collection? (3) What are the main methodological approaches used for analysing data on species and threat and to assess the exposure risk?
- At the local level for the Mediterranean Sea Region: (4) what are the main research findings and which areas have higher risk of exposure for cetaceans and sea turtles to FML?

We defined a common search strategy to answer all the questions. The key elements and the corresponding defined criteria are listed in Table 1. The definition of a priori criteria is necessary to ensure transparency and repeatability and to minimise bias (CEE 2022).

Literature search

We carried out an exhaustive systematic literature search using the SCOPUS and Web of Science databases. We curated specific search terms and synonyms according to the predefined research questions excluding terminology that was too broad (i.e., the use of the term 'marine mammals' would have also included groups of species that were not the focus of our research). Therefore, the following search terms were used for each element of the PECO framework excepted for the "output" (O) parameter:

- Population: cetacean*, dolphin*, whale*, odontocet*, mysticet*, 'sea turtle*', 'marine turtle*' AND 'marine', 'ocean*', 'sea'.
- Exposure: macroplastic*, microplastic*, 'plastic litter', 'plastic debris', 'marine litter', 'marine debris', 'anthropogenic litter', 'ocean plastic*', 'marine plastic*'.
- Comparator: risk*, 'risk assessment', 'risk evaluation', 'risk index', 'risk exposure', threat*, impact*.

The terms within each category were combined with the Boolean operator 'OR', while the categories were aggregated with 'AND'. The wildcard '*' was used to include singular/plural forms. The terms were entered for Abstract, Title and Keywords. No temporal and language restrictions were imposed. Each search was then refined by document types retaining primary literature (i.e., empirical and modelling studies) while choosing to exclude secondary literature (i.e., reviews, editorial material and early access) and grey literature (i.e., conference/proceeding papers) as the review was focused on relevant published, peer-reviewed research on the topic. The search was also refined by meso-topics, excluding those that were not related to marine science or ecology. A full list of search-related information (databases, URLs, subscribing institutions, search strings and filter options) can be found in Supplementary Table 1 (Supplementary Materials). The search result was validated by testing it against a predetermined set of 10 relevant publications found in Google Scholar, resulting in a 100% sensitivity. The benchmarking set is provided in Supplementary Table 2. The search was conducted in November 2022 resulting in a total of 550 articles; 136 duplicate articles in common between the two database results were removed, leading to a total of 415 articles (Figure 2).

Screening

The screening process was conducted in two stages: Title & Abstract and Full Text Screening (Mangano & Sarà 2017, Mangano et al. 2017, CEE 2022). In the first stage, all articles were initially assessed by title and abstract, using the inclusion criteria listed in Table 1. To check for inter-rater reliability in the inclusion criteria, a subset of the Title & Abstract (10%) that emerged from the search was independently reviewed by two authors before the screening process began and the

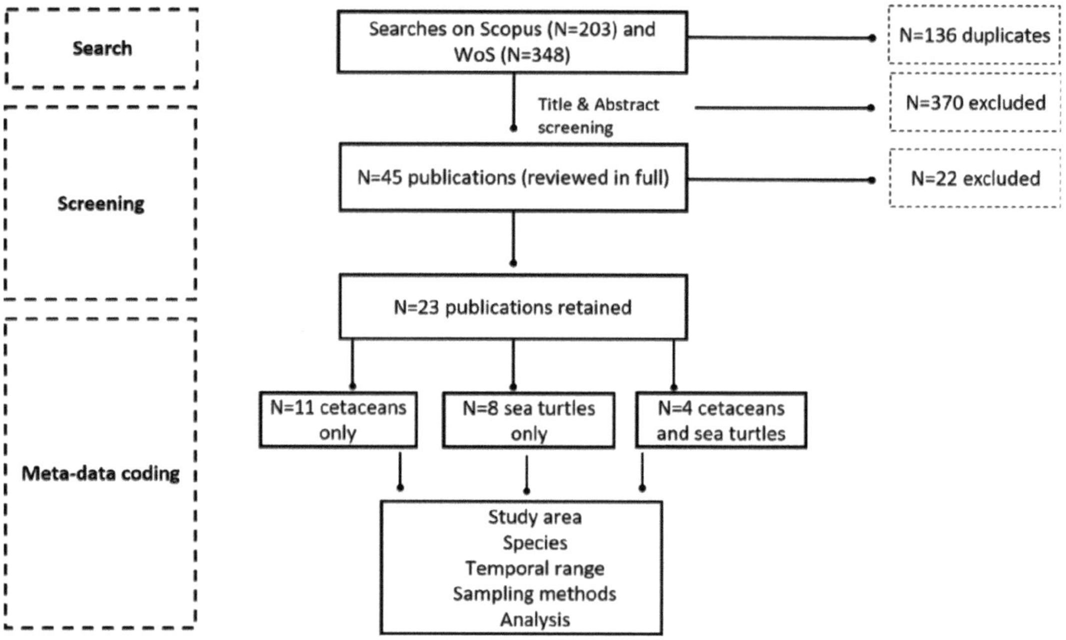

Figure 2 Flow chart outlining the systematic review process.

Cohen's Kappa test measured a perfect agreement ($k=1$, $n=30$). All full texts of the articles selected after this first step ($n=45$) were retrieved, read completely, and reviewed against the inclusion criteria (Figure 2). 22 full texts were excluded because they did not meet the specified criteria: Population ($n=6$), Exposure ($n=6$), Intervention ($n=4$) or because of the type of document (Reviews = 6), resulting in 23 articles that were included in the final review.

Meta-data extraction and coding

In order to systematically extract relevant descriptive information from the articles included in the evidence synthesis, a standard meta-data matrix was developed and agreed between the reviewers before the process of extraction began. Descriptive data were converted into a priori categories whenever possible. Studies were classified according to the following macro-categories: (1) Study (geographical) areas: indication of whether the study area is part of the Mediterranean Sea and the name of the Mediterranean sub-region (i.e., Western Mediterranean, Tyrrhenian & Sardinian-Sicilian Channels, Adriatic Sea, Central and Eastern Mediterranean, or overall Mediterranean Sea) or the non-Mediterranean area (i.e., Atlantic, Indian and Pacific Oceans or overall seas), (2) Species: cetaceans/sea turtles, indicating the name of all species studied, (3) Characteristics of FML: whether micro (<0.5 cm), meso (0.5–2.5 cm) or macro (>2.5 cm), (4) Sampling methods for data extraction/collection on species and FML and temporal range of data, (5) REA methodology: methods for species and FML data preparation (descriptive synthesis of analysis performed and inclusion in the macro-categories of occurrence points, gridded distribution, spatial generalisation of distribution, hotspot analysis, spatial distribution modelling), overlapping methods, presence/absence of a risk exposure index and its formula, whether or not the assessment considered multiple species, the temporal resolution of the REA (weekly, monthly, seasonal, annual), the number of seasons considered (1-4) and the temporal range, (6) Main results: indication of whether a risk exposure map, including seasonal maps, or no maps were produced and a description of the areas of potential high risk of exposure to FML evidenced by the 23 selected studies.

Narrative data synthesis

Data were narratively synthesised in four sections discussing key characteristics: the areas investigated; the type of datasets used on species and FML, and the methodology for the data collection; the methodology used for data analysis on species and FML and to perform the REA; the main findings from REA studies in the Mediterranean Sea. Given the limited number of resulting studies and the heterogeneous nature of the results, no statistical analysis was conducted. We provided tables and graphs to summarise data on included articles to support the narrative synthesis. As for the Mediterranean studies, areas at high risk of exposure to FML were grouped and presented in two maps, one for cetaceans and one for sea turtles. The areas were derived from the discussion sections of the 13 selected papers and from the risk exposure maps, when provided. As the REA approaches differed among studies, the synthesis maps should be considered qualitative descriptions only.

Results

The searches returned a total of 415 articles. Following the steps of inclusion/exclusion (outlined above) only 23 articles were retained and included in this review (11 on cetaceans-only, 8 on sea turtles-only, and 4 articles on both cetaceans and sea turtles). The full list of retained articles is provided in Supplementary Table 3. Analysing these 23 articles, it emerged that the risk exposure topic was considered only recently, with the first study being published in 2011 focusing on cetaceans (Williams et al. 2011), followed by another about sea turtles in 2013 (Wilcox et al. 2013, Figure 3); since 2016 risk assessment has been a constantly studied topic, with the highest number of papers published in 2022 (Figure 3).

Geographic areas investigated

Most of the studies were conducted in the Mediterranean Sea ($n=13$, 57%) and Pacific Ocean ($n=6$, 26%), spanning Canada, California, Hawaii, Galapagos, China and Australia; two studies were from the Atlantic (Darmon et al. 2017, Sá et al. 2021) and only one from the Indian Ocean (Wilcox et al. 2013, Figure 4). Only two studies were carried out at a global scale (Schuyler et al. 2016, Hoiberg et al. 2022). In the Mediterranean Sea, the western sectors were the most studied

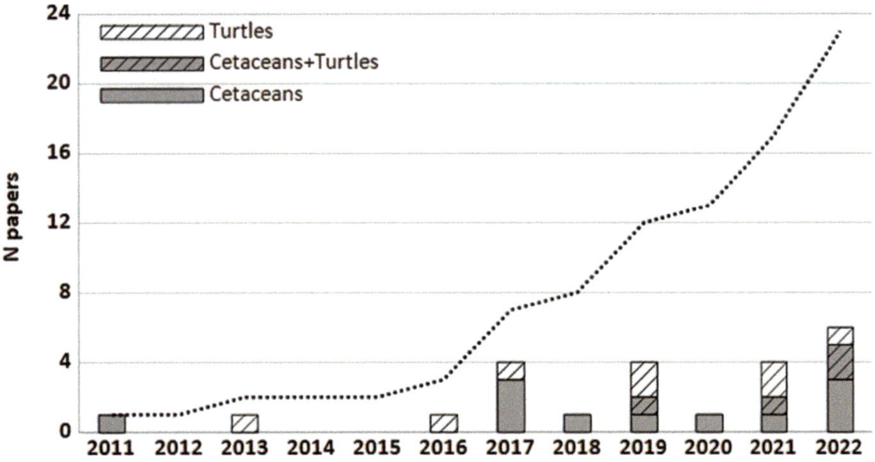

Figure 3 Temporal trend of papers about risk assessment used in this study ($n=23$), differentiated by species category. Dotted line represents the cumulative trend.

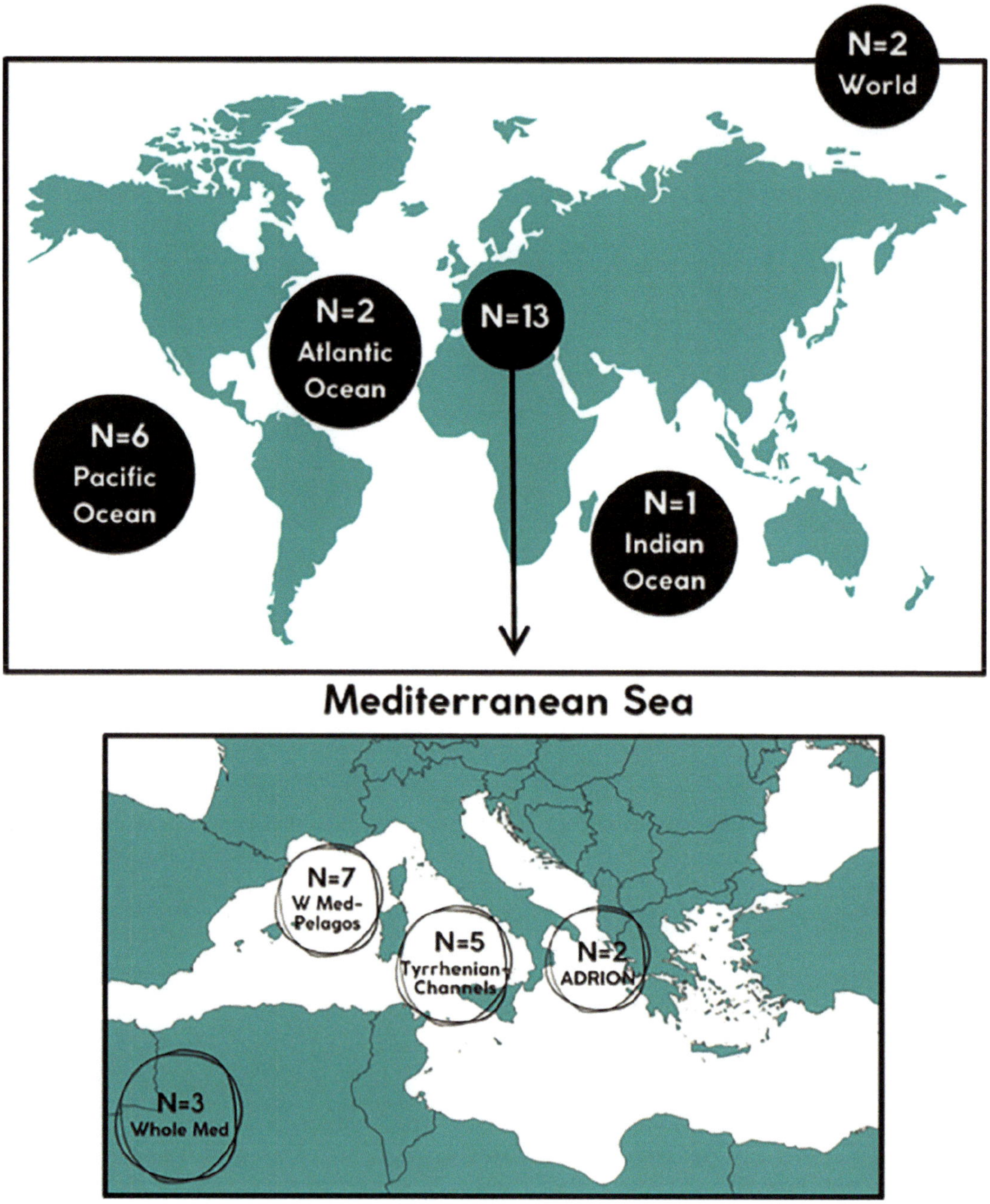

Figure 4 Global distribution of studies (upper panel, $n=23$) with a focus for the different regions of the Mediterranean Sea (bottom panel, $n = 13$).

($n=12$, 92%), including the Pelagos Sanctuary, the Tyrrhenian Sea and the Sardinian-Sicilian Channels (Figure 4). Two studies were conducted in the Adriatic-Ionian (ADRION) sub-region (Arcangeli et al. 2019, Galli et al. 2022), and three considered the whole Mediterranean basin (Compa et al. 2019, Soto-Navarro et al. 2021, Almpanidou et al. 2022).

Datasets and methodology for data collection

Fin whales (*Balaenoptera physalus*), bottlenose dolphins (*Tursiops truncatus*), striped dolphins (*Stenella coeruleoalba*) and sperm whales (*Physeter macrocephalus*) were the most studied cetaceans among the 22 cetacean species investigated for REA, while loggerheads (*Caretta caretta*) were the most studied among the sea turtle species (Supplementary Table S4). The use of data on multiple species was the most common approach ($n = 16$), while 30% of the analysed articles focused on a single species.

Datasets on FML were mostly based on the macrolitter (>2.5 cm) with 15 papers (65%) examining only this category and another including the mesolitter (0.5–2.5 cm, Hoiberg et al. 2022). Two studies considered macro and micro-litter (Critchell et al. 2019, Zhang et al. 2020), while four (17%) examined only micro-litter data (<0.5 cm). Only one study analysed all size categories (i.e., Fossi et al. 2017).

More than 60% of risk exposure assessments used field data collected at sea on species ($n = 17$) and litter ($n = 14$): the most common methodology was performing visual observations from boat surveys ($n = 12$, 52% for species; $n = 10$, 43% for litter) and in one case, all data were collected from aerial surveys (Darmon et al. 2017). In three papers (13%), field data from telemetry were used to investigate species distribution, while five articles (22%) used data on the species retrieved from available collections or repositories (e.g., OBIS, Aquamaps). In four studies (17%) data on micro-litter were collected by manta samplings. Other studies based the risk assessment on data obtained from niche or distribution models for species ($n = 3$, 13%) and hydrodynamic dispersion models for litter ($n = 8$, 35%); four papers (17%) combined field data for the species coupled with a modelled distribution of litter.

Data were collected over different temporal ranges, from 10 days to 20 years. Seven publications (30%) considered all four seasons using data collected from one (Zhang et al. 2020) to seven years (Gregorietti et al. 2021); six studies (26%) were carried out during the summer season only, ranging from few days of surveys (Fossi et al. 2017, Galli et al. 2022) to nine years averaged (Guerrini et al. 2019). Three studies (13%) considered opposite seasons (autumn/winter and spring/summer, Darmon et al. 2017, Critchell et al. 2019, Campana et al. 2022), and one study used data collected during one month in spring (Jones et al. 2021). The remaining articles used data gathered in different periods or did not specify the temporal resolution.

Approaches for REA

Methods used to analyse data on cetacean and sea turtle species and floating litter

The reviewed articles reported on a variety of methods for analysing spatial data on species and FML, the results of which are then used as data inputs for REA. For the purposes of this study, the methods were grouped into five general categories of increasing complexity, ranging from simple visualisation of "occurrence points", through "gridded distribution" and "spatial generalisation of distribution", to "hotspot analysis" or advanced "spatial distribution modelling". The first three are mutually inclusive, as the simple occurrence points (i.e., visualisation of georeferenced sighting points) can be used as input data to obtain quantitative information on gridded distribution (i.e., density/abundance of species/FML in grid cells), which in turn can be used to spatially generalise the distribution through smoothing factors (i.e., geometric features representing the area/s where the species/FML occur at different densities). For these first three categories, the methods were counted according to the higher degree of complexity. Further on, hotspot analysis can be performed on geometric features (points, lines or polygons) to infer the significance of hotspots (i.e., spatial clusters of high values), or spatial distribution modelling (SDM hereafter) can be applied to occurrence/density data of species/FML to infer distribution in unsurveyed areas. For the methods of the species distribution only, a sixth method category was included: "integrated multispecies analysis", which considers the cumulative occurrence of more than one cetacean/sea turtle species or biodiversity indices (e.g., species richness) as the starting data for REA.

Most of the studies used the same approach for the analysis either on species or on FML (i.e., occurrence points 4.3%, gridded distribution 8.7%, spatial generalisation 30.4%, hotspot analysis 8.7%, SDM 21.7%) while 35% of the studies used different approaches on species and FML.

In particular, for the analysis of cetacean and sea turtle species distribution (Figure 5A, Table 2), seven studies (30%) used the occurrence points from visual sightings or telemetry data, while the gridded distribution of species density was used in two studies (Wilcox et al. 2013, Currie et al. 2017). On the other hand, spatial generalisation of distribution was the predominant method used to determine the core area of the species distribution ($n = 10$, 43%): within this category, eight studies used the Kernel density estimation technique based on species occurrence or abundance, while two studies used the area included in the Regional Management Units (RMU) as representative of higher sea turtle occurrences (Schuyler et al. 2016, Hoiberg et al. 2022). Hotspot analysis was conducted in two studies, in particular using the G* analysis (Getis & Ord 1992) to highlight the statistical significance of the different hotspots (Arcangeli et al. 2019, Gregorietti et al. 2021). SDM

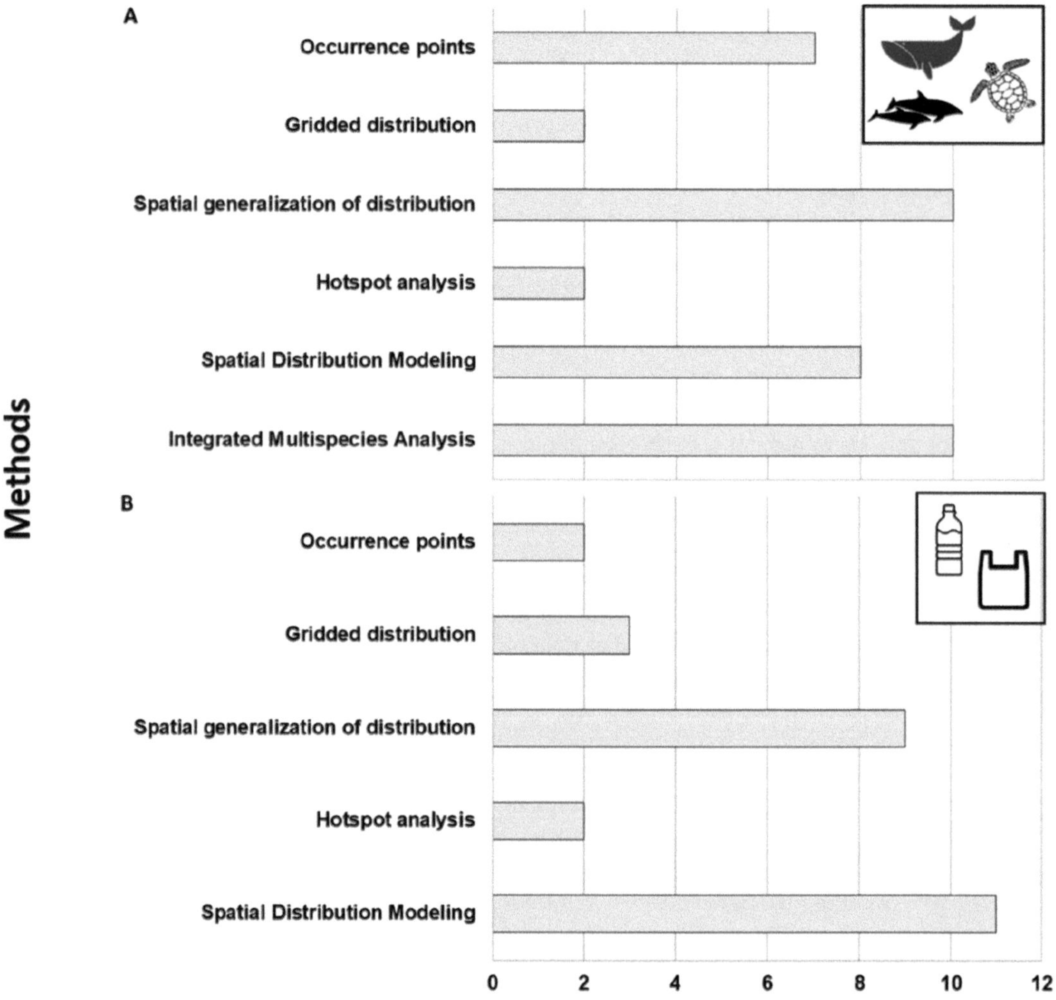

Figure 5 Methods used to analyse distribution data on cetacean and sea turtle species (A) and floating litter (B) prior to performing the REA; x-values refer to the number of articles.

Table 2 Summary of Methodological Approaches for Data Collection, Methods Used to Analyse Data on Vulnerable Species and Floating Marine Litter (FML) and for RISK Exposure Assessment (REA)

#	Authors	Type of data Collection		Methodological Approaches for Analysis		
		Species	FML	Species	FML	REA
1	Almpanidou et al. (2022)	Telemetry	No data collection (from existing model)	SDM (ensemble modeling)	SDM (Lagrangian modeling)	• Threat risk × cell in turtle foraging grounds • Map
2	Arcangeli et al. (2019)	Visual monitoring	Visual monitoring	Spatial generalisation of distribution; Hotspot analysis	Spatial generalisation of distribution; Hotspot analysis	• Aggregated raster of species and litter densities • Map
3	Atzori et al. (2021)	Visual monitoring	Visual monitoring	Spatial generalisation of distribution; SDM (GAM)	Spatial generalisation of distribution	• Risk × cell (litter density × species abundance) • Map
4	Campana et al. (2018)	Visual monitoring	Visual monitoring	Occurrence points	Spatial generalisation of distribution	• Overlap of species sightings and plastic density isopleth • Map
5	Campana et al. (2022)	Visual monitoring	Visual monitoring	SDM (GAM); Integrated Multispecies Analysis	Spatial generalisation of distribution	• Biological traits • Sensitivity index × cell (litter density × species sensitivity) • Map
6	Compa et al. (2019)	No data collection (databases)	No data collection (from existing model)	Occurrence points, Integrated Multispecies Analysis	SDM	• Litter density × binary/weighted species distribution • Biological traits • Ingestion probability estimation with GAM • Map
7	Critchell et al. (2019)	Telemetry	No data collection (from existing model)	Occurrence points	SDM (Lagrangian modeling)	• Overlap of homerange % in the relative exposure categories (litter density)
8	Currie et al. (2017)	Visual monitoring	Visual monitoring	Gridded distribution	Gridded distribution	• Risk × cell (litter density × species density) • Map

(Continued)

Table 2 (*Continued*) Summary of Methodological Approaches for Data Collection, Methods Used to Analyse Data on Vulnerable Species and Floating Marine Litter (FML) and for RISK Exposure Assessment (REA)

#	Authors	Type of data Collection		Methodological Approaches for Analysis		
		Species	FML	Species	FML	REA
9	Darmon et al. (2017)	Visual monitoring	Visual monitoring	Spatial generalisation of distribution	Spatial generalisation of distribution	• Linear distance between turtles and litter • Frequency of turtles surrounded by litter • Mean number of surrounding litter × turtle occurrence for each distance class
10	Di-Meglio and Campana (2017)	Visual monitoring	Visual monitoring	Spatial generalisation of distribution	Spatial generalisation of distribution	• Overlap of species sightings and plastic density isopleth • Map
11	Fossi et al. (2017)	Visual monitoring	Visual monitoring	SDM (FHO)	SDM (Lagrangian modeling)	• Qualitative overlap
12	Galli et al. (2022)	Visual monitoring	Manta sampling	Spatial generalisation of distribution; Integrated Multispecies Analysis	Gridded distribution	• Risk index × cell (species sensitive scores × litter density) • Map
13	Gregorietti et al. (2021)	Visual monitoring	Visual monitoring	Spatial generalisation of distribution; Hotspot analysis; SDM (biomod)	Spatial generalisation of distribution; Hotspot analysis	• Risk index × cell (litter density × species abundance) • Map
14	Guerrini et al. (2019)	No data collection (databases)	No data collection (from existing model)	SDM (FHO)	SDM (Lagrangian modelling)	• Risk index × cell (litter density × species habitat) • Map
15	Hoiberg et al. (2022)	No data collection (databases)	No data collection (from existing model)	Spatial generalisation of distribution; Integrated Multispecies Analysis	SDM (Oceanographic dispersal modelling)	• Species sensitivity distribution (litter density × potential entanglement exposure areas) • Potentially Affected Fraction of species (PAF) • Map
16	Jones et al. (2021)	No data collection (databases)	Manta sampling	Occurrence points; Integrated Multispecies Analysis	Occurrence points	• Biological traits • Species distribution × Conservation scores × Severity
17	Kahane-Rapport et al. (2022)	Telemetry	Manta sampling	Occurrence points	SDM	• Modelled litter ingestion rate

(*Continued*)

Table 2 (*Continued*) Summary of Methodological Approaches for Data Collection, Methods Used to Analyse Data on Vulnerable Species and Floating Marine Litter (FML) and for RISK Exposure Assessment (REA)

#	Authors	Type of data Collection		Methodological Approaches for Analysis		
		Species	FML	Species	FML	REA
18	Sá et al. (2021)	Visual monitoring	Visual monitoring	Spatial generalisation of distribution; Integrated Multispecies Analysis	Spatial generalisation of distribution	• Risk index × cell (litter density × species density) • Map
19	Schuyler et al. (2016)	No data collection (databases)	No data collection (drifting buoys)	Spatial generalisation of distribution; Integrated Multispecies Analysis	SDM (Lagrangian modeling)	• Risk index × cell (litter density in species RMU • Map
20	Soto-Navarro et al. (2021)	No data collection (databases)	No data collection (from existing model)	SDM; Integrated Multispecies Analysis	SDM (Lagrangian modeling)	• Risk index × cell (averaged litter density, vulnerability and species distribution) • Map
21	Wilcox et al. (2013)	Visual monitoring	No data collection (model of beached ghost nets)	Gridded distribution; Integrated Multispecies Analysis	SDM	• Risk index × cell (relative turtle density (caught) × ghost fishing effort) • Map
22	Williams et al. (2011)	Visual monitoring	Visual monitoring	Spatial generalisation of distribution; Integrated Multispecies Analysis	SDM (GAM)	• Risk index × cell (litter density × species density) • Map
23	Zhang et al. (2020)	Visual monitoring	Visual monitoring	Spatial generalisation of distribution	Spatial generalisation of distribution	• Risk index × cell (litter density × species density × coefficients) • Map

approaches were applied in eight studies (35%): for example, Generalized Additive Models (GAM) and density surface modelling were used to define areas of high presence of the target species, or the Feeding Habitat Occurrence (FHO) model was applied to trace the potential fin whale feeding areas (Fossi et al. 2017, Guerrini et al. 2019).

For FML spatial analysis (Figure 5B, Table 2), only two studies used the occurrence locations of FML as data input for the REA (Jones et al. 2021, Kahane-Rapport et al. 2022). Three studies calculated the gridded distribution of FML density instead (13%), while nine articles (39%) used a spatial generalisation of the density distribution, smoothed through the Kernel function to define areas of higher pressure of FML. Two studies performed the hotspot analysis (i.e., G* analysis) to highlight the statistical significance of hotspots (Arcangeli et al. 2019, Gregorietti et al. 2021). Nevertheless, the majority of the studies ($n = 11$, 48%) applied SDM approaches to define areas of high FML densities, mainly Lagrangian modelling.

Methods used for REA

The main principle in spatial REA is the identification of the areas of potential risk exposure for vulnerable species, by understanding the source of the pressure and the pathways by which exposure might occur. Two general approaches are commonly used for spatial REA: (1) the overlap of areas with a high density of both species and FML to identify areas with higher exposure; (2) quantification of REA through indices, calculated directly on spatially gridded data by combining species and FML occurrences/densities. Of the reviewed articles, four investigated the risk exposure by using the spatial overlap of hotspots of megafauna species and FML defining the percentage of seasonal overlap (Di-Méglio & Campana 2017, Campana et al. 2018) and describing it qualitatively (Fossi et al. 2017) or by assessing the proportion of the species' home range in areas of high ML densities (Critchell et al. 2019). The majority of the studies ($n = 19$, 78%) estimated instead the extent of risk exposure using the risk index-based method. Ten studies aggregated the spatially gridded densities of FML and the vulnerable species (Williams et al. 2011, Schuyler et al. 2016, Currie et al. 2017, Arcangeli et al. 2019, Compa et al. 2019, Zhang et al. 2020, Atzori et al. 2021, Gregorietti et al. 2021, Sá et al. 2021) or their suitable habitat (Guerrini et al. 2019). Other studies integrated the risk index with other variables, such as species vulnerability, biological characteristics or conservation status, to measure the degree of the impact (i.e., Specificity, Severity, Table 2). As for Specificity, Jones et al. (2021) considered a conservation value based on IUCN categories to calculate risk indices on sea turtles, one study calculated sensitivity values considering the presence of different marine species (Galli et al. 2022), and similarly another study considered a sensitivity index of cetacean species including biological features such as life stage (Campana et al. 2022). Other studies calculated specific risk indices, including ingestion probability (Compa et al. 2019, Soto-Navarro et al. 2021, Almpanidou et al. 2022, Kahane-Rapport et al. 2022) and turtle/cetacean entanglement (Wilcox et al. 2013, Darmon et al. 2017, Hoiberg et al. 2022).

Main findings from spatio-temporal risk exposure studies in the Mediterranean

The findings from the Mediterranean REA studies resulting from the literature search ($n = 13$) were heterogeneous in terms of species, temporal resolution and provided maps. Areas reported as being at higher-risk of exposure to FML were summarised and reported separately for cetaceans, and sea turtles.

Cetaceans

The investigations carried out in the western Mediterranean basin revealed the main risk exposure areas during summer in the Liguro-Provençal basin, including the Pelagos Sanctuary (Figure 6). Di-Méglio and Campana (2017) highlighted monthly changes in risk exposure areas for cetacean

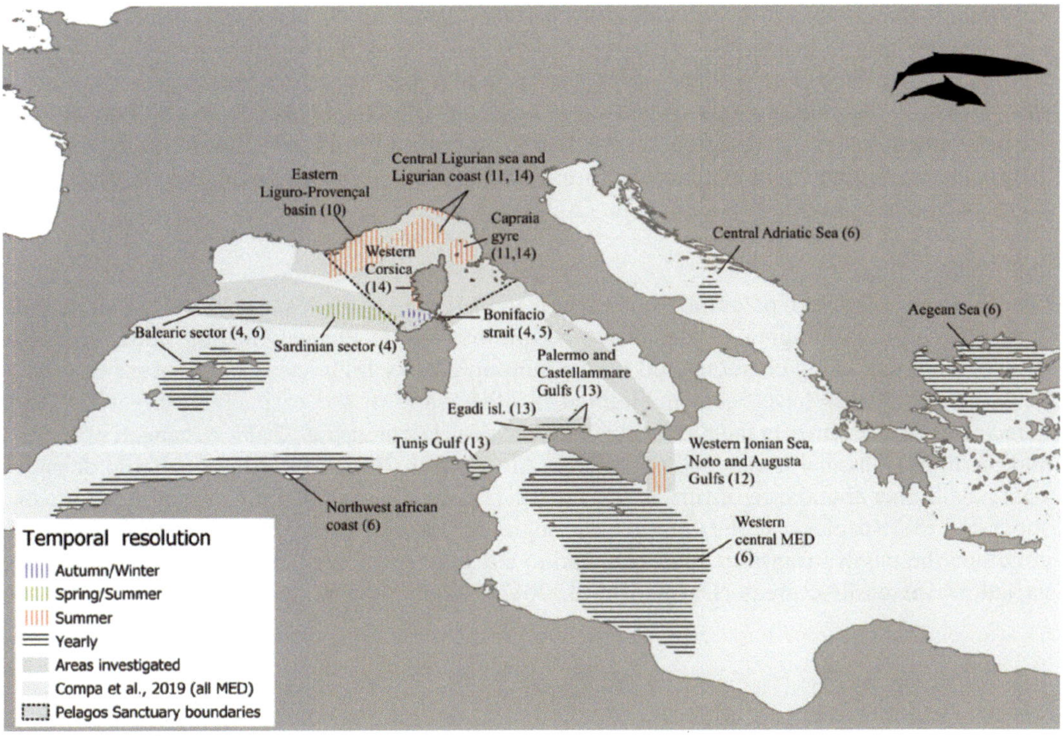

Figure 6 Synthesis map of annual/seasonal areas of higher risk of exposure to FML for different species of cetaceans emerged from the studies analysed (vertical lines coloured by seasonal resolution and black horizontal lines for yearly resolution); the numbers in brackets correspond to the studies to which each area refers, based on the numeration used in Table 2. For reference on the season and species investigated by the different studies, refer to the text. The areas investigated by the reviewed studies are indicated in grey in the figure; only Compa et al. (2019) (study N. 6) investigated the whole Mediterranean basin (light grey in figure).

species driven by a dynamic pattern of FML densities, with minimum values in the middle of the summer compared to other months (from May to September). FML resulted in overlapping by 50% with the known ranges for six cetacean species (fin, sperm and pilot whales, and Risso's, bottlenose and striped dolphins), indicating high potential of interaction, especially in the western part of the Pelagos Sanctuary. Fossi et al. (2017) showed a possible overlap of fin whale summer feeding habitats and areas of high microplastic concentration, which would increase the risk of ingestion, in the external part of both cyclonic and anticyclonic structures such as the Capraia gyre. As well, Guerrini et al. (2019) investigated the risk of fin whales to microplastic exposure via food ingestion during the summer season and found spatial and interannual variability of risk exposure patterns, with the highest values in the Central Ligurian Sea and along the Ligurian and Western Corsica coasts. At mid-latitudes in the western Mediterranean Sea, Campana et al. (2018) provided seasonal data for fin whales, dolphins and squid eaters (sperm, Cuvier's beaked and pilot whales, and Risso's dolphins), showing the most evident overlaps between high density of FML and all cetacean species in the Balearic Sea in all seasons, as also identified by Compa et al. (2019). During spring and summer, risk exposure areas for fin, sperm whales and dolphins were found in the Sardinian Sea and the coastal waters of the Bonifacio Strait (Campana et al. 2018, 2022). In the Bonifacio Strait, a generally higher risk was also identified particularly for fin whales and striped dolphins in autumn-winter, due to the offshore accumulation of FML (Campana et al. 2022). In the southern Mediterranean basin, several areas have been identified at particular risk during all the seasons, such as the African coast, the western central Med (Compa et al. 2019), the Sicilian Channel, where Gregorietti et al. (2021) identified the waters outside Palermo harbour along the coast to the Castellammare Gulf, the Egadi Islands and the Tunis Gulf as areas of risk exposure to plastic for striped and bottlenose dolphins throughout the year, albeit with seasonal variation of intensity. Other areas of exposure of cetacean species to FML have been described in the central Adriatic Sea and the Aegean Sea (Compa et al. 2019), while a specific area in the western Ionian Sea, in the Noto and Augusta Gulfs, has been highlighted during summer by Galli et al. (2022).

Sea turtles

One of the major areas of potential exposure risk to FML for sea turtles is reported by Almpanidou et al. (2022) in the foraging area enclosed within the Tunisia Plateau/Gulf of Sidra, with about 40% of its extent being characterised by medium up to very high levels of exposure (Figure 7). Seasonal differences have been highlighted in other parts of the basin, with larger risk areas detected during summer in the Sardinian-Balearic Sea (Darmon et al. 2017, Arcangeli et al. 2019) and Sardinian Channel (Arcangeli et al. 2019, Atzori et al. 2021) due to the increased density of FML and higher abundance of turtles (Figure 7). These authors in fact, investigating all seasons, suggested that species-specific foraging strategies, biological cycle, and oceanic features may condition the passive transport of debris, and to a degree sea turtles, explaining spatio-temporal variations in sensitive areas (Darmon et al. 2017, Arcangeli et al. 2019). In the middle of the

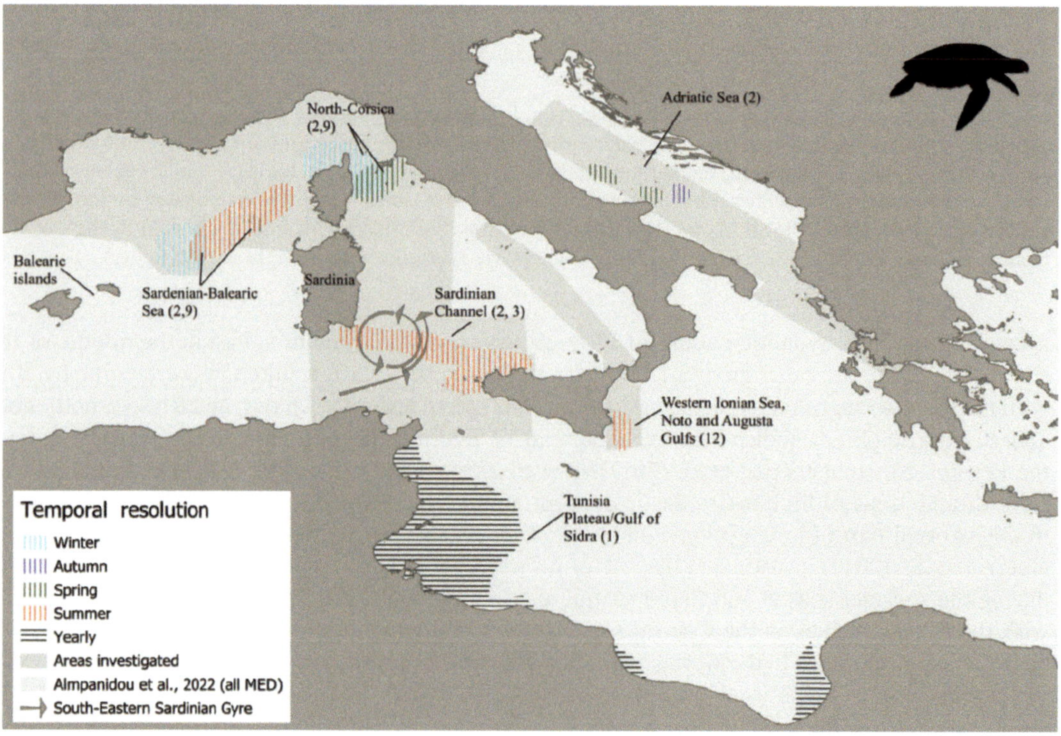

Figure 7 Synthesis map of annual/seasonal areas of higher risk of exposure to FML for sea turtles emerging from the analysed articles (vertical lines coloured by seasonal resolution and black horizontal lines for yearly resolution). The numbers in brackets correspond to the studies to which each area refers, based on the numeration used in Table 2. For reference on the season investigated by the different studies, refer to the text. The areas investigated by the reviewed studies are indicated in grey in the figure; only Almpanidou et al. (2022) (study N. 1) investigated the whole Mediterranean basin (light grey in figure).

Sardinian Channel, for example, the large South-Eastern Sardinian Gyre appears to act as a trap or convergence area for both animals and FML (Atzori et al. 2021). During winter, Darmon et al. (2017) identified two major overlaps between areas of FML concentration and loggerhead turtles in the North of Corsica and between Sardinia and Balearic islands, while Arcangeli et al. (2019) highlighted a high-risk exposure area for sea turtles in northern Corsica during spring. That study also confirmed the well-known area for loggerhead sea turtles in the Adriatic Sea, where high risk of exposure was recognised during spring as well. Finally, a localised area for microlitter exposure was described in the western Ionian Sea, in the Noto and Augusta Gulfs, during summer by Galli et al. (2022).

Discussion

Despite the relevance of the topic, and the legislative requirements, this review revealed that using risk assessment to analyse the spatial risk exposure produced by FML on highly vulnerable cetaceans and sea turtles has only been considered as recently as 2011 and by a limited number of studies (23 articles). Among the global, EU and Regional legislative frameworks that address the threat posed by the anthropogenic pressures such as marine litter on marine ecosystems (e.g., CBD, Habitats Directive, MSFD, MSPD, UNEP-MAP Barcelona Convention), since 2008 the EU MSFD specifically requires member states to assess the risk for marine species by the ML threat, connecting the Descriptor 1 (i.e., Biodiversity) to the Descriptor 10 (i.e., Marine Litter) likely giving an effective impulse to the REA studies. However, despite the increase of recent research efforts, it appears that a large gap still remains in the literature dedicated to identifying critical areas for species, hampering the possibility to identify priorities for conservation and successive effectiveness of mitigation measures.

By reviewing the available papers on spatial REA, this study offered a first insight into the available methodological approaches that focus on the initial phases of potential risk assessment (Figure 1) not requiring detailed information on true impact (Phase 3) that could be difficult to collect at sea. The information collected will potentially contribute to the standardisation of an approach in line with the precautionary principle.

Main studied areas

Most of the risk assessment studies were concentrated in European countries with few case studies carried out in the ocean near other continents (Northern Australia, China, Hawaii, Canada). This could be due to the implementation of the EU MSFD that, since 2008, gives a direct input for integrating the effect of pressures in the assessment of the status of marine waters, and thus identifying mitigation measures according to the information gathered. Within the EU, results confirmed one of the findings of Maes et al. (2019), with most of the studies concentrated in the Mediterranean Sea, and only two in the Atlantic Ocean. Our review also showed that, within the Mediterranean basin, the western Mediterranean Sea was the most investigated area accounting for a higher number of studied species, likely due to the greater research effort in this subregion especially on cetaceans (e.g., Gnone et al. 2023).

Datasets

This review revealed that only a limited number of studies were able to use empirical data collected at sea on species and ML, and almost 40% of reviewed articles based the assessment on modelling data, especially for larger scale assessments. Most studies focused on common cetacean (i.e., fin whale, bottlenose and striped dolphins) and sea turtle species (i.e., loggerhead turtle), reflecting the relatively larger availability of data, while little effort is dedicated to the other highly vulnerable cetaceans and sea turtle species. The limited availability of continuous data on abundance and distribution of species and on the amount, distribution, and composition of ML in representative geographical areas on a seasonal basis is an impediment for exhaustive REA and likely the main cause for the limited number of articles addressing this topic. For FML in particular, the high variability in FML empirical data leads to a reliance, especially on modelled approaches. However, at a global level, it could also be the case that, despite several empirical or modelled datasets available for either species or for FML, these have not yet been used for assessing exposure risk (REA) (e.g., Miranda-Urbina et al. 2015). At the EU level, the EEA Report (2020) recognises that the ranges, population sizes and suitable habitat areas of cetacean species (apart from two Annex II species) remain unknown in most Member States, which is generally due to the lack of appropriate monitoring programmes. The review of reports produced by Member States on the MSFD Descriptor 1 (i.e., Biodiversity, Palialexis & Boschetti 2021) and Descriptor 10 (i.e., Marine Litter, Ruiz-Orejón et al. 2021) also recognised the lack of data on both, species and ML, and on the overlap between spatial distribution of human activities/pressures and species distribution.

Some large-scale synoptic visual data collection programs for FML and vulnerable species (e.g., cetaceans, sea turtles) exist in the Mediterranean basin (i.e., IMAP, UNEP-MAP Integrated Monitoring and Assessment Programme; ASI, Accobams Survey Initiative; FLT, Fixed Line Transect Mediterranean Monitoring Network). These make use of standardised protocols for data collection on cetaceans, marine turtles and/or on litter (e.g., Arcangeli et al. 2020), but are not yet sufficiently comprehensive temporally or spatially, or in terms of the species considered. Telemetry is generally applied on sea turtles and was indeed used by three of the investigated articles to infer the spatial pattern of the vulnerable species. Technological advances in satellite telemetry facilitated the increasing research effort and the identification of key neritic and oceanic areas, for example, loggerhead turtle in the Mediterranean Sea (Pasanisi et al. 2022). Telemetry could also provide useful information on demographic parameters (Girard et al. 2022) and biological traits for evaluating the probability of impact. Thus, despite some existing initiatives or techniques already available, there is still a large lack of coordination and integration. Indeed, the integration of existing datasets and initiatives ensuring that continuous standardised data are collected on species and threats should be a priority to guarantee that a solid baseline of information exists, allowing long-term assessment of the effectiveness of mitigation measures.

Approach for risk exposure assessment

This review confirmed, in line with Soto-Navarro et al. (2021), that methodologies for assessing spatial risk exposure vary according to the data available on the species and ML, their resolution and the study's specific scope. Generally, they make use of modelled or empirical data collected at sea on ML and sensitive species to assess the degree of overlap between the areas of species distribution and high pressure from ML. The choice of the investigated species is crucial for assessing the risk posed by marine litter on marine biodiversity and ecosystems. Most articles examined focused on a single taxonomic group, either sea turtles or cetaceans, and 30% examined a single species. The selection of species is likely mostly driven by the availability of data. A few articles calculated the risk of exposure across multiple taxa including polychaetes, crustaceans, gastropods, cephalopods, bivalves, fishes, sharks, other reptiles, pinnipeds and seabirds in addition to cetaceans and sea turtles, making use of large-scale data repositories trying to represent marine biodiversity in general (Compa et al. 2019, Soto-Navarro et al. 2021, Galli et al. 2022). Some nature and biodiversity regulatory frameworks (e.g., Habitats Directive, HD 92/43/EEC) are mostly species-focused and centred on the conservation of single threatened cetaceans and sea turtles, requiring the identification of specific priority needs. Instead, other more global or more recent legislation (e.g., CBD, MSFD, MSPD, UNEP-MAP Barcelona Regional Sea Convention) uses the ecosystem approach, which considers species to be functional components of the local biodiversity, driving ecosystem functions and processes, and the conservation effort is focused on preserving balanced and functioning ecological communities (Hartje et al. 2003, Shepherd 2004).

Cetaceans and sea turtles are generally considered valuable umbrella species (UNEP-MAP RAC/SPA 2010, Pace et al. 2015). They are good indicators of ecological processes, as their protection has been shown to have positive effects on community structure and ecosystem functioning (e.g., Pace et al. 2015, Pennino et al. 2017). They are among the species identified as target indicators for monitoring the impacts of ML (e.g., Fossi et al. 2017, 2020, Matiddi et al. 2017). The accurate selection of multiple species sensitive to the pressure but representative of different trophic guilds and different potential interactions with the threats could support scaling up the REA from the single species to the ecosystem-level approach. To this end, a framework for a consistent and transparent selection of marine environment components (species) according to a set of scientific and practical criteria was recently suggested to identify the most representative species to be selected for REA (Palialexis et al. 2022).

Investigating the spatial overlap between hotspots of megafauna species and FML is the first step to identify areas of higher risk exposure. Few articles limited the analysis of risk exposure at this stage, or just qualitatively describing it, while most of the reviewed articles attempted to quantitatively estimate the magnitude of the exposure risk (78%). Moving from a qualitative approach towards more quantitative assessments of risk is surely required to better focus on the higher-risk contexts. The examined literature revealed that some progress is being made to quantify exposure intensity, but methodological approaches are still highly heterogeneous. Most of the studies used spatially-explicit risk indices over grid cells, quantifying the magnitude of risk exposure by multiplying the density of litter and species. Some examples are also available of studies that accounted for biological features, which increase species vulnerability, such as integrating species sensitivity scores based on biological (e.g., Campana et al. 2022) or functional (e.g., Jones et al. 2021) traits. The inclusion of vulnerability and trait analyses (e.g., integrating information on life history, morphological and behavioural characteristics of species present in assemblages to indicate aspects of their ecological functioning) into the conservation frameworks would enable predictions to be made regarding organism responses to future environmental changes and allows broader conservation actions to be selected (Miatta et al. 2021, Clark et al. 2023). Ideally, the exposure risk index equation should include data on amounts of litter and species density, and parameters that account for species vulnerability (e.g., biological trait, species richness and presence of juvenile) and information on litter characterisation correlated with the probability of impact (e.g., type, size, material and shape).

Finally, the integration of multiple pressures for REA analysis is required specifically by the MSFD and MSPD, to account for their combined effects on marine species and habitats. Recently, in addition to North and Baltic Sea, also the western Mediterranean Sea and the Adriatic Sea were recognised among the areas with the most extensive combined pressure effects within the EU marine areas, being under severe anthropogenic pressure from pollution, habitat loss and increasing disturbance due to intensive fishery and coastal activities (Korpinen et al. 2019). A framework for recognising critical areas that integrates spatio-temporal information on sensitive species and multi-sectoral sources of potential impacts is still lacking (Maes et al. 2019). The MSP Directive is designed to manage multiple uses of the seas, ensuring efficiency and sustainability of human activities. Even if not directly targeting ML, the MSPD aims to sustainably manage marine activities recognised as potential sources of ML, both from land-based activities (coast-land interaction) or sea-based (e.g., aquaculture, fishing, maritime transport and tourism). The MSPD aims to implement a framework for consistent and evidence-based decision-making. As such, it could contribute to the identification of the spatial and temporal distribution of relevant existing and future activities, and to the sustainable development of these sectors to ensure the preservation of species and the protection of the environment.

Main research findings in the Mediterranean Sea

The available studies in the Mediterranean Sea were heterogeneous in terms of area, species studied, maps provided and temporal resolution. Nevertheless, findings were in most cases consistent across studies, with some high-risk exposure areas emerging from different authors, such as the Balearic Sea (Campana et al. 2018, Compa et al. 2019) and Pelagos Sanctuary (Fossi et al. 2017, Guerrini et al. 2019, Campana et al. 2022) for cetaceans, and the Sardinian-Balearic Sea, north Corsica (Darmon et al. 2017, Arcangeli et al. 2019) and the Sardinian Channel (Arcangeli et al. 2019, Atzori et al. 2021) for sea turtles. High spatio-temporal variability of FML exposure was found in all studies, either basin-wide (Arcangeli et al. 2019, Compa et al. 2019, Soto-Navarro et al. 2021, Almpanidou et al. 2022) or at regional or local scales (e.g., Di-Méglio & Campana 2017, Campana et al. 2018, Guerrini et al. 2019, Atzori et al. 2021, Galli et al. 2022), regardless of the species considered. These findings are in line with the known variability in the distribution of either the vulnerable species and the pressure. High seasonal and inter-annual abundance and dispersal variability are indeed known for cetaceans and sea turtle species (e.g., Arcangeli et al. 2017, Arcangeli et al. 2019, Zampollo et al. 2022). Also, the accumulation/dispersion pattern of litter is highly variable in the Mediterranean basin, with a large transboundary dispersion and the absence of any formation of permanent large or local FML accumulations (i.e., Arcangeli et al. 2018, Liubartseva et al. 2018, Mansui et al. 2015, 2020, Macias et al. 2022). Marked seasonal and inter-annual variability with areas of FML accumulation was recognised mainly during summer/autumn, but with lower values indicating the gradual formation/disintegration of accumulation pattern also seen in June, October and November (Liubartseva et al. 2018). A spatial stratification for potential exposure was also identified between offshore and coastal areas (Compa et al. 2019, Arcangeli et al. 2020). In coastal areas, the risk of exposure is considered to be mainly due to the marine fauna's proximity to sources of ML, such as estuaries, ports and touristic areas, that produce higher concentrations of FML (e.g., Compa et al. 2019, Gregorietti et al. 2021, Campana et al. 2022). In offshore waters, the main drivers of variability in risk exposure were recognised in the dynamic oceanographic factors that determine FML accumulation and, partially, influence the distribution of early-stage sea turtles (e.g., Darmon et al. 2017, Atzori et al. 2021) and the species active selection of foraging grounds in areas with a high accumulation of litter (e.g., Darmon et al. 2017, Fossi et al. 2017, Campana et al. 2022).

Knowledge gaps and research needs

Despite the difficulties in gathering the needed data at an adequate spatio-temporal resolution, and the lack of a standardised methodological approach, a growing number of studies have recently made an effort to support management decisions by identifying the highest risk context in which managers should prioritise mitigation measures or conservation efforts. However, high informational heterogeneity was recognised and several priorities to improve future research efforts can be found. With this scope, the following points outline some key aspects that must be considered to improve the ability to effectively assess cetacean' and sea turtle' risk of exposure to one of the main anthropogenic pressures, i.e., marine litter (Table 3).

- The risk exposure assessment process largely relies on the availability of empirical data, although there are few data sources available. The need for data is particularly necessary to account for the high variability of either the species or the pressure. The integration of different datasets has proved to be effective for sea turtle species (Palialexis & Boschetti 2021, Girard et al. 2022). Additionally, a combined effort of research and monitoring programmes, including those responding to the different directives/regional sea conventions, are able to enhance the availability of continuous data. In general, continuous long-term monitoring programmes at appropriate spatial and temporal scales are needed to provide

Table 3 Main Knowledge Gaps Identified by the Review for Risk Exposure Assessment and Recommendations for Future Improvement

Knowledge Gaps	Research Needs
Scarcity of empirical data on ML and distribution of sensitive species	• Integration of available dataset. • Combined effort from research and monitoring programmes. • Support continuous long-term monitoring programmes.
Information biased towards the summer season	• Support year-round monitoring programmes to include important areas for the species in all seasons and ensure adaptive measures targeting the right sector and area at the right time.
Lack of studies on less common sensitive species (e.g., squid-eating species)	• Enlarge the REA to other sensitive species. • Enhance multi-target studies targeting species representative of different potential harms and ecosystem compartments.
Lack of studies outside EU and in specific EU regions	• Promote REA studies where data are available, and enhance data collection in unsurveyed regions.
Lack of standard protocol for REA, mostly based on more qualitative analysis	• Advance towards a standardisation of analytical approaches to assess trends in risk over time, and allow evaluation of the effectiveness of mitigation measures

realistic information, allow analysis of trends and support the development and design of important conservation and management measures to reduce the risk of exposure;

- In terms of temporal resolution, only a third of the reviewed papers considered all the seasons, while another third focussed on the summer season only. While present results underline that many studies are conducted during summertime, when still high risk can be found (e.g., Campana et al. 2018), seasonality plays a key role as the main driver of variability in species distribution and abundance (e.g., Arcangeli et al. 2017, Zampollo et al. 2022) as well as the composition and the source of litter change with the seasons (e.g., Mansui et al. 2015, Arcangeli et al. 2018). Thus, a bias of surveys towards summer might seriously compromise the validity of some of the REAs, as it does not account for phenology of species that might reproduce or aggregate in sensitive habitats during all seasons, when the information is missing. Therefore, studies on seasonality are crucial in understanding how adaptive measures may be implemented in the right sector and areas at the right time;
- So far, attention has been given to the most common species (e.g., fin whale, bottlenose and striped dolphin, sperm whale, loggerhead turtle), however info is especially urgently needed for the least abundant species known to be affected by litter, such as the squid-eating species (e.g., sperm whale, Cuvier's beaked whale, López-Martínez et al. 2023) or other sea turtles (e.g., green turtle, Duncan et al. 2019). The inclusion of multi-species targets, effectively selected as representative of different potential harms and ecosystem compartments, and a trait-analysis (e.g., integrating information on life history, morphological and behavioural characteristics of species occurring in groups to reveal aspects of their ecological functioning), is the main path to support a smart-implementation of a functional and ecosystem-based risk assessment approach, increasing the reliability of (bio-geo) distributional analysis and allowing prediction of organism responses to environmental changes;
- Few studies were conducted outside the EU countries. The Mediterranean basin was the most investigated among seas. However, large geographical gaps in risk assessment emerged also in the Mediterranean Sea, especially in the southern and Levantine basin and we need to increase our research efforts to provide such information;

- Methodological approaches to assess the contextual presence between ML and animals from which to derive a robust assessment of risk exposure are still highly heterogeneous. There are, however, some examples for quantitatively investigating the magnitude of risk exposure, also incorporating biological traits to account for species vulnerability. Advancement towards a standardisation of approach for analysis, limiting informational heterogeneity and uncertainty degree, would enhance the possibility to assess risk trends over time and evaluate the effectiveness of mitigation measures;
- Lastly, while this review was specifically focussed on the first two phases of risk assessment (i.e., Risk Exposure Assessment), further input should be given to integrate data and analytical approaches that include information about the potential of items to be harmful, the biological-ecological variables that we expect having an influence on the species vulnerability (e.g., life stage), the rate of entanglement/ingestion (i.e., specificity), and the severity of the impact at the individual level.

Conclusion

Spatial risk exposure assessment provides a reliable framework, in line with the precautionary principle, to prioritise conservation and mitigation measures on the highest risk settings. By summarising the methodological approaches and knowledge available to date on the indicative species of cetaceans and sea turtles, this review is a first step towards supporting the conceptualisation and standardisation of the spatial risk exposure approach that can be potentially extended to other species as well. While the review evidenced inconsistent geographical applications and high informational heterogeneity, main limits and future recommendations were identified regarding raw data availability, information bias, geographical gaps, target species selection and lack of standard protocols needed to assess trends to evaluate the effectiveness of mitigation measures. Ultimately, the study showed that a spatial-contextual approach (possibly functional trait-based) is needed to effectively support long-term year-round monitoring programmes, especially in still unsurveyed regions. Given the urgent need to conserve marine biodiversity and ecosystem functioning, standardising REA to include potential threats for marine megafauna throughout all habitats and regions cannot be postponed.

Acknowledgements

This work was firstly conceived within the MEDREGION project and was updated, integrated and finalised within the purposes of the Life CONCEPTU MARIS. The authors would like to thank the editors and the reviewers for the valuable and constructive suggestions that greatly helped improve the study.

References

Abreo, N.A.S., Aurelio Jr, R.M., Kobayashi, V.B. & Thompson, K.F. 2023. 'Eye in the sky': off-the-shelf unmanned aerial vehicle (UAV) highlights exposure of marine turtles to floating litter (FML) in nearshore waters of Mayo Bay, Philippines. *Marine Pollution Bulletin* **186**, 114489.

Almpanidou, V., Tsapalou, V., Chatzimentor, A., Cardona, L., Claro, F., Hostetter, P., Kaska, Y., Liu, W., Mansui, J., Miliou, A. & Pietroluongo, G. 2022. Foraging grounds of adult loggerhead sea turtles across the Mediterranean Sea: key sites and hotspots of risk. *Biodiversity and Conservation* **31**, 143–160.

Arcangeli, A., Campana, I. & Bologna, M.A. 2017. Influence of seasonality on cetacean diversity, abundance, distribution and habitat use in the western Mediterranean sea: implications for conservation. *Aquatic Conservation: Marine and Freshwater Ecosystems* **27**(5), 995–1010.

Arcangeli, A., Campana, I., Angeletti, D., Atzori, F., Azzolin, M., Carosso, L., Di Miccoli, V., Giacoletti, A., Gregorietti, M., Luperini, C., Paraboschi, M., Pellegrino, G., Ramazio, M., Sarà, G. & Crosti, R. 2018. Amount, composition, and spatial distribution of floating macro litter along fixed trans-border transects in the Mediterranean basin. *Marine Pollution Bulletin* **129**, 545–554.

Arcangeli, A., Maffucci, F., Atzori, F., Azzolin, M., Campana, I., Carosso, L., Crosti, R., Frau, F., David, L., Di-Méglio, N., Roul, M., Gregorietti, M., Mazzucato, V., Pellegrino, G., Giacoletti, A., Paraboschi, M., Zampollo, A., de Lucia, G.A. & Hochscheid, S. 2019. Turtles on the trash track: loggerhead turtles exposed to floating plastic in the Mediterranean sea. *Endangered Species Research* **40**, 107–121, doi:10.3354/esr00980

Arcangeli, A., David, L., Aguilar, A., Atzori, F., Borrell, A., Campana, I., Carosso, L., Crosti, R., Darmon, G., Gambaiani, D., Di-Méglio, N., Di Vito, S., Frau, F., Garcia-Garin, O., Orasi, A., Revuelta, O., Roul, M., Miaud, C. & Vighi, M. 2020. Floating marine macro litter: density reference values and monitoring protocol settings from coast to offshore. Results from the MEDSEALITTER project. *Marine Pollution Bulletin* **160**, 111647.

Atzori, F., Carosso, L., Cadoni, N., Frau, F., García Gutiérrez, M.L., De Lucia, G., Gregorietti, M., Pellegrino, G., Liubartseva, S., Lecci, R., Coppini, G., Maffucci, F., Hochscheid, S., & Arcangeli, A. 2021. Loggerhead sea turtle, Caretta caretta, presence and its exposure to floating marine litter in the Sardinia Channel and the Strait of Sicily: results from seven years of monitoring using ferry as platform of observation. *Mediterranean Marine Science* **22**(2), 422–435, doi:10.12681/mms.25579

Baulch, S. & Perry, C. 2014. Evaluating the impacts of marine debris on cetaceans. *Marine Pollution Bulletin* **80**, 210–221.

Berlino, M., Mangano, M.C., De Vittor, C. & Sarà, G. 2021. Effects of microplastics on the functional traits of aquatic benthic organisms: a global-scale meta-analysis. *Environmental Pollution* **285**, 117174.

Bucci, K., Tulio, M. & Rochman, C.M. 2020. What is known and unknown about the effects of plastic pollution: a meta-analysis and systematic review. *Ecological Applications* **30**(2), e02044.

Butterworth, A., Clegg, I. & Bass, C. 2012. *Untangled - Marine Debris: A Global Picture of the Impact on Animal Welfare and of Animal-focused Solutions*. London: World Society for the Protection of Animals.

Campana, I., Angeletti, D., Crosti, R., Di Miccoli, V. & Arcangeli, A. 2018. Seasonal patterns of floating macro-litter across the Western Mediterranean Sea: a potential threat for cetacean species. *Rendiconti Lincei. Scienze Fisiche e Naturali* **29**(2), 453–467.

Campana, I., Angeletti, D., Giovani, G., Paraboschi, M. & Arcangeli, A. 2022. Cetacean sensitivity index and threats analysis to assess effectiveness of protection measures. An example of integrated approach for cetacean conservation in the Bonifacio Bouches. *Biodiversity and Conservation* **31**(2), 517–541.

CBD (Secretariat of the Convention on Biological Diversity and the Scientific and Technical Advisory Panel). 2012. *Impacts of Marine Debris on Biodiversity: Current Status and Potential Solution*. Montreal, Technical Series No. **67**, 61 pp.

CBD (Secretariat of the Convention on Biological Diversity). 2016. *Marine Debris: Understanding, Preventing and Mitigating the Significant Adverse Impacts on Marine and Coastal Biodiversity (2016)*. Montreal, Technical Series No. 83, 78 pp.

CEE (Collaboration for Environmental Evidence). 2022. *Guidelines and Standards for Evidence Synthesis in Environmental Management*. Version 5.1 A.S. Pullin et al. (eds.). Available at www.environmentalevidence.org/information-for-authors. (Accessed December 2022).

Clark, B.L., Carneiro, A.P., Pearmain, E.J., Rouyer, M.M., Clay, T.A., Cowger, W. & Quillfeldt, P. 2023. Global assessment of marine plastic exposure risk for oceanic birds. *Nature Communications* **14**(1), 3665.

Claro, F., Fossi, M.C., Loakeimidis, C., Baini, M., Lusher, A.L., Mc Fee, W., McIntosh, R.R., Pelamattig, T., Sorcei, M., Galgani, F. & Hardesty, B.D. 2019. Tools and constraints in monitoring interactions between marine litter and megafauna: insights from case studies around the world. *Marine Pollution Bulletin* **141**, 147–160.

Coe, J.M. & Rogers, D.B. 1997. *Marine Debris: Sources, Impacts and Solution*. New York: Springer-Verlag.

Cole, C., Jones, L., McVeigh, J., Kicman, A., Syedc, Q. & Bellis, M. 2011. Adulterants in illicit drugs: a review of empirical evidence. *Drug Testing and Analysis* **3**(2), 89–96.

Coll, M., Piroddi, C., Albouy, C., Ben Rais Lasram, F., Cheung, W.W., Christensen, V., Karpouzi, V.S., Guilhaumon, F., Mouillot, D., Paleczny, M., Palomares, M.L., Steenbeek, J., Trujillo, P., Watson, R. & Pauly, D. 2012. The Mediterranean Sea under siege: spatial overlap between marine biodiversity, cumulative threats and marine reserves. *Global Ecology and Biogeography* **21**(4), 465–480.

Compa, M., Alomar, C., Wilcox, C., van Sebille, E., Lebreton, L., Hardesty, B.D. & Deudero, S. 2019. Risk assessment of plastic pollution on marine diversity in the Mediterranean Sea. *Science of the Total Environment* **678**, 188–196.

Critchell, K., Hamann, M., Wildermann, N. & Grech, A. 2019. Predicting the exposure of coastal species to plastic pollution in a complex island archipelago. *Environmental Pollution* **252**, 982–991, doi:10.1016/j.envpol.2019.06.031.

Currie, J.J., Stack, S.H., McCordic, J.A. & Kaufman, G.D. 2017. Quantifying the risk that marine debris poses to cetaceans in coastal waters of the 4-island region of Maui. *Marine Pollution Bulletin* **121**, 69–77.

Darmon, G., Miaud, C., Claro, F., Doremus, G. & Galgani, F. 2017. Risk assessment reveals high exposure of sea turtles to marine debris in French Mediterranean and metropolitan Atlantic waters. *Deep Sea Research Part II: Topical Studies in Oceanography* **141**, 319–328.

Davison, P. & Asch, R.G. 2011. Plastic ingestion by mesopelagic fishes in the North Pacific Subtropical Gyre. *Marine Ecology Progress Series* **432**, 173–180.

Derraik, J.G.B. 2002. The pollution of the marine environment by plastic debris: a review. *Marine Pollution Bulletin* **44**, 842–852.

Di-Méglio, N. & Campana, I. 2017. Floating macro-litter along the Mediterranean French coast: composition, density, distribution and overlap with cetacean range. *Marine Pollution Bulletin* **118**, 155–166.

Duncan, E.M., Arrowsmith, J.A., Bain, C.E., Bowdery, H., Broderick, A.C., Chalmers, T., Fuller, W.J., Galloway, T.S., Lee, J.H., Lindeque, P.K., Omeyer, L.C.M., Snape, R.T.E. & Godley, B.J. 2019. Diet-related selectivity of macroplastic ingestion in green turtles (Chelonia mydas) in the eastern Mediterranean. *Scientific Reports* **9**(1), 11581.

European Commission (EC). 2017. Commission Decision (EU) 2017/848 of 17 May 2017 laying down criteria and methodological standards on good environmental status of marine waters and specifications and standardised methods for monitoring and assessment, and repealing Decision 2010/477/EU. L 125/43.

European Environment Agency (EAA). 2020. State of nature in the EU Results from reporting under the nature directives 2013–2018. EEA Report No 10/2020.

Fossi, M.C., Casini, S., Caliani, I., Panti, C., Marsili, L., Viarengo, A., Giangreco, R., Notarbartolo di Sciara, G., Serena, F., Ouerghi, A. & Depledge, M.H. 2012. The role of large marine vertebrates in the assessment of the quality of pelagic marine ecosystems. *Marine Environmental Research* **77**, 156–158.

Fossi, M.C., Romeo, T., Baini, M., Panti, C., Marsili, L., Campani, T., Canese, S., Galgani, F., Druon, J.N., Airoldi, S., Taddei, S., Fattorini, M., Brandini, C. & Lapucci, C. 2017. Plastic debris occurrence, convergence areas and Fin Whales feeding ground in the Mediterranean Marine Protected Area Pelagos Sanctuary: a modelling approach. *Frontiers in Environmental Science* **4**, 167.

Fossi, M.C., Baini, M. & Simmonds, M.P. 2020. Cetaceans as ocean health indicators of marine litter impact at global scale. *Frontiers in Environmental Science* **8**, 255.

Gall, S.C. & Thompson, R.C. 2015. The impact of debris on marine life. *Marine Pollution Bulletin* **92**, 170–179.

Galli, M., Tepsich, P., Baini, M., Panti, C., Rosso, M., Vafeiadou, A., Pantelidou, M., Moulins, A. & Fossi, M.C. 2022. Microplastic abundance and biodiversity richness overlap: identification of sensitive areas in the Western Ionian Sea. *Marine Pollution Bulletin* **177**, 113550.

Getis, A. & Ord, J.K. 1992. The analysis of spatial association by use of distance statistics. *Geographical Analysis* **24**, 189–206.

Girard, F., Girard, A., Monsinjon, J., Arcangeli, A., Belda, E., Cardona, L., Casale, P., Catteau, S., David, L., Dell'Amico, F., Gambaiani, D., Girondot, M., Jribi, I., Lauriano, G., Luschi, P., March, P., Mazaris, A.D., Miaud, C., Palialexis, A., Sacchi, J., Sagarminaga, R., Tepsich, P., Tomás, J., Vandeperre, F. & Claro, F. 2022. Toward a common approach for assessing the conservation status of marine turtle species within the European marine strategy framework directive. *Frontiers in Marine Science* **9**, 1–22, doi:10.3389/fmars.2022.790733

Gnone, G., Bellingeri, M., Airoldi, S., Gonzalvo, J., David, L., Di-Méglio, N., Cañadas, A.M., Akkaya, A., Awbery, T., Mussi, B., Campana, I., Azzolin, M., Dede, A., Tonay, A.M., Monaco, C., Pellegrino, G., Tepsich, P., Moulins, A., Arcangeli, A., Labach, H., Scheinin, A.P., Mevorach, Y., Carlucci, R., Santacesaria, F.C., Chicote, C.A., Gazo, M., Tintore, B., Alessi, J., Mandich, A., Bittau, L., Diaz Lopez, B., Azzinari, C., La Manna, G., Pace, D.S., Decandia, D., Castelli, A., Nuti, S., Santoni, M.C., Verga, A., Tomasi, N., Giacoma, C., Costantino, M., Falabrino, M. & Azzellino, A. 2023. Cetaceans in the Mediterranean sea: encounter rate, dominant species, and diversity hotspots. *Diversity* **15**, 321, doi:10.3390/d15030321

Gormley, A., Pollard, S. & Rocks, S. 2011. Guidelines for Environmental Risk Assessment and Management. Green Leaves III. www.defra.gov.uk

Gregorietti, M., Atzori, F., Carosso, L., Pellegrino, G., Sarà, G. & Arcangeli, A. 2021. Cetacean distribution in the central Mediterranean sea and potential risks deriving from plastic pollution. *Marine Pollution Bulletin* **173**, 112943.

Gregory, M.R. 2009. Environmental implications of plastic debris in marine settings-entanglement, ingestion, smothering, hangers-on, hitch-hiking and alien invasions. *Philosophical Transactions of the Royal Society B: Biological Sciences* **364**(1526), 2013–2025.

Guerrini, F., Mari, L. & Casagrandi, R. 2019. Modelling plastics exposure for the marine biota: risk maps for fin whales in the pelagos sanctuary (north-western mediterranean). *Frontiers in Marine Science* **6**, 299, doi:10.3389/fmars.2019.00299

Guzik, T.J., Mohiddin, S.A., Dimarco, A., Patel, V., Savvatis, K., Marelli-Berg, F.M., Madhur, M.S., Tomaszewski, M., Maffia, P., D'Acquisto, F., Nicklin, S.A., Marian, A.J., Nosalski, R., Murray, E.C., Guzik, B., Berry, C., Touyz, R.M., Kreutz, R., Wang, D.W., Bhella, D., Sagliocco, O., Crea, F., Thomson, E.C. & McInnes, I.B. 2020. COVID-19 and the cardiovascular system: implications for risk assessment, diagnosis, and treatment options. *Cardiovascular Research* **116**(10), 1666–1687.

Hartje, V., Klaphake, A. & Schliep, R. 2003. *The International Debate on the Ecosystem Approach: Critical Review, International Actors, Obstacles and Challenges.* Germany: Federal Agency for Nature Conservation, 50 pp.

Høiberg, M.A., John, S.W. & Verones, F. 2022. Global distribution of potential impact hotspots for marine plastic debris entanglement. *Ecological Indicators* **135**, 108509.

Jones, J.S., Porter, A., Muñoz-Pérez, J.P., Alarcón-Ruales, D., Galloway, T.S., Godley, B.J., Santillo, D., Vagg, J. & Lewis, C. 2021. Plastic contamination of a Galapagos Island (Ecuador) and the relative risks to native marine species. *Science of the Total Environment* **789**, 147704.

Kahane-Rapport, S.R., Czapanskiy, M.F., Fahlbusch, J.A., Friedlaender, A.S., Calambokidis, J., Hazen, E.L., Goldbogen, J.A. & Savoca, M.S. 2022. Field measurements reveal exposure risk to microplastic ingestion by filter-feeding megafauna. *Nature Communications* **13**(1), 6327.

Korpinen, S., Klančnik, K., Peterlin, M., Nurmi, M., Laamanen, L., Zupančič, G., Popit, A., Murray, C., Harvey, T., Andersen, J.H., Zenetos, A., Stein, U., Tunesi, L., Abhold, K., Piet, G., Kallenbach, E., Agnesi, S., Bolman, B., Vaughan, D., Reker, J. & Royo Gelabert, E. 2019. Multiple pressures and their combined effects in Europe's seas. ETC/ICM Technical Report 4/2019: European Topic Centre on Inland, Coastal and Marine waters, 164 pp.

Kühn, S., Rebolledo, E.B., van Franeker, J.A., Bergmann, M., Gutow, L. & Klages, M. 2015. Deleterious effects of litter on marine life. In *Marine Anthropogenic Litter*, M. Bergmann (ed.). Cham: Springer International Publishing, **4**, 75–116, doi:10.1007/978-3-319-16510-3_4

Laist, D.W. 1997. Impacts of marine debris: entanglement of marine life in marine debris including a comprehensive list of species with entanglement and ingestion records. In *Marine Debris: Sources, Impacts and Solution*, J.M. Coe & D.B. Rogers (eds.). New York: Springer-Verlag, 99–139.

Liubartseva, S., Coppini, G., Lecci, R. & Clementi, E. 2018. Tracking plastics in the Mediterranean: 2D Lagrangian model. *Marine Pollution Bulletin* **129**, 151–162.

López-Martínez, S., Giménez-Luque, E., Molina-Pardo, J.L., Manzano-Medina, S., Arribas-Arias, H., Gavara, R., Morales-Caselles, C. & Rivas, M.L. 2023. Plastic ingestion by two cetacean groups: Ziphiidae and Delphinidae. *Environmental Pollution* **333**, 121932.

Macias, D., Stips, A. & Hanke, G. 2022. Model based estimate of transboundary litter pollution on Mediterranean coasts. *Marine Pollution Bulletin* **175**, 113121.

Maes, T., Perry, J., Alliji, K., Clarke, C. & Birchenough, S.N.R. 2019. Shades of grey: marine litter research developments in Europe. *Marine Pollution Bulletin* **146**, 274–281.

Mangano, M.C. & Sarà, G. 2017. Collating science-based evidence to inform public opinion on the environmental effects of marine drilling platforms in the Mediterranean Sea. *Journal of Environmental Management* **188**, 195–202, doi:10.1016/j.jenvman.2016.12.013

Mangano, M.C., Sarà, G. & Corsolini, S. 2017. Monitoring of persistent organic pollutants in the Polar Regions: knowledge gaps & gluts through evidence mapping. *Chemosphere* **172**, 37–45, doi:10.1016/j.chemosphere.2016.12.124

Mansui, J., Molcard, A. & Ourmières, Y. 2015. Modelling the transport and accumulation of floating marine debris in the Mediterranean basin. *Marine Pollution Bulletin* **91**, 249–257.

Mansui, J., Darmon, G., Ballerini, T., van Canneyt, O., Ourmieres, Y. & Miaud, C. 2020. Predicting marine litter accumulation patterns in the Mediterranean basin: spatiotemporal variability and comparison with empirical data. *Progress in Oceanography* **182**, 102268.

Marn, N., Jusup, M., Kooijman, S.A. & Klanjscek, T. 2020. Quantifying impacts of plastic debris on marine wildlife identifies ecological breakpoints. *Ecology Letters* **23**(10), 1479–1487.

Matiddi, M., Hochsheid, S., Camedda, A., Baini, M., Cocumelli, C., Serena, F., Tomassetti, P., Travaglini, A., Marra, S., Campani, T., Scholl, F., Mancusi, C., Amato, E., Briguglio, P., Maffucci, F., Fossi, M.C., Bentivegna, F. & de Lucia, G.A. 2017. Loggerhead sea turtles (Caretta caretta): a target species for monitoring litter ingested by marine organisms in the Mediterranean Sea. *Environmental Pollution* **230**, 199–209.

Miatta, M., Bates, A.E. & Snelgrove, P.V.R. 2021. Incorporating biological traits into conservation strategies. *Annual Review of Marine Science* **13**, 421–443, doi:10.1146/annurev-marine-032320-094121

Miranda-Urbina, D., Thiel, M. & Luna-Jorquera, G. 2015. Litter and seabirds found across a longitudinal gradient in the South Pacific Ocean. *Marine Pollution Bulletin* **96**(1–2), 235–244.

Molina, J.M.E., del Mar Chaves Montero, M., Galgani, F., Giorgetti, A., Vincia, M., Le Moignec, M. & Brosicha, A. 2019. EMODnet marine litter data management at pan-European scale. *Ocean and Coastal Management* **181**, 104930.

Moore, C.J. 2015. How much plastic is in the ocean? You tell me! *Marine Pollution Bulletin* **92**, 1–3.

Moore, K.T. & Barco, S.G. 2013. *Handbook for recognizing, evaluating, and documenting human interaction in stranded cetaceans and pinnipeds.* 510 NOAA technical memorandum NMFS.

Ory, N.C., Sobral, P., Ferreira, J.L. & Thiel, M. 2017. Amberstripe scad *Decapterus muroadsi* (Carangidae) fish ingest blue microplastics resembling their copepod prey along the coast of Rapa Nui (Easter Island) in the South Pacific subtropical gyre. *Science of the Total Environment* **586**, 430–437.

Oxford English Dictionary (OED). 2005. *'Risk'. Oxford English Dictionary.* Oxford: Oxford University Press, 3rd edition.

Pace, D.S., Tizzi, R. & Mussi, B. 2015. Cetaceans value and conservation in the Mediterranean sea. *Journal of Biodiversity & Endangered Species* **2015**, 4, doi:10.4172/2332-2543.S1.004

Palialexis, A. & Boschetti, S.T. 2021. Review and analysis of Member States' 2018 reports Descriptor 1: species biological diversity, EUR 30664 EN, Publications Office of the European Union, Luxembourg, 2021, ISBN 978-92-79-34256-4, doi:10.2760/27700, JRC124085.

Palialexis, A., Barnard, S., Smith, C.J. & Papadopoulou, N. 2022. Getting better odds: a risk-based approach for the monitoring and assessment of biodiversity within the EU Marine Strategy Framework Directive, Publications Office of the European Union, Luxembourg, 2022, doi:10.2760/600038, JRC130785.

Pasanisi, E., Chimienti, M., Blasi, M.F., Maffucci, F. & Hochscheid, S. 2022. Ocean highways in the Western Mediterranean: which are the areas with increased exposure to maritime traffic for loggerhead turtles? *Frontiers in Marine Science* **9**, 1734.

Pennino, M.G., Arcangeli, A., Prado Fonseca, V., Campana, I., Pierce, G.J., Rotta, A. & Bellido, J.M. 2017. A spatially explicit risk assessment approach: Cetaceans and marine traffic in the Pelagos Sanctuary (Mediterranean Sea). *PLoS One* **12**(6), e0179686.

Poeta, G., Staffieri, E., Acosta, A.T.R. & Battisti, C. 2017. Ecological effects of anthropogenic litter on marine mammals: a global review with a "black-list" of impacted taxa. *Hystrix, Italian Journal of Mammalogy* **28**, 253–264, doi:10.4404/hystrix-00003-2017

Ruiz-Orejón, L.F., Tornero, V., Boschetti, S.T. & Hanke, G. 2021. Marine Strategy Framework Directive - Review and analysis of EU Member States' 2018 reports - Descriptor 10: Marine litter, EUR 30665 EN, Publications Office of the European Union, Luxembourg, 2021, ISBN 978-92-76-34263-2, doi:10.2760/238367, JRC124701.

Sá, S., Bastos-Santos, J., Araújo, H., Pereira, A.T., Ferreira, M., Sarmento, P., Vingada, J. & Eira, C. 2021. Floating marine litter and their risks to cetaceans off Portugal. *Marine Pollution Bulletin* **170**, 112603.

Sakellariou, S., Sfougaris, A., Christopoulou, O. & Tampekis, S. 2022. Integrated wildfire risk assessment of natural and anthropogenic ecosystems based on simulation modelling and remotely sensed data fusion. *International Journal of Disaster Risk Reduction* **78**, 103129.

Salerno, M., Berlino, M., Mangano, M.C. & Sarà, G. 2021. Microplastics affect the fish functional traits: insight from a global meta-analysis. *Global Change Biology* **27**, 2645–2655.

Schuyler, Q.A., Wilcox, C., Townsend, K., Wedemeyer-Strombel, K.R., Balazs, G., van Sebille, E. & Hardesty, B.D. 2016. Risk analysis reveals global hotspots for marine debris ingestion by sea turtles. *Global Change Biology* **22**(2), 567–576, doi:10.1111/gcb.13078.

Shepherd, G. 2004. *The Ecosystem Approach: Five Steps to Implementation.* IUCN, Gland, Switzerland and Cambridge, UK. vi+30 pp.

Soto-Navarro, J., Jordá, G., Compa, M., Alomar, C., Fossi, M.C. & Deudero, S. 2021. Impact of the marine litter pollution on the Mediterranean biodiversity: a risk assessment study with focus on the marine protected areas. *Marine Pollution Bulletin* **165**, 112169.

Suter II, G.W. 2016. *Ecological Risk Assessment.* Boca Raton, FL: CRC Press.

Sutherland, W.J., Clout, M., Cote, I. M., Daszak, P., Depledge, M.H., Fellman, L., Fleishman, E., Garthwaite, R., Gibbons, D.W., De Lurio, J., Impey, A.J., Lickorish, F., Lindenmayer, D., Madgwick, J., Teuten, E.L. & Reddy, C.M. 2007. Halogenated organic compounds in archived whale oil: a pre-industrial record. *Environmental Pollution* **145**, 668–671.

UNEP 2009. *Marine Litter: A Global Challenge.* Nairobi: UNEP, 232. https://wedocs.unep.org/20.500.11822/7787

UNEP-MAP RAC/SPA. 2010. The Mediterranean Sea biodiversity: state of the ecosystems, pressures, impacts and future priorities. *Palaeogeography, Palaeoclimatology, Palaeoecology*, **297**(1), 37–53. Retrieved at: https://rac-spa.org/sites/default/files/doc_cop/biodiversity.pdf.

UNEP-MAP. 2016. Integrated monitoring and assessment guidance. 19th Ordinary Meeting of the Contracting Parties. Athens, Greece, 9-12 February 2016.

Vegter, A.C., Barletta, M., Beck, C., Borrero, J., Burton, H., Campbell, M.L., Costa, M.F., Eriksen, M., Eriksson, C., Estrades, A. Gilardi, K.V.K., Hardesty, B.D., Ivar do Sul, J.A., Lavers, J.L., Lazar, B., Lebreton, L., Nichols, W.J., Ribic, C.A., Ryan, P.G., Schuyler, Q.A., Smith, S.D.A., Takada, H., Townsend, K.A., Wabnitz, C.C.C., Wilcox, C., Young, L.C. & Hamann, M. 2014. Global research priorities to mitigate plastic pollution impacts on marine wildlife. *Endangered Species Research* **25**(3), 225–247.

Vighi, M., Ruiz-Orejón, L.F. & Hanke, G. 2022. Monitoring of Floating Marine Macro Litter - State of the art and literature overview. EUR **31073** EN, Publications Office of the European Union, Luxembourg, ISBN 978-92-76-52436-6, doi:10.2760/78914, JRC129261.

Werner, S., Budziak, A., van Franeker, J., Galgani, F., Hanke, G., Maes, T., Matiddi, M., Nilsson, P., Oosterbaan, L., Priestland, E., Thompson, R., Veiga, J. & Vlachogianni, T. 2016. Harm caused by Marine Litter. MSFD GES TG Marine Litter - Thematic Report; JRC Technical report; EUR 28317 EN; doi:10.2788/690366

Wilcox, C., Hardesty, B.D., Sharples, R., Griffin, D.A., Lawson, T.J. & Gunn, R. 2013. Ghostnet impacts on globally threatened turtles, a spatial risk analysis for northern Australia. *Conservation Letters* **6**, 247–254.

Williams, R., Ashe, E. & O'Hara, P.D. 2011. Marine mammals and debris in coastal waters of British Columbia, Canada. *Marine Pollution Bulletin* **62**(6), 1303–1316.

Woods, J.S., Rødder, G. & Verones, F. 2019. An effect factor approach for quantifying the entanglement impact on marine species of macroplastic debris within life cycle impact assessment. *Ecological Indicators* **99**, 61–66.

Wright, S.L., Thompson, R.C. & Galloway, T.S. 2013. The physical impacts of microplastics on marine organisms: a review. *Environmental Pollution* **178**, 483–492.

Yu, W., Zhang, T., Lu, Y., Han, F., Zhou, Y. & Hu, D. 2020. Engineering risk analysis in cold regions: state of the art and perspectives. *Cold Regions Science and Technology* **171**, 102963.

Zampollo, A., Arcangeli, A., Costantino, M., Mancino, C., Crosti, R., Pietroluongo, G., Giacoma, C. & Azzolin, M. 2022. Seasonal niche and spatial distribution modelling of the loggerhead (Caretta caretta) in the Adriatic and Ionian Seas. *Aquatic Conservation: Marine and Freshwater Ecosystems* **32**(7), 1141–1155, doi:10.1002/aqc.3815.

Zhang, D., Yu, W., Chen, B., Chen, H., Zhao, L., Wang, X., Wu, F., Hu, W. & Ma, Z. 2020. Risk assessment for marine plastic debris ingestion by Indo-Pacific humpback dolphins (*Sousa Chinensis*) in Xiamen Bay, China. *China Environmental Science* **40**(4), 1809–1818.

Supplementary Materials are provided online at: https://www.routledge.com/9781032761961